THE AVIAN EGG

THE AVIAN EGG

by

Alexis L. Romanoff, Ph. D.
PROFESSOR OF CHEMICAL EMBRYOLOGY
CORNELL UNIVERSITY

and

Anastasia J. Romanoff, B. Sc.

JOHN WILEY & SONS, INC., New York
CHAPMAN & HALL, LIMITED, London

COPYRIGHT, 1949
BY
ALEXIS L. ROMANOFF
AND
ANASTASIA J. ROMANOFF

All Rights Reserved

PRINTED IN THE UNITED STATES OF AMERICA

DEDICATED

to

the scientists whose original studies on the egg
have made it possible for the authors
to write this book

PREFACE

This book represents an attempt to compile all the facts known about the bird's egg. Heretofore, these facts have been scattered throughout the scientific literature, where they could be found only with considerable effort. Our intention in assembling them here has been to unify this piece-meal information and to make it available in one volume.

In writing this book, we found it logical and convenient to divide its contents into three sections, which successively consider the biological origins and structure of the egg, its physical and chemical properties, and its usefulness to man. We have limited our discussion to the egg before the activation of life within it, although we have tried not to lose sight of the relationship between the egg's organization and its reproductive function. We have also made an effort to point out the operation of the many factors which make no two eggs alike. The illustrations, drawn or reconstructed from various sources by the senior author, are included to supplement or clarify the descriptive material. It is unfortunate that space did not permit reference to all the publications—nearly fifteen thousand in number—that we reviewed, but the bibliography may be of value to those who wish to consult the original works mentioned.

We offer this volume to all who seek knowledge about the bird's egg, whether they be teachers, students, research workers, or merely individuals with inquiring minds. It is our hope that they will find here the answers to their questions. Above all, we feel that our undertaking will have been justified if interest in further study of the egg is stimulated. Much has been discovered about the egg, but it still offers a challenge to the investigator.

ALEXIS L. ROMANOFF
ANASTASIA J. ROMANOFF

Ithaca, New York

ACKNOWLEDGMENTS

The authors regret that there is insufficient space to express their gratitude to everyone who assisted in the preparation of this book. Some suggestions were so helpful and constructive, however, that those who offered them must be singled out. Particularly valuable advice was received from the persons named below, who reviewed various chapters in the indicated sections of the book:

Part I

Professor Arthur A. Allen, Cornell University
Dr. George W. Bartelmez, University of Chicago
Professor Frank A. Hays, Massachusetts State College
Dr. Walter Landauer, University of Connecticut
Professor B. H. Willier, Johns Hopkins University

Part II

Professor C. A. Brandly, University of Wisconsin
Dr. A. M. Bueche, Cornell University
Professor Henry B. Bull, Northwestern University
Dr. G. F. Somers, United States Nutrition Laboratory, Ithaca, N. Y.
Professor James B. Sumner, Cornell University
Dr. Alexander Zeissig, New York State Department of Health, Albany, N. Y.

Part III

Dr. E. C. Bate-Smith ⎤
Dr. J. Brooks ⎥ Department of Scientific and Industrial
Mr. H. P. Hale ⎦ Research at the University of Cambridge
Professor Clive M. McCay, Cornell University
Dr. Jeanette B. McCay, Ithaca, N. Y.
Emeritus Professor E. V. McCollum, Johns Hopkins University
Dr. Mary E. Pennington, New York, N. Y.

The senior author also deeply appreciates the freely given counsel of his colleagues and associates. He is especially grateful for the encouragement he received from Professor Emeritus James E. Rice, and for the numerous courtesies extended by Dr. J. H. Bruckner, head of the Department of Poultry Husbandry at Cornell University during the time the major part of this work was in progress.

In addition, the authors sincerely thank Marion Pearsall, Everett T.

Erickson, and Frances E. Sage for their help in putting the manuscript into its final form. Mention must also be made of the kindness and willing cooperation of many librarians, especially those of the John Crerar Library of Chicago, and of the libraries of the Field Museum of Natural History, the New York Academy of Medicine, and Harvard and Cornell universities.

Finally, the authors wish to acknowledge their indebtedness to the investigators whose original work was the source of so much of the material in this book, and to the publishers of the various journals that record the day-by-day progress of science throughout the world.

CONTENTS

Part II

BIOPHYSICOCHEMICAL CONSTITUTION

Part III

BIO-ECONOMIC IMPORTANCE

INTRODUCTORY NOTE

"I think if required on pain of death to name instantly the most perfect thing in the universe, I should risk my fate on a bird's egg." Those were the words of T. W. Higginson, in the year 1863.

The above thought was expressed during the era when scientists emphasized the study of external form and color and were occupied in accumulating and classifying specimens. In this period, both naturalists and laymen alike collected birds' eggs with almost fanatical enthusiasm. The vast collections of museums were started; and later, in the second half of the nineteenth century, it became possible for museums and various ornithological organizations in the United States and Europe to sponsor many excellent memoirs and catalogues, beautifully illustrated with colored plates of birds' eggs. Especially worthy of mention are *Coloured Figures of the Eggs of British Birds,* by H. Seebohm, published in 1896, and *Life Histories of North American Birds,* by C. Bendire, Vol. I of which appeared in 1892, and Vol. II, in 1895.

The approach of the naturalist, however, did not long satisfy scientific minds interested in the study of eggs. Approximately in the middle of the nineteenth century, great curiosity developed concerning the internal structure of the egg. The egg was then subjected to the most painstaking dissection and microscopic examination. Long before the close of the century, the gross organization and the histology of the egg were described in *The Anatomy of Vertebrates,* by R. Owen (*1866*), and in *Elements of Embryology,* by M. Foster and F. M. Balfour (*1874*). A very long series of papers on the eggshell, by W. von Nathusius (*1867–1895*), is one of the outstanding works of this period.

During these years, the foundation was laid for a thoroughly factual approach to the study of birds' eggs. The next logical step was chemical analysis. In the latter part of the nineteenth century, and the early part of the twentieth, various chemical compounds were identified in the egg and isolated from it in purified form. The true nature of the egg was glimpsed for the first time, when it was revealed how great a multitude of substances enter into its composition. For fairly recent reviews in which some attention is given to the chemistry of the egg, the reader is referred to two monumental works: *Chemical Embryology,* by Joseph Needham, published in Cambridge, England, in 1931; and *Handbuch der Eierkunde,* by J. Grossfeld, published in Berlin, Germany, in 1938.

The disclosures of the chemists served to attract attention to the egg as a rich source of experimental material, and the development of physico-chemical techniques broadened the scope of endeavor. To many workers in the pure sciences, the egg provided substances to be studied in the search for general, fundamental truths. Other theoretical scientists became interested in the egg as an organized system, composed of inter-related parts. Simultaneously, independent inquiry into the nature of the egg was pursued by investigators seeking the answers to practical questions. As a result, the egg has been closely scrutinized in the course of research in biochemistry, biophysics, embryology, bacteriology, medicine, nutrition, food chemistry, and poultry husbandry. From all directions, valuable contributions have been made to the cumulative fund of information on the egg. As the various branches of science merge and become more and more interdependent, an increasingly great coordination of effort is made possible in approaching the unsolved problems which the egg still presents.

Part I

MORPHOGENETIC EXPRESSION

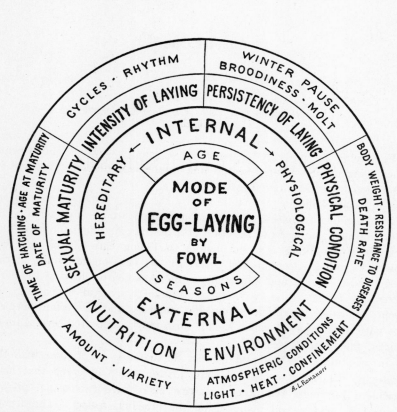

Interrelation of factors that affect the number of eggs laid by fowls.

Chapter One

MODE OF LAYING

The abundance of avian eggs depends chiefly upon the exploitation of complex hereditary and physiological factors.

Egg-laying by birds is the result of a complex natural endowment. The laying of eggs is an essential link in the reproductive cycle and is associated with the parental instinct.

The parental instinct is manifested by the bird's responses to a series of commanding impulses. These impulses lead to (1) migration; (2) courtship and mating; (3) selection of a nesting site and building of the nest; (4) egg-laying; (5) incubation, including care of the eggs by shielding, rolling, cleaning, and covering; (6) care of the young in the nest; and (7) care and education of the young after leaving the nest (*Herrick, 1907*). The bird's parental urge is cyclical in nature. It recurs annually, usually in spring and summer; and, in certain species, the pattern of behavior which it calls forth may be repeated several times during the breeding season.

In wild birds, the parental instinct fluctuates in accordance with the phase of the sexual cycle. Howard (*1914, 1920*) observed that male warblers, buntings, and other passerine birds have two sexual phases, neutral in winter, and male during the breeding season. The female passes from the neutral state of winter into a partly sexed condition and finally becomes fully sexed. During the intermediate phase, she seeks out the company of a male, appears interested in some of his advances, and may construct the outer part of one or more nests. However, it is not until she is in the fully sexed phase that she will solicit or permit coition or will construct a lining for a nest. This period begins about 3 to 6 days before the first egg is laid.

The instinctive preparation of the female for egg-laying is accompanied by definite physiological changes, which indicate that, in some parts of her body, metabolic activity is highly intensified. These physiological changes reach their maximum just before the laying of each egg (see Chapter 4, "Formation").

FECUNDITY AND EGG-LAYING

The term "fecundity" is used to describe the inherent capacity of an organism to reproduce rapidly. In higher animals, reproduction is possible only after the ovum (female gamete) is fertilized or united with the male gamete. In birds, however, "fecundity" usually refers to the ability of the female to lay eggs in large numbers, because in many species fertilization is not a necessary preliminary to egg-laying. The female pigeon, for instance, may form and lay eggs without mating (*Riddle, 1931*) or even in the total absence of the male (*Craig, 1911*). The domestic hen can also lay eggs continuously without being mated or without being stimulated to lay by the presence of a male. This biological phenomenon has been advantageously utilized by man in producing infertile eggs for food. (Infertile eggs are of more economic value than fertile, because there is no danger of loss through development of the embryo.)

SIZE OF CLUTCH

Wild birds usually lay eggs in clutches. The term "clutch" refers to the set of eggs laid for one incubation. The number of eggs in a clutch, or set, is fairly uniform in each species but varies greatly from one species to another. Some birds lay a single egg (auks, murres, ruffins, penguins); others never lay more than 2 (pigeons), or 4 (plovers); and still others lay as many as 12 to 20 eggs (partridges).

Several hypotheses have been advanced to explain the size of the clutch. It has been stated (*Fox, 1899*) that the number of eggs laid in a clutch bears a definite relationship to the amount of danger to which the species is exposed, or to its power for self-defense or escape. Another possible factor is the extent of the food supply (*Harvie-Brown and Bunyard, 1910*). Tropical species lay fewer eggs than those in temperate regions. Chapman (*1937, p. 83*) gives, among other examples, the sooty terns and bridled terns of the tropics, which lay 1 egg, as compared with the arctic common terns, which usually lay 3 eggs. Averill (*1933*), however, believes the tropical bird is at a disadvantage in feeding its young. In the temperate and arctic zones there are more hours of daylight during the breeding season than in the tropics, and consequently the northern bird has a longer time for foraging.

Most wild birds living under natural conditions lay only one clutch of eggs during the breeding season. Should disaster occur, resulting in destruction of the clutch, the bird's reproductive cycle may again commence, and the bird then lays another clutch. However, a few wild birds (such as jungle fowl, robins, and bluebirds) lay, incubate, and brood two, and sometimes three, clutches in one season. The semidomesticated and gregarious house sparrow is credited with as many as four clutches. The domestic pigeon, which spends about 1 month in incubating and tending the squabs of each brood, will sometimes lay upward of ten clutches of eggs in a year.

If a bird lays several large clutches in one season, the number of eggs in each successive clutch is usually smaller. For example, turkeys will lay about 18 eggs in the first clutch. The number of eggs laid in the second clutch averages about 12, and in the third about 10 (*Weiant, 1917*).

Some wild birds will abandon the nest if any eggs are removed from it. Many other birds will content themselves with incubating the eggs left to them. For example, it has been observed (*Davis, 1942c*) that removing eggs from the nest of the herring gull does not induce her to lay more than the normal complement of 3 eggs; nor does adding eggs restrain her from laying the usual number.

On the other hand, if all eggs but one are removed from the nest, some wild birds continue to lay eggs in an effort to achieve a nest complement. Phillips (*1887*) recorded a remarkable instance. When the eggs of a flicker were removed from the nest as soon as they were laid, the bird produced 71 eggs during 72 days. Under similar circumstances, a common wryneck laid 48 eggs in succession (*Hanke, quot. by Floericke, 1909, p. 76*), and a house sparrow, 51 (*Wenzel, 1908*). The mallard duck may lay 80 to 100 eggs (*Austin, 1908*). One mallard is known to have laid 146 eggs in 158 days (*Mouquet, 1924*).

It almost seems that birds are able to count, as Ingersoll (*1897*) expressed it, up to the proper limit of their nest complement, or, at any rate, realize instinctively when each clutch has been completed. The islanders of Jutland have taken advantage of this fact by gathering day by day the eggs of the sheldrake, which breeds in artificial burrows along the coast. Encouraging birds to lay by removing eggs from the nest is a device which has been put to particularly good use in increasing the egg production of domestic poultry. The extent to which the laying of eggs can be stimulated in this way is shown for domesticated and game birds in Table 1. A domesticated hen can lay as many as 300 eggs in a year. An exceptional record was made by a hen that laid 361 eggs during 364 days (Fig. 1).

Few domestic hens lay eggs in clutches. However, it has been observed by Harland (*1927*) that the native fowl of Trinidad still have a tendency to lay in clutches of about 12 eggs. Each clutch is regularly followed by a broody period of approximately 16 days, so that roughly 28 days are

TABLE 1

Egg-Laying by Domesticated and Game Birds *

Species	Natural Size of Clutch	Maximum Laid in a Year When Eggs Are Removed
Chickens:		
Egg-producing ⎤		⎧361
Meat-producing ⎬	11–14	60
Game or fancy ⎦		⎩60
Ducks:		
Egg-producing ⎤		⎧309
Meat-producing ⎦	14–20	⎩120
Turkeys	15–20	205
Guinea fowl	14	100
Geese	10–15	100
Peafowl	5– 9	37
Ostriches	12–15	100
Pheasants (in confinement)	10–12	104
Grouse (in confinement)	9–15	36
Quails (in confinement)	12–20	128
Mallard ducks (in confinement)	5–14	146
Pigeons	2	50
Canaries	4– 6	60

* Compiled from various sources.

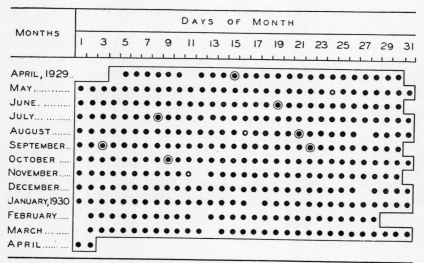

FIG. 1. Daily record of unusual egg production by a Black Orpington hen. (Courtesy of Taranaki Egg-Laying Competition Society, New Zealand.)

required to complete the cycle. Under similar environmental conditions, hens that are crosses of native and pure breeds lay multiples or sub-multiples of the basal clutch number of the native fowl.

Purebred fowls of American and European descent have no clutch habit. They lay eggs continuously for a considerable time and rarely cease laying for intervals of more than a few days. Consequently some of them never become broody. A modern 300-egg hen represents a remarkable evolution from the ancestral jungle fowl, which lays one clutch, of about 13 to 15 eggs, at a time.

NESTING HABITS

A flock of domesticated hens lays its eggs throughout the day, from about 7 o'clock in the morning until 5 o'clock in the afternoon (*Turpin, 1918*). Most hens lay during the forenoon, as shown in the tabulation.

TIME	HENS LAYING (per cent)
7 A.M. to 9 A.M.	17.7
9 A.M. to 11 A.M.	28.5
11 A.M. to 1 P.M.	27.3
1 P.M. to 3 P.M.	19.5
3 P.M. to 5 P.M.	7.0

It is interesting to observe that nearly 56 per cent of the flock lays between 9 and 1 o'clock.

Among other species, domesticated ducks, as well as mallards, lay most of their eggs very early in the morning, whereas common pigeons lay in the early afternoon (*Riddle, 1923*). A turkey hen lays at no particular time of day, although most of her eggs are usually laid in the morning.

The time spent in the nest varies greatly. Turpin (*1918*) showed that a hen, when laying, spends approximately 2 hours in the nest. The quail, on the other hand, requires only 3 to 10 minutes to deposit an egg (*Stoddard, 1931, p. 26*).

The time spent in the nest evidently increases during the laying of a clutch. Weiant (*1917*) states that a turkey hen ordinarily remains in the nest about 1.5 hours, when laying the first eggs in the clutch, but may remain there as long as 8 hours when laying the last few eggs.

BREED DIFFERENCES

Although many varieties of chickens, and some ducks, are bred specifically for egg-laying, others are considered to be game or fancy breeds (such as Bantams), and still others are raised chiefly for their meat (Brahmas, Cochins, Langshans, New Jersey Giants, etc.). The game and flesh-producing fowl are usually poor layers; they lay an average of perhaps 60 eggs a year. The only chickens that lay well are those that have been bred for egg-laying. There are several "egg-laying" breeds, but they are not equally popular. In 1943, the egg-laying chicken population of the United States was composed of these breeds in differing

Egg-Laying Breeds	Proportion of Total (per cent)
White Leghorns	34
Plymouth Rocks	25
New Hampshires	10
Rhode Island Reds	6
Crosses	9
Mixed and other breeds	16

proportion, as the table shows. White Leghorns and Plymouth Rocks together made up more than one-half of the total number.

VARIATION AMONG INDIVIDUALS

Among wild birds of the same species there is little variation in the size of the clutch. On the other hand, the number of eggs laid by domesticated birds, especially by chickens of the egg-producing breeds, is extremely variable.

Trapnest records have shown that some hens lay few eggs, or none, whereas others lay large numbers. Consequently, a flock of hens may demonstrate a wide individual variation in annual egg production, which may range from 0 to 300 eggs.

In a study of four of the most popular egg-producing breeds of fowl, made about 20 years ago, it was found that a frequency distribution curve of individual egg production was not symmetrical (Fig. 2). On the side of high production the frequency curve descended from the peak more abruptly than it ascended on the side of low production. This asymmetry in frequency distribution (skewness) is typical of all egg-producing breeds and may be still more pronounced in modern flocks that have higher egg records than those of the past.

INFLUENCE OF AGE

The life span of birds varies widely. According to Gurney (*1899*) and Krzhishkowskii (*1933, p. 243*), some birds, such as the kite, may live as long as 120 years. Others, especially small songbirds (passerine group)

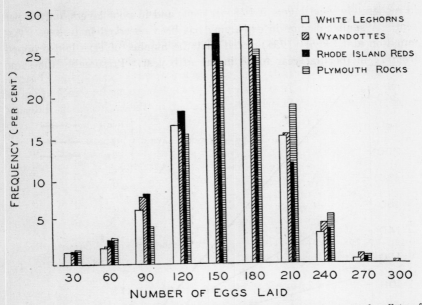

Fɪɢ. 2. Frequency distribution, according to annual egg production, of pullets of four different breeds. (After Dunn, 1924a, 1924b, 1924c, 1927.)

and chickens, live a much shorter time, not more than about 20 years. In the opinion of Gurney (*1899*), the life span usually bears an inverse relationship to the birth rate, as represented by the number of eggs in a clutch. A bird that lays 1 egg is therefore likely to live longer than a bird that lays 10 eggs.

Egg-laying is limited to the early life of the bird and, in fowls, seldom extends beyond the tenth year. During the reproductive period, however, a hen may lay many eggs. One hen, for example, is known to have laid 1515 eggs in 8 years (*Hall, 1938*).

Egg production declines gradually throughout the hen's life; it is heaviest in the pullet year and decreases during the following years. This change in the rate of egg-laying is in almost linear relationship to age (*Brody, Henderson, and Kempster, 1923; Clark, 1940; Thompson, 1942*).

It has been observed, also, that egg production declines with age more rapidly in those individuals that lay intensely during their first year than in those that are low producers from the beginning. This point is well illustrated by Fig. 3, which shows that, in high producers, the average annual decrease in the number of eggs laid is about 10 per cent, whereas in moderately good layers it is 8 per cent, and in poor layers, only 5 per cent. A similar decline in egg-laying has been observed in turkeys. Asmundson and Lloyd (*1935*) noted that the number of eggs laid dropped from 77 in the first year to 28 in the fifth year. Presumably the same

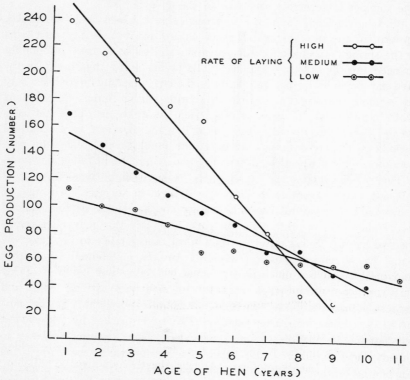

Fig. 3. Decline in egg-laying with advancing age (White Leghorns). (After Dryden, 1921; Hall and Marble, 1931.)

changes in egg production might be observed during the life of other domesticated birds.

The decrease in egg production in later life is due to the progressively diminishing metabolic activity of the organs and tissues (*Zander, Lerner, and Taylor, 1942*). The fertility of eggs is also affected by senescence; short as is the egg-laying period in the life span of birds, the period during which their eggs are fertile is even shorter. The ability to reproduce presumably declines with age at a more rapid rate in male than in female birds (*Pearl, 1917*).

LAYING SEASONS

Physiological activity in wild birds is correlated with the season of the year. The heightening of the instinct for reproduction usually occurs in spring, and egg-laying coincides with the time of general revival in nature. The summer is devoted to the care of the offspring.

Wild jungle fowl, from which domesticated chickens have descended, conform to the same general laws of nature and usually lay their eggs in the spring. In chickens, domestication and selection have profoundly changed the physiology of laying. The laying season has been lengthened far beyond the natural limits. Modern, carefully bred hens may lay without interruption during the entire year. Actual observations show, nevertheless, that the average hen does not distribute her eggs evenly throughout the seasons. She still tends to revert to the natural periodicity displayed by her ancestor, the jungle fowl. Maximum production occurs in spring and summer, and minimum in fall and winter.

The pullet reaches her reproductive stage in late summer or early fall and, under favorable conditions, will begin to lay eggs. The fall, of course, is not a natural breeding season, and egg-laying at this time of year indicates how profoundly the process of selection may alter reproductive habits. Since this is an unnatural time for her to lay eggs, the pullet has a tendency to stop laying. Autumn is followed by another unnatural laying season—winter. Some hens may not lay well during this time, because of their inability to withstand extreme cold, and especially because of the relatively few hours of daylight. Spring, normally the season of high reproductive activity, is marked by the heaviest egg-laying. During the summer, many hens show a strong tendency to brood; under natural conditions, this time of the year is spent in rearing the new generation. However, in many varieties of fowl, selected primarily for egg production, the brooding instinct is nearly absent; in other breeds, it is not usually allowed to run its normal course but is checked in its initial stages.

INFLUENCE OF LATITUDE

Seasonal changes in egg-laying are influenced by latitude, largely because of variation in the daily amount of light in different parts of the world (see Chapter 4, "Formation"). Spring is the breeding season for birds in the temperate zone of both the northern and the southern hemispheres, but in southern latitudes spring extends from September to November, rather than from March to May. The peak of egg-laying in Argentina or New Zealand, for example, therefore occurs at the time of lowest production [1] in the central part of the United States or in southern Europe (Fig. 4). In the tropics (approximately 23° north and south of the equator), egg-laying varies with the seasons in much the same manner as in the temperate zones; on the other hand, in such equatorial regions

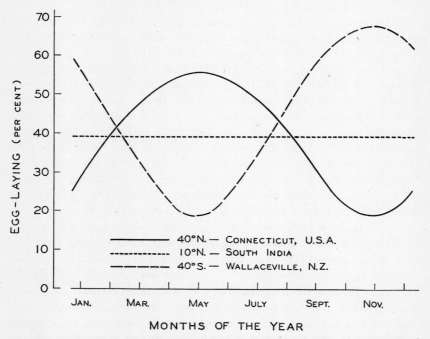

FIG. 4. Seasonal variations in egg production of hens (each with an annual record of about 150 eggs) in the temperate regions of the northern and southern hemispheres, and in an equatorial region. (From the data of Whetham, 1933.)

as southern India (about 10° N.) it is nearly uniform throughout the year (cf. Fig. 4), and seasonal changes are almost imperceptible.

CHANGES WITH ADVANCING AGE

Not only does the annual number of eggs decrease as the bird grows older, but the seasonal character of egg-laying also becomes accentuated (Fig. 5).

During the first year, the pullet, after reaching her full sexual development, begins to lay eggs early in the fall. She may continue to lay well

[1] The rate of production is usually expressed as the percentage of the egg production which would be possible each month, if 1 egg were laid each day.

through winter, spring, and summer. In the following fall, her rate of laying begins to decline and reaches its lowest point in early winter. This period of infrequent laying corresponds to the period of physiological rest.

During the second year and thereafter, the hen begins to lay in the spring and continues to lay through the summer. In the fall, her rate

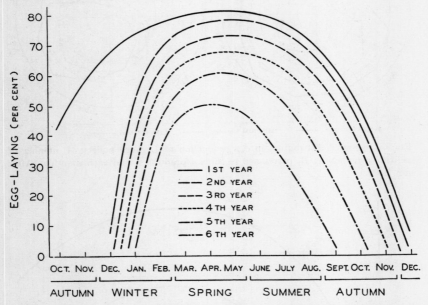

FIG. 5. Smoothed curves of seasonal variations in egg-laying during first 6 years of the hen's life. (From data on a selected group of fowls with records of 250 eggs each during the first year.)

of laying decreases again, and the second period of lowest production thus comes in winter. With advancing age, laying begins later in each successive year and terminates earlier; consequently, an old hen lays eggs only during the spring and summer.

RATE OF LAYING

Seasonal changes in egg-laying depend upon the relative fecundity of the hen and are least marked in prolific birds. The hens that lay the greatest number of eggs during each season also tend to rest for the shortest time in the fall and winter (Fig. 6). Hens of low fecundity lay chiefly during the spring and summer, and their period of rest is of long duration. Figure 7 shows how the curve of the annual egg production

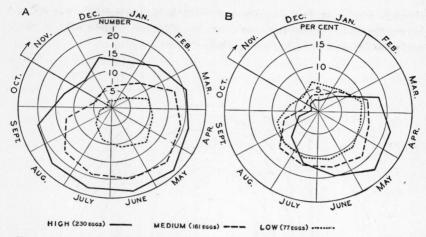

FIG. 6. Seasonal distribution in egg production of Leghorn pullets: *A,* number of eggs laid each month; *B,* percentage of annual production laid each month. (After Dunn, 1927.)

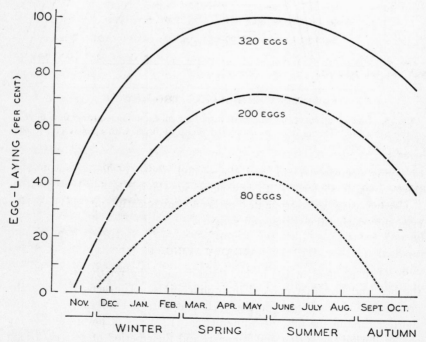

FIG. 7. Seasonal variation in egg-laying by hens with different records of annual production. (After Ball and Alder, 1917, and unpublished data at Cornell University.)

varies in width as well as height, according to the rate of egg-laying. Advancing age considerably modifies the seasonal fluctuations in egg-laying, particularly in the prolific bird (Fig. 8).

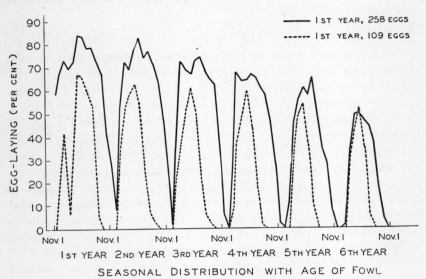

Fig. 8. Effects of advancing age on seasonal variation in egg-laying by hens of high and low fecundity. (From unpublished data at Cornell University.)

FACTORS INFLUENCING EGG-LAYING

Egg-laying is subject to the influence of various internal and external factors. It is, in large measure, controlled by the individual bird's heredity; it also is largely dependent upon her physiological efficiency and metabolic activity. These, in turn—although inherited to a certain extent—may be profoundly affected by many environmental circumstances.

The hen, therefore, not only must possess the inherent capacity to lay well; she must, in addition, be free of physical defects and disease and be well fed and well housed. She cannot maintain a high level of egg production if her functional balance is disturbed, for example, by infection, insufficient food, lack of certain dietary essentials, inadequate amounts of fresh air and sunlight, or exposure to sudden weather changes.

HEREDITARY TRAITS

The domestic fowl's fecundity is inherited through a very complex mechanism, involving the transmission of several independent characters or traits. Many attempts have been made to segregate these traits in the offspring of experimental matings, and a number of theories have sought to explain the manner in which Mendel's laws operate in the inheritance of egg-laying capacities.

Mendel's Law of Inheritance

The first suggestion that the hen's productivity depended upon heredity was made by Pearl (*1912a*). After a careful analysis of available records, he advanced the theory that high and low egg production in winter are influenced by three genetic factors and that the egg record of a hen, by itself, is not a sound basis upon which to breed for fecundity. A few years later, however, Goodale (*1918*) pointed out that Pearl's explanation was inadequate and suggested that egg-laying capacities depend on a number of more or less independent internal factors. Consequently, the operation of different sets of factors may produce similar results. It follows, therefore, that each factor must be studied separately, both from genetic and physiological standpoints. Hurst (*1921*) modified Goodale's hypothesis and enlarged previous concepts of the inheritance of fecundity by indicating that, in order to obtain maximum egg production, it is necessary to achieve the proper combination of dominant and recessive traits in the same individual.

These theories were further elaborated by Hays (*1924–25; 1927; 1932b; 1944b*), who demonstrated the importance of several additional physiological factors. Other studies, especially statistical analyses on chickens (*Knox, Jull, and Quinn, 1935; Lerner and Taylor, 1937, 1943*) and on turkeys (*Asmundson, 1938*), indicate that all known physiological factors affecting egg-laying are subject to some hereditary influences.

Much valuable information has been obtained regarding the mechanism of the transmission of egg-laying capacity from generation to generation. However, it has been found impossible to prove that a single specific hereditary factor (gene) is responsible for fecundity, and it is generally thought that many hundreds of genes are in operation.

PHYSIOLOGICAL FACTORS

The ability to lay eggs requires normal body structure and involves numerous metabolic processes. To lay a large number of eggs, the hen must be an efficient physiological machine for the rapid transformation of raw feed into finished eggs. Although she does not begin to lay until she is mature, preparation for laying starts early in her life (see Chapter 4, "Formation").

The number of eggs that a hen lays annually depends upon her age at sexual maturity, the rate at which she lays, her ability to lay continuously, and her physical condition.

Sexual Maturity

Sexual maturity is manifested by the laying of the first egg. Almost all wild birds of the northern hemisphere achieve sexual maturity in April, May, or early June, when they are approximately a year old. The process of domestication has accelerated the rate at which a number of birds attain maturity. The chicken, for example, usually starts to lay in the fall or early winter, when she is about 6 months old. There is considerable individual variation, however, in the age at which pullets mature. In addition, the time required for the bird to reach maturity, and consequently the date when she lays her first egg, vary according to the season when she is hatched (*Hays, Sanborn, and James, 1924; Knox and Bittenbender, 1927; Upp and Thompson, 1927; Jeffrey and Platt, 1941*).

The chicken's total egg production during her first year depends to some extent upon her age at sexual maturity. Early sexual maturity is ordinarily associated with intense reproductive activity and therefore with heavy egg-laying. However, if birds are hatched at unnatural seasons, they may mature very rapidly and yet lay few eggs.

TIME OF HATCHING

Upp and Thompson (*1927*) hatched chicks every 15 days throughout the year and found that those hatched in spring and early summer matured more slowly (and consequently weighed more at maturity) than those hatched during the fall and winter (Fig. 9). Riddle (*1931*) observed that doves and pigeons became sexually mature in about 180 days if hatched in the months of September to January, inclusive, but required about 300 days to reach maturity if hatched in other months of the year.

Pullets hatched in the spring produce the largest number of eggs, although they require the longest time to mature (*Buss, 1919; Kempster and Henderson, 1922; Berry and Walker, 1927; Davidson, McCrary, and Card, 1946*). The highest autumn egg production may be achieved by birds that are hatched in February and begin to lay in July, but these birds may stop laying for a time in November and December and go through a partial molt (Fig. 10). On the other hand, birds hatched as late as June or July may not begin to lay before January. In either case, the actual laying period for the first year is considerably shortened, and

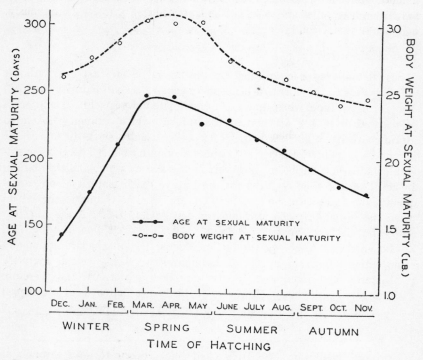

FIG. 9. Variation in age and body weight, at the time of sexual maturity, among birds hatched at different seasons. (After Upp and Thompson, 1927.)

a high record of production is not attained. The tabulated data (*Jeffrey and Platt, 1941*) show the relationship between hatching time and annual egg production.

MONTH HATCHED	ANNUAL PRODUCTION (number of eggs)
January	148
April	173
June	166
September	143
November	113

From the biological point of view, it appears that under ordinary flock management there is no particular advantage in hatching before April 1.

Birds hatched at the natural time will mature in October and will begin to lay in the fall without resting and molting. Such birds are very likely to have good records of egg-laying during the first year. However, under modern methods of management, which employ enriched diets and artificial illumination, it is sometimes advantageous to hatch chickens either very early in the spring (*Thompson, Philpott, and Page, 1936*) or late in the summer, in order to produce eggs unseasonally.

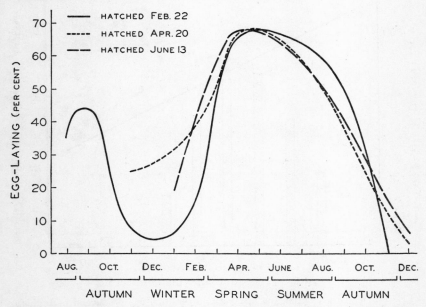

FIG. 10. Seasonal variations, during the first year, in the egg-laying of Leghorns hatched at different dates. (From the data of Buss, 1919.)

AGE AT SEXUAL MATURITY

The average age at maturity is not the same in different species of birds, such as doves, pigeons, and chickens. This fact indicates that genetic factors influence the rate of development. Chickens of various breeds, however, mature at approximately the same age. Leghorn pullets hatched in springtime usually begin to lay when they are about 6.5 months old (*Knox, 1930a*), as do Rhode Island Red pullets (*Hays, 1934*), and those of certain other varieties. Pullets of heavy breeds mature when they are about 1 month older, unless the flock has been selected for early egg-laying.

Chickens of the same breed, hatched simultaneously, do not become sexually mature at the same age. There may be more than 6 months' variation in the time required for sexual development. A few pullets become mature at the age of 4 months; others do not reach maturity until they are a year old (Fig. 11). The enormous differences in age at sexual maturity naturally result in individual variations in egg production during the first year (cf. Fig. 11). Except for a few premature individuals that lay intermittently, the first birds to mature lay the

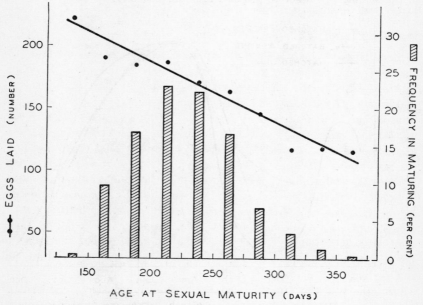

Fig. 11. Variation, during the first year, in the total egg production of Leghorns maturing at different ages. (After Hervey and Decker, 1926; Buster, 1927; Kempster, 1927; and Knox, 1930a.)

largest number of eggs. Birds that mature late invariably lay the fewest eggs.

It is common knowledge that feeding influences the rate of sexual development. Birds fed intensively mature rapidly and, according to Prentice, Baskett, and Robertson (*1930*), may reach their normal mature weight in 135 days, whereas birds fed an ordinary diet require 186 days to attain the same weight. These authors indicate that it is the mineral rather than the protein content of the ration that chiefly affects the rate of development. However, it has been demonstrated that maturity can be delayed by feeding insufficient protein, or vegetable protein alone (*Gutteridge, Pratt, and O'Neil, 1944*).

Schönberg and Ghoneim (*1946*) demonstrated that administration of stilbene (a synthetic estrogenic compound) can hasten development. The treated birds began to lay eggs at the age of 114 days, whereas the control group did not lay until 146 days old.

The use of additional illumination during the growing period can bring about sexual maturity at an earlier age (*Svetosarov and Streigh, 1940; Callenbach, Nicholas, and Murphy, 1944*). The effect of light on egg-laying will be discussed more fully later.

DATE OF LAYING FIRST EGG

Since chickens mature at different ages, there is a wide variation in the date when the first egg is laid. If birds are hatched from February to May—as is usually the case—some may begin to lay as early as July, and others not until March, that is, 8 months later.

From numerous studies (*Kempster, 1927; Maw and Maw, 1928; Knox, 1930b; and others*) it is evident that the date on which the first egg is laid provides a basis for the prediction of a bird's total egg production. Analysis of the data presented by Kempster (*1927*) shows that this date is especially valuable as an aid in foretelling fall and winter production.

Pullets that start to lay as early as July usually produce more eggs in the fall (up to November 1) than from November 1 to March 1. When the laying season begins too soon, a pause, known as fall or winter molt, materially reduces the egg record in winter. Pullets that start to lay in September, October, or November produce few eggs in the fall but perform well in winter and during the remainder of the laying year. During the first year of production, their records are highest. Pullets that begin to lay as late as February or March, however, produce few or no eggs in winter and make relatively poor records throughout the remainder of the laying year.

It has therefore been generally concluded that the optimum date for a pullet to lay her first egg is approximately November 1 (Fig. 12). Birds that begin to lay at this time usually show the highest and most economical production of eggs during the first, and sometimes the following, laying years.

Intensity of Laying

The rate at which a hen lays, or the intensity of her laying, vitally affects her total egg production. The rate may be rapid or slow, uniform or variable. It is determined by the number of successive days on which an egg is laid and by the length of the intervals between periods of laying.

The number of eggs laid successively may be conveniently called a cycle of laying (*Patterson, 1916; Atwood, 1929b*).   These cycles have no relation to the clutches laid by wild birds; nevertheless, they have been called "clutches" in many publications.   The frequency with which these cycles recur may be called the laying rhythm (*Goodale, 1915; Patterson, 1916; Warren, 1930a*).

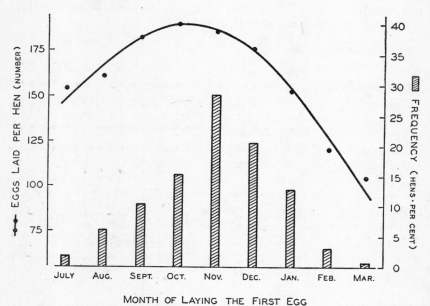

FIG. 12.   Variation in the total first-year egg production of birds that began laying in different months.   (After Kempster, 1927.)

Many domesticated hens have a tendency to lay in rhythmic cycles (*Hays, 1938*).   This regularity in laying habit remains unchanged throughout the pullet year and persists into the second year of production (*Atwood, 1929b*).   A bird may lay every other day, 2 days out of 3, or 3 out of 4; but, whatever the length of the cycle, the rhythm remains fairly constant (Fig. 13).   However, cycles and rhythm are imperfect and scarcely discernible when the rate of egg production is low.

The length of the cycle is due to the maintenance of a uniform interval between eggs.   This interval is usually somewhat longer than 24 hours, and therefore each successive egg is laid a little later in the day.   Eventually, the hour falls very late in the afternoon, and the egg is not laid.

The cycle is thus terminated. The eggs of a 4-egg cycle, for example, may be laid at 9 A.M., 11 A.M., 1 P.M., and 3 P.M. on succeeding days (*Atwood, 1929a*). The eggs of long cycles, however, are usually laid at approximately the same time each day, except on a few days at the beginning and the end of the cycle. The longest known laying cycle is accredited to a White Leghorn hen that laid an egg every day for 235 days (*Robinson, 1930*). Instances have been cited (*Drew, 1907*) of hens

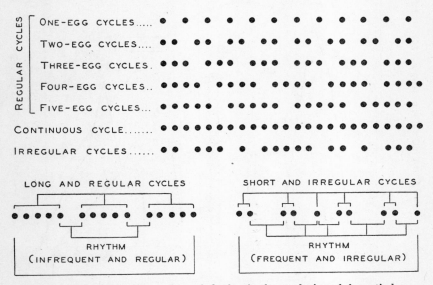

Fig. 13. Diagram showing cycles and rhythm in the egg-laying of domestic hens.

having laid 2 normal, fertile eggs in 1 day. Usually 1 egg was delivered in the morning and the other in the afternoon. Within 1 month, one hen made five deliveries of 2 eggs in a single day and failed to lay on only 3 days; in other words, she laid 33 eggs during 31 days.

It has been generally thought that the egg which is scheduled to be laid late in the day is held in the oviduct overnight and deposited early in the morning, thus becoming the first egg in a new cycle. However, it is more probable that the intervals between eggs are determined by ovulation intervals (see Chapter 4, "Formation") rather than by oviducal factors. The interval is characteristic of the functioning ovary; even hens that do not lay because of pathological conditions may still ovulate and visit the nest regularly (*Goodale, 1918*). Scott and Warren (*1936*) demonstrated that each succeeding egg of a cycle can be felt in the oviduct within 5 hours after the oviposition of the preceding egg, but that the first egg in the cycle cannot be detected until nearly 20 hours after the laying of the last egg of the previous cycle. A delay in ovulation is thus indicated as causing the end of the cycle.

It is interesting to note that throughout the year the ovulation interval of the ringdove evidently varies in accordance with the changing number of daylight hours. This bird lays the first egg of its 2-egg cycle in the

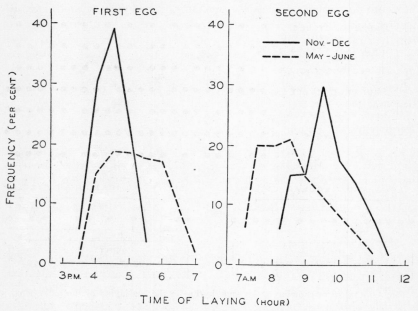

FIG. 14. Seasonal variation in the time of day at which the ringdove lays the first and second eggs of its cycle. (After Schooley and Riddle, 1944.)

afternoon, and the second egg in the morning of the second day following. Ovulation occurs during the night. As the length of the days increases in the spring, the first egg is laid later and the second egg earlier (Fig. 14).

Intensity of laying is not entirely independent of environmental conditions. The poor weather associated with certain seasons tends to reduce the intensity of laying and consequently the mean length of the cycle. Riddle (*1925a*) presents data which show that the common pigeon lays the greatest percentage of 1-egg clutches during January and February, and the smallest percentage during July and August. Since the characteristic clutch of the pigeon consists of 2 eggs, it seems probable that weather conditions can adversely affect this bird's rate of laying and consequently reduce the mean clutch size.

Warren (*1930a*) demonstrated that disturbing the hen, as by transferring her to new living quarters, may result in shortening the cycle and in lengthening the intervals between cycles. Ovulation may even cease if the hen is subjected to sufficiently adverse conditions (*Stieve, 1918*). Many internal and external factors may have an unfavorable influence on normal reproductive processes. On the other hand, environmental conditions sometimes stimulate egg-laying in such a manner as to lengthen the cycle. For example, artificial illumination may do so when it is used to increase the daily hours of exposure to light (*Byerly and Moore, 1941*).

As will be seen in the following discussion, total egg production also depends upon the persistency with which a hen lays.

Persistency in Laying

A hen's relative persistency in laying is indicated by her ability to produce eggs more or less continuously throughout the year. Persistency may be measured by the extent to which laying is interrupted by winter pause, broodiness, and molting.

WINTER PAUSE

Many of the pullets that begin to lay in the fall stop laying for a period of a few days to 2 months, or more, during the winter. This nonlaying period is called "winter pause" and usually occurs between November 1 and March 1 (Fig. 15).

The duration of the winter pause may depend upon environment as well as upon inheritance (*Hays, 1924–25*). Many environmental factors that affect the length of the pause are probably beyond the control of the breeder. On the other hand, a winter pause may sometimes be brought on by disease, abrupt changes of feed, or moving to new quarters.

The winter pause may or may not constitute a continuous period of nonlaying; usually there is intermittent egg-laying between several short or long pauses (*Goodale, 1918*). The long pauses are those exceeding 10 days in length, and the short pauses are those lasting for 2 to 10 days. The pauses that occur at frequent intervals may be called "multipauses." Long pauses occur chiefly in December and January, and to a lesser extent in February; but short pauses are distributed in a fairly uniform manner throughout the winter.

High-producing pullets may not pause at all in winter. After they begin to lay, their egg production increases rapidly without interruption until it reaches a maximum rate, at which it continues for a period of several months during the spring and summer. In modern, rigidly selected flocks, there is no pronounced winter pause even among hens more than a year old.

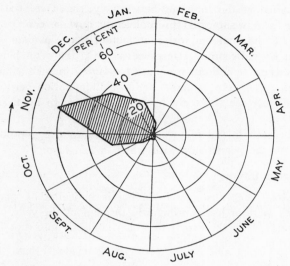

Fig. 15. Average percentage of the flock that undergoes a pause in laying each month. (After data collected on Wyandottes, Rhode Island Reds, and Plymouth Rocks, by Dunn, 1924a, 1924b, 1924c.)

BROODINESS

Under natural conditions, the ancestor of the domestic hen laid only 25 to 30 eggs during a year, that is, two clutches. After laying the first clutch, she spent considerable time in hatching and rearing her chicks. Then, if time and weather permitted, she laid another clutch of eggs and hatched and brought up a second family.

Domestication has produced profound changes in the fowl's habits. Through selective breeding, adequate feeding, and special methods of management, the hen, instead of laying only a few eggs and hatching them herself, has become a veritable egg-making machine with little or no time for rest or the rearing of a family.

In spite of the fact that the natural tendency to incubate eggs has been very largely bred out, it still recurs cyclically in domesticated fowl. Consequently all breeds of fowls exhibit broodiness in some degree. The Asiatic or meat breeds are all intensely broody; the American and many European egg-laying breeds show considerable broodiness; and the Mediterranean egg-laying breeds, although almost nonbroody, always produce some broody individuals.

Each broody period, if allowed to run its course, lasts about 20 days, a length of time corresponding to the chicken's natural incubation period. Usually, when a hen becomes broody, egg-laying ceases. There is, therefore, a 3 weeks' loss in production during each recurrence of broodiness. The manner in which broodiness may decrease the number of eggs laid annually by chickens (*Hays and Sanborn, 1926*) and by turkey hens (*Parker and Barton, 1945*) may be seen in the table.

	ANNUAL EGG RECORD	
TIMES BROODY	CHICKENS	TURKEYS
0	181	71
1	178	56
2	157	53

The tendency toward broodiness is considered to be inherited (*Bateson, 1902; Hurst, 1905; Goodale, Sanborn, and White, 1920; Punnett and Bailey, 1920; Hurst, 1921; Roberts and Card, 1934*). Because of selective breeding for nonbroodiness, the average number of days that each hen spends in broodiness per year (*Kirkpatrick and Card, 1917*) is now relatively small. From the figures given, it is evident that Leghorns (a

BREED	ANNUAL BROODINESS (days)
White Leghorn	4.0
Plymouth Rock	26.0
Wyandotte	27.4
Rhode Island Red	39.5

Mediterranean breed) are almost nonbroody, whereas Plymouth Rocks, Wyandottes, and Rhode Island Reds (American breeds) lose, on an average, about 1 month out of the year.

Within each breed, individual variability in the tendency to become broody is also very great. Some hens do not brood at all and lay eggs continuously throughout the spring and summer, whereas others have as many as thirteen broody periods in a season (*Hays and Sanborn, 1926*). Presumably about 60 per cent of all hens of the popular American breeds are broody, as compared with 10 per cent among Mediterranean breeds (*Hannas, 1920*).

Broodiness may be induced by several environmental factors, such as the presence of chicks (*Réaumur, 1749*), the accumulation of eggs in the nest (*Punnett, 1923, p. 180*), high temperature (about 90° F.), or darkness (*Burrows and Byerly, 1938*). It has also been shown that injection of prolactin, an anterior pituitary hormone, induces broodiness in doves and pigeons (*Riddle, Bates, and Lahr, 1935; Riddle, 1938a*). Certain steroids, among them testosterone (*Riddle and Lahr, 1944*), may also do so, although these substances are not normally involved in evoking the broody instinct.

The resumption of egg-laying after very prolonged broodiness may be still further delayed by molting.

MOLT

Molting, or the shedding and replacing of the feathers, is a natural physiological phenomenon in birds. In general, it is similar to the shedding of hair by an animal.

The time and duration of complete molt vary considerably with the individual bird. The majority of fowls shed feathers and acquire new ones during 6 to 12 months of the year, but the peak of the true molt usually occurs in fall or in winter. Some hens begin to molt in August, others not until November; but all complete their molt by February or March (*Marble, 1930a*).

The onset of the molt is associated with a decline in the functional activity of the reproductive organs. When the molt begins, egg-laying ceases, and there is a marked regression in the secondary sexual characters, such as the comb, for example (*Greenwood, 1936*). Molting therefore corresponds to a period of physiological rest.

It has been observed (*Rice, Nixon, and Rogers, 1908; Marble, 1930a; Hays and Sanborn, 1930*) that the most persistent layers begin their molt latest and require the shortest time to shed their old feathers and grow new ones. Moreover, a few birds that do not exhibit complete molt may lay without lengthy interruptions for 2 or more years. Consequently, there is a close relationship between the duration of molt and the fall, winter, and annual egg production of both hens and pullets. Marble (*1930a*) found that the average annual egg production for hens beginning to molt in September was 178 eggs, whereas for those beginning in November it was 228 eggs. Similarly, pullets laying an average of 140 eggs go through a long molting period; those laying 220 eggs, a late, short one, or none at all (*Goodale, Sanborn, and White, 1920; Marble, 1930a*).

It also has been noted that early and prolonged molting can be induced by starvation (*Rice, Nixon, and Rogers, 1908*), or by an abrupt decrease in the daily amount of artificial light (*Lesher and Kendeigh, 1941; Larionov, 1941*). Derivatives of thyroxine induce molting in chickens and cause depigmentation of the feathers (*Zavadovski, Titaiev, and Faierniark, 1929*). Iodine, although it does not change the onset or duration of the molt, may hasten the loss of the feathers (*Klein, 1933*). Methionine has practically no effect on the length of the molting period (*Taylor and Russell, 1943*). In thyroidectomized and castrated pigeons, the molting process may deviate very slightly from normal (*Woitkewitsch, 1940*).

Physical Condition

The ability of the hen to lay a large number of eggs depends to a great extent upon her inherent physical constitution and vigor. However good her inherited capacity for intense and persistent egg-laying, it will be realized only if she is a bird of superior vitality. If she is not, the strenuous demands of sustained and rapid egg production may adversely affect her general health, lower her resistance to disease, and shorten her life.

The importance of the hen's physical condition is revealed by several relationships. Egg-laying is affected by the bird's body weight and by the presence of disease. The mortality rate in the flock also appears to be correlated with productive capacity.

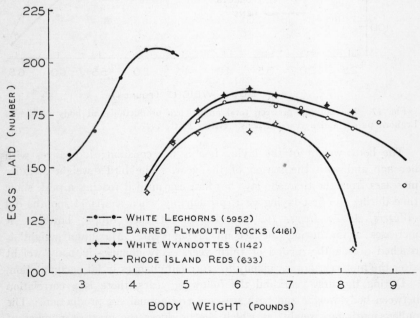

Fig. 16. Correlation between total annual egg production and body weight in four breeds of fowl. (After Anonymous, 1929a.)

BODY WEIGHT

Body weight is a criterion of good health, especially in the hen. An underweight bird cannot endure long and persistent laying, and an excessively fat hen also stops laying. It is therefore important that the hen's weight be maintained at the optimum.

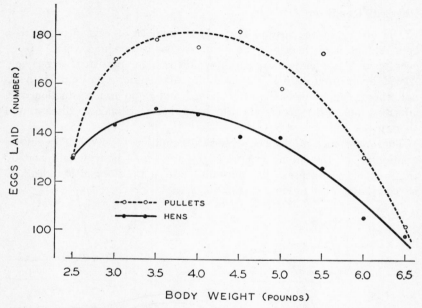

FIG. 17. Relationship between total annual egg production and body weight of Leghorn pullets and 2-year-old hens. (After Platt, 1927.)

The body weight of the laying hen is not constant; it changes with her age and with the season of the year. The bird's weight steadily increases from the time she lays her first egg until it reaches a peak some time during the first laying year, presumably when she is 11 months old (*Waters, 1937; Hays, 1939*). Her weight then drops; however, it increases again during the second laying year. Maximum weight is reached during the second or third year. Any changes in body weight after the first year appear chiefly to represent changes in fat accumulation.

During the first year and the following years, there is a correlation between body weight and both seasonal and annual egg production. The pullet's optimum weight, at which she produces the greatest number of eggs (*Platt, 1927*), is, for the Leghorn, 4.0 to 4.5 lb.; for the Rhode

Island Red, 5.5 to 6.0 lb.; and for the Plymouth Rock and the Wyan-
dotte, 6.0 to 6.5 lb. (Fig. 16). It is interesting to observe that there
is greater variation in egg production among pullets of different body
weight than among 2-year-old hens of different weight (Fig. 17).

Seasonal changes in weight are very regular and presumably of identical character in all breeds of fowl. In general, a bird weighs the

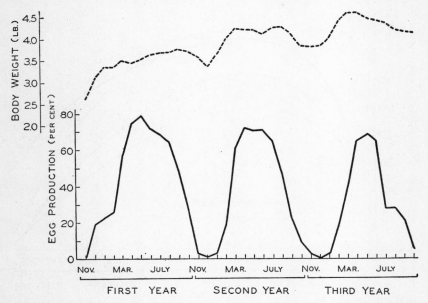

Fɪɢ. 18. Relationship between monthly egg production of a flock of Leghorns
and seasonal changes in average body weight during 3 consecutive years. (After
Atwood and Clark, 1930b.)

most at the time of heavy egg production during her first and second
laying years and often during the following years (Fig. 18). Body
weight is usually lowest when egg-laying is at a minimum, that is, in
the winter.

PRESENCE OF DISEASES

The relationship between the presence of a diseased condition in a
bird and her rate of egg-laying cannot be easily ascertained. However,
it is reasonable to assume that various factors that may bring about an
unhealthy state in the organism can diminish egg-laying.

For example, egg-laying may decrease when birds are infected with
bacillary white diarrhea (caused by *Salmonella pullorum*). In one flock

of pullets, the average annual production of positive reactors to the agglutination test for *Salmonella pullorum* was 160 eggs, whereas that of nonreactors was 222 eggs (*Asmundson and Biely, 1928, 1930*). A difference of 62 eggs per year is significant. Moreover, the egg-laying performance of the nonreactors was considerably better than that of the reactors throughout all seasons of the year (Fig. 19).

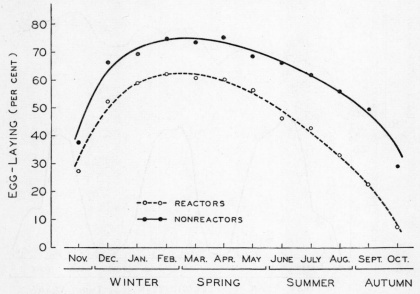

FIG. 19. Egg-laying by reactors and nonreactors to the test for bacillary white diarrhea (infection with *Salmonella pullorum*) in a flock of 689 fowl. (After Asmundson and Biely, 1930.)

"Pullet disease," an acute disease of the kidneys and liver of young chickens, has been shown to curtail egg production to a great extent. During a period of 6 weeks, one flock of chickens afflicted with this disease was found to lay an average of about 2 eggs per bird (*Scott, Jungherr, and Matterson, 1944a*).

It has also been observed (*Hinshaw, Jones, and Graybill, 1931*) that infectious laryngotracheitis (inflammation of the larynx and trachea) may cause a noticeable drop in egg-laying. In one instance, the disease reached its maximum about 15 days after its outbreak, as indicated by the peak of the mortality curve, but laying was not resumed at the normal rate until more than 30 days later (Fig. 20).

In some instances, a low rate of egg-laying may be the result of an infection contracted during the growing period. Some diseases, such as chronic coccidiosis, will retard growth and delay sexual maturity. It has been shown (*Mayhew, 1934*) that artificial inoculation of young birds with coccidia organisms can eventually reduce the number of eggs laid during the first year of production by more than 19 per cent.

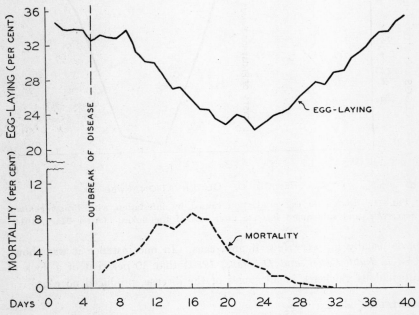

FIG. 20. Egg-laying and mortality rate in a flock of fowl, subsequent to outbreak of disease (infectious laryngotracheitis). (After Hinshaw, Jones, and Graybill, 1931.)

It has also been found that hens infested with the oviduct fluke, *Prosthogonimus macrorchis*, may lay only about one-tenth of the number of eggs laid by hens in good health (*Macy, 1934*). Infestation with the cysts may be followed by a sudden drop in egg production within a week, which is the time required for the parasites to mature. Since the flukes are usually lost between the third and the fifth week, production may recover subsequently (Fig. 21). Infestation has been noted to have similar effects on the egg-laying of ring-necked pheasants (*Macy, 1940*).

One of the most serious avian diseases is tuberculosis. Tuberculous

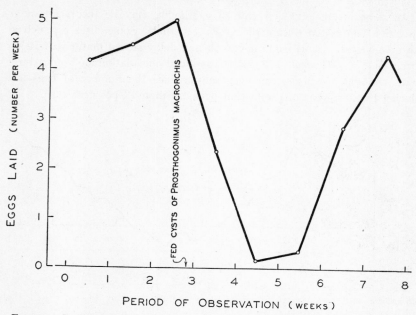

Fig. 21. Decline in egg production caused by infestation with *Prosthogonimus macrorchis,* and subsequent increase after loss of the flukes. (After Macy, 1934.)

birds usually lay very few, if any, eggs. In one instance it was shown (*Fitch, Lubbehusen, and Dikmans, 1924*) that 30 per cent of a flock of tuberculous chickens, in all stages of the disease, produced no eggs.

MORTALITY RATE

The mortality rate in a laying flock bears a definite relationship to the rate of egg production. It is apparent that the intense metabolic activity necessary for the production of many eggs may eventually have some ill effects, and may even reduce the length of the hen's life.

Harris (*1926, 1927*) observed that birds that died during their first year laid at a lower monthly rate than birds that survived (Fig. 22). The inferior performance of short-lived birds may be attributed to several interrelated factors. Decreased egg production may be the result of low vitality, which indirectly causes the death of the hen. It is also possible that illness, as distinct from an inherently poor constitution, reduces her egg-laying capacity, and shortens her life. On the other hand, death is sometimes the direct result of the conditions and circumstances that cause hens to lay many eggs. This is indicated by the

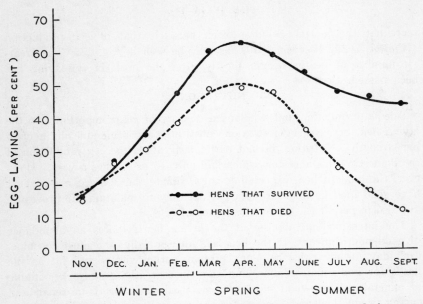

FIG. 22. Egg-laying performance of birds that survived their first laying year, compared with the egg production of those that died during the same period of their lives. (After Harris, 1927.)

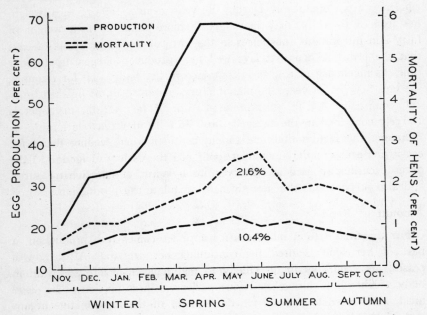

FIG. 23. Seasonal changes in egg-laying and in the incidence of mortality in two flocks of fowl. (After Winton, 1929, and Weaver, 1930.)

fact that the mortality curve often parallels that of egg production (*Winton, 1929; Weaver, 1930*), as may be seen in Fig. 23. The greatest number of deaths occurs shortly after the peak in egg-laying has been passed.

NUTRITION

Adequate nutrition and a balanced diet are of great importance to the laying hen, because egg formation constitutes a considerable drain upon her strength. Nutrition did not particularly affect the egg production of the domestic hen's ancestor, which laid only 20 to 30 eggs a year. However, the modern hen, expected to approximate the 300-egg mark, must have good food in amounts sufficient not only to maintain life processes, but also to permit the continuous laying of eggs.

The intensified metabolism of the laying hen requires an abundance of good nourishment. If the hen lays an egg daily, she must consume, each day, nearly twice as much protein, carbohydrate, and fat as she needs for body maintenance, and she must greatly increase her intake of minerals. These food substances are used, first, for tissue repair and energy production, and, second, for the formation of eggs. A deficiency in the amount or variety of food, therefore, directly affects egg-laying.

Egg production by the domestic hen is a remarkable example of efficiency in an organism. During the year, even a 150-egg bird, weighing from 3 to 4 lb., may manufacture more than her own weight of fatty and nitrogenous substances in the form of the 18 or 19 lb. of eggs that she lays. A hen that lays well may produce, in eggs, up to four times as much dry matter as is contained in her body. If, for example, she lays 300 eggs a year, she manufactures about 38 lb. of finished product, comprising 4.6 lb. of protein, 4.1 lb. of fat, 4.2 lb. of minerals (largely contained in the eggshell), and 25.1 lb. of water.

Among the factors that one should consider when feeding flocks for egg-laying are: (1) the amount of food; (2) the variety of food; (3) the drugs occasionally used as laxative, prophylactic, or curative measures, and also preparations that are given to stimulate egg production.

Amount of Food

For the first 6 to 8 months after a pullet comes into production, a part of her food is used for continuing her growth. After growth ceases, her food contributes chiefly to maintenance of life and gain in body weight; but, during egg-laying periods, a large portion of her food must be diverted to provide for the activity of the reproductive organs and for the deposition of nutrients in the egg.

It has been determined that the individual requirements for nonlaying hens of various breeds are from 70 to 80 gm. of food per day. When in full egg production, the same birds each require from 125 to 135 gm. of food daily. It is therefore evident that more than one-half the food is used for maintenance, and that the remainder, about 55 gm., is used for the formation of the egg. The production of 200 or 300 eggs per year would require about 22 to 33 lb. of extra food.

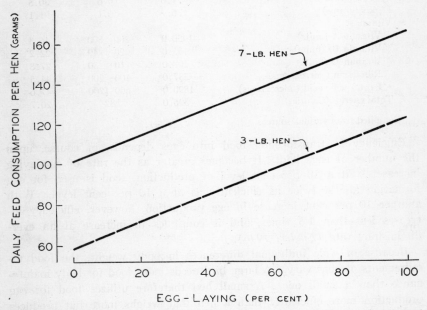

Fig. 24. Relationship between feed consumption and egg production of 3-lb. and 7-lb. hens. (After Byerly, 1941.)

Figure 24 indicates the extent to which food consumption rises as the percentage monthly egg production increases. The illustration also shows that a 7-lb. hen requires a proportionately greater amount of food than a 3-lb. hen laying at the same rate.

The biological efficiency of the hen in transforming raw food materials into an egg has been calculated to be about 54 per cent (*Brody, Funk, and Kempster, 1938*); that is, it is equivalent to the efficiency of the cow in producing milk.

How efficiently the hen converts various organic substances of her food into egg materials is shown in Table 2. It may be seen that the recovery of lipids and vitamins in the egg is especially good.

TABLE 2

EFFICIENCY OF THE HEN IN CONVERTING ORGANIC NUTRIENTS OF THE FEED
INTO EGG NUTRIENTS *

	Nutrients in 135 Gm. of Feed	Nutrients in 51.6 Gm. of Egg Contents	Recovery of Nutrients (per cent)
Total organic matter, dry (grams)	126.0	13.2	10.5
Protein (grams)	21.6	6.6	30.5
Lipids (grams)	6.7	6.1	91.1
Vitamins:			
Vitamin A (units)	1080.0	200– 800	46.3
Vitamin B_1 (units)	135.0	20– 40	22.2
Vitamin D (units)	108.0	10– 50	27.8
Riboflavin (micrograms)	337.0	100– 200	44.5
Pantothenic acid (micrograms)	1890.0	600–1200	47.6
Total energy (Calories)	396.0	82	20.7

* Compiled from various sources.

Efficiency in transforming food into eggs depends, of course, upon the number of eggs laid. It becomes greater as the rate of egg-laying increases. At a 20 per cent level of production, food is used for egg formation almost twice as efficiently as at a 10 per cent level. With another 10 per cent increase in egg production, however, efficiency increases less than 1.5 times, and it continues to increase at an ever-diminishing rate (*Engler, 1936*).

Because of wide individual differences in body weight, the food requirements of hens vary. A large bird needs more food for body maintenance than a small one. A small hen therefore utilizes food for egg production more efficiently than a hen that weighs more but produces the same number of eggs per year. Big birds, however, may compensate for this difference by laying larger eggs.

Efficiency in converting food into eggs is obviously of economic importance. If her production is less than 100 or 150 eggs annually, the domestic hen is a financial loss.

The amount of food consumed each month by the hen varies throughout the year (Fig. 25) and is greatest during the periods of highest egg production. Changes in food consumption are also associated with fluctuations in body weight, since the hen normally loses weight during a long period of laying and regains it at times of rest.

Heywang (*1940*) limited a group of heavily laying birds to 75 per cent of the amount of food consumed by another group and noted that the egg production of the restricted group markedly decreased. It is evident that, to obtain uninterrupted egg-laying, the food supply must be continuous. One day of food deprivation may reduce the rate of laying. A brief period of starvation may cause the complete cessation of egg-laying, followed by a molt.

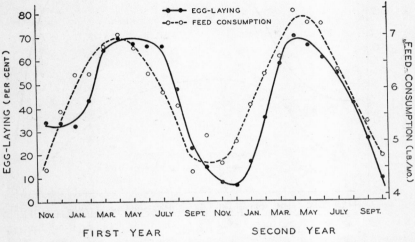

FIG. 25. Seasonal changes in egg-laying and feed consumption of 600 Leghorns. (After Lewis, Hannas, and Wene, 1919.)

Variety of Food

To encourage the hen to lay more eggs, her diet must include the essential nutrients. It must contain an adequate quantity of proteins, carbohydrates, fats, water, minerals, and vitamins.

Most natural foods contain all these nutrients, but not in the same proportions. Unless carefully compounded, the hen's diet may be deficient in any one of the essential ingredients; and such a deficiency may lower her reproductive activity and therefore her rate of laying.

The source of the ingredients used is also of considerable importance. Proteins, for example, may be of animal or vegetable origin. In feeding for egg production, it has been found (*Romanoff, 1926; and others*) that protein concentrates of animal origin, such as meat scrap, tankage, fish meal, or milk products, are more effective than the vegetable proteins (cottonseed meal, soybean-oil meal, peanut meal). Usually a combination of animal and vegetable proteins is best.

A lack of animal protein, as well as an inadequate amount of any protein (*Zöllner, 1932; and others*), in the diet of the laying hen will reduce the number of eggs she lays (Fig. 26). On the other hand, too

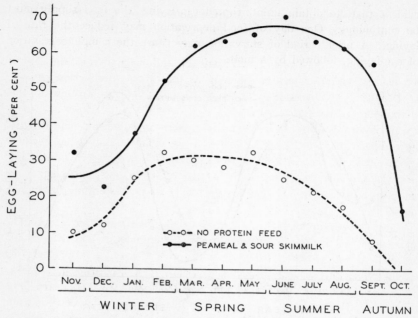

Fig. 26. Effect of dietary deficiency of protein on rate of egg-laying by Leghorns.
(After Parkhurst, 1924.)

liberal a use of animal proteins, besides being wasteful, may have some
detrimental effect on the organism.

The table shows how various proportions of animal protein in the
hen's diet may affect egg production (*Kempster, 1924*).

Protein Concentrate (per cent)	Eggs Laid (number)
0	61
5	105
10	112
15	118
20	120
25	117
34	126

The importance of supplying the hen with sufficient protein of good
quality is also indicated by the fact that the hen is apparently unable
to synthesize such amino acids as arginine, histidine, isoleucine, leucine,
lysine, methionine, phenylalanine, threonine, tryptophane, and valine
(*Munks, Robinson, Beach, and Williams, 1945*). Since each egg that she

lays contains these amino acids in large amounts (see Chapter 6, "Chemical Composition"), her diet must obviously serve as the source of these amino acids in the egg proteins.

In addition, it has been found that the sulfur balance of the laying hen usually becomes negative as soon as the rate of production exceeds 50 per cent (*Holman, Taylor, and Russell, 1945*). Protein is the chief dietary source of sulfur.

Carbohydrates and fats are usually present in sufficient quantities in the ordinary diet of the laying hen and are deficient only when feeding is restricted. A diet of low fat content tends to depress egg production (*Russell, Taylor, and Walker, 1941*), but it does not cause laying to cease abruptly, as would the total absence of an essential nutritional factor.

Water is very important to the laying hen. Although water accounts for nearly two-thirds of the body weight in animals, a loss of only 10 per cent of water, through dehydration and excretion, usually results in serious physical disorders. A hen's daily loss of water exceeds this level. Each egg laid requires about 38 gm. of water. When a hen is fed only on solid food, the water supplied with grains and the water derived from the oxidation of the digested food (metabolic water) are insufficient in quantity to compensate for the water lost. It has been observed that the laying hen drinks at least 150 gm. of water daily (*Henderson, 1945; and unpublished data at Cornell University*). As the figures show, there is a definite correlation between her egg production and her water intake. (These figures have been corrected for 10 per cent loss through surface evaporation in containers.)

ANNUAL EGG PRODUCTION	DAILY INTAKE OF WATER (grams)
179	146
216	174
230	192
243	200

It has also been demonstrated that, if a flock of hens is supplied with a limited amount of drinking water and allowed to drink during only 2 hours each day, egg-laying is suddenly reduced by about 50 per cent (*Quisenberry, 1915–16*).

The demand for minerals is perhaps the greatest that is made on the laying hen. Unlike a cow producing milk, the hen cannot easily draw upon her body stores of minerals to compensate even temporarily for the deficiencies in her diet. The hen's body could supply certain essential minerals in sufficient quantities for only a very limited number of eggs (Table 3).

TABLE 3

NUMBER OF EGGS THAT COULD BE PRODUCED FROM MINERALS STORED IN THE HEN'S BODY *

Essential Minerals	Minerals in the Hen's Body (grams)	Minerals in an Egg (grams)	Number of Eggs That Could Be Produced
Calcium	21.115	1.975	11
Phosphorus	12.298	0.115	107
Sulfur	3.902	0.114	34
Chlorine	2.588	0.088	29
Sodium	1.708	0.073	23
Potassium	4.196	0.067	63
Magnesium	0.998	0.027	37
Iron	0.162	0.002	81

* After Halnan (*1936*).

The minerals for which there is greatest demand are calcium, sodium, chlorine, manganese, and sulfur. In particular, deficiencies of calcium, sodium, and possibly chlorine may be limiting factors in egg-laying. Calcium and phosphorus (*Miller and Bearse, 1934*) are the most essential minerals for egg formation.

It has often been demonstrated (*Buckner and Martin, 1920; Buckner, Martin, and Peter, 1923; Tully and Franke, 1935*) that the lack of calcium in a utilizable form in the daily diet of a laying hen causes a

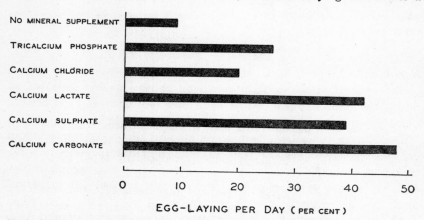

FIG. 27. Daily rate of egg-laying, as affected by various sources of dietary calcium. (After Buckner, Martin, and Peter, 1928.)

marked decrease in egg production (Fig. 27). The best source of calcium appears to be calcium carbonate, in such natural forms as limestone or oyster shell.

There is some evidence that manganese-deficient diets may somewhat decrease egg-laying by chickens (*Gallup and Norris, 1939*). The influence of iodine on egg production has been studied extensively (*Klein,*

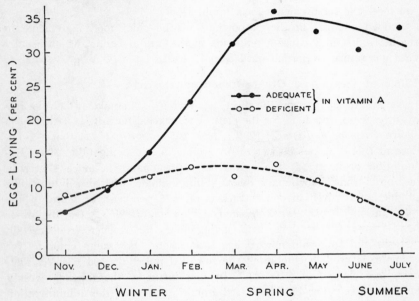

Fig. 28. Effect of dietary deficiency of vitamin A on egg-laying by fowl. (After Payne and Hughes, 1933.)

1933; Zaitschek, 1934; Asmundson, Almquist, and Klose, 1936; Lee, Hamilton, and Henry, 1936), but without conclusive results. The amount of this element in the egg, however, may be considerably increased by raising the level of iodine in the hen's diet (see Chapter 6, "Chemical Composition").

The vitamin content of the hen's ration also affects the production of eggs. Lack of vitamins, or poor assimilation of them, results in a decrease in the number of eggs laid. However, when egg production declines because of vitamin deficiency, it may be returned to normal levels by supplementary, vitamin-rich feeding.

Among the vitamins known to influence egg-laying are: vitamin A, as shown in Fig. 28 (*Sherwood and Fraps, 1932; Rubin and Bird, 1942*);

vitamin B_6, or pyridoxine (*Cravens, Sebesta, Halpin, and Hart, 1943, 1946*); and vitamin D (*Martin, Erikson, and Insko, 1930*). Vitamin C is not essential in the hen's ration. Ordinarily, alfalfa (particularly the leafy portion) and yellow corn are the feeds that supply vitamin A. Common grains contribute adequate amounts of vitamin B. Enough sunshine at all seasons, or such substitutes as ultraviolet light, cod-liver oil, or certain synthetic preparations irradiated with ultraviolet light, can meet the requirements for vitamin D.

Sudden changes in the proportions of the chief food ingredients in the hen's diet may cause a reduction in the rate of laying, which usually does not return to normal until 8 to 10 weeks later (*O'Neil, 1946*).

Effect of Drugs and Other Preparations

Among drugs known to affect egg-laying is kamala, which is frequently given with food for the control of internal parasites. It has been shown (*Atwood and Clark, 1929; Maw, 1934; Smith, 1943*) that a 1-gm. dose of this drug per bird causes a decline in egg-laying (Fig. 29). Resumption of laying at the normal rate presumably is much slower in old birds than in younger ones. Sulfanilamide will also inhibit egg production if given in concentrations of 0.25 per cent, or more, of the total ration (*Scott, Jungherr, and Matterson, 1944b*). On the other hand, some preparations help laying birds to recover from certain diseases and to regain normal egg production. Potassium chloride, for example, is used in combating the so-called pullet disease (*Scott, Jungherr, and Matterson, 1944a*). In one instance, the average weekly number of eggs per bird increased from 2 to 4 after the administration of potassium chloride.

Numerous attempts have been made to increase egg-laying by feeding hormonal tissues, such as thyroid (*Crew and Huxley, 1923; Cole and Hutt, 1928; Winchester, 1939*), pituitary (*Clark, 1915; Pearl, 1916; Simpson, 1920, 1923; Gutowska, 1931*), ovary (*Vaček, 1935*), and corpus luteum (*Pearl, 1916*), but frequently without apparent success. Protamone, a protein of high thyroid activity, seems to improve egg production when fed throughout the year at the level of 5 to 10 gm. per 100 lb. of feed (*Turner, Irwin, and Reineke, 1945*), and there is evidence that the seasonal nature of egg-laying may be due in part to a decreased secretion of thyroxine during the summer (*Turner, Kempster, Hall, and Reineke, 1945*). Furthermore, injection of anterior pituitary extract (*Noether, 1931; Phillips, 1943*) or follicular hormone (*Asmundson, 1931a*) may produce a temporary stimulation of the ovary, but it is usually followed by cessation of laying.

Schönberg and Ghoneim (*1946*) studied the influence of feeding various synthetic estrogenic compounds (triphenylethylene, stilbestrol, triphenylchloroethylene, and stilbene). Of these, only stilbene caused an increase in egg production.

Cow manure, because of its content of an androgenic substance, may cause a marked decrease in egg production when ingested by fowl. It

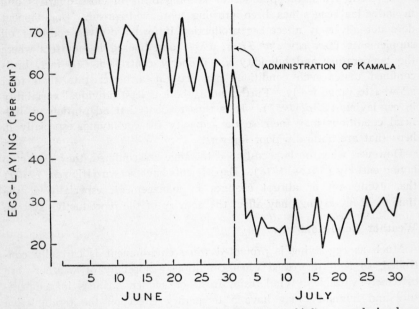

Fig. 29. Effect of a single dose of kamala (1 gm. per bird) on egg-laying by a flock of 165 Leghorns. (After Atwood and Clark, 1929.)

is therefore not advisable to allow laying hens free access to a range shared by cattle. A test of 48 weeks' duration showed that a group of fowl laid a significantly smaller number of eggs when fed a good practical all-mash diet to which was added 8 per cent dried cow manure of high androgenic potency (*Whitson, Titus, and Bird, 1946*).

ENVIRONMENT

In becoming domesticated, chickens and other birds have adapted themselves to the unnatural circumstances imposed by man. The hen, native to equatorial regions, thrives very well in temperate zones, largely because man has provided surroundings that promoted acclimatization. Her greatly increased capacity for egg-laying is one of the many results of the care she has received. The fact that the physical elements of her environment may affect her egg production, both favorably and unfavorably, may clearly be seen by the influence of weather conditions and various methods of management.

Confinement

In recent years, the practice of keeping birds in close quarters and in individual cages has been growing more widespread. Confinement does not adversely affect egg production if fowl are given cod-liver oil supplements (*Godfrey and Titus, 1934*) and housed in quarters where the temperature and humidity are properly controlled. In fact, birds confined under such conditions may lay at a better rate than fowl allowed to range freely. Furthermore, there is less individual variability in egg-laying (*Bird, 1937*). It is thus evident that adaptation to artificial conditions may increase the capacity for egg-laying, especially in hens that are naturally poor foragers.

However, when birds accustomed to range are confined, they may stop laying entirely (*Stieve, 1918*). Experiments have shown (*Warren, 1930a*) that excitement or abrupt changes in management, especially at the time of heavy laying, may affect the activity of the reproductive system.

Weather Conditions

Much as egg-laying is promoted when environment is carefully controlled, so is it depressed when conditions—especially of weather—are extremely variable. Wind, rain, frost, or sudden changes in temperature and humidity may have a temporary effect on the reproductive activity of the hen (*Fronda, 1935*), either by restricting exercise, causing physical discomfort, lowering vitality, or reducing food consumption.

In temperate climates, the onset of cold weather early in winter brings a decline in the egg production of birds housed in ordinary buildings. As soon as the weather is stabilized, the rate of laying improves. Thereafter, it varies according to changes in temperature and in humidity. Weather conditions therefore have considerable influence on both winter (*C. Smith, 1930*) and annual egg production (*Bruckner, 1935*).

Variations in barometric pressure presumably have little immediate effect on egg-laying (*Card, 1917; Lapland, 1924*). It has been observed, however, that intermittent housing of birds in mountain regions (6000 ft. above sea level) may result in more sustained egg-laying throughout the year and in an increase in total egg production (*Vezzani, 1939*).

Artificial Light

In 1907, Schafer suggested that the bird's sex cycle was related to the annual variation in the hours of daylight. His observations were later corroborated by the results of various experiments on wild birds (*Bissonnette, 1932a; Rowan, 1928, 1936*). In the temperate zones, there is clearly a correlation between the seasonal variation in the hen's egg

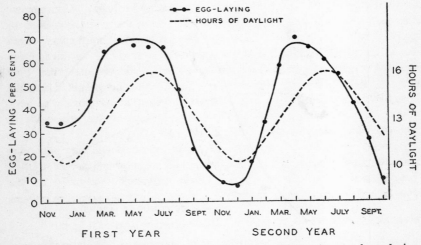

FIG. 30. Parallel seasonal changes in hours of daylight and rate of egg-laying. (After Lewis, Hannas, and Wene, 1919.)

production and the changes in the length of the daylight period throughout the year (Fig. 30).

Increases in the daily amount of light have so stimulating an effect on the hen's reproductive system that it is possible to promote a high rate of egg-laying in winter by augmenting natural daylight with artificial illumination (*Dougherty, 1922; Gutteridge, Bird, MacGregor, and Pratt, 1944; Nicholas, Callenbach, and Murphy, 1944*). A greater improvement can be obtained in the performance of hens than in that of pullets, since hens usually begin to lay much later in winter (*Curtis, 1920*). The initial gain in production, however, is usually followed by an eventual decline, although it appears that a high rate of laying can be maintained by alternating 14-hour periods of light with 12-hour periods of darkness, in synchronization with the hen's usual 26-hour ovulation cycle (*Byerly and Moore, 1941*). Continuous light for 24 hours a day stimulates pullets as well as hens, with the result that total winter egg production may be greatly increased, and its peak shifted from March and April to November and December (Fig. 31). The total number of eggs laid annually by the turkey can be almost doubled by continuous light (*Milby and Thompson, 1941*). It therefore appears that the bird's inherent egg-laying potentialities are not always realized; but the tendency toward an increase in the mortality rate observed in flocks exposed to artificial

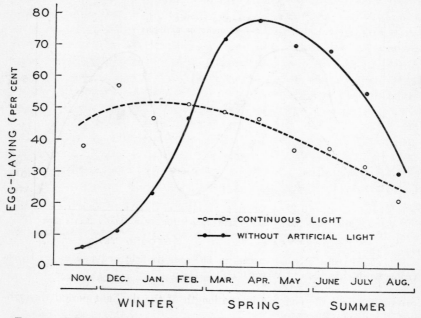

Fɪɢ. 31. Effect of continuous artificial lighting on seasonal fluctuations in egg-laying by fowl. (After Penquite and Thompson, 1933.)

light (*Ogle and Lamoreux, 1942*) also indicates that the limits of physiological endurance can be reached.

Various experiments with artificial illumination have demonstrated the profound effect that light may have upon the bird's reproductive system. In one instance, the time of laying was shifted from day to night by reversing the hours of light and darkness (*Scott and Warren, 1936*); the hen required only about 60 hours to adjust to the drastic change. Additional light each day has been found to advance the breeding season of doves (*L. Cole, 1933*) and grouse (*Clark, Leonard, and Bump, 1937*) by 1 month, and to induce sexual maturity 2 months earlier than usual in turkeys (*Scott and Payne, 1937*). Continuous light has been found valuable for hastening sexual development in late-hatched and slow-maturing pullets (*Kennard and Chamberlin, 1931; Ebbell, 1935*), and for preventing early winter molt (*Kennard and Chamberlin, 1931*). The sudden discontinuance of artificial light tends to throw fowl into a molt (*Kable, Fox, and Lunn, 1928; Hall, 1946*).

The spectral quality of the light is important. Ultraviolet light apparently has no effect on sexual activity (*Rowan, 1936*). Male starlings are stimulated only by light of wave lengths between 0.58 and 0.68 micron, or blue-green to red (*Burger, 1943*). There are, quite possibly, species differences in sensitivity, since English sparrows are much more responsive to red light than green (*Ringoen, 1942*). The intensity of the light is also a factor (*Bissonnette, 1931; Asmundson, Lorenz, and Moses, 1946*). The extent to which the bird's reproductive system is affected by the amount and character of the diurnal illumination is indicated by many observations on pigeons, doves, and wild birds (*Riddle, 1931; Rowan, 1931; Bissonnette, 1931, 1932a, 1932b, 1933b; Bissonnette and Chapnick, 1930*), as well as on chickens (*Bissonnette, 1933a*).

According to the most widely accepted theory, light acts indirectly upon the reproductive organs, both male and female (*Benoit, Grangaus, and Sarfati, 1941*), by first stimulating the pituitary through the nervous system (*Benoit, 1935a*). The initial stimulus may apparently be received through the skin as well as the eyes, for testicular growth has been found to occur in sparrows when the eyes were covered and a defeathered area of the skin was exposed to light (*Ivanova, 1935*).

Artificial Heat

There is little conclusive evidence in favor of heating the domestic hen's quarters during the cold winter months (*Graham, 1930; C. Smith, 1930*). Sudden changes in temperature and in humidity may adversely affect the rate of egg-laying, particularly in low-producing birds (*Willham, 1931*); however, cold, of itself, is not primarily responsible for decreasing production. In dry houses, well ventilated but not drafty, there is no great danger of a reduction in egg-laying caused by exposure or physical discomfort.

On the other hand, it has been shown that high temperatures tend to curtail the number of eggs laid by English sparrows (*Kendeigh, 1941*).

EFFECT OF SELECTIVE BREEDING

The intensive egg-laying of modern domesticated birds is largely the result of selective breeding as well as of improved feeding, housing, and management. Breeding for high egg production may be accomplished by (1) flock selection, (2) recording the performance of individual birds, and (3) progeny testing.

Flock selection is the simplest system of selective breeding. It is based on the doctrines "like produces like," and "breed the best to get the best" (*Pearl, 1912b*). In the practice of this system, selection is made for health, longevity, body type, and other characteristics related to egg-laying.

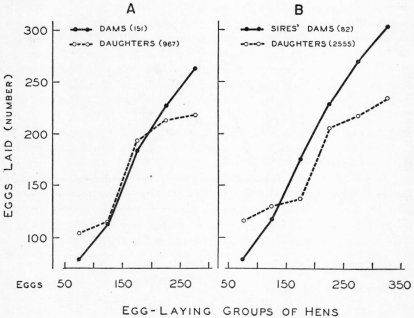

Fɪɢ. 32. Annual egg-laying records of pullets, compared with the first-year egg-laying records of *A*, their dams, and *B*, their sires' dams. (After Hall, 1935.)

In selecting breeders on the basis of their *records of performance*, consideration is given chiefly to individual egg-laying records. The egg production of each bird is determined by trapnesting, or other means, either during a few months—preferably in the winter (*Lerner and Taylor, 1940*)—or during the entire year.

In *progeny testing*, complete records of performance in egg-laying, as well as other useful data, are required for several generations on both sides, the dam's and the sire's. The value of selection on this basis has been demonstrated by Hall (*1935*), who showed that the egg production of the dam or of the sire's dam bears a significant relationship to the pullet's production (Fig. 32).

An analysis of the records of egg-laying contests, and of egg-production statistics accumulated by agricultural colleges throughout the world during the past 50 years, shows that a great improvement has been made in the average annual number of eggs laid per fowl. In Fig. 33, it is evident that the initial effect of selection on egg-laying may be very marked. As selective breeding continues, egg production increases at a

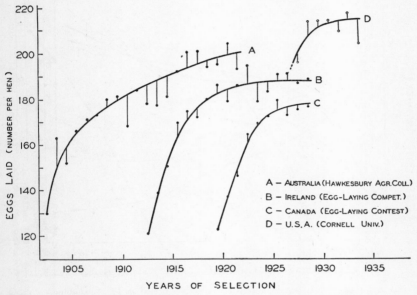

Fig. 33. Influence of selective breeding on the average egg production of fowl in *A,* Australia, *B,* Ireland, *C,* Canada, and *D,* the United States. (After Houghton, 1935; Hennerty, 1930; Taylor, 1930; and Hall, 1935, respectively.)

diminishing rate and may finally reach an almost stationary level. The final level of annual egg production varies, however, from one locality to another, largely because selection is not practiced everywhere to an equal extent or with equal success. In 1940, for example, the hen's yearly egg production differed considerably from one continent to another, as the data show.

	ANNUAL EGG PRODUCTION
CONTINENT	PER HEN
Europe	113
North America	111
Asia	83
South America	82
Africa	49

During the process of selection, many of the characteristics associated with egg-laying undergo great changes as egg production increases. Sexual maturity is attained at an ever-earlier age. The winter cycle becomes considerably longer, and the winter pause grows shorter, as winter production increases. Broodiness progressively declines in the flock, and

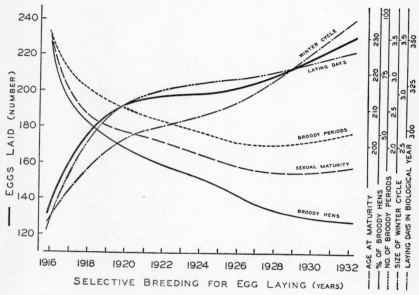

FIG. 34. Changes produced by selective breeding for egg-laying.

1, Decrease in age at sexual maturity; 2, reduction in the number of broody hens in the flock; 3, reduction in the average number of broody periods for each hen; 4, increase in the length of the winter laying cycle; and 5, increase in the number of days from the date of laying the first egg to the beginning of the molting period in the following fall, that is, increase in the length of the biological year. (After Hays and Sanborn, 1934.)

there are fewer broody periods per hen. Persistency in laying is greatly improved (Fig. 34).

The improvements that have been brought about by selective breeding cannot be accredited to genetic factors alone; they can be attributed also to simultaneous betterment in environmental (zootechnical) conditions, such as housing and feeding (*Petrov, 1935*). Unquestionably both inheritance and environment have played their parts in increasing the hen's egg production.

In spite of the fact that the pullet's egg production was enormously increased between 1920 and 1940, there are still some birds in every

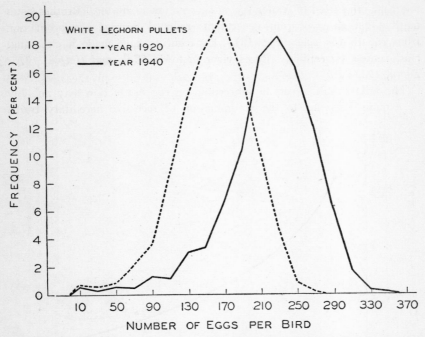

Fɪɢ. 35. Frequency distribution of White Leghorn pullets according to egg production in 1920 and in 1940. (After Dunn, 1924a, 1924b, 1924c; and New York State Egg-Laying Test, respectively.)

flock that do not lay well (Fig. 35). Selection must still be practiced if a further improvement in egg production is to be attained.

RESULTS OF INBREEDING

Inbreeding, or mating of blood relatives, has been practiced for many years both to obtain uniformity in flocks and to establish newly developed strains and breeds. At present it is doubtful whether close inbreeding within a flock of an already established breed is an advantageous method of achieving high egg production. In fact, inbreeding of closely related birds, even of the best laying stock, tends to produce inferior egg-layers; and, if the practice is continued, egg production decreases with each new generation.

The experiments of Dunn (*1923*), Goodale (*1927*), Dunkerly (*1930*), Jull (*1933*), and Hays (*1924, 1929b, 1934, 1935*) show how inbreeding reduces egg-laying. These investigators found that, as inbreeding continued, there was a consistent increase in the age at sexual maturity, and that the period of winter laying became much shorter. Brother-sister matings for three or four generations reduced average egg production from 231 to 154 eggs (Fig. 36). Experiments have made it clear that close inbreeding rapidly reduces vigor in fowls (*Cole and Halpin, 1916; Dunn, 1923; Hays, 1924*).

The unfavorable effects of inbreeding on egg production may possibly be explained as due to the accumulation of recessive hereditary traits.

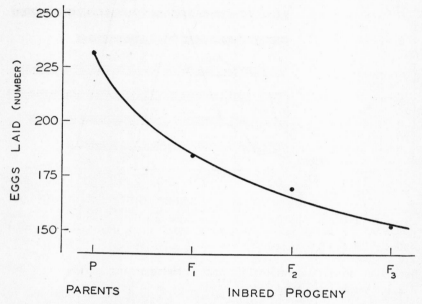

Fɪɢ. 36. Effects of close inbreeding on egg-laying through 3 generations. (After Hays, 1934.)

RESULTS OF CROSSBREEDING

Crossbreeding is the mating of birds of different breeds or of unrelated individuals from independently bred strains of the same breed. The egg-laying record of the offspring obtained from either type of mating may be superior to that of one or both of the parental lines (*Warren, 1930b; Jull, 1933; Hays, 1935*). The improvement in egg production that may result from crossing certain breeds of poultry is shown in Fig. 37. The increase in egg-laying capacity observed in crossbred chickens is evidently one of the indirect results of the well-known stimulation that hybridization often produces. Crossbred domestic animals and hybrid plants commonly surpass their progenitors in size, vigor, and viability. Unfortunately, however, the crossbreeding of fowl sometimes accentuates the tendency toward broodiness (*Warren, 1930b; Knox and Olsen, 1938*). Crossbred chickens, furthermore, are useless for breeding purposes, as their desirable traits are not consistently transmitted to succeeding generations.

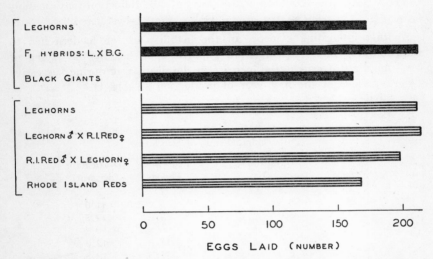

FIG. 37. Egg-laying by the hybrids obtained from crossing various breeds of fowl. (After Warren, 1927, 1930b.)

PHYSICAL ENDURANCE VERSUS EGG-LAYING

Heavy egg-laying demands a superior physical constitution and intense metabolic activity. The burden that it places on the hen can progressively undermine her health and decrease her resistance to diseases.

Constitutional deterioration actually has been observed in all flocks highly selected for egg-laying. The mortality rate for the entire first laying year, which was about 4 per cent 20 years ago, has increased in recent years to more than 20 per cent and, in some instances, to as much as 50 per cent (*Platt, 1936; Hutt, 1938; Hutt, Cole, and Bruckner, 1941*). For example, the following increases in the annual mortality rate of laying birds have been noted in England (*Anonymous, 1939*) in chickens and in ducks. These changes have some economic significance, since commercial producers of eggs cannot operate profitably when the adult mortality rate exceeds 20 per cent.

| Years | Mortality Rate (per cent) | |
	Chickens	Ducks
1922–25	5.4	. . .
1925–28	6.4	1.8
1928–31	8.9	2.9
1931–34	15.5	3.7
1934–37	20.5	7.3
1937–40	18.1	7.5

Another interesting evidence of the inability of the modern hen to withstand the physiological strain imposed upon her by the laying of large numbers of eggs is the increase in the death rate during successive 4-week periods of the first year of production (Table 4).

TABLE 4

Changes in the Death Rate of Pullets During Successive Egg-Laying Periods *
(Based on initial entry of 2273 birds)

Four-Week Periods	Death Rate (per cent)	Four-Week Periods	Death Rate (per cent)
1	2.5	7	8.5
2	5.5	8	11.0
3	5.0	9	12.5
4	6.0	10	10.5
5	7.0	11	12.0
6	6.0	12	13.5

* From the records of the 1937–39 laying trials at the Harper Adams Agricultural College (*Anonymous, 1939, p. 315*).

From the records of progeny testing, it also becomes increasingly evident that a high adult mortality rate is characteristic of individual families. Entries in egg-laying tests over a period of many years indicate that there is an enormous variation in longevity and health among birds of different inbred flocks (Fig. 38). The mortality rate in some flocks has consistently been very high; in others, relatively low. This situation has focused attention on the importance of rigid selection for health and resistance to disease, as well as for egg-laying, in order to combine productivity and vigor.

METHODS OF ESTIMATING EGG PRODUCTION

The most accurate method of determining the annual egg production of individual birds is by trapnesting them throughout the year. However, this procedure is so laborious that often it is not feasible.

Since a definite relationship exists between total annual production and the rate of production during various seasons of the year, a hen's egg record in any month of the year may serve as a more or less accurate index of her past and probable future production (*Thompson, Philpott, and Page, 1936*). The egg-laying record for October, or for November and December (*Dryden, 1921; Lerner and Taylor, 1940*), has been

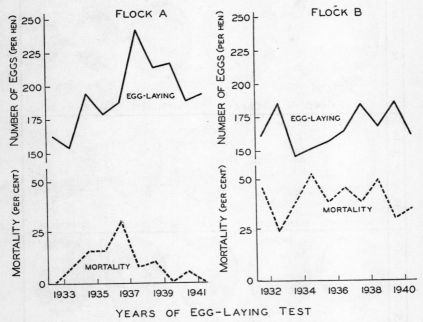

Fig. 38. Adult mortality rate and egg production in two flocks of hens over a number of years. (Courtesy of New York State Egg-Laying Test.)

found to be especially valuable. The rate of laying, and to some extent past and future production, can also be roughly estimated by observing rudimentary pigmentation (as on beak and shanks), the extent of molt, and other aspects of external appearance frequently associated with the hen's laying activity (*Blakeslee, Harris, Warner, and Kirkpatrick, 1917*). These indications of egg-laying, however, cannot be used successfully in evaluating egg production during the pullet year (*Lerner, 1942*).

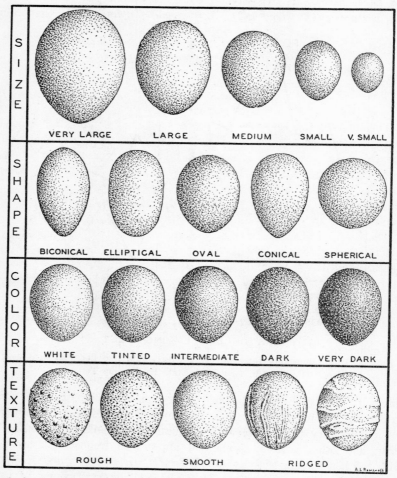

Variations in size, shape, color, and shell texture of hen's eggs. (Reduced about two and one-half times.)

Chapter Two

EXTERNAL CHARACTERISTICS

The avian egg is presented in an attractive and practical container.

The intact avian egg possesses various external qualities by which it may be identified. Among these attributes are its size and shape, and the color and texture of its shell. There is considerable variation in these characteristics among the eggs of all birds, both wild and domesticated. External differences not only serve to distinguish the eggs of various species but also are observable in eggs laid by individuals of the same species, or even by the same bird.

The external characteristics of the egg vary because the reproductive functions of the bird are influenced by numerous hereditary, physiological, and environmental factors. The operation of these factors, particularly as modified by domestication, is a subject of interest, to consider which is the purpose of the present discussion.

SIZE OF THE EGG

Among the eggs of different species of birds, there is an enormous range in size (Fig. 39). In the past, topography, prevailing climatic conditions, water supply, vegetation, type and amount of available food, and various other environmental factors have profoundly influenced birds' nesting and feeding habits, body size and structure, evolutionary status, and physiological processes. In turn, the size of the egg has been affected.

The size of the eggs laid by a number of different species is shown in Table 5, where egg size is expressed in terms of weight. (Weight provides a basis of comparison which is more convenient than dimensions or volume.)

Because of its great bulk of nutrients, the bird's egg is a very large reproductive cell, both absolutely and in relation to the size of the parent. There is obviously a general relationship between the size of the egg and the size of the bird (Fig. 40). However, the largest species of birds lay relatively the smallest eggs (*Heinroth, 1922; Huxley, 1927*).

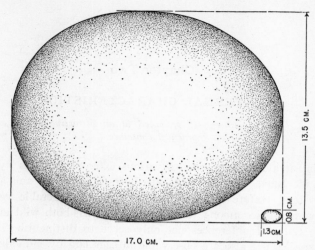

Fig. 39. Comparison of the dimensions of eggs laid by existing species: egg of ostrich (*Struthio camelus*), weighing 1400 gm., and egg of ruby-throated hummingbird (*Trochilus colubris*), weighing less than 1 gm. One-half natural size. (Based on authors' data.)

TABLE 5

ᴀᴘᴘʀᴏxɪᴍᴀᴛᴇ Eɢɢ Wᴇɪɢʜᴛ ᴏғ ᴀ Fᴇᴡ Rᴇᴘʀᴇsᴇɴᴛᴀᴛɪᴠᴇ Sᴘᴇᴄɪᴇs ᴏғ Bɪʀᴅs

Species	Egg Weight (grams)
Aepyornis (extinct)	12,000
Ostrich	1,400
Swan	285
Embden goose	215
Canada goose	135
Peafowl	90
Domestic turkey	85
Domestic duck (Pekin)	80
Wild turkey	75
Muscovy duck	70
Mallard duck	60
Leghorn fowl	58
Guinea fowl	40
Ringnecked pheasant	32
Ruffed grouse	18
Pigeon	17
Bobwhite quail	9
Canary	2
Hummingbird	0.5

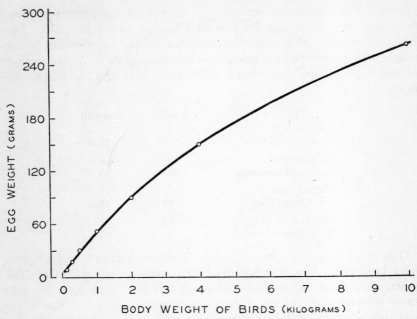

Fig. 40. The relationship between egg weight and the bird's body weight. (Based on data of Huxley, 1927.)

When 427 species are divided into four groups according to size, it is clear that relative egg weight diminishes as body weight increases.

NUMBER OF SPECIES	BODY WEIGHT (grams)	EGG WEIGHT / BODY WEIGHT
164	2– 180	⅑
177	400– 1,500	1/15
80	2,600–12,000	1/28
6	20,000–90,000	1/55

A few birds represent exceptions. The kiwi of New Zealand, which weighs about 2000 gm., lays an egg equal to almost one-fourth of its body weight. The egg of the British species of cuckoo is not one-ninth, but one-thirtieth, of the bird's body weight of about 100 gm. It has been suggested that the cuckoo's parasitic habit of depositing eggs in the nests of smaller species, or of carrying them there in her bill, has caused an adaptation of egg size (*Ingersoll, 1897*).

Egg size is largely a breed characteristic and has been determined by many events in the history of the species. Domestication has shown particularly well how extrinsic forces may modify egg weight. Good nutrition and a favorable environment, as well as the process of selection, have been responsible for a notable increase in the size of the eggs laid by a number of species. An indication of the improvement in egg weight which domestication has accomplished is shown in the comparison of the eggs of some domesticated birds with those of their wild prototypes.

Species	Egg Weight (grams)	Increase (per cent)
Red jungle fowl	40	
Dark Brahma fowl	68	70.0
Canada wild goose	135	
Embden White goose	215	59.3
Mallard duck	57	
Pekin White duck	83	45.6
Wild turkey	75	
White Holland turkey	85	13.3

The greatest increase in egg weight has been attained in fowls that have been under domestication the longest time, presumably over 3000 years. The least increase is observed in turkeys, which were domesticated very recently. Much of the improvement in egg size is due to the fact that selection has been rather consistently made for larger body size.

VARIATION WITHIN THE SPECIES

Variation in egg size within a single species is often not less striking than it is among different species. The domestic chicken may be offered as an example, particularly as numerous statistics are available on this bird. The eggs of Dark Brahmas, for instance, are more than twice as heavy as those of Japanese Bantams. The average egg weights of representative breeds of fowl are presented in Table 6. It may be seen that the eggs of most of the heavily laying breeds weigh from 45 to 64 gm., and that the eggs of the large meat-producing breeds are at the upper extreme of size and those of the small game birds at the lower.

The values shown in Table 6, however, are not representative of all isolated flocks. For example, the weight of White Leghorn eggs varies enormously from country to country, as the figures show.

Weight of Leghorn Eggs (grams)

Country	First Year	Second Year	Investigator
Bulgaria	57.7	59.3	Tabakoff (*1939*)
Germany	49.7	54.4	Lauprecht (*1939*)
Netherlands	59.0	62.5	Tukker (*1930*)
Sweden	55.0	58.5	Axelsson (*1934*)
United States	53.2	56.8	Clark (*1940*)

TABLE 6

COMPARATIVE WEIGHT OF EGGS OF SOME REPRESENTATIVE BREEDS OF FOWL

Breeds of Fowl	Average Egg Weight (grams)
Dark Brahma	68.4 †
Plymouth Rock ⎱ Black Minorca ⎰	63.9 †
Andalusian	63.4 *
Light Brahma	62.0 ‡
Langshan ⎱ Ancona ⎰	61.4 *
Orpington	60.1 *
Rhode Island Red	59.3 *
Leghorn ⎱ Wyandotte ⎰	58.1 *
Black Hamburg	57.1 *
Cantonese	43.2 ¶
Bantam	37.7 §
Japanese Bantam	30.8 ‖
Jungle fowl (possible progenitor)	29.1 *

* Averaged by the authors from various sources.
† Gilbert (*1891*).
‡ Dryden (*1899*).
§ Purvis (*1921*).
‖ Walther (*1914*).
¶ Fronda and Clemente (*1934*).

In view of the variation in egg weight which exists even within the same breed, it is necessary to study the data from each inbred flock separately.

INDIVIDUAL VARIATION

The size of the eggs laid by one individual may differ widely from those laid by another of the same species and breed. On a single day, for example, one flock of Leghorn pullets laid eggs ranging in weight from 41 to 77 gm., although most eggs fell within the limits of 53 to 61 gm. (Fig. 41). The average annual egg weight of some individuals in a flock may be as low as 47 gm., in others as high as 69 gm. (Fig. 42).

The eggs laid by a single bird may also vary considerably in size. In the clutches of wild birds, there is frequently at least 1 egg larger or smaller than the average. The eggs of domesticated birds may be of very diverse size. Hens, for example, occasionally lay extremely small, yolkless eggs which weigh only a few grams; they also may lay unusually large eggs which contain two or three yolks; and one completely double egg has been found that weighed nearly 200 gm. (*Romanoff and Hutt,*

1945). Such cases represent definite aberrations (see Chapter 5, "Anomalies"), but structurally perfect eggs also vary in size over almost as great a range. Normal eggs have been found that weighed as little as 13 gm. (*Pearl and Curtis, 1916*) and as much as 113 gm. (*Kohmura, 1931*)

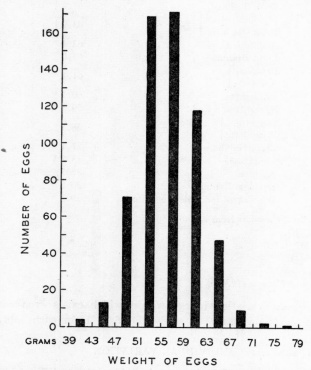

Fɪɢ. 41. Frequency distribution of egg weight. Eggs laid by 605 Leghorn pullets on February 1. (Based on data of Romanoff, unpublished.)

and 117 gm. (*Romanoff, unpublished*). Although exceptionally large or small eggs might be expected to appear either at the beginning or the end of a laying period, they are, on the contrary, most often laid during the time of heavy egg production, when egg size is most uniform.

In general, however, egg size is a distinctive characteristic of the individual hen. Each bird more or less consistently lays eggs that vary in weight within fairly narrow limits. Figure 43 makes a graphic comparison of the egg weight records of two hens and shows how relatively constant is the size of the individual's eggs.

Each hen's eggs vary in weight in a manner typical of the individual. This fact becomes apparent when the weights of all the eggs laid by a number of hens are plotted on frequency distribution curves. The curves are usually of three distinct types (*Hadley and Caldwell, 1920*), as shown in Fig. 44. Some curves are symmetrical; but they reveal that the eggs

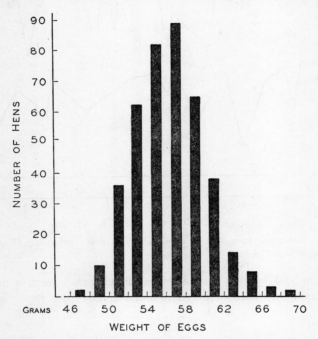

Fig. 42. Frequency distribution of 413 Leghorn pullets according to the average annual weight of their eggs. (Based on data of Romanoff, unpublished.)

of different hens neither vary around the same mean weight (cf. Fig. 44-*A*) nor deviate from the mean weight to the same extent (cf. Fig. 44-*B*). Other curves are asymmetrical, with skewness toward heavy or light weight (cf. Fig. 44-*C*). Occasionally, a curve is bimodal, with two peaks or modes (cf. Fig. 44-*D*).

The wide variation in weight exhibited by eggs leads to the conclusion that egg size is the result of many complex biological phenomena.

VARIATION WITH AGE OF THE BIRD

The average size of the eggs laid by an individual bird changes during successive years of laying. Hens' eggs are smallest during the first, or

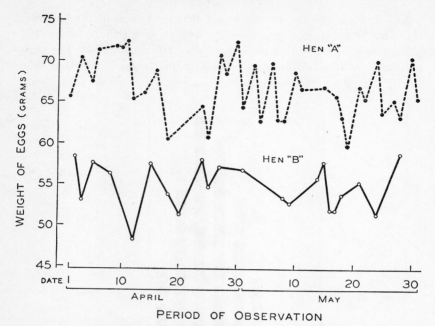

FIG. 43. Constancy in weight of eggs laid by each of two hens. (Based on the authors' unpublished data.)

pullet, year and increase in size thereafter until they reach a maximum, which may be attained in different flocks at any time from the second

TABLE 7

CHANGES IN AVERAGE EGG WEIGHT WITH AGE OF FOWL

	White Leghorns		Plymouth Rocks
Laying Year	Clark (*1940*) (grams)	Unpublished data, Cornell University (grams)	Hadley and Caldwell (*1920*) (grams)
1	53.2	55.0	59.5
2	56.8	58.1	60.2
3	56.5	58.3	64.1
4	56.0	58.4	61.8
5	54.1	57.8	58.9
6	53.7	57.6	59.2
7	52.8	56.9	58.8
8	50.4	55.7	
9	49.6	54.5	
10	47.4	53.3	
11		52.4	

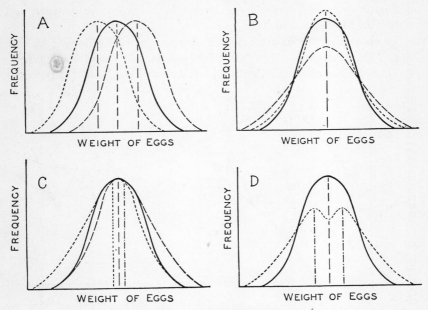

FIG. 44. Various types of frequency distribution curves for egg weight.

A, symmetrical, varying around different mean weights. *B,* symmetrical, with different degrees of deviation from the mean. *C,* asymmetrical, with skewness toward high and low weight. *D,* bimodal, with two peaks. In *B, C,* and *D,* a curve of type *A* (heavy line) is shown for comparison. (Based on observations made by Hadley and Caldwell, 1920.)

to the fourth year (Table 7). Subsequently, the size of the egg declines as the fowl's age advances.

The increase in the size of the egg from the first to the second year has been found to be greater in birds that reach sexual maturity at an early age than in late-maturing birds not well selected for egg-laying.

SEASONAL CHANGES

Observations show that the weight of the chicken egg is not uniform throughout the year but varies from month to month in a very definite manner. However, pullets' eggs and mature hens' eggs exhibit quite different seasonal variations in size (Fig. 45). Pullets' eggs are smallest in the autumn, at the time production begins (*Féré, 1898a*). As the young birds continue to lay, their eggs gradually grow larger, reaching maximum size in February or March. Egg weight then declines somewhat until midsummer but subsequently increases until it approximates that of 2-year-old hens' eggs in November or December. Eggs laid by all fowl 2 or more years old are largest in December or January and smallest in May, June, or July.

The above conclusions are based on observations on Leghorns (*Atwood, 1914, 1923; Benjamin, 1920; Clark, 1940*), Rhode Island Reds (*Hays, 1930a, 1944a*), and Plymouth Rocks (*Hadley and Caldwell, 1920*), but it

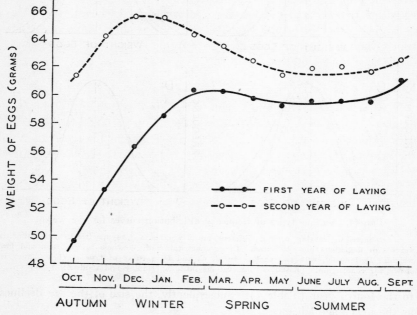

FIG. 45. Seasonal variation in egg weight. Eggs of 163 White Leghorns in their first and second years of laying. (After Tukker, 1930.)

may fairly be assumed that the general trend of seasonal variation in egg weight is similar in other breeds of fowl.

HEREDITARY INFLUENCES

In spite of the fact that size is a fundamental biological attribute of the egg, there is still much conjecture regarding the laws governing the inheritance of this characteristic. Possibly the relative lack of knowledge is due in part to the difficulty encountered in performing experiments with animals of such complex genetic constitution as domesticated birds. In addition, hens exhibit a very high metabolic activity during egg-laying and are affected by a multitude of internal and external factors. It is therefore not easy to separate the hereditary from the nonhereditary influences or to interpret correctly the results of any investigation into the exact mode of transmission of the trait for egg size.

Several experiments performed on domesticated fowl (*Benjamin, 1920; Hurst, 1921; Kopeć, 1924*) indicate that the small egg is dominant to the large one. In other words, if birds are not selected for large egg size, succeeding generations have a tendency to lay smaller and smaller eggs, and thus to approach the egg size of their wild ancestors.

However, the reversion to smaller egg size is not a rapid one. The trait of laying large eggs has become fairly permanent, and consequently many hens consistently lay such eggs.

Family Traits

It is well known that egg weight varies from one family of birds to another; some families lay large eggs, others, small. The trait of laying eggs of a specific size is obviously transmitted from one generation to the next (*Hays, 1940*). Large eggs possibly result from the cumulative action of two or more genes (*Hays, 1937a*).

The direct relationship between the egg weights of dams and their daughters is indicated by the accompanying figures (*Axelsson, 1934*).

	FAMILY TRAIT	
	LARGE EGG (grams)	SMALL EGG (grams)
Dams	61.33	56.03
Daughters	57.73	55.30

These data also show the tendency for progeny to produce somewhat smaller eggs than their parents.

Results of Inbreeding and Crossbreeding

Inbreeding, within reasonable limits, does not necessarily decrease the size of the eggs laid by any strain of birds which has not been selected for large eggs (*Waters, 1945c*); yet, as the degree of inbreeding increases, an adverse effect upon egg size is usually manifested. Waters (*1941*) inbred six families over a period of 9 years and eventually noted a rather significant change in average egg weight. Other observa-

EGG WEIGHT (grams)		DECREASE
IN 1926	IN 1935	(per cent)
56.3	52.0	7.6

tions indicate that, under certain conditions, inbreeding may cause egg weight to decrease much more rapidly, as shown below (*Dunn, 1923*).

GENERATION	EGG WEIGHT (grams)
Parental	56.3
First	55.0
Second	52.5

In another flock, the percentage of birds whose eggs averaged at least 52 gm. in weight, from the commencement of laying to January 1, was found to diminish as follows through successive inbred generations (*Hays, 1934*).

GENERATION	PER CENT OF FLOCK
Parental	96.30
First	84.62
Second	61.54
Third	61.54

The extent to which sire and dam are respectively responsible for the weight of their daughters' eggs is not definitely known, nor can it be said whether choice of a sire should be based upon the weight of his dam's or his sisters' eggs. It has even been maintained (*Waters, 1945a*) that egg size is not influenced by the male parent and that, in breeding for large eggs, it is necessary only to select dams that lay such eggs. On the other hand, it is also believed (*Olsen and Knox, 1940; Hays, 1941; Hutt and Bozivich, 1946*) that both parents contribute substantially to determination of egg weight in their offspring.

Influence of Mass Selection for Egg Size

Although the inheritance of egg size is not entirely understood, there have been many practical demonstrations of the manner in which breeding from carefully selected males and females can increase (within limits) the average size of the eggs laid by a flock. The tabulated data constitute an illustration (*Olsen and Knox, 1940*). Selection must be con-

GENERATION	AVERAGE EGG WEIGHT (grams)
First	54.7
Second	55.7
Third	56.8
Fourth	58.5
Fifth	59.7

stantly practiced, since each succeeding generation inevitably produces a few birds that lay eggs of smaller size than desired.

PHYSIOLOGICAL FACTORS

Since the egg is the product of the functional activity of the reproductive organs, it can be seen that the diversity in the weight of the eggs laid by a flock of hens is due in part to the great constitutional variation among individuals. Birds are very unequal in the rate at which they mature, in the intensity and persistency with which they lay, and in their natural stamina; and the size of their eggs differs accordingly.

Sexual Maturity

The first steps in reproduction, both in animals and in plants (*Pearl, 1909*), are likely to be atypical. Perhaps in agreement with this general biological law, the pullet's first egg is her smallest. The second is considerably larger, and those that follow gradually increase in weight and more nearly approach the size which will later characterize the individual's eggs. The trend is the same, whether the bird lays large eggs or small (*Féré, 1898a; Pearl, 1909*).

	EGG WEIGHT (grams)		
EGGS	HEN A	HEN B	HEN C
First	37.9	48.6	53.2
Second	45.8	55.1	60.0
Third	46.8	55.2	62.9
Fourth	47.4	55.3	63.0

The size of the first egg gives a fairly good indication of the relative size of the eggs that the bird will lay in the future. If the first egg is large, the bird usually continues to lay large eggs. There is, therefore, a close relationship between the weight of the first egg and the average annual egg weight, as the accompanying data indicate (*Hadley and Caldwell, 1920*).

WEIGHT OF FIRST EGG (grams)	AVERAGE ANNUAL EGG WEIGHT (grams)
49.2	57.7
54.0	58.8
59.6	63.0 '
62.6	67.8

These figures also show that, in general, a bird whose first egg is small eventually increases her egg size—as shown by average annual egg weight—to a greater extent than a bird whose first egg is large. The increase in weight may vary from a fraction of 1 per cent to over 30 per cent.

It has been suggested (*Maw and Maw, 1932*) that, if a bird's eggs are to weigh about 56 gm., each, before the end of the pullet year, the average weight of her first 10 eggs must be 47.5 gm. Hays (*1929a*) states that pullets (Rhode Island Reds) should lay eggs averaging 52 gm. in weight during November in order to attain an egg weight of 56.7 gm. by March.

To a large extent, the average weight of an individual bird's eggs depends upon her age and body weight at sexual maturity.

AGE AT SEXUAL MATURITY

The younger the bird when she lays her first egg, the smaller that egg will be. Early-maturing birds continue to lay comparatively small eggs throughout the first year, and sometimes the following years, whereas late-maturing pullets lay relatively large eggs at the start and thereafter (*Jull, 1924b; Parkhurst, 1926; Heuser and Norris, 1934*). The relationship between age at sexual maturity and the average size of the first 10 eggs is illustrated in Fig. 46.

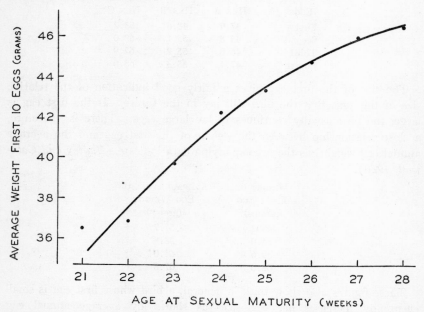

FIG. 46. Graph showing the manner in which the average weight of the first 10 eggs varies according to the bird's age at the time of sexual maturity. (After Heuser and Norris, 1934.)

BODY WEIGHT AT SEXUAL MATURITY

Since body weight in pullets usually depends to a large extent upon age, those birds that mature when relatively young weigh less than those that do not begin laying until they are somewhat older. The smallest eggs consequently are produced by the birds which are not only youngest but also lightest, as shown in the table (*Heuser and Norris, 1934*).

Body Weight at Sexual Maturity (grams)	Average Weight of First 10 Eggs (grams)
1400	39.1
1600	41.8
1800	42.7
2000	46.2
2200	48.0

Among various breeds, the growth rate from hatching to maturity, and consequently the body weight at maturity, are very different, although the age at maturity is more or less the same (see Chapter 1, "Mode of Laying"). The heaviest Asiatic breeds, such as Brahmas, grow much

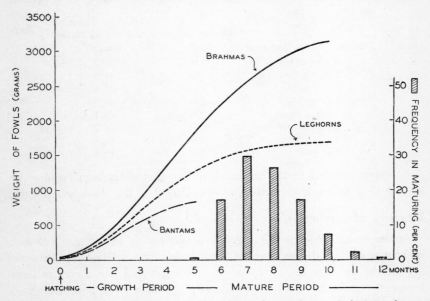

FIG. 47. Frequency distribution in age at sexual maturity; and the growth rate of three breeds of fowl, from time of hatching to maturity. (After Hervey and Decker, 1926; Kempster, 1927; Hays, 1932a; Waters, 1931, 1934; and Jaap, Penquite, and Thompson, 1943.)

more rapidly and have a much greater body weight at maturity than the lightest Mediterranean breeds, such as Leghorns, or the small fancy breed, the Cornish Bantam (Fig. 47). At maturity, a Light Brahma pullet weighs about 3200 gm., a Plymouth Rock about 2500 gm., a Leghorn about 1800 gm., and a Bantam only about 600 gm. These breed differences in body weight at the time laying begins are correlated with breed differences in egg weight.

Increase in egg size throughout the first year is associated with gain in body weight. Early onset of laying may retard growth, and as a result the pullets that mature first are the last to reach their full body weight and therefore their full egg size (*Jull, 1924b*). Body weight at the commencement of laying is accordingly a factor upon which annual egg weight depends (*Lippincott, 1921; Lippincott, Parker, and Schaumburg, 1925; Robertson, 1931; Callenbach, 1934*).

SEASON WHEN HATCHED

The age at which pullets begin to lay, and therefore the size of their eggs, varies according to the season in which the birds are hatched. This fact is indicated in the table (*Upp and Thompson, 1927*). It is evident

SEASON WHEN HATCHED	AGE AT LAYING (days)	AVERAGE WEIGHT OF FIRST 12 EGGS (grams)
Early winter	156	36.4
Late winter	185	41.3
Early spring	236	48.6
Late spring	234	48.9
Early summer	229	48.1
Late summer	214	45.7
Early fall	196	41.8
Late fall	179	40.4

that pullets hatched in the natural breeding season (spring) require the longest time to reach maturity and also that they lay the heaviest eggs.

Intensity of Laying

An important objective in breed improvement is to increase egg size without reducing the number of eggs laid. The relationship between egg weight and the intensity, or rate, of egg-laying has therefore been frequently studied. Many contradictory opinions have been expressed on the subject. On the whole, however, there is enough evidence to warrant the conclusion that small eggs are most likely to be laid during periods of heavy egg production.

It has been observed (*Hadley and Caldwell, 1920; Atwood and Clark, 1930a; Tukker, 1930; Wijk and Ubbels, 1930; and many others*) that, when the curves of egg production and mean egg weight are plotted on monthly ordinates (Fig. 48), the weight maximum is roughly coincident with the maximum of egg production during the pullet year. In other words, during the first year of laying, egg weight and egg production

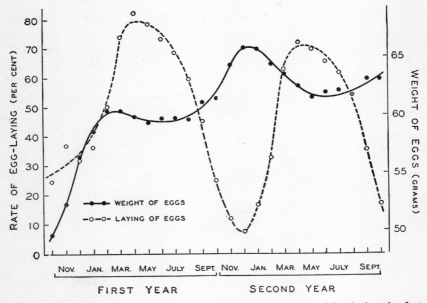

FIG. 48. Seasonal changes in the rate of laying and in egg weight during the first and second years. Based on monthly averages of 163 White Leghorns in the Netherlands. (After Tukker, 1930.)

increase simultaneously. During the second and following laying years, however, the largest eggs often appear in winter, when egg production is at its minimum.

Sometimes no correlation is evident between the rate of production and egg size, probably because of the simultaneous influence of a number of other factors upon which egg size depends. For example, many modern flocks have been bred both for high production and large egg size (*Bennion and Warren, 1933b*). Furthermore, it has been pointed out (*Marble, 1930b*) that the smallest eggs are laid not only by the birds of the highest yearly production but also by those whose production is lowest (Fig. 49). The small size of the poor layer's eggs may be the result of the bird's lack of vigor and her failure to attain maximum egg weight, whereas the low weight of the heavy producer's eggs is probably due to her lengthy laying cycle.

Egg weight also varies to a significant extent according to the position of the egg in the cycle and the time of the day when the egg is laid.

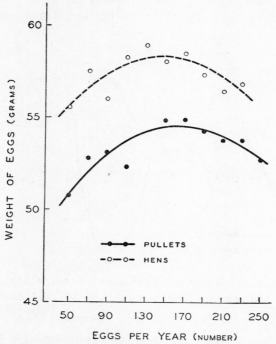

FIG. 49. Relationship between average egg weight and total annual egg production of pullets and hens. (After Marble, 1930b.)

POSITION OF THE EGG IN THE CYCLE

The weight of the chicken egg changes throughout the laying cycle (*Féré, 1898a; Curtis, 1914a; Atwood and Weakley, 1917; Atwood, 1926, 1929b; Bennion and Warren, 1933b*). There is a general tendency for the first egg of the cycle to be the heaviest, and for the succeeding eggs to decrease gradually in size, as shown in Table 8. However, the weight of the successive eggs may not decrease with perfect regularity, and often the last egg is somewhat heavier than the egg immediately preceding it. In short cycles, the decline in weight is, of course, more rapid than in long cycles. The longer the cycle, the smaller is the average decrease in the weight of each egg within the cycle, but the

TABLE 8

EGG WEIGHT IN RELATION TO THE POSITION IN THE CYCLE *

Length of Cycle (number of eggs)	Weight of Eggs		Total Decrease per Cycle (grams)	Average Decrease per Egg (grams)
	First Egg (grams)	Last Egg (grams)		
2	51.2	50.1	1.1	1.1
3	53.7	52.3	1.4	0.7
4	54.5	53.0	1.5	0.5
5	53.2	51.6	1.6	0.4
6	54.8	53.1	1.7	0.3
7	55.4	53.8	1.6	0.3
8	54.0	52.1	1.9	0.3

* After the data of Bennion and Warren (*1933b*).

greater is the total decrease from the first egg to the last (*Bennion and Warren, 1933b*). Egg weight sometimes declines throughout the cycle to as great an extent at the beginning of the first laying year as during the following years (*Curtis, 1914a*). Often, however, the tendency for the pullet's eggs to grow larger during the initial stages of laying may somewhat offset the trend toward diminishing weight within the cycle (*Bennion and Warren, 1933b*).

Egg weight, largely determined by the bird's physiological efficiency, probably diminishes throughout the cycle because the fowl is unable to build up enough material to form a series of full-sized eggs. It is also possible that, as the cycle advances, the strain on the hen's reproductive organs is sufficient to lower her capacity for converting available material into eggs (*Curtis, 1914a*).

TIME OF DAY WHEN THE EGG IS LAID

When a hen lays in cycles of medium length, the first egg is usually laid in the morning and the succeeding eggs at ever later hours, so that the last egg of the cycle is laid late in the afternoon (see Chapter 1, "Mode of Laying"). After resting for 1 or more days, the bird starts a new cycle by again laying the first egg early in the day.

Since the first egg of the cycle is generally the heaviest, the largest eggs are usually laid in the morning (*Atwood, 1927, 1929b; Funk and Kempster, 1934; Pritzker, 1940*). Atwood (*1927*) recorded the time of the day at which about 5000 eggs were laid by 186 Leghorn pullets over a period of 39 days. The eggs were weighed individually. Table 9 shows how the average weight of the eggs declined during the laying day, which was divided into six periods. The decrease was fairly large

TABLE 9

RELATIONSHIP BETWEEN EGG WEIGHT AND TIME OF DAY WHEN LAID *

Periods	Average Weight of Eggs (grams)	Changes in Egg Weight (grams)
Before 9 A.M.	55.22	
		−0.68
From 9 to 10 A.M.	54.54	
		−0.49
From 10 to 11 A.M.	54.05	
		−0.21
From 11 A.M. to 12 M.	53.84	
		−0.10
From 12 M. to 2 P.M.	53.74	
		+0.11
From 2 to 5 P.M.	53.85	
Total		−1.37

* After the data of Atwood (*1927*).

from the first to the second period, and significant, though smaller, from the second to the third period. During the two following periods it was much less; and in the last period egg weight increased slightly. The total decrease in weight during the day was 1.37 gm., or about 2.5 per cent. It is therefore evident that the size of the egg is correlated with the time of the day at which the egg is laid.

Persistency of Laying

Our discussion, thus far, has chiefly considered egg size during periods of fairly continuous production. Certain consistent variations in egg size are also associated with laying pauses of more than a few days' duration.

PAUSES IN LAYING

Laying may be interrupted for some time by the winter pause, the occurrence of which depends upon genetic factors (*Goodale, 1918; Hays, 1924–25, 1930b*), or by a period of brooding or molting. When laying is resumed, the first eggs are not of standard size. The weight of the first egg laid after a pause may be 2 to 4 gm. less than the average weight of the first eggs in the last three cycles preceding the pause (Fig. 50). Usually, the normal size of the egg is recovered with the second or third egg after the resumption of laying.

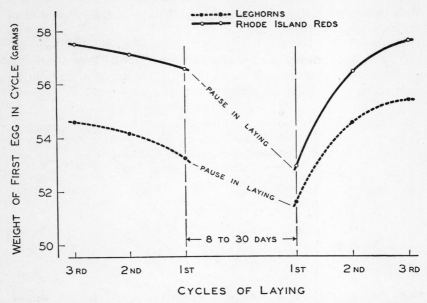

Fig. 50. Changes in egg weight after a pause in laying. (After Bennion and Warren, 1933b.)

Physical Condition

The activity of the reproductive organs is unquestionably influenced by the hen's general physical condition. If the bird's vitality is lowered, egg weight may be affected.

Physiological disturbances may be indicated by changes in the size of the egg even before they are revealed by the behavior of the bird. Curtis (*1914a*) cites the case of a fowl which, for no apparent reason, laid 90 consecutive eggs of smaller size than those she laid during the previous year. Eventually, it could be seen that she was sick. An autopsy showed various pathological conditions which were obviously of long standing. Disease may thus result in a decrease in egg size rather than in the cessation of laying.

BODY WEIGHT

The physiological factors that are responsible for the bird's continued growth after sexual maturity are also responsible for the associated increase in egg weight. Egg size reaches a definite value with the termination of the hen's somatic development. Within the species, the mature birds of greatest body weight usually lay the largest eggs, and birds of lighter weight lay smaller eggs.

The correlation between body weight and egg weight has been frequently demonstrated in studies of domesticated fowl. Presumably, the relationship is linear in adult birds (Fig. 51), although it may not be so in young birds during the period of growth (*Marble, 1930b*).

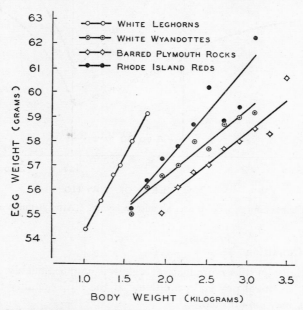

Fɪɢ. 51. Relationship between body weight and egg weight in four breeds of domesticated fowl. (After Robertson, 1931.)

NUTRITION

It is logical to assume that there might be some relationship between the bird's diet and the weight of her eggs. However, since egg weight depends upon many concomitant factors, it is extremely difficult to evaluate the direct influence of nutrition. It is therefore possible to make only a few definite statements regarding the effect of the ration on egg size.

Kind and Amount of Food

Size is a fairly permanent and fundamental characteristic of the individual's eggs; it is less readily influenced by dietary regime than is the rate at which eggs are laid. For this reason, changes in the quantity or quality of the ration tend to affect egg production before modifying egg weight. For example, restriction of a well-balanced diet to 50 per cent of the normal amount may curtail the number of eggs laid yet have no adverse effect on their size (*Bennion and Warren, 1933a; Heywang, 1940*). However, there is evidence that nutritional factors determine egg weight to a certain extent.

Of all dietary constituents, proteins probably have the greatest influence on egg weight. They materially contribute to the production of large eggs. The effect of a protein deficiency can be readily demonstrated. In one instance, decreasing the amount of protein in the diet of pullets from 21 per cent to 12 per cent resulted in a reduction in average egg weight from 53.8 gm. to 52.9 gm. (*Hendricks, 1934*). Proteins of animal origin are superior to those of vegetable origin, but best results are obtained by feeding combinations of both types. Various milk products give good results and are especially valuable for enhancing the effect of meat products, as shown in the tabulation (*Romanoff, 1926*).

Type of Protein	Average Egg Weight (grams)
Liquid milk	56.2
Condensed milk	55.7
Dried milk	54.7
Meat products	54.6
Meat and milk products	56.6

The form in which calcium is supplied also appears to affect egg weight. Calcium phosphate, for example, seems to be poorly assimilated. When it is the sole source of calcium in the diet, egg weight declines (*Massengale and Platt, 1930*). It is probable that a deficiency in vitamin D tends to reduce egg weight, since an increase in egg size often follows the correction of deficiency by treatment with cod-liver oil. However, the administration of cod-liver oil has no effect on egg size when there is no vitamin D deficiency.

Inclusion of iron salts in the hen's ration is said to result in an increase in egg size (*Tangl, 1939*).

Drugs and Other Preparations

The influence of certain drugs on the size of the egg is fairly evident. A single 1.0-gm. dose of kamala, which is commonly used for the eradication of tapeworm in fowl, often causes a significant decrease in egg size during the days following treatment. The maximum effect, which may be a 10 to 15 per cent decline in weight, is usually observed in about 8 days (*Atwood and Clark, 1929; Maw, 1934*). Normal size is then rapidly regained (Fig. 52). There is no definite agreement as to whether the reduction in the size of the egg is due to diminution of the yolk or of the albumen, and therefore it is not possible to state whether the drug chiefly affects the ovary or the oviduct.

Hens consistently respond to the daily feeding of desiccated thyroid by laying eggs of smaller size than usual. Thyroid is effective within a

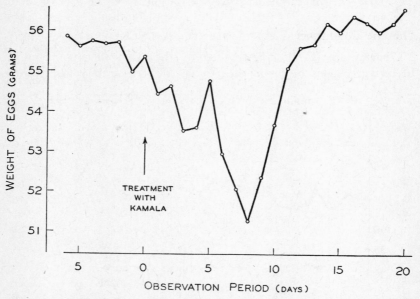

FIG. 52. Effect of kamala (1.0 gm. per bird) on average weight of eggs laid by 165 Leghorns in their third year. (After Atwood and Clark, 1929.)

very few days when given in quantities as small as 0.3 gm. (*Asmundson, 1931a, 1931d; Schmidt, 1932; Chrzaszcz, 1935*). It causes a relatively greater reduction in the size of the yolk than in the amount of the albumen. Rhiodine, an organic iodine compound, when administered in amounts of only 0.04 gm. per day, may result in more than a 10 per cent decrease in average egg weight within 11 days (*Peano and Pissaro, 1935*).

Féré (*1897*) observed that a stupefying dose of morphine (about 0.7 gm. per kilogram of body weight) caused a hen to lay an egg of smaller size than usual on the day the drug was given. She did not lay again for 4 days. The first 2 or 3 eggs which she laid upon resuming production were also smaller than those preceding the morphine intoxication.

Very little experimental work has been done on the effect of hormonal preparations on egg size. However, it has been reported that neither estrogen (follicular hormone) nor anterior pituitary extract has any apparent influence (*Asmundson, 1931a; Parkhurst, 1934*).

ENVIRONMENT

Among the external factors whose possible influence on egg weight has been investigated are climatic conditions, confinement, and artificial light.

Climatic Conditions

Climates are chiefly differentiated by their prevailing temperatures. In certain regions of the world, notably the temperate zones, there is

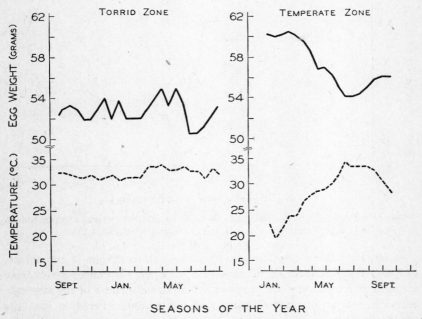

Fig. 53. Seasonal changes in environmental temperature and in average egg weight in Laguna, Philippine Islands (torrid zone), and in Kansas, U.S.A. (temperate zone). (After Warren, 1939.)

great annual fluctuation in temperature. In the same regions, egg weight has been observed to vary seasonally. This fact suggests the possibility that temperature is a factor in determining egg size, although it should be remembered that the effect of one aspect of climate on the animal organism may be modified by the effect of another aspect.

Temperature. Egg weight records accumulated in various latitudes, extending from the equator to Scotland, indicate that extremely low temperatures have little effect upon the size of the egg. However, there appears to be a consistent decrease in egg weight whenever the maximum temperature exceeds 21° C. for a few consecutive days. In cold climates, where the temperature seldom remains above this level for any length

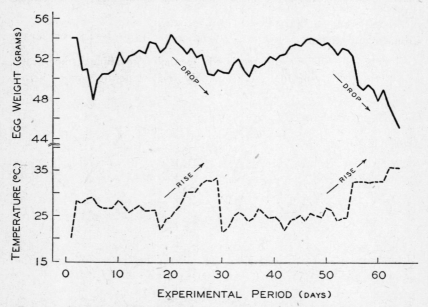

Fig. 54. Changes in egg weight induced experimentally by changes in room temperature. The average weight of eggs laid by 32 hens declines as the room temperature rises, and vice versa. (After Bennion and Warren, 1933a.)

of time, there is little decline in the size of eggs laid in summer. In tropical or semitropical regions, such as the Philippine Islands, temperature varies within a rather narrow range throughout the year, and egg weight is also fairly constant. The greatest variation in egg size is noticeable in the latitudes where there is also the greatest seasonal fluctuation in temperature (Fig. 53).

In general, undue heat appears to have an adverse effect on egg size (*Heywang, 1946*). An abrupt reduction in egg weight, of the magnitude of 15 to 20 per cent, has been produced experimentally by subjecting hens to increased environmental temperature (*Bennion and Warren, 1933a*), and it was observed that there was a relatively slow recovery in size after normal temperatures were restored (Fig. 54). Wild birds also seem to require moderate temperatures for several days previous to egg formation, in order to achieve maximum egg size (*Kendeigh, 1941*).

Confinement

In recent years, the trend has been increasingly toward the partial or complete confinement of laying fowl. The practice offers obvious advantages, not only in space-saving, but also in permitting control of various environmental conditions to a degree not otherwise possible. If birds in confinement are adequately supplied with vitamin D, or if their houses are equipped with the type of window that transmits ultraviolet light, egg size is not appreciably affected; sometimes, in fact, it may be slightly increased (*Godfrey and Titus, 1934*). There is probably a general improvement in efficiency of food utilization when the bird's physical activity is restricted.

Artificial Light

The use of artificial illumination to lengthen the period of daylight has no apparent effect on the weight of chicken or turkey eggs (*Parkhurst, 1933; Wilcke, 1939*).

SHAPE OF THE EGG

Birds' eggs differ considerably in shape. Although many are truly ovate, some are nearly spherical, and others are elongated. Eggs may be almost equally pointed or rounded at both ends, or they may taper sharply from the large end to the small. The eggs laid by different birds of the same species resemble each other in shape, but they are not identical; nor are all the eggs of a particular bird exactly alike.

Aristotle (384–322 B.C.) believed that the cock hatched from the more pointed chicken egg and the hen from the rounder type. Early in the nineteenth century, a number of naturalists argued that the egg's contours indicated the general body form of the bird that would develop within it. Somewhat later, advocates of the theory of natural selection made elaborate attempts to show that, through adaptation, the eggs of different birds had assumed the shape most likely to insure the survival of each species in its particular environment. Pearl and Surface (*1914b*) found, however, that dimensional relationships may vary as much in the eggs of a wild species as in those of the domestic hen, upon which the forces of natural selection obviously do not operate. It is now generally agreed that physiological factors are largely responsible for the diversities in the form of the egg.

CAUSATION OF SHAPE

The egg would tend to be a sphere if it were not subjected to external forces while still in a plastic condition. As early as 1772, Günther stated that the ovoid contour of the egg was the result of pressure exerted by the oviducal muscles. Essentially, his theory is still valid.

The walls of the oviduct contain two layers of muscle, the inner circular layer, which moves the egg forward, and the outer longitudinal layer, which expands the oviduct (see Chapter 4, "Formation"). Differences in the degree of coordination between these muscles probably account for many of the minor variations in the shape of the egg. According to Curtis (*1914a*), if the circular fibers are contracted behind the egg while the oviduct is expanded ahead of it, the egg will meet little resistance, and will be long and narrow; as resistance increases, the egg becomes shorter and broader. (Ryder, in 1893, similarly explained that the egg's shape was generated by the interaction between force and resistance, but he erroneously assumed that the blunt end of the egg advanced first down the oviduct.) Another factor that must be considered is the relation of the size of the egg to the caliber of the oviduct. Asmundson (*1931c*) suggests that an elongated egg might be formed if a large volume of albumen were secreted and then forced through a narrow isthmus, and Thompson (*1943, p. 939*) points out that a small egg could pass with ease through a large oviduct and would probably tend to be round.

Opinions differ as to where in the oviduct the egg is given its form. Szielasko (*1905*) stated that the egg was shaped in the uterus, which he considered to be merely a passive mold. Pearl (*1909*) showed that the egg is ovoid before it reaches the uterus, but he believed that the muscular activity of the uterine walls determined the particular form of each individual egg. Curtis (*1914a*) concluded that the process of shaping the egg occurred continuously throughout the whole oviduct. According to Asmundson (*1931c*), the contours of the egg are established in the albumen-secreting part of the duct and in the isthmus but may be altered afterward in the uterus. Harper and Marble (*1945*) suggest that restrictions imposed by the dorsal ligament, which supports the oviduct, exert some influence on the shape of the egg while the egg is still in the isthmus.

EGG SHAPE INDEX

The numerous variations in the contour of individual eggs obviously cannot be expressed in mathematical terms. However, the shape of the egg can be approximately indicated by the ratio between length and breadth. The first attempt to describe egg shape by means of a formula based on the two diameters was made in 1870 by Reichenow, who studied wild birds' eggs. At present, the shape index is commonly employed. This value is obtained by dividing the transverse diameter of the egg by the length [1] and multiplying the result by 100. The shape index is independent of absolute size and varying contours; a relatively long and narrow egg of any size will have a low index, and a short and broad egg, whether large or small, will have a high index.

Other formulae attempt to express the shape of the egg with extreme accuracy. One is based on four measurements of diameter, taken at equidistant levels (*Asmundson, 1931c*), and another on two measurements of diameter, made at a certain distance from either end (*Serebrovsky and Serebrovsky, 1926*).

SPECIES CHARACTERISTICS

In general, the shape of the egg is a recognizable species characteristic. Although the eggs of many wild birds are ovoid, like those of the chicken,

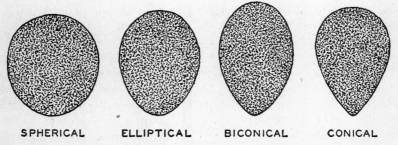

SPHERICAL   ELLIPTICAL   BICONICAL   CONICAL

FIG. 55. Variation in the shape of wild birds' eggs; one-half natural size. (After Seebohm, 1896.)

Spherical: egg of the hooded merganser (*Mergus cucullatus*). Elliptical: egg of the common bittern (*Botaurus stellaris*). Biconical: egg of the great crested grebe (*Podicipes cristatus*). Conical: egg of the golden plover (*Charadrius pluvialis*).

other species lay eggs that diverge widely from this type. For example, almost spherical eggs are peculiar to the owl, the titmouse, the kingfisher, the penguin, and the hooded merganser. The eggs of the grebe, the pelican, the cormorant, and the bittern are nearly alike at either extremity and are therefore described as biconical or ellipsoidal. Those of the curlew, the plover, and the sandpiper are conical, tapering sharply from the broad end (Fig. 55).

[1] The outline of the egg may be projected on paper by a tracing device known as a copyscope.

The average indices for several species are given in the table.

Species	$\text{Index} \left(\dfrac{\text{Breadth}}{\text{Length}} \times 100 \right)$
Hooded merganser	90
Ostrich, snowy owl	80
Ruddy sheldrake	70
Eagle, cormorant	60
Guillemot	50

BREED AND FLOCK VARIATION IN FOWLS

Flocks of the same breed of chicken differ greatly in average egg shape index; and within a single flock there may be enormous variation. In one instance, it was found that the shape indices of 450 eggs laid in one day by Barred Plymouth Rock pullets ranged from 60.0 to 85.9 (*Pearl and Surface, 1914b*). Similarly, the authors noted index values from 63.1 to 81.7 in the eggs laid by a flock of 262 Leghorn hens (Fig. 56).

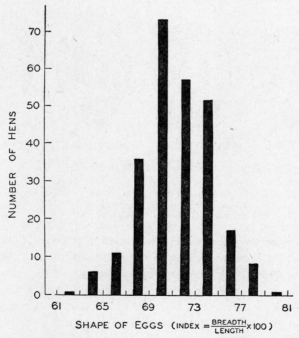

Fig. 56. Frequency distribution of 262 Leghorn hens according to the shape of their eggs. (After the data of Romanoff, unpublished.)

INDIVIDUAL VARIATION

The individual hen lays eggs that are more or less uniform in contour and shape index (Fig. 57). Pearl and Surface (*1914b*) concluded that, in passing from the race to the individual, the variability of the egg's length is reduced by 32 per cent and that of its breadth by 41 per cent.

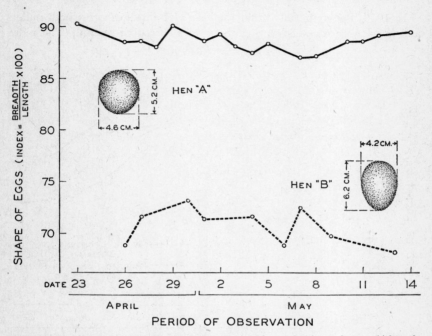

F꜐g. 57. Constancy in the shape of the eggs laid by each of two hens. (After the data of Romanoff, unpublished.)

Breadth is usually the more constant dimension. This circumstance is not surprising, in view of the fact that the egg originates in a tube whose lumen is limited in extensibility. By statistical methods, Curtis (*1914a*) determined the comparative variability in the length and breadth of a large number of eggs, with the following results:

Dimension	Coefficient of Variation
Length	3.32
Breadth	2.38

Occasionally a hen may be found whose eggs vary to a greater extent in breadth than in length. However, unusual variability in one dimension is generally correlated with unusual variability in the other (*Pearl and Surface, 1914b*). The eggs of some fowl may be extremely and consistently variable and may even show more diversity among themselves than is found among the eggs of the entire flock.

SEASONAL VARIATION

In 1920, Benjamin noted that the eggs of White Leghorns, both pullets and hens, were roundest during the natural breeding season. The index

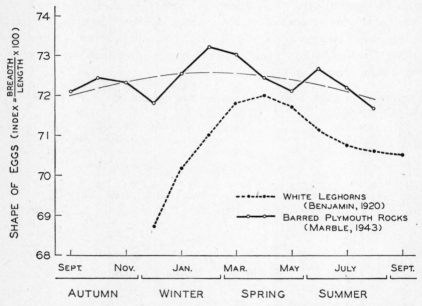

FIG. 58. Seasonal changes in the shape of the hen's egg, as observed from 1911 to 1919 and from 1936 to 1940. (After Benjamin, 1920, and Marble, 1943.)

value increased from 69 in December (the first month of production at that time) to 72 in March and subsequently declined to 70.5 in September, when laying ceased. On the other hand, nearly 25 years later, Marble (*1943*) was unable to demonstrate significant seasonal fluctuations in the shape of eggs laid by individual Barred Plymouth Rock hens, or by the flock as a whole. His data show only a slight trend toward rounder eggs at the same time of the year during which Benjamin noted the highest index (Fig. 58). Flock and breed differences may be responsible for the discrepancy between these observations, as may be, also, the innovations in poultry management which were introduced in the intervening quarter of a century. Selection for egg production, the use of artificial illumination, new methods of feeding, and other recently adopted practices may have modified the hen's reproductive physiology sufficiently to eliminate seasonal variation in egg shape.

INFLUENCE OF HEREDITY

In the inheritance of egg shape, neither the round egg nor the long egg appears to possess a clear-cut dominancy. If dam and sire's dam lay eggs of identical type, the progeny lay eggs of the same shape. However, when parents are derived from strains which produce the extremes of egg shape, the eggs of the offspring are intermediate in form (*Benjamin, 1920; Marble, 1943*). These observations are summarized in the tabulated data (*Marble, 1943*).

Egg Index of Sire's Dam	Egg Index of Dam	Egg Index of Progeny
76.8 (round)	77.4 (round)	76.2 (round)
65.8 (long)	67.2 (long)	69.4 (long)
67.7 (long)	77.2 (round)	73.0 (intermediate)
77.6 (round)	68.2 (long)	74.8 (intermediate)

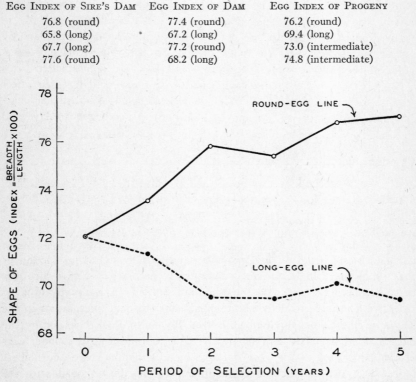

Fig. 59. Two egg shape lines in Plymouth Rocks, established through selection. (After Marble, 1943.)

Chickens may be bred for a specific egg shape with comparative ease. Figure 59 shows the success with which two different egg shape lines were established as the result of 5 years of selection (*Marble, 1943*). In the course of this experiment, it was observed that the average shape index of the progeny more closely approached that of the parent in the round-egg line than in the long-egg line.

There are indications that a limited number of genes is involved in the inheritance of egg shape, but there is not sufficient evidence to establish a single-gene hypothesis (*Marble, 1943*).

PHYSIOLOGICAL FACTORS

Changes in the general physiological condition of the reproductive system, and particularly in the muscular tone of the oviduct, are presumably responsible to some extent for the variability in the form of the individual bird's eggs. Physiological factors probably account for the fact that a wild bird occasionally lays an egg that differs markedly in shape from the others in the clutch. For example, of 3 eggs laid by a russet-barred thrush, the index of 1 was 60.9, whereas the average index of the other 2 was 71.2 (*Ingersoll, 1910*). In domesticated birds as well, variations in functional efficiency are manifested by their effect on the shape of the egg. A pullet usually does not establish her characteristic egg shape until some time after she has begun to lay, and the shape of the egg is noticeably affected by pauses.

Commencement of Laying

The first eggs laid by a pullet are likely to be atypical in shape, as in other characteristics. Apparently the deviation is not necessarily always in the same direction. Pearl (*1909*) reported the extreme case of a pullet whose first egg was almost abnormally long and narrow, and of lower shape index (48.8) than any egg that she laid during the remainder of her life. As she continued to lay, her eggs became progressively rounder, at first rapidly, and later at an ever more gradual rate (Fig. 60). On the other hand, the data given by Benjamin (*1920*) consistently indicate that pullets' eggs are somewhat rounder than hens' eggs, especially during the first part of the laying year.

Cycle of Laying

The first egg of a cycle is usually longer and more narrow than the second egg, but no significant trend is discernible thereafter in the cycle.

Marble (*1943*) found the average shape index of the first and second egg of the cycle to be 72.1 and 73.0 respectively.

Pause in Laying

The first egg laid after a pause of 7 or more days is also longer and narrower than the last egg preceding the pause; normal shape may not

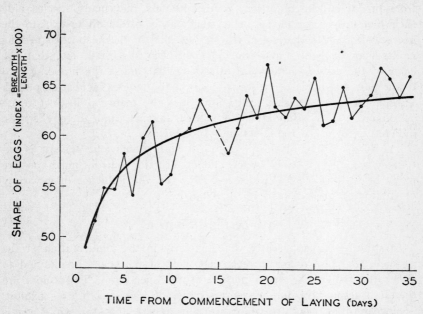

Fig. 60. Changes in egg shape after commencement of laying. Eggs of a Barred Plymouth Rock pullet. (After Pearl, 1909.)

be attained until the fifth egg. Data indicate that about 75 per cent of pauses are associated with changes in egg shape of this nature (*Marble, 1943*).

COLOR OF THE EGG

The great variety and subtle beauty of eggshell tints have always been interesting to ornithologists and laymen alike. The appearance of the egg has a certain esthetic appeal, to which eggshell color inevitably contributes.

There has been much speculation regarding the reasons why birds lay eggs that are white or colored, light or dark, unadorned or profusely speckled. According to one theory (*Darwin, 1868; McAldowie, 1886*), eggshell pigmentation represents a natural adaptation for shielding the egg contents from the sunlight. Many birds nesting in the darkness of holes or hollow trees lay white or light-colored eggs, and the eggs of tropical species are often of deeper tone than those of birds whose habitat is less sunny. Exceptions are very numerous, however. Adaptive processes are also considered responsible for the laying of spotted eggs, which blend with the surroundings and are thus supposedly concealed from predators. In objection, it may be said that many species hide their blotched or speckled eggs so well that it is difficult to explain the necessity for protective color, unless it existed in the past when nesting habits may have been different. It has been suggested, too, that the color of the eggs is a powerful attraction that draws the brooding bird to the nest and holds her there throughout the tedious incubation period. It is argued that eggshell color intensifies the maternal instinct much as plumage color stimulates the mating instinct, and thereby serves, indirectly, to insure reproduction (*Abercrombie, 1931*).

The biochemistry of the pigments in both plants and animals is extremely complex. It is largely under genetic control, much as are many other physiological processes. However, the influence of a variety of environmental factors is not precluded.

VARIATION IN COLOR

In spite of the apparently wide diversity in the pigmentation of the eggshell, the spectroscope reveals that most of the colors are limited to either the red or the green portion of the spectrum. Two colors are therefore responsible for the entire range of hues which eggs exhibit.

Since reptiles' eggs are white, we may assume that the white egg is the primitive type, but it is not possible to demonstrate a logical sequence in the development of other colors. It is possible that white eggs occasionally represent a reversion to the ancestral color.

The eggs of many birds are marked with spots, flecks, blotches, streaks, or fine lines, superimposed on white, or on almost infinitely varied tones of the two basic colors (Fig. 61). It may also be noted that the ground color itself appears far from uniform in distribution when it is viewed through the microscope (Fig. 62). Small surface irregularities on the shell bear pigment in greatly different concentrations.

The eggs of the domestic hen may be white, yellow, or many shades of brown; one breed lays blue-green eggs. Sometimes extremely small, dark flecks are present on the shell, especially if it is brown (*Hays, 1937a*). The extensively varied tones of brown eggs almost defy classification. Benjamin (*1920*) remarked on the impossibility of matching eggs with any other colored surface and eventually improvised his own standard of comparison from empty eggshells. He was able to identify fifty colors, from chalk white to dark chocolate brown; but he found seventeen of these, ranging from white through "cream tinted" to "brown tinted," sufficient for practical use. Some investigators have devised

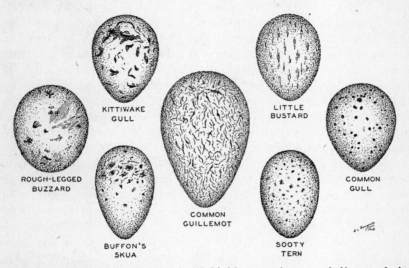

FIG. 61. Superficial markings on wild birds' eggs; about one-half natural size. (After colored plates of Seebohm, 1896.)

Blotches, of varied form and distribution: eggs of rough-legged buzzard (*Archibuteo lagopus*), kittiwake gull (*Larus tridactylus*), and Buffon's skua (*Stercorarius buffoni*). Fine lines: egg of common guillemot (*Alca troile*). Streaks: egg of little bustard (*Otis tetrax*). Flecks, evenly distributed: egg of sooty tern (*Sterna fuliginosa*). Spots, unevenly distributed: egg of common gull (*Larus canus*).

numerical systems for measuring eggshell color (*Kopeć, 1922; Axelsson, 1932*). Others have employed the Ridgeway Color Standards (*1912*), from which eight to ten distinctive shades of pink, buff, cinnamon, and "vinaceous cinnamon" are usually chosen to designate egg color. It is obvious that the subjective element is strong in judging eggshell tint.

RACIAL CHARACTERISTICS

Among wild birds, eggshell color is typical of the species. The eggs of some birds, however, resemble those of closely related species.

Among domestic fowl, the color of the egg is more or less peculiar to the breed, although tinted eggs occasionally appear in breeds that ordinarily lay white eggs. Of the four races officially recognized in the

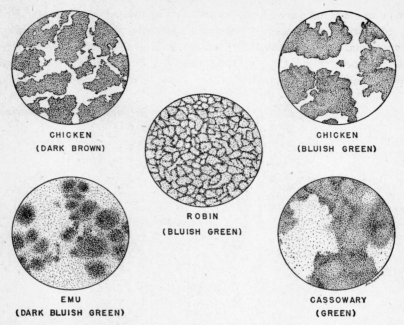

Fig. 62. Uneven distribution of pigment on eggshells of several different species of bird; magnified about 30 times. (Drawn from specimens in the authors' collection.)

United States, the Mediterranean (comprised of Leghorns, Minorcas, Anconas, Black Spanish, and Blue Andalusian) alone lays white eggs; the other three races—Asiatic, English, and American—lay tinted eggs, with the exception of two or three breeds. Cochin China hens lay eggs that range from bright yellow to dark yellow, speckled with fine red dots. Langshans lay dark yellow eggs, Brahmas, reddish yellow (*Hasterlik, 1916*). The eggs of continental European breeds are predominantly white, save for those of a very small number native to Belgium, Holland, and France. The Araucana of South America lays light bluish green eggs.

The color of the chicken egg often assumes economic importance, as there are numerous local prejudices in favor of certain shell tints.

INDIVIDUAL VARIATION

Among wild birds, only occasionally is an individual found whose eggs deviate widely from the normal color. Among domestic hens, also, each bird lays eggs that are more or less consistently of the same shade.

However, the successive eggs of certain hens may show noticeable differences in color intensity. Individual variability in eggshell pigmentation is more likely to occur at the beginning of the laying season than later (*Axelsson, 1932*).

SEASONAL VARIATION

In color intensity, the eggs of all breeds of fowl may vary during the course of the laying year. Eggs are darkest either in early winter

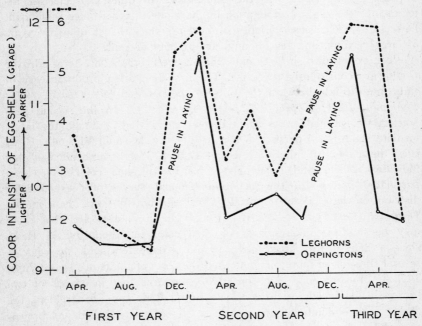

Fɪɢ. 63. Seasonal variation in intensity of eggshell color. Eggs of Orpington and Leghorn fowl, observed in the temperate zone of the northern hemisphere (at Pulawy, Poland). (After Kopeć, 1926a.)

or in spring (*Benjamin, 1920; Kopeć, 1926a; Axelsson, 1932; Hall, 1944*), as shown in Fig. 63.

It is probable that the apparent relationship between season and shell pigmentation is a false correlation. Data from all sources indicate that darkest shell color is usually observed in a pullet's first eggs, or in eggs laid at a slackening rate or immediately following a pause. Although Kopeć (*1927*) believed that temperature changes influenced shell color, it is more likely that pigment accumulates during periods of nonproduction or infrequent laying and becomes depleted when egg-laying is intense (*Needham, 1931a, p. 1377*).   The phenomenon may be analogous to the disappearance of body pigments during continuous laying.

HEREDITARY FACTORS

The general hereditary nature of shell pigmentation is revealed by the manner in which egg color is identified with species and breed. The transmission of shell color in the domestic chicken seems to be on a less complex genetic basis than the transmission of other characteristic attributes of the egg. Investigation of the mode of inheritance is difficult, however, because most varieties of fowl are the product of many years of crossbreeding, and few, if any, are of pure descent.

Whenever brown-shelled and white-shelled breeds are crossed, the results are consistently the same. The first generation produces eggs of tints intermediate between the tints of the parents' eggs (*Hurst, 1905; Benjamin, 1920; Kopeć, 1922; Warren, 1930c*).   In the second generation, the original colors again become segregated, although the intermediate shades still persist (*Kopeć, 1922*), as shown in Fig. 64. However, the manner of inheritance is not so simple as these phenomena may indicate.   Major and minor genes for brown shell color have been postulated because of the fact that bimodal curves of eggshell color distribution have been obtained in certain crossbreeding experiments (*Punnett, 1923, pp. 164–173*).   In addition, it is suspected that there may be multiple determinants of shell color in Rhode Island Reds, since a hen of this breed often produces daughters that lay eggs of several tints, some quite different from the color of the dam's eggs (*Hays, 1937a*).   Furthermore, Axelsson (*1932*) suggests that in crosses of two brown-shelled breeds, such as Rhode Island Reds and Barnevelders, the darker shell is dominant, whereas in crosses of white-shelled and brown-shelled breeds, the white shell is incompletely dominant.   There is also evidence that, in crossbreeding, eggshell color inheritance is sex-linked to the sire (*Kopeć, 1926b; Warren, 1930c; Axelsson, 1932; Hall, 1944*).

It is of interest to note that the blue egg behaves as a simple dominant to eggs of other colors (*Punnett, 1933*).   In crossbreeding, blue combines with various shades of brown to produce greens and olives.

Xenia.   At one time, it was thought that a hen tended to lay eggs resembling, in color, those of her mate's breed, if hen and cock were of two breeds characterized by different eggshell pigmentation.   This theory was given credibility by the observations of various German biologists (*von Nathusius, 1879; Holdefleiss, 1911; von Tschermak, 1915*).   However, Duerden (*1918*) was unable to demonstrate modification in the egg's color upon mating two different breeds of ostrich, and it is now considered quite unlikely that mating can in any way alter the pigment-

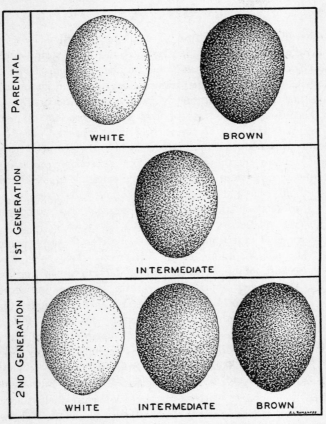

Fig. 64. Effect of crossbreeding on egg color. First-generation birds from a cross between a white-shelled and a dark-shelled breed produce eggs of an intermediate tint. Segregation (multifactorial) occurs in the second generation. (Modified from Punnett and Bailey, 1920.)

secreting function. Apparently the normal periodical variations in the color of the egg were responsible for the formerly prevalent belief in the ability of the male to make a direct impression on eggshell tint (*Kopeć, 1927; Axelsson, 1932*).

CONSTITUTIONAL AND FUNCTIONAL FACTORS

It is possible that there is a relationship between eggshell color and the changes in functional efficiency which are associated with advancing age or which occur during a laying cycle.

AGE OF THE BIRD

It is rather difficult to obtain evidence in support of the theory that wild birds lay darker eggs as they grow older (*Newton, 1893, p. 185*). However, the shell pigment of hens' eggs has been observed to increase gradually from the second to the sixth year (*Benjamin, 1920*), as egg production declines (see Chapter 1, "Mode of Laying").

POSITION OF THE EGG IN THE CLUTCH

Within a single clutch, there may be great variation in the color intensity of a wild bird's eggs (*Newton, 1893, p. 186*), particularly when there are 5 or 6 eggs in the clutch. On the other hand, the clutch of some birds, such as the golden eagle, comprises only 2 eggs, yet 1 is often light, the other dark. There appears to be no correlation, however, between the depth of color and the order in which the eggs are deposited.

It has not been established that there is a definite trend in color intensity throughout the domestic hen's laying cycle.

SHELL TEXTURE

Eggs show great variety in the surface characteristics of their shells. Some are exceedingly glossy, others very dull; some are smooth, some rough. Textures intermediate between the extremes are common.

SPECIES VARIATION

Among wild birds, shell texture is typical of the species. For example, the eggs of the tinamou reflect the light as if they had been polished, and those of the catbird are also unusually glossy. Dull, chalky eggs identify the ibis and the megapode. The eggs of certain ducks are noticeably oily to the touch, and those of the emu are coated with a calcareous film of considerable thickness. Ostrich eggs are pitted with large pores, and cassowary eggs bear pigmented granulations uniformly distributed over the shell surface.

The shell of the domestic hen's egg may be mat or fairly lustrous. Its many variations in surface texture make it difficult to designate abnormalities. Ridges and furrows are frequently present. Small "sandy" deposits of calcareous material are often found at one end or the other and are occasionally distributed over the entire shell (see frontispiece to this chapter). On brown eggs, minute superficial deposits are sometimes noticeable as white spots.

INFLUENCE OF HEREDITY

Among domestic fowl, glossy and chalky shells appear to be hereditary to some extent, as certain strains within breeds may lay eggs predominantly of one type or the other. There is some evidence that the chalky egg is dominant within the breed. However, it is possible that the glossy egg may be dominant when two different breeds are crossed (*Taylor and Lerner, 1941*).

Daughters tend to resemble dams with respect to the smoothness or roughness of their eggshells. Hays (*1937a*) noted that the trait of laying smooth eggs was exhibited by 79 per cent of the daughters of hens that laid such eggs, but by only 52 per cent of the daughters of hens that laid "sandy" eggs.

EFFECT OF NUTRITION AND DRUGS

The influence of nutrition on shell texture is still a subject of investigation. Experiments have not demonstrated conclusively what deficiency in the diet, if any, can be held accountable for the numerous abnormalities that occur in the nature of the shell surface (see Chapter 5, "Anomalies"). However, French (*1935a*) found that salt exerted a very adverse effect when fed in amounts exceeding 2 per cent of the total food intake. The authors have observed that rancid cod-liver oil in the diet causes the eggshell to lose its natural glossiness and to become detectably rough to the touch.

It has also been noted that sulfonamides produce marked irregularities in the surface of the eggshell (*Genest and Bernard, 1945*). When sulfanilamide, soluseptazine, neoprontosil, or sulfapyridine is given at the level of 0.3 to 0.5 per cent of the dry mash, granular deposits are formed on the shell. The deposits caused by these drugs differ somewhat in appearance (Fig. 65). The administration of other sulfonamides, such as sulfathiazole, sulfaguanidine, and sulfamerazine, tends to result in the formation of small pits in the shell. It is suggested that the effect of the latter group of drugs is due to the inhibition of certain enzymes.

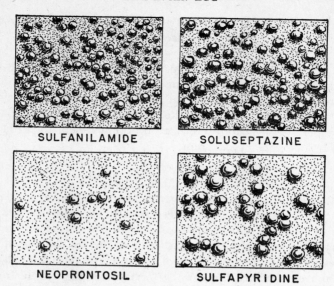

SULFANILAMIDE   SOLUSEPTAZINE

NEOPRONTOSIL   SULFAPYRIDINE

Fɪɢ. 65. Appearance of granular deposits formed on the eggshell as the result of feeding four different sulfonamides; magnified about 15 times. (After Genest and Bernard, 1945.)

PHYSIOLOGICAL AND ANATOMICAL FACTORS

Shell texture is produced by glandular secretion. The egg is glossy, for example, when an abundant deposit of organic material (cuticle) is made on the surface of the shell. Glandular activity depends in part upon certain aspects of anatomical structure, such as the size and number of secretory cells. For this reason, the individual's eggs tend to resemble one another in shell texture. Any radical departure from the usual type of shell may ordinarily be considered as indicating a change in physiological condition. Since rough or "sandy" texture is most likely to appear at the end of a prolonged period of high egg production, it probably indicates that the cells supplying calcific material are showing the effect of sustained egg-laying (*Hays, 1937a*). Minor variations in texture are sometimes observable even within the laying cycle.

Cycle of Laying

In the opinion of Berg (*1945*), the relative smoothness of the shell varies according to the egg's position in the cycle. He notes that the first egg of any cycle is usually the smoothest, and that the shell becomes progressively rougher as the cycle advances. In cycles of more than 3 eggs, however, he observes that the last 1 or 2 eggs are again smoother than those immediately preceding them.

Anatomical Obstructions

When the isthmus is made narrower by surgical operation, the eggs laid subsequently may bear ridges at either end, as if the egg membranes were wrinkled at the time of shell deposition (*Asmundson, 1931c*). It has never been demonstrated whether or not naturally occurring ridges are due to analogous obstructions.

MATHEMATICAL DETERMINATION OF SIZE

Although it is possible to determine the egg's weight, volume, and surface area by direct methods, it is obviously desirable to have mathematical formulae for calculating them from a few simple measurements. Such formulae are useful, for example, in estimating the original weight of museum eggs, or of eggs that have grown lighter because of the evaporation of moisture (see Chapter 10, "Preservation"). It is frequently more convenient, also, to calculate the volume and surface area of an egg than to ascertain these values directly without laboratory facilities.

However, the great variability with which the egg deviates from the perfect ellipsoid makes it impossible to calculate its weight, volume, and surface area with complete accuracy. The only available mathematical method for describing the egg's shape is by means of the shape index, which has been previously discussed.

THE STANDARD EGG

In formulating equations for the mathematical appraisal of the egg, or in making computations, a standard of reference is valuable. For this purpose, the egg of most typical configuration and dimensions may be chosen. In outline, it is an asymmetrical ellipse, or Cassinian oval, of which one end is somewhat blunter than the other. It is delineated by a smooth curve, without irregularities; and a cross section at any level is a perfect circle.

The standard hen's egg, with several self-explanatory dimensions, is

Weight	58.0	gm.
Volume	53.0	cc.
Specific gravity	1.09	
Long circumference	15.7	cm.
Short circumference	13.5	cm.
Shape index	74	
Surface area	68.0	sq. cm.

illustrated in Fig. 66. The data given above make its description more complete. Standard length and breadth measurements for the eggs of some common birds may be found in Table 10.

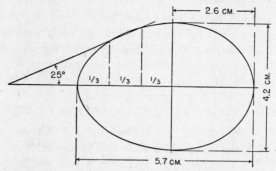

Fig. 66. A standard, or ideal, chicken egg. (Based on numerous measurements made by the authors.)

TABLE 10

Average Dimensions of Eggs of Some Common Domesticated and Game Birds *

Species	Length (centimeters)	Width (centimeters)
Ostrich	17.0	13.5
Swan	11.3	7.4
Goose (Embden)	8.7	6.1
Goose (Canada)	8.4	5.9
Peafowl	6.8	5.1
Turkey (Bourbon Red)	6.7	4.7
Turkey (Holland)	6.6	4.7
Duck (Pekin)	6.6	4.8
Duck (Muscovy)	6.2	4.5
Duck (Runner)	6.1	4.5
Duck (mallard)	5.9	4.2
Chicken (heavy)	5.8	4.4
Chicken (medium)	5.7	4.2
Chicken (light)	4.4	3.3
Jungle fowl	4.8	3.5
Guinea fowl	4.9	3.7
Pheasant (ringnecked)	4.4	3.5
Grouse (ruffed)	3.9	2.9
Pigeon (common)	3.7	2.8
Quail (bobwhite)	3.0	2.3
Canary	2.1	1.1
Hummingbird	1.3	0.8

* From the records of the authors.

ESTIMATION OF WEIGHT

The original weight of the museum egg can be estimated, approximately, by filling the blown shell with water, sealing the hole, and weighing the egg. The specific gravity of the egg contents is so close to that of water that the results of this method are accurate enough for general purposes. Bergtold (*1929*) filled the empty eggshell with chloroform of known density, instead of water, and developed the following equation:

$$X = \frac{W - S}{A} \times B + S,$$

where X is the original weight of the egg, W the weight of the egg when filled with chloroform, S the weight of the empty shell, A the specific gravity of chloroform, and B the assumed specific gravity of the egg. A second equation that he devised also takes into consideration the specific gravity of the egg, as well as its length (L) and breadth (B):

$$W = \tfrac{11}{21}(LB^2)S,$$

where S is the specific gravity of the fresh egg. After the substitution of 1.075 and 1.043, respectively, for the specific gravity of the eggs of precocial birds (specifically, chickens) and of fourteen species of altricial birds, the equation becomes, for precocial birds,

$$W = 0.5632LB^2,$$

and for altricial birds,

$$W = 0.5463LB^2.$$

He believed that his equations were accurate within a 5 per cent margin of error.

Other equations for calculating the egg's weight are based on the dimensions of the egg, usually both the long and the short diameter. Hoxie (*1887, 1890*), however, neglected breadth in his formula. Since the length of the egg is more variable than the breadth, his error was 5 to 20 per cent. Pearl and Surface (*1914b*), in devising a number of equations, succeeded in reducing this error to less than 3 per cent by taking both length and breadth into account and using several constants of variation.

Since the shell, in the eggs of each species, constitutes a certain constant percentage of the egg's total weight, Schönwetter (*1932*) proposed the following equation:

$$W = \tfrac{1}{2}(LB^2 + w),$$

in which W is the weight of the egg, L its length, B its breadth, and w the weight of the shell. In the eggs of small altricial birds, the weight of the shell may be only 5 per cent of the total weight, but in the eggs of precocial birds it is 10 to 14 per cent of the total (see Chapter 3, "Structure").

ESTIMATION OF VOLUME

The volume of the egg may be obtained by measuring the volume of water that it displaces when submerged.

Purely mathematical estimations of the egg's volume have been based chiefly upon the close resemblance of the egg to the prolate spheroid. After correcting for a standard error, determined by observation, it is possible to calculate the egg's volume by the equation for computing the volume of the prolate spheroid. Pearl and Surface (*1914b*) adjusted the equation to compensate for an average 2.2 per cent error, so that, instead of

$$V = \frac{\pi LB^2}{6} \quad \text{or} \quad V = 0.5236LB^2,$$

it became:

$$V = \left(\frac{\pi LB^2}{6}\right) - 0.022\left(\frac{\pi LB^2}{6}\right).$$

Worth (*1940*), however, considered the error to be 15 per cent and revised the formula accordingly:

$$V = 0.85\left(\frac{\pi LB^2}{6}\right).$$

A number of constants have been used (*Asmundson and Baker, 1940; Baten and Henderson, 1941*). Szielasko (*1904*) and Grossfeld (*1933a*) claim that an error of only 1.1 per cent is made in using the equation $V = 0.519LB^2$. Similarly, by using the constant 0.526 for hens' eggs of various sizes and shapes, an error of less than 2 per cent has been obtained (*Romanoff and Koshkin, unpublished*).

If the weight is known, and represented by y, the volume may be calculated by the equation $V = 0.933y$, according to Baten and Henderson (*1941*). The accuracy of this equation may be increased somewhat by using one constant for precocial birds and another for altricial birds (*Romanoff and Koshkin, unpublished*). Thus it becomes, for precocial birds:

$$V = 0.913W,$$

and for altricial birds:

$$V = 0.959W,$$

where W is the original weight of the egg.

ESTIMATION OF SURFACE AREA

The surface area of the egg has been determined by covering it with adhesive tape, the area of which is measured after removal (*Murray, 1925*).

Mathematically, the egg's surface area may be estimated by the equation (*Saija, 1899; Schönwetter, 1932*):

$$S = \pi \frac{BP}{2},$$

where S is the surface area, B the breadth, and P the long circumference.

Sometimes, the egg's surface is simply considered as the two-thirds power of the volume. The equation for the volume of the prolate spheroid may therefore be modified as follows, to apply to the egg:

$$S = k \left(\frac{\pi L B^2}{6} \right)^{2/3}.$$

The constant, k, has been given various values, 4.63 (*Dunn and Schneider, 1923*), 5.07 (*Murray, 1925*), and 4.83 (*Romanoff and Koshkin, unpublished*).

After calculating the egg's volume by the equation $V = \frac{\pi}{6} (LB^2)$, and determining the actual volume by the specific-gravity method, substitutions may be made in the following equation (*Marshall and Cruickshank, 1938*):

$$A = S \left(\frac{V}{V'} \right)^{2/3},$$

where A is the actual surface of the egg, S the calculated surface of a prolate spheroid of the same size as the egg, V the egg's actual volume, and V' the calculated volume. It is said that this method is accurate within 1 per cent, if the difference between V and V' is not over 4 per cent.

Several other very useful and accurate equations for calculating the egg's surface area have been suggested by Romanoff and Koshkin (*unpublished*). The constants employed in these equations were determined from observations on eggs of the standard 58-gm. weight. However, the equations are applicable to an egg of any size. The smallest error is given by the equations based on volume:

$$S = kV^{2/3} = 4.831V^{2/3},$$

and on breadth and length:

$$S = k\pi(B^2L)^{2/3} = 3.138(B^2L)^{2/3}, \quad \text{where } k = 0.999.$$

An average error of 2.4 per cent is given by the equations based on weight:

$$S = kW^{2/3} = 4.558W^{2/3};$$

on both circumferences (C_B and C_L):

$$S = \frac{C_B C_L}{k\pi} = \frac{C_B C_L}{3.010}, \quad \text{where } k = 0.958;$$

and on center of gravity $\left(\dfrac{L + B}{2}\right)$:

$$S = \pi\,\frac{B\left(\dfrac{L + B}{2}\right)}{k} = 3.279B\left(\frac{L + B}{2}\right), \quad \text{where } k = 0.958.$$

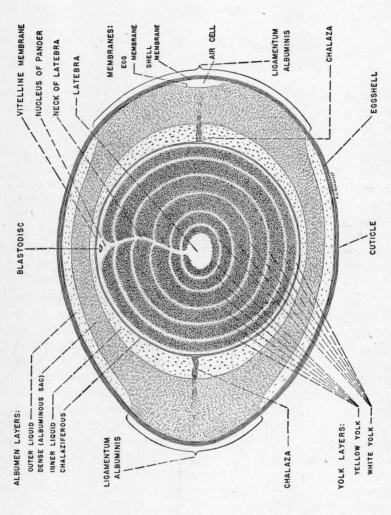

VITELLINE MEMBRANE

NUCLEUS OF PANDER

NECK OF LATEBRA

LATEBRA

MEMBRANES:

EGG MEMBRANE

SHELL MEMBRANE

AIR CELL

LIGAMENTUM ALBUMINIS

CHALAZA

EGGSHELL

BLASTODISC

CUTICLE

ALBUMEN LAYERS:

OUTER LIQUID

DENSE (ALBUMINOUS SAC)

INNER LIQUID

CHALAZIFEROUS

LIGAMENTUM ALBUMINIS

CHALAZA

YOLK LAYERS:

YELLOW YOLK

WHITE YOLK

Structure of the hen's egg, shown by a section through the long axis.

Chapter Three

STRUCTURE

The avian egg is an architectural marvel.

The bird's egg is one of the most complex and highly differentiated reproductive cells. Its structural elements are arranged with great precision, and their total organization is essential to the specific function of each part. The morphology of the egg is inseparably linked to its biological significance, its potential ability to maintain the continuity of life.

Fundamentally, the egg is comprised of a minute center of life, about which are accumulated relatively enormous amounts of inanimate food substances, the whole enclosed in protective structures. Given the proper combination of circumstances, the living fraction of the egg is activated and transforms the nonliving mass into an organism capable of independent existence.

GENERAL DESCRIPTION

The interdependence of structure and function is rarely more manifest than in the bird's egg. Deep in the interior of the egg, the living cells are clustered in a small area, the blastoderm (analogous to the blastodisc in the infertile egg), which is in intimate contact with the yolk and its great stores of food. Surrounding these most essential elements is the albumen, an elastic, shock-absorbing, insulating semisolid with a large content of water. Together, the yolk and the albumen furnish readily accessible materials in sufficient amount to sustain the life of the embryo for an extended period of time—a period of less than 2 weeks for the wren, but of almost 8 weeks for the emu. Finally, the egg is invested with multiple external coverings which shield the embryo from physical harm, provide for the respiratory exchange of gases, and conserve the food and water supply.

PROPORTIONS OF COMPONENT PARTS

The egg's total weight is not always distributed in the same way among the three chief component parts, which are yolk, albumen, and shell (with

113

membranes). All the eggs of a species, however, are of more or less similar proportional composition.

SPECIES DIFFERENCES

When the eggs of birds are grouped according to the relative amounts of the yolk and albumen, they fall naturally into two classes. Eggs in which the yolk makes up 30 to 40 per cent of the total weight belong without exception to precocial birds; eggs in which the yolk constitutes a lower percentage of the total weight (between 15 and 20 per cent) belong to altricial species (Table 11).. Of the parts of the egg, the yolk has the greatest food value. A relatively large yolk assures a fairly

TABLE 11

PROPORTIONAL PARTS OF BIRDS' EGGS *

Species	Weight of Egg (grams)	Proportional Parts		
		Albumen (per cent)	Yolk (per cent)	Shell (per cent)
Precocial birds:				
Ostrich	1400	53.4	32.5	14.1
Emu	710	52.2	35.0	12.8
Goose	200	52.5	35.1	12.4
Turkey	85	55.9	32.3	11.8
Duck	80	52.6	35.4	12.0
Chicken	58	55.8	31.9	12.3
Guinea fowl	40	52.3	35.1	12.6
Pheasant, ringnecked	32	53.1	36.3	10.6
Partridge	18	50.8	37.0	12.2
Plover	15	50.7	40.8	8.5
Average	..	52.9	35.2	11.9
Altricial birds:				
Golden eagle	140.0	78.6	12.0	9.4
Buzzard	60.0	76.8	14.0	9.2
Dove	22.0	72.4	18.1	9.5
Pigeon	17.0	74.0	17.9	8.1
Jay	8.5	68.1	26.6	5.3
Starling	7.0	78.6	14.3	7.1
Robin	2.5	70.3	24.2	5.5
Hedge sparrow	2.0	72.5	21.6	5.9
Golden-crested wren	1.0	71.0	24.1	4.9
Hummingbird	0.5	69.7	25.3	5.0
Average	...	73.2	19.8	7.0

* Compiled from various sources, including Tarchanoff (1884), Friese (1923), Groebbels and Möbert (1927), Romanoff (unpublished), and others.

advanced stage of development in the young at hatching; but in species that lay small-yolked eggs, the young are helpless nestlings.

In addition, most altricial birds lay eggs that have relatively thin shells as well as small yolks. In proportion to total egg weight, the shells are considerably heavier in the eggs of precocial birds.

By listing the eggs of various species of altricial or of precocial birds by size, in descending order, it can be seen that egg weight and the percentage of shell diminish simultaneously; the yolk, however, constitutes an ever-larger proportion of the total weight of the egg (cf. Table 11). The smaller the egg, the greater, relatively, is its surface area. A large surface area and a thin shell tend to permit rapid loss of heat. As if to compensate, the embryo is provided with larger amounts of food substances of high energy value, in the form of yolk.

Values for the Chicken Egg

By weight, the hen's egg is roughly six parts albumen, three parts yolk, and one part shell. The quantitative relationships of the average egg's component parts are given in Table 12.

TABLE 12

PROPORTIONAL COMPOSITION OF THE AVERAGE HEN'S EGG

Parts of the Egg	Actual Weight (grams)		Relative Weight (per cent)	
Albumen:	32.9		55.8	
Outer fluid layer		7.6		23.2
Middle dense layer		18.9		57.3
Middle fluid layer		5.5		16.8
Chalaziferous layer		0.9		2.7
Yolk	18.7		31.9	
Shell with membranes:	6.4		12.3	
Shell		6.2		96.9
Shell membranes		0.2		3.1
Whole egg	58.0		100.0	

The actual and relative weights of the egg's structural elements, especially of the shell (*Curtis, 1914b; Jull, 1924a; Asmundson, 1931d; Olsson, 1936*), may deviate rather widely from the values given in Table 12. In fact, there is a prevailing lack of uniformity in the proportional composition of eggs, even of the eggs of a single individual.

INFLUENCING FACTORS

Diverse relative amounts of the major component parts are found in eggs laid at various seasons by hens of different strains, ages, and productivity, subjected to a variety of environmental conditions, and fed sundry diets. The relative proportions of the egg's components are modified by many of the same influences that affect the total weight of the egg (see Chapter 2, "External Characteristics"). This is not surprising, in view of the relationship existing between the percentages of the parts and the weight of the entire egg.

Egg Weight

When hens' eggs are arranged in a series according to weight, the trend in their percentage composition is much the same as that already seen

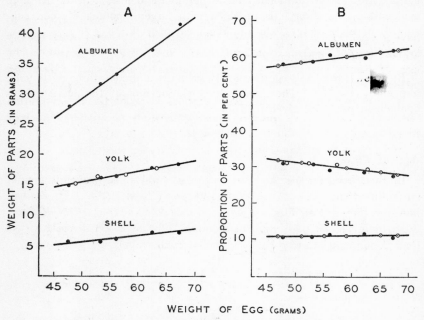

FIG. 67. Relationship between the weight of the Leghorn egg and *A,* the actual, and *B,* the proportional, amounts of its major components (albumen, yolk, and shell). (After the data of Atwood and Weakley, 1917, in solid dots; Jull, 1924a, in circles; and Olsson, 1936, in circles with dots.)

upon similarly arranging the eggs of altricial and precocial birds. The larger chicken eggs generally contain greater absolute amounts of all component parts (Fig. 67), and relatively less yolk (*Atwood and Weakley, 1917; Olsson, 1936*) and more albumen, than the smaller eggs (*Tso, 1925a; Olsson, 1936*). On the other hand, the relative amount of shell is the same in hens' eggs of all sizes.

If the hen's diet is deficient, especially in vitamin D (*Olsson, 1936*), there may be greater than normal variability in the proportional composition of the egg. If the correlation between egg weight and the relative amounts of albumen and yolk is not evident (*Krizenecky, 1934*), lack of control of some dietary factor may be suspected.

Breed Differences

Among chickens, there are no detectable breed differences in the degree of variability in the weight of the egg's major components. However, in a particular breed, it is possible for the relative weight of some part to be consistently different from the average for the species. Eggs from certain flocks of Plymouth Rocks, for example, have been observed to have a smaller percentage of shell than those from several other heavily laying breeds (*Taylor and Martin, 1928; Hall, 1939*). The yolk in the eggs of Rhode Island Reds may be disproportionately small (*Meszaros, 1934*).

It must be pointed out that the variations in proportional composition observed in eggs laid by different breeds of fowl are no greater than those found in eggs laid by different flocks of the same breed.

Age of Fowl

In their proportions, the components of pullets' eggs exhibit greater variability than those of mature hens' eggs (*Olsson, 1936*). In general, however, there is a smaller percentage of yolk (*Asmundson, 1933a; Olsson, 1936*) and a larger percentage of albumen (*Olsson, 1936*) in pullets' eggs than in the eggs of older birds, as shown in the tabulation.

	YOLK (per cent)	ALBUMEN (per cent)	SHELL (per cent)
Young hens:			
Asmundson (*1933a*)	27.4	61.6	11.0
Olsson (*1936*)	28.0	61.1	10.9
Older hens:			
Olsson (*1936*)	29.8	59.6	10.6

During the first few months after laying begins, the relative amount of yolk in the pullet's egg gradually increases, whereas that of the shell decreases rather rapidly. The percentage of albumen remains the same (Fig. 68). At the end of the first laying year, the proportional composition of the egg has reached a fairly stationary level.

As the total egg weight increases throughout the first year (see Chapter 2, "External Characteristics"), the amount of yolk increases absolutely

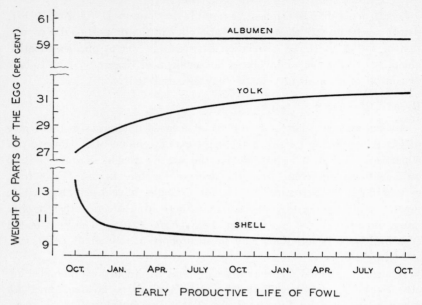

FIG. 68. Changes in the proportional composition of the egg during the hen's early productive life. Based on monthly averages of twelve birds from the commencement of laying to the end of the second laying year. (After Curtis, 1914b.)

(Fig. 69) as well as relatively. The increments in yolk size are quite large at first and then gradually become smaller.

Seasonal Changes

Of all the egg parts, the yolk shows the least seasonal fluctuation in actual weight. In the winter, the yolk is proportionally at its smallest. With the approach of the breeding season, it increases in absolute size. By the time it has attained its maximum, total egg weight and the actual amount of albumen and shell have diminished. Consequently, the relative increase in the amount of yolk is also considerable (*Jull, 1924a; Philpott, 1933*).

The albumen shows the greatest seasonal variation in proportional amount (*Curtis, 1914b*). During the winter, the albumen increases in weight more rapidly than the yolk. At the beginning of the breeding season, while the yolk is growing larger in absolute amount, the albumen remains constant and therefore suffers a proportional loss. A further loss, both proportional and actual, is sustained during the summer, when the amount of albumen and shell and the weight of the egg decrease, and the yolk remains the same size or becomes only slightly smaller.

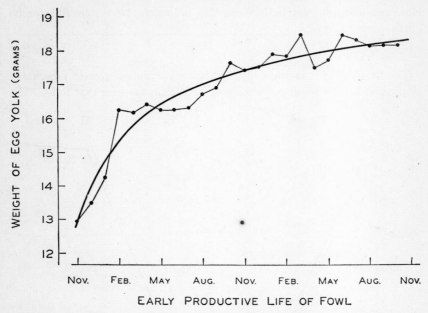

FIG. 69. Increase in the weight of the yolk during the early productive life of the fowl. Based on monthly average of eggs from twelve Barred Plymouth Rocks. (After Curtis, 1914b.)

The weight of the shell declines, both absolutely and proportionately, during the warmer months of the year. The tabulated data indicate the correlation between decrease in shell weight and increase in average maximum temperature (*Heywang, 1946*). An artificially produced increase in the temperature of the hen's environment demonstrates the

	SHELL WEIGHT	
TEMPERATURE (°C.)	ACTUAL (grams)	PROPORTIONAL (per cent)
19	5.6	8.8
23	5.4	8.5
32	4.9	8.0
39	4.2	7.1

effect of undue heat upon the proportional composition of the egg. The egg and all its component parts decrease in actual size; the yolk gains relatively, but the albumen and shell suffer a proportional loss, the shell to the greater extent (*Bennion and Warren, 1933a*). Of all the components of the egg, the shell shows the greatest variability in amount.

However, many factors besides seasonal influences are responsible for its fluctuations in weight.

Laying Cycle

The first egg of a laying cycle invariably contains a greater proportion of albumen, and a lower percentage of yolk, than the eggs that follow. In

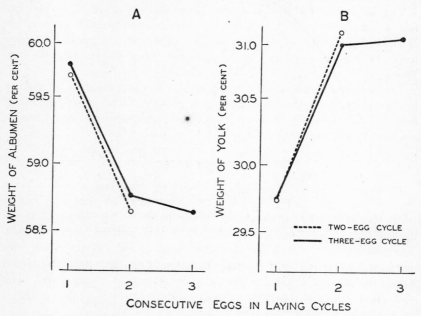

Fig. 70. Changes in proportional amounts of *A*, albumen, and *B*, yolk, in successive eggs of a cycle. Based on averages for 232 two-egg cycles and 59 three-egg cycles. (After Philpott, 1933.)

proportional composition, the second egg differs markedly from the first. Its total weight and its content of albumen are so much less that its yolk, although somewhat smaller in actual size, is relatively much heavier. The albumen of the second egg decreases proportionally as well as absolutely (Fig. 70). In the succeeding eggs, there is a gradual decline in weight, and percentage composition tends to become more constant as minimum size is approached. The albumen shows the greatest actual and proportional decrease (*Philpott, 1933*).

Nutrition

Certain dietary factors affect the proportional composition of the egg. For example, the percentage of shell decreases when there is insufficient calcium in the diet (*Buckner, Martin, and Peter, 1928*). For this reason, a reduction in total feed consumption in hot weather may diminish shell weight by limiting calcium intake.

The amount of the shell may also be reduced by a deficiency in vitamin D, since vitamin D is necessary for proper utilization of calcium. It is largely because of its effect on the shell that a vitamin deficiency produces unusual variability in the egg's percentage composition (*Olsson, 1936*).

The feeding of desiccated thyroid results in the formation of slightly heavier shells. On the other hand, the growth of the yolk is retarded, and the yolk therefore becomes actually and relatively smaller (*Asmundson and Pinsky, 1935*).

THE BLASTODISC

Location. When a fresh egg is opened by removing a sizable piece of shell and shell membrane from the upper side, an opaque, circular, white spot is usually visible on the surface of the yolk. This spot in the unfertilized egg is called the *blastodisc*, or, in older writings, the *cicatricula*. The protoplasm of which the blastodisc is composed and the white yolk material immediately surrounding and underlying the blastodisc are of lighter weight and lower density than the remainder of the yolk. The yolk therefore turns to bring the blastodisc uppermost, unless there is unusually great tension on the chalazae.

External Appearance. The entire blastodisc measures 3 to 4 mm. in diameter. Its central portion, or the germinal area, is encircled by a more opaque collar, the periblast. These two regions are divided by the inner periblastic ring. The margin of the blastodisc may merge into the surrounding yolk so that the blastodisc's periphery, known as the outer periblastic ring, may be somewhat indefinite. The periblast may have a mottled appearance, due to the presence of vacuoles called lacunae (*Kosin, 1944*). The lacunae are characteristic of the infertile egg, since they are usually very few, or totally lacking, in the fertilized egg. The blastodisc in the infertile egg also differs in size and shape from the blastoderm [1] in the fertile egg. A small circular area with many lacunae usually indicates an unfertilized egg. A larger, oval, homogeneous area is evi-

---

[1] The blastodisc of the unfertilized egg should not be confused with the homologous structure, the *blastoderm*, in the fertilized egg.

dence of development, which, except for some parthenogenetic segmentation (*Bartelmez and Riddle, 1924; Kosin, 1945*), does not occur without fertilization.

MICROSCOPIC STRUCTURE

Blastodisc (in Unfertilized Egg)

For microscopic examination of the blastodisc, sections are made through it perpendicular to the surface of the yolk. The center of the

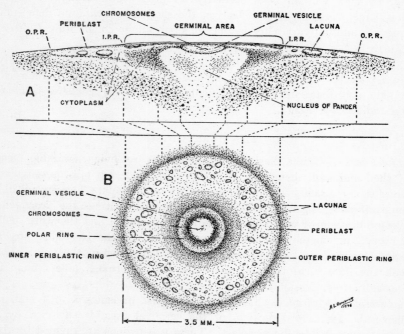

FIG. 71. Structure of the blastodisc (in unfertilized egg) showing *A*, perpendicular section, and *B*, surface view. (Drawn from the authors' specimens, and from several sources, including Oellacher, 1872; Kionka, 1894; Barfurth, 1896; Bartelmez, 1912; and Kosin, 1944.)

blastodisc is occupied by a roughly lens-shaped mass about 0.5 mm. in diameter and less than 0.1 mm. thick (*Sonnenbrodt, 1908*), which lies directly beneath the vitelline membrane. This mass is all that remains of the germinal vesicle, or nucleus, of the egg, since the nuclear membrane breaks down shortly before the yolk is released from the ovary. The chromosomes lie bunched in the center of the nuclear mass in the metaphase of the final stage of maturation. A relatively thin, disk-shaped layer of cytoplasm extends from the germinal vesicle to the outer peri-blast. An extremely thin layer of cytoplasm continues over the re-mainder of the yolk's surface, beneath the vitelline membrane (Fig. 71).

Immediately surrounding the central protoplasmic mass, consisting of nuclear material and a little cytoplasm, is the polar ring, composed of granules of white yolk. About the polar ring is a larger ring of deeply staining white yolk, which in cross section appears as a wedge with the acute angle outward. Its outer border constitutes the inner periblastic ring. Continuous with the lower surface of the wedge, a very thin collar of white yolk extends beneath the entire broad margin of the blastodisc to the outer periblastic ring. Underlying the blastodisc and the collar of wedge granules is a mass of white yolk in the form of an inverted cone, called the *nucleus of Pander*. Thus the entire underside of the blastodisc is invested with white yolk (cf. Fig. 71).

Blastoderm (in Fertilized Egg)

At the time of laying, the germinative portion of the fertilized egg, the *blastoderm,* is composed of many cells, disposed in two layers. The central portion of the blastoderm is separated from the underlying yolk by the subgerminal cavity (Fig. 72). Cell division has been most active along the long axis of the embryo; consequently, the blastoderm has as-sumed an oval shape. It has also perceptibly increased in size, so that its average diameter is now 4.41 mm. (*Edwards, 1902*).

Elongation in the direction of the embryonic axis reveals the egg's latent bilaterality (*Bartelmez, 1912, 1918*), previously undetectable. Although there is some variation in the direction of the embryonic axis, it generally deviates 20 degrees from the short axis of the egg. At this stage of development, the anterior end of the embryo is nearer the pointed end of the egg.

Two regions are easily distinguishable in the blastoderm. These are the transparent *area pellucida,* in the center; and the *area opaca,* a wide, oval collar which forms a margin (cf. Fig. 72). Together these areas give rise to all the embryonic and extraembryonic structures of the developing egg. The oval *area pellucida,* which averages 2.51 mm. in diameter, gives rise to the embryo proper; the *area opaca,* to structures that function only during development.

THE YOLK

The yolk is the most important part of the egg. It cradles the blasto-derm, from which the embryo arises; it also contains the mass of nutri-tive materials that support embryonic development.

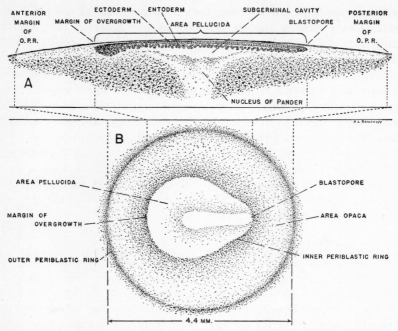

F IG. 72. Structure of the blastoderm (in fertilized egg) showing *A,* perpendicular section, and *B,* surface view. (Drawn from the authors' specimens, and from various sources, including Duval, 1884.)

EXTERNAL CHARACTERISTICS

The yolk, or *vitellus,* of the bird's egg is an almost spherical, yellow to orange body, located near the center of the egg, and enclosed in a delicate, elastic, lustrous membrane.

Location. In some newly laid eggs, the yolk is firmly held in a central position. In others, it is surrounded by a relatively large amount of liquid albumen and is consequently less restricted in movement. The chalazae tend to support the yolk along the long axis of the egg. Since the chalaza at the pointed end of the egg is usually the more firmly attached of the two, it permits less latitude for movement than the chalaza at the blunt end. The yolk can rotate about the egg's long axis until the chalazae become tautly twisted. When a fresh egg is spun on its long axis and stopped abruptly, the yolk turns through at least one more revolution, according to Thomson (*1859, p. 63*).

Immediately after the egg is laid, the yolk, of greater specific gravity than the albumen, tends to sink below the center of the egg. Later, however, the albumen becomes more concentrated, because of loss of water to the yolk and to the surrounding atmosphere. The yolk then tends to rise. The extent to which the yolk shifts its position depends on the amount of fluid envelope in which it floats.

Shape. The shape of the yolk has erroneously been considered spherical. Careful measurements of the yolk of the chicken egg disclose the fact that the vertical diameter through the poles (the upper of which is occupied by the blastodisc) is considerably shorter than the diameter in the direction of the long axis of the egg. The side of the yolk toward the

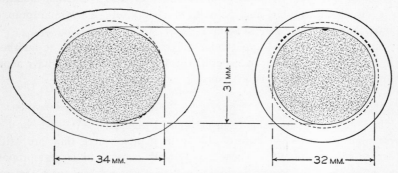

FIG. 73. Diagram showing the shape of the yolk in the newly laid chicken egg. (After Romanoff, 1943a.)

sharp end of the egg is slightly pointed. The surface bearing the blastodisc in its center is somewhat flattened (Fig. 73).

Size, Area, and Volume. In eggs from a flock of Leghorns, the average vertical, transverse, and long diameters of the yolk were found to be 31 mm., 32 mm., and 34 mm., respectively (*Romanoff, 1943a*). An average of these three values may be taken as the diameter. When one-half the diameter is substituted for r, the radius, in the following formulae,

$$(1) \ 4\pi r^2 \quad \text{and} \quad (2) \ \tfrac{4}{3}\pi r^3,$$

one arrives at close approximations of (1) the area and (2) the volume of the yolk. The average area of the yolks mentioned above is thus found to be 32.2 sq. cm., and the volume, 17.1 cc.

The volume of one day's deposition of yolk, which is represented by two successive rings, one white and one yellow, can be calculated by the method of Asmundson and Pinsky (*1935*). In this equation r_1 is equal to one-half the outside diameter of the outer of the two rings, r_2 to one-half the inside diameter of the inner of the two rings:

$$V = \tfrac{4}{3}\pi r_1^3 - \tfrac{4}{3}\pi r_2^3.$$

Yolk Color

It is possible for the yolks of eggs to vary considerably in color, from a very pale yellow to a dark and brilliant orange. For convenience, yolk color is usually classified merely as "light," "medium," or "dark." It may be more accurately evaluated by employing various color standards. On the scale devised by Ridgeway (*1912*), for example, yolk pigmentation ranges from 27 for light cream to 13 for dark orange red. In addition, a "color rotor" has been especially designed for grading yolks (*Heiman and Carver, 1935*). It consists of twenty-four watch crystals mounted on a circular turntable and painted in shades of yellow and orange. Numerically, its scale ranges from 1 (light) to 24 (dark).

Consumer preference in yolk color differs from one locality to another, and from time to time. In general, the intermediate shades have more appeal than those at either extreme.

CAUSES OF VARIATION

The yolk derives its color from carotenoid pigments. At least sixty of these pigments occur in nature, chiefly in plants, where they are often obscured by chlorophyll. Those in egg yolk consist, in largest proportion, of alcohol-soluble xanthophylls, and, to a lesser extent, of petroleum ether-soluble carotenes and cryptoxanthin (see Chapter 6, "Chemical Composition").

The intensity of the yolk's yellow color varies primarily according to the bird's consumption of carotenoid pigments (*Parker, 1927a*). Feeds differ considerably in their content of these pigments, which, for example, are lacking in most cereal grains except yellow corn. The concentration of the carotenoids in plants changes throughout the growing season and usually decreases after harvesting. The tabulated data indicate the diversity in the amounts of yellow pigments in some common feeds (*Peterson, Hughes, and Payne, 1939*).

Hens differ in their ability to assimilate carotenoid pigments from their diet, and also in their appetite for the feeds containing them. As a result, a flock often lays eggs in which the yolks cover a wide range of color, even when all birds receive identical rations. When choice is eliminated, as by feeding mashes alone, yolk color may still vary among the eggs of different birds, because individuals are not alike in their physiological capacity to transpose pigments from the feed to the egg yolk (*Hunter, Van Wagenen, and Hall, 1936; Peterson, Hughes, and Payne, 1939*).

| | PIGMENTS |
FEED	(milligrams/100 gm. dry material)
Carrots (yellow)	94.9
Oat grass	87.4
Wheat grass	78.7
Oat-grass silage	43.7
Fresh green alfalfa	26.7
Dehydrated alfalfa	10.4
Fresh green corn plant	9.2
Sudan grass	5.8
Alfalfa hay	0.7–4.5
Timothy hay	0.8–1.9
Yellow corn	0.28
White corn	Trace

Nutrition. As early as 1903, Stewart and Atwood noted that more deeply colored yolks were laid when white corn was replaced by yellow in the hen's diet. In 1905, Dryden observed a similar darkening of yolks upon feeding alfalfa. Later, Parker, Gossman, and Lippincott (*1926*) demonstrated that the peak of the distribution curve of yolk pigmentation could be shifted to a darker color by increasing the amount of greens or yellow corn, or both, in the flock's diet (Fig. 74). In addition, their results clearly showed how wide a variation in color may normally be expected in the yolks from any flock, regardless of the type of ration. The close relationship between yolk color and the actual amount of greens consumed by the bird was also revealed. Yolk colors corresponding to Ridgeway grades of 19 and 14 were respectively associated with the daily consumption, per bird, of 20 gm. and 45 gm. of greens.

The various carotenoid pigments are transferred from the feed to the yolk with different efficiency. Xanthophylls are best utilized (see Chapter 6, "Chemical Composition"). In addition, pigments from certain sources, notably yellow corn, are more efficiently transferred than those from other sources (*Hughes and Payne, 1937*).

A striking effect on yolk color is produced by feeding pimiento peppers (see Chapter 6, "Chemical Composition"), which contain the carotenoid pigment, capsanthin (*Brown, 1934*). Dark red yolks of very objectionable hue may result from the daily consumption of only 1.0 gm. of dried peppers per bird. Desirably bright yellow yolks may be obtained by reducing this quantity of pimiento by one-half (*Morgan and Woodroof, 1927*).

Yolk color quickly responds to a change in the carotenoid pigment content of the ration. How soon yolk pigmentation is affected depends, in part, upon the rate at which eggs are being laid. The color of the yolk may become abruptly darker or lighter with the first egg produced after modification of the diet (*Albright and Thompson, 1935*). Depending upon the quantity and source of the pigments added to the ration or withdrawn, the maximum effect may be observed with the second to the eleventh egg laid after the pigment content of the diet is

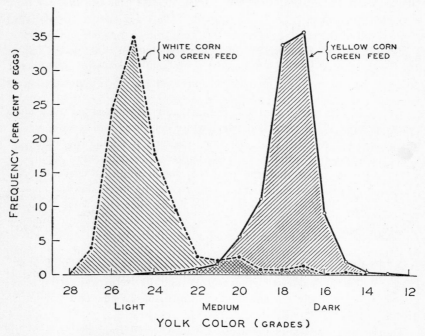

FIG. 74. Distribution of yolk color (standards of Ridgeway, 1912) in eggs from two flocks of chickens fed different rations. (After Parker, Gossman, and Lippincott, 1926.)

changed (*Titus, Fritz, and Kauffman, 1938*). In general, yolk pigmentation decreases more gradually than it increases (*Heiman and Wilhelm, 1937b*). An exceptionally long time is required for yolk color to return to its original intensity after the feeding of pimiento peppers is discontinued.

Because the yolk is built up in successive concentric layers from the center outward, it is possible for a sudden change in the carotenoid content of the bird's diet to cause the production of yolks that are not of the same color intensity throughout. The outer layers thus may be darker or lighter than the interior of the yolk.

The yolk of the new-laid egg occasionally shows an olive-green tinge.

There is no evidence that greenish yolks differ from yellow yolks in any respect other than color. The green pigmentation of the yolk in the so-called "grass egg" has been variously attributed to the ingestion of certain weeds (see Chapter 6, "Chemical Composition"), acorns (*Olsen, 1942b*), grass silage (*Bolton and Common, 1941*), and molasses oat-grass silage (*Gish, Payne, and Peterson, 1940*). Grass eggs are distinct from eggs that develop olive yolks in storage. The color change in held eggs is the result of the reaction between the ferric iron of the yolk and gossypol, a pigment derived from malvaceous plants in the hen's diet (see Chapter 10, "Preservation"). Even when fed in capsule form daily for 42 days, gossypol has been found to have no effect on the yolk pigmentation of new-laid eggs (*Thompson, Albright, Schnetzler, and Heller, 1932*), apparently because the ferric ions in the fresh egg are insufficiently dissociated to enter into a visible reaction (*Swensen, Fieger, and Upp, 1942*).

Effect of Dyes. Artificial colors may be imparted to the yolk by feeding birds various oil- and alcohol-soluble dyes, which become incorporated into the yolk fats (*Gage and Gage, 1908; Rogers, 1912; Riddle, 1911; Gage and Fish, 1924; Warren and Conrad, 1939*). From 0.160 mg. to 0.659 mg. of Sudan Red B, for example, may appear in the yolk when the hen receives 100 mg. of the dye daily (*Grossfeld and Kanitz, 1937*).

A few dyes and their corresponding effects on yolk pigmentation are given in the tabulation (*Denton, 1940*). The dyed yolk material is

Dye	Color Imparted to Yolk
Oil Red O	Red
Hexyl Blue	Blue
Alizarin Cyanine Green	Green
Alizarol Purple SS	Green
Oil Brown D	Magenta
Comassie Fast Black	Pink

simultaneously deposited on a number of ovarian yolks of various degrees of maturity, and therefore of various sizes. A single feeding of dye results in the appearance of a single artificially colored layer in every yolk laid subsequently for some time. In a cross section of the yolk, the dyed material is first seen as a peripheral ring. In each succeeding egg, the dyed ring is closer to the center of the yolk, and the periphery of the yolk is of normal color. If the dye is administered daily without interruption, the superficial layer of dyed yolk merely increases in thickness with every egg as the normally pigmented central portion diminishes in diameter, until, eventually, the entire yolk is artificially colored (as shown very diagrammatically in Fig. 75). When the dye is discontinued, yolk substance of normal pigment content is again deposited over the surface of all the immature dyed yolks present in the ovary. The two colors, natural and artificial, are now reversed in relative position. The dyed central portion grows smaller in each successively laid yolk, until the yolk's original homogeneous color is restored (cf. Fig. 75).

DEPOSITION OF PIGMENT IN YOLKS OF SUCCESSIVE EGGS

DISAPPEARANCE OF PIGMENT FROM YOLKS OF SUCCESSIVE EGGS

FIG. 75. Diagrams showing the distribution of the fat-soluble dye, Sudan III, in the yolks of a series of eggs laid during continuous administration of the dye (upper row), and after discontinuing the dye (lower row). (Based on the observations of Gage and Fish, 1924, and Henderson and Wilcke, 1933.)

Other Variables. When hens are confined and their rations controlled throughout the year, their yolks show no significant trend in color from month to month (*Van Wagenen and Hall, 1936*). In those parts of the world where green vegetation is of seasonal occurrence, yolk pigmentation varies periodically, if birds are permitted to forage at all times. After a winter of confinement on a diet regulated for the production of pale yolks, birds are sometimes given sudden access to free range in the spring. An almost immediate deepening of yolk color then occurs. Yolks may become progressively darker for at least 2 weeks. A return to the original regime restores the light shade within a comparable period of time (Fig. 76). Because of the ease with which yolk pigmentation may be controlled, the yolks of market eggs are fairly uniform in color throughout the year, in accordance with consumer demand.

There is little or no evidence for or against the existence of breed characteristics in yolk color. Kaufman and Baczkowska (*1938*) observed that Polish Greenlegs laid more deeply pigmented yolks than Leghorns or Rhode Island Reds, but only when the flocks were on free range.

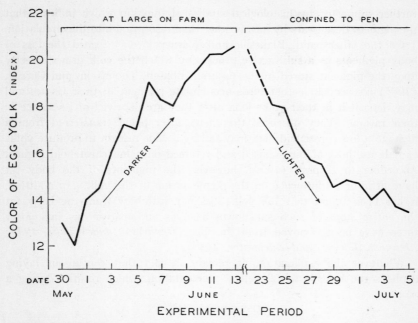

Fig. 76. Temporary increase in yolk color intensity (according to the scale of Heiman and Carver, 1935) during a period when confined birds were allowed access to green pasture. (After Snyder, 1945.)

RELATION OF YOLK PIGMENT TO BODY PIGMENT

The yellow color of the hen's beak, shanks, body fat, and (in some breeds) ear lobes fades during extended periods of intensive egg production. This phenomenon has been used for many years as a basis upon which to judge the laying capacity of fowl. As yet, it is not definitely known to what metabolic process the loss of pigment may be attributed.

Palmer and Kempster (*1919*) have demonstrated that loss of body color is not necessarily correlated with egg-laying, as it may be produced in cockerels by entirely removing the carotenoid pigments from the diet. Furthermore, female birds, if raised on carotenoid-free rations, deposit no yellow pigment in their body tissues upon being given carotenoid-rich feeds for the first time; yet they begin to lay yellow-yolked eggs. It has therefore been suggested that deposition of pigment is an excretory process which may take place either through the egg or through the surface of the body, but more readily by the former route. This view is further supported by histological studies of the skin which indicate that the pigment disappears by means other than resorption into the circulation.

On the other hand, Blakeslee and Warner (*1915*) regard the loss of body pigments as a subtractive process by which the yolk draws directly upon the pigment stored in the tissues. Bohren, Thompson, and Carrick (*1945*) also conclude that birds accumulate pigment in their tissues and may deposit it in their yolks long after they are deprived of its source in their ration. They observed that hens, after being transferred from a high- to a low-carotenoid diet, may fail by a wide margin to produce yolks as pale as those of birds raised and retained on a low-carotenoid regime. According to Kupsch (*1934*), however, the pigments of the body fat have little or no influence on the pigmentation of the yolk. In addition, it has been found that the yolk does not withdraw from the body fat the entire store of such substances as dyes and malvaceous fats, after these have been removed from the diet (*Henderson and Wilcke, 1933; Almquist, Lorenz, and Burmester, 1934*).

Whatever the explanation, it remains a fact that continuous laying depletes the hen's external yellow pigmentation, which returns during a pause.

GROSS ORGANIZATION

When the membrane surrounding the yolk is broken, a viscous fluid of seemingly homogeneous nature oozes out. However, there is actually great diversity and segregation of material in the yolk. By the simple procedure of hard-boiling the egg, the separate elements may be fixed in their original position. If the boiled egg is cut through the polar axis, the complex arrangement of the yolk becomes apparent.

Latebra

In the center of the yolk is an approximately spherical core about 6 mm. in diameter. This is the *latebra*. It contains a particularly fluid type of white vitellin which is not always completely hardened by boiling. The latebra constitutes about 0.6 per cent of the entire yolk (*Spohn and Riddle, 1916*). Extending from the latebra toward the blastodisc is a vase-shaped accumulation of white yolk, the *neck* of the latebra, whose outer portion bells out beneath the blastodisc to form the *nucleus of Pander*, or *nucleus cicatriculae*. This expanded portion constitutes less than 0.5 per cent of the whole yolk.

Stratification of Yolk

A concentric layer of yellow yolk in the form of a spheroid encloses the latebra. About the spheroid of yellow yolk is a very thin layer of white yolk. The entire yolk mass outside the latebra and neck is composed of similarly alternating yellow and white layers. An extremely thin white stratum immediately underlies the envelope that invests the vitellus. The layers in the interior of the yolk are interrupted and deformed by the neck of the latebra, as are the outer strata by the nucleus of Pander and the blastodisc. There are usually about six layers each of yellow and white yolk in eggs produced by hens that lay daily (Fig. 77). More strata are found when the rate of ovulation is lower.

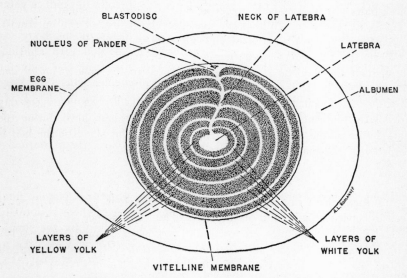

FIG. 77. Diagrammatic representation of the stratification of the yolk in a newly laid chicken egg. (After Romanoff, 1943a.)

The white layers of yolk are the narrower. They measure only 0.25 to 0.4 mm. in depth and contribute but little to the thickness of the various pairs of strata. The white yolk, exclusive of that in the latebra, the neck of the latebra, and the nucleus of Pander, constitutes only 3 to 4 per cent of the total yolk material (*Spohn and Riddle, 1916*). It differs in composition from yellow yolk, as it contains less fat and pigment (see Chapter 6, "Chemical Composition"). The successive layers of white yolk mark the ever-changing boundary of the ovum and record its daily growth.

The broader layers of yellow yolk vary in thickness, averaging about 2 mm. (*Romanoff, 1943a*). The tabulated data show the thickness of each successive pair of yellow and white strata, and the volume of the

PAIR OF STRATA	THICKNESS (millimeters)	VOLUME (cubic centimeters)
First	0.52	0.10
Second	1.07	0.29
Third	2.83	1.16
Fourth	2.80	3.19
Fifth	2.67	5.02
Sixth	1.56	4.06

material in each pair, proceeding from the center outward. The first three pairs are progressively thicker. The next two are slightly thinner, the last markedly so. With each successive layer, except the outermost, an increasingly great amount of yolk material is deposited.

When the fat and pigment content of the hen's diet does not vary throughout the day, no stratification is visible in the yolk (*Conrad and Warren, 1939*). It is only when pigment-containing feeds are restricted to a short period in the day, or when the feeds regularly given every day are low in pigments, that strata of white yolk are laid down during the dark hours of the early morning. At that time, the blood of the hen is depleted of the materials necessary for yellow yolk formation.

MICROSCOPIC STRUCTURE

A suspension of raw yolk material, when viewed in the microscope, is seen to consist of spherules of varying size and appearance (Fig. 78). In yellow yolk, the globules are comparatively large, from 0.025 to 0.150 mm. in diameter (*Foster and Balfour, 1874; Wiley, 1908*). The largest masses are found near the center of the yolk (*Moran, 1925*) and are

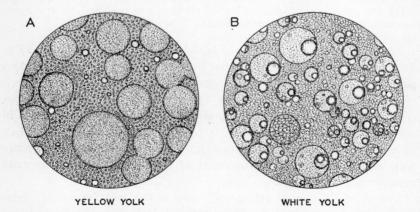

YELLOW YOLK　　　　　　　　WHITE YOLK

FIG. 78. Microscopic structure of *A*, yellow yolk, and *B*, white yolk, in the chicken egg. Magnified about 150 times. (After Wiley, 1908.)

filled with numerous highly refractive granules. The granules may possibly be protein, since they are not extractable with fat solvents (*Foster and Balfour, 1874*). In white yolk, the spherules are smaller, approximately 0.004 to 0.075 mm. in diameter, and contain one or more spherical refractive bodies of various sizes, suspended in clear fluid.

Vitelline Membrane

The yolk of the newly laid egg is enclosed in a thin, pliable envelope, known as the *membrana vitellina,* or yolk membrane. The structure and

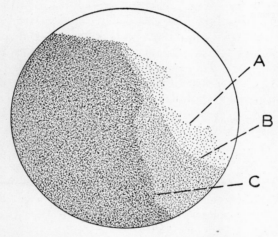

Fig. 79. Surface view of vitelline membrane, showing three layers: *A,* inner layer. *B,* middle layer. *C,* outer layer. Magnified 60 times. (After Moran and Hale, 1936.)

thickness of the vitelline membrane are still under discussion. In its natural state, it is possibly about 0.024 mm. thick (*Needham, 1931b*). Histological techniques probably cause it to swell or shrink, particularly during fixation and staining. By one method, it was found to consist of three layers (Fig. 79), the middle of which was equal in thickness to the other two combined. It was suggested that the middle layer was composed of keratin, and the others of mucin (*Moran and Hale, 1936*). A somewhat different technique revealed only two strata, a thin collagen layer within a much thicker mucin layer, which together totalled 0.048 mm. in thickness (*McNally, 1943*).

THE ALBUMEN

Surrounding the yolk, and enclosed by the external coverings of the egg, is a clear material of yellowish tint (Fig. 80), constituting 60 per cent of the weight of the egg. This is called the albumen (*albus,* white), or egg white, because of its appearance after coagulation. The yellowish cast is due to a pigment, ovoflavin (*Kuhn, György, and Wagner-*

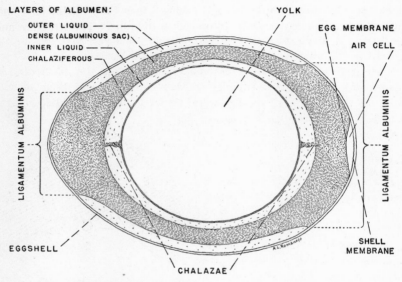

FIG. 80. Gross structure of the albumen of a newly laid chicken egg. (After Romanoff, 1943c.)

Jauregg, 1933). Occasionally, egg albumen may be pink (*Almquist and Lorenz, 1932b; Schroeder, 1933*) or slightly green (*Fronda and Clemente, 1934*). Certain ingredients of the diet, not commonly used, may cause these unusual tints.

GROSS ORGANIZATION

The uniformity of the albumen is only apparent. A variable amount of it, usually about 40 per cent, is a viscous liquid; the remainder is gelatinous and semisolid. The whole is organized in a definite manner (cf. Fig. 80).

Chalazae

Along the egg's long axis, a ropy structure of cloudy appearance spirals from the yolk into the albumen at either end of the egg. These two structures are the *chalazae,* so called from the Greek for "tubercles," or, according to Bartelmez (*1918*), "hailstones." The chalazae are composed of numerous fine, mucinlike fibers. At the proximal end of each chalaza, the fibers are firmly attached over the surface of the yolk, in its equatorial region. At the distal end, the fibers are interlaced with similar fibers in the albumen. Of the two chalazae, the cloacal, which extends toward the sharp end of the egg, is the longer and the larger; its fibers are disposed in two heavy strands, twisted counterclockwise on each other. The infundibular chalaza, at the blunt end of the egg, consists of a single strand of fibers, twisted clockwise and less firmly moored in the albumen than those of the cloacal chalaza.

When the egg is turned, the yolk rotates until the blastodisc is uppermost. The chalazae become twisted and taut, and their tension pulls the yolk nearer to the geometric center of the egg. The chalazae thus serve to stabilize the position of the yolk.

Layers of Albumen

The whole body of the albumen is disposed in four concentric layers.

CHALAZIFEROUS LAYER

The fibers of the chalazae continue in a slightly spiral course over the surface of the yolk. They form a matted, fibrous capsule, inseparable from, and embedded in, a thin layer of dense albumen immediately about the vitelline membrane and sometimes closely united with it. Together, the fibrous capsule and the envelope of dense albumen constitute the chalaziferous layer, or the *membrana chalazifera.* This entire layer, including the chalazae, constitutes only 2.7 per cent of the total volume

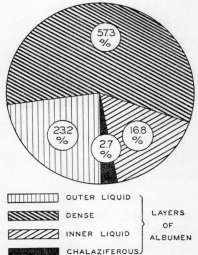

Fig. 81. Relative volume of each layer of albumen in the hen's egg. (From the data of Romanoff and Sullivan, 1937.)

of the albumen (Fig. 81) and weighs less than 1 gm. The weight of the two chalazae alone is 0.15 gm. (*Burmester and Card, 1939*).

INNER LIQUID LAYER

The yolk and the chalaziferous layer, moored in the approximate center of the egg by the chalazae, float in a fluid, viscous albumen. This is the inner liquid layer of the albumen. It constitutes 16.8 per cent of the total albumen (cf. Fig. 81). Mucin fibers are almost totally absent from this layer.

MIDDLE DENSE LAYER

The middle dense layer of albumen, often termed the albuminous sac, surrounds the inner liquid albumen. The albuminous sac constitutes 57.3 per cent, by volume, of the entire albumen (cf. Fig. 81). It is a heavy, plastic envelope capable of maintaining its shape to some degree. The numerous semisolid mucin fibers found in this layer provide a structural framework which contains liquid albumen dispersed in its interstices. The albuminous sac forms a cushion for the protection of the yolk and is sufficiently firm to provide anchorage for the chalazae, whose ends are embedded in it.

The albuminous sac is attached to the inner shell membrane at each extremity of the egg by a *ligamentum albuminis,* first described by Tredern (*1808*). The two ligaments are formed by the penetration of numerous mucin fibers of the albumen into the membrane. The average area of the attachment at the cloacal or sharp end of the egg has been found to be 2.54 sq. cm., that at the infundibular or blunt end, 10.2 sq. cm. (*Romanoff, 1943c*).

OUTER LIQUID LAYER

Surrounding the albuminous sac, except where the ligaments are attached, lies the outer liquid layer of albumen. It constitutes 23.2 per cent of the total volume of the albumen (cf. Fig. 81). Like the inner liquid layer, the outer liquid layer is a viscous fluid containing few mucin fibers.

PROPORTIONAL AMOUNTS OF LAYERS

Only the average value has been given above for the percentage of the entire albumen which each layer constitutes. In eggs taken at random (*Almquist and Lorenz, 1933a*), the distribution of the albumen among the three major layers is remarkably variable. In the chicken egg, the middle dense layer may constitute from 30 per cent to 80 per cent of the total volume of the albumen, the outer liquid layer from

10 to 60 per cent, and the inner liquid layer from 1 to 40 per cent (Fig. 82). The degree of variability differs from one strain of fowl to another. For example, in eggs laid by three different flocks, the proportion of the dense layer varied within the limits shown in the tabulation.

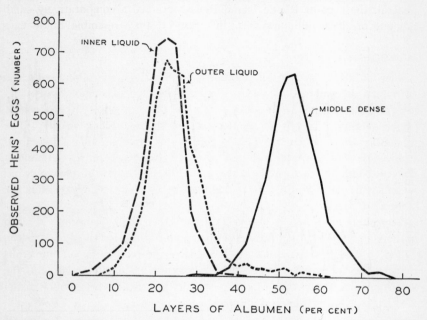

Fig. 82. Variability in the proportional amounts of the three major layers of egg albumen. (After Van Wagenen and Wilgus, 1935.)

	Dense Albumen (per cent)
Investigator	
Almquist and Lorenz (*1933a*)	35–57
Van Wagenen and Wilgus (*1935*)	30–78
Halnan and Day (*1935*)	32–86

The least variability in the proportional composition of the albumen is found in the eggs of the individual bird (*Holst and Almquist, 1931b*). The constancy with which a hen may lay eggs with a fairly uniform proportion of dense albumen is shown in Fig. 83. In the same flock, one hen may lay eggs in which the percentage of dense albumen fluctuates only slightly around an average of 43.5 per cent, and another hen may produce eggs in which it is never far from an average of 79.3 per cent (*Lorenz, Taylor, and Almquist, 1934*). On the other hand, there are some individuals in whose eggs the percentages of the different layers of albumen are quite variable (*Hunter, Van Wagenen, and Hall, 1936; Munro, 1938*).

In studying breed differences in the proportional composition of the albumen, Knox and Godfrey (*1934*) and Hall (*1939*) concluded that, in general, the eggs of Rhode Island Reds contained a comparatively small amount of dense albumen. In this respect, they resemble Silkie eggs

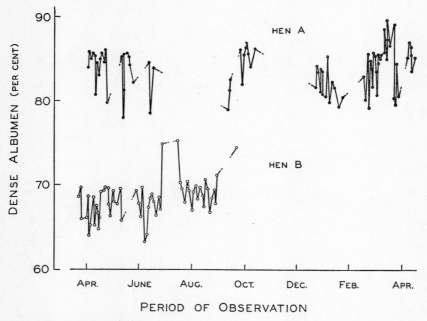

Fig. 83. Constancy in the percentage of dense albumen in eggs laid by each of two hens. (After Halnan and Moran, 1937.)

(*Gordon, 1940*). Both Barred Plymouth Rock (*Hall, 1939*) and Light Sussex eggs (*Gordon, 1940*) tend to contain a high percentage of dense albumen.

The dense albumen content of the egg seems to be an inherited characteristic (*Knox and Godfrey, 1938*), which is probably controlled by multiple factors (*Van Wagenen, Hall, and Wilgus, 1937*). By selection over a period of 5 years, Knox and Godfrey (*1940*) were able to develop a strain of Rhode Island Reds in whose eggs the dense albumen content was almost 20 per cent higher than in the eggs of the original stock. At the same time, the percentage of firm albumen was reduced slightly in the eggs of a collateral line of the same origin (Fig. 84). Crossbreeding results in progeny that lay eggs of intermediate type, although the influence of the dam is usually more apparent than that of the sire (*Lorenz and Taylor, 1940*).

Environmental conditions, particularly temperature, have a considerable effect on the firm albumen content of eggs. High temperature

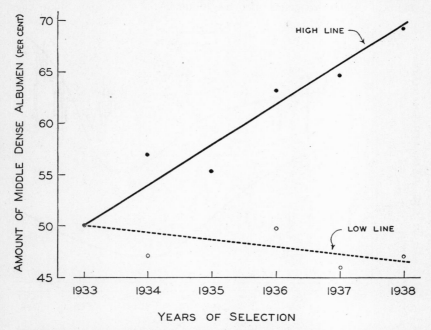

FIG. 84. Results of selective breeding for high and low percentage of dense albumen in the egg, as shown in two lines of Rhode Island Reds. (After Knox and Godfrey, 1940.)

may cause a reduction in the amount of firm white immediately after the egg is laid (*Lorenz and Almquist, 1936; Skoglund and Tomhave, 1941*). Possibly for this reason, summer eggs have a comparatively low percentage of firm albumen (*Hunter, Van Wagenen, and Hall, 1936*). It has been noted, however, that, in the eggs of both pullets and hens, there is a trend for the percentage of dense albumen to decrease from October to midsummer (*Knox and Godfrey, 1938*), as shown in Fig. 85.

The percentage of firm white is independent of the size of the egg (*Lorenz and Almquist, 1936*) and of the rate of production (*Knox and Godfrey, 1938*). Various investigators have been unable to demonstrate that the quantity of dense albumen contained in the egg is influenced by such factors as the type or amount of protein, green feed, or mineral supplement in the bird's diet (*Card and Sloan, 1935; Sowell and Morgan, 1936; Ringrose and Morgan, 1939*).

Few studies have been made of variations in the percentage of thin or liquid albumen. However, the proportion of the liquid albumen is usually reciprocal to the proportion of dense albumen.

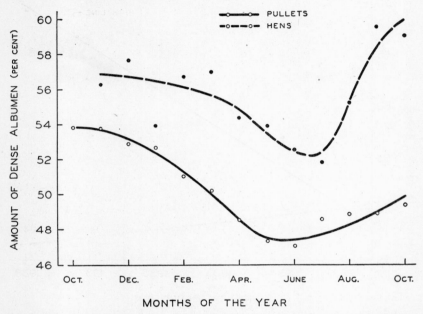

Fɪɢ. 85. Seasonal changes in the proportional amount of dense albumen in eggs of pullets and hens. (After Knox and Godfrey, 1938.)

MICROSCOPIC STRUCTURE

The albumen is composed of a framework of fine, semisolid fibers enmeshing liquid material. The majority of the fibers is found in the dense albumen, which owes its semblance of form solely to its fibrillar structure. The outer boundary of the albuminous sac is often sharply demarcated by a zone of long and closely packed fibers (*R. K. Cole, 1938*). Within the dense layer, stratification is visible macroscopically in the heat-coagulated and appropriately stained albumen (*Remotti, 1929; Moran and Hale, 1936*). Microscopic examination reveals that the strata, which are approximately 1.0 mm. wide, consist of compact layers of numerous fibers alternating with layers in which the fibers are few and scattered

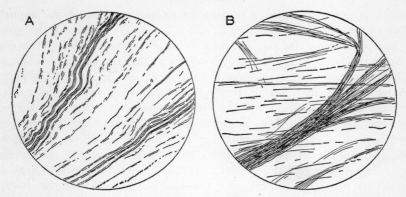

Fɪɢ. 86. Microscopic views of dense layer of egg albumen, showing fibrillae that enmesh the liquid portion.

A, magnified about 40 times (after Moran and Hale, 1936). *B,* magnified about 400 times (after Almquist and Lorenz, 1932b).

(Fig. 86-*A*). High magnification shows that the fibers are rather intricately interwoven (cf. Fig. 86-*B*).

THE MEMBRANES

A two-ply, parchmentlike envelope, shaped like the egg, fits snugly about the egg contents. The two layers of this envelope comprise the shell membranes.

GROSS ORGANIZATION

The inner, or egg, membrane, the *membrana putaminis* (*putamen,* Latin for pod or husk), surrounds the albumen. The membrane is in contact with the outer liquid albumen in all but the polar regions of the egg, where some of the mucin fibers of the albuminous sac penetrate the membrane to form the *ligamenta albuminis* (cf. Fig. 80). The outer surface of the inner membrane is firmly cemented to the inside of the outer membrane, except in a small area, usually at the blunt end of the egg. In this area, the space between the two membranes is occupied by the air cell, which will be discussed later.

The outer or shell membrane, the *membrana testae* (*testa,* Latin for shell), lies between the inner membrane and the shell. The fibers of the outer surface of the shell membrane are so firmly embedded in the inner surface of the shell that it is difficult to detach the membrane without tearing it.

In the hen's egg, the membranes appear chalk white. On closer examination, however, they are found to be slightly pinkish, due to the presence of a very small amount of porphyrin pigment (*Klose and Almquist, 1937*).

The membranes are pliable when moist but become brittle when dry (*Romankewitsch, 1932*). Although very thin, they are remarkably tough. Their toughness compensates for the brittleness of the shell. It is interesting to note that the relatively thin-shelled eggs of certain species contain thicker shell membranes than the comparatively thick-shelled eggs of other species. The shell membranes constitute but 0.63 to 0.83 per cent of chicken, silver pheasant, and ringnecked pheasant eggs, in which the shells are fairly sturdy. On the other hand, the membranes constitute 1.61 and 2.61 per cent of turkey and quail eggs, the shells of which are proportionally thinner (*Asmundson, Baker, and Emlen, 1943*).

On an average, the membranes in the hen's egg weigh 0.36 gm., equivalent to 0.6 per cent of a 58-gm. egg.

Thickness

The thickness of the two shell membranes in the eggs of different species of birds is roughly related to the size of the egg (*Asmundson, Baker, and Emlen, 1943*). Their relationship is shown in the table

Species	Thickness of Membranes (millimeter)
Ostrich	0.200
Swan	0.165
Chickens:	
Brahma	0.092
Leghorn	0.065
Bantam	0.050
Quail	0.067
Zebra finch	0.005

(*Romanoff, unpublished*), in which the species are arranged in descending order of egg size. It may be seen that the actual thickness of the membranes decreases as the size of the egg diminishes.

The combined thickness of the shell membranes may vary in eggs laid by birds of the same species. The greatest variation is probably found in chicken eggs (*Stewart, 1935*), because of their wide range in size. The eggs of the three breeds of chickens listed here in the table differ markedly in size, as well as in the thickness of their membranes.

The membranes are not of uniform thickness from one end of the egg to the other. They are thickest at the blunt end (*Ferdinandoff, 1931, p. 32*).

Region	Thickness of Membranes (millimeter)
Pointed end	0.057
Equatorial region	0.065
Blunt end	0.069

Various values have been observed for the thickness of each of the two shell membranes measured separately, probably because of the shrinking or swelling of the membranes under various treatments. In a study of Leghorn eggs, Hays and Sumbardo (*1927*) found the thickness of the outer membrane to be 0.0505 mm., or three times that of the inner membrane. The thickness of each membrane and the ratio between the two are given, for several species, in the table (*Romanoff, unpublished*).

Species	Outer (millimeter)	Inner (millimeter)	Ratio O/I
Ostrich	0.12	0.080	1.5
Swan	0.11	0.055	2.0
Turkey	0.09	0.017	5.3
Chickens:			
Brahma	0.07	0.022	3.0
Leghorn	0.05	0.015	3.0
Bantam	0.04	0.010	4.0
Quail	0.06	0.008	7.5

There is a general relationship between the thickness of each membrane and the size of the egg. However, the ratio of the thickness of the outer membrane to that of the inner appears to be entirely unrelated to the size of the egg. In hens' eggs, the ratio is surprisingly constant, considering the divergence in egg size.

MICROSCOPIC STRUCTURE

The shell membranes are composed mainly of protein fibers felted together and are strengthened by an albuminous cementing material. This material is found between the two membranes and in the interstices among the fibers, particularly those of the inner membrane. On its inner surface, where it is in immediate contact with the albumen, the egg envelope is smooth; its outer surface, beneath the shell, appears frayed and tattered (Fig. 87), even when the shell has been carefully dissolved with weak acid (*Szuman, 1925*). According to Hays and Sumbardo (*1927*), both membranes contain pores, which are more numerous in the inner membrane. In the outer membrane, the openings may

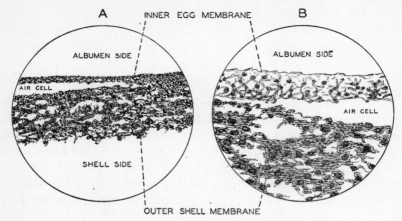

FIG. 87. Transverse sections of the egg and shell membranes at the margin of the air cell. (After Romankewitsch, 1932.)

A, magnified about 100 times. B, magnified about 500 times.

be as large as 0.028 mm. (*von Wittich, 1851*). However, the passage of gases or liquids through the membranes is mainly by osmosis or diffusion.

The outer membrane, next to the shell, has three major layers, which are fairly distinct and which can be teased apart (Fig. 88). Each layer is composed of numerous netlike strata. The outermost of the three major layers is formed of large, unbranched keratin fibers, flattened like ribbons (*Thomson, 1859*), the diameter of which may be from 0.002 to 0.005 mm. (*Moran and Hale, 1936*), or even up to 0.01 mm. (*Romanke-*

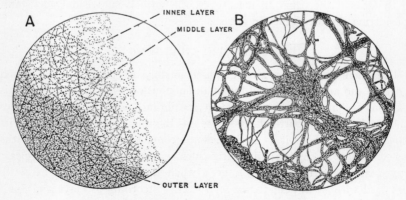

FIG. 88. Structure of outer shell membrane.

A, three layers; magnified about 50 times (after Moran and Hale, 1936). B, individual fibers; magnified about 500 times (after Romankewitsch, 1932).

witsch, 1932). Many of them adhere so firmly to the shell as to remain attached to it when the membrane is forcibly removed. The fibers lie, for the most part, parallel to the surface. Knots are present on the sides of the fibers, particularly where they cross to form a network (cf. Fig. 88). The fibers forming the deeper layers are successively finer and may be but 0.008 mm. in diameter (*Moran and Hale, 1936*). They apparently consist of mucin.

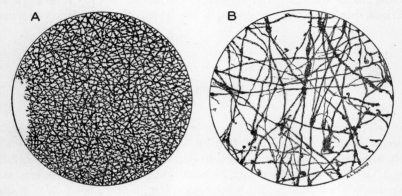

Fɪɢ. 89. Structure of inner egg membrane.

A, magnified about 100 times (after Moran and Hale, 1936). *B,* magnified about 500 times (after Romankewitsch, 1932).

The inner membrane, next to the albumen, is divided with great difficulty into two somewhat indistinct layers (*Moran and Hale, 1936*). This membrane is made up of very fine branching fibers of keratin and mucin, which form a network both perpendicular and parallel to the surface (Fig. 89). Because of the manner in which these fibers are interwoven, the membrane is surprisingly durable for a structure only 0.0148 mm. thick (*Hays and Sumbardo, 1927*). The interstices between the fibers are filled with albuminous cementing material. The smoothness of the surface fronting the albumen (cf. Fig. 89) is due to a thick coating of this cement.

THE AIR CELL

The egg, at the moment it is laid, contains no air cell; the membranes and the shell are filled to capacity. Soon, however, the air cell appears as a small circular space, usually at the blunt end of the egg. It is situated between the inner and outer shell membranes, which remain in contact with the albumen and shell, respectively. In shape, the air cell resembles a double convex lens. The curvature of its outer boundary is the same as that of the shell; the opposite side may be less convex if the air cell is small, but it approaches the curvature of the outer boundary if the air cell is large. The air cell is sometimes called the air space or the air chamber.

The size of the air cell varies according to the permeability of the egg-shell, the age of the egg, and the conditions of temperature and humidity to which the egg is subjected. The size of the air cell is also a function of the size of the egg. The average size of the air cell in "fresh eggs" of various species of birds is shown in the table. The correlation between

SPECIES	DIAMETER OF AIR CELL (centimeters)
Ostrich	4.0
Goose	2.0
Chicken (Leghorn)	1.5
Grouse	1.0

air cell size and the size of the egg is discernible in the eggs of a single species, such as the chicken (*Olsson, 1936*). The diameter of the air cell, although 1.5 cm. in the Leghorn egg, is 1.3 cm. in the smaller (42.23 gm.) Los Baños Cantonese egg 24 hours old. In the Cantonese egg, the average height of the air cell is 0.15 cm. Romijn and Roos (*1938*) found the average volume of the cell in a group of Leghorn eggs to be 0.4 cc.

The rapidity with which the air cell forms depends upon the rate at which the egg is cooled after it is laid. If the air temperature is low, the cell may appear within as short a time as 2 minutes. In a warm, moist environment, several hours may be necessary for its formation. In most chicken eggs, it appears in 6 to 60 minutes. Immediately on appearance, it is 0.5 to 0.9 cm. in diameter (*Meharliscu, 1933*) and 0.1 to 0.2 cc. in volume, depending on the size of the egg (*Moran and Piqué, 1926*). After 2 hours, it is 1.3 to 2.5 cm. in diameter.

It is generally thought that the formation of the air cell is the result of the different rates at which the shell and the egg contents contract as the new-laid egg cools from the temperature of the bird's body. Immediately after it is laid, the egg starts to lose heat and continues to do so until its temperature approaches that of the surrounding air. The rigid shell, entirely different in chemical composition from the semisolid contents of the egg, contracts only to an insignificant extent. The initial size of the air cell therefore represents the decrease in the volume of the egg contents, less the slight decrease in the capacity of the shell.

There is considerable evidence in support of the above explanation of the origin of the air cell. For example, upon refrigerating the egg, the air cell enlarges; correspondingly, it becomes smaller when the egg is warmed (*Almquist, 1933*). The formation of the air cell does not depend upon the presence of either the chalazae or the yolk (*Lataste, 1924*). In composition, the gas within the air cell is a mixture similar to air (*Romijn and Roos, 1938*). In addition, a structure analogous to the air cell was experimentally produced in the egg many years ago by Coste (*1847*), in demonstrating the permeability of the shell and outer shell membrane to oil. He tightly ligated an oviduct, *in situ,* at either end of the egg which it contained. He then removed the ligated segment of oviduct, immersed it in oil, and cut a window in its wall, so that a small area of the eggshell was exposed. After a time, he removed the egg. Beneath the portion of the shell that had been in contact with the oil, he found a small oil bubble between the shell membranes. By varying the position of the window in succeeding experiments, he was able to change the location of the oil bubble.

The normal position of the air cell at the blunt end of the egg appears to bear a relationship to its function of supplying air to the embryo at the time when pulmonary respiration is initiated. The head of the embryo lies directly beneath the air cell. If the air cell is not at the blunt end of the egg, the chick may die of asphyxia within the shell. The physical factors determining the air cell's location are not altogether clear. Generally, but not invariably, the pores in the eggshell are more numerous at the large end of the egg than at the small end. According to von Wittich (*1851*), the meshwork of the outer shell membrane contains little albuminous cement and unusually large interstices (0.028 mm. in diameter) in the region of the air cell, so that penetration and accumulation of air is facilitated. Szuman (*1925*), however, was not able to detect any peculiarities in the histological structure of the shell membrane in the vicinity of the air cell, whether this was at the blunt end of the egg or elsewhere.

After its formation, the air cell increases in size as the evaporation of moisture causes the volume of the egg contents to decrease. For this reason, the size of the air cell is frequently taken as an index of the quality and age of the egg (see Chapter 9, "Food Value," and Chapter 10, "Preservation").

THE EGGSHELL

The shell of the bird's egg is a relatively smooth, hard, calcareous coat, attached to the outer of the two membranes. So firm is the attachment that shell and membrane can be separated only with difficulty.

The shell of the hen's egg is translucent when the egg is laid but

becomes opaque as soon as it dries.   By transmitted light, most eggs
appear uniformly translucent.   Occasionally, an egg shows numerous
bright flecks in its shell.   These are due to the presence of aggregations
of protein which retain moisture better than other parts of the shell and
pass light more readily (*Holst, Almquist, and Lorenz, 1932b; Almquist
and Burmester, 1934*).

THICKNESS

Eggshells vary greatly in thickness, which is determined by numerous
factors.   In general, however, the thickness of the shell is correlated with
the size of the egg and the size of the bird.   The shell, whose breaking
strength depends largely upon thickness (see Chapter 7, "Physicochemical
Properties"), must be strong enough to support the weight of the parent,
yet fragile enough to crack easily when the young bird hatches.

Species Characteristics

Because the thickness of the shell·is characteristic of the species, it
is useful in taxonomic classification (*von Nathuşius, 1871a*).   At one extreme is the paper-thin shell of the hummingbird's egg, and at the other
the container, ⅙ inch thick, which held the 3-gal. egg of the flightless,
extinct *Aepyornis*.   The table shows the relationship between average
shell thickness and the size both of the bird and of the egg.   Species are
listed in descending order of body size.

	EGG WEIGHT	SHELL THICKNESS	
SPECIES	(grams)	(millimeters)	INVESTIGATOR
Aepyornis	12,000	4.40	Schönwetter (*1930*)
African ostrich	1,400	1.95	von Nathusius (*1885*)
Australian swan	700	0.69	Romanoff (*unpublished*)
Holland turkey	80	0.41	Romanoff (*unpublished*)
Chicken:			
Cochin China	65	0.36	von Nathusius (*1882a*)
Leghorn	58	0.31	Romanoff (*1929a*)
Bantam	38	0.26	von Nathusius (*1882a*)
Ringnecked pheasant	32	0.26	Romanoff, Bump, and Holm (*1938*)
Quail	9	0.13	Romanoff, Bump, and Holm (*1938*)
Australian finch	1	0.09	Romanoff (*unpublished*)
Hummingbird	0.5	0.06	Romanoff (*unpublished*)

Since the weight of the egg is correlated with its volume, so also is the
thickness of the shell (*Olsson, 1936*).   The direct linear relationship between volume and shell thickness is especially evident in the chicken egg
(Fig. 90).

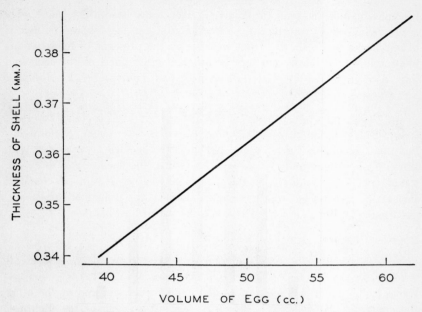

Fɪɢ. 90. Relationship between the volume of the hen's egg and the thickness of the shell. (After Olsson, 1936.)

Individual Variation

Shell thickness is not always the same in eggs laid by wild birds of the same species. It is even more variable in chicken eggs, not only because they are laid in larger numbers, but also, no doubt, because of the many influences to which hens have been subjected throughout centuries of domestication.

In the eggs laid by a flock of hens, the range in shell thickness may be so great that eggshells from some individuals are 50 per cent thicker than those from other individuals. For example, in one flock of ninety-five White Leghorns (*Romanoff, 1929a*), three birds laid eggs with shells as thin as 0.26 mm., and two produced shells 0.38 mm. thick. The eggs of thirty-one hens bore shells of the average thickness of 0.32 mm. (Fig. 91). Two flocks of the same breed may produce eggshells of quite dissimilar average thickness, and it is therefore difficult to demonstrate breed differences in this characteristic.

Among the eggs of the individual hen, variability in shell thickness is limited (Fig. 92). Normally, each bird secretes a fairly constant amount of shell material on her eggs.

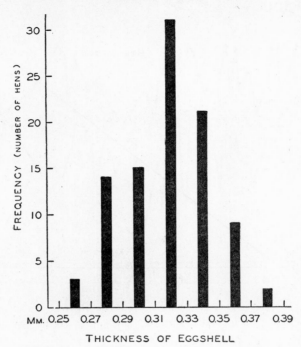

F<small>IG</small>. 91. Frequency distribution of ninety-five White Leghorns according to thickness of their eggshells. (Based on data of Romanoff, 1929a.)

Factors Influencing Shell Thickness

The thickness of the eggshell is influenced by many of the same factors that affect the egg in other respects. Among these are the season of the year, the hen's nutrition, her physiological efficiency, and probably her hereditary endowment.

HEREDITY

The thickness of the individual hen's eggshells is a manifestation of her calcium metabolism. Her relative efficiency in assimilating and secreting calcium and other minerals involved in shell formation apparently comes under hereditary control to some extent.

Fowl tend to show family resemblances in eggshell thickness. Variation in average shell thickness from one flock to another may be due in part to accentuation of family differences by a certain degree of inbreeding, since a flock often represents a single strain, or a few closely related strains.

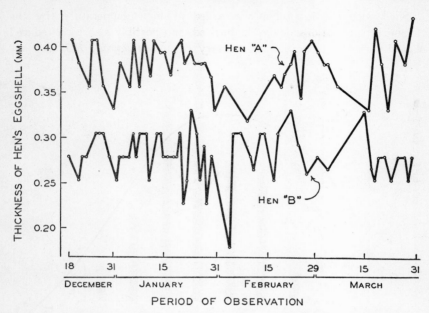

Fig. 92. Constancy in the thickness of the shell in eggs laid by each of two hens. (After data of Romanoff, 1929a.)

Taylor and Lerner (*1939*) succeeded in establishing two lines of hens, "thick-shelled" and "thin-shelled," but they were unable to demonstrate clear-cut genetic factors. Hereditary influences apparently account for the inability of some families to produce thick shells even under the most favorable conditions (*Taylor and Martin, 1928; Hays, 1937b*).

SEASONAL VARIATION

It is generally known that eggshells are of greatest and most uniform thickness in winter and become thinner during the spring and summer. This fact is well illustrated in Fig. 93, which shows that shell thickness decreases continuously from March to September.

Seasonal decline in thickness of the shell is probably correlated with rise in prevailing temperature (*Wilhelm, 1940*). Warren and Schnepel (*1940*) obtained thinner eggshells almost immediately after experimentally increasing the environmental temperature from 20° C. to 32.5° C. (Fig. 94). A recovery in shell thickness occurred after a subsequent decrease in temperature. They also observed that high humidity accentuated the effects of high temperature, and that the hens' feed consumption at 32.5° C. was 27 per cent less than at 20° C.

Shell secretion is probably retarded at high temperatures by the diminished intake of calcium in the feed, and possibly also by a reduced capacity of the blood stream to carry calcium (*Conrad, 1939*).

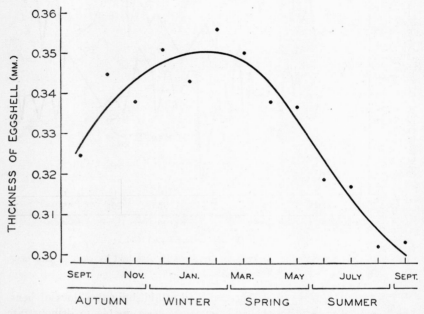

FIG. 93. Seasonal changes in average thickness of the shells of Leghorn pullets' eggs. (After Wilhelm, 1940.)

NUTRITION

Shell thickness is influenced by the amounts of several nutritional elements contained in the bird's diet, especially calcium, phosphorus, and vitamin D, and probably manganese. However, the capacity of hens to utilize these elements for shell deposition is extremely variable and no doubt depends upon individual differences in the physiological processes governing assimilation, secretion, and excretion.

Hens must receive supplementary calcium in their ration in order to produce eggshells of adequate thickness (*Collier, 1892; Buckner and Martin, 1920*). The effect of increased calcium intake upon mean shell thickness is shown in the table below (*Evans, Carver, and Brant, 1944b*).

For shells of good thickness, the optimum level of calcium in the diet is 1.8 per cent (*Norris, Heuser, Ringrose, and Wilgus, 1934*) to 2.5 per cent (*Evans, Carver, and Brant, 1944b*). The feeding of excessive

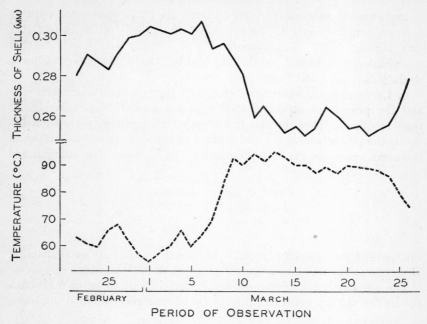

FIG. 94. Influence of environmental temperature on the thickness of eggshells produced by a flock of twenty-five Leghorn hens. (After Warren and Schnepel, 1940.)

CALCIUM IN DIET (per cent)	SHELL THICKNESS (millimeter)
1.0	0.302
3.0	0.347

amounts of calcium results in a decrease in the efficiency with which it is utilized (*Gutowska and Parkhurst, 1942b*) and may even have a deleterious effect, especially upon uniformity of shell thickness (*Tyler, 1946*) and egg production.

The hen does not assimilate calcium equally well from all sources. Calcium carbonate, of which the eggshell chiefly consists, is the most readily utilized form, especially when fed as oyster shell.

Because of the intimate relationship between calcium and phosphorus metabolism, an adequate amount of phosphorus must be given to insure good shell formation. In general, a concentration of 0.75 to 0.85 per cent of the diet is sufficient. In addition, vitamin D, of well-known importance in the metabolism of both calcium and phosphorus, must be supplied at the proper level.

Manganese, which is present in the egg in extremely minute quantities,

is important in mammalian reproduction. In birds, manganese appears
to contribute indirectly to the formation of eggshells of adequate thickness (*Lyons, 1939*). Experiments of a preliminary nature indicate that
the addition of iodocasein to the diet also favors the production of heavy
shells (*Gutteridge and Pratt, 1946*).

In England, it has been noted that both the incidence of cracked eggs
and the proportion of thin-shelled eggs are greater in certain regions
than in others. Investigation has revealed an apparent correlation between the percentage of cracked shells and the type of soil prevailing in
the areas where the eggs are produced (*Coles, 1938*). The strength and

Type of Soil	Cracked Eggs (per cent)
Sandy	12.5
Chalky	10.5
Clay	3.0

thickness of the eggshell is possibly influenced by the varying availability
of certain substances in the vegetation grown in different soils.

The administration of various sulfonamides at the level of 0.3 to 0.5
per cent of the dry mash has a distinctly adverse effect on shell thickness,

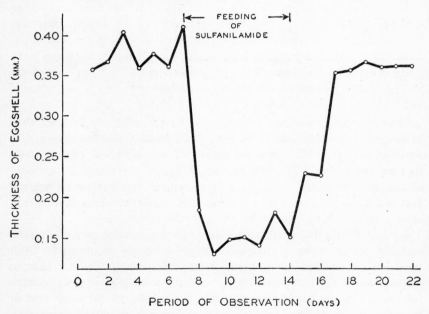

Fig. 95. Decrease in thickness of eggshell during administration of sulfanilamide
to hens, at the level of 0.3 per cent of dry mash. (After Genest and Bernard, 1945.)

as indicated in the tabulation *(Genest and Bernard, 1945)*. Figure 95 shows how promptly the production of thin shells follows the first dose of sulfanilamide and also how rapidly normal shell thickness is again

DRUG	DECREASE IN SHELL THICKNESS (per cent)
Sulfanilamide	61
Soluseptazine	43
Neoprontosil	41
Sulfapyridine	24

established after the drug is discontinued. Other sulfonamides, such as sulfathiazole, sulfaguanidine, sulfadiazine, and sulfamerazine, have no noticeable inhibitory effect on shell secretion.

PHYSIOLOGICAL FACTORS

The physiological efficiency of the individual hen's shell-secreting glands apparently varies. Often, as a result, the shell material is not deposited uniformly over the surface of the egg. In addition, shell thickness usually changes throughout the laying cycle.

In 2-egg cycles, the shell of the second egg is thicker than that of the first egg; in cycles of 3 or more eggs, the shells of the first and last eggs are the thickest *(Wilhelm, 1940; Berg, 1945; Romanoff, unpublished)*. Those of the intervening eggs are progressively thinner, except when the cycle consists of 5 or more eggs. The next to the last egg then bears a shell somewhat thicker than that of the egg immediately preceding it (Fig. 96).

Since the eggs with the thickest shells are laid after the longest intervals of time, it appears that the relative thickness of the shell depends to some extent upon the time available for the accumulation of shell-forming substances. The presence of an egg in the uterus is probably necessary to stimulate the storage of calcareous material; otherwise the first egg in the cycle would be disproportionately thick, because it is laid after a pause *(Berg, 1945)*.

Because hens differ in the capacity to store and secrete shell-forming material, all eggshells are not of uniform thickness from end to end. Variability in shell thickness is greater at the poles of the egg than in the equatorial region. Observations on eggs with greater than average thickness have indicated that the shell is heaviest at the sharp end of the egg *(Olsson, 1936)*. During an investigation of the breaking strength of nearly 4000 eggs, Romanoff *(1929a)* noted that shells fall into three classes: (1) shells of average and uniform thickness; (2) shells thinner than average, in which the blunt end is the thicker; and (3) shells thicker than average, in which the pointed end is the thicker (Fig. 97). As the average thickness of the entire shell becomes greater, the thickness of the pointed end increases rapidly and linearly, but increase in the thickness of the blunt end lags (cf. Fig. 97).

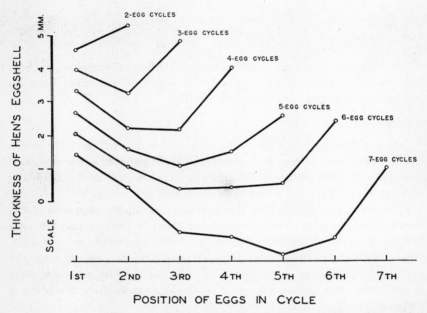

FIG. 96. Changes in the thickness of the shell of the hen's egg, throughout cycles from 2 to 7 eggs in length. Average shell thickness in the first egg of each cycle is 0.35 mm. (After Berg, 1945.)

These findings may be interpreted as indicating that shell-secreting activity, if normal, remains constant during the time that the egg is in the uterus. Consequently, the shell is of uniform thickness. Greater or less than normal efficiency in shell formation, on the other hand, is accompanied by an ever-changing rate of secretion. If the rate of shell deposition continually diminishes, the shell will be of greatest thickness at the leading end of the egg. If the rate of shell secretion continually accelerates, the other end of the egg, of course, receives the thickest deposit of shell.

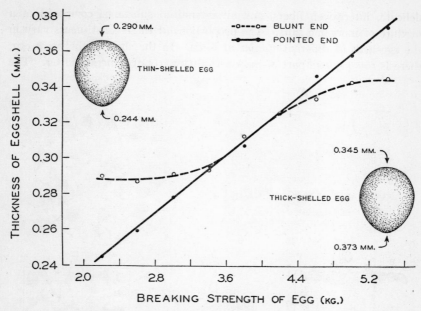

FIG 97. Relationship between average thickness of the entire eggshell (as shown by breaking strength) and shell thickness at either end of the egg. Based on observations of about 4000 eggs. (After data of Romanoff, 1929a.)

GROSS ORGANIZATION

To fulfill its many functions, the eggshell must meet several requirements. It must be sufficiently strong and rigid to withstand the weight of the adult bird. It must be porous enough to permit the respiration of the embryo and yet compact enough to prevent the entrance of microorganisms and the escape of too much moisture. In addition, the shell must contain various inorganic salts in amounts sufficient to satisfy the greater part of the mineral requirements of the developing embryo.

The shell is gracefully curved to fit the taut egg membranes, over which the secretions of the uterus pour out and harden, much like cement (see Chapter 4, "Formation"). The domed architecture of the shell contributes to its strength; it utilizes the principle of the arched stone bridge, designed to bear extraordinary weight on its convex surface. Additional strength is given to the shell by the radial orientation of crystals in its outer surface. If the crystals were parallel to the surface, the shell might scale and thus be weakened.

The bird's eggshell is not a homogeneous structure. It is compounded of two widely different materials: an organic matrix, or framework, of

delicate, interwoven fibers, and an interstitial substance composed of a
mixture of inorganic salts. The proportions of these constituents vary in
the eggshells of different species of birds. In the shell of the hen's egg,
there is roughly one part of matrix to fifty parts of mineral matter. The

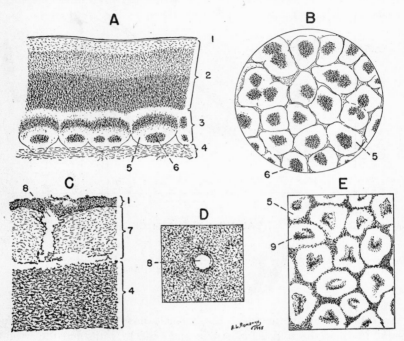

FIG. 98. Histological structure of the shell of the hen's egg.

A, radial section through the shell; magnified about 100 times (after von Nathusius, 1882b).
B, tangential section through the mammillary layer of the shell, viewed from the outer surface;
magnified about 100 times (after Stewart, 1935). *C*, radial section through eggshell after
decalcification; magnified about 200 times (after Almquist, 1933). *D*, tangential section
through the organic matrix material of the decalcified shell, cutting across a pore; magnified
about 200 times (after Almquist, 1933). *E*, a portion of the inner surface of the shell after
removal of the organic material by ashing; magnified about 100 times (after Ferdinandoff,
1931, p. 27).

The numbers indicate: 1, cuticle; 2, spongy layer; 3, mammillary layer; 4, shell membrane;
5, mammilla (mammillary knob); 6, protein matrix material forming the core of the mammilla; 7, protein matrix of shell; 8, pore in the decalcified eggshell; 9, space in a mammilla
remaining after the organic core has been removed by burning.

organic matrix is a collagen-like protein (*Almquist, 1934*); the minerals,
mainly carbonates and phosphates of calcium and magnesium. Of the
latter substances, calcium carbonate is by far the most plentiful. In
mineral composition, the eggshell is strikingly similar to limestone (see
Chapter 6, "Chemical Composition").

MICROSCOPIC STRUCTURE

The matrix and the minerals of which the shell is composed are apparently secreted and deposited simultaneously. The matrix is a meshwork of fine fibers and granules, and the minerals are laid down in its interstices. The matrix, represented by the granulation in Fig. 98-*A*, is far more

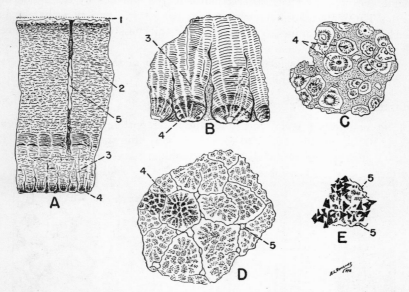

FIG. 99. Structure of *Aepyornis* egg (extinct species). (After von Nathusius, 1871b.)

A, cross section through entire shell; magnified about 15 times. *B*, section showing mammillae; magnified about 50 times. *C*, tangential section through mammillae, close to the ends of the knobs; magnified about 50 times. *D*, tangential section through mammillae, with clearer delineation of the matrix columns in two mammillae; magnified about 50 times. *E*, tangential section through a single mammilla; magnified about 100 times.

The numbers indicate: 1, eroded surface of the shell; 2, spongy layer; 3, individual mammilla of mammillary layer; 4, matrix nucleus; 5, pore canals.

plentiful in the deeper portion of the shell; three distinct gradations in layer 2 may be noted in this illustration. The greater share of the magnesium compounds and the phosphates are found in the inner portion of the shell, whereas the outer portion is nearly pure calcium carbonate. Toward the inside of the shell the minerals are amorphous or noncrystalline, with no definite orientation. Toward the outer surface, the calcium is in the form of calcite crystals, oriented in such a manner that their long axes are perpendicular to the surface (*Kelly, 1901; Mayneord, 1927; Nilakantan, 1940*). Here, the matrix is so scarce that it causes little interference with crystal formation and orientation. For this reason, the outermost strata of the shell are the hardest and most compact. When the shell is broken, the cleavage is roughly radial, although it may deviate laterally in the inner portion of the shell.

In Fig. 98-*C* is shown a radial section through the shell and shell membrane of a hen's egg, after removal of the minerals by the action of acid. The matrix is intact, although it has shrunk until it is barely as thick as the membranes. (Normally, the thickness of the shell is three times that of the membranes.)

There are two main layers in the shell proper of the bird's egg. These are the inner, or mammillary, layer, and the outer, or spongy, layer.

Mammillary Layer

The mammillary layer constitutes the lesser portion of the shell (Figs. 99-*A*, 100-*A*; cf. Fig. 105-*A* to *I*), save in very thin-shelled eggs (cf. Fig. 105-*I*) such as the swallow's. The mammillary layer rests on the outer surface of the shell membrane and is partially embedded in it.

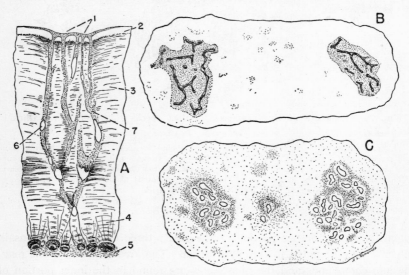

Fig. 100. Structure of the shell and the pore canal system of the ostrich egg; magnified about 35 times. (After von Nathusius, 1868.)

A, radial section showing: 1, multiple pores of a single canal system; 2, cuticle; 3, spongy layer; 4, mammillary layer; 5, a small portion of the shell membrane still adhering to the shell; 6, pore canals intersected by the plane of the section; 7, paths of the pore canals. *B*, surface view of a portion of a shell stained with Carmine Red. The heavily staining areas (represented by stippling) are matrix-filled depressions where the mouths of the pores open into countersunk, branched grooves (heavily shaded). *C*, tangential section just below the cuticle, through two groups of pore canals.

The mammillary layer is composed of numerous roughly conical knobs or mammillae (cf. Figs. 98-*A*, 99-*B*, 100-*A*, 101-*A*), which are oval to circular in cross section (cf. Figs. 98-*B*, 99-*D*, 101-*C*). The mammillae are tightly compressed, side by side, in a single stratum. Their broad, domed tops are somewhat flattened or indented at the sides where they

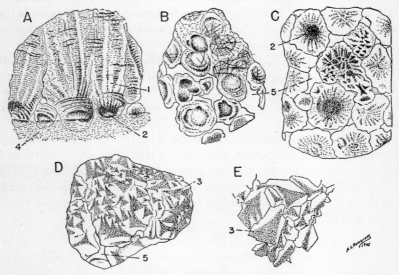

FIG. 101. Histological structure of the ostrich eggshell. (After von Nathusius, 1868.)

A, radial section, cutting several mammillae lengthwise; magnified 100 times. Note the sunburst form in which the matrix granules are deposited, the rays emanating outward and upward. *B,* tangential section through the lower portion of the mammillary layer; magnified 100 times. The plane of section is not quite parallel to the plane of the shell and therefore cuts into the shell membrane at the top of the figure. *C,* tangential section through the mammillary layer (at a somewhat higher level than in the previous figure), cutting across the mammillae where the opaque matrix deposit is densest; magnified 100 times. Four mammillae are shown in detail. Note the sunburst arrangement of the matrix granules. *D,* tangential section through the upper portion of the mammillae; magnified 100 times. Note the wedge-shaped matrix rays. *E,* same as *D,* but magnified 160 times.

The numbers indicate: 1, mammilla; 2, granular matrix body; 3, wedge of matrix granules; 4, shell membrane; 5, air space.

fit against each other and are cemented together to form a rigid, though fragile, foundation for the outer portion of the shell. The diameter of each mammilla diminishes toward the base, which terminates within the surface of the shell membrane. Although the knobs fit snugly at the top, air spaces remain between the tapered bases. These irregularly shaped spaces communicate with each other to form an uninterrupted network of air channels throughout the lower portion of the layer. At frequent intervals, very small canals issue outwardly between the tops of the mammillae (cf. Fig. 99-*D*) from the intermammillary ventilation system.

A mammilla consists of noncrystalline minerals concentrically aggregated about granular matrix material. If the organic matrix is removed by ashing, only the minerals remain (cf. Fig. 98-*E*). A nucleus of matrix is centrally located near the base of the mammilla (cf. Fig. 99-*A* and *C*). Additional matrix granules may be grouped in various patterns. In the shell of the *Aepyornis* egg, they radiate upward and outward in columns, wedge shaped in cross section (cf. Fig. 99-*B*, *D*, and *E*). In the ostrich egg (cf. Fig. 101-*A*, *B*, *C*, *D*, and *E*), wedge-shaped rays of granules proceed outward and upward in sunburst pattern from the nucleus near the base. In the chicken egg (cf. Fig. 98-*A*), the matrix, exclusive of the nucleus, is in the form of a convex cap beneath the calcareous dome of the mammilla.

In the eggs of each species, the shape and size of the mammilla and the arrangement of the matrix within it are characteristic. The range in size is so small in the eggs of a single species that von Nathusius (*1871a; 1874*) found the cross-sectional area of the mammilla useful in distinguishing closely related groups of birds. This measurement is given for several species in Table 13.

TABLE 13

Size of Mammillary Bodies (Mammillae) in the Eggshells of Various Species of Birds *

Species	Area of Cross Section (square millimeter)
Domestic hen	0.0072–0.0163
Geese:	
Domestic goose	0.021 –0.024
Anser cinereus	0.023 –0.025
Anser segetum	0.0145–0.015
Anser cygnoides	0.014
Cereopsis novae hollandiae	0.015
Doves:	
Domestic pigeon	0.011 –0.014
Columba turtur	0.0072–0.0074
Columba palumbus	0.0083–0.0098
Columba oenas	0.011 –0.012
Columba livia sera	0.013 –0.014
Crows:	
Corvus corone	0.0094–0.0103
Corvus cornix	0.0055–0.0071

* Calculated from data of Blasius (*1867*) on the chicken egg and von Nathusius (*1871a, 1874*) on eggs of other species.

In the hen's egg, the thickness of the mammillary layer is about 0.11 mm., or approximately one-third that of the entire shell. The height of a single mammilla corresponds to the thickness of the mammillary layer and the diameter is 0.096 to 0.144 mm. (*Blasius, 1867*).

Spongy Layer

This layer is superimposed on the mammillary layer and is firmly cemented to its uneven surface. Except in the eggs of some small birds (cf. Fig. 105-*I*), the spongy layer constitutes the greater part of the shell (*Gadow, 1891, p. 876*). Contrary to its name, this layer is very compact, although numerous microscopic canals traverse its entire depth at irregular intervals. Only when the shell is decalcified with acid (cf. Fig. 98-*C*) does it appear spongy. Its deposition about the fragile mammillary layer confirms the shape of the egg and gives rigidity and strength to the shell.

Although the spongy layer is added concentrically, there is no visible indication of stratification in its mineral structure. However, examination by X-rays (*Herzog and Gonell, 1925*) shows that, toward the surface, the shell becomes progressively denser, and a greater share of the mineral is crystalline.

The matrix content of the spongy layer is not uniformly distributed, as shown by the application of protein stains. The matrix fibers are usually most plentiful in the region nearest the mammillary layer (cf. Figs. 98-*A*, 105-*C*, *D*, *F*, *G*, and *H*). In the eggs of some water birds, such as the pelican and the gull (cf. Fig. 105-*B* and *E*), matrix is absent in narrow horizontal and vertical bands, so arranged as to give the appearance of masonry.

Protein staining of the chicken eggshell reveals three fairly distinct strata in the spongy layer which differ in their content of matrix material (cf. Fig. 98-*A*). Within each stratum, however, the distribution of matrix fibers is fairly uniform. They are plentiful in a wide stratum adjoining the mammillary layer, sparse in a narrow zone at the surface of the shell, and moderately abundant in the middle zone of intermediate thickness.

PORES

In the bird's egg, there are minute, oval-to-circular openings in the surface of the shell. These are the pores, which are the mouths of the pore canals. The pores vary in size; the largest are barely visible to the naked eye. Their location is indicated by shallow indentations in the shell surface, somewhat unevenly spaced.

Number of Pores. The pores in an eggshell are so numerous, and counting them is so laborious, that little information exists concerning the numbers present in the shells of different species. The average number of pores in the shell of the hen's egg is somewhat more than 7500 (*Rizzo, 1899; Penionzhkevich, 1933; Romanoff, unpublished*). The goose's egg, because it is larger, surpasses the hen's egg in total number

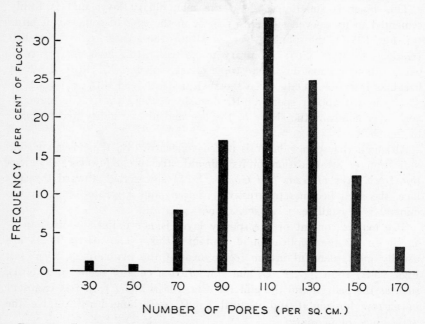

Fig. 102. Frequency distribution of a flock of 258 hens according to the average number of pores per square centimeter in their eggshells. (After unpublished data of Romanoff.)

of pores, but it has fewer pores per square centimeter of shell surface (*Penionzhkevich, 1933*); the duck's egg has more per square centimeter than the hen's egg (*Kato and Ko, 1938*). In eggs laid by the different hens of a flock, however, the number of pores per square centimeter of shell varies widely (Fig. 102). Similarly, the total number of pores per shell appears to be as variable among the eggs of a single strain or breed of chickens as among the eggs of the entire species. It is only among the eggs of the individual hen that shell porosity is fairly constant (*Almquist and Holst, 1931; Romanoff, 1943b*). Black and Tyler (*1944*) found that the first egg of the cycle usually lost weight less rapidly than any of the others (see Chapter 10, "Preservation"), and therefore may have fewer pores, or smaller ones, than the eggs that follow it. The relatively low porosity of the first egg is apparent even when the calcium supplement in the hen's diet is insufficient.

Distribution of Pores. The pores are distributed unevenly over the surface of the shell. On the hen's egg, they are most numerous either in the equatorial region or at the blunt end and sparsest on the pointed end, as indicated in the tabulation.

NUMBER OF PORES (per square centimeter)

BLUNT END	EQUATORIAL REGION	POINTED END	INVESTIGATOR
149	131	90	Rizzo (*1899*)
92.7	150.3	31.4	Tishima (*1934*)
77.6	106.9	16.0	Tishima (*1934*)

In 793 eggs from 260 hens, Romanoff (*unpublished*) found that, at the blunt pole of the egg, the pores were more numerous in the center than at the periphery of a circular area of 1.0 cm. radius. At the sharp pole of the egg, the pores were fewer in the center than at the periphery of a circular area of the same size. These findings are summarized below.

REGION	PORES (per square centimeter)
Blunt end:	
Center	125.6
Periphery	106.1
Pointed end:	
Center	73.7
Periphery	113.4

The normal distribution of the pores in the shell of the hen's egg is shown in Fig. 103. The abundance of pores in the center of the blunt end, where the air cell is located, allows air to pass more readily into the air cell. Thus, respiration is facilitated during the latter part of the incubation period, when the head of the embryo is turned in this direction.

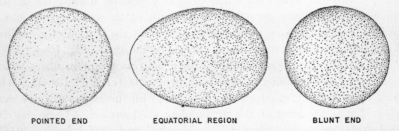

POINTED END   EQUATORIAL REGION   BLUNT END

FIG. 103. Typical distribution of pores on the hen's egg (as indicated after dyeing the plaques over the pore openings). (Drawn from the authors' prepared specimens.)

Occasionally, there may be portions of the surface in which pores are few or lacking, usually in a sizable area at the pointed end or in a smaller area at the blunt end.

Size and Shape of Pores. Although the shape of the pores is characteristic of the eggs of a particular species, the range in the size of the pores is wide in a single egg. The dimensions of the smallest and the largest pores in the eggs of several species of birds are shown in the table (*von Nathusius, 1868*). It may be seen that the size of the pores,

SPECIES	LARGEST PORE (millimeter)	SMALLEST PORE (millimeter)	SHAPE
Ostrich	0.050	0.020	Circular
Swan	0.042 x 0.038	0.029 x 0.026	Oval
Turkey	0.055 x 0.037	0.037 x 0.031	Oval
Duck	0.036 x 0.031	0.014 x 0.012	Oval
Hen	0.029 x 0.022	0.011 x 0.009	Oval
Pheasant	0.014 x 0.012	0.013 x 0.010	Oval
Auk	0.090 x 0.057	0.011 x 0.006	Oval
Gull	0.016 x 0.013	0.011 x 0.010	Oval

though variable, is correlated to a certain extent with the size of the egg laid by each species. The auk's egg, as shown in the data given here, is a notable exception.

In the shell of the hen's egg, the size of the pores is especially variable. Although the largest, as shown above, may be only 0.022 mm. by 0.029 mm. (*von Nathusius, 1868*), the smallest may sometimes measure 0.038 by 0.054 mm. (*von Wittich, 1851*).

Microscopic Structure

The pore canals traverse the spongy layer approximately at right angles to the surface of the shell and form connecting passages between the exterior of the shell and the network of air spaces in the mammillary layer. Each pore canal is of smallest bore at the base of the spongy layer; from this point the canal gradually widens toward its mouth, where the diameter is greatest. The mouth of the canal opens into a narrow, branched groove which lies lengthwise at the bottom of a shallow, irregularly oval indentation on the surface of the shell (Fig. 104).

The entire pore system is filled with a matrix of protein fibers which stain in the same manner as the shell membrane (cf. Fig. 104-*A*). The fibers close the pore canal and its mouth and fill the groove and the oval depression to the level of the surrounding shell surface. These matrix-filled areas are called "plaques" by Marshall and Cruickshank (*1938*).

The pore canals are single, or unbranched, in the shells of most birds' eggs, including that of the enormous, extinct *Aepyornis* (cf. Fig. 99-*A*). In the eggshell of the extinct moa (*Dinornis crassus*) (*von Nathusius, 1871b*) and of the ostrich (*von Nathusius, 1868*), the canals are branched. From a single pore canal, several canals radiate upward through the spongy layer and twine about each other (cf. Fig. 100-*A* and *C*). They

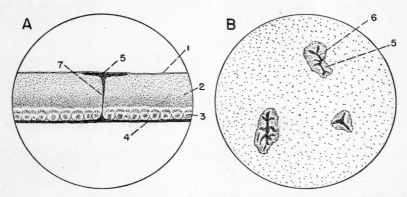

FIG. 104. Pore in shell of hen's egg (semidiagrammatic), stained with Methylene Blue; magnified about 50 times. (Modified from Ferdinandoff, 1931, p. 24.)

A, radial section of the shell through a pore. The canal leads inward from the pore, diminishing in diameter as it goes, and passes directly through the spongy layer to the underlying mammillary layer, where it connects with the air spaces between the tips of the mammillae. The organic matrix material in the grooved mouth of the pore and in the pore canal stains deeply, as does the shell membrane. *B*, surface view of the shell. The grooved and scalloped mouths of the pores are countersunk and filled with matrix material indistinguishable from that in the pore canals.

The numbers indicate: 1, cuticle; 2, spongy layer; 3, mammillary layer; 4, membranes; 5, plaque; 6, groove; 7, pore canal.

terminate at the surface of the shell as a group of pores in a single depression (cf. Fig. 100-*B*).

The pore canal in the shell of the hen's egg is unbranched (cf. Figs. 98-*C* and *D,* 104-*A*). Its length depends on the thickness of the spongy layer it traverses. In a shell of the average thickness of 0.31 mm., the pore canal is therefore about 0.20 mm. long. In diameter, a single canal may measure 0.013 mm. at its mouth and 0.006 mm. at its inner end (*Haines and Moran, 1940*).

Some canals may end blindly before reaching the surface of the shell. Abrasion of the shell surface opens these canals and markedly increases the porosity of the shell (*Bryant and Sharp, 1934*).

THE CUTICLE

On the external surface of most birds' eggs there is an extremely thin, transparent coating of protein, probably mucin (*Moran and Hale, 1936*). This outermost covering of the egg is the cuticle. It is very securely attached to the shell and is continuous over the entire surface, including the mouths of the pores. Although the cuticle is a closed envelope, containing no visible openings, it is permeable to gases. The nature of its

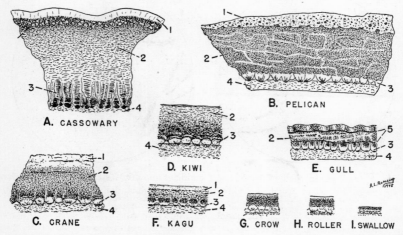

FIG. 105. Cross sections of eggshells of various species of wild birds; all magnified 25 times.

A, cassowary (after von Nathusius, 1869). B, pelican (*Pelicanus crispus*), (after von Nathusius, 1869). C, crane (*Grus cinerea*), (after von Nathusius, 1871b). D, kiwi (*Apteryx Mantelli*), (after von Nathusius, 1871b). E, gull (after von Nathusius, 1868). F, kagu (*Rhinochetus rufescens*), (after von Nathusius, 1871b). G, crow (*Corvus cornix*), (after von Nathusius, 1878). H, roller (*Coracias garrula*), (after von Nathusius, 1878). I, swallow (*Hirundo riparia*), (after von Nathusius, 1878).

The numbers indicate: 1, cuticle; 2, spongy layer; 3, mammillary layer; 4, membranes; 5, pigment bands; 6, pigment granules.

surface determines whether the eggshell appears smooth and lustrous or dull and chalky, as in the crane's egg (cf. Fig. 105-*C*). The membranous cuticle of the chicken's egg resembles "bloom" (*Szielasko, 1913*). (See Chapter 2, "External Characteristics.")

In thickness, the cuticle is the most variable of all the egg's coverings. Figure 105 shows that the cuticular material is thick on the eggs of the cassowary (*A*), the pelican (*B*), and the crane (*C*); thin on the kiwi's egg (*D*); and lacking on the eggs of the gull (*E*) and some small birds. As a hard glaze, it coats the ostrich egg to a depth of 0.036 mm. (*von*

Nathusius, 1868). On the domestic duck's eggshell, it is only 0.003 mm. thick (*von Nathusius, 1894*).

On some eggs, such as those of the pelican (cf. Fig. 105-*B*) and the quail (cf. Fig. 107), the cuticular material varies markedly in thickness. On the quail's egg, it is 0.008 to 0.0195 mm. thick, depending on the amount of shell pigmentation, as will be explained later. Cuticle is usually deposited on the hen's egg to a depth of 0.005 to 0.010 mm. It is

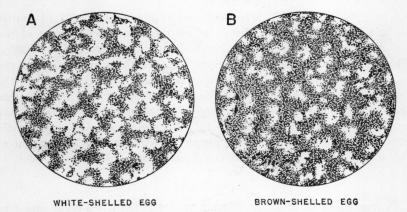

WHITE-SHELLED EGG   BROWN-SHELLED EGG

FIG. 106. Surface views of the cuticle on white and dark brown shells of chicken eggs, after impregnation with silver salts; magnified about 20 times. Only the outer layer of the cuticle contains granules stainable with silver. These granules are more numerous in the depressions above the mouths of pores but are not confined to these areas. The granules are lacking in the unstained areas. (After Romankewitsch, 1934.)

A, cuticle of white shell. The granules stainable with silver are relatively few, and the unstained intervening areas are large. *B,* cuticle of brown shell. The granules in the outer layer of the cuticle are very numerous, the unstained areas small.

present in greatest quantity in the small concave areas about the mouths of pores.

Although the cuticle appears to be homogeneous, it contains two distinct layers. Treatment of the shell with weak acid (*Dickie, 1849*) releases carbon dioxide and lifts the outer layer free from the inner, which remains attached to the shell. Staining (*Romankewitsch, 1934*) reveals that the outer layer possesses a microscopic structure characteristic of the species.

When the cuticle of the hen's egg is impregnated with silver salts, the thicker outer layer appears granular, particularly in the neighborhood of the pores (Fig. 106). The granules are fewer in white-shelled (*A*) than in brown-shelled eggs (*B*). In the ostrich egg, this layer is a meshwork

0.024 mm. thick, consisting of fine fibers (*Gadow, 1891, pp. 876–877*). Whatever the nature of the outer cuticular layer, the details of its structure contribute a characteristic appearance to the particular egg.

Minute fat droplets can be demonstrated in the outer cuticular layer by fat stains (*Romankewitsch, 1934*). Measurable amounts of fats have been found on some duck eggs (*Pritzker, 1941*).

In spotted eggs, the inner cuticular layer contains pigment granules. Otherwise, it is clear and structureless.

The so-called plaques (*Marshall and Cruickshank, 1938*) of matrix material in the mouths of the pores must be considered a part of the cuticle because they are firmly united with the inner layer of the cuticle. They are of different origin, however. Their staining properties are the same as those of the shell membranes and of the matrix threads in the pore canals (cf. Fig. 104-*A*). When Methylene Blue is injected into the egg and allowed to seep by capillarity through the pore canals, the roughly oval plaques are stained deeply, but the intervening areas remain unstained. The total area of the plaques constitutes only 1.5 per cent of the area of the shell (*Marshall and Cruickshank, 1938*).

PIGMENTS

Pigments derived from the coloring matter of red blood corpuscles (see Chapter 4, "Formation," and Chapter 6, "Chemical Composition") are found in all the coverings of birds' eggs, including the membranes, in which they are present in minute quantities (*Klose and Almquist, 1937*). Usually they are most plentiful in the cuticle and the superficial layers of the shell, where they are visible and therefore determine the egg's color. Shells that appear pure white nevertheless contain pigments.

The pigments of the egg coverings may be classified, according to their location (*Gadow, 1891, pp. 878–885*), as shell pigments or cuticular pigments. The ground color of the egg is determined by the shell pigment. The cuticular pigment, when present, is unevenly distributed over the shell's surface as darker superficial markings.

Shell pigment, of a single color or a mixture of colors, is intermingled with the calcium salts of the shell. It is present only in traces in the mammillary layer, with few exceptions. Usually the bulk of it is contained in the spongy layer, particularly in the outermost strata (cf. Fig. 105-*A*). Its amount decreases gradually from the outer to the inner portion of the spongy layer. The gradation is the same at every point in the shell. However, in the shells of a few birds, such as the gull, the pigment is concentrated in bands deep within the spongy layer (cf. Fig. 105-*E*). In rare cases, the true color of a pigmented shell is concealed by a very thin, pigment-free layer of calcium salts deposited on the surface.

Cuticular pigment is deposited as granules in the deeper portion of the cuticle. The intensity of pigmentation depends upon the thickness of this deposit, as shown in Fig. 107. Pigment granules are scarce or lacking in cross sections *C* and *D*, each taken from an area of the quail's egg where the ground color shows through and the cuticle is thin. In section *A*, taken from a dark spot, the cuticle is thick and contains a broad band

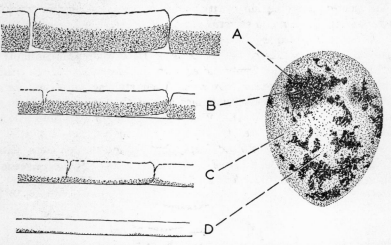

FIG. 107. Cross sections of the cuticular layer on the eggshell of the common European quail (*Coturnix communis*); magnified about 600 times. (After von Nathusius, 1894.)

Uneven pigmentation, or blotching, is due to variations in the depth to which pigment granules are deposited in the lower portion of the cuticle. The cuticle is thickest, and the pigment band widest, in heavily pigmented areas. However, the external, unpigmented portion of the cuticle is of fairly uniform thickness over the entire surface of the egg.

A, a deep band of pigment (0.0125 mm. thick) in a thick portion of the cuticle from a bluish brown blotch. The entire cuticle is 0.0195 mm. thick, the unpigmented portion, 0.007 mm. thick. *B* and *C*, successively thinner cuticle, containing smaller amounts of pigment, from progressively lighter areas on the egg. *D*, Cuticle, 0.007 mm. thick, with little or no pigment. (Picture of the quail's egg, 1.3 times natural size, after Seebohm, 1896.)

of granules in its inner stratum. In areas of intermediate color intensity, as at the edge of a blotch (section *B*), the pigment band is of moderate depth. The cuticle undulates gently according to the varying thickness of its pigmented portion.

In the eggs of a few species, where cuticle is sparse or absent, the markings are borne on the shell itself and tend to be localized at one end (*Giersberg, 1922*). In areas of irregular outline, the films of pigment are tightly affixed to the shell and even penetrate the mouths of the pores. The color of hens' eggs is due entirely to shell pigments, since superficial pigment is present only in traces, usually undetectable by ordinary vision.

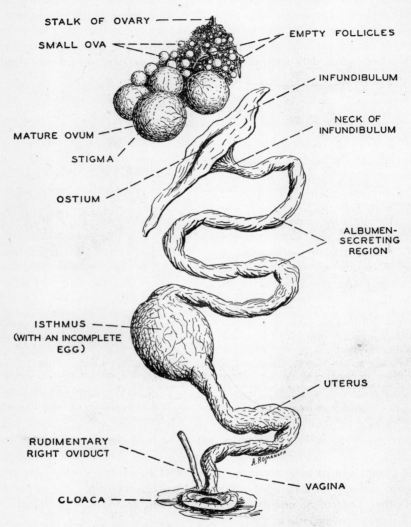

STALK OF OVARY

SMALL OVA

EMPTY FOLLICLES

INFUNDIBULUM

NECK OF
INFUNDIBULUM

MATURE OVUM

STIGMA

OSTIUM

ALBUMEN-
SECRETING
REGION

ISTHMUS
(WITH AN INCOMPLETE
EGG)

UTERUS

RUDIMENTARY
RIGHT OVIDUCT

A. ROMANOFF

VAGINA

CLOACA

Reproductive organs of the hen.

Chapter Four

FORMATION

The avian egg is the product of a remarkable physiological factory.

Through the centuries, classical works have described in detail the development of the embryo within the hen's egg, but little attention has been given to the processes by which the egg itself is formed. Only recently has interest been shown in the physiology and biochemistry of egg formation.

The egg is built up gradually over a considerable period of time. The production of eggs by the hen is not unlike mass production in a factory. Raw materials, in the form of food, must be taken in and processed. Many organs and systems aid in converting the raw materials into the various substances that become incorporated into the egg. Some of these constituents are formed and deposited long before the egg is laid. Yolk, for example, is gradually accumulated in several ova simultaneously; and the ovary may contain many ova in various stages of completion. The oviduct acts as the final assembly room where the albumen and the egg envelopes are added.

The laying hen, constantly diverting large quantities of life-sustaining substances from her body to the egg, must be a bird of great stamina and vigor. Whatever her hereditary capacity for reproduction, it will not be realized unless her reserves of strength are such that egg formation does not constitute an excessive strain upon her; otherwise egg production ceases, or the hen's health is impaired. An inherent resistance to disease and a naturally strong constitution are essential.

ORGANS OF EGG FORMATION

Egg formation consists of two distinct processes: (1) the growth and maturation of the germ cell; and (2) the deposition of nonliving materials—yolk, albumen, and egg envelopes. The organs in which the egg is formed are the ovary and oviduct. In the ovary, which contains the germ cells, the yolk of the egg is produced. The oviduct supplies, in succession, the albumen, two shell membranes, and the shell.

Asymmetry. Although two symmetrically placed ovaries and two oviducts are formed during the early embryonic development of the female chick, normally only the left ovary and oviduct develop fully. They are destined to carry on the reproductive functions of the bird. The cortex does not form on the right ovary; and the right oviduct, with few exceptions, ultimately retrogresses and persists only as a rudiment. The left oviduct in the 4- to 7-day-old female chick embryo is already distinctly larger than the right (*Gruenwald, 1942*). In the 9-day-old embryo, asymmetry in size, form, and structure is pronounced (*Greenwood, 1925*). At the time of hatching, the right ovary, as compared with the left, is diminutive and appears to be degenerating (*Romanoff, 1933*).

However, the frequency with which the right ovary persists in adult females of certain species of birds is very high, as shown in the tabulation.

SPECIES	NUMBER OF BIRDS	PERSISTENT RIGHT OVARY (per cent)	INVESTIGATOR
Hawk	50	66.0	Gunn (*1912*)
Hawk	48	47.8	Picchi (*1911*)
Ringdove	200	23.5	Riddle (*1925b*)
Common pigeon	200	9.5	Riddle (*1925b*)
Robin	21	4.8	Riddle (*1925b*)
English sparrow	22	4.5	Riddle (*1925b*)

These investigators did not specify whether or not the persistent right ovary possessed a cortex. When the cortex is absent, the organ compares more nearly in structure with a testicle. The presence of the cortex is necessary for the production of ova.

Ordinarily, the right oviduct remains infantile (*Finlay, 1925*); its size is not correlated with that of the right ovary, when the latter persists (*Stanley and Witschi, 1940*). The injection of estrogen into the incubating egg on or before the fourth day of development tends to prevent the retrogression of the right oviduct (*Domm and Davis, 1941*).

Sex Reversal. The results of removing the functional left ovary indicate that a fundamental difference exists between the two gonads of the female bird. After sinistral ovariectomy, the right gonad grows and develops into a smooth body similar to a testis, or male gonad (Fig. 108), particularly when the cortex is entirely lacking. A sperm duct often appears. The microscopic structure of the gonad is also testicular. Leydig cells, resembling those present in the interstitial tissue of normal testes, are found in the transformed ovary (*Benoit, 1932*). If ovariectomies are performed on young chicks, the right gonad, in a small percentage of the birds, may later produce sperm (*Domm, 1929*). As the bird approaches maturity, male hormones are secreted, as manifested by the development of such male secondary characteristics as large comb, typical male plumage, and crowing.

The phenomenon of sex reversal may indicate the possibility that both sets of reproductive organs, male and female, were originally present in

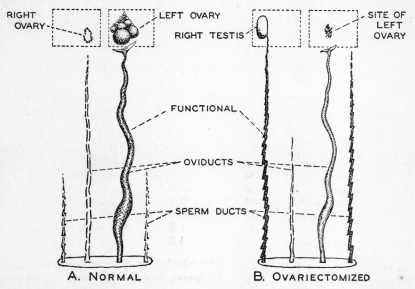

FIG. 108. A diagram showing sex reversal in the fowl after surgical removal of the left ovary.

A, in a normal hen: the left ovary is functional; the right ovary is diminutive, or degenerated. *B,* in a hen several months after removal of the left ovary: the right ovary has grown, but, instead of developing into an ovary, it has developed as a testis; the oviducts are atrophic, and a sperm duct has appeared on each side.

every individual, as they are in some of the lower forms of animal life. During the long evolutionary process, a division of labor in reproduction may have occurred, with the result that the organs of one or the other sex retrogress during early development. Possibly the growth of one set of organs and the degeneration of the other may also have been associated with the necessity for accommodating an increasingly large and complex egg.

THE OVARY

The ovary is known as the *ovarium,* or egg holder. To the naked eye, it appears very much like a bunch of grapes. Each of the spherical ova is loosely attached to the ovary by a stem, or peduncle, which is embedded in the connective and vascular tissues of the ovary. These tissues also attach the ovary to its stalk. They form the foundation of the ovary.

Since eggs are produced in succession, it is not surprising to find that the ovarian ova are not of uniform size. Because the ova differ in relative maturity and in the length of time they have spent in active growth, their size ranges from microscopic to that of the large, yolk-filled, fully mature ovum ready to leave the ovary.

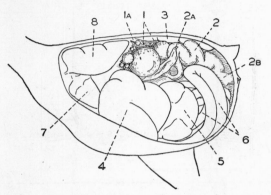

Fig. 109. General location of the ovary and the oviduct, in relation to other organs in the body cavity of the hen.

The numbers indicate: 1, the ovary; 1a, the mature ovum; 2, the oviduct; 2a, the infundibulum, or anterior end of the oviduct; 2b, the uterus, or posterior end of the oviduct; 3, the left kidney; 4, the liver; 5, the gizzard; 6, the intestines; 7, the heart; 8, the left lung. (Drawn from a dissected hen.)

The color of the individual ovum is also an indication of its stage of development and maturity. As they increase in age and size, the ova become more intensely colored. The minute, youngest ova are gray or grayish white. The more mature ova have a yellowish cast, which indicates that yolk substance is beginning to form. The fully ripened ovum, with its vast accumulation of yolk, is the familiar bright yellow seen in newly laid eggs.

Location. The single ovary that develops to maturity is situated in the sublumbar region of the abdominal cavity (Fig. 109). It lies at the anterior end of the left kidney and slightly to the left of the midline of the body. The heavy stalk, or base, of the ovary is attached to the dorsal wall of the abdominal cavity by a fold of peritoneum, called the *mesovarium.* This fold keeps the ovary in position, suspended in the body cavity.

The neighboring organs in the abdominal cavity so surround the ovary as to enclose it in what Curtis (*1910*) calls the ovarian pocket. When the yolk is released from the ovary, it may be received directly by the active oviduct, or it may fall into the ovarian pocket. The gentle movements of the organs bordering the pocket assist in guiding the yolk into the oviduct and decrease the possibility of its going astray.

Variation in Size. The size, as well as the weight, of the hen's ovary varies remarkably, depending on whether the reproductive system is in a functional condition or in a state of inactivity.

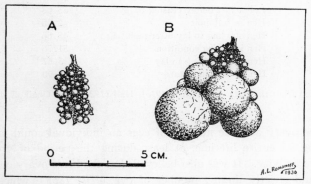

Fig. 110. Size and shape of the hen's ovary: *A*, in the inactive state. *B*, in the active state.

During prolonged laying, the ovary attains its maximum size and weight; it becomes from ten to fifteen times smaller during the period of rest (Fig. 110). The size of the ovary, therefore, is regulated mainly by the relative reproductive activity of the bird.

The age at which a bird reaches sexual maturity is characteristic of the particular species. In the domestic chicken, sexual maturity is reached at 5 or 6 months of age (see Chapter 1, "Mode of Laying"). The ovary of the pullet, after her first egg is laid, is almost six times as large as that of a hen during a nonlaying period.

Table 14 illustrates the changes in the weight of the chicken's ovary from the age of 1 day to the mature, laying stage. The weight of the ovary drops sharply during periods of reproductive inactivity, such as molting and brooding.

Number of Ova

During the nineteenth century, the number of ova contained in the bird's ovary was a subject of great interest to scientists. It was generally thought that a direct correlation should exist between the number of

TABLE 14

CHANGES IN THE WEIGHT OF THE HEN'S OVARY *

Age and Condition of the Bird (Single-Comb White Leghorn)	Weight of Ovary (grams)
Day-old chick	0.03
Three-month-old pullet	0.31
Four-month-old pullet	2.66
Five-month-old pullet	6.55
Pullet after laying first egg	38.00
Pullet just before ceasing to lay	33.63
Hen in full molt	2.98
Hen starting to lay after molt	48.66
Hen in laying condition	51.76
Hen after ceasing to lay	3.67
Hen in brooding state	2.90

* Compiled from the data of Wilkins (*1915*), Hall (*1926*), and Chaikoff, Lorenz, and Entenman (*1941*).

ova in the ovary and the number of eggs an individual could produce, if not during its entire lifetime, at least during the period of greatest reproductive activity. It was also believed that the hen's ovary contained a definite and strictly limited number of ova (*His, 1868*), which Geyelin (*1865*) stated to be 600. A hen would thus have a reproductive period of less than 2 years if she laid an egg every day. Actually, a modern hen can lay many more than 600 eggs during her lifetime. On the other hand, domestic fowls were not so systematically selected for intensive laying during Geyelin's time as at present; perhaps for this reason, the number of visible ova he observed was less than one-third the average number found today (Fig. 111). Even the ovary of domestic waterfowl now contains more than 1000 visible ova. The effects of selection become

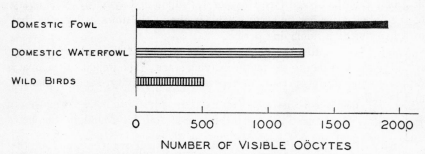

FIG. 111. Number of visible oöcytes in the ovary in various species of birds. (From the data of Pearl and Schoppe, 1921.)

obvious when the ovary of a domesticated bird is compared with that of a wild bird, in which about 500 ova are seen (cf. Fig. 111).

The visible ova in the hen's ovary have been counted by Pearl and Curtis (*1912*) and Pearl and Schoppe (*1921*). The largest number found in the examination of thirty-six birds was 3605; the smallest, 586. The number of visible oöcytes in the ovary of several breeds of chicken may be seen in Table 15, which not only gives an indication of the

TABLE 15

NUMBER OF VISIBLE OÖCYTES IN THE OVARY OF VARIOUS BREEDS OF FOWL *

Visible Oöcytes

Breeds of Fowl	Discharged Follicles	Under 1 Mm. in Diameter	From 1 Mm. to 1 Cm. in Diameter	Over 1 Cm. in Diameter	Total Number of Oöcytes
White Leghorn	87	2268	124	5	2484
F₁ Crossbred	46	2257	64	7	2374
Barred Plymouth Rock	31	1483	88	6	1608
Cornish Indian Game	54	1323	167	6	1550
Light Brahma	28	1074	119	9	1230

* From the data of Pearl and Schoppe (*1921*) and Romanoff (*1931*).

enormously increased reproductive potentialities of the modern hen but also shows the effect of the selective breeding of the Leghorn for high egg production.

However, the hen lays far fewer eggs during her lifetime than there are oöcytes visible in the ovary. The best laying record at present is a little more than 1500 eggs, fewer than half the maximum number of oöcytes found in an ovary by Pearl and Schoppe (*1921*). Egg production probably is limited by factors other than the capacity of the ovary to produce ova—possibly by the capacity of the organs and tissues to supply the materials needed to complete the egg.

As the hen advances in age and continues to lay, progressively fewer oöcytes can be seen in the ovary (*Hansemann, 1913*). Theoretically, the total number of eggs laid should correspond to the number of ruptured follicles, since these are formed upon the discharge of ova from the ovary. However, such is not the case, because the ruptured follicles normally are resorbed. The largest number of ruptured follicles found in the hen's ovary by Pearl and Schoppe (*1921*) was 87.

It is evident that resorption of ruptured follicles is relatively very rapid. In the pheasant, during the laying period, Romanoff (*1943a*) found that the follicle corresponding to the most recent ovulation weighed 0.17 gm., whereas the fifth previous follicle weighed only 0.007 gm.

In hens, 1 week after ovulation, the postovulatory follicle has regressed so as to be barely visible to the naked eye. After a month, it has practically disappeared in both chickens and pigeons (*Davis, 1942b*).

THE OVIDUCT

The oviduct (from the Latin, *ovum,* egg; *ductus,* canal) is the tube through which the egg passes. However, in the bird, its function is far more complex than merely to provide a means for transport. The oviduct must apply to the yolk all the structures necessary to complete the egg. While the yolk traverses it, the oviduct secretes and applies several structures in succession: the albumen, two shell membranes, and the calcific shell, which in some species is pigmented. These elements comprise two-thirds of the weight of the egg.

Location. The single oviduct of the bird is a long, convoluted, intestinelike organ with extremely elastic walls. It is capable of great changes in size. In the laying hen, it occupies a large portion of the left half of the abdominal cavity. The diameter of the oviduct varies throughout its length. In some places, the canal is narrow and the musculature of the walls is thin; in other portions, where the muscles and glands in the walls are better developed, the passageway is large. The upper end of the oviduct—the infundibulum—opens into the body cavity near the ovary but is not attached to it. The lower end opens into the cloaca.

Supporting Structures—Ligaments. The oviduct is supported in the body cavity by two structures, the dorsal and ventral ligaments (Fig. 112).

The dorsal ligament suspends the oviduct from the dorsal body wall above the abdominal air sac. The ligament is a thin, fan-shaped curtain, thrown into folds by the convolutions of the oviduct, to the upper edge of which it is attached. The ventral ligament hangs from the lower side of the oviduct into the abdominal cavity. Although it is more muscular than the dorsal ligament, the ventral ligament is likewise thin and veil-like. The anterior end extends forward to the region of the ovary and assists in guiding the ova into the infundibulum of the oviduct. The caudal portion terminates as a solid muscular cord which divides into bundles of fibers; these pass on to the ventral and ventrolateral sides of the anterior portion of the vagina (*Kar, 1947*). Both ligaments are very elastic and flexible and, according to Curtis (*1910*), permit considerable movement of the oviduct, especially its infundibular portion.

When the oviduct is functionally active, the ligaments are well supplied with blood. In the region of the oviduct, the large blood vessels break up into branches which pass forward and backward on the oviduct, anastomosing with other branches either in the ligaments or on the walls of the oviduct (*Curtis, 1910*).

Variation in Size. The size of the oviduct in the mature domestic hen depends on the bird's age and the state of the oviduct's functional

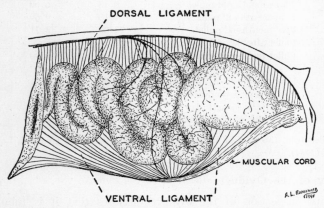

FIG. 112. Location of dorsal and ventral oviducal ligaments in the hen. (Drawn from a dissected specimen.)

activity. The elasticity of the walls of the oviduct, and of the ligaments that support it, permits great extremes of dilation and contraction.

The dimensions of the oviduct in mature hens, laying and nonlaying, have been measured by several observers. In nonlaying hens, the oviduct may be from 11 cm. (*Kaupp, 1918*) to 18 cm. long (*Sacchi, 1887; Sturm, 1910*), and from 0.4 mm. to 7.0 mm. in diameter, depending on the portion measured. When the bird is at the height of reproductive activity, the dimensions of the oviduct not only are greater but also vary over a much wider range. The length ranges from 37 cm. (*Cushny, 1902*) to 86 cm. (*Surface, 1912*), with an average of 64.8 cm.; the width ranges from 0.6 cm. to 10.0 cm. The data in Table 16 show that, when the hen is in full laying condition, the oviduct may be from twelve to twenty times heavier and from two to four times longer than during nonlaying periods.

To evaluate the significance of these enormous changes in size, it is necessary to examine the anatomical organization and the microscopic structure of the oviduct and the chemical nature of its secretions.

TABLE 16

CHANGES IN WEIGHT AND LENGTH OF THE HEN'S OVIDUCT *

	Oviduct	
Age and Condition of the Bird (Single-Comb White Leghorn)	Weight (grams)	Length (centimeters)
Day-old chick	.....	0.45
Three-month-old pullet	0.18	6.60
Four-month-old pullet	1.10	9.69
Five-month-old pullet	22.00	32.21
Pullet after laying first egg	77.20	67.74
Pullet just before ceasing to lay	74.43	64.90
Hen in full molt	4.20	16.92
Hen starting to lay after molt	75.50	67.84
Hen in laying condition	78.12	68.80
Hen after ceasing to lay	5.43	29.85
Hen in brooding state	5.70	29.85

* Compiled from the data of Wilkins (*1915*), Hall (*1926*), and Chaikoff, Lorenz, and Entenman (*1941*).

Anatomy of the Oviduct

A casual inspection of the oviduct discloses that it is tubular and varies considerably in diameter throughout its length. Closer examination shows that this variation in diameter may be used as a basis for dividing the organ into regions.

MORPHOLOGICAL DIVISIONS

Structurally, the oviduct of the hen may be divided into five more or less distinct regions, each having fairly specific physiological functions in egg formation. These regions are (1) the infundibulum, or funnel; (2) the albumen-secreting region;[1] (3) the isthmus, or shell membrane-secreting region; (4) the uterus, or shell-secreting region; and (5) the vagina. The dimensions of these divisions in the active and inactive oviduct are shown in Table 17.

The two structural features of the oviduct most essential to egg formation are (1) the muscular layers which lend support to the oviduct and propel the egg, and (2) the glandular epithelial lining.

The muscular tissue is arranged in two layers, the outer and inner, the fibers of which, respectively, run longitudinally and circularly. The muscle layers impart a certain degree of firmness to the wall of the

[1] This term is considered more specific than "magnum," which has recently come into use to designate this region of the oviduct.

TABLE 17

Size of Various Sections of the Hen's Oviduct in Active and Inactive Periods *

Sections of the Oviduct	Size of Oviduct			
	Active Stage		Inactive Stage	
	Length	Width	Length	Width
	(centi-meters)	(centi-meters)	(centi-meters)	(centi-meters)
Infundibulum, or funnel	7.0	8.6	2.4	. . .
Albumen-secreting portion	33.6	1.7	5.4	0.8
Isthmus	8.0	0.9	2.2	0.4
Uterus, or shell-secreting portion	8.3	2.9	2.4	1.2
Vagina	7.9	0.9	3.0	0.4
Total	64.8	. . .	15.4	. . .

* Averaged from the data of Blasius (*1867*), Sacchi (*1887*), Giacomini (*1893*), Sturm (*1910*), Surface (*1912*), Giersberg (*1922*), Buckner, Martin, and Peter (*1925*), Romanoff (*1931*), and Warren and Scott (*1935a*), on active oviduct; and of Sacchi (*1887*), Sturm (*1910*), Kaupp (*1918*), and Giersberg (*1922*), on inactive oviduct.

oviduct. The outer longitudinal layer of smooth muscle shortens or lengthens the oviduct, and the inner circular layer changes the oviduct's diameter. Coordination of the muscular action of both layers results in peristalsis, which serves to advance the egg in the oviduct.

In contrast to the smooth peritoneal covering of the outside of the oviduct, the epithelium forming the lining is a series of ruts and wrinkles (Fig. 113). It is made up of two elements: (1) the outer mucous membrane, and (2) an inner ciliated epithelium.

The mucous membrane is thrown into numerous primary folds, or ridges, which continue with little interruption throughout the oviduct. The size of the folds varies from region to region; the larger the folds, the rougher the inner oviducal surface. The area of the secretory surface increases in proportion to the amount of folding. The mucous membrane and its slimy secretions form a soft, resilient cushion for the delicate yolk. As the yolk passes, it is rotated gently by the slight spiral direction of the ridges (cf. Fig. 113).

Throughout the length of the oviduct, the innermost lining consists of the ciliated epithelium, which covers the ridges in the mucous membrane. Where the epithelium is intensely developed, it forms secondary and tertiary folds. In the mucous and epithelial lining of the oviduct are the glands that secrete all the structures of the egg, except the yolk.

The Infundibulum. The most anterior portion of the oviduct opens into the body cavity near the ovary by way of an elliptical aperture, the

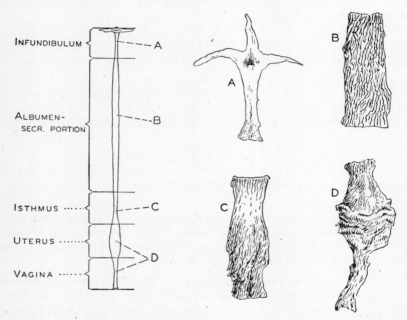

Fig. 113. Gross anatomy of the hen's oviduct. Portions of the oviduct, slit longitudinally and laid open, in order to show the folding of the mucous membrane.

A, the infundibulum. B, the albumen-secreting region. C, the transitional zone, between the albumen-secreting region and the isthmus. D, the expanded uterus and the narrow vagina. Reduced 4 times. (After Surface, 1912.)

infundibulum, for which the anatomical term is *ostium tubae abdominale.* The walls of the oviduct expand and become thinner at the ostium to form the flared lips of the funnel. The lips are greatly extended in the dorsoventral direction and are connected with the anterior borders of the dorsal and ventral ligaments of the oviduct. In the laying hen, the funnel is about 8.6 cm. in diameter. The ligaments afford considerable freedom of movement to the infundibulum, which is thereby enabled to clasp the follicle of the ovary and receive the mature ovum at the time of ovulation.

The funnel is supplied with a net of capillaries and lymphatic vessels. The muscle fibers are sparse and are arranged in scattered bundles, rather than in definite layers.

The flaring funnel converges sharply into a rather narrow tube, 2 to 4 cm. long. This tube is the neck of the infundibulum. It is the most thin-walled and delicate portion of the oviduct. The two distinct layers of muscles (*Surface, 1912*) are very thin and well embedded in connective tissue.

The inner surface of the infundibulum is covered with longitudinal folds of mucous membrane. These folds are low and small, in comparison with those in other portions of the oviduct. However, they increase in height throughout the tubular portion of the funnel. In the posterior part, where the transition is made to the albumen-secreting region, the folds suddenly become ridges. In the epithelium covering the large primary folds in the mucous membrane are numerous smaller secondary folds (*Surface, 1912*).

The Albumen-Secreting Region. This region (sometimes called the *convoluted glandular portion*) is named according to its physiological function and constitutes the major portion of the oviduct. In the laying hen, it attains an average length of 33.6 cm. Most of the coils and convolutions of the oviduct are in this region. The walls are markedly thicker than those of the infundibulum. Although the muscle layers are somewhat heavier, the thickness of the walls is due mainly to the intense development of the glands.

The size and number of ridges or folds in the lining of the tube reach their maximum in this portion (cf. Fig. 113). According to Giersberg (*1922*), there are fifteen to twenty-two principal folds, with additional secondary and tertiary folds. The primary folds attain a height of 4.5 mm. and a thickness of 2.5 mm., or about five times the dimensions of those in other regions. Some of the folds are continuous with the folds of the infundibulum and take a slightly spiral course. During phases of active secretion, the lining of this portion of the oviduct is either milky white or luminous gray.

The albumen-secreting region is sharply divided from the next portion of the oviduct, the isthmus. The boundary between the two regions appears as a narrow ring, devoid of glands and extending around the tube. It is easily distinguished by its translucency. In this area, the folds decrease in height to about 1.5 mm.

The Isthmus. This is a short section, about 8 cm. in length. The duct is narrower here than in the albumen-secreting region. The muscle layers, particularly the circular layer, are thicker and firmer, as are the walls. The folds of the inner surface are not so high as in the preceding portion, and they are arranged longitudinally rather than spirally. The mucous membrane is yellowish brown and is darker than in the other portions of the duct.

The narrow isthmus merges gradually into the wider uterus, without any definite boundary or striking anatomical differences. As the tube enlarges,

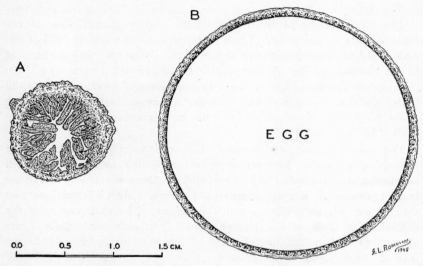

Fig. 114. Cross section of the uterine portion of the oviduct of the jackdaw, showing its natural extensibility.

A, the appearance without the egg. *B*, the same region at the time of passage of the egg. (After Stieve, 1922, with some modifications.)

the primary folds of the lining are flattened. The lesser folds become thin, leaflike plates covering the inner surface.

The Uterus. The length of this portion, about 8.3 cm., is comparable with that of the isthmus; but the uterus appears much shorter because of its expanded, saclike form.

The walls are thick, and the musculature, especially in the longitudinal layer, is well developed. Although the glands resemble those in the other regions, the folds of the lining are lower and are interrupted by secondary folds running diagonally and transversely. The mucous membrane is a characteristic pink.

The uterus is extremely extensible, as shown in Fig. 114. During the passage of a nearly completed egg, the diameter of the uterine tube increases about three times.

The uterus opens into the short terminal portion of the oviduct, the vagina. The boundary between the two regions is formed by a strong sphincter muscle.

The Vagina. The walls of the oviduct constrict, as they pass the sphincter muscle, to form the vagina. This consists of a narrow, muscular tube extending to the cloaca. The length of this region is about 7.9 cm. Longitudinal ridges and folds are present in the lining of the vagina, but they are low and narrow. The muscle layers are well developed, particularly the circular layer, which is several times thicker than in any other portion of the oviduct. The egg arrives in the vagina fully developed, covered with membranes and shell, and ready to be expelled. It is a solid object upon which the muscles of the vagina may act.

The vagina and the intestine both open into the cloaca; the vaginal opening is just dorsal to the rectum, through which the intestine empties.

Histology of the Oviduct

The wall of the oviduct contains seven layers of tissue. From exterior to interior of the oviduct, these are: (1) a thin, serous epithelium, which covers the outside of the oviduct and is continuous with the lining of the body cavity; (2) a layer of longitudinal muscle fibers, which are continuous with the musculature of the ventral and dorsal ligaments of the oviduct; (3) a layer of connective tissue containing the larger blood vessels; (4) a layer of circular muscle fibers; (5) a second layer of connective tissue richly supplied with capillaries; (6) a layer of highly glandular mucous membrane in large or small folds, depending on the degree of development of the glands; and (7) an internal lining of ciliated epithelium (Fig. 115).

The several egg envelopes are secreted and added successively by the glandular epithelium of the various regions as the egg is advanced through the oviduct by muscular action. The thickness of the secretory epithelium at the different levels in the oviduct is generally greatest where there is most active secretion of egg-forming substances, that is, in the posterior portion of the albumen-secreting region and in the uterus (Fig. 116). The vaginal epithelium is exceptional in that it forms a deep layer yet contributes little, if anything, to the egg.

The height of the epithelium in the albumen-secreting region is greatest just prior to the passage of a yolk. At the extreme caudal end of this region, the average goblet cell height is 0.034 mm. (*R. K. Cole, 1938*).

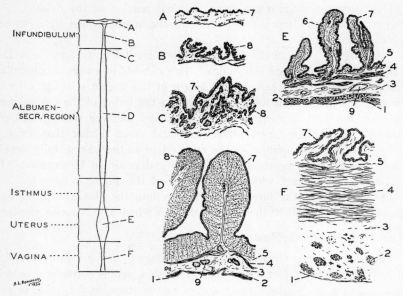

FIG. 115.   Transverse sections of various regions of the wall of the oviduct of a laying hen.

   A, the lips of the infundibulum.   B, the neck of the infundibulum, showing primary and secondary folding of the epithelium.   C, the region of transition from the infundibulum to the albumen-secreting region (muscular layers are not shown).   D, the albumen-secreting region, showing the thick layer of tubular glands.   E, the walls of the uterus.   F, the walls of the vagina.   Magnified 20 times.

   The numbers indicate: 1, peritoneal membrane or layer; 2, longitudinal muscle fibers; 3, connective tissue; 4, circular muscular fibers; 5, inner layer of connective tissue; 6, thick layer of convoluted tubular glands; 7, epithelium of the duct; 8, ducts of convoluted glands; 9, blood vessels.   (After Surface, 1912; somewhat modified.)

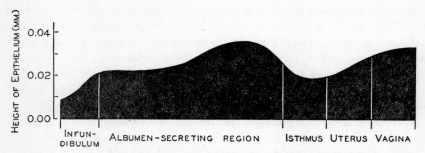

FIG. 116.   Height of the epithelial cells in different regions of the oviduct.   (From the observations of Bradley, 1928.)

THE SECRETORY LINING

According to Surface (*1912*), Giersberg (*1922*), and Bradley (*1928*), the lining of the oviduct is made up of three kinds of secretory cells: (1) ciliated cells of the columnar type; (2) nonciliated cells of the goblet type; and (3) nonciliated cells not of the goblet type. The first two types of cells are found in the epithelium, and those of the third type,

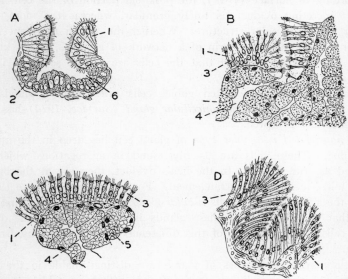

Fig. 117. Sections of typical glands from the four regions of the hen's oviduct. *A*, the infundibulum. *B*, the albumen-secreting region. *C*, the isthmus. *D*, the vagina. Magnified 500 times.

The numbers indicate: 1, ciliated cells, columnar type; 2, nonciliated cells, unicellular gland type; 3, goblet cells, interspersed among the ciliated epithelial cells; 4, tubular glands; 5, fluid albumen; 6, glandular groove. (After Surface, 1912.)

which form the tubular glands, are found in the mucous membrane (Fig. 117).

The *ciliated cells* are tall and contain finely granular cytoplasm. In the individual cell of this type, the nucleus is large and usually oval. It lies slightly toward the free, or distal, end of the cell. The free end of the cell is profusely furnished with strong cilia which extend into the lumen of the oviduct. The function of these hairlike appendages is to sweep the secretory products onward.

The ciliated cells in some regions of the oviduct are likely to be much distorted, because they are crowded by the goblet type of nonciliated cells with which they alternate.

The *nonciliated goblet type* of epithelial cell found in the inner lining of the hen's oviduct is considered glandular in function. In addition to having no cilia, it has other structural peculiarities.

The nucleus in this type of goblet cell differs in two respects from that in the ciliated cell: (1) it lies well toward the base, or proximal end, of the cell; (2) it is more nearly spherical.

According to Surface (*1912*), the proximal portion of the cell, in the actively secreting oviduct, is finely granular, whereas the more distal portion has an alveolar structure. Whether the alveoli represent secretion, or cytoplasm in the form of a network of granules, is not known. It can certainly be assumed that the cells are full of secretion in the active oviduct.

Functionally, the nonciliated goblet cells are primitive glands, for which the descriptive name *unicellular gland* (*Surface, 1912*) has been quite generally accepted.

The *nonciliated nongoblet type* of gland cell lies deep in the mucous membrane. These cells are usually found in aggregations which are regarded as true glands. In the hen's oviduct, they are at the bottom of the folds, in a tubular arrangement. For this reason, Surface (*1912*) named the aggregations of cells *tubular glands*. Differential staining indicates that the function of these glands is different from that of the goblet cells. The existence of this difference was confirmed by Bradley (*1928*).

To summarize, two types of glands are distinguishable in the hen's oviduct: (1) the *unicellular glands* of the epithelium, which are composed of nonciliated goblet cells, and which alternate with ciliated columnar cells (Fig. 118-*A*); and (2) the *tubular glands* (cf. Fig. 117) found in the folds of the mucous membrane.

The Glands of the Infundibulum. In the anterior portion of the infundibulum there are no folds in the mucous membrane; this region is lined with ciliated cells only. In the posterior part, where the folds begin, unicellular glands or goblet cells (*Bradley, 1928*) appear, alternating with ciliated cells, as in the rest of the oviduct. In the depressions between the shallow folds in the posterior part of the infundibulum are secretory cells typical of tubular glands. These folds, the "glandular grooves," or "gland pouches" (*Surface, 1912*), represent the initial phase in the development of the tubular gland (cf. Fig. 117-*A*). They are more shallow than the true tubular gland.

The Glands of the Albumen-Secreting Region. In passing from the neck of the infundibulum to the albumen-secreting region, the primary folds of the mucous membrane suddenly become taller and broader and attain their maximum development. The secondary folds also become deeper and more numerous (cf. Fig. 115-*D*).

In this region, all the glandular elements, and particularly the tubular glands, attain their greatest development, functionally as well as structurally (*Loos, 1881; Giacomini, 1893; Cushny, 1902; Giersberg, 1921*). The tubular glands branch and coil in an extremely complex manner, and it is difficult to trace the course of a single gland (*Surface, 1912*). The glands lie buried in the connective tissue below the mucous membrane and communicate with the lumen of the oviduct by short ducts

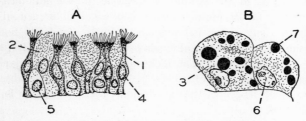

FIG. 118. Gland cells from the albumen-secreting region.

A, unicellular glands alternating with ciliated cells. *B,* cells from a tubular gland. Magnified 1000 times.

The numbers indicate: 1, ciliated cell; 2, nonciliated cell of the goblet time; 3, cell of a tubular gland; 4, 5, and 6, nuclei; 7, granules. (After Surface, 1912.)

opening into the bottom of the folds. The connective tissue surrounding the glands is richly supplied with capillaries.

The tubular glands are composed of nongoblet, nonciliated cells (cf. Fig. 117-*B*). In the active secretory phase, they appear distorted because of crowding. The cytoplasm of the cells is filled with fine, dark granules (cf. Fig. 118-*B*); these, according to Giersberg (*1922*), have angular projections which cause the cytoplasm to appear alveolar. A substance that stains deeply is scattered between the granules in such a manner that the cytoplasm resembles a spongioplasm.

The unicellular glands, or goblet cells, are as numerous as, and much broader than, the ciliated cells in this region. They are larger than the goblet cells of the funnel region. They may so compress the ciliated cells that the epithelium appears to be a continuous mucin-secreting surface.

As to the secretory activity of the goblet cells, there seems to be a difference of opinion. Surface (*1912*) says, ". . . the secretion from these glands can often be seen pouring out in little streams." According to Giersberg (*1922*) and Bradley (*1928*), the secretion is extruded in the form of droplets which disintegrate into fine granules immediately upon being released. At the time they are extruded, the droplets give the staining reaction characteristic of mucin or mucoid. After the drop-lets disintegrate, the staining reaction of the secretion is lost.

The secretion of the tubular glands is at first granular, but after a short time it loses its granular nature and becomes fluidlike.

The secretions of the two types of gland become mixed in the lumen of the oviduct and lose their identity. It has not been definitely deter-mined whether these secretions are the same substance or different sub-stances which combine to form the albumen.

The Glands of the Isthmus. Tubular glands and unicellular goblet-type glands are present in the isthmus as well as in the albumen-secreting region. Surface (*1912*) found little histological difference in the two regions, except that, in the isthmus, the tubular glands do not form so thick a tissue layer as in the albumen-secreting region. Also, according to Bradley (*1928*), the granular inclusions in the tubular glands of the isthmus are the larger (cf. Fig. 117-*C*). These glands are absent in the narrow band that separates the two regions, and their place is taken by connective tissue.

Some histological differences in the glands of the two regions were first noted by Giacomini (*1893*), who compared glands in the same phase of activity. His observations were confirmed by Giersberg (*1922*) and Bradley (*1928*).

Although the unicellular glands of the isthmus are similar to those in the funnel, they are not so tall nor nearly so broad as those in the albumen-secreting region, and they contain less secretion. Bradley (*1928*) found that the unicellular glands of the isthmus and the albumen-secreting region did not react in the same manner to staining. These differences are explained by the fact that the two regions secrete different parts of the egg.

The Glands of the Uterus. In the uterus, as in the previous regions, there are tubular glands and unicellular goblet-type glands (cf. Fig. 115-*E*). In this region, which furnishes the shell, the glands are so slightly differ-entiated that little indication is given of their highly specialized function.

Goblet-type unicellular glands alternate with ciliated epithelial cells and are similar in size, form, and staining reaction to the goblet-type glands of the isthmus.

The tubular glands of this region differ from those in the isthmus in that (1) the cytoplasm is more finely granular (Fig. 119); (2) the secre-tory granules in the cells are finer, more diffuse, and fewer, and they are confined to the edges of the cell, particularly toward the lumen; and (3) the nuclei are more nearly spherical and nearer the center of the cell. The gland is more shallow and sometimes lacks a true duct; it begins to lose its identity as a true tubular gland.

The Glands of the Vagina. In the vagina, ciliated and nonciliated cells similar to those in previous regions form the epithelial lining, which is the only important secretory element in this region. Both types of cell are narrower but usually taller than comparable cells in other portions of the oviduct (cf. Fig. 117-*D*).

Although the mucous membrane is richly folded, the ridges are low and narrow and are supported by connective tissue which obliterates the

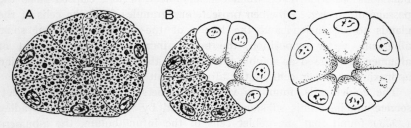

FIG. 119. Transverse sections of tubular glands from the isthmus and the uterus and from the junction of these two regions; magnified 1000 times.

A, gland from the isthmus, composed of cells containing large granules. *B,* gland from the transitional region, composed of cells characteristic of both the isthmus and the uterus. *C,* gland from the uterus, composed of cells containing small granules. (After Bradley, 1928.)

layer of tubular glands usually found there. Numerous aggregations of cells similar to gland cells are found in the depressions between the folds. Since the egg is complete when it arrives in the vagina, these cells probably contribute little or nothing to egg formation. According to Giersberg (*1922*), tubular glands are present in the vagina, although Bradley (*1928*) states that "the vagina has no glands in the ordinary sense of the word."

The folds in the mucosa disappear on passing through the sphincter into the cloaca.

Chemistry of the Oviduct

The oviduct does not differ greatly in chemical composition from other soft tissues. Analyses for ordinary tissue constituents of the hen's oviduct (*Common, 1938*) yielded the results shown in the tabulation.

CONSTITUENT	PER CENT
Moisture	76.0
Protein	18.6
Fat	4.3
Ash	1.1

The secretion of certain substances that enter into egg formation is localized in definite regions of the oviduct; for example, the secretion of protein occurs in the albumen-secreting region, that of minerals chiefly in the uterus. The detection of these materials has been attempted by various methods: differential staining (*Bradley, 1928; Conrad and Scott, 1942b*); analyses of isolated portions of the oviduct (*Buckner, Martin, and Peter, 1925*); and microincineration (*Richardson, 1935*).

Conrad and Scott (*1942b*) found, by staining, that the tubular glands of the albumen-secreting region, without exception, were filled with albuminous secretion, even when considerable time had elapsed since the passage of an egg. Albumen prosecretory granules filled the cells of the glands to distension, and the lumen of the oviduct contained some albumen. After oviposition, the hourly increment in the amount of albumen (expressed as soluble nitrogen) in the albumen-secreting region approximates 4 per cent of the albumen content of the egg.

The goblet cells, found in varying numbers throughout the oviduct, secrete mucin. This substance is detected by several stains. Bradley (*1928*) found that the goblet cells of the infundibulum, the albumen-secreting region, and the isthmus stained deeply, thereby revealing an abundant content of mucin. Goblet cells in the uterus stained slightly, indicating their relative lack of mucin.

During shell formation, the glands of the uterus provide more than 2 gm. of calcium, as well as other minerals. A complete explanation of the mode of secretion has not yet been made. Richardson (*1935*) found, by microincineration, that the uterine epithelium contained more ash than did deeper-lying tissues; the ciliated cells in contact with the egg were particularly rich in minerals. Analyses made by Buckner, Martin, and Peter (*1925*) showed that the uterus contained no more calcium and phosphorus than other portions of the oviduct (Table 18), and that the uterine tissues in the laying hen were only slightly richer in calcium than those in the nonlaying hen. In fact, the entire oviduct contains very little calcium (*Common, 1938*). It is evident, therefore, that calcium is not stored in the oviduct in preparation for eggshell formation but is removed from the circulation and deposited as shell during the egg's sojourn in the uterus. The fact that the phosphatase content of the uterus is also low (*Gutowska, Parkhurst, Parrott, and Verburg, 1943*) is an additional indication that the uterus excretes rather than mobilizes calcium.

The production of avidin (see Chapter 9, "Food Value") is restricted to the albumen-secreting region of the oviduct (*Fraps, Hertz, and Sebrell,*

TABLE 18

MINERAL CONTENT OF VARIOUS REGIONS OF HEN'S OVIDUCT *

	Mineral Content of Dried Material	
Regions of the Oviduct	Calcium (per cent)	Phosphorus (per cent)
In laying hen:		
Albumen-secreting region	0.06	0.40
Isthmus	0.31	0.52
Uterus (shell-secreting region)	0.17	0.52
In nonlaying hen:		
Albumen-secreting region	0.06	0.56
Isthmus	0.16	0.67
Uterus (shell-secreting region)	0.04	0.50

* From the data of Buckner, Martin, and Peter (*1925*).

1943). Apparently, avidin is not secreted unless the oviduct is in the laying condition.

HYDROGEN-ION CONCENTRATION

The pH of the internal environment of the oviduct has a significant bearing on fertilization and is therefore of biological interest. In the vesicles of the cock, where the reaction is slightly alkaline (pH 7.3), the spermatozoa are motile (*Buckner and Martin, 1928*). In an increasingly acid medium, their activity declines and motility eventually ceases.

According to Buckner and Martin (*1928*), the lining of the anterior portion of the oviduct, down to the end of the albumen-secreting region, is of slightly acid reaction (pH 6.4). Posterior to this point, the mucosa of the oviduct is somewhat more acid, as shown in the tabulation. (In

SECTION OF OVIDUCT	pH
Infundibulum	6.4
Albumen-secreting region	6.4
Isthmus	5.8
Uterus	5.8
Vagina	5.9

the nonlaying hen, the various regions of the oviduct differ very little in acidity.) Although the sperm is more active in alkaline solutions, it is able to traverse the entire length of the oviduct, where, in certain regions, the acidity approaches the limit beyond which the sperm cannot remain active.

Although very slightly acidic when secreted, the albumen becomes somewhat alkaline when the egg is in the isthmus or the uterus (pH 7.4).

FORMATION OF THE YOLK

A large part of the yolk of the bird's egg consists of nonliving material. In proportion, the amount of living substance, or protoplasm, is very small.

The formation and deposition of yolk and the differentiation, growth, and maturation of the germ cell are different yet interdependent processes. Together they result in the mature ovarian ovum filled with yolk.

The germ cell originates early in the embryonic development of the bird. A series of orderly changes in the germ cell continues until, eventually, ovulation occurs. The whole process is termed oögenesis and is divided into three stages: (1) differentiation of the primordial germ cells, followed by (2) multiplication, and (3) growth and maturation. During

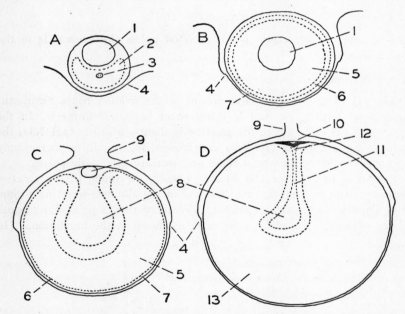

Fig. 120. Cross-sectional diagrams showing the development of the oöcyte in the pigeon.

A, a primordial follicle, 0.05 mm. in diameter; magnified about 400 times. *B,* an oöcyte 0.6 mm. in diameter; magnified about 50 times. *C,* an oöcyte 3 mm. in diameter; magnified about 10 times. *D,* a mature ovarian egg 20 mm. in diameter; magnified about 3 times.

The numbers indicate: 1, germinal vesicle; 2, yolk crescent; 3, centrosphere; 4, cloacal end of stigma; 5, central protoplasm; 6, peripheral spherule zone; 7, peripheral protoplasm; 8, latebra; 9, stalk of follicle; 10, blastodisc; 11, neck of latebra; 12, nucleus of Pander; 13, yolk. (After Bartelmez, 1912.)

The first three stages correspond roughly to the same stages in the chicken; but the ovum in the final stage is at least 32 mm. in diameter in the chicken, rather than 20 mm., as in the pigeon.

the first growth phase and the first of the maturation stages, the germ cell is called a *primary* and a *secondary oöcyte,* respectively. The young primary oöcyte is largely protoplasmic, with comparatively little yolk. It is during the last oöcyte stage that the major portion of true yolk is accumulated (Fig. 120).

ORIGIN AND HISTORY OF THE GERM CELL

The primordial germ cells are distinguishable very early in the bird's embryonic development, during the primitive streak stage. After 12 hours of incubation, they are found in the entoderm, at the anterior margin of the blastoderm (*Goldsmith, 1928*). They migrate by ameboid movement into the mesoderm, after the latter has extended into this region of the blastoderm between the two original germ layers. With the formation of blood vessels, the germ cells are carried by the blood to all parts of the 27-hour embryo. They remain in the blood stream until about 43 hours of incubation. At this time, they begin to congregate in the epithelium which lines the dorsal side of the body cavity in the region where the gonads will develop. A few hours later they are absent from the blood stream (*Swift, 1914*).

Throughout this period, the germ cells are about 0.016 mm. in diameter; their nuclei are about 0.01 mm. in diameter (*Swift, 1914*). The germ cells may be differentiated from all other types of cell by their greater size, by the presence near the nucleus of a darkly staining diploid centrosome, and by a Golgi apparatus larger than in other cells (*Woodger, 1925*). Like other cells in the embryo, the germ cells contain large yolk granules.

In the 5-day-old embryo, the germ cells have not changed greatly either in size or appearance, although the yolk granules are now fewer and smaller. However, the germinal epithelium, previously invaded by the germ cells, has now thickened and is budding and growing downward into the medulla of the gonad (*Swift, 1915*). These downgrowths of epithelium carry with them the primordial germ cells. Together they comprise the sexual cords, and their development makes possible the definite determination of the sex of the embryo on the following day. A second downgrowth, forming the cortex of the ovary, follows.

From the eighth to the eleventh day of incubation, the germ cells in the ovary of the embryo multiply rapidly and also decrease in size. The germ cells are now called *oögonia.* The centrosomes in the oögonia become obscured by the Golgi apparatus, which enlarges and accumulates fat about itself to form the *yolk-attraction* sphere, or the *vitelline nucleus of Balbiani* (*d'Hollander, 1902*).

Through an increase in cytoplasm, the oögonia enlarge and become primary oöcytes. Very little additional growth occurs up to the time the chick hatches, although by then the oöcyte has again attained a size comparable to the primordial germ cell when first seen. Many nuclear changes take place in the oöcyte during the later stages of embryonic development, and further changes take place as the oöcyte accumulates yolk. The walls of the germinal vesicle disintegrate about 24 hours prior to ovulation (*Olsen, 1942a; Olsen and Fraps, 1944*). The final phase of maturation occurs when and if the egg is fertilized. The relationships between stage of development and age are shown in Table 19.

TABLE 19

ORIGIN, TRANSFORMATION, AND GROWTH OF OÖCYTES IN FOWL *

Age of Fowl	Transformation of Oöcyte	Condition of Nucleus
Embryonic	Primordial cells (location):	
12 hours	In entoderm	Resting
24 hours	In splanchnic mesoderm	Resting
46 hours	In blood vascular system	Resting
55 hours	In gonad region	Resting
	Germ cells (location and condition):	
6 days	In germinal epithelial cords of female	Resting
9–11 days	Rapid division	Mitotic changes
12–14 days	Growth	Resting
15 days	Beginning of first stage of maturation	Resting
16–18 days	First stage of maturation	Meiosis
19 days	First stage of maturation	Leptotene stage
20 days	First stage of maturation	Zygotene stage
21 days	First stage of maturation	Pachytene stage
		Diplotene stage
Postembryonic	Ova (condition and size):	
2 days	Beginning of follicle formation	Resting
4 days	0.01–0.02 mm. diameter	Resting
10 days	Accumulation of deutoplasm	Resting
	Formation of definite follicular ring	Resting
21 days	0.03–0.07 mm. diameter	Resting
37 days	Follicle condensing	Resting
42 days	0.04–0.08 mm. diameter	Resting
65 days	Beginning of second stage of maturation	Renewed activity
70 days		Pachytene stage
90 days	0.04–0.10 mm. diameter	Resting
180 days	Rapid accumulation of yolk material	Resting
Mature	4.0–6.0 mm. diameter	Disappearance of membrane
	30.0–35.0 mm. diameter	
	Second maturation stage	Metaphase
	Ovulation of the largest ovum	

* Based on the data of van Durme (*1914*), Swift (*1914*), Goldsmith (*1928*), and others.

Deposition of Yolk

Yolk deposition is a complex process which begins early in the life of the oöcyte and which depends to some extent upon the physiological efficiency of the bird.

In the primordial germ cell from which the oöcyte arises, the nucleus, or germinal vesicle, occupies the center of the cell, with the Golgi apparatus arranged in a thin crescent about one side of it (*Brambell, 1926*). Small, rodlike bodies, the mitochondria, are scattered throughout the cytoplasm.

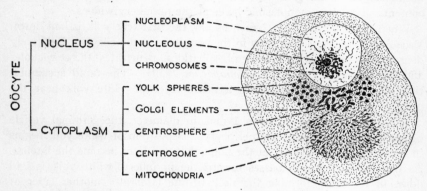

Fig. 121. Arrangement of various elements in the oöcyte (about 0.1 mm. in diameter) of the chick. (After Brambell, 1926, and others.)

In the oögonia, the mitochondria form a dense layer about the yolk-attraction sphere.

In the oöcyte, the central mitochondria condense to form the mito-chondrial cloud at one side of the nucleus. The Golgi apparatus becomes associated with the centrosphere and enlarges. The nucleus is displaced somewhat to one side. The central mitochondria fuse to produce the first center of fat formation in the oöcyte. The elements composing a typical oöcyte are shown in Fig. 121.

Toward the end of incubation, the yolk-attraction sphere and the fat globules break up, and the centrosphere disappears. Fat globules and the larger mitochondria of the oöcyte then combine with material from the cytoplasm and form a fatty layer a short distance from the periphery. Both central and peripheral fat layers become vacuolated. This stage corresponds to the first phase of yolk formation mentioned by Marza and Marza (*1935*).

Further increase in the size of the oöcyte results from the accumulation of materials passed to the oöcyte from the follicle, which begins to form about each oöcyte soon after the chick hatches. The Golgi bodies in the follicle cells divide and are extruded into the cortical layer of the oöcyte. Later they act as centers for the aggregation of yolk substance. After the vitelline membrane forms, the extrusion of Golgi bodies ceases. Materials then either diffuse directly through the membrane into the oöcyte or are passed through the pores of the membrane by means of the intercellular bridges of the follicle cells (*Paladino, 1891*).

During the first 6 weeks of the chick's postembryonic life, the growth of the oöcyte is due mainly to a slow increase in cytoplasm and fluid content. The clear cytoplasmic layer at the periphery widens.

When the chick is about 2 months old, the first true yolk is laid down. Vacuoles of liquid protein arise in the periphery of the oöcyte. Relatively denser protein, combined with phosphatides, penetrates these vacuoles and forms compact globules (*Konopacka, 1933*). The rapid accumulation of these vacuoles produces a dense layer of white yolk near the periphery.

When the oöcyte approaches 1 mm. in diameter, the germinal vesicle begins to migrate toward the periphery. Deposition of white yolk begins in the center of the oöcyte, at the point that will later become the latebra. As the germinal vesicle passes toward the periphery, white yolk is laid down in its wake, forming the neck of the latebra. Another layer of white yolk is laid down at the periphery. Periodicity in yolk deposition is thus established.

The increase in the diameter of the oöcyte from 1 mm. to 5 mm. is due chiefly to the increase in fluid content and the accumulation of yolk in the center; the cytoplasmic layer at the margin grows progressively thinner.

As the final stage, or third phase (*Marza and Marza, 1935*), of growth is approached, the first layer of yellow yolk is deposited about the latebra, followed by the deposition of a layer of white yolk. According to Konopacka (*1933*), white yolk is transformed into yellow yolk by the penetration of fat into the vacuoles of white yolk. According to Marza and Marza (*1935*), yellow yolk is formed directly from substances in the oöcyte without passing through the white-yolk stage. Large amounts of carotenoid pigments (*Conrad and Scott, 1938*) are chiefly responsible for the color of the yellow yolk.

At first the concentric layers of yolk are very thin. During the last 4 days before ovulation, when the greater part of the yolk is deposited (*Liebermann, 1888; Gage and Fish, 1924*), a layer of yellow yolk may attain a thickness of 2 mm., although the white-yolk layer is only 0.4 mm. thick (*Romanoff, 1943a*). The yellow yolk is laid down during the day, up to midnight; the white, during the remainder of the night (see Chapter 3, "Structure").

Stages of Growth of the Oöcyte. The elliptical oöcyte of the newly hatched chick is comparable in size—0.01 to 0.02 mm. in diameter (*Brambell, 1926*)—with the primordial germ cell first seen in the 12-hour embryo. In the meantime, the germ cell has undergone many changes. It has been carried into the ovary by the epithelial cords; it has multiplied, to form the relatively small oögonia; the oögonia have grown, mainly by an increase in protoplasm and water, to produce primary oöcytes; and lipid vacuoles have begun to appear in the center of each cell, about the yolk-attraction sphere.

The increases in size that follow during the bird's postembryonic life are due to a small increase in the protoplasmic elements and to a tremendous increase in the yolk content. The growth of the oöcyte may be divided into stages determined by (1) the type of material deposited in the yolk (*Marza and Marza, 1935*) or by (2) the rate of increase in size (*Romanoff, 1931*).

During the 6 weeks immediately after hatching, the oöcyte of the chick enlarges approximately fourfold (*Brambell, 1926*), mainly by an increase in cytoplasm and water content. However, some deutoplasm (fat) begins to accumulate in the center of the oöcyte in the 10-day-old chick (*Goldsmith, 1928*).

These growth processes continue slowly until just before the bird reaches sexual maturity. The oöcyte is then about 1 mm. in diameter. Some phospholipids, but very few neutral fats, have been accumulated up to this time.

While the oöcyte grows from 1 mm. to 3 mm. in diameter, the character of yolk deposition changes. Vacuoles, similar to those already present in the center, appear in the periphery and become white yolk (*Konopacka, 1933*). White yolk is laid down in the latebra during this phase. When the oöcyte reaches a diameter of 3 mm., the first phase of growth is ended (*Romanoff, 1931*).

During the next phase of growth, the oöcyte increases from 3 mm. to 6 mm. in diameter. This stage is characterized by the slow deposition of alternate layers of white and yellow yolk about the latebra as a center.

During the final phase of growth, layers of yellow and white yolk are deposited very rapidly. After the oöcyte reaches a diameter of 6 mm., its rate of growth accelerates sharply. In chickens, the large diameter of the yolk increases from 6 mm. to 35 mm. in a period of 6 days.

The several largest oöcytes in an ovary, when arranged in order of weight, yield interesting data on the rate of their growth. Each ovum

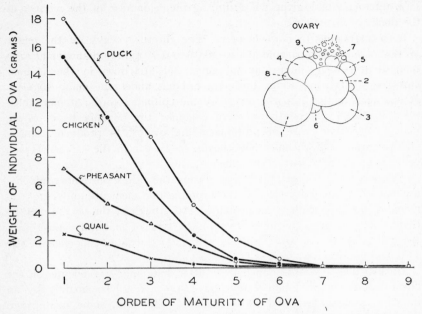

Fig. 122. Weights of individual ova during the final period of growth in ovary of duck, chicken, pheasant, and quail. (After Romanoff, 1943a.)

may well represent a developmental stage. A bird in full reproductive condition ordinarily lays no more than one egg a day. The difference in volume or weight of any two ova in the series is then a measure of the growth rate. It may be seen, in Fig. 122, how greatly the weight of the oöcyte increases in a number of birds just prior to ovulation. The table

Ova	VOLUME (cubic centimeters)	INCREMENTS (cubic centimeters)
1	0.4	
		0.6
2	1.0	
		1.0
3	2.0	
		2.4
4	4.4	
		3.5
5	7.9	
		4.4
6	12.3	
		5.1
7	17.4	
		6.3
8	23.7	

(*Chomkovic, 1928*) gives the daily increments in the volume of ova in the duck.

It is interesting to note, however, that the increments in the weight of the ovum, when calculated as percentages of the total weight, are smallest at the beginning and the end of the growth period and largest during the midperiod. The accompanying data show the actual and relative increments in the weight of the chicken's ovum (*Romanoff, 1943a*). It

Weight of Successive Ova (grams)	Increments (grams)	Increase (per cent)
0.2		
	0.1	50
0.3		
	0.3	100
0.6		
	1.7	284
2.3		
	3.3	144
5.6		
	5.2	93
10.8		
	4.2	39
15.0		

thus becomes apparent that the growth of the ovum follows the same course as the growth of multicellular organisms.

The Deposition of Chemical Substances. The oöcyte of the chick, at the time of hatching, is composed mainly of protoplasm high in protein and fluid content. Lipid vacuoles are just beginning to form about the yolk-attraction sphere. As the peripheral fatty layer is laid down, the lipid content of the oöcyte increases slowly. Before periodicity in yolk deposition is established, vacuoles of dilute protein appear in the periphery, phosphatides combined with dense protein penetrate the vacuoles, and finally a limited amount of fat is added.

As maturity is approached, several of the largest oöcytes successively enter the most rapid phase of growth. The formation and deposition of the principal components of yolk, the phospholipids and the protein, ovovitellin, are accelerated. Both the rate and the mode of synthesis of these phosphorus-containing substances have been traced by the use of radioactive phosphorus. During the last 4 days before ovulation, ovovitellin is formed and stored in the oöcyte more rapidly than phospholipids (*Chargaff, 1942b*).

During the embryonic and early postembryonic periods, the growth of the oöcyte is due mainly to the accretion of protein and water (Table 20).

TABLE 20

CHEMISTRY OF YOLK FORMATION

Phases of Growth of Ova

	Diameter from 2.0 to 6.0 Mm. (per cent)	Diameter from 6.0 to 35 Mm. (per cent)	Investigator
Water	86.0	45.0	Romanoff (*1931*)
Proteins	44.4	35.1	
Neutral fats	23.1	42.3	
Phospholipids	11.0	12.7	Spohn and Riddle (*1916*)
Organic extractives	8.6	3.9	
Ash	12.9	6.0	

The absolute quantities of all the components of the oöcyte increase throughout growth, but the percentage of water decreases because of the addition of solids. The percentage of solids in the ovarian ovum more than doubles in the last 7 days before ovulation (Fig. 123). In spite of the addition of large amounts of ovovitellin, the percentage of protein falls, since fats and phospholipids are accumulated in even greater

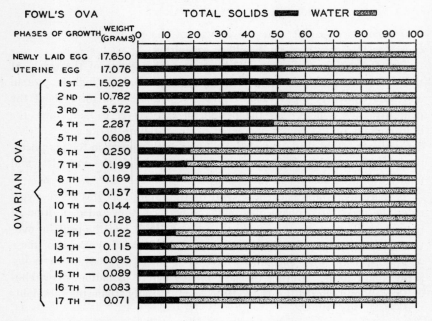

FIG. 123. Proportions of total solids and water in fowl's ovum, at various stages of growth. (After the data of Romanoff, 1931.)

quantities. Part of the increase in ash accrues from the phosphorus in the ovovitellin and in the phospholipids.

After ovulation, the ovum absorbs some liquid on entering the oviduct (cf. Fig. 123). There is a further slight dilution of the yolk during the addition of the albumen.

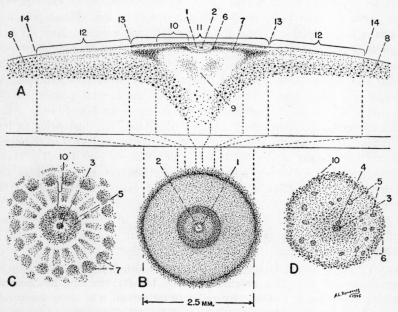

FIG. 124. Structure of blastodisc in the pigeon egg, approximately at the time of fertilization. (Rearranged from the drawings and photographs of Bartelmez, 1912, and Harper, 1904.)

A, parasagittal section of the blastodisc; magnified about 35 times. *B,* surface view of the blastodisc. *C,* horizontal section through the central portion of the blastodisc as far as the inner edge of the wedge granules; magnified 100 times. *D,* horizontal section through the nucleus (showing only the restricted area inside the polar ring), just before the fusion of the male and female pronuclei which constitutes fertilization; magnified 400 times.

The numbers indicate: 1, germinal vesicle; 2, chromatin; 3, egg pronucleus; 4, sperm pronucleus; 5, accessory sperm nuclei; 6, polar granules; 7, wedge granules; 8, periblastic granules; 9, nucleus of Pander; 10, polar ring; 11, inner periblastic area; 12, outer periblastic area; 13, inner periblastic ring; 14, outer periblastic ring.

The Blastodisc.

The blastodisc (Fig. 124), consisting of nucleus and cytoplasm, can be traced back to the primordial germ cell. As the final stage of rapid accumulation of yolk approaches, the nucleus and some cytoplasm begin to migrate from the center of the oöcyte to the periphery.

On coming to rest beneath the vitelline membrane of the yolk, the nucleus becomes lens shaped (cf. Fig. 124-*A*). In its path of migration is a vase-shaped accumulation of white yolk. The bulbous portion in the center of the yolk is the latebra. The thinner portion is the neck of the latebra. The flaring end of the neck is the nucleus of Pander, which consists of an extension of the underside of the blastoderm (cf. Fig. 124-*A*).

A large part of the thin outer layer of cytoplasm, which extends over the surface of the yolk, is also displaced toward the animal pole and forms a lens-shaped envelope about the germinal vesicle. Next to it is a circle of polar granules, surrounded by a collar of wedge granules tapering toward the germinal vesicle (cf. Fig. 124-*A* and *B*). Underlying the collar and blastodisc is a cone of finely granular white yolk extending to the nucleus of Pander. The appearance of the central portion of the blastodisc of the unlaid, unfertilized egg is shown in Fig. 124-*C*. After the penetration of sperm cells into the blastodisc, but just prior to fertilization (the fusion of the pronucleus of the sperm with that of the egg), the central portion of the blastodisc appears as in Fig. 124-*D*. Many sperm nuclei that do not take part in fertilization are lodged in the peripheral region. They may initiate some development, but ultimately they degenerate.

The Follicle and Its Function.   In the ovary of the chick, soon after hatching, a thin ring of cells forms about each oöcyte (Fig. 125-*A*). These cells are derived from the epithelium that covers the gonad of the very young embryo; they eventually give rise to the ovarian follicle. Ten days after the chick hatches (cf. Fig. 125-*B*), the ring has become distinct (*Goldsmith, 1928*). By cell multiplication, it is soon many cell layers deep (cf. Fig. 125-*C*). Its depth increases to 19 μ in the oöcyte 1 to 3 mm. in diameter (*Marza and Marza, 1935*). Some of the cells stain deeply and eventually disintegrate to form intercellular cement. As the oöcyte grows and the pedicle forms, the follicle is accordingly stretched thin, and its cells are flattened (cf. Fig. 125-*D*).

During the three phases of yolk formation, the depth of the follicular epithelium increases and then diminishes as the ovum grows, as follows (*Marza and Marza, 1935*).

DIAMETER OF OVUM	DEPTH OF FOLLICULAR EPITHELIUM
(millimeters)	(microns)
0.05– 2.0	4–20
2.0 – 6.0	20– 7
6.0 –35.0	7– 4

In addition, there is formed about the yolk a compact capsule of thin cells which border on several less compact outer layers of cells. These are the *theca interna* and *theca externa* of the follicle, respectively. The less compact theca externa borders on the stroma of the ovary, which is well supplied with blood vessels. Blood vessels enter the follicle through the pedicle and extend into all portions of the follicle, except for a narrow band, the *stigma,* opposite the pedicle.

Early in the development of the oöcyte, the cells of the follicle extrude Golgi bodies into the cytoplasm of the oöcyte (*Brambell, 1926*). These

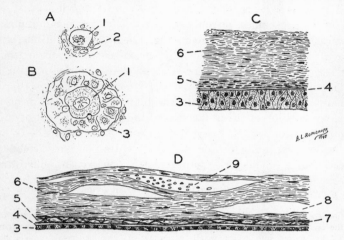

FIG. 125. Stages of development of the follicle in the chicken.

A, germ cell from the ovary of the chick 2 days after hatching, showing the follicle cells beginning to form a ring. *B,* an ovum from the ovary of a 10-day-old chick; the follicle cells have formed a definite ring around the ovum, and supporting cells can be seen outside the ring. *C,* a portion of the follicle of an ovum about 4 mm. in diameter. *D,* a portion of the follicle of a mature ovum (about 35 mm. in diameter), just before ovulation. All magnified about 500 times.

The numbers indicate: 1, germ cell; 2, follicle cell; 3, follicular epithelium; 4, membrana propria; 5, theca interna; 6, theca externa; 7, capillaries; 8, blood vessel; 9, blood cells. (*A* and *B* after Goldsmith, 1928; *C* and *D* after Novak and Duschak, 1923.)

bodies break up, and the fragments act as centers of yolk formation. When the vitelline membrane has formed, the follicle cells cease their extrusion of Golgi bodies.

Later, when the most rapid deposition of yolk is taking place, the follicle (Fig. 126-*A*) passes the materials necessary for yolk formation from the blood to the oöcyte. The follicle may possibly elaborate some of the materials; but it more probably acts as a selective ultrafilter and merely passes the nutritive substances across the intercellular bridges which its cells maintain with the cytoplasm of the oöcyte. At first, the follicle permits the passage chiefly of neutral fats; then of proteins and water; and, finally, in the last phase of growth, of phospholipids. Thus the permeability of the follicle seems to change three times during the development of the ovum (*Marza and Marza, 1935*).

When the yolk is full grown, the follicle loses its significance as a nutritive organ but may assume the function of the hormonal regulation of the egg-laying cycle (*Rothchild and Fraps, 1944*).

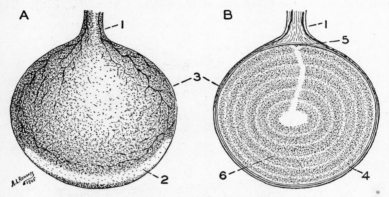

FIG. 126.   Diagrams of a full-sized ovarian ovum of the fowl.

A, external view.   *B*, cross section.

The numbers indicate: 1, pedicle, the connection between the ovum and the ovary; 2, stigma, the area devoid of blood vessels, where the follicle ruptures at the time of ovulation; 3, follicle, the ovisac into which vessels penetrate; 4, vitelline membrane; 5, blastodisc; 6, alternating layers of yellow and white yolk. (Drawing from prepared specimens.)

A cross section of the full-grown follicular ovum (cf. Fig. 126-*B*) shows the orientation of the blastodisc and the structural arrangement of the yellow and white yolk.

Atresia.   At ovulation, the follicle ruptures and the ovum escapes. The resorption of the ruptured follicle, known as *follicular atresia,* is a normal occurrence after ovulation. Eventually, all that remains of the follicle is a connective tissue scar (*Davis, 1942a*). Many small follicles, with their contained ova, cease developing and also undergo atresia. The degeneration of more mature follicles, or *bursting atresia,* is common in birds (*Rowan, 1930; Davis, 1942b*). The follicles of maturing ova, containing considerable yolk, may burst on the side attached to the ovary, so that the yolk flows into the ovarian stroma (Fig. 127). The follicle becomes wrinkled and thickened. The yolk material is engulfed by swarms of white corpuscles and eliminated from the ovary by the blood stream.

The Vitelline Membrane.   At the time the yolk enters the last growth phase, a clear, thin, noncellular membrane appears between the outer edge of the oöcyte and the follicle. This is the vitelline membrane, or *zona radiata,* so called because of the numerous striations in it, which are actually small canals.

After the follicle has formed about the oöcyte, and before the rapid deposition of yolk begins, the outer and inner layers of the vitelline membrane are formed, respectively, of intercellular cement from the follicle and cytoplasm from the oöcyte. At first, the membrane is very thin and has numerous perforations through which strands of cytoplasm ex-

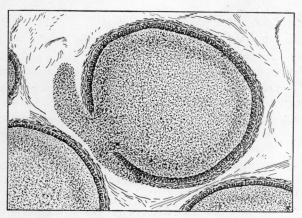

Fig. 127. Rupture of the follicular wall and extrusion of yolk into the adjacent stroma, in bursting atresia. (Drawn from photomicrograph of Rowan, 1930.)

tend from the follicle cells to the peripheral cytoplasm of the oöcyte. As the membrane thickens, the pores become canaliculae.

The vitelline membrane is very elastic, providing a highly extensible envelope to accommodate the ever-greater amounts of yolk deposited within it. As the rapid deposition of yolk increases the size of the oöcyte, a space, filled with lymph, arises between the vitelline membrane and the follicle. This is the perivitelline space. The oöcyte is now free to rotate within the follicle, and can orient itself with respect to gravity. The heavier vegetal pole sinks, and the lighter animal pole rises. Eventually, so much yolk accumulates that the oöcyte, suspended by the pedicle, hangs into the body cavity. The blastodisc (or animal pole) then comes to rest below the point of attachment of the pedicle to the follicle (cf. Fig. 126-*B*).

After the vitelline membrane has formed, the materials for yolk formation must pass through it to reach the oöcyte. Materials, either particulate or in solution, may pass across the intercellular bridges which extend through the pores of the membrane; soluble materials may pass directly through the membrane. Both means of transfer probably take place; but, when the yolk approaches maturity, the pores disappear, and the passage of materials must depend on the permeability of the membrane. According to Marza and Marza (*1935*), the permeability changes three times, as in the follicle, to permit passage of the different types of materials that are deposited in the oöcyte during the three periods of growth.

ACTIVATORS AND SUPPRESSORS OF THE GROWTH OF THE OÖCYTE

Hormones. Reproduction in general, and the growth of the oöcyte in particular, are intimately bound up with the bird's endocrine system. The regulating hormones of the anterior pituitary gland, and the sex hormones produced by the ovary, establish the rhythms for the growth of the oöcyte and the production of yolk. Other endocrine secretions modify the processes either directly, by acting on the ovary, or indirectly, by acting on the pituitary or by affecting the bird's general physiological activity.

One of the secretions of the anterior pituitary, the follicle-stimulating hormone, causes rapid growth of follicles (*Koch, 1934*). Phillips (*1943*) observed that, when an extract of the anterior pituitary was injected into laying hens, the ovary was stimulated, and several oöcytes grew rapidly.

This gonadotrophic hormone, besides stimulating ovarian activity, also causes the production of ovarian hormones that prepare the oviduct to receive the ovum. The ovarian hormone in turn acts on the pituitary to reduce the production of the follicle-stimulating hormone (*Riley and Fraps, 1942*). Thus a balance of secretory activity is established. When the follicular hormone of the ovary is injected into a bird in which follicles are developing, there is a repression of follicular growth, because the amount of gonadotrophic hormone produced by the pituitary is diminished (*Asmundson, 1931a; Ringoen, 1940*). In fact, injection of 4 to 8 mg. of follicular hormone has been observed to cause permanent cessation of egg formation in pheasants and subsequent appearance of male plumage characteristics (*Champy-Coujard, 1945*).

Effects similar to those of the gonadotrophic hormone of the pituitary are produced by a hormone found in the serum of pregnant mammals. When this hormone is injected into hens, the rate of yolk production is increased about 50 per cent (*Fraps, 1940*). Although this hormone increases the weight of the ovary (*Asmundson, Gunn, and Klose, 1937*), it does not cause ripening of follicles (*Asmundson and Wolfe, 1935*).

The thyroid gland, also controlled by the pituitary, acts mainly in regulating the rate of metabolism. When thyroid (or a thyroid iodine preparation) is administered to laying hens, the size of the ripened yolk is decreased (*Asmundson, 1931a; Asmundson and Pinsky, 1935*), probably because the thyroid secretion, by increasing the metabolic rate, hastens the excretion of egg-forming materials (*Riddle, Hollander, Mc-Donald, Lahr, and Smith, 1945*).

Prolactin, another hormonal secretion of the anterior pituitary, decreases the production of the follicle-stimulating hormone. The ovary atrophies, and the follicles fail to ripen (*Riddle, 1938a*).

External Factors. The reproductive cycle in birds is easily affected by many environmental factors. High temperatures may cause a reduction in the size of the mature yolk (*Bennion and Warren, 1933a; Kendeigh, 1941*), but they do not affect the yolk so much as the other parts of the egg. The ovary is less sensitive to environmental conditions than is the remainder of the reproductive system.

Light is the most effective single external factor controlling the reproductive cycle in the higher animals. It has been established through the work of many investigators (*Warren and Scott, 1936; Rowan, 1938; Ringoen and Kirschbaum, 1939; Benoit, Grangaus, and Sarfati, 1941; Kendeigh, 1941*) that the ovary is stimulated when the bird is exposed to an increased amount of light. This stimulation is the indirect result of an increased production of gonadotrophic hormone by the pituitary.

In species that breed seasonally, the ovary's state of development is related to the amount of light to which the bird is exposed (*Rowan, 1938; Kendeigh, 1941*). In birds, such as chickens, that lay eggs more or less continuously throughout the year, the influence of light upon ovarian function is still apparent (see Chapter 1, "Mode of Laying"). Svetosarov and Streigh (*1940*), for example, caused pullets to mature 1 month sooner than usual by the use of additional illumination.

Mechanical Injury. Ovarian injury is almost an everyday occurrence in the laying hen. As the mature yolk separates from the ovary, the follicle is ruptured. Since the follicle, after its degeneration, is finally replaced by connective tissue, the ovary is constantly accumulating scars.

These normally acquired injuries apparently stimulate the production of oöcytes. The effect of scar tissue is shown by experiments in which surgical damage to the ovary increased the number of visible oöcytes by 33 to 68 per cent (*Pearl and Schoppe, 1921*). However, damaging or partially removing the ovary does not hasten the sexual maturity of the hen (*Hutt and Grussendorf, 1933*), in spite of opinions to the contrary (*Steggerda, 1928; 1931*).

Effects of Chemicals and Drugs. The administration of various chemical substances may have a pronounced influence on egg formation, especially ovarian function. Quinine sulfate, when given to laying ringdoves, reduces the size of the yolk (*Riddle and Anderson, 1918*). The inhalation of alcohol has a similar effect on the size of the pigeon's ovum, and the effect persists for a considerable time after treatment is stopped (*Riddle and Basset, 1916*). The feeding of small amounts of thyroid iodine also reduces the growth rate of the ovum (*Asmundson and Pinsky, 1935*), but the normal growth rate is restored a few days after iodine feeding is discontinued. The administration of the vermifuge, kamala, results in a temporary decrease in the weight of the ovum during a period of about 2 weeks (*Maw, 1934*). Iodine diminishes the yolk's content of lecithin (*Bonnani, 1912*).

OVULATION

The liberation of the ovum from the ovary by the bursting of the follicle is the most momentous event in the reproductive cycle of the bird. It is truly a miniature birth. When it occurs, there is initiated a train of physiological events that result in the completed egg. Although the act of ovulation has often been observed (*Coste, 1847; Bartelmez, 1912; Pearl and Surface, 1914a; Phillips and Warren, 1937*), a complete explanation of all the attendant phenomena has yet to be made.

Ovulation usually follows about a half hour after the previously formed egg is laid (*Warren and Scott, 1935a*). While a completed egg is in the posterior part of the oviduct, the anterior part is at rest. After oviposition, the anterior end of the oviduct is suffused with blood and becomes very active. The infundibulum advances, clasps the ovary as if to swallow it, and then retreats. More often than not, it engulfs the largest follicle protruding into the body cavity from the ovary.

Ovulation occurs through a bandlike area in the follicle, known as the *stigma*. As the follicle forms, blood vessels enter it from the ovary by way of the pedicle. They penetrate all portions of the follicle and become prominent and visible, except in the stigma. The stigma is a compact layer of smooth muscle extending over a large part of the circumference of the follicle on the side opposite the pedicle (cf. Fig. 126-*A*). The stigma bears a relation to the bilaterality of the egg and to the orientation of the embryo (*Bartelmez, 1912*), should the latter develop. It is an area of weakness in the follicle.

A few minutes before ovulation occurs, the muscles of the stigma contract and produce tension on the follicle. The tension is hardly perceptible, but it is enough to compress the blood vessels, which pale and become indistinct. Shortly, a small rupture appears at one end of the stigma, presumably because of the internal pressure produced by the contracting muscles of the stigma. The soft ovum bulges out. Instantaneously the follicle splits along the entire length of the stigma, beginning at the point of rupture. The ovum falls into the ovarian pocket (*Curtis, 1910*), an irregular cavity formed about the ovary by the surrounding organs.

The infundibulum continues to advance, swallow, and retreat, partially engulfing the ovum, then releasing it. This activity may continue for half an hour before the ovum is entirely within the oviduct.

Why the stigma of the follicle ruptures is an unsettled question. Some early investigators thought that it was ruptured by the pressure exerted on the follicle by the infundibulum (*d'Hollander, 1905; Sonnenbrodt, 1908*); yet ovulation continues to occur after removal of the oviduct (*Pearl and Curtis, 1914*). It was also believed that internal pressure from the accumulation of yolk or follicular fluid caused the rupture (*Holl, 1890; Bartelmez, 1912*). However, when the follicle is about to rupture, ovulation cannot be induced either by increased tension on the follicle or by injection of fluid into the yolk (*Phillips and Warren, 1937*). It is thought that ovulation is ultimately under hormonal control, as is much of the reproductive process.

Hormonal Control of Ovulation

In nature, reproduction often bears a relationship to regular changes in environmental conditions, such as seasonal variations in weather, or in the length of the day. These changes cause cycles of secretion in the endocrine system. In consequence, during heightened secretory activity, the reproductive organs of the adult or maturing bird enlarge, the germ cells ripen, and the entire system is brought into a state of tension. Only a slight stimulus is now necessary to start the sequence of hormonal secretion that sets the reproductive processes in motion.

In many species, nervous stimuli initiate changes in the endocrine system, upon the activity of which depends the entire pattern of reproduction. To the hen, psychic influences are not of prime importance, since she can begin and continue to lay in complete isolation. However, in the pigeon, mating provides the stimulus for ovulation, which follows in about 8 days (*Harper, 1904*). When 2 eggs are present in the nest, the pigeon becomes broody, ovulation stops, and no more eggs are laid. However, if the eggs are removed from the nest, ovulation may recur (see Chapter 1, "Mode of Laying").

The pituitary, the master gland that regulates reproduction, is located beneath and is in contact with the midbrain. Stimuli are transmitted to the pituitary over nerve pathways connecting it to centers in the brain. The pituitary responds by secreting the hormones that control reproduction. There is experimental evidence that the stimulation of the pituitary by light occurs in this manner (*Benoit, 1935b*).

Two secretions of the forepart of the pituitary are instrumental in bringing about ovulation: (1) the follicle-stimulating hormone and (2) the luteinizing hormone. The development of the oöcyte is to some extent independent of the pituitary hormones, but the influence of the follicle-stimulating hormone is responsible for the final period of rapid growth (*Bates, Lahr, and Riddle, 1935*). When the follicle is full grown, ovulation is accomplished under the control of the luteinizing hormone (*Fraps, Riley, and Olsen, 1942; Fraps, Olsen, and Neher, 1942*).

The presence of a formed egg in the uterus prevents ovulation by inhibiting the release of the luteinizing hormone from the pituitary. On the other hand, premature expulsion of the egg does not hasten ovulation. Ovulation intervals seem to be determined largely by the rhythm with which the ova mature (*Warren and Scott, 1935a*). This, in turn, is under pituitary control.

The most recently ruptured follicle also influences the rate of ovulation in the hen, by regulating the time at which the next egg is expelled. If the follicle is removed, oviposition and the subsequent ovulation are both delayed (*Conrad and Scott, 1938; Rothchild and Fraps, 1944*). Injections of anterior pituitary extract, however, may overcome the effect of removing the ruptured follicle (*Neher and Fraps, 1946*).

Ovulation in hens tends to occur in the forenoon. Hens begin ovulating very early in the morning, and by noon most of them have ovulated. The tabulation (*Warren and Scott, 1936*) shows the relative frequency of ovulation during different periods of the day, up to early afternoon.

TIME	OVULATIONS (per cent)
2 A.M. to 5 A.M.	13.4
5 A.M. to 8 A.M.	27.5
8 A.M. to 10 A.M.	35.8
10 A.M. to 2 P.M.	23.3

Stimulation and Inhibition of Ovulation

Ovulation is not only the most important event in the bird's reproductive cycle; it is also the most easily affected by both external and internal influences.

If it is due as night is approaching, ovulation in the hen may be delayed until the following morning (*Shibata, 1933; Conrad and Scott, 1938*). When the hen is exposed to continuous illumination, the laying of the egg and the subsequent ovulation can take place at any time of the day or night (*Warren and Scott, 1936*). Light does not initiate ovulation, but darkness delays it.

The extreme heat of late summer decreases the rate of ovulation (*Brody, 1921*). Egg production then declines.

When the hen is disturbed as she is about to lay, the egg is retained (*Patterson, 1910*), and ovulation is delayed. Not only is ovulation interrupted when the hen is sufficiently disturbed, but degenerative changes are produced in the nuclei of the larger ova (*Stieve, 1918*). The degeneration extends to the younger ova if the unfavorable conditions persist.

The retention of the egg seems to result from psychological tension. The stimulus that leads to the delay in ovulation is presumably the uterine distension caused by the retained egg. However, no nervous pathway has been found leading from any part of the oviduct to the ovary; apparently, therefore, ovulation is not inhibited directly. Conceivably, the stimulus could alter the rhythm of ovulation by affecting first the central nervous system and the bird's psychological state, and then the ovary or the glandular system.

Ovulation occurs at the normal rate even when the oviduct or any of its parts has been removed. It appears, therefore, that stimuli from the oviduct, although able to delay ovulation when an egg is retained in the uterus, are not instrumental in initiating ovulation.

When it became known that reproduction was under the control of endocrine glands, experimental efforts to accelerate egg production followed as soon as hormonal preparations were made available. At first, some confusion arose as to the specific effects of the various hormones, the administration of which often suppressed ovulation or otherwise disturbed normal processes. As improvement was made in purifying and standardizing the preparations, investigation yielded more definite results. The influence of several hormones, originating in the pituitary gland, the reproductive system, and the adrenals, is now fairly clear.

Injection of a pituitary extract, prolan, accelerates ovulation in the hen (*Koch, 1934*). In more recent experiments, various fractions extracted from the anterior pituitary have been used separately and in combination, in order to determine the specific effect of each. The follicle-stimulating hormone, in combination with the luteinizing hormone, causes premature ovulation (*Fraps, Riley, and Olsen, 1942*), especially if the luteinizing hormone is given as a preliminary treatment. If the luteinizing hormone is injected while the last egg of the cycle is in the uterus, the egg is expelled 3 to 6 hours prematurely. The first ovulation of the next cycle occurs simultaneously, about 24 hours prematurely (*Fraps, 1942*).

When injected into hens, progesterone, formed by the interstitial cells of the ovary, causes premature ovulation of ripe follicles (*Fraps and Dury, 1943*); the effectiveness of this preparation, however, seems to depend on the mode of its injection. In pigeons, ovulation, except of the ovum of the first egg, is prevented by progesterone (*Riddle, Bates, Wells, Miller, Lahr, Smith, Dunham, and Opdyke, 1941*). Progesterone, combined with the follicular ovarian hormone, induces the formation of avidin in the oviduct of immature female chicks. Neither preparation is effective when used alone (*Hertz, Fraps, and Sebrell, 1943*). Hormonelike steroids from the ovary and adrenals cause ovulation of ripe follicles and extensive atresia in less mature ones (*Dunham and Riddle, 1942*).

The first recorded inhibition of ovulation in hens by hormonal means was by the use of corpus luteum substance (*Pearl and Surface, 1914a*). Ovulation may also be delayed a few hours to several days by the injection of nonspecific, proteinous, hormone-free substances, such as ovalbumin, casein, peptone, whole dried egg albumen, or desiccated tissues from the cow or chicken. Follicular atresia does not occur, however. These substances possibly suppress the release of the gonadotrophic hormone of the anterior pituitary (*Fraps and Neher, 1945*).

Abnormal and Multiple Ovulation

Double-yolked and even triple-yolked eggs are occasionally produced (*Curtis, 1915a*); more rarely, a hen lays a daily succession of double-yolked eggs. The inclusion of two or three yolks in the same egg may result from a number of causes (see Chapter 5, "Anomalies"), one of which is a slight disturbance in the ovulatory cycle (*Conrad and Warren, 1940*). Two yolks may even be ovulated almost simultaneously.

The rate of ovulation may be increased by the proper use of certain hormone preparations. Multiple ovulations result if hens are given injections of anterior pituitary-like hormone from pregnant-mare serum for several days before the injection of luteinizing hormone. Fraps and Riley (*1942*) found that an average of more than three ovulations could be produced in a bird by one such treatment.

THE SECRETION OF ALBUMEN

The secretion of albumen and its deposition on the yolk are largely the function of the anterior portion of the oviduct. Like many other aspects of reproduction in the hen, the development of the oviduct, and its preparation to receive the mature ovum, are directly controlled by the hormones of the ovary (*Asmundson, Gunn, and Klose, 1937; Herrick, 1944*). As the time of ovulation approaches, the individual gland cells and the glandular folds enlarge, and the ducts and cells become filled with secretory products.

Most of the proteins that contribute to the albumen are formed and accumulated at a fairly regular rate in the tubular glands in the anterior portion of the oviduct, during the intervals between successive eggs (*Conrad and Scott, 1942b*). The albumen is deposited in roughly concentric layers about the yolk as it passes (see Chapter 3, "Structure"). When the egg arrives in the uterus, most, if not all, the protein constituents of the albumen have been acquired (Fig. 128).

Chalaziferous Layer. The first coat of albumen is applied to the naked yolk as it traverses the posterior portion of the infundibulum (*Scott and Huang, 1941*). The firm, mucinlike albumen is deposited as a sievelike sheet and forms the inner portion of the chalaziferous layer.

The Middle Dense Layer. The mucin-coated yolk continues down the oviduct, rotating slowly on the spiral folds in the lining of the albumen-secreting region. The middle and posterior portions of this region contribute the largest share of the albumen content of the egg (*Asmundson and Jervis, 1933*).

Mucin and a more liquid albumen are secreted at the same time by the highly glandular mucosa and the epithelium of the albumen-secreting region. The mucin forms an intricate meshwork of microscopic fibers (*R. K. Cole, 1938*), and the liquid albumen is dispersed in the meshes. The concentric deposition of the two substances builds about the yolk a thick, plastic envelope, capable of maintaining its form to some extent.

The Inner Liquid Layer. Together, the yolk and the albuminous sac rotate slowly as they are propelled through the oviduct. The rotation twists the microscopic mucin fibers in the inner layers of the albumen and draws them taut (*Conrad and Phillips, 1938*). The liquid albumen dispersed in the meshes is squeezed out to form the inner liquid layer (*Almquist, 1936*) in which the yolk floats free. The yolk is now able to orient itself; the lighter animal pole rises.

The Chalazae. The rotation twists the fine mucin fibers of the inner layers of the albuminous sac in opposite directions at the two ends of the egg. The fibers become tightly stretched about the yolk and are spun into larger threads (*Conrad and Phillips, 1938; Conrad and Scott, 1938*) which form the chalazae. The ends of the chalazae become extended into the albuminous sac at the poles of the egg, as dense albumen continues to be deposited.

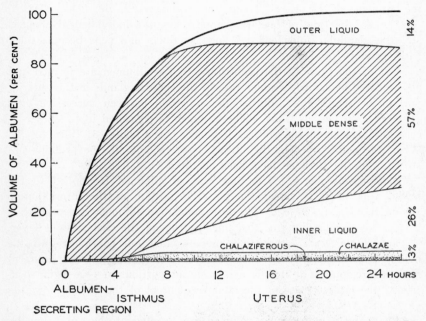

Fig. 128. Rate of increase in the volume of the entire albumen as the egg passes through the various regions of the oviduct, and the concomitant changes in the proportional amounts of the albumen layers. (Constructed from the data of Pearl and Curtis, 1912, and Burmester, 1940.)

The formation of the chalazae continues throughout the entire time the egg is in the isthmus (*Asmundson and Burmester, 1936*) and during the first part of the egg's sojourn in the uterus (*Scott and Huang, 1941*).

The Outer Liquid Layer. As it enters the isthmus, where the shell membranes are acquired, the egg is composed of the yolk, surrounded by the chalaziferous, inner liquid, and dense layers of the albumen. Very little of the liquid albumen that will form the outer liquid layer is present (*Asmundson, 1931b*). Up to this time, only 40 to 50 per cent of the total amount of albumen has been acquired by the egg, leaving 50 to 60 per cent to be added in the isthmus and uterus (cf. Fig. 128).

The anterior part of the albumen-secreting region of the oviduct probably elaborates the albuminous material of the outer liquid layer, since resection of this portion of the oviduct causes a reduction in the amount of the outer liquid layer (*Asmundson and Burmester, 1936*). Some of the liquid albumen is added to the egg in the isthmus while the membranes are being formed. When the egg first arrives in the uterus, the shell membrane is still very porous. During the first 8 hours in the uterus, liquid material passes readily through the membrane into the egg (*Burmester, Scott, and Card, 1940*), and the volume of the albumen now approximates that of the albumen in the completed egg. After the membranes are fully formed and the shell begins to be deposited, the volume of albumen continues to increase, due to the absorption of inorganic salts and water (*Hansen, 1933; Beadle, Conrad, and Scott, 1938; Burmester, 1940*) rather than absorption of proteins (*Scott, Hughes, and Warren, 1937*). The material that is added to the albumen in the uterus is comparable in composition to the uterine secretion, which is practically devoid of protein and consists mainly of a water solution of inorganic salts, rich in potassium and bicarbonate ions (*Hughes, 1936; Beadle, Conrad, and Scott, 1938*).

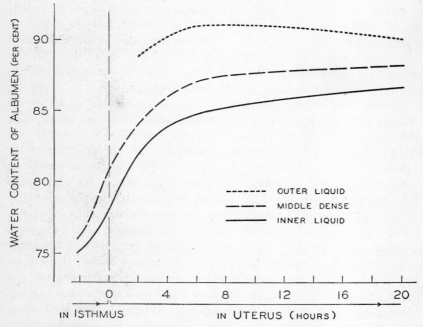

FIG. 129. Changes in water content of various layers of egg albumen while in the isthmus and the uterus of the hen's oviduct. (After Burmester, 1940.)

As the egg leaves the albumen-secreting region, the solids content of the entire albumen is almost double that of the albumen in the finished egg. The inner liquid layer is somewhat more concentrated than the middle dense layer. During passage through the isthmus and uterus, each of these layers becomes greatly diluted, the middle dense layer to a somewhat larger extent than the inner liquid layer. The outer liquid layer loses water during the later stages of egg formation, as the other two layers are diluted (Fig. 129). In the completed egg, the percentage of solids decreases from the inner to the outer regions of the albumen (*Romanoff, 1929b*) as a result of the manner in which water diffuses into the egg during formation.

The oviduct is more sensitive than the ovary to factors that adversely affect egg formation. The amount of albumen in the egg is seriously diminished by excessive temperatures (*Bennion and Warren, 1933a*) or by the administration of quinine (*Behre and Riddle, 1919*).

FORMATION OF EGG INTEGUMENTS

Shell Membranes

As the egg reaches the end of the albumen-secreting region, it consists of the yolk surrounded by the chalaziferous layer and the dense albuminous sac. As soon as the egg noses into the isthmus, the shell membrane begins to be deposited on the forward end. Eggs, thus partially covered on one end when laid, have been reported by various authors (*Coste, 1847; Asmundson, 1931b; Burmester, 1940*).

The stimulus of the egg's presence causes the tubular glands of the isthmus to secrete a granular, keratinlike material into their ducts. The granules swell by taking up water and coalesce (*Giersberg, 1921*) into viscous strands. The sticky fibers, on issuing from the apertures of the numerous tubular glands, become closely felted into a lacelike membrane (*Richardson, 1935*) and are applied to the surface of the albuminous sac. By the time it has fully entered the isthmus, the egg is already enclosed in the very thin, tight-fitting inner membrane, composed of fine fibers. The egg pauses for a short time (*Richardson, 1935*) and then proceeds slowly onward (*Burmester, 1940*), as the glands of the entire isthmus contribute the materials for the outer membrane. This structure is made up of coarser, more loosely interwoven fibers than those of the inner membrane, although the fibers are formed in the same manner. Microscopically, the relatively finer, more compact inner shell membrane, and the loosely woven outer membrane, can easily be distinguished (see Chapter 3, "Structure").

The membranes fit the egg snugly when they are first applied. As the egg traverses the isthmus, the membranes expand. When the end of the isthmus is reached, they are quite loose (*Conrad and Scott, 1938*).

As the egg passes through the isthmus, the amount of shell membrane increases (*Asmundson and Burmester, 1936*). The removal of a part of the isthmus results in a reduction in the amount of shell membrane (*Asmundson and Jervis, 1933*). Furthermore, material placed in the oviduct at the end of the isthmus is not furnished with shell membranes (*Pearl and Curtis, 1914*). This evidence definitely indicates that the membranes of the egg are formed by the isthmus.

The deposition of a small amount of oöporphyrin gives the membrane a pale pink hue (*Klose and Almquist, 1937*).

Eggshell

The egg, loosely attired in its shell membranes, passes from the isthmus to the uterus. The uterus is stimulated to produce its watery secretion. The membranes, fairly permeable at first, permit the water and some of the dissolved salts of the uterine secretion to enter the egg and thus to dilute the albumen. The egg becomes plump. Again the membranes fit snugly and provide a more or less firm surface on which the shell can form.

The uterus is clearly responsible for shell formation. Eggs laid through a fistula posterior to the uterus have fully formed shells (*Asmundson, 1931b*). The removal of part of the uterus results in a reduction in the amount of shell (*Asmundson and Jervis, 1933*) and in the calcium and protein content of the shell (*Asmundson and Burmester, 1938*).

The deposition of calcium salts, the major constituents of the shell, proceeds slowly at first (*Burmester, 1940*). Extremely small granules appear on the surface of the shell membrane. The granules increase in size, mainly by the accretion of calcium salts. A small amount of protein matrix material, however, is intermingled with the salts, particularly near the center of the concretions. Eventually the concretions become solid, knoblike objects, somewhat pointed at the base. These are the mammillae. Collectively, they constitute the mammillary layer (Fig. 130-*A*). The basal portion of each mammilla is embedded among the fibers of the shell membrane and is thus firmly attached. The broad distal ends of the mammillae may be in contact with each other, but minute spaces remain about the conical tips above the membrane. These spaces are part of an extensive ventilation system which, in the completed egg, permits gaseous exchange.

A collagenlike protein (*Almquist, 1934*) is secreted in the fore portion
of the uterus and deposited on the mammillary layer in the form of fibers.
Simultaneously, crystals of calcium salts, in a more homogeneous form
than before, are slowly deposited in the interstices of this fibrous framework, or matrix. Thus the formation of the spongy layer (cf. Fig. 130-*B*
and *C*) is begun. This layer will comprise almost two-thirds of the shell

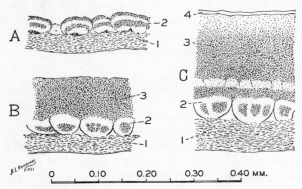

Fig. 130. Transverse sections through the shell of the hen's egg at various stages of
formation, seen by reflected light.

A, with bare mammillary layer only. *B,* with mammillary and spongy layers. *C,* completed shell.

The numbers indicate: 1, shell membranes; 2, mammillary layer; 3, spongy layer; 4, cuticle.
(After von Nathusius, 1893.)

(*Conrad and Scott, 1938*). The secretion of the matrix is nearly complete when the egg reaches the middle portion of the uterus, but most of
the calcific material is yet to be added.

Burmester (*1940*) determined the amounts of calcium salts and protein
deposited in the shell and the shell membranes of the egg in various parts
of the isthmus and uterus. His results are reproduced here.

	Protein	Calcium (CaCO₃)
Region of Oviduct	(gram)	(grams)
Isthmus:		
Anterior	0.38	. . . . . .
Mesial	0.111	0.0002
Posterior	0.153	0.0007
Uterus:		
Anterior	0.161	0.011
Mesial	0.213	1.99
Posterior	0.222	5.24

Shell material is deposited slowly during the first 3 hours after the egg
arrives in the uterus. By the end of that time, a more uniform and

higher rate of deposition has been attained. During the final 17 hours, the weight of the shell increases in proportion to the time the egg has remained in the uterus (*Burmester, 1940*). The amount of shell secreted is influenced by the size of the yolk and the amount of albumen to be covered (*Asmundson, 1933b*).

Crystals of calcite, interspersed among the collagenlike fibers of the matrix, continue to be deposited, until the outer (or spongy) layer of the shell is materially thicker than the inner layer. The spongy layer (*Landois, 1865*), particularly the outer stratum, is firm and dense, despite its name. Only when the calcium salts are dissolved out does its spongy texture become evident.

Pores are formed in places where the eggshell, in the process of formation, is in contact with the uterine epithelium. They extend through the spongy layer and open into slight depressions in the surface of the shell. According to von Nathusius (*1893*), matrix threads are included in the crystallizing masses of calcium salts as the latter are deposited. The inner end of the radial canal, which is formed about the matrix threads, communicates with the intermammillary spaces. The regularity in the spacing of the pores suggests that the arrangement of the protein strands must be predetermined in some way (*Marshall and Cruickshank, 1938*). Shrinkage of the matrix threads possibly opens the pores for gaseous exchange.

The secretion of calcium has not been localized to any special portion or gland of the uterus. The tissues in the uterus of the heavily laying hen contain less calcium than those in the remainder of the oviduct (*Buckner, Martin, and Peter, 1925*). Since more than 5 gm. of calcium carbonate is deposited on the egg in about 20 hours, some source of calcium other than the uterus itself is indicated. The calcium in the blood of the laying hen rises to a level two or three times that in the blood of the nonlaying hen (*Hughes, Titus, and Smits, 1927*). During shell formation, the blood supply to the uterus is greatly increased. Presumably, calcium salts in solution pass from the blood into the cells of the uterine glands, which secrete the calcareous solution into the lumen of the uterus. The calcium leaves no trace of its place of secretion or route of passage, except for a slight increase in the inorganic salt content of those uterine cells that come into contact with the egg (*Richardson, 1935*).

Although the calcium in the shell is found preponderantly as the carbonate (*Buckner, Martin, and Insko, 1930*), this salt is so insoluble as to preclude the secretion of the shell in this form (*Conrad and Scott, 1938*). However, if there were an excess of carbonate ions, the calcium might exist as soluble bicarbonate, to dissolve which comparatively little fluid would need to be secreted. Although the major portion of the calcium required for the shell is probably secreted as bicarbonate, some may be in the form of phosphate, chloride, and calcium-protein complex.

The deposition of the shell is probably aided by the activity of enzymes (*Buckner, Martin, and Insko, 1930*), especially by that of carbonic anhydrase. This enzyme, by catalyzing the reaction $CO_2 + H_2O \rightleftarrows H_2CO_3$ (*Meldrum and Roughton, 1933*), may influence the rate at which the carbonate anion of calcium carbonate is produced. The uterine epithelium, which appears to be the tissue most directly concerned in the secretion of shell materials (*Richardson, 1935*), has a significantly greater carbonic anhydrase activity than the remaining oviducal tissues (*Common, 1941*). Further evidence of the importance of carbonic anhydrase in eggshell production is provided by the fact that the feeding of sulfanilamide, said to be a specific inhibitor of this enzyme (*Mann and Keilin, 1940*), results in the formation of very thin and characteristically pitted shells, and occasionally in the entire absence of the shell (*Benesch, Barron, and Mawson, 1944; Gutowska and Mitchell, 1945*). Among other enzymes that may influence shell deposition is phosphatase (*Laskowski, 1934; Common, 1936b; Peterson and Parrish, 1939; Gutowska, Parkhurst, Parrott, and Verburg, 1943*). Phosphatase may be largely responsible for the transfer of the calcium cation to the shell. The abnormally thin, rough shells that appear subsequent to the administration of some of the sulfonamides are thought to be due to the inactivation of this enzyme, or some enzyme other than carbonic anhydrase (*Bernard and Genest, 1945*).

When subnormal body temperature of several hours' duration is induced in hens, the production of soft-shelled eggs is likely to occur. Since the rate of the egg's passage through the uterus is unaffected, it therefore appears that the secretion of shell material is inhibited, possibly by inactivation of carbonic anhydrase (*Sturkie, 1946*).

Shells slightly heavier than usual are formed when hens are fed thyroid (*Asmundson and Pinsky, 1935*).

The Secretion of Pigment

In colored eggs, oöporphyrin pigments (*Fischer and Kögl, 1923*), whose source is the hemoglobin of the red blood corpuscles, are deposited during shell formation as uniform ground color. After the shell is complete, further pigment may be added as spots, streaks, blotches, or flecks (Fig. 131).

The ground pigments are produced by a series of transformations (*Sorby, 1875; Liebermann, 1878; Krukenberg, 1883; Giersberg, 1922*).

The capsules of worn-out red corpuscles rupture, and the hemoglobin is released, dissolved in the blood, and transformed into hematin. The hematin is transformed by mesothelial cells, particularly by those in the liver, into bile pigments of various colors—red, yellow, blue, brown, and black. These are carried by the blood to the uterus and there secreted by glands of the uterine mucosa into the lumen of the uterus while the shell is being formed.

In various species of birds, different bile pigments are secreted; and pigments are deposited neither in the same concentration, nor at the same

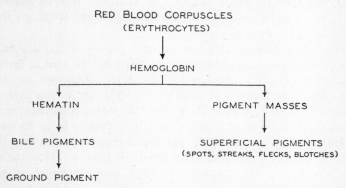

FIG. 131. Origin of the ground and superficial pigments of the bird's eggshell.

stage in shell formation. There is little or no pigmentation in the inner layer of the hen's eggshell. Beginning slowly, the secretion of ground pigment proceeds at an ever-increasing rate until the shell is completed. During the last 5 of the 20 hours the egg spends in the hen's uterus, 75 per cent of the total pigment is deposited (*Warren and Conrad, 1942*).

In the interval between the completion of the shell and the expulsion of the egg, the shells of many birds' eggs receive additional (superficial) pigmentation, distributed over the surface in an irregular pattern characteristic of the species. If a further protective coating (cuticle) is added around the shell, the material from which this coat is formed carries with it pigment masses from the lumen of the uterus. Otherwise the masses adhere directly to the shell surface (*Giersberg, 1922*).

Like the ground pigments, the superficial pigments are derived from the hemoglobin of the red corpuscles, but more directly (*Giersberg, 1922*). Red cells are forced out among the tissues when the uterine wall is suffused with blood. Hemoglobin from these erythrocytes is gathered up by wandering mesoderm cells (lymphocytes) and transformed into pigment masses (cf. Fig. 131). These cells, bearing pigment, pass through

the uterine epithelium into the lumen of the uterus where they disintegrate. The pigment masses, thus freed, remain suspended in the sticky
secretion in the lumen and coalesce. The masses may adhere to the
shell surface on coming in contact with it, or, as noted above, they may
be applied with the mucilaginous material that forms the final envelope
about the shell.

Superficial pigments are added shortly before the egg is laid. When
turkey eggs are surgically removed from the uterus a short time before
they are due to be expelled, the speckles are found to be so recently
applied that they are easily smudged in handling (*Warren and Conrad,
1942*). The eggs of some species are laid before the spots are fixed, so
that streaks form.

At one time or another, all the posterior regions of the oviduct, from
the isthmus to the vagina, inclusive, as well as the cloaca, have been
considered responsible for pigment formation. Although the isthmus
supplies a minute amount of pink color to the egg membranes (*Klose
and Almquist, 1937*), it is unlikely that this region could contribute the
relatively vast amounts of pigment to a shell that is to be formed at a
lower level in the oviduct. Fully pigmented eggs have been laid through
a fistula at the end of the uterus (*Asmundson, 1931b*). Thus the regions
beyond the uterus, the vagina and cloaca, are eliminated as sites of
pigment formation.

The amount of pigment in the shells of eggs removed from the uterus
is roughly proportional to the amount of time the egg has spent in that
region of the oviduct (*Warren and Conrad, 1942*). All evidence points
to the uterus as the site of pigment secretion.

The Cuticle

The lower portion of the oviduct applies a thin, protective film of transparent material to the surface of the shell (cf. Fig. 130-*C*). This coat,
the cuticle, is formed in the interval between the completion of the
shell and the expulsion of the egg. In chemical composition, it is similar
to the egg membranes.

When mineral deposition is ended, those basal cells in the uterine
epithelium that are in contact with the shell begin to secrete a granular
material comparable to the cuticle in staining properties (*Richardson,
1935*). The granules coalesce and adhere to the shell, to the exposed
matrix fibers, and to the matrix material in the mouths of the pores. In
birds that lay speckled eggs, the first cuticular material secreted picks up
pigment masses and particles from the uterine lumen. These pigment
masses are carried with the cuticle and deposited simultaneously on the

shell. Where pigment is most plentiful, cuticular material is laid down in greatest depth. A layer of unpigmented cuticle of fairly uniform thickness is finally deposited over the entire shell surface (*von Nathusius, 1894*). The cuticle of the hen's egg, which contains no visible pigment masses, is evenly distributed and varies little in thickness.

The cuticle of a newly laid egg, whether visibly pigmented or not, contains porphyrins, which give off a red fluorescence in ultraviolet light. When the oviduct of a laying hen is cut open and irradiated, only the uterine portion gives a red fluorescence (*Furreg, 1931*). This evidence and the staining experiments of Richardson (*1935*) indicate that the uterus elaborates the material from which the cuticle is formed.

RATE OF PASSAGE THROUGH THE OVIDUCT

In hens that lay on successive days, the ovulation of each yolk ordinarily occurs within a half hour after the previous egg is laid (*Warren and Scott, 1935a*). A quarter of an hour later, the yolk may already have been engulfed by the infundibulum and started on its tortuous journey through the oviduct.

Once enclosed by the infundibulum, the yolk travels rapidly through this first portion of the oviduct. Of the total time required for the yolk's passage through the oviduct, less than 2 per cent is spent in the infundibulum, which constitutes 5 per cent of the entire length of the oviduct. After 20 minutes, the yolk has entered the albumen-secreting region (*Warren and Scott, 1935b*).

With but little slowing of pace (Fig. 132), the yolk proceeds through the albumen-secreting portion of the oviduct, gathering albumen about itself as it passes. Only 13 per cent of the entire time is spent in this region, which makes up about 60 per cent of the total length of the oviduct. In the posterior portion of the albumen-secreting region, where most of the firm albumen is acquired, the speed of passage diminishes (*Burmester, 1940*); it was found by Warren and Scott (*1935a*) to average 2.3 mm. per minute in five hens.

The egg may pause after entering the isthmus (*Richardson, 1935*) or continue on its way at a steadily decreasing rate (cf. Fig. 132) while its first protective coats, the shell membranes, are being applied. The isthmus, constituting 15 per cent of the oviduct, is traversed in 5 per cent of the total time. The average speed attained in the isthmus is 1.4 mm. per minute (*Burmester, 1940*).

At the isthmo-uterine junction, the progress of the egg becomes slower. The egg then moves onward at a fairly uniform rate throughout the uterus (cf. Fig. 132). This portion of the oviduct approximates the isthmus in length, yet the egg spends about 80 per cent of the total time of passage in it (*Warren and Scott, 1935b*). Comparatively little time is necessary for the fully formed egg to traverse the vagina.

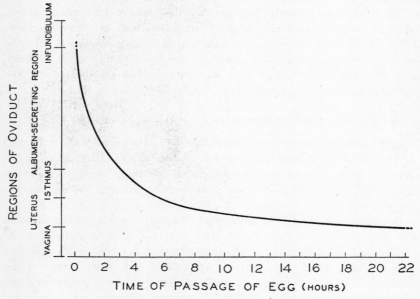

Fig. 132. Rate of passage of the egg through the secretory portions of the hen's oviduct.

The average time the egg requires to pass through each region of the oviduct, and the percentage of its total sojourn spent at each level, are indicated below:

REGION OF OVIDUCT	DURATION OF STAY HOURS	MINUTES	PROPORTION OF TOTAL TIME (per cent)
Infundibulum	..	20	1.4
Albumen-secreting region	3	0	12.8
Isthmus	1	10	5.0
Uterus	19	0	80.8
Vagina	Presumably very brief		....

TIME INTERVAL IN LAYING

Over a number of days, a laying hen will produce an egg each day, cease laying for a day or so, and begin laying again. The production of a series of eggs without interruption is known as a *laying cycle* (see Chapter 1, "Mode of Laying"). Hens that produce heavily pause for a single day only, at long intervals, whereas poor producers pause for a longer time and more frequently. Delay in ovulation contributes to the lengthening of the pause between laying cycles (*Warren and Scott, 1935b*).

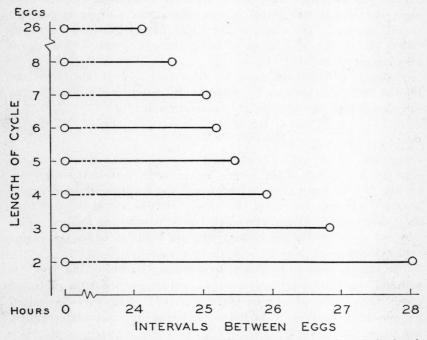

FIG. 133. Relationship between the number of eggs in the cycle and the length of the interval between successive eggs. (After the data of Atwood, 1929b.)

As shown in Fig. 133, there is a correlation between the number of eggs in a cycle and the length of the interval between the eggs. In 2-egg cycles, the second egg is laid somewhat more than 4 hours later in the day than the first egg (*Atwood, 1929a; Heywang, 1938*), possibly because of the delaying influence of the hormone secreted by the ruptured follicle of the first egg (*Fraps, 1946*). The longer the cycle, the more nearly the interval between eggs approaches 24 hours. In a single cycle, the interval shortens toward the middle of the cycle and lengthens toward the end. The interval between the last 2 eggs laid before a pause is usually the longest (*Atwood, 1929a; Heywang, 1938*).

In an effort to discover the factors that influence the length of the interval between eggs, Warren and Scott (*1935a*) found that, in sixteen hens, 14 to 75 minutes elapsed between the laying of an egg and the ovulation of the next yolk. This variability is too slight to account for differences of several hours in the length of the interval. Variations in the rate of the egg's passage through the albumen-secreting region of the oviduct and the isthmus are also of negligible importance. On the other hand, the egg's uterine period may be several hours longer in one bird than in another. The length of the laying interval, therefore, is determined chiefly by the number of hours the egg spends in the uterus (*Warren and Scott, 1935b*).

The cycle is terminated, however, by a delay in ovulation (*Scott and Warren, 1936*), rather than by retention of the egg in the uterus for an extended period (see Chapter 1, "Mode of Laying").

MECHANISM OF LAYING

For the egg, the vagina is a heavily muscled avenue of escape to the outside. The length of the vagina approximates that of the egg. According to Wickmann (*1896*), the uterus, containing the finished egg, is prolapsed through the vagina and cloaca, and the egg is dropped from the cloacal orifice. The uterine wall everts, unwrapping the egg, which does not come into actual contact with the walls of either the vagina or the cloaca.

The vagina seems to be under the voluntary control of the hen. Its powerful musculature probably contracts voluntarily about the prolapsed uterus, the eversion of which is thereby assisted. Eggs may be retained for considerable periods if conditions for laying are unfavorable (*Patterson, 1910; Stieve, 1918*).

Orientation of the Egg in the Oviduct

The orientation of the egg in the oviduct has been a matter of discussion since the first recorded observation by Aristotle. Interest in this subject became especially keen early in the nineteenth century. Experimental evidence now indicates that most eggs, possibly 90 per cent (*Olsen and Byerly, 1932*), form in the oviduct with the sharp, or cloacal, end toward the posterior of the hen.

However, not all eggs are laid sharp end foremost, contrary to the statement of Purkinje (*1825*). Probably at least 20 to 30 per cent are laid blunt end first (*Olsen and Byerly, 1932*) and therefore must of necessity be rotated end for end while passing down the posterior portion of the oviduct. By the simple expedient of marking the egg as it neared the vent and recording the orientation when laid (*Wickmann, 1896; Olsen and Byerly, 1932*), it has been shown that some eggs actually are reversed in the uterus.

On some occasions the egg, before entering the vagina, deforms the thin-walled uterus in such a way that a bulge, or blind sac, is formed. The sharp end of the egg extends into the blind sac. The greatest portion of the egg lies below the inner orifice of the vagina (Fig. 134-*A*).

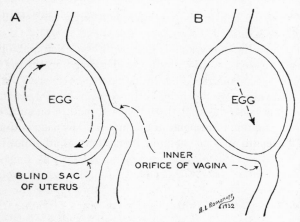

Fig. 134. Diagrammatic explanation of the turning of the egg in the uterus before laying.

A, retention of the egg in a deep blind sac on the posterior wall of the uterus, so that the egg must be turned to be laid. *B*, the normal straightforward course of the egg. (Modified after Curtis, 1916.)

Force applied to the egg by the contraction of the uterine musculature rotates the egg so that the blunt, or infundibular, end leads as the egg is expelled.

It is not known whether or not end-to-end rotation anterior to the uterus accounts for those eggs that are found in the uterus with the blunt end caudad.

PHYSIOLOGY OF FORMATION

The freeing of the mature yolk from the ovary, and its engulfment by the infundibulum, are the most extraordinary phenomena in the formation of the bird's egg. Striking changes, visible and invisible, in the reproductive organs and in the bird's general functional activity anticipate ovulation by many days.

When the pullet is preparing to lay her first egg, or the hen the first egg of a new cycle, her blood becomes greatly enriched in the materials that will be transformed and incorporated into the egg. The extra energy required for these extensive transformations is derived from an increase in the blood's content of glucose and is associated with a rise in the metabolic rate.

MORPHOLOGICAL CHANGES IN THE OVARY AND OVIDUCT

The ovary of the hen in the resting condition is very small, weighing only 2 to 8 gm. (*Chaikoff, Lorenz, and Entenman, 1941*). Of the numerous ova present in the ovary, one is destined by position and advantage in blood supply to reach maturity first. As this ovum enters the final stage of rapid yolk accumulation, the ovary increases sharply in size and weight. Meanwhile, yolk is being deposited in several smaller ova of various sizes, at such rates that these ova will mature on successive days.

The oviduct is quick to respond to the increased ovarian activity. From the dormant weight of 2 to 20 gm., it enlarges rapidly until it weighs

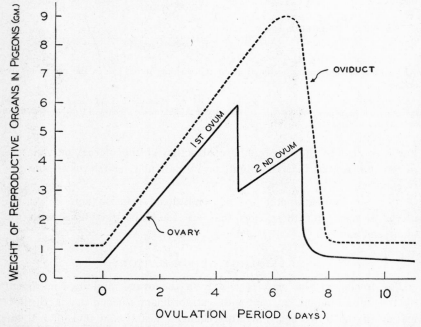

FIG. 135. Changes in weight of ovary and oviduct during the reproductive cycle of the pigeon. After the ovulation of the second ovum, the weight of the oviduct suddenly drops to that of the oviduct in the inactive state. (After Riddle, 1942a.)

40 to 60 gm. in the actively laying hen (*Chaikoff, Lorenz, and Enten-man, 1941*). The glands of the oviduct become taller, the secretory cells become turgid with secretory products, and secretion appears in the ducts. The oviduct is now ready to receive the mature ovum and to contribute, in orderly sequence, the various materials that form the coverings of the yolk.

How rapidly the oviduct of the pigeon responds to changes in the ovary is strikingly shown in Fig. 135. This species normally lays only 2 eggs in a clutch. In all, only 8 days are required for the preovulatory growth of the reproductive system, ovulation, and the laying of both eggs. When the second ovum is released, the ovary suffers a sudden loss in weight. Within a day, the weight of the oviduct has also diminished to that of the oviduct in the nonlaying bird.

The ovary-oviduct relationship in the hen is comparable with that in the pigeon. There are a few dissimilarities. The number of eggs in a cycle is greater and more variable. The oviduct may not regress immediately after the last egg of the cycle, even though a new cycle does not begin at once. Ova may continue to develop after egg-laying ceases, but

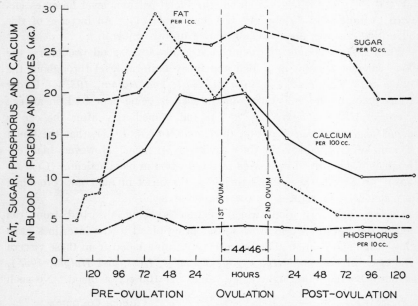

FIG. 136. The levels of fat, sugar, calcium, and phosphorus in the blood of pigeons and doves, before, during, and after ovulation. (After Riddle and Honeywell, 1924, on sugar; Riddle and Reinhart, 1926, on calcium; Riddle and Burns, 1927, on fat and phosphorus; and Riddle, 1938b, 1942b.)

at a diminishing rate. These ova degenerate while still in the ovary, or they are ovulated into the body cavity and their yolk subsequently resorbed (*Chaikoff, Lorenz, and Entenman, 1941*). There is mobilization of egg-forming materials and stimulation of the oviduct as long as ova are produced, no matter what their fate.

CHANGES IN THE CHEMISTRY OF THE BLOOD

With the onset of laying, the concentration of lipids (*Warner and Edmond, 1917*), carbohydrates (*Hayden and Sampson, 1931*), and minerals (*Benjamin and Hess, 1933; Heller, Paul, and Thompson, 1934*), including magnesium (*Charles and Hogben, 1933*), increases in the hen's blood. Similar changes have been observed in the blood of pigeons and doves (*Riddle, 1927*) about 5 days before the first ovulation of the cycle (Fig. 136), in association with a 5 to 10 per cent increase in the basal metabolic rate (*Riddle, Smith, and Benedict, 1933*).

CALCIUM

In the nonlaying hen, the blood plasma contains about 10 mg. of calcium per 100 cc. Just before ovulation, the blood calcium level becomes elevated to more than 20 mg. per 100 cc. (Table 21). An increase of this magnitude requires an abundant source of this mineral.

When large amounts of calcium are available to good producers, the blood calcium remains at a high and fairly uniform level throughout the laying cycle (*Deobald, Christiansen, Hart, and Halpin, 1938*). Under ordinary conditions, it is only when a shell is being deposited on an egg (*Knowles, Hart, and Halpin, 1935*), and immediately after the egg is laid (*Macowan, 1932; Knowles, Hart, and Halpin, 1935*), that the blood calcium level is reduced in the heavily laying hen. However, high environmental temperatures may cause a reduction in blood calcium (*Conrad, 1939*). When the hen ceases to lay at the onset of molt and brooding, the calcium soon returns to the nonlaying level.

During shell formation, when calcium is being absorbed from the food in the intestines, the arteries that bring blood to the intestines have a lower calcium content than the veins returning blood from these viscera (*Buckner, Martin, and Hull, 1930; Knowles, Hart, and Halpin, 1935*), as shown by the following average values per 100 cc. of blood. No such

BLOOD VESSEL	CALCIUM CONTENT (milligrams)
Anterior mesenteric artery (to viscera)	15.3
Anterior mesenteric vein (from viscera)	21.8

TABLE 21

LEVEL OF VARIOUS BLOOD CONSTITUENTS IN LAYING AND NONLAYING HENS

Amounts in Blood Plasma

Constituent	Nonlaying (milligrams per 100 cc.)	Laying (milligrams per 100 cc.)	Increase (per cent)	Investigator
Lipids:				
Total lipid	379.0	1347.0	255	Warner and Edmond (*1917*), Lorenz, Entenman, and Chaikoff (*1938*)
Fatty acids	345.0	1387.0	302	Lorenz, Entenman, and Chaikoff (*1938*)
Cholesterol	102.0	123.0	21	Warner and Edmond (*1917*), Lorenz, Entenman, and Chaikoff (*1938*)
Proteins:				
Serum proteins	5540.0	8970.0	62	Rochlina (*1934a*)
Glucose	171.1	179.2	5	Hayden and Sampson (*1931*)
Calcium:				
Total calcium	12.8	21.5	68	Russell, Howard, and Hess (*1930*), Laskowski (*1933*, *1934*), Paul (*1934*), Roepke and Hughes (*1935*), Taylor and Russell (*1935*), Feinberg, Hughes, and Scott (*1937*)
Filterable calcium	6.0	5.9	−2	Correll and Hughes (*1933*), Taylor and Russell (*1935*)
Phosphorus:				
Total phosphorus	11.8	35.6	202	Laskowski (*1934*), Roepke and Hughes (*1935*)
Inorganic phosphorus	4.3	5.4	26	Peterson and Parrish (*1939*)
As total lipid	14.0	28.5	104	Lawrence and Riddle (*1916*)
As lecithin	7.5	17.0	127	Lawrence and Riddle (*1916*)

difference exists in the blood of the nonlaying hen (*Buckner, Martin, and Hull, 1930*).

Some of the calcium in the blood is in the form of inorganic salts, which are filterable and immediately available. The remainder, or bound calcium, is mainly in combination with the proteins of the blood. The level of filterable calcium in the mature hen, whether she is laying or not, remains fairly stable (cf. Table 21). The difference between the values for laying and nonlaying hens is apparently due to variations in the amount of bound calcium.

Although the calcium for the shells of a series of eggs must ultimately come from the diet, some is supplied by temporary drafts on the body reserves (*Common and Hale, 1941*). The hen's skeleton contains 98 per cent of the calcium in her body. As the hen continues to lay, her bones suffer a considerable loss in weight. Since only 1 gm. of calcium can be deposited in the bones in 1 day, the hen that lays daily can draw from her body reserves less than half of the calcium needed for the egg. These intermittent withdrawals must be replaced by calcium from the diet (*Common, 1936a*).

When the calcium in the hen's diet is restricted, ovulation is less frequent or may cease altogether, and progressively thinner eggshells are produced (*Deobald, Lease, Hart, and Halpin, 1936*). When the diet contains only 1 per cent of calcium, the hen's bones soften to such an extent that they cannot support her weight (*Evans, Carver, and Brant, 1944b*).

Briefly, then, the calcium for shell formation eventually comes from the diet. The amount of calcium retained daily—about 50 per cent of intake (*Common, 1943*)—is usually insufficient for shell formation. During the short, intensive period of utilization, when the egg is in the uterus, the calcium deficit is made up by a draft on body calcium, primarily on that of the bones. When the deficit is not made up from the diet, the skeleton therefore grows lighter as long as egg-laying continues. During intervals of nonlaying, or between periods of shell formation, calcium is again deposited in the bones. This has been noted especially in the pigeon in the prelaying period (*Kyes and Potter, 1934; Bloom, Bloom, and McLean, 1941; Riddle, Rauch, and Smith, 1944*). The formation of medullary bone, both in male and in nonlaying female pigeons, has been induced experimentally by injections of estrogen (*Pfeiffer and Gardner, 1938*). Ringoen (*1945*) observed similar medullary bone formation in the English sparrow (*Passer domesticus*) and bobwhite quail (*Collinus virginianus*).

PHOSPHORUS

As calcium is mobilized before egg formation, the phosphorus in the blood of the hen also increases greatly. The phosphorus level in the laying hen's blood is elevated to three times that in the blood of the nonlaying hen (cf. Table 21). The simultaneous increase in these minerals occurs several days before ovulation, at the time yolk is being rapidly produced. The enrichment of the blood in these substances is maintained throughout the period of egg production (Fig. 137).

Most of the phosphorus used in egg formation is incorporated into phospholipids and phosphoprotein (ovovitellin). A small amount is deposited in the shell as calcium phosphate. A minute quantity is used in the formation of albumen.

The rate at which phosphorus is utilized in forming the various parts of the egg, as demonstrated by the injection of radioactive phosphorus, is shown in Fig. 138. Available phosphorus, in the form of inorganic salts, is immediately deposited in the shell. The small amount used in

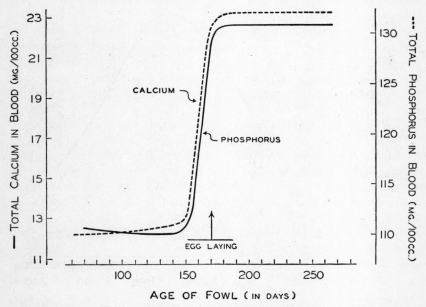

FIG. 137. Increase in the total calcium and total phosphorus in the whole blood of the fowl at the time of sexual maturity, when egg-laying begins. (After the data of Heller, Paul, and Thompson, 1934.)

albumen formation is incorporated more slowly. Several days are necessary for the fabrication of phospholipids and ovovitellin. In the hen, the phosphorus content of the blood remains elevated throughout the cycle of laying, since yolk production is continuous during the entire period. In the pigeon, the blood phosphorus returns to the normal level before ovulation, after most of the phosphorus has been utilized in yolk formation.

The level of phosphorus in the blood serum of laying and nonlaying hens differs greatly. Precisely at the time of greatest calcium utilization, the phosphorus in the blood is increased (*Feinberg, Hughes, and Scott, 1937*). Generally, variations in these two minerals are in the

same direction and are fairly proportional, irrespective of the amount
of calcium in the diet (*Deobald, Christiansen, Hart, and Halpin, 1938*).

With the onset of laying, the excretion of phosphorus in the feces is
increased (*Russell and McDonald, 1929*). If the calcium in the diet is
deficient, phosphorus is excreted more rapidly than usual during the time

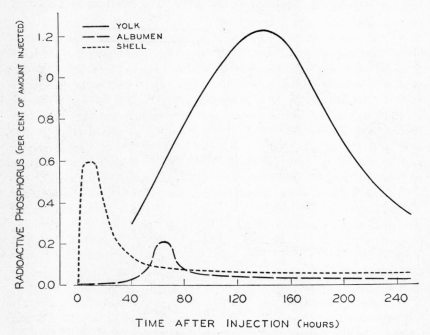

Fɪɢ. 138.   The rate of deposition of radioactive phosphorus in shell, albumen, and
yolk.   (After Lorenz, Perlman, and Chaikoff, 1943.)

of shell formation (*Common, 1936a*). The increase in phosphorus in the
blood plasma and in the feces, at the time that the skeleton is losing
weight, is evidence that egg formation draws on the body reserves of
phosphorus. In the heavy producer, the phosphorus of the bones, along
with the calcium, must be promptly replaced by the diet. When there is
no deficiency of calcium, a diet containing 0.8 per cent of phosphorus
is adequate to restore the minerals withdrawn from the bones during
egg-laying (*Evans, Carver, and Brant, 1944a*). The ratio of the minerals in the newly deposited bone may differ from the original ratio
(*Common and Hale, 1941*).

LIPIDS

Great quantities of phospholipids, fats, and cholesterol are required for the formation of yolk; insignificant amounts enter into the formation of the albumen, shell membranes, and eggshell. With the onset of maturity in the fowl, the phospholipids and fats in the blood plasma increase sharply; the cholesterol increases only moderately (*Lorenz, Entenman, and Chaikoff, 1938*).

In laying hens, the average blood lipids are at a level several times higher than in nonlaying fowls (cf. Table 21). A close relationship exists between the level of the blood lipids and the size of the oviduct (*Entenman, Lorenz, and Chaikoff, 1938*). When the oviduct is in full reproductive condition, the blood lipid values are the highest. When regression is complete, the blood lipids are reduced to the nonlaying level (*Chaikoff, Lorenz, and Entenman, 1941*). The condition of the oviduct is due directly to the activity of the ovary; the heightened blood lipid level may well be due to this activity also. At most, the oviduct can be only a mediator of the blood lipid level and may act as an intimate regulator.

The rise in blood lipids occurs simultaneously with the increase in the mineral content of the blood. A large part of the phosphorus used in egg formation is incorporated into the lipids of the yolk. By employing radioactive phosphorus, Hahn and Hevesy (*1937*) found that the phospholipids are formed in the liver. In the hen, fats are accumulated there during the prelaying period. Both fat and the newly formed phospholipid are carried to the ovary by the blood (*Heller, Paul, and Thompson, 1934; Lorenz, Entenman, and Chaikoff, 1938*) and deposited in the ovum.

The production of the yolk's full complement of phospholipid is a slow process. In the pigeon, the simultaneous elevation of blood phosphorus and fats (cf. Fig. 136) 2 or 3 days before ovulation provides the materials for synthesis at a sufficiently early time.

PROTEINS

The proteins in the yolk and albumen of the egg are derived from the proteins of the hen's blood serum. Coincident with ovulation, the day before laying, the blood serum protein increases by more than 50 per cent (cf. Table 21). Shortly after laying, it returns to a level below that found in the nonlaying hen (*Rochlina, 1934b*).

The formation of yolk protein is a process requiring about 10 days. The synthesis of ovovitellin draws upon the large quantities of phosphorus in the blood and, in fact, probably occurs in the blood stream, since a protein resembling ovovitellin can be demonstrated in the blood of the laying hen. After the protein has been produced, the blood carries it to the ovary (*Roepke and Bushnell, 1936*).

The soluble proteins of the albumen accumulate at a constant rate in the glands of the oviduct and are stored there until deposited in the egg; mucin, however, appears to be formed in the goblet cells as needed (*Conrad and Scott, 1942b*). Those components of the albumen that contain phosphorus are probably produced by a synthetic process which requires from 24 to 54 hours (cf. Fig. 138) for completion (*Lorenz, Perlman, and Chaikoff, 1943*).

CARBOHYDRATES

A hen that lays daily must ingest a large quantity of raw materials, mobilize her body reserves, transform into numerous complex egg-forming substances the materials from diet and body, and, finally, rebuild her body reserves. These processes must take place at a rapid rate.

The energy necessary for maintaining normal body processes is derived mainly from carbohydrates. The additional energy required for egg formation is provided by a rise in the glucose content of the hen's blood before the first ovulation of a laying cycle (*Hayden and Sampson, 1931*). The metabolic rate remains elevated during the entire cycle and for a short period after the hen ceases to lay (*Winchester, 1940*).

HORMONAL CONTROL OF REPRODUCTION

A delicate balance in the functioning of the entire glandular system is necessary for the maintenance of normal bodily functions in the higher animals. Before and throughout periods of reproduction, this balance is altered to meet the increased demands on the organism. Seasonal and other environmental influences (*Ringoen, 1940*) bring about rhythmic alterations in the activity of the anterior pituitary gland, which regulates the activity of the reproductive organs. In the modern hen, the reproductive system is maintained in a highly functional state for long periods of time, because of the abundant production of pituitary hormones.

The anterior pituitary produces two hormones that are essential to the growth of the ovary and the growth and maturation of the ova. These are the follicle-stimulating and luteinizing hormones (Fig. 139). The effect of the follicle-stimulating hormone can be seen in Fig. 140. When the hen is given injections of pituitary extract, the larger yolks in the ovary are stimulated to mature simultaneously (cf. Fig. 140-C). The anterior pituitary-like hormone of pregnant-mare serum stimulates all the ova to grow rapidly (cf. Fig. 140-*B*), so that the gradation in size is less apparent than in the normal ovary (cf. Fig. 140-*A*).

Under the influence of the pituitary, the follicles of the more rapidly growing ova produce a hormone, estrin, which stimulates the oviduct.

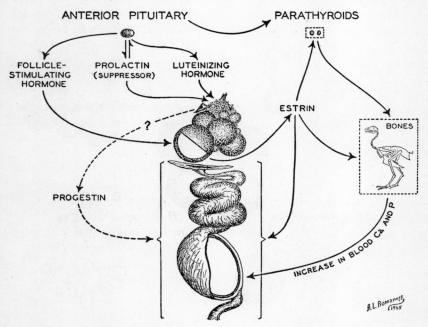

Fɪɢ. 139. Schematic representation of hormonal control of reproduction in the chicken.

In turn, the oviduct grows, and its glands become highly secretory. Large doses of the follicular ovarian hormone, injected into female chicks from the eighteenth to the fortieth day of life, may result in enlargement of the oviduct to forty-eight times normal size (*Herrick, 1944*).

Estrin produces other profound changes in the general physiology of the hen. It mobilizes from the food, and from the body reserves, the building materials needed in egg formation, as well as the metabolic fuel necessary to transform them.

With ovulation approaching, the follicle becomes more mature and secretes estrin at a greater rate. The formation, deposition, and storage of egg-forming materials proceed at a faster pace. When the fabrication of the materials approaches completion, ovulation occurs, and the final assembly of the egg in the oviduct follows.

Within 2 days after ovulation, the ruptured follicle begins to atrophy (*Davis, 1942a, 1942b*). That the ruptured follicle assumes an endocrine function is questionable. In mammals, the ruptured follicle is transformed into the corpus luteum, which secretes a hormone, progestin; this

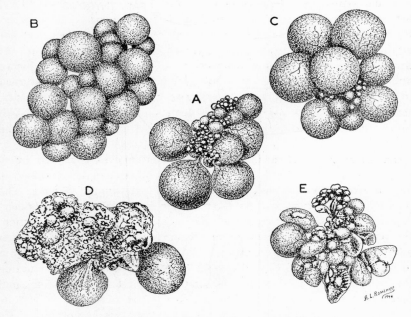

FIG. 140. Various conditions of the ovary of the fowl.

A, normal ovary. *B* and *C,* ovaries stimulated by injections of pregnant-mare serum and pituitary gonadotrophin, respectively. *D,* adenocarcinomatous ovary. *E,* ovary from bird infected with pullorum disease. (After Rettger, Kirkpatrick, and Jones, 1914; Olson and Bullis, 1942; and Phillips, 1943.)

hormone aids in maintaining the oviduct in a reproductive condition. There is some doubt that progestin is produced in birds (*Riddle and Schooley, 1944*). However, when the most recently ruptured follicle is removed from the ovary of the hen, laying may be delayed several hours or even days (*Rothchild and Fraps, 1944*). The ruptured follicle may therefore elaborate a hormone that affects the time of laying.

All rapidly growing follicles secrete estrin in proportion to their state of maturity. Their additive secretions maintain the oviduct in a highly functional state, permitting continuous egg formation with only occasional short interruptions.

Because the weight of the oviduct and the level of lipids in the blood are closely related, the secretion of the growing follicle, by controlling the development of the oviduct, is indirectly responsible for the mobilization of lipids.

The enrichment of the blood in minerals is an immediate response to increased ovarian activity before ovulation. In the pigeon, when the first and second ova begin to grow rapidly 5 and 7 days, respectively, before the first ovulation, the minerals in the blood increase sharply. By referring to Figs. 135 and 136 one may see that the oviduct does not attain its maximum size until the approach of the second ovulation. At that time, the phosphorus and calcium, as well as the fat, have long since attained their maxima and are either declining or have already returned to the resting level. The rapid decrease of all the egg-forming substances in the blood may be due to depletion as the eggs are formed, rather than to diminished hormonal stimulation up to the time of the second ovulation.

The mobilization of calcium and phosphorus involves a complex interplay of hormonal effects. The resting level of calcium in the blood is directly controlled by the secretion of the parathyroid glands. The parathyroids, however, are controlled by the master gland, the pituitary. As pituitary activity increases in the female bird during reproduction, the parathyroids are stimulated to greater activity, and blood calcium and phosphorus levels are elevated. The growing follicles secrete estrin in increasing amounts. This hormone (either directly, or by stimulating the parathyroids) may cause the simultaneous release of calcium and phosphorus from the skeleton into the blood (cf. Fig. 139).

When gonad-stimulating hormones from the anterior pituitary are injected into pigeons after removal of the pituitary, the blood calcium level rises (*Riddle and Dotti, 1936*). The injection of female sex hormones produces a like result even when the parathyroids are removed (*Riddle, Rauch, and Smith, 1945; Riddle and McDonald, 1945*). It is clear that the secretions of the ovary, as well as of the parathyroids and the anterior pituitary, are directly or indirectly involved in the mobilization of calcium.

In the pigeon, when the second egg is laid, the anterior pituitary releases a hormone, prolactin. This hormone decreases the production of gonad-stimulating hormones by the anterior pituitary (*Riddle, 1938a*). The ovary atrophies; and, soon afterward, the oviduct shrinks to the resting size (cf. Table 16). Because it depresses ovarian secretion by inhibiting the gonad-stimulating hormones, prolactin indirectly decreases the calcium in the blood (*Riddle and Dotti, 1936*).

Among the many effects of prolactin is the production of broodiness in birds (*Riddle, Bates, and Lahr, 1935*), and the stimulation of the crop glands of the pigeon to secrete the nutritious "milk" necessary for the rearing of the young (*Riddle, Bates, and Dykshorn, 1933*). In some races of fowl—Leghorns, for example—the capacity to respond to prolactin by becoming broody has been almost lost. The net result is greater productivity with fewer periods of rest.

DISEASES OF THE HEN'S REPRODUCTIVE SYSTEM

The most critical time in the life cycle of animals is the reproductive period. The great fecundity of the hen renders her particularly susceptible to a multiplicity of diseases. High mortality among heavily laying birds seems inevitable. A disproportionate number of deaths is traced directly to diseases of the reproductive system (Table 22).

TABLE 22

Occurrence of Various Diseased Conditions in the Reproductive System of Domestic Chickens

Diseased Condition	Number of Autopsies	Incidence (per cent)	Investigator
Ovary:			
Ruptured yolk	2282	5.0	N. Y. State Vet. Coll. (Quot. by *Ogle, 1938*)
Tumors	880	3.4	Curtis (*1915b*)
Hemorrhage	2645	0.4	Brunett (*1925*)
Oviduct:			
Impacted	2645	9.8	Brunett (*1925*)
Cystic	363	0.3	Brunett (*1925*)
Tumorous	880	0.9	Curtis (*1915b*)
Abnormal	2282	0.09 ⎱	N. Y. State Vet. Coll. (Quot. by *Ogle, 1938*)
Infected	2282	0.09 ⎰	

Many deaths due directly to pathological conditions in other portions of the body may also be the indirect result of the extraordinary physiological strain placed upon the system by intensive egg-laying. Diseased reproductive organs were found in 17 per cent of more than 7000 hens autopsied by Goss (*1941*).

The oviduct appears to be more susceptible to disease than the ovary. In egg formation, the oviduct undergoes great mechanical stress. Opening to the outside as it does, it furnishes an avenue of entry for infectious diseases. Impaction, or obstruction, of the oviduct is the most frequent reproductive disorder. Undoubtedly, many cases of obstruction may be due to mechanical failures of the oviduct, but some result from irritation caused by infection.

The ovary is more likely to be a site for tumors than the oviduct. Ruptured yolk, not in itself a disease, is a condition often associated with death from various causes. Many ovarian diseases do not originate in the ovary but are due to bacterial infection carried into the body cavity by an infected, irritated oviduct.

The bacteria most commonly found in hens suffering from reproductive disorders are much the same as those found in birds that die of other ailments. Bacterial cultures of various organs, made at autopsy, show the following frequency of occurrence of the organisms isolated (*Moore and Marten, 1944*).

ORGANISM	FREQUENCY (per cent)
Staphylococcus, aureus and *albus*	44.0
Escherichia coli	38.2
Bacillus subtilis	10.2
Salmonella species	2.6
Pasteurella avicida	1.3
Streptococcus viridans	0.6
Hemolytic *Streptococcus*	0.6
Others	2.5

DISEASES OF THE OVARY

Ruptured Yolk. Irritation, inflammation, or obstruction, due directly or secondarily to bacterial invasion, may cause reverse peristalsis in the oviduct. Contaminated yolks or other materials ·may thus be returned to the body cavity. Infection of the ovaries may follow, and the ova may become soft, misshapen, and easily ruptured. Various inflammatory diseases of the body cavity, including "pullet disease," may affect the ovary secondarily and cause softening of the immature ova. Ruptured yolk may occur in combination with fowl paralysis (*Moore, 1940*). Peritonitis and death may result from the rupture of a yolk (*Hutt, Cole, and Bruckner, 1941*), although yolk material, if uncontaminated by bacteria, is quickly resorbed from the body cavity (*Hoffman, 1932*).

Occasionally, solid concretions of yolk are found in the body cavity. The concentric arrangement of the material in these masses indicates that they are accumulated over a period of time. They may become sufficiently large to crowd the organs in the abdomen.

Tumors. In the fowl, the reproductive organs are second only to the blood-forming organs in susceptibility to tumors. Over 90 per cent of all neoplasms, except leucotic tumors, were found in the reproductive system in more than 7000 hens autopsied by Goss (*1941*). By far the majority of these were located in the ovary.

Tumors of the ovary are most frequently adenocarcinomas. An ovary with a typical neoplasm of this kind is shown in Fig. 140-*D*. Fibrosarcomas, carcinomas (*Goss, 1941*), folliculomas (*Murisier, 1928*), and ovarian cysts (*Laurie, 1912*) occur less often. Cysts may be multiple or single and may vary from pin-point to quart size.

The degeneration of ovarian tissue caused by neoplasms is sometimes followed by a proliferation of testicular tissue, particularly in the rudimentary right ovary (*Murisier, 1928*). In rare cases, two testes are produced (*Reed and Martin, 1933*). The hormones secreted by the newly formed testicular tissue may cause enlargement of the comb, and the growth of spurs and of plumage of the male type. In many instances of spontaneous sex reversal, a tumor of the ovary is found (*Friedgood and Uotila, 1941*), although masculinization of hens may also be due to other atrophic diseases of the ovary (*Crew, 1923b*).

Few of the neoplastic growths in the reproductive system are fatal. The usual result is cessation of egg formation.

Hemorrhage. Slight hemorrhages from the ovary are not unusual, as indicated by the blood spots occasionally found in eggs laid during periods of heavy production. As the stigma of the follicle tears to permit the escape of the yolk, capillaries may be ruptured. The blood released is enclosed with the yolk in the coverings of the egg. Ordinarily, such hemorrhages have no effect on the health of the bird or the edibility of the egg.

In a few instances, bleeding from the ovary into the body cavity may be extensive and may contribute to death. Abortion of immature yolks, accompanied by bleeding, is said to be caused by fright. Injury to hens may result in extensive bleeding into the yolks (*Laurie, 1912*). Bleeding may contribute to the contents of cysts in the ovary. Vitamin K deficiency, which extends the clotting time of the blood, may be associated with hemorrhage.

Atrophy. When an ovary atrophies, the larger ova first become wrinkled and misshapen, and later the smaller ova are affected in succession. The appearance of the ovary is characteristic, whatever the cause of the atrophy. The atrophic ovary shown in Fig. 140-*E* is typical; in this case, the cause was pullorum disease.

In normal laying hens, the ovary occasionally undergoes temporary atrophy during periods of rest or molting. The larger ova may suffer "bursting atresia" (*Davis, 1942a*), an apparently normal process. Extensive follicular atresia may be produced when laying hens are closely confined or when they are subjected to unaccustomed disturbances (*Stieve, 1918*) or to surgical operations (*Rothchild and Fraps, 1945*).

In old age, the ovary becomes atrophied after the last laying period and remains in that state. If the hen is given injections of anterior pituitary extract or pregnant-mare serum during a period of heavy production, the ovary, although stimulated initially, may eventually become injured and atrophic (*Phillips, 1943*).

Atrophy Due to Bacterial Infection. As mentioned previously, infections of the oviduct usually cause hyperirritability, which may produce reverse peristalsis. The infected contents of the oviduct are returned to the body cavity. Along with the other viscera, the ovary may become infected. Inflammation of the ovary is usually followed by nonspecific atrophy, and by discoloration and darkening of the maturing ova. Although this condition is associated with pullorum disease (cf. Fig. 140-*E*), it may be caused by other infections of the ovary. In pullorum disease, the infection of the ovary is usually permanent (*Rettger, Kirkpatrick, and Card, 1919*).

In isolated instances, gangrene may be present in the hen's ovary. The ova in such cases are darkened and fragile, and the contained yolk becomes decomposed.

Occasionally, tuberculosis may spread from other organs to the ovary. In one instance, atrophy of tuberculous origin preceded complete sex inversion (*Crew, 1923a*). Fowl typhoid produces pathological changes in the ovaries of chicks and mature hens (*Beach and Davis, 1927*).

DISEASES OF THE OVIDUCT

Impaction (obstruction) of the oviduct is the most common pathological condition found in the laying hen; it is a frequent cause of death (cf. Table 22). The oviduct may become obstructed by an egg too large to pass (especially in pullets, before the oviduct has attained its full size); by hardened feces in the cloaca; by tumors in the oviducal wall; by scar tissue in the posterior portion of the oviduct, at the site of a previous rupture; or by strangulation of the oviduct. Overexertion, fright, or close confinement may cause paralysis of the oviduct, and eggs and egg materials then accumulate (*Laurie, 1912*) in a mass that becomes too large to expel. The presence of such a mass usually causes irritation and stimulates secretion, so that the mass continually increases in size as the secreted materials harden around it. A cyst may eventually form, although oviducal cysts are comparatively rare. A large one, filled chiefly with yolk, is shown in Fig. 141.

Bacterial infection of the oviduct probably is indirectly responsible for most cases of oviducal obstruction. Although the oviduct is normally

free of bacteria (*Rettger, 1913*) because of the germicidal properties of
the secreted albumen (*Horowitz, 1903*), it is nevertheless possible for
infection to spread to it from the cloaca or from a diseased ovary.   The
oviduct may then become inflamed and hyperirritable.   Chronic constriction at the site of severe inflammation may result in the accumulation of

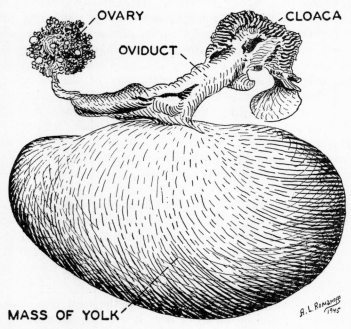

FIG. 141. An oviducal cyst from a domestic hen. The cyst measured 8 cm. by
more than 15 cm., weighed about 630 gm. without the attached structures, and contained a mass of yolk. (After Reese, 1919.)

yolk, albumen, or broken shell.   In addition, peristalsis may be reversed
in an irritated, infected oviduct.   Hiden (*1921*) reports finding a fully
formed egg impacted in the upper portion of the oviduct, where it had
evidently been carried by antiperistalsis; the egg was encased in a mass
of mingled yolk and hardened albumen.

   Reverse peristalsis in a diseased oviduct may also carry infected
material to the body cavity.   Infection of the ovary and other viscera,
and eventual peritonitis, commonly follow.

   Obstruction of the oviduct may lead to other pathological conditions.
An impacted oviduct may rupture under the stress of extreme muscular
effort to expel an egg.   If the rupture occurs in the anterior portion of

the oviduct, it may heal; but if it occurs in the posterior portion, bacteria may be admitted into the body cavity, and peritonitis may then result. Straining to pass an egg may also cause prolapse of the cloaca. Unless the

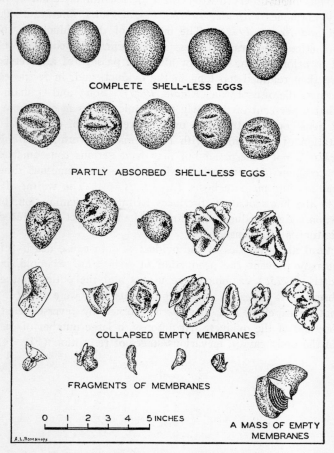

COMPLETE SHELL-LESS EGGS

PARTLY ABSORBED SHELL-LESS EGGS

COLLAPSED EMPTY MEMBRANES

FRAGMENTS OF MEMBRANES

0 1 2 3 4 5 INCHES

A MASS OF EMPTY MEMBRANES

A.L.ROMANOFF

FIG. 142. A collection of shell-less eggs in various stages of resorption, found in the body cavity of a hen with an incomplete oviduct. The total estimated weight of the eggs is about 600 gm. (After Curtis, 1915c.)

condition is corrected, the wall of the cloaca becomes perforated, and septicemia follows.

Any marked departure from normal oviducal function may not only damage the oviduct but also result in abnormal eggs. For example, anti-peristalsis is presumably a factor in the formation of double-yolked eggs (see Chapter 5, "Anomalies"). Hyperactivity of the oviduct may cause an egg to be propelled so rapidly that only an incomplete, fragile shell can form. Premature expulsion of the egg, due to overactivity of the oviduct, has been observed after the experimental injection of posterior pituitary hormones (*Riddle, 1921b; Burrows and Byerly, 1940, 1942; Burrows and Fraps, 1942*). The muscles of the oviduct, on the other hand, sometimes perform with less than normal efficiency, especially at the end of a long laying cycle, or after the passage of an oversized egg. Eggs are then retained too long in the oviduct, and, in fertile eggs, the embryos may die of asphyxia (*Riddle, 1923*). Boley and Graham (*1939*) found that 6 per cent of the early embryonic deaths in eggs from a flock of 1372 hens could be traced to oviducal retention of eggs.

Infestation of the oviduct with the trematode worm, *Prosthogonimus macrorchis,* is not widespread but may have serious consequences. The fluke is common only in regions where there are numerous dragonflies, which act as intermediate hosts (*Macy, 1934*). The worm, after it is lodged in the hen's oviduct, produces irritation, inflammation, bleeding, and anemia. Egg production declines. If the oviduct is ruptured, death from peritonitis often ensues.

Structural abnormalities of the oviduct seldom cause death (cf. Table 22) but may prevent the production of eggs. For example, in a hen that Curtis (*1915c*) examined, the uterus and the outlet to the cloaca were absent. As a result, only membranous eggs were formed. These could not be expelled, and they were returned by reverse peristalsis to the upper end of the oviduct. At autopsy, a large number of them were found in the body cavity, in various stages of resorption (Fig. 142).

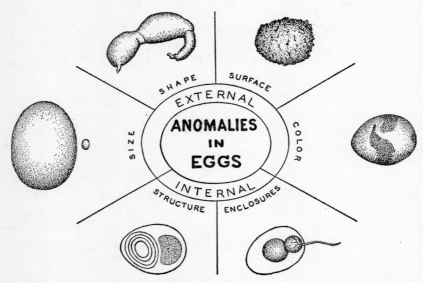

Examples of anomalous conditions found in the chicken egg. Size: Giant eggs, after Clinton (*1927*); dwarf eggs, after Szuman (*1926*). Shape: After Rayer (*1849*) and Davaine (*1860*). Surface: After Durski (*1907*). Color: From authors' collection of eggs. Structure: After Curtis (*1916*). Enclosures: After Cleyer (*1682*) and Davaine (*1860*).

Chapter Five

ANOMALIES

The avian egg is sometimes a great oddity.

The normal avian egg is the product of the coordinated function of reproductive organs that are anatomically perfect. Occasionally, these organs perform deficiently, because of physiological disturbance, morphological abnormality, or injury. Any temporary or permanent impairment in the efficiency of the reproductive system may result in malformations of the egg.

It is remarkable that, of the large number of eggs laid, especially by domestic hens, only a few exhibit abnormalities. On the other hand, it is inevitable that anomalies in eggs should occur, since throughout nature freaks are produced with a certain degree of regularity.

There has always been considerable popular interest in abnormal eggs. In the past, the mystery of their origin evoked a folklore of fantastic explanations, but superstition has now been largely dispelled by the findings of science. As the result of numerous investigations, the causes of many types of aberrant eggs are fairly clear.

In the past, attempts were made to classify abnormal eggs systematically. Davaine (*1860*) considered as one type all anomalies of the yolk, blastodisc, and vitelline membrane; and, as another type, all anomalies that involved any of the structures surrounding the yolk. On the other hand, Parker (*1906*) divided deviations from the normal egg into three varieties, according to their origin in either an abnormal ovary, an abnormal oviduct, or both.

Abnormalities are classified here according to their effect on the internal and external appearance and morphology of the egg. Aberrations in size, shape, shell surface, shell pigmentation, and internal structure are discussed, as well as the enclosures of foreign objects within the egg. The most probable origin of each anomaly is considered, in so far as it is possible to do so. In many instances, the immediate cause of malformation may be clear; yet the more fundamental reason for it may be obscure.

EXTERNAL ABNORMALITIES

External abnormalities of the egg are readily recognized. Deviations from the normal in size, form, shell surface, or coloring of the egg are immediately noticed, even by the casual observer. For this reason, anomalies that affect the external characteristics of eggs have long been matters of interest and curiosity to scientist and layman alike.

SIZE

Eggs laid by birds of the same species, or even by the same individual, are by no means uniform in size (see Chapter 2, "External Characteristics"). It is therefore difficult to define normal and abnormal eggs on the basis of their size. In general, however, it may be said that chicken eggs weighing more than 80 gm., or less than 35 gm., may be considered as deviating from normal and can be classified as giants or dwarfs, respectively.

In the past, it was believed that abnormally large or small eggs marked the beginning or end of a laying period (*Féré, 1898b*), but it is now known that this is not the case. Eggs of unusual size may appear unexpectedly at any point in a series of normal eggs, as illustrated in Fig. 143.

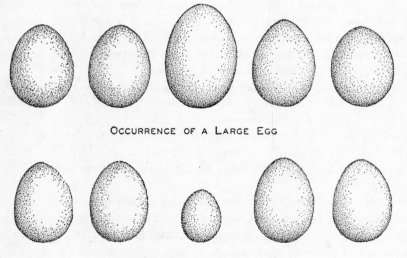

OCCURRENCE OF A LARGE EGG

OCCURRENCE OF A SMALL EGG

FIG. 143. The occurrence of large and small eggs in the laying cycle of the hen. The 5 eggs in each group were laid on consecutive days. (After Warner and Kirkpatrick, 1916.)

Giants

The largest chicken egg on record is probably one weighing 320 gm., which is on display in the Museum of Curiosities of the Pasteur Institute in Paris. This egg is approximately five and one-half times as large as the average hen's egg, which weighs 58 gm. (see Chapter 2, "External Characteristics"). The White Leghorn that produced this giant egg usually laid eggs averaging 80 to 100 gm. (*Anonymous, 1936a*).

Few excessively large eggs attain such great size. One 227-gm. egg was laid by a Rhode Island Red (*Brown, 1939*); and the recorded weight of another egg is 204 gm. (*Henneguy, 1911*). Among unusually large eggs which have been reported are one weighing 198.5 gm. (*Clinton, 1927*), and seven ranging in weight from 170.1 gm. to 192.9 gm. (*Romanoff and Hutt, 1945*). Many eggs weighing slightly more than 100 gm. have been found.

Birds other than chickens have also been known to lay abnormally large eggs. The average runner duck egg weighs about 67 gm., but occasionally one weighs as much as 100 or 106 gm. (*Sumulong, 1925*), or even 113 gm. (*Bauer, 1895*). Extremely oversized pigeon eggs have been observed (*Levi, 1941, p. 176*). Riddle (*1921a*) noted several instances of twin ringdoves and pigeons hatching from abnormally large eggs. The weights of the normal and the large eggs he found are compared in the tabulation.

	Normal Eggs (grams)	Large Eggs (grams)
Ringdove	8.1	10.4
Common pigeon	15.8	20.6

The reasons for the production of giant eggs are not completely understood, although the bird's hereditary propensities and individual peculiarities are no doubt partially responsible. Of more than 200,000 eggs laid by 1820 hens during 20 months, 89 were unusually large; and, of these large eggs, 99 per cent were laid at the time of heaviest production (*Warner and Kirkpatrick, 1916*).

Dwarfs

At the opposite extreme in size from giant eggs are abnormally small, or dwarf, eggs. Since early times, many tiny but externally normal eggs have been found. Perhaps the smallest hen's egg on record is one which weighed only 1.29 gm. (*Szuman, 1926*), or 98 per cent less than the average. Figure 144 shows how this egg would compare with a giant egg.

Dwarfs are more common than all other types of abnormal eggs except those with two yolks. Warner and Kirkpatrick (*1916*) found about 5 dwarfs per 10,000 eggs; Pearl and Curtis (*1916*), about 9.

For many centuries, unusually small eggs have been regarded super-stitiously and have been given many nicknames. It was once believed that miniature eggs were laid by cocks (hence their common name of

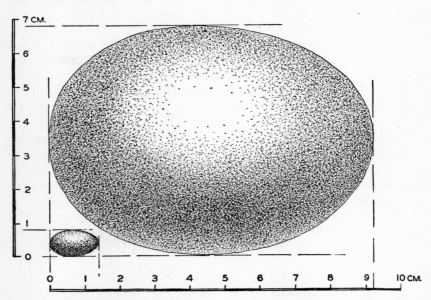

Fig. 144. A giant egg, weighing 198.5 gm., compared with a dwarf egg, weighing 1.3 gm.; both laid by hens. (After Clinton, 1927; and Szuman, 1926, respectively.)

"cock eggs"), and that they would hatch into a serpent, the basilisk, whose breath or look was fatal. Dwarf eggs have also been called "wind eggs" or "witch eggs." They were thought by some people to have the power to bewitch one's enemies; by others, they were considered omens of good luck. According to one superstition, if a dwarf egg were thrown over a building, a wish made while the egg was still in the air would come true.

Externally, dwarf eggs differ from normal eggs in both size and shape. Two shapes appear to be especially distinctive of dwarf eggs: cylindrical, and prolate-spheroidal, which most resembles true egg shape. Cylindrical dwarfs are usually much smaller than the prolate-spheroidal dwarfs (*Pearl and Curtis, 1916*). The average shape index (or breadth-length ratio multiplied by 100) for each type is as follows.

BREADTH/LENGTH × 100

| Prolate-spheroidal | 80 |
| Cylindrical | 53 |

Dwarfs have been noted a number of times among the eggs of birds other than domestic fowl; and undoubtedly many are laid and never observed. Occasionally pigeons are known to produce very small, round eggs (*Levi, 1941, p. 176*). Carpenter (*1906*) reported a hummingbird's egg measuring only 0.53 cm. by 0.74 cm., as compared with the usual

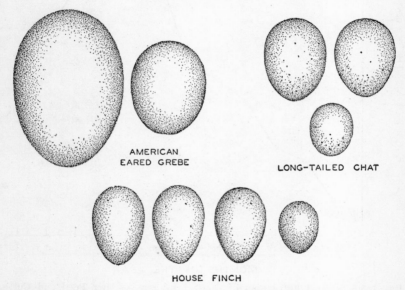

AMERICAN
EARED GREBE

LONG-TAILED CHAT

HOUSE FINCH

FIG. 145. Occurrence of small eggs in the clutches of wild birds. (After Ingersoll, 1910.)

dimensions of 0.84 cm. by 1.27 cm. A single small egg has been found in the clutch of such wild birds as the long-tailed chat, the house finch, and the American eared grebe (Fig. 145). Dwarfs have also been reported among the eggs of pheasants, lapwings, robins, song thrushes, gulls, and ducks (*M'William, 1927*).

The occurrence of dwarf eggs is only partially explained even today. The investigations made since the beginning of the century have resulted in only a few definite conclusions concerning the production of dwarf eggs.

Among the eggs of the domestic fowl, dwarfs occur most frequently in the spring and early summer. Over 70 per cent of all dwarf eggs appear from March to July (Fig. 146). Most of them are laid by pullets just coming into production. Figure 147 indicates how much greater is the tendency to lay small eggs during the first year of production than during subsequent years (*Pearl and Curtis, 1916*). No significant pause in egg production nor change in egg size precedes or follows the occurrence of a dwarf egg.

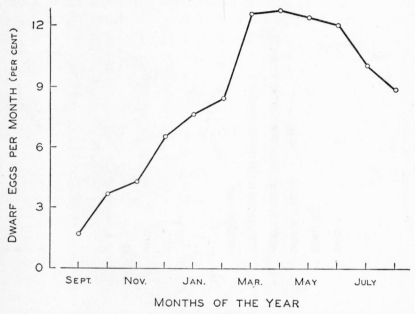

Fig. 146. Seasonal distribution of dwarf hens' eggs. (After Pearl and Curtis, 1916.)

Heredity apparently plays no part in the production of dwarfs. As a rule, individual hens exhibit no tendency to be habitual dwarf producers. Instead, dwarf eggs occur as isolated phenomena. Pearl and Curtis (*1916*) found that, of 200 birds that laid dwarf eggs, 89 per cent laid but 1 such egg, 7.5 per cent laid 2, and only 3.5 per cent laid more than 2. Warner and Kirkpatrick (*1916*) observed that 94.1 per cent of all hens produced only 1 dwarf each.

Experimental surgical resection of the anterior portion of the albumen-secreting region of the oviduct has yielded interesting results (*Burmester and Card, 1939*). Of thirty-one operated hens, twenty-five survived and laid eggs. Six of the birds laid normal eggs, but nineteen of them each laid at least 1 dwarf egg. Seven of the nineteen hens laid dwarf eggs when they first came into production after the operation and later laid normal eggs. The mean weight of the normal eggs produced by these hens was about 53 gm., but the dwarf eggs they produced averaged only 18 gm. When autopsies were performed, two of the hens that had laid

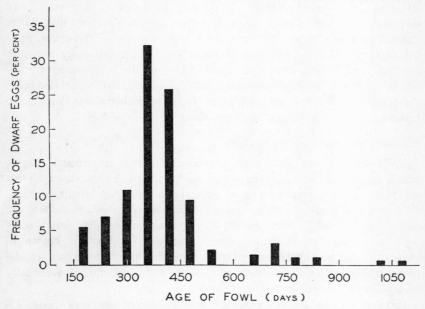

FIG. 147. Frequency distribution of dwarf eggs, according to the age of the hens producing them. (After Pearl and Curtis, 1916.)

dwarf eggs were found to have a noticeable constriction in the oviduct at the place of operation.

From the foregoing evidence, it becomes apparent that dwarf eggs are the result of a temporary disturbance or accident, rather than of a permanent abnormality or disorder of the reproductive organs. It can be seen that, the greater the activity of the reproductive organs, the more numerous are the opportunities for a dwarf to occur.

Infection of the posterior end of the funnel or the anterior portion of the albumen-secreting region, with consequent incomplete constriction of the oviduct, may be a factor causing dwarf eggs (*Pearl and Curtis, 1916; Burmester and Card, 1939*). It also seems likely that ovulation precedes the formation of a yolkless dwarf egg, since autopsies on birds that had just laid yolkless dwarf eggs revealed a large empty follicle in the ovary of each bird, and a yolk lying in the body cavity (*Pearl and Curtis, 1916*).

FORM

Eggs vary greatly in shape, to such an extent that there is no rigid standard for normal contour. They may be rounder, narrower, more elliptical, or more conical than the true ovate form, without being anomalous (see Chapter 2, "External Characteristics"). However, radical departures in the direction of roundness, or an extremely pyriform contour, may be considered abnormal. Two of the most common variations in shape, prolate-spheroidal and cylindrical, are associated with dwarf eggs. Many other deviations from normal contour have been reported, including definite malformations.

Unusual Shapes of Single Eggs

Fully formed, complete eggs occasionally occur in unusual shapes, which have been variously described as elongated, elliptical, or almost spherical; or as resembling a pear, a spindle, a kidney, a biscuit, a retort, or a screw (*Landois, 1878*). In some instances, an oddly shaped egg is the only one of its sort produced by a hen; on the other hand, some hens habitually lay misshapen eggs. One bird, at least, is known to have produced peculiarly long, narrow, somewhat pyriform eggs when she began laying; each successive egg was shorter and broader, until eventually normal shape was achieved (*Pearl, 1909*).

Peculiar shapes are not confined to the eggs of domestic hens. Pigeons' eggs also show a similar range in shape from round to pyriform, with extremes that are definitely abnormal (*Levi, 1941, p. 176*). A sausage-shaped egg laid by a water hen was reported by M'William (*1927*).

Truncated Eggs. Occasionally eggs possess a normal hard shell and are of normal shape, save that one side appears truncated, as if a portion had been sliced off. In some instances, the truncated surface is wrinkled and slightly convex, rather than flat (*von Nathusius, 1869*). It is of interest to note that the flattened portion of the egg is generally toward the pointed, rather than the blunt, end of the egg. Several typical truncated eggs are shown in Fig. 148.

The formation of truncated eggs is presumably effected by any internal pressure on the egg during the early stages of shell formation in the uterus. Scott (*1940*) observed that when an egg is retained beyond the time it would normally be laid, the succeeding egg may become pressed against it and, if not yet invested with a hard shell, may thus acquire a flattened surface at the point of contact.

Constricted Eggs. An egg sometimes becomes constricted at one end during formation and therefore appears to consist of two separate

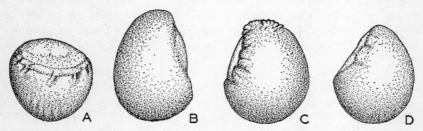

FIG. 148. Various forms of truncated eggs. (*A*, after Anonymous, 1937b; *B* and *D*, drawn from specimens in the authors' collection; and *C*, after von Nathusius, 1869.)

portions, a main body, which is nearly normal in size and shape, and a smaller portion, which seems to have been pinched from one end of the main part (Fig. 149). In some instances, the extension is in line

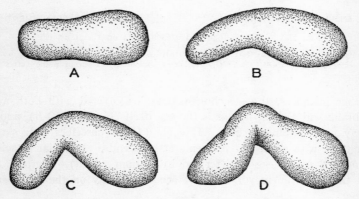

FIG. 149. Hens' eggs, constricted and bent to various degrees.

A, constricted, but not bent (after Landois, 1878). *B*, constricted, slightly bent (after Panum, 1860). *C*, constricted, markedly bent (after Hargitt, 1912). *D*, constricted, bent, and twisted (after Chidester, 1915).

with the principal part (*Landois, 1878*). In other cases, the smaller portion is at an angle to the larger (*Panum, 1860; Kunstler, 1907; Hargitt, 1912*), and sometimes the two parts appear to have been twisted after the angle formed between them (*Chidester, 1915*).

A constricted egg probably results when some of the egg materials are displaced from the main body of the egg. This displacement may be the result of a constriction of the oviduct (*Chidester, 1915*) or of unusual muscular activity of the oviduct (*Asmundson, 1931e*).

Eggs with Attachments. Some abnormally shaped eggs have attachments at one end, or both ends. Such attachments are of several types. In some cases, a series of membranes may be joined to an otherwise normal, hard-shelled egg, as shown in Fig. 150-*A* (*Jakob, 1916*).

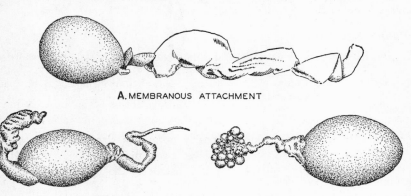

A. MEMBRANOUS ATTACHMENT

B. SHELLED ATTACHMENTS   C. TISSUE ATTACHMENT

FIG. 150. Various forms of attachments on the hen's egg.

A, membranous attachment (after Jakob, 1916). *B*, hard-shelled attachments at both ends of the egg (after Davaine, 1860). *C*, attachment composed of ovarian tissue (after Parker and Kempster, 1940).

In other instances, the attachments may be covered with both a shell membrane and a shell, just as is the egg itself (cf. Fig. 150-*B*), and may or may not contain albumen. Davaine (*1860*) describes 2 eggs of the latter type, with shelled attachments at both ends. A similar attachment on one end of an egg is described by Panum (*1860*). Curtis (*1916*) notes that attachments of this sort occur generally on the blunt end of the egg.

Eggs with long stalks, or attachments, may be laid subsequent to surgery. After undergoing two operations for the insertion of artificial "yolks," a hen laid an egg 5.8 cm. long, with an attachment that was 6.8 cm. long and covered with shell membrane and a small amount of shell (*Asmundson, 1931e*).

Attachments are apparently caused by the displacement of a portion of one component part of the egg, rather than by a change in the shape of all component parts, as in constricted eggs. Part of the albumen may be displaced if the oviduct is constricted or if the oviducal muscles function improperly (*Curtis, 1916; Asmundson, 1931e*).

Sometimes the attachments are composed of tissue, rather than membranous or calcareous egg materials (cf. Fig. 150-*C*). In one case, a portion of the ovary, containing approximately twenty-five ova, was attached to a normal egg by a stalk and a large mass of clotted blood. No albumen or shell had been deposited over the tissues. Tissue attachments are the result of an extremely pathological condition, such as an ovarian hemorrhage (*Parker and Kempster, 1940*).

Bound Eggs

Another type of abnormality, of which there are a number of recorded cases, consists of 2 eggs, of equal or different size, bound together by a narrow connection; this connection may be membranous or covered with a shell (*Baer, 1845; Landois, 1878*). Various forms of bound eggs have been observed (Fig. 151). Sometimes the connection is between the sharp ends of the eggs; in one such instance, both eggs were soft-shelled and had only a thin deposit of calcareous material over the entire shell membrane (*Hargitt, 1899*). The total length of these 2 eggs and their connecting portion was 11.3 cm. An egg weighing 39.7 gm., bound to another weighing 28.2 gm., was described by Dürigen (*1922, p. 299*). Another bound egg, composed of 2 eggs of different sizes, is reported by Weimer (*1918*).

Two or more eggs may be bound together directly, without a connecting segment (cf. Fig. 151-*D*). In such instances, 1 egg appears to have settled as a cap over the other; or there may be only a slight constriction between them (*Landois, 1878*).

Several groups of 3 small, hard-shelled eggs, attached to each other in various ways, were reported by Landois (*1878*) to have been laid by a very old hen. In each case, 1 egg of the group was much larger than the others (Fig. 152). The eggs were either clustered about a point (cf. Fig. 152-*A*), arranged in a row (cf. Fig. 152-*B*), or irregularly attached (cf. Fig. 152-*C*).

Bound ducks' eggs have been described (*Kummerlöwe, 1931*); but among wild birds, the production of bound eggs appears to be a very great rarity, since no eggs of this type have been reported.

Bound eggs are difficult to explain. Connections between only 2 eggs possibly result when the eggs are so close together in the oviduct that the glands are stimulated to secrete albumen into the lumen between the eggs. Membranes, followed by shell, are deposited not only on the eggs but also on the connecting strand of albumen (*Hargitt, 1899*). When

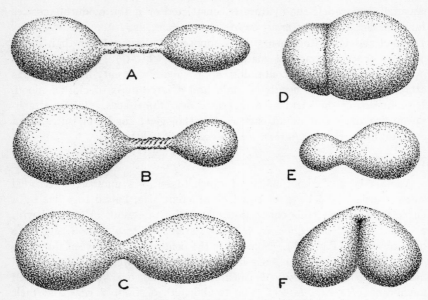

FIG. 151. Various forms of bound eggs (double eggs).

A and *B*, bound eggs with long connections (after Landois, 1878, and Baer, 1845, respectively). *C* and *E*, bound eggs with short connections (after Weimer, 1918, and Kummerlöwe, 1931, respectively). *D*, 2 eggs overlapping each other (after Landois, 1878). *F*, bound eggs with short, bent connection (after Landois, 1878).

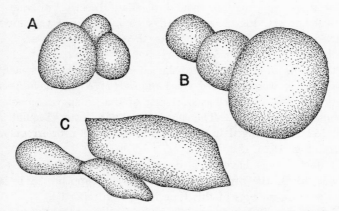

FIG. 152. Bound eggs in groups of three (triple eggs); natural size. (After Landois, 1878.)

A, eggs arranged about a central point. *B*, eggs connected in a row. *C*, eggs irregularly attached.

3 small eggs are bound together, they probably form separately and accumulate in the uterus before the shell is deposited. There they come in contact with each other in random arrangement, which becomes fixed as the shell material hardens.

Freak Eggs

Occasionally, hens produce amazing monstrosities, so malformed that the term "egg" is of questionable fitness, in spite of the presence of all

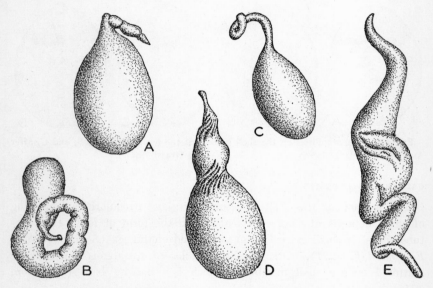

FIG. 153. Unusual malformations of the hen's egg. (*A*, after Brown, 1910; *B*, after Roux, 1923; and *C*, *D*, and *E*, after Landois, 1878.)

the egg components, including the shell. Such freaks may have long, twisted appendages (Fig. 153). Several abnormal eggs of this type are included in a collection made by Landois (*1878*). An egg that tapers to an elongated coil is described by Roux (*1923*).

SHELL SURFACE AND STRUCTURE

Surface Irregularities

Not uncommonly, the shell of an otherwise normal egg presents unusual features. It may appear wrinkled or ridged, sometimes over its entire surface, sometimes over only a portion of its surface. The ridges may follow a spiral course, or cross and recross in every direction (Fig. 154), or be arranged in various patterns. Occasionally, a single ridge extends around the equatorial region of the egg, so that the two halves of the shell appear to overlap. In other instances, a portion of the shell may be missing, and the shell membrane partially exposed. In extreme cases, the entire shell may be absent.

Ridges on the shell are not peculiar to the chicken egg, since a partridge egg with spiral ridges has been reported (*M'William, 1927*).

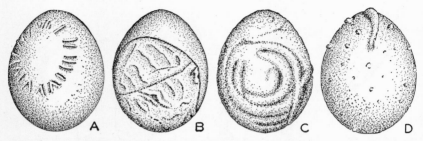

FIG. 154. Abnormalities on the shell surface of the hen's egg.   (*A, B,* and *C,* after Anonymous, 1937a, 1937b; *D,* after Parona and Grassi, 1877.)

CALCAREOUS DEPOSITS

The surface of the shell sometimes displays irregularities consisting of small deposits of calcareous material in the form of folds, tiny protuberances, or small extensions at either end of the egg (*Kunstler, 1907; Ferdinandoff, 1931*). Occasionally, nodules of calcareous material are scattered over a shell membrane which is otherwise bare (*Patterson, 1911*).

An egg in the authors' collection, illustrated in Fig. 155, bears superficial collections of calcareous material on its shell. When viewed under the microscope, these patches resolve themselves into individual granules, some nearly round, others elongated, which bear a striking resemblance to pebbles lying on a seashore. The granules are easily detached from the surface of the egg.

Von Nathusius (*1869*) described a somewhat similar condition in a turkey egg (Fig. 156). In this egg, many nodules of calcareous material, containing the same proportions of lime and organic matter as the shell itself, are present on the shell surface, embedded in the cuticle. Microscopically, these deposits are of various forms: in some, there are concentric layers; in others, two separate deposits overlap; and in many, the overlapping arrangement of layers is so complex that the actual

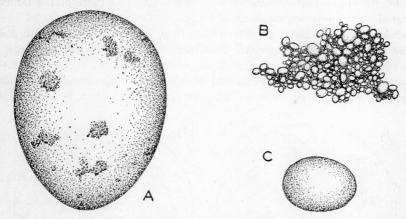

FIG. 155. Deposits of calcareous granules over the surface of a hen's egg. (Drawn from the original specimen.)

A, general view of the egg, showing aggregations of calcareous granules. *B,* a single aggregation of granules; magnified about 10 times. *C,* usual appearance of a large granule about 0.5 mm. long; magnified about 50 times.

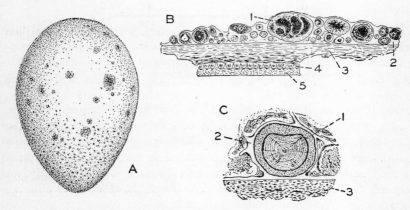

FIG. 156. Abnormal aggregations of granules on the surface of a turkey's egg. (After von Nathusius, 1869.)

A, general view of eggshell, showing granular aggregations. *B,* cross section through a large group of granules; magnified about 25 times. *C,* cross section through a small granule; magnified about 150 times.

The numbers indicate: 1, individual granules; 2, cuticle in which the granules are embedded; 3, spongy layer of eggshell; 4, mammillary layer of eggshell; 5, shell membrane.

structure of the mass cannot be discovered. Around most of the larger granules are numerous smaller nodules.

Sometimes the shell, complete with cuticle, is covered by an additional thin layer of calcific material. This rare condition is illustrated in Fig. 157-*A*. A lightly pigmented shell of average thickness, 0.31 mm. (*Romanoff, 1929a*), is entirely covered with a deeply pigmented layer 0.05 mm. thick. This egg was probably retained by the hen for a consid-

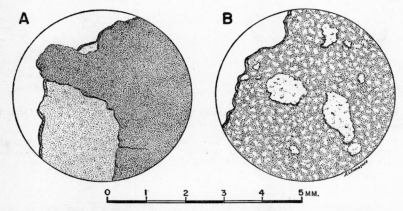

FIG. 157. Calcareous deposits on the surface of otherwise normal shells of hens' eggs. (Drawn from specimens in the authors' collection.)

A, uniform, darkly pigmented calcareous layer, deposited about a well-formed, lightly pigmented shell, complete with cuticle. *B*, small islands of unpigmented calcareous material, attached firmly to the surface of a dark brown eggshell.

erable time after the completion of the first shell (*Steggerda and Hollander, 1944*). The pigment glands, which are stimulated slowly (see Chapter 4, "Formation"), expended little pigment during its formation and did not exhaust their pigment stores. The thin secondary deposition of shell material was made before the calcium in the hen's body was restored to a normal level; the pigment glands, however, were still in a secretory condition and yielded their residual pigment.

Islands of calcareous material, irregular in size, shape, and distribution, may be attached firmly to the surface of an otherwise normal shell. Such a case is shown in Fig. 157-*B*. The shell, 0.29 mm. thick (slightly thinner than average), is deeply pigmented. The islands, only 0.02 mm. thick, are unpigmented. They probably represent the delayed secretion of isolated groups of calcium glands that had not previously contributed to the shell. The pigment glands were depleted, having just furnished a large amount of pigment, and failed to contribute to the islands.

In some cases, unpigmented islands may be so numerous as to form a more or less continuous layer, which conceals most or all of the normal pigmentation (*Steggerda and Hollander, 1944*).

"Glassy" Texture. Some eggshells are thin and fragile and emit a musical tinkle when tapped. Before the candle, they present a mottled appearance, because of the collection of excessive amounts of moisture in the organic matter that lies in the pores. Some hens have a greater tendency than others to produce "glassy" shells (*Almquist and Burmester, 1934*).

PROMINENCES DUE TO DEFECTIVE SHELL STRUCTURE

The surface of the hen's eggshell may be granular in nature, either because of bulges in the shell itself, or because of isolated granular masses that project above the surrounding shell surface. In either case,

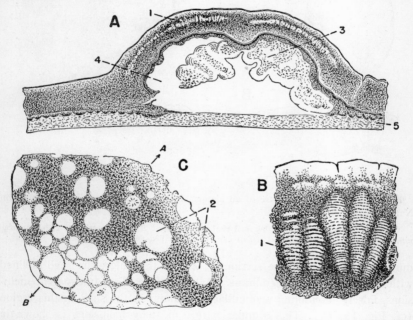

FIG. 158. Structure of an area of abnormal shell on the hen's egg. (After von Nathusius, 1868.)

A, radial section through a bulge on the eggshell; magnified about 40 times. B, radial section through a transparent layer of calcareous inclusions in the bulge; magnified about 300 times. C, tangential section through the edge of the transparent layer, showing the nature of the granular inclusions; magnified about 190 times. The axis of the section is inclined so that A is toward the surface and B is deeper in the shell.

The numbers indicate: 1, transparent calcareous layer; 2, transparent granules; 3, albumen; 4, space within bulge; 5, shell membrane.

the local structure of the shell is abnormal. The former type of abnormality is illustrated in Fig. 158, where a bulge, somewhat less than 2.0 mm. in diameter and 0.5 mm. in height, is shown to have formed over a small mass of albumen adhering to the shell membrane. (The space shown within the bulge is due to shrinkage of the mass of albumen

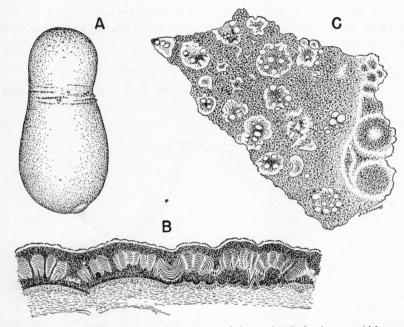

FIG. 159. The shell structure of an abnormal (constricted) hen's egg. (After von Nathusius, 1869.)

A, external appearance of the abnormal egg; natural size. *B,* radial section of eggshell; magnified 50 times. *C,* horizontal section of eggshell; magnified 50 times.

during preparation of the section.) Within the bulge, the shell is lined with a very thin shell membrane. A layer of peculiar, transparent inclusions, falsely resembling mammillae, are found in the shell in this region (cf. Fig. 158-*B*). The granular nature of these inclusions is shown in Fig. 158-*C*.

The same type of anomalous shell structure is the cause of the roughened surface of the hen's egg shown in Fig. 159. This egg is an abnormally small one, constricted and wrinkled about the equator. The shell surface at the pointed end is granular (cf. Fig. 159-*A*) and rough (cf. Fig. 159-*B*). As in the egg described above, the shell contains a layer of inclusions. A horizontal section through the most superficial part of the shell reveals that these inclusions, again, are spherical and transparent granules (cf. Fig. 159-*C*).

"CHECKS"

The eggshell may become cracked during the course of its deposition in the uterus of the hen. The cracks may then be sealed by the secretion of additional calcific material, although visible grooves and ridges mark the lines of fracture. The shell is obviously weakened.

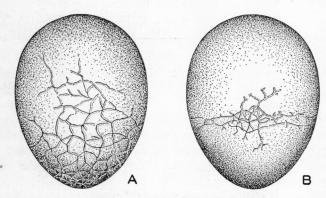

FIG. 160. Most common forms of body-checked eggs laid by hens. (Drawn from the authors' specimens.)

A, check at sharp end of the egg. *B*, check in the equatorial region of the egg.

"Checks," or "body checks," may occur in any area of the shell (Fig. 160), but they are most frequent at the ends, especially the sharp end. Checks in the equatorial region sometimes completely encircle the egg.

The cause of checked shells is not definitely known. It has been suggested that they are formed because the bird becomes unduly excited or is handled roughly.

Shell-less Eggs

Shell-less or soft-shelled eggs, which are covered with little or no shell substance, are quite frequently produced by domestic birds, especially during periods of heavy production (*Hewitt, 1939*). Shell-less eggs have also been obtained experimentally after resection of a portion of the uterus (*Asmundson, 1933b*), or as a result of feeding sulfa drugs (Fig. 161), especially sulfanilamide (*Genest and Bernard, 1945*).

Shell-less eggs may be otherwise normal; however, they are frequently abnormal, both in size and shape. Dwarf eggs are sometimes shell-less

(*Pearl and Curtis, 1916*).   Many of the bound, or connected, eggs are also without shells (*Hargitt, 1899*).

Shell-less eggs are known to have been laid not only by hens, but also by pigeons (*Levi, 1941, p. 176*).

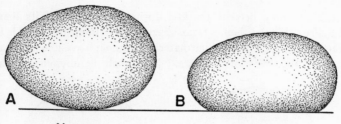

NORMAL   SHELL-LESS

FIG. 161.   Appearance of a shell-less egg when placed on a flat surface.   *A*, normal egg.   *B*, shell-less egg, laid by a hen that had been given sulfanilamide.   (After Benesch, Barron, and Mawson, 1944.)

Eggs with insufficient or incomplete shells are no doubt of fairly common occurrence, but they are so fragile as to be destroyed before they are discovered.   Microscopic views of an abnormally thin shell from a turkey egg are shown in Fig. 162.   The formation of the shell was obviously interrupted before even the first layer was completed.   The shell membrane is visible in both views (cf. Fig. 162-*A* and *B*) between the knobs, or mammillae, which are in the early stages of formation.   The

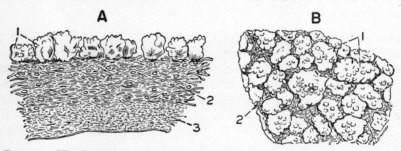

FIG. 162.   Histological structure of a thin-shelled turkey egg; magnified about 100 times.   (After von Nathusius, 1869.)

A, radial section.   *B*, outer surface view.

The numbers indicate: 1, rudimentary knobs of the mammillary layer; 2, outer shell membrane; 3, inner egg membrane.

egg was expelled prematurely, possibly because the bird became disturbed, or perhaps because of some irritation in the oviduct.

At times, birds lay naked yolks, without albumen, shell membrane, or shell.   Hewitt (*1931*) reports such an abortion.   Hutt (*1939*) describes a fully developed ovum that was laid, still in its unruptured ovarian follicle, without a trace of albumen or shell.

Causes. The immediate cause of shell-less eggs is either a failure of the glands in the shell-secreting portion of the oviduct, or violent peristalsis which hurries the egg through this region before a shell can be formed (*Hewitt, 1939*). Experiments by Buckner, Martin, Pierce, and Peter (*1922*) show that calcium starvation is not necessarily a causative factor in the production of shell-less eggs, as was once thought.

PIGMENTATION

As a rule, the majority of hens' eggs are either white or brown in color, with the exception of a few which are greenish blue. Although

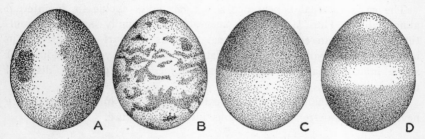

Fig. 163. Abnormalities in the distribution of pigment on brown-shelled chicken eggs. (*B* and *C*, after Anonymous, 1937b and 1937a, respectively; *A* and *D* drawn from specimens in the authors' collection.)

eggs may be of various color intensities, the entire surface of each egg is normally of a uniform shade. However, sometimes the pigment is distributed unevenly (Fig. 163). In some eggs, spots of color may be scattered at random over the surface. In others, there may be sharply demarcated areas of heavy pigmentation, so that part of the egg is darker than the remainder. Usually the regions of different color intensity encircle the egg as bands, but occasionally their direction is longitudinal.

Depigmented regions may occur on otherwise uniformly colored hens' eggs. The pigment seems to be extremely superficial and easily rubbed off (*Steggerda and Hollander, 1944*).

Failure to secrete the pigment spots that characterize the eggs of certain wild birds is more frequent than the occurrence of speckles on eggs that are normally unadorned (*Ingersoll, 1910*). However, Carpenter (*1907*) found some spotted eggs among those of such birds as the green-backed goldfinch and the lazuli bunting, both of which ordinarily lay eggs of solid color. M'William (*1927*) has observed elaborate flourishes of color on the eggs of gulls and oyster catchers, although these eggs are normally marked with spots, rather than streaks.

INTERNAL ABNORMALITIES

A great many abnormalities of the egg involve more than the egg's appearance. Indeed, many defects are not apparent until the egg is opened and its contents closely examined.

Internal anomalies of the egg may affect the size, structure, or arrangement of the egg's component parts, in a great variety of ways. In addition, eggs that are structurally perfect may contain inclusions or foreign bodies not normally present.

STRUCTURE

The egg's external appearance sometimes gives a clue to an internal abnormality. Thus, a large egg suggests an unusually large yolk, multiple yolks, or an egg within an egg, just as a dwarf egg suggests an unusually small yolk or the complete absence of a yolk. Oddities of shape indicate that the egg's internal elements are probably disposed unnaturally.

On the other hand, some eggs, externally normal in all respects, are nevertheless of extremely anomalous internal structure.

Size of Yolk

Great variation in the size of the whole egg, from dwarf to giant, has been pointed out previously. Examination of the interior of the egg shows that variation in the size of the yolk does not cover nearly so wide a range.

As might be expected, dwarf eggs are apt to have yolks considerably smaller than the normal average of 18.7 gm. (see Chapter 3, "Structure"). It is common enough to find small eggs with 7- and 8-gm. yolks; and, in one series of 30 dwarf eggs, the average yolk weight was found to be 10.3 gm. (*Romanoff, unpublished*). An extremely small but complete yolk, found in a dwarf egg (*Pearl and Curtis, 1916*), weighed only 4.3 gm., or 77 per cent less than normal.

Abnormally large yolks are relatively rare. Yolks with more than one blastodisc are the largest on record. One such yolk weighed 30.1 gm., or nearly 70 per cent more than normal (*Curtis, 1915a*).

Multiple Yolks

The occurrence of two yolks in the hen's egg is a common phenomenon. The presence of more than two yolks is rare, and there is no record of any egg with more than three complete yolks.

An increase in the number of yolks is usually associated with an increase in the total weight of the egg (Fig. 164). Pearl (*1910*) demonstrated that yolkless eggs are generally the smallest, and triple-yolked eggs, the largest.

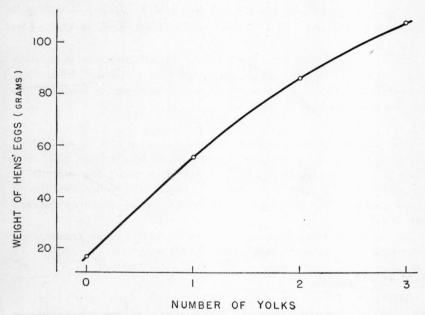

Fig. 164. Relationship between the weight of the egg and the number of yolks contained in the egg. (After Pearl, 1910.)

DOUBLE YOLKS

Of all the abnormalities in eggs, the most common is the phenomenon of two yolks. Eggs containing two yolks generally deviate slightly from normal in size and form; they tend to be longer, larger, and proportionally narrower than eggs with only one yolk. The two ends of double-yolked eggs are apt to be nearly the same size and rounded rather than pointed; often there is a definite, depressed ring around the middle of the shell (*Broca, 1861; Parker, 1906; Curtis, 1914c*). Double-yolked eggs contain all the component parts of normal eggs in greater absolute amounts (Fig. 165). The increase in the weight of the albumen, however, is not equal to that in the total yolk weight; and the shell suffers a proportional loss (*Curtis, 1914c; Asmundson, 1931e*).

The phenomenon of two yolks in one egg has attracted popular and scientific attention since the time of Aristotle, who stated that twin chicks hatched from eggs with double yolks. Comments on the occurrence of double-yolked eggs were particularly frequent during the last century. Dugès (*1839*) reported 3 double-yolked eggs laid on successive days by the same hen. Rayer (*1849*) described a large, 108-gm. egg containing two yolks connected by a small, chalaza-like ligament. Harvie-Brown (*1868*) mentions another hen that laid 3 double-yolked eggs in 6 days and then laid 2 single-yolked eggs the size of ducks' eggs. Féré

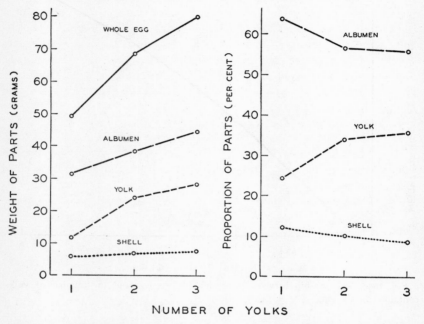

Fɪɢ. 165. Relationship between the number of yolks in the egg and the actual weights and proportional amounts of the egg's component parts. (After Curtis, 1914c.)

(*1898a*) found that by drugging hens with atropine sulfate he could force them to lay double-yolked eggs.

Descriptions in recent years, more complete and accurate than those of the past, have supplied full details of the internal structure of double-yolked eggs. In addition, attempts have been made to ascertain the factors responsible for the formation of eggs with two yolks.

Not all double-yolked eggs are exactly alike in their internal structure; and, in fact, eggs with two yolks are of three distinct types (*Curtis, 1915a*). In eggs of Type I, the yolks have separate vitelline membranes, but a common set of surrounding structures (cf. Fig. 167-*A*). In eggs of Type II, each yolk has a separate vitelline membrane and a separate inner chalaziferous layer of albumen, but the two yolks share the chalazae, albuminous sac, membranes, and shell (cf. Fig. 167-*B*). In eggs of Type III, there is a separate, complete set of internal structures around each yolk, and only the membranes and the shell are shared in common (cf. Fig. 167-*C*). Type II double-yolked eggs occur most frequently, and in this class belong the majority of eggs with two yolks described by various observers.

Examples of double-yolked eggs of Types I and III have been noted in the Philippine Islands. A Cantonese-native hen laid a 70-gm. double-yolked egg of Type III (*Fronda, 1924*), in which the albuminous sacs were entirely separate and there was a single set of shell membranes. Sumulong (*1925*) found 2 eggs of Type I weighing 66.4 gm. and 64.5 gm. respectively. A number of Type I double-yolked eggs are also described by Asmundson (*1931e*).

In a double-yolked egg, each yolk usually weighs slightly less than the average yolk in a normal egg laid by the same hen (*Warren and Scott, 1935a; Romanoff, unpublished*). However, the two yolks of double-yolked eggs are not necessarily of the same size. In 11 double-yolked eggs examined by the authors, and in 25 examined by Warren and Scott (*1935a*), the average weights of the two yolks were as tabulated. Some-

	WEIGHT (grams)	
	AUTHORS' DATA	WARREN AND SCOTT (*1935a*)
Larger yolk	16.5	11.9
Smaller yolk	15.7	11.4
Difference	0.8	0.5

times the difference in the size of the two yolks is much more pronounced than it was in these eggs. In a double-yolked egg described by Parker (*1906*), one of the yolks was approximately spheroidal in shape and measured 3.4 cm. by 3.0 cm.; the other yolk was slightly flattened on one side and measured only 2.7 cm. by 2.1 cm. A similar disparity in the size of the two yolks has been noted by the authors (Fig. 166).

Although double-yolked eggs are usually large, Curtis (*1915a*) reports an interesting double-yolked egg that weighed only about 20 gm. and contained two small yolks, one much smaller than the other. The two yolks had separate vitelline membranes which had fused at the point of contact, so that there was a communication between the two yolk cavities.

Pullets coming into production lay the largest proportion of double-yolked chicken eggs. Over 53 per cent of all double-yolked eggs occur during the first 60 days of production (*Conrad and Warren, 1940*). In-

vestigators, from time to time, have estimated the rate at which eggs
with double yolks may be expected to appear. Féré (*1899*) stated that
the phenomenon occurred once in every 2000 eggs, but Curtis (*1915a*)
found 1 egg of this type in every 531 eggs produced.

Double yolks are not limited to the eggs of the domestic hen. Ducks
are especially apt to lay double-yolked eggs, which deviate from the
normal shape much as do double-yolked chicken eggs and are proportionately oversized (*Sumulong, 1925*). One excessively large double-
yolked duck egg weighed over 113 gm., as compared with the normal

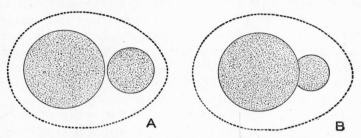

Fig. 166. Yolks of unequal size in double-yolked eggs.

A, yolks are well separated (drawn from a photograph in the authors' collection). *B,* yolks
are fused (after Curtis, 1915a).

average of 67 gm. (*Bauer, 1895*). The yolks of this egg together weighed
43 gm., the albumen 59 gm., and the shell 11 gm. Among pigeons, certain
individual birds apparently tend to lay double-yolked eggs year after
year (*Levi, 1941, p. 176*). Pigeon eggs with two yolks are usually about
twice the size of normal pigeon eggs and are equally rounded at both
ends. Double yolks have also been found in goose eggs (*Panum, 1860;
von Nathusius, 1895*) and in the eggs of the swan, the common sparrow,
and the lark (*Valenciennes, 1856*).

Twin chicks and twin pigeons have been observed to hatch from
double-yolked eggs. In one instance, nine sets of twin chicks were
obtained from the double-yolked eggs laid by a single hen (*Bernard,
1850*).

Sometimes a single-yolked egg may falsely appear to have two yolks,
as the result of the rupture of an extremely weak vitelline membrane.
It is possible, also, for a single yolk to become constricted and take on
a double appearance, probably because of a constriction in the oviduct
(*Davaine, 1860; Parker, 1906*).

Causes. Various theories have been advanced to explain the inclusion
of two yolks in the same egg and the differences between the three dis-

tinct types of double-yolked eggs. There is considerable evidence that certain individuals tend to lay eggs with double yolks more frequently than others, but it has not been demonstrated that the propensity is inherited.

The internal structure of the double-yolked egg varies according to the level in the oviduct at which the two yolks come together (Fig. 167).

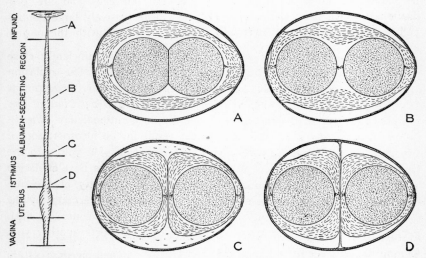

Fɪɢ. 167. Various forms of double-yolked chicken eggs, and the probable meeting place of the yolks in the oviduct.

A, Type I double-yolked egg, in which the yolks are in close contact (after Curtis, 1915a). *B,* Type II double-yolked egg, in which each yolk has its own chalaziferous layer of albumen (after Curtis, 1915a). *C,* Type III double-yolked egg, in which the yolks have separate albuminous sacs (after Curtis, 1915a). *D,* a type of doubling in which the yolks are separated by shell membranes; actually, it is a double egg (after von Nathusius, 1869).

Obviously, the yolks must meet before reaching the isthmus, if they are to be included in the same shell. If they are in contact while in the region of the infundibulum, the egg will be Type I; but if the meeting occurs in the upper part of the albumen-secreting region, Type II will result. Type III is formed when the first yolk reaches the isthmus before being joined by the second yolk. Sometimes the two yolks remain separated until the posterior portion of the isthmus is reached and each has acquired a complete set of accessory structures, except for the shell. The result is a double egg within a single shell, rather than a double-yolked egg (cf. Fig. 167-*D*).

The chances that two yolks might meet in the longest section of the oviduct, the albumen-secreting region, are obviously the greatest. As may therefore be anticipated, eggs of Type II predominate (*Curtis, 1915a*). The frequency with which yolks may be expected to meet in the various regions of the oviduct compares favorably with the observed incidence of each type, which is as follows.

Type	Frequency (per cent)
I	16
II	71
III	13

There are several reasons why two yolks can be in sufficiently close proximity to be enclosed in a single set of egg envelopes. The ovulation of two ova at one time, due either to their simultaneous development (*Parker, 1906; Little, 1924; Warren and Scott, 1935a; Conrad and Warren, 1940*) or to the retention of one for an additional day (*Curtis, 1915a; Conrad and Warren, 1940*), might result in the incorporation of two yolks in the same egg. So, too, might ovulation of a yolk into the body cavity, and its delayed recovery at the time of the next ovulation (*Curtis, 1915a*). Then again, an abnormally low tonicity of the oviducal muscles might delay one yolk in its passage until overtaken by the next yolk. Another possibility is the reversal of a yolk's course by antiperistalsis, so that a yolk traveling up the oviduct meets another yolk on its way down (*Curtis, 1915a; Asmundson, 1931e*).

Experiments have been performed in recent years to test the validity of these explanations. Warren and Scott (*1935a*) observed that the yolks of double-yolked eggs are often almost equal in weight, and that the weight of each yolk in a double-yolked egg is fairly comparable to that of the yolk in the normal egg laid by the same hen. They therefore concluded that two yolks can be ovulated simultaneously. Conrad and Warren (*1940*) determined statistically how many of the pauses occurring before or after the production of double-yolked eggs might be considered as signifying the delay of a yolk in its passage. Their results indicate that 65 per cent of double-yolked eggs are due to simultaneous development of two ova, 25 per cent to premature ovulation of one of the two ova, and 10 per cent to retention of one yolk in the body cavity. These figures are supported by an additional study of 11 double-yolked eggs dyed by successive injections of Sudan III into the hen's blood stream, so that the relative age of each yolk was shown by dyed layers corresponding to the location of the follicular wall at the time of each injection. However, 1 of these 11 eggs suggested the additional possibility that occasionally very rapid ovulation may occur, with the production of an extra ovum over a period of a few days.

In one instance, autopsy of a hen that habitually laid double-yolked eggs revealed a singular condition of the ovary. A number of follicles were attached to the main body of the ovary by elongated "suspensoria." Many follicles appeared to be compound, and some were observed in the process of budding (Fig. 168). Weakness of the tunica albuginea appar-

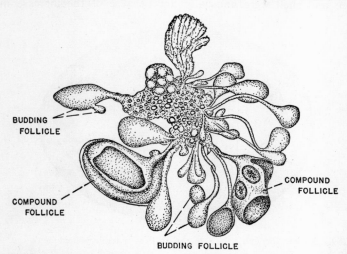

FIG. 168. Abnormal ovary of a hen that laid double-yolked eggs. (After Glaser, 1913.) The follicles are attached by suspensoria, some of which are more than 3.5 cm. in length. In addition, there are several budding and compound follicles.

ently permitted the local outgrowth of the ovarian stroma. The establishment of identical blood supply possibly led to the simultaneous development of the ova in the double follicles (*Glaser, 1913*).

TRIPLE YOLKS

Eggs containing three yolks are extremely rare. A typical example is shown in Fig. 169. In 6 years, only 3 triple-yolked eggs were found at the Maine Agricultural Experiment Station (*Curtis, 1914c*); on this basis, it was estimated that only one bird in 1000 ever lays a triple-yolked egg. Perhaps 1 egg in every 5000 or 6000 has three yolks. Triple-yolked eggs are almost always produced by pullets less than 6 months old.

Valenciennes (*1856*) described several eggs, normal in shape and size, that contained three yolks, each in a separate vitelline membrane. He stated that triple-yolked eggs were found only five or six times a year in the Paris market, where, at that time, more than 140,000,000 eggs were handled annually.

Pearl (*1910*) made note of a large (87.1 gm.) but not abnormally shaped egg with three yolks. These yolks were in contact with each other but had separate vitelline membranes. Although this egg contained normal amounts of thick and thin albumen, the proportions of albumen and shell were unusually low because of the increase in yolk

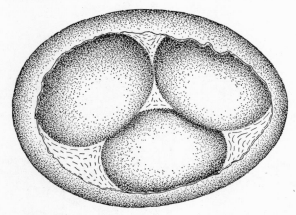

Fig. 169. A view through a hole in the shell of a triple-yolked hen's egg. (After Pearl, 1910.)

weight. The egg was produced by a pullet that had previously laid only 4 eggs, 3 normal eggs and 1 soft-shelled. After laying the triple-yolked egg, the pullet paused for 2 days and thereafter produced normal eggs.

Two other triple-yolked eggs (*Curtis, 1914c*) were not excessively large; they weighed 79.6 and 77.4 gm. respectively. Another (*Clark, 1932*) weighed no more than a normal egg, but possessed one-third more yolk. This egg was laid by a bird that had been vaccinated against fowl pox 6 days previously; possibly some physiological disturbance, caused by vaccination, contributed to abnormal egg formation.

Only once has it been reported that a hen laid more than 1 triple-yolked egg (*Clinton, 1927*).

Triple-yolked eggs are probably formed in much the same way as double-yolked eggs. It is possible that triple yolks, usually found in the eggs of young pullets, are often the result of delay in establishing regularity in the rhythm of ovulation (*Pearl, 1910*).

Multiple Blastodiscs

Very rarely, a single yolk possesses two blastoderms. The development of twins from single-yolked eggs, a very infrequent occurrence, is apparently the result of this phenomenon. Single-yolked pigeon eggs have been known to produce twins (*Levi, 1941, p. 176*).

Dareste (*1877, p. 10*) described double blastoderms. He noted that in some cases the blastoderms are very close together, even to the point of being in contact with each other, whereas in other cases they are widely separated on the surface of the yolk (Fig. 170). Double blasto-

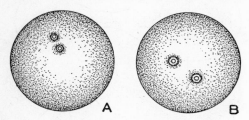

FIG. 170. Double blastodiscs.

A, in close proximity to each other (after Davaine, 1860). *B*, separated more widely (after Curtis, 1915a).

derms have been reported by many others (*Flourens, 1835; Thomson, 1844; Baer, 1845; Serres, 1860; Panum, 1860; Chatin, 1887; Nalbandov, 1942*). In 1 egg with two blastoderms, the observed proportions of the component parts were as follows (*Curtis, 1915a*). It is interesting

	AMOUNTS (per cent)
Yolk	36.6
Albumen	54.1
Shell	9.3

to note that, in relative amount, this unusually large yolk compares favorably to the total yolk mass in double-yolked eggs (*Curtis, 1915a; Romanoff, unpublished*); and, in fact, it was possibly formed by the fusion of two yolks, as indicated by the presence of a thin white line between the two blastodiscs. The occurrence of two blastodiscs on the same yolk, as the result of doubling of oöcytes within the ovary, appears to be extremely unusual in chicken eggs (*Curtis, 1915a*).

Structure of Constricted and Bound Eggs

Constricted and bound eggs are generally unusual in internal structure as well as external appearance (Fig. 171).

In origin, bound eggs appear closely related to double-yolked eggs (*Curtis, 1915a*), as shown in Fig. 171-*A*. In the instance illustrated, each of the 2 eggs contains a yolk. Although the eggs are of different sizes, the yolks are nearly equal and are connected by strands of chalazae extending through the connecting band. The difference in the size of these 2 eggs is due to their unequal content of albumen. Had the yolks been only slightly closer together in their course through the oviduct,

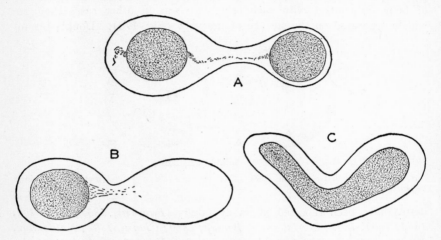

Fig. 171.   Internal structure of bound and constricted eggs.

A, bound eggs, both containing yolks (after Hargitt, 1899). *B*, constricted egg, containing a single yolk (after Weimer, 1918). *C*, constricted egg, containing a constricted yolk (after Chidester, 1915).

they would undoubtedly have become incorporated into a double-yolked egg.

Constricted eggs usually possess only one yolk, which may be confined to one part of the egg (*Panum, 1860; Weimer, 1918*), as shown in Fig. 171-*B*. Occasionally, however, the yolk is misshapen and elongated and conforms to the external contours of the egg (cf. Fig. 171-*C*).

Double Eggs

The presence of a more or less complete egg within another egg is comparatively rare. However, double eggs have attracted a great deal of attention since the time of Aristotle, and much has been written about the so-called *ovum in ovo*. Double eggs are distinct from multiple-yolked eggs in that the enclosed egg is surrounded by a shell, or at least a membrane, and usually contains some albumen.

Perhaps the earliest written description of an egg within an egg is that given by Albertus Magnus (*1250*), who mentions an egg with two shells. Other early writers who discussed double eggs were Harvey (*1651*), Perrault (*1666–1699*), and Elsholtii (*1675*). In 1877, Parona and Grassi described an abnormal egg of this type and made an extensive review of the literature on double eggs. Within a few years, additional descriptions and reviews appeared (*Schumacher, 1896; Supino, 1897; Hargitt, 1899, 1912; Herrick, 1899; Parker, 1906*).

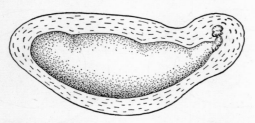

FIG. 172. Double egg, in which both enclosed and enclosing eggs show identically abnormal shape. (After Kunstler, 1907.)

Since eggs of this type are infrequent, a series of them is a great rarity. Simmonds (*1889*) quotes the following from the *Birmingham Mercury* of May 9, 1857: "A half-breed Cochin China hen belonging to Mr. Campbell of Darlaston laid eleven eggs of extraordinary size. Seven of them contained eggs enclosed in them. Around one weighing seven ounces a third shell had started to form."

Double eggs may appear in a variety of shapes and sizes. Externally, the egg may be either larger or smaller than normal; the enclosed egg is generally smaller than normal. Either egg may be shell-less. Double eggs are often abnormal in shape. In some, identical abnormalities in shape may appear in both the enclosed and the enclosing egg (Fig. 172).

The descriptions of double eggs reveal that four general types occur: (1) a normal and complete egg containing a second complete egg; (2) a complete egg containing a yolkless egg, which may or may not have a shell; (3) a yolkless egg containing a complete egg; and (4) a yolkless egg containing another yolkless egg (Fig. 173).

COMPLETE EGG WITHIN COMPLETE EGG

Complete eggs enclosed within complete eggs are relatively rare (cf. Fig. 173-*A*). Mention of complete double eggs was made by Jung (*1671*), Cleyer (*1682*), and Rivaliez (*1683*). Davaine (*1860*) reviewed

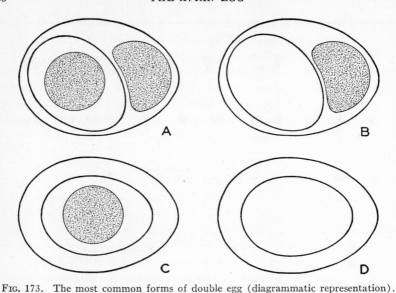

FIG. 173. The most common forms of double egg (diagrammatic representation).

A, a complete egg within another complete egg. *B,* a yolkless egg within a complete egg.
C, a complete egg within a yolkless egg. *D,* a yolkless egg within another yolkless egg.

thirty-eight instances of *ovum in ovo,* only four of which, however, were complete double eggs.

A typical example of a complete double egg is described by Patterson (*1911*). The egg was normal in shape, very thin-shelled, and only slightly above average size. Inside lay a yolk and a hard-shelled egg. Complete double eggs have also been described in recent years by Hargitt (*1912*), Curtis (*1916*), von Frankenberg (*1928*), and Asmundson (*1933b*).

Although some double eggs of this type are of only average size (*Brites, 1933*), most of them are abnormally large and sometimes weigh more than 200 gm. (*Henneguy, 1911*). The weights of the two double eggs reported by Roberts and Card (*1929*) are typical:

	WEIGHT (grams)	
	EGG 1	EGG 2
Entire egg	171	164
Enclosed egg	62	60

The production of no less than 10 complete double eggs by one hen (*Romanoff and Hutt, 1945*) within 3 months (Fig. 174) is indeed remarkable. None of these double eggs was preceded by an egg on the previous day, but 2 of them were followed by eggs on the next day.

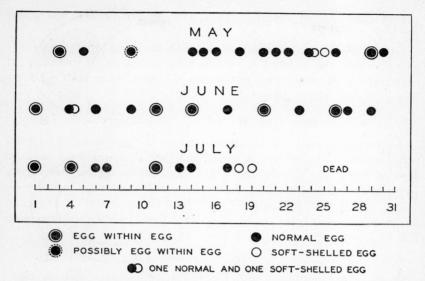

Fig. 174. Chart showing a White Leghorn hen's egg-laying record for 3 months, during which time she laid 10 double eggs. (After Romanoff and Hutt, 1945.)

Ordinarily, there was a 2-day interval both before and after each double egg. The tabulated data give the average weights and the proportional amounts of the component parts of both the double and single eggs laid by this hen.

	Proportions (per cent)
Double egg (182 gm.)	
Enclosing egg (107 gm.)	
Albumen	74
Yolk	18
Shell	8
Enclosed egg (75 gm.)	
Albumen	68
Yolk	25
Shell	7
Single egg (70 gm.)	
Albumen	66
Yolk	28
Shell	6

YOLKLESS EGG WITHIN COMPLETE EGG

Double eggs of the second type—a yolkless egg enclosed in a complete egg—are fairly numerous. To this class belong the double eggs described by such investigators as von Haller (*1768*), Rayer (*1849*), Herrick

(*1899*), Féré (*1902*), Parker (*1906*), O'Donaghue (*1911*), and others (cf. Fig. 173-*B*).

Curtis (*1916*) reports several yolkless eggs contained in larger complete eggs. In one, the enclosed egg was unusually complex in structure. It consisted of four concentric membranes separated from each other by layers of clear, thick albumen.

COMPLETE EGG WITHIN YOLKLESS EGG

A complete normal egg sometimes appears to have acquired an extra set of egg envelopes, without a second yolk, and thus represents the third type of double egg (cf. Fig. 173-*C*). Many double eggs of this variety have been described, first by Bartholin (*1661*) and more recently by Hargitt (*1899, 1912*), Kunstler (*1907*), and Bujard (*1917*). The tabulation gives the weights of the component parts in a typical specimen, one of several observed by Curtis (*1916*):

	YOLK (grams)	ALBUMEN (grams)	SHELL (grams)	TOTAL (grams)
Enclosing egg	..	57	8	65
Enclosed egg	23	6	..	29
Total weight	..	..	..	94

A double egg described by Justow (*1927*) also belongs to this class. Examination revealed that its weight was distributed as follows:

	PROPORTIONS (per cent)
Double egg (155 gm.)	
Enclosing egg (101 gm.)	
Albumen	90
Shell	10
Enclosed egg (54 gm.)	
Albumen	63
Yolk	26
Shell	11

A chicken laid 3 double eggs of this type after an operation on the uterus (*Asmundson, 1933b*). In all 3, the shell membranes were normal, but the shells were very thin and had rough, irregular surfaces.

YOLKLESS EGG WITHIN YOLKLESS EGG

The least common form of double egg is the fourth type, in which neither the enclosed nor the enclosing egg has a normal, complete yolk (cf. Fig. 173-*D*). Housset (*1785*) and Bujard (*1917*) comment on this anomaly. Curtis (*1916*) describes, in detail, 5 or 6 double eggs that may be placed in this class. In some specimens, no yolk at all was visible in either the enclosed or the enclosing egg; in others, however, small drops of yolk substance were found.

OTHER FORMS

More unusual forms of double eggs are illustrated in Fig. 175. There are known instances of an egg enclosing 2 eggs (*Curtis, 1916; Asmund-*

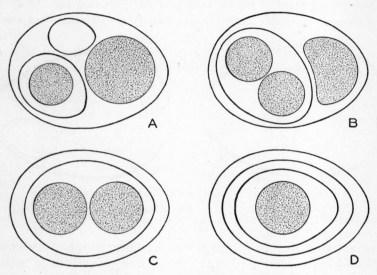

FIG. 175. Less frequent forms of egg multiplicities.

A, an egg with 2 enclosed eggs, one of which is yolkless. *B,* single-yolked egg enclosing double-yolked egg. *C,* yolkless egg enclosing double-yolked egg. *D,* an egg with three complete sets of envelopes.

son, 1933b). In some double eggs, the enclosing egg is double-yolked, and the enclosed egg may or may not contain a yolk (*Panum, 1860; Clinton, 1927; Brown, 1939*). An unusual egg described by Davaine (*1860*) had three complete sets of egg envelopes.

In one instance (Fig. 176-*A*) the enclosed egg was said to lie within the yolk of the enclosing egg (*Herrick, 1899*). Such an inclusion within the yolk would necessitate rupture of the vitelline membrane. The actual condition of the vitelline membrane was possibly not observed, since the egg had been hard-boiled before the abnormality was discovered. Another very odd egg was noted by Gruvel (*1902*); in this, a small egg was enclosed between the inner and outer layers of the shell membrane (cf. Fig. 176-*B*).

Partial Doubling. An egg may occasionally be incompletely double; that is, the enclosing egg may surround only a portion of the enclosed egg (*Curtis, 1916*). For example, the blunt end of the egg may be capped by a second egg. In 1 egg of this type, the enclosing cap was filled with albumen and possessed a long stalk, also filled with albumen, which was folded down along the side of the egg. Shell material covered both portions, which, externally, appeared to be a single unit.

Many descriptions of double eggs have been omitted here because of incomplete or unreliable data. Unfortunately, double eggs often are not

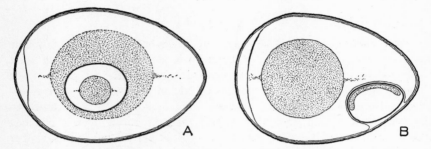

FIG. 176. Two cases of most unusual doubling in the hen's egg.

A, a small complete egg within yolk (after Herrick, 1899). For explanation see text. *B,* a small egg between the two layers of the shell membrane (after Gruvel, 1902).

discovered until they have been broken for domestic use. Sometimes a portion of the contents are lost before any observations can be made or the arrangement of the component parts noted accurately.

Domestic hens have produced most of the observed specimens of double eggs. However, it is known that double eggs have been laid by the jungle fowl (*Panum, 1860*), the duck (*Bonnet, 1883; von Frankenberg, 1928; Kummerlöwe, 1931*), the goose (*Rayer, 1849; Bonnet, 1883; Hahn, 1930*), the turkey (*Panum, 1860; Hargitt, 1899*), the pigeon (*Levi, 1941, p. 176*), the red-backed shrike (*Landois, 1892*), the ostrich (*Hargitt, 1899*), and the emu (*Clinton, 1927*).

AXIAL RELATIONSHIPS IN DOUBLE EGGS

It is of interest to note the axial relationships in double eggs. In many instances, it is impossible to distinguish the blunt from the pointed end of the outer egg. When the two ends are differentiated, the orientation of the inner egg is usually the same as that of the outer (*Chobaut, 1897; Parker, 1906; Curtis, 1916*). Sometimes, however, the long axis of the enclosed egg lies at an oblique angle to the long axis of the enclosing egg

(*Patterson, 1911; Romanoff and Hutt, 1945*). In compound eggs with 2 contained eggs, the long axes of the enclosed eggs are occasionally parallel to each other; such a relationship indicates that the 2 eggs were side by side when they passed through the isthmus (*Curtis, 1916*).

When a yolk is present in the enclosing egg, its position may vary. Parker (*1906*), Patterson (*1911*), and Curtis (*1916*) found that the yolk lay in the blunt end of the large outer egg, and the enclosed egg toward the sharp end. Exactly the opposite orientation was observed by Pick (*1911*) in 1 double egg, by Brites (*1933*) in another, and by Romanoff and Hutt (*1945*) in no fewer than 4 eggs.

SEASONAL OCCURRENCE OF DOUBLE EGGS

Since there are records of only a relatively small number of double eggs, it would be unwise to make definite statements concerning their seasonal occurrence. It may be significant, however, that most of the double eggs described in the past were laid in the winter or spring, between December and April (*Parker, 1906; Patterson, 1911*); those observed by Romanoff and Hutt (*1945*) were produced by a single hen between May and July (cf. Fig. 174). It will be remembered that other abnormalities, such as dwarf eggs and multiple-yolked eggs, occur with greatest frequency in the spring and summer.

ORIGIN OF DOUBLE EGGS

There has been much speculation concerning the origin and formation of double eggs. It is apparent that double eggs are the result of a combination of normal and abnormal functioning of the reproductive system. Parker (*1906*) considered double eggs to be caused by dysfunction either of the oviduct or of both the oviduct and the ovary. It is obvious that the doubling of the egg must occur in the oviduct, whether or not the egg receives a normal yolk.

According to a theory first proposed by Panum (*1860*) and later supported by others (*Chobaut, 1897; Rabaud, 1902; Henneguy, 1911; Justow, 1927; Hahn, 1930*), doubling is the result of retention of an egg in the uterus until overtaken by a second egg; the 2 eggs then become enclosed within a single set of envelopes. The occurrence of reverse peristalsis is considered unlikely. However, the not infrequent discovery of fully formed eggs in the body cavity shows that eggs can be returned even after they have traversed almost the entire length of the oviduct (*Curtis, 1916; Asmundson, 1931e*).

A second theory, proposed by Davaine (*1860*) and supported by many others (*Schumacher, 1896; Herrick, 1899; Parker, 1906; Curtis, 1916;*

Asmundson, 1931e, 1933b; Romanoff and Hutt, 1945), suggests that antiperistalsis, or a similar process, forces an egg from the posterior end of the isthmus back up the oviduct to the albumen-secreting region. There the egg encounters another, and together they return to the posterior end of the oviduct and acquire an additional set of egg envelopes. It is evident that the enclosed egg of a double egg must have attained the posterior portion of the oviduct, in order to have acquired a shell membrane and a shell; furthermore, the enclosed egg could not lie within the albumen of the intact outer egg without having been returned to the albumen-secreting portion of the duct (*Romanoff and Hutt, 1945*).

Although it is not thoroughly understood how antiperistalsis is produced, it is conceivable that the normal sequence of contraction and relaxation of the oviducal muscles might be disturbed. A series of contractions, displaced slightly toward the egg's cloacal end, could easily reverse the movement of the egg. According to Warren and Scott (*1935a*), the entire oviduct is affected when an ovum is released from the ovary and engulfed by the infundibulum, and even the muscles of the uterus contract. An egg, if present in the uterus at this time, could be forced up the oviduct. The intervals that preceded the 10 double eggs reported by Romanoff and Hutt (*1945*) undoubtedly indicate that each of the primary eggs was retained in the uterus until after the ovulation of the second egg. The presence of a dwarf egg in the uterus at the time of ovulation is also possible and is not necessarily indicated by a laying interval, since the production of dwarfs is often independent of the normal rhythm of ovulation.

It is difficult to account for the failure of the oviducal glands to be stimulated by an egg on its return passage up the oviduct. If activation of the glands occurred, the enclosed eggs of double eggs would possess a set of reversed egg envelopes; this phenomenon has never been observed.

Hargitt (*1912*) believed that the glands might become exhausted from the downward passage of the egg and be unable to secrete additional material over a returning egg. This explanation is inadequate, since the glands are able to secrete excessive amounts of albumen when double- and triple-yolked eggs are formed (*Curtis, 1916*). Patterson (*1911*) and others have suggested that the oviduct is polarized to such an extent that the glands respond only to a stimulus from the normal direction.

It is more probable that, when an egg is returned, it traverses the oviduct so rapidly that there is insufficient time for the glands to produce their secretions (*Curtis, 1916; Romanoff and Hutt, 1945*). The great speed of the completed egg's reverse passage is indicated by the fact that the succeeding yolk is often met in the upper portion of the oviduct soon after having been released from the ovary.

The type of double egg that is formed depends upon how far antiperistalsis carries the egg that is destined to be enclosed. The presence of a yolk in the outer egg indicates that the inner egg was forced to the anterior part of the oviduct, where it met a newly released ovum. If the outer egg is yolkless, the first egg probably traversed only a portion of the oviduct in the reverse direction. Whether the inner egg is complete or yolkless is determined, of course, by its original structure.

Structure of Dwarf Eggs

In structure, dwarf eggs may exhibit a variety of aberrations. Both the yolk and the albumen may deviate from normal.

Yolk. In most dwarf eggs, the yolk is absent or incomplete. Warner and Kirkpatrick (*1916*) found no yolk at all, or only fragments of yolk, in more than 90 per cent of all the abnormally small eggs that they examined. The incidence of various conditions of the yolk, as noted by Pearl and Curtis (*1916*), is shown in the tabulation. These investigators

Condition of Yolk	Incidence (per cent)
Without vitelline membrane	51
Completely absent	35
Complete	10
With ruptured vitelline membrane	4

also observed that the yolk was absent in all cylindrical dwarf eggs and in 89 per cent of the prolate-spheroidal type.

Dwarf eggs containing yolks are in general longer, broader, and heavier than those without yolks; their average weight is approximately 26 gm., whereas yolkless dwarfs weigh only about 17 gm. The average weight of the yolk in dwarf eggs is 4 gm. Most dwarf eggs weighing less than 9 or 10 gm. are yolkless, but the proportion of yolkless eggs decreases steadily among dwarfs of increasingly greater weight (Fig. 177). Some yolkless dwarf eggs, however, may attain relatively large size. Many eggs of this type weigh more than 20 gm.; Asmundson (*1931e, 1939*) observed several weighing between 22 and 24 gm. Some may weigh as much as 27 to 33 gm. Dwarfs that are heavier than 33 gm. usually contain yolks (cf. Fig. 177).

Usually, when no yolk is present, a foreign object is found in the center of the dwarf egg and appears to have served as a nucleus around which the egg materials were secreted. Such objects as a bit of hardened albumen, a clot of blood, or a fragment of yolk substance have been found. The shape of the nucleus appears to have considerable influence on the final shape of the egg. A globular nucleus stimulates the formation of the prolate-spheroidal egg; cylindrical dwarfs are usually formed around somewhat longer objects. Most dwarf eggs tend to be round,

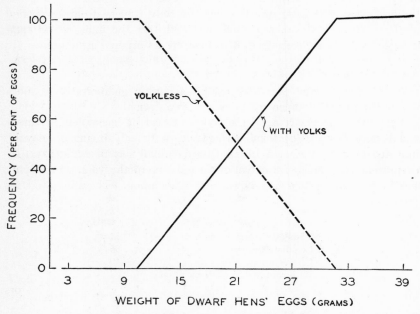

FIG. 177. Frequency with which yolks are present and absent in dwarf eggs of various weight classes. (From the data of Pearl and Curtis, 1916.)

since the oviduct does not exert so great an elongating force on a small object as on a large one (*Pearl and Curtis, 1916*).

Dwarf eggs may be formed around artificial "yolks" inserted into the albumen-secreting region of the oviduct. In one instance, after the insertion of a prune, a dwarf egg with an elongated stalk was laid (*Asmundson, 1931e*). A few fibers of the prune were found in the egg; these had apparently acted as a nucleus. Autopsy later showed the remainder of the prune still lodged in the oviduct.

An unusual dwarf egg, described by Benoit and Courrier (*1933*), was laid by a fowl during the process of sex inversion from female to male. The nucleus of this egg consisted of a mass of spermatoid material. Around this core had been secreted normal albumen, shell membranes, and shell.

Autopsy revealed that the right ovary of this bird was well developed, whereas the left ovary was in a state of atrophy. Male hormones were apparently beginning to gain dominance, but the aggregation of spermatoid material had stimulated the glands of the oviduct to normal function.

Albumen. Most dwarf eggs have a tendency to contain a larger than normal proportion of dense albumen. In specific gravity, the albumen of dwarf eggs may vary greatly, but the maximum of 1.2 (*Pearl and Curtis, 1916*) is well above the maximum of 1.04 found in normal eggs (see Chapter 7, "Physicochemical Properties").

Shell. The shell of the dwarf egg is sometimes partly or wholly lacking. When present, it shows a normal variation in thickness.

Proportional Composition. In dwarf eggs with small yolks or no yolks at all, the proportional amounts of the egg components deviate widely from normal; the percentage of albumen—especially the dense portion—is naturally large (*Burmester and Card, 1939*). This fact is clearly indicated by the data given in the table (*Asmundson, 1931e, 1939*). In a representative series of yolkless dwarf eggs, averaging 18 gm. in weight, the mean weights and percentages of the component parts were as tabulated.

	WEIGHT (grams)	PROPORTION (per cent)
Albumen	16.0	88.9
Chalazae	0.1	0.6
Dense	12.3	76.9
Liquid	3.6	22.5
Shell	2.0	11.1

Even in dwarf eggs containing complete yolks, the proportional composition is abnormal. Figure 178 shows the relationship between the weight of dwarf eggs and the proportional amounts of yolk, albumen, and shell. This illustration shows that the yolk, which is nearly 30 per cent of the normal egg, varies from 4 per cent of the total weight, in the smallest dwarfs, to approximately 23 per cent, in the largest. The shell, which, in normal eggs, remains fairly constant at 12 per cent, constitutes from 13 to 20 per cent of the dwarf egg; the albumen constitutes approximately 65 to 75 per cent of the dwarf egg, as compared with the average of 56 per cent in the normal egg (cf. Table 12, Chapter 3, "Structure").

Causes. Evidence discussed previously in this chapter indicates that dwarf eggs are produced only when the ovary is in active laying condition. Mechanical stimulation of the glands of the oviduct by artificial "yolks," or foreign bodies, results in the production of dwarf eggs only at times when yolks are being matured and ovulated normally by the ovary. Although many objects have been found as nuclei of dwarf eggs, it is significant that fully 65 per cent of all dwarfs contain at least a particle of yolk matter (*Pearl and Curtis, 1916*).

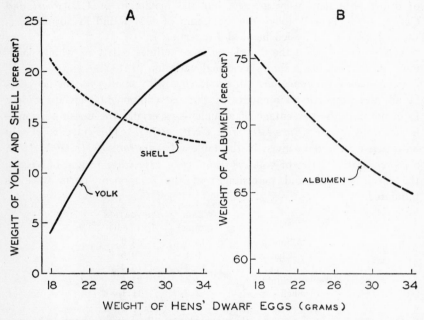

Fig. 178. Relationship between weight of dwarf eggs and proportional amounts of *A*, yolk and shell, and *B*, albumen. (After the data of Pearl and Curtis, 1916.)

The presence of an incomplete yolk in a dwarf egg probably indicates that a normal yolk was produced and ovulated but, except for a small portion, failed to reach the oviduct. The yolk may have been ruptured and the greater part of its contents extruded into the body cavity, eventually to be absorbed by the visceral peritoneum. A yolkless dwarf egg may result if a normal yolk starts down the oviduct and initiates the secretion of albumen but is broken up afterward and extruded into the body cavity (*Pearl and Curtis, 1916*).

Yolk Defects

Various abnormalities of the yolk have been observed. Yolks may be ruptured, fragmentary, misshapen, or discolored.

RUPTURED YOLK AND YOLK FRAGMENTS

Incomplete or fragmentary yolks, the incidence of which is highest in dwarf eggs, are also associated with other malformations, or they may be the egg's only abnormality. In double eggs, either the enclosed or the enclosing egg may possess an abnormal, rudimentary, or ruptured yolk. For example, in a double egg described by Parker (*1906*), the yolk substance of the outer egg was scattered around the inner egg (Fig. 179-*A*), probably because the vitelline membrane had been ruptured.

Another double egg (*Curtis, 1916*) contained a normal yolk, about which was curved an elongated, membranous egg. In the enclosed egg,

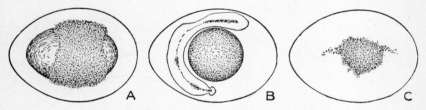

FIG. 179. Examples of ruptured yolk and yolk fragments.

A, yolk mass surrounding a small inner egg (after Parker, 1906). *B*, yolk fragments enclosed in a membrane (after Curtis, 1916). *C*, ruptured yolk (after Pearl and Curtis, 1916).

yolk material was arranged in droplets along a central axis, like beads on a string (cf. Fig. 179-*B*).

As noted by Pearl and Curtis (*1916*), a normal yolk may be replaced by a large mass of yolk material which lacks a vitelline membrane (cf. Fig. 179-*C*) and which may take a variety of shapes. Disproportionately large masses of this type may be present in eggs of normal size.

As previously mentioned, the vitelline membrane of a yolk may rupture and, by causing a constriction, give the yolk the appearance of being double.

ABNORMAL SHAPE

The yolk may assume a variety of abnormal forms. It may be biconcave or biconvex, or constricted at some point and twisted into two or more segments (*Davaine, 1860*). If the constriction is slight and not well defined, the result is an elongated yolk in the shape of a dumbbell (*Parona and Grassi, 1877*), as shown in Fig. 180-*A*. Invaginations may give the yolk the shape of a crescent (*Curtis 1916*) or a triangle (cf. Fig. 180-*B* and *D*).

When the constriction in the yolk is clearly defined, the yolk may present the false appearance of being double (*Parker, 1906*). The "fused" yolks described by Harvey (*1737*) and Dareste (*1877*) are probably of this type.

If the yolk is constricted in more than one place, the result may be a twisted amorphous mass, such as the unusual yolk found by Schumacher (*1896*) in the enclosed egg of a double egg (cf. Fig. 180-*C*).

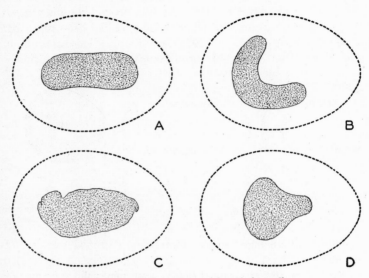

FIG. 180. Various deformities of the yolk.

A, flattened yolk (after Parona and Grassi, 1877). *B*, invaginated yolk (after Curtis, 1916). *C*, misshapen yolk (after Schumacher, 1896). *D*, triangular yolk (after the authors' records).

Causes. The occurrence of oddly shaped yolks is probably due to more than one cause. A weak vitelline membrane may rupture and allow the yolk material to assume a pseudo-doubled appearance (*Parker, 1906*). Davaine (*1860*) considered yolk shape to be dependent to some extent on the shape of the oviduct. A constricted oviduct might cause the yolk, and indeed the entire egg, to become elongated and constricted.

SPOTTING OF THE YOLK

Normally, the surface of the yolk is of uniform color. Occasionally, however, dark spots of various sizes are unevenly distributed over it (Fig. 181). Schaible, Davidson, and Moore (*1936*) found that surface spotting is correlated neither with the rate of production, the size of the yolk, nor the position of the egg in the cycle; nor could they demonstrate that individual birds consistently produced spotted yolks, although it was noted that one hen laid several eggs with a distinctive S-shaped spot.

Another abnormal condition, rarely found in fresh eggs, is mottling of the yolk (*Almquist, 1933*). Mottled yolks differ from spotted yolks in origin.

Occasionally, the yolk displays an opaque, whitish blemish which may cover from one-tenth of the surface to the entire yolk. Although the edibility of the egg is in no way altered, the yolk appears rotten (*Jeffrey, 1945*). This particular type of blemish extends more deeply into the yolk than spots. It is produced more frequently by some breeds of fowl than others and is most pronounced in eggs laid by certain strains of Rhode Island Reds.

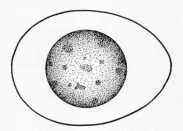

FIG. 181. A spotted yolk. (Drawn from a collection of colored slides at Cornell University.)

Causes. Spotting on the yolk's surface is apparently caused by uneven pressure exerted on the yolk as it passes through the oviduct. The condition may be duplicated by the application of slight pressure to small areas on the yolk of an opened egg. An imperfection in the follicle or oviduct, whether temporary or permanent, might imprint itself on the yolk by displacing the superficial layer of light yolk and exposing the dark yolk beneath (*Schaible, Davidson, and Moore, 1936*).

Mottling is apparently due to inability of the vitelline membrane to prevent the penetration of protein from the albumen. Chemical analysis has revealed that the protein-fat ratio is higher than normal in mottled areas (*Almquist, 1933*). The blemished yolk is of unknown origin, although possibly it is associated with a weak vitelline membrane.

Absence of Albumen

The complete absence of albumen in an avian egg has rarely been noted. Liégeois (*1859*) describes a double egg consisting of a small egg with a normal yolk, but no albumen, enclosed within a yolkless egg. Such an anomaly may have been made possible either by a temporary inhibition in the physiological activity of the albumen-secreting region of the oviduct, or by very rapid passage of the yolk through this region. The latter explanation is the most plausible.

Malposition of the Air Cell

In fresh eggs, there is seldom displacement of the air cell from its usual location at the blunt end of the egg. However, at times the air cell may be found at the sharp end or at one side. In rare instances, the air space appears to be within the inner egg membrane, so that it moves freely around the albumen to whatever part of the egg is upper-most (*Almquist, 1933*).

The shape of the air space occasionally varies from its normal round-ness and may assume a long, narrow, spiral form, as in a hen's egg observed by the authors (Fig. 182).

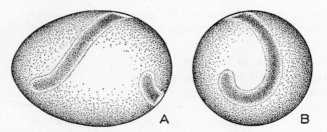

Fɪɢ. 182.   Unusual form of air cell found in a hard-boiled chicken egg.   (From the records of the authors.)

A, side view.   *B*, end view.

In double eggs and double-yolked eggs, the air space is nearly always shifted to the side of the egg (*von Nathusius, 1869; Kunstler, 1907; Romanoff and Hutt, 1945*). This change in position has been observed not only in abnormal chicken eggs but also in double-yolked goose eggs (*von Nathusius, 1895*).

INCLUSION OF FOREIGN SUBSTANCES

Many eggs that are perfect in structure and organization must never-theless be classed as abnormal, because, in the course of their formation, extraneous material has been incorporated in them. In some cases, the inclusions originate in the bird's body; in others, they are derived from the environment.

Blood Clots

Blood clots are often detected when fresh eggs are candled (see Chap-ter 9, "Food Value"). They appear as a dark red spot or streak, usually adhering to the yolk but sometimes apparently floating freely in the albumen. When similar, but somewhat browner, spots are observed, the term "meat spot" rather than "blood spot" is usually used. For many years, blood spots and meat spots were believed to be two separate phenomena; but recent studies have shown that most, if not all, so-called meat spots are in reality blood clots in various stages of decomposition (*Nalbandov and Card, 1944*).

Blood clots vary greatly in size, shape (Fig. 183), and frequency. It is commonly assumed that from 2 to 4 per cent of all chicken eggs, selected at random, contain blood spots (*Nalbandov and Card, 1944*).

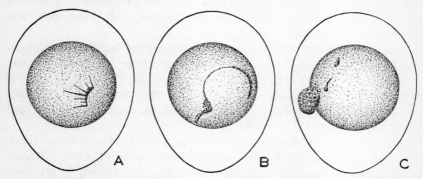

FIG. 183. Blood streaks (*A* and *B*) and a blood clot (*C*) observed in chicken eggs. (After Nalbandov and Card, 1944; and the authors' observations.)

Blood spots and clots are very common in the eggs of heavy breeds, such as Rhode Island Reds, and relatively rare in the eggs of White Leghorns. Quinn and Godfrey (*1940*) noted that, of all the blood spots they observed, 62 per cent occurred in the eggs of Rhode Island Reds, and only 4 per cent in White Leghorn eggs. Lerner and Smith (*1942*) found that there is also seasonal variation in the incidence of blood clots, which increase in frequency after the first of April, during the period of heavy laying.

The incidence of blood spots is not the same in eggs laid by hens of all ages. Nalbandov and Card (*1944*) found that, on an average, 44 per cent of all hens laid eggs with blood spots during their first year, 26 per cent during their second year, and 22 per cent during their third. As hens grow older, most of them produce blood spots less frequently, but about 35 per cent do so more often.

Causes. Until recently, most of the explanations for the occurrence of blood spots and clots have been based on speculation rather than experimental data. Aristotle believed that blood clots were the result of premature expulsion of the yolk. Fabricius (*1600*), like Aristotle, considered the presence of blood clots as ample proof of the theory that yolks are formed directly from blood.

Later theories explained blood spots as the result of ovarian bleeding at the time of ovulation or as the consequence of frightening the bird. However, it is now known that bleeding at the time of ovulation is so rare and so slight that it could not possibly be responsible for all the blood spots found in eggs. It is also unlikely that fright causes hemorrhages and blood spots. In one experiment (*Jeffrey and Pino, 1943*), a dog was allowed to chase the flock for 5-minute periods several times a day, yet no significant increase in the incidence of blood spots was observed.

It is now thought that blood spots are caused by intrafollicular bleeding (*Nalbandov and Card, 1944*). A hemorrhage may occur at any point in the follicle, and blood may adhere to the yolk surface or accumulate in the pedicle. If it forms in the pedicle, the clot does not necessarily become attached to the yolk but may remain free and eventually be engulfed in the albumen.

Meat spots were formerly thought to consist of glandular tissue sloughed off from the walls of the oviduct (*von Nathusius, 1867*). Recent observations, however, have revealed that they are actually blood spots which have undergone a certain amount of chemical change. Microscopic studies (*Burmester and Card, 1938*) have shown that meat spots are masses of protein material and that all but about 3 per cent of them contain red blood cells. Usually, no other cellular tissue is present, except, rarely, a few bacteria. However, degenerating blood spots have been found to contain macrophages which are apparently derived from fibroblasts of the parent tissue rather than from lymphocytes (*Lucas, 1946*).

Nalbandov and Card (*1944*) have demonstrated that blood spots may be transformed into meat spots, probably by the liberation of hematin from oxyhemoglobin. The change is favored by an alkaline medium and a high environmental temperature, but it requires some time to take place. Meat spots therefore are probably derived from blood clots formed by an intrafollicular hemorrhage several days before ovulation. If a blood clot forms and later disintegrates completely, the albumen may take on a bloody appearance.

It is still unknown why certain individuals and breeds produce eggs with blood and meat spots with greater frequency than others. To what extent heredity and environment are responsible for these egg defects has not been determined. Most investigators are convinced that some hens are congenitally subject to intrafollicular hemorrhages (*Quinn and God-*

frey, 1940; Nalbandov and Card, 1944; Lerner and Smith, 1942). Some relationship has also been noted between diet and the occurrence of blood spots. The incidence of blood spots is reduced when hens are allowed to run free on the range (*Nalbandov and Card, 1944*). It also is relatively low in eggs laid by birds that do not survive their first year (*Lerner, 1946*).

Worms and Parasites

In addition to blood and meat spots, which may be considered as foreign bodies, many other objects have been observed in eggs. In the past, various forms of animal life, from insects to the embryo of "the serpent basilisk," and "a human-faced monster with hair and beard of serpents," have allegedly been found in the egg.

There have been many reports of worms in eggs, although the accuracy of some of these accounts has been questioned. It has been claimed that all so-called worms are probably nothing more than masses of hardened albumen, or similar concretions. However, in some cases, the authenticity of the reports seems well established.

In several instances, flukes have been observed in the albumen of newly laid eggs. These are usually members of the genus *Prosthogonimus,* a type of trematode which matures in the oviduct and *bursa Fabricii* of hens and ducks. Eggs are particularly apt to contain flukes in China and Japan, where conditions are frequently unsanitary, and where many hens are infested with *Prosthogonimus japonicus* and *Prosthogonimus ovatus* (*Wu and Noyes, 1928; Du and Williams, 1930; Khaw, 1930*). In Europe, *Prosthogonimus pellucidus* and *Prosthogonimus cuneatus* are occasionally found in eggs (*Wolfhügel, 1906; Khaw, 1930*). Authenticated discoveries of flukes in hens' eggs are reported by Landois (*1882*), Linton (*1887*), and Braun (*1901*). Inclusion of flukes in eggs is also described by Hanow, Purkinje, Eschholz, and Schilling, whom Davaine (*1860*) quotes. In the Great Lakes region of the United States, infestation with the oviduct fluke is also quite prevalent (*Kotlan and Chandler, 1925; Lakela, 1931*). In America, the species most commonly found is *Prosthogonimus macrorchis,* which is closely related to, but distinct from, *Prosthogonimus pellucidus* (*Macy, 1934*). Macy (*1934*) fed cysts of *Prosthogonimus macrorchis* to a hen; 43 days later the hen laid an egg that contained a living fluke in the albumen.

Roundworms have also been found in eggs; undoubtedly, some of the "serpents" observed in the past were in reality worms. Landois (*1882*) and Pellegrino (*1927*) each reported the inclusion of the roundworm, *Heterakis inflexa,* in an egg. Henry (*1928*) describes an egg in whose shell a roundworm was embedded in such a way as to produce a visible welt (Fig. 184). He mentions similar instances described by d'Aldrovandi (*1642*), Cleyer (*1682*), Montius (*1757*), and Pavesi (*1893*). Recently, Hall (*1945b*) reported the presence of a roundworm, *Ascaridia lineata,* in the thin albumen of an egg (cf. Fig. 184). The roundworm *Ascaridia columbae* (*Heterakis columbae*) is sometimes found in the eggs of pigeons (*Levi, 1941, p. 310*).

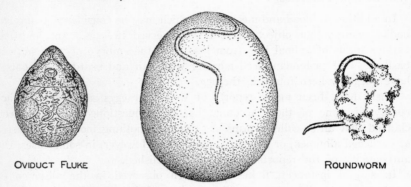

OVIDUCT FLUKE   ROUNDWORM

FIG. 184. Parasites and worms found in birds' eggs.

On the left, an oviduct fluke, *Prosthogonimus macrorchis,* sometimes found in the albumen of the hen's egg; magnified about 4 times. (After Macy, 1934.)

In the center, a hen's egg with a roundworm (*Ascaridia*) embedded in the shell. (After Henry, 1928.)

On the right, a roundworm, *Ascaridia lineata,* found in a chicken egg and shown here embedded in a mass of albumen coagulated by preservative; magnified about 2 times. (After Hall, 1945b).

It is probable that roundworms, which are commonly found in the intestinal tract, are able to migrate from the cloaca into the oviduct, where they may become included in an egg. Once the parasite is surrounded by the albumen of a forming egg, it becomes a part of the egg and is covered by the shell membranes and shell. If the worm penetrates to the albumen-secreting region of the oviduct, it is incorporated into the albumen; if it reaches only the shell-secreting region, it becomes embedded in the shell (*Parker, 1906; Henry, 1928*).

Other Foreign Bodies

In addition to the types of inclusion already mentioned, a wide variety of miscellaneous objects, both organic and inorganic, have been found in eggs. Some of these obviously originate from the environment. Perrault (*1675*) describes a metal pin found in a hen's egg as being "covered with a whitish crust and resembling a frog's thigh bone; under this crust the pin was black and slightly rusted." Bonnet (*1883*) mentions such inclusions as insects, pebbles, bits of sand, feathers, horsehair, and even a coffee bean. Bits of fecal matter are also found occasionally in eggs; these are conceivably forced up into the oviduct from the cloaca (*Hewitt, 1939*). Inanimate foreign bodies of extraneous origin are probably picked up when part of the hen's oviduct is everted as an egg is laid, and are then passed up the oviduct by means of antiperistaltic contractions.

Concretions of egg materials may become engulfed and included in otherwise normal eggs. Sometimes concretions are formed in the oviduct

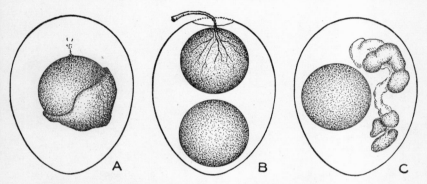

FIG. 185. Reconstructed pictures of various enclosures in the hen's egg.

A, tissue sac surrounding part of the yolk (after Davaine, 1860). *B,* a fully formed follicle within the egg (after Davaine, 1860). *C,* five small yolks and a fragment of coagulated albuminous material in the egg (reconstructed after Panum, 1860).

by unusual secretion of the glands. At other times, an obstruction may cause an egg to break up. Remnants of the egg may then cling to the walls of the oviduct, gradually harden, and, after accumulating layers of secretions, eventually find their way into eggs (*Durant and McDougle, 1943*). An enclosure of this type, described by Panum (*1860*), consisted of five yolks, or pieces of yolk, connected by a twisted mass of hardened albumen (Fig. 185-*C*). Eggs containing concretions are also described by Duhamel (*1826*), Leblond (*1834*), and Davaine (*1860*).

In several instances, an entire follicle (intrafollicular ovum) has been found in an egg (*Cleyer, 1682; Laboulbène, 1859*), as shown in Fig. 185-*B*. An egg with such an inclusion, described by Hutt (*1946*), lacked shell on the blunt end when brought into the laboratory. The stalk of the follicle was thought to have protruded from the blunt end.

Occasionally, bits of the ovary may be torn off and become embedded in, or attached to, an egg (cf. Fig. 185-*A*). Inclusions of ovarian tissue are described by Bailly (*1838*), Laboulbène (*1859*), Davaine (*1860*), Chatin (*1883*), von Nathusius (*1895*) and Durski (*1907*).

Part II

BIOPHYSICOCHEMICAL CONSTITUTION

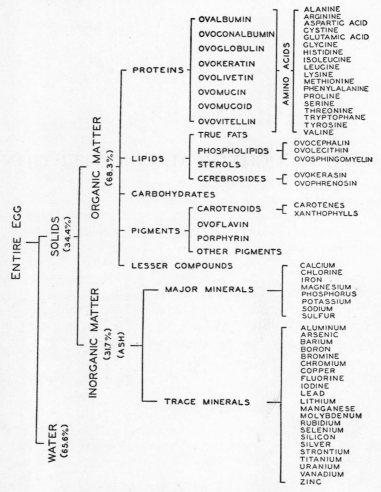

Chemical constituents of the entire chicken egg.

Chapter Six

CHEMICAL COMPOSITION

The avian egg is a chemical storehouse.

The bird's egg is composed of the substances that form the basis of all animal life. It is the egg's function to produce a new living organism, whose needs are supplied by the egg alone. The egg therefore contains most of the varied materials of life in amounts that are sufficient to insure the growth and development of the embryo.

Like animal tissues, the egg in large measure consists of water. Its solid portion is composed of proteins, fats, carbohydrates, and minerals, and also small amounts of many fairly simple organic compounds. All these chemical constituents are distributed among the egg's component structures in a very specific manner.

Since it is the product of metabolic processes which take place within the individual organism, the bird's egg is subject to minor variations in composition. The absolute and proportional amounts of the egg's numerous ingredients may be influenced to a certain extent by the heredity and environment of the bird in whose body the egg is formed. In particular, the nature of the bird's diet is responsible for the variability in the egg's chemical constitution.

GENERAL COMPOSITION

The eggs of different species are made up of somewhat dissimilar quantities and proportions of the many substances they contain. Eggs laid by a flock of domestic fowl, however, may be fairly alike, if the birds are genetically homogeneous and maintained under well-controlled management. The following paragraphs give the average composition of the egg, obtained from the results of analyses of the entire egg and its contents and integuments.

THE ENTIRE EGG

The entire avian egg (contents with egg coverings) has a large content of water. Its solids consist of both organic compounds—proteins, lipids

(or fatty substances), and carbohydrates—and inorganic matter, or minerals. The average amounts of these chemical components found in the
entire chicken egg are approximately as tabulated.

	AMOUNT (grams)
Total	58.0
Water	38.1
Solids	19.9
Organic matter	13.6
Proteins	7.0
Lipids	6.1
Carbohydrates	0.5
Inorganic matter	6.3

The percentages of these chemical components are shown in Fig.
186-*A*. Various chemical analyses of the entire egg (*Baudrimont and*

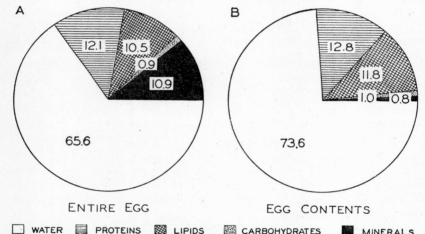

FIG. 186. Average percentages of the major groups of chemical compounds in *A*,
the entire chicken egg, and *B*, the egg contents alone.

*St. Ange, 1847; Lebbin, 1900; Willard and Shaw, 1909; Langworthy,
1917*) indicate that these percentages vary widely in eggs laid by birds
of the same species.

Because of species variation in the relative amount of eggshell, which
contains most of the inorganic matter, the mineral content of some birds'
eggs may be proportionately higher or lower than that of the hen's
egg. For instance, it is higher in the guinea fowl's egg and lower in the
plover's. The eggs of all altricial birds, such as the robin and the sparrow,
are very thin-shelled and hence have a considerably lower percentage of

minerals. The percentages of other chemical constituents may differ also, but in much smaller degree.

Elementary Composition

The principal elements that enter into the chemical compounds of the egg are carbon, oxygen, nitrogen, hydrogen, phosphorus, and sulfur.[1] The relative amounts of these chemical elements in the dry substance of the average chicken egg (*Barrow, 1890*) are as listed.

	RELATIVE AMOUNT (per cent)
Carbon	53
Oxygen	20
Nitrogen	15
Hydrogen	7
Phosphorus	4
Sulfur	1

Of all the elements composing the egg, oxygen alone exists (at least in part) in uncombined form. All the others appear as compounds, most of which are of a very complex nature.

EGG CONTENTS

The chief ingredient of the egg contents is water; it constitutes about 75 per cent of the total weight. Next in relative amount are the organic compounds. These are mainly proteins and lipids (about 12 per cent of each), except for small amounts of carbohydrates (about 1 per cent). Inorganic matter constitutes about 1 per cent of the egg contents. These percentages are shown in Fig. 186-*B*. In addition to these major groups of chemical substances, there are smaller quantities of other compounds, which are discussed below.

The average amounts of the major substances in the contents of the chicken egg are as tabulated.

	AMOUNT (grams)
Total	51.6
Water	38.0
Solids	13.6
Organic matter	13.2
Proteins	6.6
Lipids	6.1
Carbohydrates	0.5
Inorganic matter	0.4

[1] Of the ninety-six known elements, at least thirty-five have been identified in the bird's egg.

The percentage composition of the egg contents varies somewhat among the eggs of different species of birds (Table 23). The eggs of water fowl (ducks and geese) contain less water and more fat than the eggs of

TABLE 23

PROPORTIONAL AMOUNTS OF MAJOR CHEMICAL CONSTITUENTS IN THE EGG CONTENTS
OF VARIOUS BIRDS

Chemical Constituent	Land Birds					Waterfowl (Anseriformes)	
	Galliformes			Others			
	Chicken (51.6 gm.) *	Turkey (71.6 gm.)	Guinea Fowl (32.9 gm.)	Plover (14.0 gm.)	Ostrich (865.0 gm.)	Duck (66.6 gm.)	Goose (177.0 gm.)
	per cent	per cent	per cent	per cent	per cent	per cent	per cent
Water	73.6	73.7	72.8	74.7	74.0	69.7	70.6
Solids	26.4	26.3	27.2	25.3	26.0	30.3	29.4
Organic matter	25.6	25.5	26.3	24.3	24.6	29.3	28.2
Proteins	12.8	13.1	13.5	11.8	12.2	13.7	14.0
Fats (lipids)	11.8	11.7	12.0	11.7	11.7	14.4	13.0
Carbohydrates	1.0	0.7	0.8	0.8	0.7	1.2	1.2
Inorganic matter	0.8	0.8	0.9	1.0	1.4	1.0	1.2

* The figures in parentheses indicate the average weight of the egg contents of each species.

References

CHICKEN: Prévost and Morin (*1846*); Payen (*1863*); Commaille (*1874*); König and Farwick (*1876*); Lebbin (*1900*); Langworthy (*1901*); Mansfeld (*1901*); Greshoff, Sack, van Eck, and Bosz (*1903*); Welmanns-Köln (*1903*); Forbes, Beegle, and Mensching (*1913*); von Czadek (*1916*); van Meurs (*1923*); Ferdinandoff (*1931, p. 39*); Weinstein (*1933*).
TURKEY: Langworthy (*1901*); Hepburn and Miraglia (*1937*).
GUINEA FOWL: Langworthy (*1901*).
PLOVER: König and Krauch (*1878*); Langworthy (*1901*); Greshoff, Sack, van Eck, and Bosz (*1903*).
OSTRICH: Anonymous (*1918*); Avenati-Bassi and Bravo (*1925*).
DUCK: Commaille (*1874*); Langworthy (*1901*); Hepburn and Katz (*1927*); Weinstein (*1933*); Danilova and Nefedova (*1935*).
GOOSE: Langworthy (*1901*); Hepburn and Katz (*1927*).

land birds, probably because of the greater heat requirements of the developing embryo. These requirements are high for two reasons. The nests of water fowl are usually situated near a body of water or in a swamp, and the bottom of the nest is cool. In addition, water fowl have a slightly higher body temperature, and possibly a higher rate of metabolism, than land birds.

Distribution of Constituents between Yolk and Albumen

The chemical constituents of the egg contents are not uniformly distributed between yolk and albumen. The albumen, the volume of which is twice that of the yolk, contains three times as much water and only half as much solid matter. Although the yolk and albumen are approximately equal in their content of proteins, carbohydrates, and minerals, the yolk contains practically all the lipids.

Egg Yolk

The composition of the yolk is much more complex than that of any other part of the egg and differs greatly from that of the albumen. Lipids form the largest part of the organic solid matter in the yolk; proteins and a very small complement of carbohydrates comprise the remainder. The yolk is also relatively rich in minerals and pigments, as well as in vitamins (see Chapter 9, "Food Value").

The yolk of the fresh chicken egg contains the following approximate amounts of the chief constituents. The percentages of these components are shown in Fig. 187-*A*.

	AMOUNT (grams)
Total	18.7
Water	9.1
Solids	9.6
Organic matter	9.4
Proteins	3.1
Lipids	6.1
Carbohydrates	0.2
Inorganic matter	0.2

There is considerable variation in the percentage of solids in the yolk. Figure 188 shows that the solids were found to range from 50.5 per cent to 54.5 per cent in the yolks of 94 chicken eggs, with an average value of 52.5 per cent (*Romanoff, unpublished*).

The proportions of the various chemical constituents of the yolk vary in the eggs of different species (Table 24). The eggs of altricial birds,

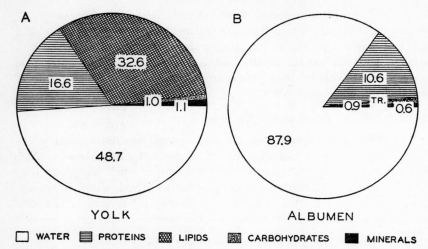

FIG. 187. Average percentages of the major groups of chemical compounds in A, the yolk, and B, the albumen of the chicken egg.

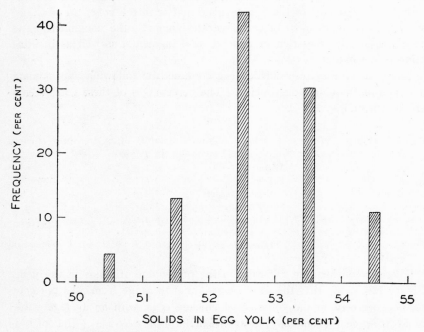

FIG. 188. Frequency distribution of 94 fresh chicken eggs according to the percentage of solids in the yolk (mean value, 52.5 ± 0.11 per cent). (After Romanoff, unpublished.)

TABLE 24

PROPORTIONAL AMOUNTS OF MAJOR CHEMICAL CONSTITUENTS OF THE EGG YOLK OF VARIOUS BIRDS

Chemical Constituent	Praecoces						Altrices
	Galliformes				Anseriformes		Pigeon (3.9 gm.)
	Chicken (18.7 gm.) *	Jungle Fowl (12.4 gm.)	Turkey (27.4 gm.)	Guinea Fowl (13.0 gm.)	Duck (26.2 gm.)	Goose (66.8 gm.)	
	per cent	per cent	per cent	per cent	per cent	per cent	per cent
Water	48.7	48.7	48.3	49.2	44.8	43.3	55.7
Solids	51.3	51.3	51.7	50.8	55.2	56.7	44.3
Organic matter	50.2	50.5	50.4	49.8	54.0	55.1	43.3
Proteins	16.6	15.3	16.3	16.0	17.7	18.0	12.4
Fats (lipids)	32.6	34.2	33.2	33.0	35.2	36.0	29.7
Carbohydrates	1.0	1.0	0.9	0.8	1.1	1.1	1.2
Inorganic matter	1.1	0.8	1.3	1.0	1.2	1.6	1.0

* The figures in parentheses indicate the average weight of the egg yolk of each species.

References

CHICKEN: Gobley (*1847*); Parke (*1868*); König and Krauch (*1878*); Stützer (*1882*); Lebbin (*1900*); Langworthy (*1901*); Carpiaux (*1903a*); Willard and Shaw (*1909*); Pennington (*1909–10*); Kojo (*1911*); von Czadek (*1916*); Iljin (*1917*); Plimmer (*1921*); van Meurs (*1923*); Thomson and Sorley (*1924*); Mitchell (*1932*); Weinstein (*1933*); Danilova and Nefedova (*1935*).
JUNGLE FOWL: Spohn and Riddle (*1916*).
TURKEY: Langworthy (*1901*); Hepburn and Miraglia (*1937*).
GUINEA FOWL: Langworthy (*1901*).
DUCK: Langworthy (*1901*); Plimmer (*1921*); Hepburn and Katz (*1927*); Danilova and Nefedova (*1935*); Weinstein (*1933*); Shibata (*1936*).
GOOSE: Langworthy (*1901*); Hepburn and Katz (*1927*).
PIGEON: Spohn and Riddle (*1916*).

such as pigeons, contain a smaller percentage of protein and lipids, and a much larger percentage of water, than those of precocial birds, such as chickens and ducks. This difference in composition may in part explain the more advanced development of precocial birds at the time of hatching. The young of the pigeon, whose eggs contain relatively little structural material (proteins) and a small energy supply (fats), are hatched in a comparatively helpless state.

Not only the entire egg contents, but also the egg yolks, of Galliformes differ from those of Anseriformes in water and fat content. The yolks of the turkey, the guinea fowl, the chicken, and the chicken's possible progenitor, the jungle fowl, contain relatively more water and less fatty substance than those of the duck and goose.

WHITE AND YELLOW YOLK

The difference between the alternating layers of white and yellow yolk (see Chapter 3, "Structure") is more fundamental than mere pigmentation, for white and yellow yolk are also markedly dissimilar in composition. White yolk consists of about 86 per cent water (*Romanoff, 1931*) and, according to Spohn and Riddle (*1916*), 4.6 per cent proteins and 3.5 per cent lipids; yellow yolk contains considerably less water (45.4 per cent) and more proteins and lipids (15.0 per cent and 36.4 per cent, respectively). Small amounts of carbohydrates and ash are present in approximately equal proportions in both white and yellow yolk.

These figures indicate that the white yolk, which is formed slowly, has a chemical composition approximating that of living, growing tissues. The yellow yolk, which is stored nearly twenty-six times more rapidly than the white yolk, departs widely in composition from animal tissues.

It should be pointed out that the amount of white yolk is relatively small; it does not exceed 5 per cent of the whole yolk. Furthermore, the white yolk is almost inseparable from the yellow. In fact, the above analyses of white yolk were made on very immature ova (which contain white yolk only), and hence it does not necessarily follow that the white yolk in the fully developed egg is of identical composition.

Because white yolk constitutes such a relatively small proportion of the entire yolk and is so difficult to separate from yellow yolk, most chemical analyses of the yolk have been made on the whole yolk body.

Egg Albumen

The egg albumen consists chiefly of water, of which it is the developing embryo's principal reservoir. The albumen's main organic constituent is protein; other components are small amounts of carbohydrates and minerals, and negligible amounts of lipids. In the chicken egg, these amounts are as tabulated. The relative amounts are shown in Fig. 187-*B*.

	AMOUNT (grams)
Total	32.9
Water	28.9
Solids	4.0
Organic matter	3.8
Proteins	3.5
Lipids	Trace
Carbohydrates	0.3
Inorganic matter	0.2

The percentage of solids in the albumen is subject to considerable variation among the eggs laid by birds of the same species. Figure 189 shows that the solids content of the albumen of 538 chicken eggs ranged from 8.5 per cent to 14.5 per cent, with an average value of 11.5 per cent. The whites of eggs laid by a single bird, however, are usually fairly uniform in their proportions of water and solids. On the other hand, the average percentages of the various constituents of the albumen are much

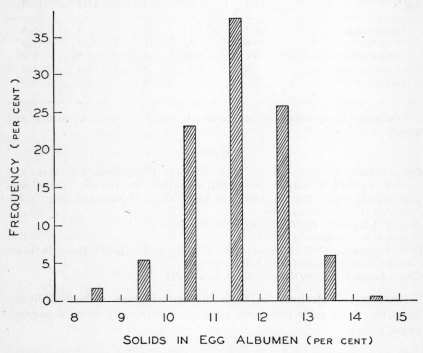

FIG. 189. Frequency distribution of 538 1-day-old chicken eggs according to the per cent of solids in the albumen (mean value, 11.5 ± 0.03 per cent). (After Almquist and Lorenz, 1933a.)

the same in the eggs of all species of precocial birds, as shown in Table 25. This lack of species difference may explain why it is possible, in the

TABLE 25

PROPORTIONAL AMOUNTS OF MAJOR CHEMICAL CONSTITUENTS IN THE EGG ALBUMEN OF VARIOUS BIRDS

Chemical Constituent	Galliformes			Anseriformes	
	Chicken (32.9 gm.) *	Turkey (44.2 gm.)	Guinea Fowl (19.9 gm.)	Duck (40.4 gm.)	Goose (110.2 gm.)
	per cent	per cent	per cent	per cent	per cent
Water	87.9	86.5	86.6	86.8	86.7
Solids	12.1	13.5	13.4	13.2	13.3
Organic matter	11.5	12.8	12.6	12.4	12.5
Proteins	10.6	11.5	11.6	11.3	11.3
Fats (lipids)	0.03	0.03	0.03	0.08	0.04
Carbohydrates	0.9	1.3	1.0	1.0	1.2
Inorganic matter	0.6	0.7	0.8	0.8	0.8

* The figures in parentheses indicate the average weight of the egg albumen of each species.

References

CHICKEN: Lebbin (*1900*); Langworthy (*1901*); Carpiaux (*1903a*); Greshoff, Sack, van Eck, and Bosz (*1903*); Willard and Shaw (*1909*); Kojo (*1911*); Rakuzin (*1915*); von Czadek (*1916*); Iljin (*1917*); Plimmer (*1921*); van Meurs (*1923*); Thomson and Sorley (*1924*); Mitchell (*1932*); Weinstein (*1933*).
TURKEY: Langworthy (*1901*); Hepburn and Miraglia (*1937*).
GUINEA FOWL: Langworthy (*1901*).
DUCK: Langworthy (*1901*); Plimmer (*1921*); Hepburn and Katz (*1927*); Weinstein (*1933*); Danilova and Nefedova (*1935*); Shibata (*1936*).
GOOSE: Langworthy (*1901*); Hepburn and Katz (*1927*).

laboratory, to substitute duck for chicken albumen without seriously interfering with early embryonic development (*Loisel, 1900; Romanoff, unpublished*).

VARIOUS LAYERS OF ALBUMEN

It has been established that the concentric layers of fresh egg albumen differ in chemical composition (*Romanoff, 1929b; 1943c; Moran, 1937d*). The percentage of water in the four layers—outer liquid, middle dense, inner liquid, and chalaziferous—varies as tabulated.

SPECIES	OUTER LIQUID	MIDDLE DENSE	INNER LIQUID	CHALA-ZIFEROUS	INVESTIGATOR
Chicken	88.8	87.6	86.4	84.3	Romanoff (*1929b, 1943c*), Almquist and Lorenz (*1933a*), Moran (*1937d*), and Romanoff and Sullivan (*1937*)
Pheasant	89.0	88.0	86.3	85.3	Romanoff (*1943c*)
Quail	88.1	87.2	85.8	84.9	Romanoff (*1943c*)
Duck	87.4	86.7	85.8	84.3	Romanoff (*1943c*)
Average	88.3	87.4	86.1	84.7	

Since the dry matter of the albumen consists chiefly of protein, the percentage of protein therefore increases from the outermost to the innermost layer. So, also, does the percentage of minerals (*Romanoff, 1943c*).

EGG INTEGUMENTS

The composition of the egg integuments—shell and shell membranes— is quite different from that of the egg contents.

Eggshell

The eggshell, including the outer gelatinous covering or cuticle, contains a large proportion of inorganic salts, some organic material, and very little water. The average chicken eggshell has approximately the following composition:

	AMOUNT	
	GRAMS	PER CENT
Totals	6.1	100.0
Water	0.1	1.6
Solids	6.0	98.4
Organic matter	0.2	3.3
Proteins	0.2	3.3
Lipids	Trace	(0.03)
Inorganic matter	5.8	95.1

Shell Membranes

The shell membranes, which make up about 4 or 5 per cent of the entire egg covering (*Almquist and Burmester, 1934*), consist almost wholly of protein, except for very small amounts of water and traces of minerals. The estimated average composition of the membranes of the chicken egg is as follows:

	AMOUNT	
	GRAM	PER CENT
Totals	0.30	100.0
Water	0.06	20.0
Solids	0.24	80.0
Organic matter	0.21	70.0
Proteins	0.21	70.0
Inorganic matter	0.03	10.0

WATER OF THE EGG

By far the greater part of the egg is water, which is distributed among the component structures of the hen's egg as shown in the tabulation.

	AMOUNT OF WATER	
	GRAMS	PER CENT
Yolk	9.1	23.9
Albumen	28.9	75.9
Shell with membranes	0.1	0.2
Total	38.1	100.0

From these data, it can be seen that the albumen contains three times as much water as the yolk, and that the shell, with its membranes, contains only traces of water.

Table 26 compares the percentage of water in precocial birds' eggs (yolk and albumen, individually and combined) with that in altricial birds' eggs. The comparatively higher proportion of water in the entire contents of altricial birds' eggs is due to the significantly larger amount in the yolk and the slightly greater amount in the albumen. The yolks in the eggs of ducks and geese have considerably less water than those in the eggs of any other species shown in Table 26; and, as a result, the water content of these birds' egg is the lowest.

BIOLOGICAL SIGNIFICANCE OF WATER IN THE EGG

The water in the egg is of great biological significance. Since water holds in solution widely different substances—salts, proteins, carbohydrates, and, to some extent, even fats (emulsified)—it makes possible the chemical activities of the egg. The solvent power of water is the most important chemical factor that enables life to develop within the fertilized egg.

TABLE 26

AMOUNT OF WATER IN THE YOLK, ALBUMEN, AND CONTENTS OF VARIOUS BIRDS' EGGS *

Species of Bird	Yolk (per cent)	Albumen (per cent)	Egg Contents (per cent)
Precocial birds:			
Land fowl:			
Chicken	48.7	87.9	73.6
Jungle fowl	48.7	....	....
Turkey	48.3	86.5	73.7
Guinea fowl	49.2	86.6	72.8
Pheasant	50.8	86.6	....
Quail	51.8	86.6	....
Average	49.6	86.8	73.4
Water fowl:			
Duck	44.8	86.8	69.7
Goose	43.3	86.7	70.6
Average	44.1	86.8	70.2
Altricial birds:			
Pigeon	55.7	....	....
Dove	57.9	89.2	83.0
Starling	56.6	89.9	77.4
Raven	57.1	90.2	....
Corncrake	58.0	89.5	....
Gull	....	89.1	....
Nightingale	....	89.1	80.0
Thrush	....	90.1	....
Canary	....	89.0	....
Finch	....	89.4	....
Honey buzzard	....	89.3	....
Yellowhammer	....	....	81.4
Blackbird	....	....	80.1
Carrion crow	....	....	83.3
Average	57.1	89.5	80.9

* Compiled from various sources, including the data of Tarchanoff (*1884*), Mörner (*1912*), Behre and Riddle (*1919*), and Groebbels (*1927*).

In addition, its high specific heat enables water to take up the heat liberated by the vital activities of the developing egg, and to give it off gradually. This property of water prevents violent temperature changes, which might lead to irreversible denaturation at high temperatures, or to retarded chemical reactions at low temperatures. Pure water is a fairly good electrical insulator and has a high dielectric constant. It is a good ionizing medium and, by permitting intimate contact between dissolved substances, facilitates many types of chemical reactions, such as hydrolysis and biological oxidation and reduction. The high surface tension of water is also essential for the conservation of cell boundaries.

All these properties of water obviously provide a favorable medium within the egg for various physical and chemical changes. Water also can effect the necessary transfer of chemical compounds from one part of the egg to another.

SOLIDS OF THE EGG

The entire avian egg (egg contents, together with the shell and shell membranes) contains about 34 per cent of solids (dry matter). In the egg contents alone, there is about 26 per cent of solid matter.

The distribution of solids in the entire chicken egg and in the egg contents is as follows:

	ENTIRE EGG		EGG CONTENTS	
	GRAMS	PER CENT	GRAMS	PER CENT
Yolk	9.6	48.2	9.6	70.6
Albumen	4.0	20.1	4.0	29.4
Shell with membranes	6.3	31.7	...	....
Total	19.9	100.0	13.6	100.0

ORGANIC CONSTITUENTS

The complexity of the organic matter in the bird's egg is of particular interest because of the egg's function both as a reproductive cell and as an article of food in the human diet. Various proteins, lipids, carbohydrates, and pigments are found in abundance. In addition to these major groups of chemical substances, many other organic compounds are present, some in exceedingly small amounts.

The average amounts of organic matter in the various parts of the entire chicken egg are as follows:

	AMOUNT OF ORGANIC MATTER	
	GRAMS	PER CENT
Yolk	9.4	69.1
Albumen	3.8	27.9
Shell	0.3	2.2
Shell membranes	0.1	0.8
Total	13.6	100.0

PROTEINS OF THE EGG

Proteins are found in every part of the avian egg. They are present, in a dissolved state, in the albumen; in the yolk body, where they form complex combinations with fats; in the vitelline membrane; and in the shell and shell membranes. However, the chief sources of protein in the egg are the yolk and the albumen; about 44 per cent of the protein is in the yolk, and a little more than 50 per cent is in the albumen. Only a negligible amount is found in the shell and the shell membranes.

The various parts of the average chicken egg contain the following amounts of protein:

	AMOUNT OF PROTEIN	
	GRAMS	PER CENT
Yolk	3.1	44.3
Albumen	3.5	50.0
Shell	0.15	2.1
Shell membranes	0.25	3.6
Total	7.0	100.0

The proteins in the egg are divided into two classes: simple proteins and conjugated proteins (those found in combination with other groups, such as sugars and phosphorus). In the albumen, simple proteins are predominant; the more complex forms are present in the yolk.

Yolk Proteins

To date, two proteins have been isolated from the yolk: ovovitellin and ovolivetin. The ratio of ovovitellin to ovolivetin is approximately 4:1. The average amounts of each in the chicken egg are 2.4 gm. and 0.7 gm., respectively.

Ovovitellin is one of the phosphoproteins, or proteins that contain phosphorus; in fact, ovovitellin contains about one-third of the entire amount of phosphorus present in the yolk. In contrast to ovovitellin, ovolivetin contains little phosphorus, but it has a relatively high proportion of sulfur—approximately one-third of the total sulfur content of the yolk.

HISTORY OF THE ISOLATION OF YOLK PROTEINS

The proteins of the yolk were studied long before those of the albumen. Ovovitellin has been known since the work of Dumas and Cahours (*1842*) and Gobley (*1846*); but probably the most exhaustive work was done by Osborne and Campbell (*1900*).

Apparently the first investigators to recognize the existence of two proteins in the egg yolk were Valenciennes and Frémy (*1854*). Their results were confirmed by those of Gross (*1899*); but the two proteins were first actually separated by Plimmer (*1908*), who named the second one *livetin*. It appears that little was known about ovolivetin, however, until Kay and Marshall (*1928*) isolated it in a fairly pure form.

Attempts have been made to isolate a third protein from the yolk (*Schenck, 1932*); this was named *ovovitellomucoid* by Onoe (*1936*) because of its apparent mucoid nature. However, its chemical identity is still doubtful. It has also been claimed by Piettre (*1936*) that four proteins can be detected in the yolk by the use of his acetone method.

PROTEINS OF THE VITELLINE MEMBRANE

In 1888, Liebermann separated the vitelline membrane from the yolk and found that it contained keratin. Later, Moran and Hale (*1936*) suggested that the different staining properties of the three layers of the vitelline membrane indicate that the center layer is composed of keratin, and that the two outer layers are mucinous.

Albumen Proteins

Egg albumen is composed of five different proteins, one of which, ovalbumin,[2] constitutes about 75 per cent of the total albumen. The other proteins are ovomucoid (13 per cent), ovomucin (7 per cent), ovoconalbumin (3 per cent), and ovoglobulin (2 per cent). These proteins are classified into two groups: simple proteins (ovalbumin, ovoconalbumin, and ovoglobulin), and glycoproteins (ovomucoid and ovomucin). As the name implies, the glycoproteins contain a carbohydrate radical.

Albumen proteins from the eggs of species other than the chicken have in a few instances been given specific names. Panormoff (*1905a*) isolated columbin in dove eggs, and anatin and anatidin in duck eggs. Worms (*1903*) identified corvin, corvinin, and corvinidin in crow eggs. These proteins probably correspond in composition to some of the known proteins of chicken eggs.

HISTORY OF THE ISOLATION OF ALBUMEN PROTEINS

The long controversy over the identification of the albumen proteins is indicative of the complexity of these materials and the difficulties encountered in analyzing them. In fact, the number of different protein

[2] Ovalbumin, or egg albumin, the simple protein, should not be confused with albumen, which is the white of the egg.

fractions isolated has varied from one to eight, or more, depending upon the methods used for their isolation and identification (see Chapter 7, "Physicochemical Properties").

The earliest workers apparently thought egg albumen contained a single protein. Since the middle of the past century, however, individual investigators have usually identified from three to five different albumen proteins (Table 27). On the basis of the most reliable studies, especially

TABLE 27

HISTORY OF ISOLATION OF PROTEINS FROM THE ALBUMEN OF THE CHICKEN EGG
(As shown by representative studies)

Number of Proteins	Proteins Isolated	Investigator
1	A single protein.	The earliest workers
3	One soluble and two insoluble.	Béchamp (*1873*)
4	Two coagulable and two noncoagulable.	Gautier (*1874*)
5	Alpha-, beta-, and gamma-albumins, and alpha- and beta-ovoglobulins.	Corin and Bérard (*1889*)
3	(Identified by heat coagulation.)	Hewlett (*1892*)
3	In order of solubility: ovomucoid, egg albumin, and ovomucin.	Eichholz (*1898*)
4	Mucin, crystallizable ovalbumin, noncrystallizable conalbumin, and noncoagulable ovomucoid.	Osborne and Campbell (*1900*)
3	Two albumins and euglobulin (instead of mucin).	Langstein (*1901*)
4 (7)	Globulin (separated into ovomucin, dysglobulin, euglobulin, and pseudoglobulin), ovalbumin, conalbumin, and ovomucoid.	Obermayer and Pick (*1902*)
3	Globulin, ovomucoid, and crystallizable albumin (studied by means of the gold number).	Schulz and Zsigmondy (*1903*)
3 (4)	Ovoglobulin, ovalbumin, and ovomucoid. The last two contain three distinct antigens (shown by anaphylactic reaction).	Wells (*1911*)
5	Ovoglobulin, ovomucin, crystallizable albumin, noncrystallizable conalbumin, and ovomucoid. (All contain distinct antigens.)	Hektoen and Cole (*1928*)
5	Globulin, mucin, albumin, conalbumin, and mucoid.	Sörensen (*1934*)
5 (8)	Egg albumin (A_1 and A_2), conalbumin, ovomucoid, globulin (G_1, G_2, and G_3), and mucin(?) (by electrophoresis).	Longsworth, Cannan, and MacInnes (*1940*)

those of Sörensen, it is now agreed that egg albumen contains five major proteins. It was once thought that crystalline ovalbumin was the most homogeneous of all protein substances. However, electrophoretic studies by Tiselius and Eriksson-Quensel (*1939*) indicate that it is not such a simple and typical protein as was formerly believed. Further investigation by Longsworth, Cannan, and MacInnes (*1940*) showed that both ovalbumin and ovoconalbumin could be separated into two fractions, and ovoglobulin into three; thus nine proteins were identified in albumen. Using electrophoresis, Frampton and Romanoff (*1947*) found ten protein components in chicken egg albumen.

A few attempts have been made to determine the distribution of some of the individual proteins in the various layers of egg albumen. The proportional amounts were found (*Hughes and Scott, 1936*) to be as follows:

	OUTER LIQUID (per cent)	DENSE (per cent)	INNER LIQUID (per cent)
Ovalbumin	94.4	89.2	89.3
Ovoglobulin	3.7	5.6	9.6
Ovomucin	1.9	5.2	1.1

These data suggest that the albumen layers differ in protein content. These results have been substantiated by electrophoretic studies (*Frampton and Romanoff, 1947*).

Shell-Membrane Proteins

It was determined some time ago (*Lindwall, 1881b; Liebermann, 1888*) that the organic portion of the shell membranes consisted almost entirely of a protein, the chemical composition of which agreed very closely with that of keratin. Analysis by Calvery (*1933a*) indicated that the shell membranes contained pure keratin. However, Moran and Hale (*1936*) later found that, of the three layers of the outer (or shell) membrane, the layer next to the shell contains keratin, whereas the other two layers apparently consist of mucin. The inner, or egg, membrane contains both keratin and mucin in both its layers.

The ovokeratin of the shell membranes is distinguished by its high percentage of sulfur. Relatively, ovokeratin contains from one and one-half to three times as much sulfur as the various proteins of the albumen.

Eggshell Proteins

It has been shown (*von Nathusius, 1868*) that there is protein in the mineral portion of the shell. The protein apparently serves as a matrix in which the inorganic salts are embedded. According to Almquist (*1934*), the matrix protein may be classified as a *collagen,* which is the type of protein found in bone and cartilage. This protein exhibits the general properties of the class of simple proteins known as *albuminoids.*

The outer covering of the shell, the cuticle, contains some protein, which has been identified as a mucin (*Moran and Hale, 1936*).

Composition of Egg Proteins

The various proteins of the egg are highly complex compounds. Knowledge of their organization has grown with the development of each new technique for the study of proteins.

Early in the nineteenth century, elementary analytical methods were used in the attempt to differentiate egg proteins. At the same time, by subjecting proteins to hydrolysis, their general chemical nature was revealed, and, more significantly, the various amino acids that form the protein molecule were discovered.

Although our knowledge of the molecular structure of proteins is still limited, sensitive modern techniques indicate that some of the earlier conceptions are subject to doubt. Nevertheless, it is of general interest to gather into a simple and comprehensive form all the rather disorganized data accumulated through the years. The following discussion is therefore presented as a review of the most important studies of the composition of egg proteins that have been made to date.

ELEMENTARY COMPOSITION

Investigation of the various egg proteins began with the quantitative analysis of their composition. The usual constituent elements of proteins were found to be present in much the same relative amounts as in proteins from other sources. Egg proteins contain about 51 per cent carbon, 24 per cent oxygen, 15 per cent nitrogen, 7 per cent hydrogen, and somewhat lower percentages of sulfur and phosphorus (Table 28). The amount of sulfur is relatively larger in the proteins of the shell and its membranes than in the other egg proteins. Also, there is presumably a higher percentage of phosphorus in the albumen proteins than in the yolk proteins.

The total amount of nitrogen (one of the most important elements of the protein molecule) is distributed among the various proteins of the entire chicken egg as tabulated.

PROTEIN	NITROGEN (per cent of total)
Albumins	49.8
Vitellins, etc.	42.5
Keratins	4.1
Free amino acids	3.6
Total	100.0

TABLE 28

ELEMENTARY CONSTITUTION OF THE PROTEINS OF THE CHICKEN EGG *

Egg Protein	Carbon (C) (per cent)	Oxygen (O) (per cent)	Nitrogen (N) (per cent)	Hydrogen (H) (per cent)	Sulfur (S) (per cent)	Phosphorus (P) (per cent)
Yolk:						
Ovovitellin	51.2	23.6	16.1	7.1	1.0	1.0
Ovolivetin	....	....	15.2	...	1.8	0.2
Albumen:						
Ovalbumin	52.4	22.9	15.3	7.1	1.3	1.0
Ovoconalbumin	52.2	23.0	16.0	7.0	1.7	...
Ovoglobulin	51.7	24.3	15.1	7.1	?	1.7
Ovomucin	50.7	25.8	14.5	6.7	...	2.3
Ovomucoid	48.6	29.1	12.7	7.0	2.3	0.3
Shell membrane:						
Ovokeratin	50.4	22.9	16.1	6.7	3.9	...
Eggshell:						
Ovalbuminoid	....	....	16.6	...	3.8	...

* Compiled from the data of Gobley (*1846*) and Osborne and Campbell (*1900*) on ovovitellin; Plimmer (*1908*) and Kay and Marshall (*1928*) on ovolivetin; Osborne (*1899*), Osborne and Campbell (*1900*), Langstein (*1903, 1907*), and Bywaters (*1909*) on ovalbumin; Osborne and Campbell (*1900*) on ovaconalbumin; Langstein (*1903*) on ovoglobulin; Osborne and Campbell (*1900*) on ovomucin; Zanetti (*1897*), Milesi (*1898*), Osborne and Campbell (*1900*), Langstein (*1903*), and Bywaters (*1909*) on ovomucoid; Lindwall (*1881a*), Liebermann (*1888*), and Mörner (*1901–02*) on ovokeratin.

CHARACTERIZATION BY HYDROLYSIS

Egg proteins were next classified according to the distribution of nitrogen in the products of their hydrolytic decomposition. By the method of Hausmann (*1899, 1900*), a protein is hydrolyzed into four fractions, in which the nitrogen is determined as ammonia, humin, basic, and nonbasic nitrogen. By the Van Slyke (*1911, 1915*) method, not only the amide and humin nitrogen but also the nitrogen of four amino acids (cystine, arginine, histidine, and lysine) are estimated; finally, the nitrogen in the filtrate from the bases is differentiated into amino nitrogen and nonamino nitrogen (Table 29).

Attempts have also been made to identify various chemically bound constituents of egg proteins, such as moisture, nitrogen, carbohydrates, and ash. The percentages of these substances in the proteins of the chicken egg, as determined by a number of analyses, are given in Table 30. This table shows that egg proteins are composed of approximately 7 per cent water, 16 per cent nitrogen, from 0.3 to nearly 15 per cent carbohydrates, and less than 1 per cent ash.

TABLE 29

DISTRIBUTION OF NITROGEN IN THE PROTEINS OF THE CHICKEN EGG *

| | Albumen | | Yolk | | Egg Contents (per cent) | Shell Membrane (Ovo-keratin) (per cent) |
	Ov-albumin (per cent)	Total Protein (per cent)	Ovo-vitellin (per cent)	Total Protein (per cent)		
By Hausmann's method:						
Ammonia N	1.4	....	1.4	....	1.4	3.5
Humin N	0.3	....	0.5	....	0.9	....
Basic N	....	....	....	....	26.3	....
Nonbasic N	....	....	....	....	64.1	....
Total					92.7	
By Van Slyke's method:						
Amide N	1.3	8.3	9.2	9.0	8.7	....
Humin N	0.9	2.0	1.5	2.0	0.9	2.7
Arginine N	10.0	11.7	16.6	14.5	14.5	....
Histidine N	3.3	0.2	1.8	3.1	1.8	....
Lysine N	8.2	10.1	11.1	9.4	8.9	....
Cystine N	....	....	1.2	1.6	0.8	20.9
Amino N	76.3	77.0	55.3	79.2	61.4	72.9
Nonamino N	....	1.6	3.3	....	2.7	4.2
Total	100.0		100.0		99.7	

* Compiled largely from the data of Rudd (*1924b*), Calvery (*1932*), and Calvery and Titus (*1934*) on ovalbumin; Calvery and White (*1932*) on ovovitellin; Stary and Andratschke (*1925*) and Calvery (*1933a*) on ovokeratin; and Calvery (*1929*) on the whole egg.

The determination of nitrogen distribution by these methods contributes little towards the characterization of the different egg proteins, except, perhaps, ovokeratin. Real progress in the chemistry of egg proteins began with the isolation and identification of the individual amino acids.

AMINO ACIDS

The formulation of Fischer's hypothesis that the protein molecule is composed of amino acids greatly stimulated work on the isolation of amino acids from a variety of proteins, including those of the egg. Of more than twenty amino acids identified to date, eighteen have already been found in the egg proteins. There is reason to believe that additional amino acids remain to be discovered, both in egg and other proteins, since, under the best conditions, analysis fails to account for nearly 25 per cent of the protein substance.

TABLE 30

RELATIVE AMOUNTS OF CHEMICALLY BOUND CONSTITUENTS IN THE PROTEINS OF THE CHICKEN EGG *

(Protein is calculated as nitrogen × 6.25)

Chemical Constituent	Albumen					Yolk		Shell Membrane
	Albumins			Glycoproteins		Ovo-vitellin (2.4 gm.)	Ovo-livetin (0.7 gm.)	Ovo-keratin (0.2 gm.)
	Ov-albumin (2.6 gm.)†	Ovocon-albumin (0.1 gm.)	Ovo-globulin (0.1 gm.)	Ovo-mucin (0.2 gm.)	Ovo-mucoid (0.5 gm.)			
	per cent	per cent	per cent	per cent	per cent	per cent	per cent	per cent
Moisture	6.4	7.0	....	7.7	7.7	7.0	....	8.6
Nitrogen	15.3	16.0	15.1	14.5	15.7	16.1	15.2	16.2
Carbo-hydrates	1.7	2.8	4.0	14.9	9.2	0.2	4.0	....
Ash:	0.1	....	....	0.7	1.2	0.4	0.6	0.1
Phos-phorus	0.1	....	....	....	....	0.9	0.1	....
Sulfur	1.5	....	....	1.4	1.7	1.0	1.8	3.8

* Compiled from various sources, including the data of Osborne and Harris (*1903*), Calvery (*1932, 1933a*) Calvery and White (*1932*), Jukes (*1933*), Sörensen (*1934*), and Chargaff (*1942a*).

† The figures in parentheses indicate the approximate amounts of the various proteins in the egg.

Determination of the exact amounts of the individual amino acids is extremely difficult, both because of the imperfections in analytical methods, and because of the variation in amino acid content of individual eggs. Consequently, there is a wide range in the values obtained, even in the more recent analyses. By using the most reliable data, it is nevertheless possible to arrive at average values for the amounts of amino acids in the egg proteins. Figure 190 shows the proportion of the total amino acid content that each amino acid constitutes in the chicken egg. The abundance of leucine, which accounts for about 18 per cent of the entire amount of amino acids, is most striking.

Table 31 indicates the relative amounts of the different amino acids in each protein of the chicken egg. The major yolk protein, ovovitellin, is high in glutamic acid, leucine, arginine, and lysine. The major albumen protein, ovalbumin, is predominantly glutamic acid, leucine, alanine, and aspartic acid. The shell-membrane protein, ovokeratin, consists chiefly of arginine, cystine, glutamic acid, and leucine.

This table shows that the various egg proteins can be differentiated by the number and kind of amino acids that they contain. The arrangement

TABLE 31

(Calculated to 16.0 per cent nitrogen)

Amino Acid	Percentage of Total Protein							
	Yolk		Albumen					Shell Membrane
	Ovo-vitellin	Ovo-livetin	Oval-bumin	Ovo-conal-bumin	Ovo-glob-ulin	Ovo-mucin	Ovo-mu-coid	Ovo-kera-tin
Alanine	0.7	6.0	8.3	...	...	...	...	3.5
Arginine	8.6	5.8	5.4	5.1	4.7	...	5.6	12.9
Aspartic acid	1.0	3.0	7.1	...	...	...	1.8	3.4
Cystine	1.2	3.2	1.2	3.4	...	(4.6)*	6.2	12.7
Glutamic acid	12.4	6.8	15.7	...	...	...	2.0	9.1
Glycine	0.8	...	2.0	...	...	...	...	3.9
Histidine	1.9	1.4	1.8	2.5	1.4	...	4.0	4.2
Leucine	10.0	10.6	12.5	...	...	...	4.0	7.4
Lysine	5.9	5.5	5.1	6.4	5.7	...	1.6	5.2
Methionine	2.9	2.4	5.0	...	...	...	1.7	...
Phenylalanine	1.5	2.0	5.2	...	...	...	4.0	...
Proline	3.3	2.2	4.8	...	...	...	2.4	3.9
Serine	0.5	...	1.2	...	...	...	...	...
Threonine	4.9	...	3.5	...	...	...	...	...
Tryptophane	1.4	1.7	1.7	5.7	4.1	...	2.2	2.7
Tyrosine	5.1	5.1	4.3	4.9	4.2	...	4.7	3.3
Valine	2.1	9.8	5.5	...	...	...	...	1.1

* Based on only one determination.

Selected References

Ovovitellin: Abderhalden and Hunter (*1906*); Hugounenq (*1906*); Levene and Alsberg (*1906*); Osborne and Jones (*1909*); Calvery and White (*1932*); Calvery and Titus (*1934*); Marlow and King (*1936*).

Ovolivetin: Kay and Marshall (*1928*); Jukes and Kay (*1932b*); Jukes (*1933*); Block (*1934*).

Ovalbumin: Abderhalden and Pregl (*1905*); Hugounenq and Morel (*1907*); Levene and Beatty (*1907b*); Osborne, Jones, and Leavenworth (*1909*); Skraup and Hümmelberger (*1909*); Calvery (*1932*); Vickery and Shore (*1932*); Calvery and Titus (*1934*); Chibnall, Rees, and Williams (*1943*).

Ovoconalbumin: Osborne, Leavenworth, and Brautlecht (*1908*); Jones, Gersdorff, and Moeller (*1924*); Block (*1934*).

Ovoglobulin: Fürth and Deutschberger (*1927*); Block (*1934*).

Ovomucin: Young (*1937*).

Ovomucoid: Folin and Denis (*1912*); Komori (*1926*); McFarlane, Fulmer, and Jukes (*1930*).

Ovokeratin: Abderhalden and Ebstein (*1906*); Plimmer and Rosedale (*1925*).

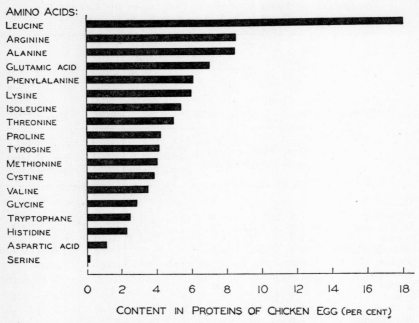

AMINO ACIDS:

CONTENT IN PROTEINS OF CHICKEN EGG (PER CENT)

Fig. 190. Percentages of the various amino acids in the total amino acid content of the proteins of the chicken egg.

of the amino acids within the molecule is also important in determining the nature of a protein. Since the protein molecule is often built up of hundreds of amino acid residues, the complexity of the problem of protein structure is apparent. For example, Bergmann and Niemann (*1937*) reported 288 amino acid residues in the molecule of ovalbumin, in part as follows: cystine, 72; histidine, 72; tyrosine, 36; arginine, 24; lysine, 24; methionine, 24; aspartic acid, 18; and glutamic acid, 8. Bernhart (*1940*) has since estimated the total number as 310, and, according to Bull (*1941a*), it is 354.

OTHER NITROGENOUS SUBSTANCES

Numerous nitrogenous substances other than proteins are present in the egg (see "Other Organic Compounds," p. 351). A small nonprotein nitrogen fraction (about 12 mg.) has been identified in the egg contents. Its composition is shown in the accompanying data (*Needham, 1931a, p. 344*), which bring out the interesting fact that the basic fraction constitutes the greatest part of the nonprotein nitrogen.

Nonprotein Nitrogen	Amount (per cent)
Basic N	88.8
Ammonia N	4.2
Free amino acid N	7.0
Creatinine N	Trace
Total	100.0

Influence of the Bird's Diet on the Protein Content of Eggs

The wide variation found in the protein content of the egg cannot be attributed solely to differences in analytical methods. It is difficult to show, however, that this variation is the direct result of dissimilarities in the diets of laying birds, in spite of the fact that nutrition is one of the most important factors that influence the chemical composition of the egg.

There is evidence that the distribution of nitrogen in the pigeon's egg can be altered significantly by diet (*Gerber and Carr, 1930*), but the amino acid composition of the proteins in the chicken egg apparently remains unchanged when different protein supplements are fed to the hen (*McFarlane, Fulmer, and Jukes, 1930; Calvery and Titus, 1934*). Biological experiments, perhaps more sensitive than chemical analysis, show indirectly that the nature of the protein in the bird's diet nevertheless affects the egg's chemical composition. The embryonic mortality rate is higher in eggs laid by hens on vegetable-protein diets than it is in eggs laid by hens that receive proteins of animal origin (*Byerly, Titus, and Ellis, 1933*). Among the vegetable proteins, those of hemp, the soybean, and wheat are apparently superior to those of corn and oats, when judged by the embryonic development of the pigeon (*Gerber and Carr, 1930*). McFarlane, Fulmer, and Jukes (*1930*) observed that the hatchability of eggs from hens fed a fat-free cod-liver-meal supplement was greatly superior to that of eggs from hens fed tankage.

Although studies of hatchability indicate that the type of protein consumed by the bird influences the composition of the egg, chemical analyses do not offer definite evidence as to whether it is the proteins or other constituents of the egg that are affected. It is therefore obvious that generalizations concerning the correlation between the bird's diet and the protein content of the egg must be avoided until more conclusive data become available.

Biological Significance of Proteins in the Egg

Egg proteins are used by the developing embryo chiefly for growth, and only to a lesser extent for maintenance. These proteins must contain amino acids in sufficient quantity and variety to supply the embryo with the necessary materials and energy for tissue building.

Some of the amino acids are used to varying extent for the construction of body cells in the growing embryo; others are quickly deaminized to form urea. In addition, certain amino acids are capable of being converted into a simple form of carbohydrate, glucose.

Egg proteins, indispensable for the growth of the chick embryo, contain the amino acids that are essential for the growth of all vertebrates (*Jukes and Kay, 1932a*). The high biological value of egg proteins, and especially of the yolk proteins, makes the egg an especially important source of nutrients for animals and man.

LIPIDS OF THE EGG

"Lipids" is a general term applied to fats and all fatlike substances. In the egg, the lipids are present as true fats, and as compounds containing, in addition to fat, phosphorus, nitrogen, and sugar.

All lipids of the avian egg, except for traces that occur in the albumen and shell,[3] are concentrated in the yolk. Of the 6.2 gm. of lipids in the chicken egg, about 99 per cent are in the yolk (*Romanoff, 1932*).

Although little is known about the nature of the lipids in the albumen, those in the yolk have been studied extensively. Many of the yolk lipids are extremely difficult to isolate in a pure state, and much confusion consequently prevails regarding their exact chemical composition. The results of numerous investigations, however, afford full justification for

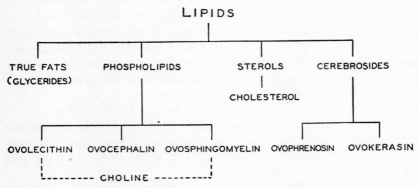

Fig. 191. Lipid constituents of the yolk of the chicken egg.

[3] Appreciable amounts (up to 4 per cent) are found in the shell of some ducks' eggs, particularly near the surface (*Pritzker, 1941*).

classifying yolk lipids with similar compounds found in animal tissues, since they possess many nearly identical characteristics.

Yolk Lipids

The lipids of the yolk, which constitute about 33 per cent of the yolk (wet weight), are very complex in structure. They consist not only of glycerides, or true fats, but also of phospholipids, sterols, and cerebrosides. Figure 191 outlines the different lipids found in the egg yolk.

Of the total lipids, the true fats constitute about two-thirds, and the phospholipids, sterols, and cerebrosides, about one-third. The approximate amounts of these substances in the yolk of the chicken egg are shown in the tabulation.

	AMOUNT	
	GRAMS	PER CENT
True fats	3.8	62.3
Phospholipids	2.0	32.8
Sterols	0.3	4.9
Cerebrosides	Trace	....
Total	6.1	100.0

The most abundant fatty acids in the egg are palmitic, oleic, and linoleic. The first of these is a saturated acid; the last two are unsaturated. Saturated and unsaturated acids are differentiated by the ability of the latter to become oxidized upon exposure to the air.

ELEMENTARY CONSTITUTION

An elementary analysis of the various egg-yolk lipids reveals both the similarities and the differences in their chemical structure (Table 32). Of particular interest are the wide variations in the proportions of carbon and oxygen, and in the ratio of nitrogen to phosphorus (in the phospholipids). These analyses are important chiefly for their historical interest, but they also fully confirm the nature of the compound lipids.

TRUE FATS

The fats of the yolk contain mixtures of glycerides. These substances are composed of glycerol (glycerine) and several fatty acids. In each glyceride, one molecule of glycerol is combined with three fatty acid radicals.

The workers who first studied the yolk fat obtained widely different results, presumably because of unreliable techniques. However, more recent analyses by Grossfeld (*1933b*), Cruickshank (*1934*), and Riemen-

TABLE 32

ELEMENTARY CONSTITUTION OF LIPID COMPOUNDS IN THE YOLK OF THE CHICKEN EGG *

Lipid Compound	Carbon (C) (per cent)	Hydrogen (H) (per cent)	Nitrogen (N) (per cent)	Phosphorus (P) (per cent)	Nitrogen / Phosphorus (molecular ratio)	Oxygen (O) (per cent)
Phospholipids:						
Ovolecithin	65.0	11.4	2.0	4.0	1:1	17.6
Ovocephalin	59.5	9.7	1.7	3.7	1:1	....
Ovosphingomyelin	63.3	10.9	2.9	3.7	2:1	....
Cerebrosides:						
Ovophrenosin ⎫ Ovokerasin ⎭	69.3	11.2	1.9	0.0	...	....
Sterol:						
Ovocholesterol	83.7	11.9	...	...	...	4.1

* Compiled from the data of Stern and Thierfelder (*1907*) and McLean (*1909*) on ovolecithin; Levene (*1916*) on ovosphingomyelin; Levene and West (*1917*) on ovophrenosin and ovokerasin; and Barbieri (*1907*) on ovocholesterol.

schneider, Ellis, and Titus (*1938*) agree essentially as to the relative amounts of the more abundant fatty acids found in the yolk fat. The tabulated percentages represent average values obtained from the data of all the above investigators.

FATTY ACIDS	AMOUNT (per cent)
Oleic	50
Palmitic	27
Linoleic	11
Stearic	6
Palmitoleic ⎫ Linolinic ⎪ Clupanodonic ⎪ Myristic ⎭	6

The saturated acids, palmitic, stearic, and myristic, constitute 34 per cent of the total fatty acids; the unsaturated acids, oleic, linoleic, and others, make up 66 per cent. To the young embryo not yet able to desaturate the saturated fatty acids, the presence of unsaturated acids in the egg may be essential.

PHOSPHOLIPIDS

The phospholipids of the yolk have been identified as three distinct substances: ovolecithin, ovocephalin, and ovosphingomyelin. A fourth substance, ovocuorin (*Erlandsen, 1907*), has also been isolated, but its chemical identity is still doubtful.

All phospholipids contain phosphorus, nitrogen, and fatty acids. In ovolecithin and ovocephalin, the nitrogen-phosphorus ratio is 1:1, but the two compounds differ with respect to the kind of nitrogen-containing base that is attached to the phosphorus. The ratio of nitrogen to phosphorus in ovosphingomyelin is 2:1. The number and kind of fatty acids vary in each phospholipid.

Table 33 shows the composition of the phospholipids in the yolk of the chicken egg.

TABLE 33

DISTRIBUTION OF CHEMICAL CONSTITUENTS IN THE PHOSPHOLIPIDS OF THE YOLK OF THE CHICKEN EGG *

			Per Cent
Ovolecithin	Fatty acids	Oleic	43
		Palmitic	32
		Clupanodonic	13
		Linoleic	8
		Stearic	4
		Arachidonic	Traces
	Glycerophosphoric acid	{ Glycerol, Phosphoric acid	
	Base	Choline	
Ovocephalin	Fatty acids	{ Stearic, Cephalinic (arachidonic) }	69.2
	Glycerophosphoric acid	{ Glycerol	11.3
		Phosphoric acid	12.0
	Base	Aminoethyl alcohol	7.5
Ovosphingomyelin	Fatty acids	{ Lignoceric, Cerebronic, Stearic	43
	Phosphoric acid		23
	Bases	{ Choline, Sphingosine }	34

* Based chiefly on the data of Cruickshank (*1934*) and Riemenschneider, Ellis, and Titus (*1938*) on ovolecithin; Levene and West (*1916*) on ovocephalin; and Levene (*1916*) on ovosphingomyelin.

Ovolecithin. The yolk of the chicken egg contains more than 1.6 gm. of lecithin. Lecithin constitutes about 8.6 per cent of egg yolk, which thus ranks next to brain and nerve tissue in its content of this substance.

When isolated from the yolk, lecithin appears as a yellowish, waxy solid, with a peculiar odor. On contact with the air, it deepens in color.

Ovolecithin has been found to contain several fatty acids, both saturated (palmitic and stearic) and unsaturated (oleic, clupanodonic, and linoleic) (cf. Table 33). There is a larger proportion of unsaturated acids than of saturated acids; the relative amounts are about 64 and 36 per cent, respectively (*Riemenschneider, Ellis, and Titus, 1938*).

It has been suggested also (*Yokoyama, 1934*) that lecithin from chicken egg yolk consists of about 75 per cent beta-lecithin, and that the remaining portion is alpha-lecithin. Alpha-lecithin yields oleic, isopalmitic, and clupanodonic acids, in the ratio 72:26:2. Beta-lecithin contains oleic and isopalmitic acids only.

Other constituents obtained by the hydrolysis of ovolecithin are glycerol and phosphoric acid, which form glycerophosphoric acid (entirely in beta- form), and choline, a nitrogenous base. Of about 0.2 gm. of choline in egg yolk (*Nottbohm and Mayer, 1933*), approximately 75 per cent is present in ovolecithin, and the next largest portion in ovosphingomyelin.

According to Masuda and Hori (*1937*), lecithin constitutes different proportions of the total phospholipids in the eggs of various species of birds. Analyses of dried egg yolk showed that lecithin made up about 69 per cent of the phospholipids in the hen's egg, 75 per cent in the duck's, 54 per cent in the quail's, and 82 per cent in the peahen's. These variations in the amount of lecithin cannot easily be explained.

Ovocephalin. About 0.4 gm. of ovocephalin is present in egg yolk. In the chicken egg, ovocephalin represents about 25 per cent of the total phospholipids (*Kaucher, Galbraith, Button, and Williams, 1943*). In many respects, ovocephalin closely resembles ovolecithin, but it is differentiated by its insolubility in alcohol. When hydrolyzed, ovocephalin yields two fatty acids (stearic and cephalinic), glycerophosphoric acid, and aminoethyl alcohol (colamin), a base (cf. Table 33). In properties and composition, ovocephalin is identical to the cephalin of brain, kidney, and liver (*Levene and West, 1916*).

Ovosphingomyelin. Ovosphingomyelin constitutes only 2 or 3 per cent of the total phospholipids of the chicken egg (*Kaucher, Galbraith, Button, and Williams, 1943*). Sphingomyelin differs from lecithin and cephalin chiefly by the absence of glycerol in its molecule. Ovosphingomyelin contains at least two (*Levene, 1916*) and possibly three (*Merz, 1930*) fatty acids (lignoceric, cerebronic, and stearic), as well as phosphoric acid, choline, and another nitrogenous base, sphingosine (*Levene,*

1916). The bases equal about 34 per cent, and the fatty acids 43 per cent, of the entire ovosphingomyelin molecule (cf. Table 33).

STEROLS

One of the most important sterols, cholesterol, is found in egg yolk. Chicken egg yolk contains about 1.6 per cent (or 0.3 gm.) of cholesterol, duck egg yolk somewhat more (*Gaujoux and Krijanowsky, 1932; Levy and Georgiakakis, 1934*). About 84 per cent is present in the free state, and the remainder in an esterified form. The ratio between free and total cholesterol varies in different layers of the yolk; more free cholesterol is found in the yellow than in the white yolk, especially the white yolk in the nucleus of Pander (*Marza and Marza, 1932*). The cholesterol content of the egg is very constant and varies only slightly among the eggs of different species of birds (*Miyamori, 1934*).

Part of the cholesterol in the egg is derived from the animal food consumed by the bird, and part is synthesized during the process of yolk formation. Its abundance in nervous tissue, and its chemical relationship to bile acids and sex hormones, indicate the importance of its presence in the egg.

CEREBROSIDES

Traces of two cerebrosides, ovophrenosin and ovokerasin, have been isolated from egg yolk (*Levene and West, 1917*). They differ from the phospholipids because they contain neither phosphoric acid nor glycerol. Instead, they contain galactose, a sugar, and sphingosine, a nitrogenous base, and are classified as glycolipids. They are optically active; ovophrenosin is dextrorotatory, and ovokerasin is levorotatory. Ovophrenosin closely resembles ovokerasin in chemical composition, except for some differences in fatty acid content (*Bloor, 1943, p. 20*). The presence of a sugar and fatty acids in the same molecule is perhaps important in the metabolism of fats.

Influence of Diet on the Lipid Composition of the Yolk

In eggs laid by hens that receive ordinary rations, the composition of the yolk fat varies only slightly from egg to egg, since the hen is able to synthesize fat from carbohydrates. Some investigators have found little or no change in the fat content of the egg yolk, even upon experimental modification of the hen's diet. Buckner, Insko, Martin, and Harms (*1939*) observed that a low-starch ration altered the reserve body fat in amount and composition to a greater extent than the yolk fat, which was hardly affected.

However, most experimental results have shown that the constituents of the yolk lipids, especially the fatty acids, are influenced by the hen's diet. When hens are fed a low-fat ration, the egg fats contain more saturated fatty acids than normal, as shown by lower iodine number (*Russell, Taylor, and Walker, 1941*). An increase in saturation, both of the neutral fats and the lecithin, may also occur when hens are given lipid-free diets (*McCollum, Halpin, and Drescher, 1912; Terroine and Belin, 1927*). On the other hand, diets high in unsaturated fatty acids lead to changes in the neutral fat of the yolk lipids, but they do not affect the lecithin (*Henriques and Hansen, 1903*).

The investigations of Cruickshank (*1934*) show that, in the synthesis of egg fat, the hen exercises a selective power. The degree of saturation may be considerably decreased by the ingestion of fats, such as hempseed and linseed oils, that have a high content of unsaturated fatty acids; and the proportions of the component fatty acids in the egg fats may be altered. The ingestion of large amounts of saturated acids, however, has little effect on the normal composition of the mixed fatty acids of the yolk fat.

Oil from cottonseed meal can be deposited in the yolk (*Almquist, Lorenz, and Burmester, 1934*). When the hen is fed fish meals of inferior grade (containing a high percentage of free fatty acids) unsaturated fatty acids appear in the yolk, which acquires a fishy flavor. Furthermore, prolonged feeding of fish oils increases the degree of unsaturation of the yolk fats (*Cruickshank, 1939*).

The egg's content of cholesterol apparently cannot be changed by minor variations in the hen's ration (*Cross, 1912; Leveque and Ponscarme, 1913*). A diet very rich in cholesterol, however, may increase the amount of cholesterol in the egg 18 to 23 per cent (*Dam, 1928, 1929*). The possible maximum limit of this substance in the egg has not yet been determined.

Biological Significance of Lipids in the Egg

Lipids, found in abundance in the egg yolk, are present in every living cell. As nutritive substances, they are indispensable to the life and growth of the chick embryo. The neutral fats act as the most important source of energy for the embryo; and the compound lipids, notably the phospholipids, are essential structural constituents of the body cells. The phospholipids may also be intermediate in the fat metabolism of the embryo.

The presence of choline in the egg is especially interesting. Choline is found in the free state in bile and blood, in small amounts. Physiologically, it is very active in lowering the blood pressure, stimulating the contractions of the small intestines, and preventing the accumulation of fat in the liver.

The composition of the egg lipids is perhaps one of the factors that affect the hatchability of the egg. Variations in lipid content therefore are of considerable biological interest.

CARBOHYDRATES OF THE EGG

The egg contains relatively small quantities of carbohydrates. The total amount averages about 0.5 gm. (*Sakuraji, 1917; Needham, 1927; Donhoffer, 1933*), of which nearly 75 per cent is in the albumen. Carbohydrates are found in the free form and in chemical combination with proteins and fats. The tabulation shows the proportion of the yolk, albumen, and total egg contents that each form of carbohydrate constitutes.

	AMOUNT OF CARBOHYDRATE (per cent)		
	IN YOLK	IN ALBUMEN	IN EGG CONTENTS
Free carbohydrates (glucose)	0.7	0.4	0.6
Combined carbohydrates	0.3	0.5	0.4
Total	1.0	0.9	1.0

Free carbohydrate is usually found as glucose; the combined, as mannose and galactose. Combined carbohydrates are present in the phosphoproteins, phospholipids, and cerebrosides of the yolk, and in the glycoproteins and other simple proteins of the albumen.

Yolk Carbohydrates

According to Fränkel and Jellinek (*1927*), the protein-bound carbohydrates in the yolk are polysaccharides of the mannose-glucosamine type. The accompanying table gives the percentages of carbohydrates (expressed as glucose) that Jukes (*1933*) found in ovovitellin and ovolivetin.

YOLK PROTEIN	CARBOHYDRATE (per cent)
Ovovitellin	2.0
Ovolivetin	4.0

The phospholipid, ovolecithin, and the cerebrosides (ovophrenosin and ovokerasin) contain the sugar, galactose (*Stern and Thierfelder, 1907*). Sakuraji (*1917*), Idzumi (*1924*), and Donhoffer (*1933*) claim that some sugar, in the form of glycogen (about 9 mg.), is present in both the yolk and the albumen of fresh eggs.

Albumen Carbohydrates

The nature of the carbohydrate group in egg albumen has been studied by numerous investigators. Among the albumen proteins, ovalbumin has been the most frequently and thoroughly studied.

Fränkel and Jellinek (*1927*) isolated from ovalbumin a carbohydrate that they believed was a disaccharide composed of glucosamine and mannose. Levene and Mori (*1929*), however, considered this compound to be a trisaccharide, containing one glucosamine residue and two mannose residues. Later, Levene and Rothen (*1929*) concluded that the carbohydrate of the ovalbumin molecule is composed of four trisaccharide units. Neuberger (*1938*) obtained a polysaccharide consisting of two molecules of glucosamine, four molecules of mannose, and one molecule of an unidentified nitrogenous constituent.

The nature of the carbohydrate of other albumen proteins is similarly indefinite. In ovomucoid, for example, Zuckerkandl and Messiner-Klebermass (*1931*) regarded the carbohydrate as a prosthetic polysaccharide, a mixture of glucosamine and mannose. Sörensen (*1934*) detected one galactose and three mannose residues. Masamune and Hoshino (*1936*) claim that the prosthetic group consists of an equimolecular mixture of mannose and acetylglucosamine.

Sörensen (*1934*) found that, in ovoconalbumin, the carbohydrate moiety was apparently formed of three parts of mannose and one part of galactose; in ovomucin, of equal parts of mannose and galactose; and in ovoglobulin, of mannose alone. The following proportions of carbohydrates were observed in the various proteins of egg albumen:

	CARBOHYDRATE	
	KIND	AMOUNT (per cent)
Simple Proteins:		
Ovalbumin	Mannose	1.7
Ovoconalbumin	Mannose and galactose	2.8
Ovoglobulin	Mannose	4.0
Glycoproteins:		
Ovomucin	Mannose and galactose	14.9
Ovomucoid	Mannose and galactose	9.2

Influence of Diet on the Carbohydrate Content of Eggs

No experimental data are available regarding the influence of the hen's diet on the carbohydrate content of her eggs. However, Biasotti (*1932*) observed a decrease in the glucose content of the egg when hens were given behenic acid. The amount of glucose was reduced to a greater extent in the yolk than in the albumen.

Biological Significance of Carbohydrates in the Egg

The egg resembles other substances of animal origin in containing very small amounts of carbohydrates as compared with plant tissues. The carbohydrates enter into the structure of new tissues in the developing embryo and assist in the functional activities of the organism.

PIGMENTS OF THE EGG

Pigment substances are present in all parts of the egg, but the chemical nature of the various pigments differs greatly. The largest amounts of pigments are found in the yolk (0.4 mg.) and in the albumen (0.03 mg.); the other parts of the egg contain only traces.

Yolk Pigments

The yolk, although it is the part of the egg richest in pigment material, nevertheless contains only about 0.02 per cent of pigments. Yolk pigments are classified as lipochromes and lyochromes.

LIPOCHROMES

Lipochromes, which are fat soluble, constitute the larger proportion of yolk pigments. These pigments belong to the group known as *carotenoids,* which are present in abundance in various plants. Carotenoid pigments are usually transferred to the yolk from carotenoid-rich plants consumed by the hen. The most common of these plants are yellow corn and various grasses. Grasses are rich in carotene and lutein; corn contains chiefly zeaxanthin.

Carotenoids are the red, orange, and yellow pigments of the chloroplast (chlorophyll granule). Their molecules are long and contain a large number of conjugated double bonds, which, with other patterns of unsaturation, are the chemical basis of color in organic compounds.

Many carotenoids have been isolated from plants, but only a few of them are found in the yolk (*Peterson, Hughes, and Payne, 1939*). These may be divided into two groups according to their composition: carotenes and xanthophylls (Fig. 192).

In addition to the two carotenes (alpha- and beta-) and the three xanthophylls (cryptoxanthin, lutein, and zeaxanthin), traces of other carotenoid pigments have been found in the egg yolk. These are: lycopene (*Kuhn, Winterstein, and Lederer, 1931*), flavoxanthinlike compounds, and neoxanthin (*Strain, 1936*).

Chemical Structure. Carotene ($C_{40}H_{56}$) is a highly unsaturated hydrocarbon. It has been found in yolk in two isomeric forms known as alpha- and beta-carotene (*Strain, 1935*). It is insoluble in water, acid, or alkali but is soluble in chloroform and ether.

Xanthophylls usually contain at least two hydroxyl groups. These pigments are closely related structurally to carotene. Cryptoxanthin ($C_{40}H_{55}OH$) is a monohydroxy derivative of beta-carotene; and lutein and zeaxanthin ($C_{40}H_{56}O_2$) are isomeric dihydroxy derivatives of alpha- and beta-carotene, respectively.

Xanthophylls, unlike carotenes, dissolve readily in alcohol and ethyl ether, but only slightly in petroleum ether. Cryptoxanthin, however, is

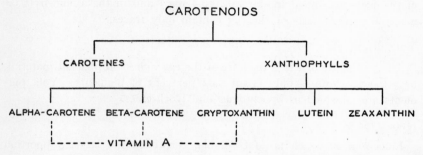

Fig. 192. Carotenoid pigments of the yolk of the chicken egg.

difficult to distinguish from beta-carotene, and also from zeaxanthin. Lutein can be differentiated from zeaxanthin by its greater solubility in boiling methanol (*Peterson, Hughes, and Payne, 1939*).

The color values of the various carotenoids are somewhat different (*Kuhn and Brockmann, 1932*); the xanthophylls, especially lutein and zeaxanthin, are approximately twice as intense in color as the carotenes.

Amounts in the Yolk. Most of the carotenoid pigments in the yolk are in the form of xanthophylls; the usual ratio of carotenes to xanthophylls is 1:10 (*Sjollema and Donath, 1940*). The average amounts of these pigments in the yolk are given in the tabulation.

Pigments	Amount (milligram)
Carotenes:	
Alpha-carotene	Traces
Beta-carotene	0.03
Xanthophylls:	
Cryptoxanthin	0.03
Lutein	0.1
Zeaxanthin	0.2

Carotenoid pigments are related to vitamin A. Beta-carotene is potentially twice as active as alpha-carotene. Of the xanthophylls, only cryptoxanthin is considered to be convertible into vitamin A, but it, too, is only one-half as potent as beta-carotene.

LYOCHROMES

There is at least one lyochrome, or water-soluble pigment, in egg yolk; this is ovoflavin (*Karrer and Schöpp, 1934*). It is orange-yellow but fluoresces yellowish green. Egg yolk contains approximately 0.13 mg. of ovoflavin (*Snell and Strong, 1939; Norris and Bauernfeind, 1940*). Chemically, it is identical with riboflavin ($C_{17}H_{20}N_4O_6$). Ovoflavin functions in the prosthetic groups of the yellow respiratory enzymes, as indicated by the work of Adler and von Euler (*1934*).

OTHER PIGMENTS

It has been claimed that protoporphyrin can be detected in the yolks of chicken, duck, and sea-gull eggs by means of its fluorescent spectrum (*Gouzon, 1934*). The amount in the chicken egg was found to be less than 2γ (*van den Bergh and Grotepass, 1936*).

Numerous pigments may be found in the yolk when the hen has access to certain foods the pigments of which are easily transferred. Unusual yolk pigments of dietary origin are discussed below.

Albumen Pigment

Egg albumen presumably contains only one pigment, the water-soluble ovoflavin, first studied by Kuhn, György, and Wagner-Jauregg (*1933*) and Ellinger and Koschara (*1933*). The average amount of this pigment in the albumen is about 0.07 mg. (*Norris and Bauernfeind, 1940*).

Shell-Membrane Pigment

The shell membranes sometimes have a distinct pink tinge, caused by a pigment that has been identified as a porphyrin (*Klose and Almquist, 1937*). A faint purple pigment, not a porphyrin, was found in the membranes by Brooks (*1937*).

Eggshell Pigments

The colors of birds' eggshells are due to two major pigments, red-brown and blue-green, which were first identified with the bile pigments—the brown with bilirubin, and the green with biliverdin (*Wicke, 1858*). The shell pigments are now considered to be more closely related to the hemoglobin of the blood than are the bile pigments.

The red-brown pigment of the eggshell has been named oöporphyrin (*Fischer and Kögl, 1923, 1924; Fischer and Müller, 1925; Bierry and Gouzon, 1939*). This pigment may easily be converted into hematoporphyrin, which shows a red luminescence in ultraviolet light, of wave length of 3660 angstrom units, generated by a quartz-mercury lamp (*Hulsebosch, 1927*). Hematoporphyrin is probably related to hematin, which is a decomposition product of the hemoglobin of the blood.

The brown eggshells of the domestic fowl are rich in oöporphyrin. According to Voelker (*1940*), oöporphyrin is present in the shell of white eggs (at least those of the chicken, duck, goose, pigeon, and owl) when they are laid, but it is quickly destroyed by light. Oöporphyrin was also found in the eggshells of many other species, including a number of passerine birds.

The blue-green pigment of some highly colored eggshells was named oöcyan by Sorby (*1875*). The chemical properties of this pigment indicate that it is a substance containing three pyrrole rings (*Lemberg, 1931*). According to Tixier (*1945*), the green pigment of the emu egg is the methyl ester of biliverdin IXa.

Cuticular Pigment

The pigment of the cuticle was first detected by its red fluorescence in ultraviolet light (*Derrien, 1924; Tapernoux, 1930*). On the basis of these observations, Furreg (*1931*) identified it as a porphyrin.

Influence of the Bird's Diet on the Pigments of the Egg

The ability to deposit pigments in the egg is inherited by the bird. Nevertheless, the intensity and, to some extent, the quality of color are influenced by the nature of the food consumed. This is much more evident in the yolk than in any other part of the egg.

Deeply colored yolks contain excessive amounts of xanthophylls. These pigments are commonly found in some natural feeds and are fairly easily transferred from the diet to the egg. After a hen has been fed yellow corn, which is rich in xanthophylls, the amount of these pigments (lutein and zeaxanthin) in the yolk increases to an extent that depends on the amount of corn fed (Fig. 193).

The bird has a selective mechanism for the separation of carotenoid pigments. She preferentially utilizes and deposits xanthophylls. Consequently, rather large amounts of xanthophylls and only small amounts of carotenes are found in the yolk. The ratio of carotenes to xanthophylls may be as high as 1:100 in eggs from hens fed large amounts of yellow corn (*Gillam and Heilbron, 1935*).

It is clear that the amounts of the vitamin-A-active carotenoids (the carotenes and cryptoxanthin) in the yolk are not so readily influenced by diet as are the xanthophylls. By adding yellow corn and alfalfa to a ration low in carotenes, Sjollema and Donath (*1940*) increased the amount of carotene to about 0.040 mg. per yolk, from the original level of 0.015 mg. The maximum increase, to a level of 0.194 mg. per yolk, has been

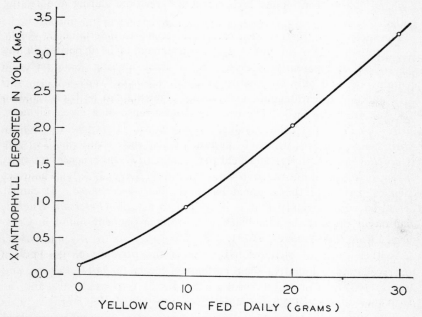

FIG. 193. Relationship between the quantity of yellow corn fed to the hen each day and the amount of xanthophyll pigments (lutein and zeaxanthin) deposited in the yolk of the egg. (After Peterson, Hughes, and Payne, 1939.)

attained by means of a diet containing large amounts of yellow corn alone (*Hughes and Payne, 1937*).

On the other hand, the administration of large amounts of vitamin A (in the form of high-potency shark-liver oil—approximately 20,000 I.U. per gram) resulted in a marked and progressively greater suppression in the content of carotenoid pigment in the egg yolk (*Deuel, Hrubetz, Mattson, Morehouse, and Richardson, 1943*), as shown in the tabulation.

VITAMIN A SUPPLEMENT (International Units/pound)	CAROTENOIDS IN YOLK (γ/gram)
0	32.3
1,000	29.4
15,000	25.3
60,000	12.9
200,000	8.4

These supplements, however, significantly increased the levels of vitamin A in the yolks (see Chapter 9, "Food Value").

According to studies by Radeff (*1935*), eggs produced after a nonlaying interval of several days possess a greater amount of carotenoids than those laid before the interval. This indicates that the pigments become depleted during continuous laying and are restored during the resting period.

The molecular structure of a carotenoid seems to determine its ability to be transferred to the yolk. For a carotenoid to be deposited in the egg yolk, it is apparently necessary that at least one ring of the molecule contain one, and only one, hydroxyl group (*Brown, 1938*).

In addition to carotenoids, many other pigments can be transferred to the yolk from the feed. The occasional occurrence of a green pigment, in the so-called olive-colored yolks, is well known. It is found especially when hens have free access to certain plants, such as shepherd's-purse (*Capsella bursa-pastoris*) and field pennycress (*Thlaspi arvense*) (*Payne, 1925*), or to an alfalfa range in winter months (*Berry, 1938*). The exact chemical nature of this pigment has never been ascertained. It cannot be stated with certainty that it is related to astasin (*Peterson, Hughes, and Payne, 1939*), the blue-black water-soluble pigment found in Crustacea (*Kuhn and Lederer, 1933*).

Another cause of olive-colored yolks is the presence of gossypol, a toxic yellow dye found in some cottonseed meals (*Schaible, Moore, and Moore, 1934*). Cottonseed meal in the diet is responsible also for the development of a reddish discoloration in the yolk during storage. This reddish pigment is due to the presence in cottonseed of a substance other than gossypol (*Lorenz, 1939*).

The orange-red color that appears in the yolk when hens are given pimiento pepper is due to the presence of the carotenoid pigment, capsanthin (*Brown, 1938*).

Certain fat-soluble dyes, such as Sudan III (*Gage and Gage, 1908*), Hexyl Blue, Oil Red, Alizarol Purple, Oil Brown, and others (*Denton, 1940*) are readily transferable to the yolk.

It has been observed (*Bilek, 1936*) that a greater intensity of yolk color may be associated with an increased content of cholesterol.

The pigments of the albumen are much less readily affected by diet than those of the yolk. The intensity of the yellowish green pigment of the albumen, however, is associated with the amount of riboflavin in the diet (*Heiman, 1935*). Lorenz (*1939*) found that cottonseed meal in the diet caused a reddish tint to appear in the albumen during storage.

OTHER ORGANIC COMPOUNDS

In addition to the well-known organic constituents of the egg, many other organic compounds have been reported in the fresh egg. However, the term "fresh egg" (see Chapter 10, "Preservation") has been very loosely used, especially in chemical literature. A considerable amount of analytical work on the chemical composition of the egg has been done without regard for the egg's age, fertility, or state of development. Consequently, some of the substances found in so-called fresh eggs may well be either the intermediate products of organic decomposition, or the products of the enzymatic activity usually associated with aging and fertility. Fertile eggs, and especially preheated fertile eggs, may contain the endogenous products of early embryonic metabolism, or the forerunners of new organic substances which are synthesized in large amounts later in development.

Table 34 lists the amounts of various organic compounds isolated from the hen's egg. Well represented are the decomposition, or endogenous, products of proteins (uric acid, creatinine, and purine nitrogen), of carbohydrates (lactic acid), and of phospholipids (choline), as well as substances synthesized during embryonic development (inositol, ethyl alcohol, and the pigment, melanin), or found in plant and animal tissues (plasmalogen, estrogen, and ergosterol, and the plant hormone auxin, the presence of which was contested by Berrier, in 1939).

The bird's egg also contains pigments, vitamins, and a variety of enzymes, which are discussed elsewhere.

The egg is a reproductive cell that fulfills its function in isolation from the maternal environment. Communication with the external surroundings is limited to the respiratory exchange of gases through the shell. Since all the metabolic processes associated with embryonic development occur within the confines of the eggshell, it is natural that various products of metabolism and synthesis should be detectable in the contents of the fertile egg after only a short period of incubation, or even before incubation. The identification of a number of substances in the egg may therefore probably be attributed to the fertility of the eggs that were examined.

INORGANIC CONSTITUENTS

A wide variety of minerals is contained in the egg. The greatest part of the egg's mineral content is associated with organic matter, but a small proportion is also found in the inorganic form. Minerals are essential components of the protoplasm of living cells. The presence of minerals in the dissociated form, as electrolytes, is also of importance in regulating not only the water content of various parts of the egg, but also the chemical reactions that take place in the body fluids and tissues of the developing embryo.

TABLE 34

OTHER CHEMICAL COMPOUNDS FOUND IN THE CHICKEN EGG

Chemical Compound	Yolk (milligrams)	Albumen (milligrams)	Egg Contents (milligrams)	Investigator
Acetaldehyde	....	....	0.02	Takahashi (*1937*)
Acetone	....	....	0.08	Takahashi (*1937*)
Avidin	....	0.02	0.02	György and Rose (*1942*)
Auxin	....	....	57 P.U.*	Robinson and Woodside (*1937*)
Choline	200	....	240	Sharpe (*1924*), Nottbohm and Mayer (*1933*)
Citric acid	2.9	0.33	....	Thunberg (*1941*)
Creatine	....	....	Negligible	Sendju (*1927*), Stepp, Feulgen, and Voit (*1927*)
Creatinine	....	....	Present	Kojo (*1911*), Salkowski (*1911*)
Ergosterol	....	....	0.5	Menschick and Page (*1932*)
Estrogen	0.03–0.04	....	....	Riboulleau (*1938*, *1940–41*), Marlow and Richert (*1940*)
Ethyl alcohol	0.79	....	3.7	Stepanek (*1905*), Aoki (*1925*), Needham, Stephenson, and Needham (*1931*)
Glutathione	....	....	Present	Talenti (*1935*)
Guanidine	....	....	30.0	Burns (*1916*)
Inositogen	....	2.1	....	Rosenberger (*1908*)
Inositol	3.0–4.0	0.003	3.0–16.0	Starkenstein (*1911*), Needham (*1924*)
Lactic acid	1.9–3.0	2.5	6.6	Tomita (*1921*), Capraro and Fornaroli (*1939b*)
Melanin	Present	....	....	Needham (*1931a, p. 1381*)
Phytin	....	....	(?)	Plimmer and Page (*1913*)
Plasmalogen	Rich	Slight	....	Stepp, Feulgen, and Voit (*1927*)
Purine nitrogen	....	....	1.6	Mendel and Leavenworth (*1908*)
Uric acid	....	....	8.0	Takahashi (*1937*)

* P.U. stands for plant units.

The bulk of the inorganic matter, about 94 per cent, is present in the shell. The remaining 6 per cent is distributed equally between the yolk and the albumen.

The table below shows the distribution of inorganic matter (or ash) in the chicken egg.

The mineral elements constituting the inorganic portion of the egg may be conveniently divided into two groups, major and trace elements.

	AMOUNT OF INORGANIC MATTER	
	GRAMS	PER CENT OF TOTAL
Eggshell with membranes	5.9	93.6
Yolk	0.2	3.2
Albumen	0.2	3.2
Total	6.3	100.0

MAJOR MINERAL ELEMENTS

Only a few mineral elements are present in relative abundance in the egg. Among these are calcium, phosphorus, potassium, sodium, chlorine, magnesium, iron, and sulfur. The largest amounts of certain minerals, especially calcium, are found in the eggshell, but the greatest variety is present in the egg contents.

Eggshell Minerals

About 95 per cent of the eggshell consists of inorganic matter. The shell of the hen's egg contains 5.9 gm. of ash, of which salts of calcium form more than 98 per cent. The remainder is made up of small amounts of phosphorus and magnesium salts, and perhaps traces of iron and sulfur.

The approximate quantities of the minerals (as elements) in the eggshell are as tabulated. In Fig. 194-*A,* the amounts of the three most

	AMOUNT (grams)
Calcium (Ca)	2.21
Magnesium (Mg)	0.02
Phosphorus (P)	0.02
Iron (Fe)	Trace
Sulfur (S)	Trace
Total	2.25

abundant eggshell minerals are shown as percentages of the shell's total content of mineral elements. It should be noted that the mineral elements themselves exist in combined form and constitute only about 38 per cent of the total mineral-containing material of the eggshell.

Little variation is found in the relative amounts of the inorganic compounds in the eggshells of different species. The major mineral compounds—calcium and magnesium carbonates and tricalcium phosphate—are shown in Table 35 as percentages of the total ash in the shells of several birds' eggs.

Calcium is the chief mineral element in the shell membranes. According to Plimmer and Lowndes (*1924*), the shell membranes of the chicken egg contain about 0.2 mg. of calcium.

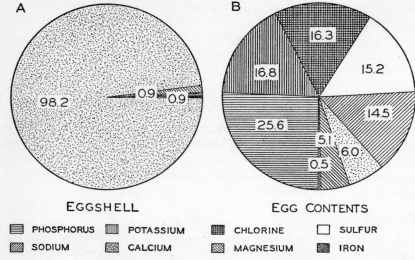

Fig. 194. Percentages of major mineral elements constituting the total mineral content of *A*, the shell, and *B*, the contents of the chicken egg.

TABLE 35

Relative Amounts of Major Mineral Compounds in Total Ash of the Avian Eggshell *

| | Major Minerals | | |
Species of Bird	Calcium Carbonate (per cent)	Magnesium Carbonate (per cent)	Tricalcium Phosphate (per cent)
Chicken	98.43	0.84	0.73
Pheasant	97.88	1.43	0.69
Duck	98.60	0.88	0.52
Goose	98.76	0.49	0.75
Kingfisher	98.84	0.44	0.72
Ostrich	97.37	1.88	0.75

* Compiled largely from the data of Wicke (*1856, 1863*) on eggs of all the above species; also Langworthy (*1901*) on duck and goose eggs; Balland (*1881*) and Torrance (*1923*) on ostrich eggs; and Buckner and Martin (*1920*) and Almquist and Burmester (*1934*) on chicken eggs.

Yolk Minerals

Egg yolk contains about 0.2 gm., or 1 per cent, of ash. The yolk is the source of many of the minerals used in the formation of various embryonic tissues, including bone, although the shell contributes a portion of the calcium for the skeletal structures.

Among the most abundant minerals is phosphorus, which accounts for a considerable proportion of the yolk's total mineral content. Most of the phosphorus occurs in organic combination and is found principally in the form of phospholipids, especially lecithin. According to Plimmer and Scott (*1909*), more than 61 per cent of the total phosphorus in the yolk is contained in lecithin.

Much smaller quantities of calcium, magnesium, chlorine, potassium, and sodium are present, as well as minute amounts of sulfur and iron. Nearly all the iron is presumably in inorganic form, although there are also indications that it may be in organic combination with ovovitellin (*Saha and Mazumder, 1945*). Sulfur, on the other hand, is found almost entirely in combination with organic substances; ovolivetin alone contains about one-half of it.

The approximate amounts and percentages of the major mineral elements in the yolk (18.7 gm.) of the chicken egg are shown in the tabulation.

	AMOUNT	
	GRAM	PER CENT
Phosphorus (P)	0.110	0.588
Calcium (Ca)	0.027	0.144
Magnesium (Mg)	0.024	0.128
Chlorine (Cl)	0.023	0.123
Potassium (K)	0.021	0.112
Sodium (Na)	0.013	0.070
Sulfur (S)	0.003	0.016
Iron (Fe)	0.002	0.011

White yolk contains eight times as much ash as yellow yolk (*Spohn and Riddle, 1916*). This fact acquires significance because of the high ratio of inorganic to organic substances found in embryonic tissues.

Albumen Minerals

The amount of mineral material in the albumen is about the same as that in the yolk: approximately 0.2 gm. Proportionally, the albumen contains slightly more than one-half as much mineral matter (0.7 per cent) as the yolk (1 per cent). Sulfur, potassium, sodium, and chlorine are present in the greatest amounts; phosphorus, calcium, and magnesium are present in much smaller amounts; and iron is found only in traces. It is believed that the calcium is bound to ovalbumin through an —OH linkage (*Abels, 1936*).

The actual and relative amounts of the major mineral elements in the albumen (32.9 gm.) of the chicken egg are as tabulated.

	AMOUNT	
	GRAM	PER CENT
Sulfur (S)	0.064	0.195
Potassium (K)	0.055	0.167
Sodium (Na)	0.053	0.161
Chlorine (Cl)	0.051	0.155
Phosphorus (P)	0.006	0.018
Calcium (Ca)	0.004	0.012
Magnesium (Mg)	0.003	0.009
Iron (Fe)	0.0003	0.0009

The presence of large amounts of inorganic sulfur in the albumen is responsible for the formation of dark stains of silver sulfide on silverware that comes in contact with the egg contents.

Distribution of Major Mineral Elements in the Egg Contents

The percentages of the major elements constituting the total mineral complement of the egg contents are shown in Fig. 194-B. From this figure, it is evident that, of all minerals, phosphorus is present in greatest abundance (25.6 per cent). There are smaller, almost equal, amounts of potassium, chlorine, sulfur, and sodium (from 14.5 to 16.8 per cent), still smaller quantities of calcium and magnesium (6.0 and 5.1 per cent respectively), and about 0.5 per cent of iron.

DISTRIBUTION OF PHOSPHORUS IN THE EGG CONTENTS

According to Plimmer and Scott (1909) and Masai and Fukutomi (1923), the phospholipids (especially ovolecithin) contain about 64 per cent of the total amount of phosphorus. The next largest portion of the phosphorus, about 28 per cent, is found in the phosphoproteins. There is some water-soluble organic and inorganic phosphorus, and possibly some nuclein phosphorus. The total amount of phosphorus in the chicken egg, averaged from the analyses of the above investigators, is divided as follows among the various forms of combined phosphorus:

	AMOUNT (per cent of total)
Phospholipid P	64.0
Phosphoprotein P	27.6
Other organic (water-soluble) P	5.4
Inorganic (water-soluble) P	2.0
Nuclein P	1.0

DISTRIBUTION OF SULFUR IN THE EGG PROTEINS

It is of some interest to note the relative amounts of sulfur in the various proteins of the egg contents. On a percentage basis, the distribution of sulfur is as follows:

	AMOUNT (per cent of total)
Yolk:	
Ovovitellin	22.5
Ovolivetin	11.1
Albumen:	
Ovalbumin	29.8
Ovoconalbumin	17.0
Ovomucoid	11.1
Ovoglobulin	6.4
Ovomucin	2.1

TRACE MINERAL ELEMENTS

In addition to the major minerals, the egg normally contains many other mineral elements in minute amounts. About 7 mg. of trace elements are present in the entire hen's egg, distributed as follows:

	AMOUNT (milligrams)
Yolk	3.8
Albumen	2.8
Shell	0.1

The importance of these so-called trace elements as activators of various reactions in living tissue makes their presence and distribution in the egg of great interest.

Table 36 gives the amounts (often so small that they can be detected only by spectroscopic analysis) of the trace mineral elements found in the chicken egg. From this table, it would appear that most of the fluorine, iodine, copper, zinc, aluminum, and manganese are present in the yolk, whereas most of the boron is in the white. A few trace elements are also found in the shell.

VARIATION IN MINERAL CONTENT OF EGGS

The amounts of the mineral elements in the egg are extremely variable. In the eggs laid by a single hen, the variation in mineral content is considerable; it is even greater in the eggs laid by a flock. Both hereditary and environmental factors are important. Seasonal influences are

TABLE 36

TRACE MINERAL ELEMENTS OF THE CHICKEN EGG *

Element	Yolk (milligram)	Albumen (milligrams)	Shell (milligram)
Aluminum (Al)	0.02 –0.09	Trace –0.01	......
Arsenic (As)	0.001–0.005	0.0004–0.001	0.009–0.023
Barium (Ba)	Present	Present	Present
Boron (B)	0.0002	0.04	......
Bromine (Br)	0.055–0.226	Trace	......
Chromium (Cr)	Occasionally present	Occasionally present	Occasionally present
Copper (Cu)	0.05 –0.30	0.02	Present
Fluorine (F)	0.016–0.146	0.0008–0.0137	......
Iodine (I)	0.003–0.008	0.0008–0.0019	0.001–0.003
Lead (Pb)	0.036–0.29	0.046 –0.188	0.006–0.096
Lithium (Li)	......	......	Trace
Manganese (Mn)	0.004–0.018	0.0014	0.000012
Molybdenum (Mo)	Occasionally present	0.0038–0.004	0.007–0.015
Rubidium (Rb)	Present	Occasionally present	......
Selenium (Se)	Occasionally present	Occasionally present	......
Silicon (Si)	0.55 –0.62	0.28 –1.14	Present
Silver (Ag)	......	Occasionally present	Occasionally present
Strontium (Sr)	Present	Present	Present
Titanium (Ti)	0.0084	0.0028	Present
Uranium (U)	Present	Present	Trace
Vanadium (V)	0.0037	0.0079	Present
Zinc (Zn)	0.7 –1.0	0.007	Occasionally present

* Compiled from numerous sources, including the spectroscopic studies of Drea (*1935*), Bell (*1938*), and Press (*1941*).

Selected References

ALUMINUM: Wolff, Vorstmann, and Schoenmaker (*1923*); Smith (*1928*); Berg (*1925*); Underhill, Peterman, Gross, and Krause (*1929*).
ARSENIC: Gautier (*1900*); Bertrand (*1903*); Fellenberg (*1930*).
BARIUM: Drea (*1935*).
BORON: Bertrand and Agulhon (*1913*).
BROMINE: d'Ambrosio (*1933*); Purjesz, Berkesy, and Gönczi (*1933*); Tenconi (*1934*).
CHROMIUM: Drea (*1935*); Press (*1941*).
COPPER: Fleurent and Lévi (*1920*); McHargue (*1925*); Elvehjem, Kemmerer, Hart, and Halpin (*1929*); Lindow, Elvehjem, and Peterson (*1929*); Loeschke (*1931*); Erikson and Insko (*1934*).
FLUORINE: Nicklès (*1856*); Tammann (*1889*); Purjesz, Berkesy, Gönczi, and Kovács-Oskolas (*1934*); Phillips, Halpin, and Hart (*1935*); Dahle (*1936*).
IODINE: Fellenberg (*1923*); Hercus and Roberts (*1927*); Jaschik and Kieselbach (*1931*); Scharrer and Schropp (*1932*); Straub (*1933*); Almquist and Givens (*1935*); Chrzaszcz (*1935*); Leone (*1936*); Westgate (*1937*).
LEAD: Bishop (*1929a*); Bishop and Cooksey (*1929*); Kogan (*1940*).
LITHIUM: Bishop (*1929b*); Press (*1941*).
MANGANESE: Pichard (*1898*); Bertrand and Medigreceanu (*1913*); McHargue (*1924*); Berg (*1925*); Lyons and Insko (*1937 a, b*); Gallup and Norris (*1939*); Lyons (*1939*).
MOLYBDENUM: Mankin (*1928*); Meulen (*1932*).
RUBIDIUM: Drea (*1935*).
SELENIUM: Moxon (*1937*); Moxon and Poley (*1938*).
SILICON: Poleck (*1850*); Rose (*1850*); Iljin (*1917*); Bell (*1938*).
SILVER: Drea (*1935*).
STRONTIUM: Drea (*1935*); Press (*1941*).
TITANIUM: Bell (*1938*).
URANIUM: Bishop (*1929*); Drea (*1935*); Hoffmann (*1943*).
VANADIUM: Bell (*1938*).
ZINC: Birckner (*1919*); Bertrand and Vladesco (*1922*); Morris (*1940*).

also of some significance. The mineral content of the hen's diet, however, is the most important single factor that influences the amount of certain minerals in the egg.

Influence of Diet on Mineral Content

The concentration of both major and trace minerals in the contents and the shell of the egg depends upon the mineral composition of the hen's diet.

MAJOR ELEMENTS

As shown above, calcium is the chief constituent of the eggshell. The total output of calcium by a hen that lays 200 eggs a year is about 0.5 kg. (approximately 1 lb.). It is obvious that large amounts of calcium must be provided in the hen's diet, in order to supply calcium not only for the eggshell, but also for deposition in the yolk and albumen.

The quantity of calcium in the hen's ration has a marked influence on eggshell characteristics, although there is no evidence that it modifies the calcium content of either the yolk or the albumen. During a period of heavy egg production, restriction of the hen's intake of calcium carbonate, either as such, or in the form of oyster shell, appreciably decreases the ash content of the fat-free bones (*Deobald, Lease, Hart, and Halpin, 1936*). It is the calcium phosphate of the bones that is utilized (*Russell and McDonald, 1929*), as shown by increased phosphorus excretion (*Common, 1933*), and by a rise in the inorganic phosphorus of the blood (*Feinberg, Hughes, and Scott, 1937*), during shell formation. The calcium withdrawn from the bones must, of course, be replaced eventually by calcium assimilated from the diet. When the supplement of calcium is below the optimum, the shell becomes thinner, although its percentage composition remains constant (*Buckner, Martin, and Peter, 1923*).

The ratio of calcium to phosphorus in the egg cannot be modified to any great extent by diet. It has been demonstrated, however, that the presence of vitamin D supplements in the hen's diet tends to increase the amount of both calcium and inorganic phosphorus in the egg yolk (*Erikson, Boyden, Insko, and Martin, 1938; Erikson, Boyden, Martin, and Insko, 1938*).

It is still unknown whether or not diet can alter the amounts of magnesium, potassium, sodium, chlorine, or sulfur in any part of the egg. Different quantitative determinations of these elements diverge so widely—often by more than 100 per cent—that analytical errors are precluded, and one must suspect the influence of such factors as local variations in nutrition and degree of inbreeding.

It is still questionable whether or not the iron content of the egg can be changed by feeding (*McFarlane, Fulmer, and Jukes, 1930*). In the early literature, it was reported that an increased amount of iron in the feed could increase the iron content of eggs. Results of three analyses for iron in the contents of eggs from hens on normal and experimental diets are given in the tabulation.

AMOUNT OF IRON (milligrams)

NORMAL RATION	WITH IRON SUPPLEMENT	INVESTIGATORS
1.7	2.1	Loges and Pingel (*1900*)
0.6	1.1	Hoffmann (*1901*)
1.6	2.6	Hartung (*1902*)

However, in the opinion of Cruickshank (*1941*) and other workers, these increases may not be significant, since the normal variation in the iron content of the egg is very high. Copper supplements in the diet of the hen apparently do not affect the amount of iron deposited in the egg (*Cunningham, 1931*), in spite of the importance of copper in the metabolism of iron. The fact that individual hens tend to lay eggs containing fairly uniform amounts of iron suggests that heredity may determine the level of iron in the egg (*Schaible, Davidson, and Bandemer, 1944*).

TRACE ELEMENTS

The amount of bromine, fluorine, iodine, manganese, or selenium in the egg can be increased by giving the element to the hen, orally or parenterally. For example, the repeated injection of sodium bromide (25 mg.) raised the level of bromine to 0.32 mg. in the yolk, and to 0.2 mg. in the albumen (*Purjesz, Berkesy, and Gönczi, 1933*).

A measurable increase in the fluorine content of eggs (to about 0.06 mg.) was observed after the addition of phosphate rock, which is high in fluorine, to the rations of laying hens (*Phillips, Halpin, and Hart, 1935*). Similarly, the egg's fluorine content rose to 0.046 mg. when the hen was given 0.03 mg. of sodium fluoride intravenously every fifth day (*Purjesz, Berkesy, Gönczi, and Kovács-Oskolas, 1934*). Fluorine is always found in the egg in the acetone-insoluble portion of the fatty substances of the yolk. This suggests that fluorine is deposited in combination with the complex yolk lipids.

It is a well-established fact that the iodine content of the egg is dependent upon, and more or less proportional to, the iodine intake of the bird (*Hercus and Roberts, 1927; Simpson and Strand, 1930; Scharrer and Schropp, 1932*). When hens receive large amounts of iodine in the diet, their eggs may contain 0.8 mg. to 2.8 mg. of iodine (*d'Ambrosio, 1933; Straub, 1933*). Experiments indicate that both organically combined and inorganic iodine are equally available for transfer to the egg. In one instance, it was found that, by adding to the hen's diet such marine products as fish meal or oyster shell, the iodine content of the egg could be increased to 0.06 mg.; this increase, however, was not closely proportional to the increase in the iodine content of the ration (*Almquist and Givens, 1935*). The results of another study showed that the administration of iodine, in the form of dried kelp, iodized linseed meal, or sodium or potassium iodide, increased the iodine content of the egg 75 or 150 times, according to the daily levels—2 mg. or 5 mg. per bird —at which iodine was fed (*Wilder, Bethke, and Record, 1933*). Intravenous injections of 0.15 gm. of sodium iodide can likewise increase the amount of iodine in the egg to as much as 0.4 mg. (*Berkesy and Gönczi, 1933b*).

In a study of the effect of iodine supplement in the diet, Straub (*1932*) found that the iodine content of the egg increased rapidly after iodine had been added to the diet, but that the maximum value was not attained until 10 days or 2 weeks later (Fig. 195). A similar length of time was required for iodine to return to the normal level after iodine feeding was discontinued.

In normal eggs, the iodine is present chiefly in the yolk. However, it was found (*Scharrer and Schropp, 1932*) that, although the yolk's iodine content showed the greatest elevation (almost twenty times above normal) upon addition of iodine to the hen's ration, it was also possible to increase the amount of iodine in the albumen more than eight times, and in the shell one and one-half times. It has been suggested that the iodine combines with the fatty acids, particularly the oleic acid of the yolk.

In Taramahi, New Zealand, Hercus and Roberts (*1927*) found that eggs laid in summer (December and January in the southern hemisphere) contained nearly four times as much iodine as eggs laid at other times of the year. Seasonal variation in the iodine content of the egg was apparently correlated with a parallel fluctuation in the amount of iodine in the plants that served as food.

The manganese content of the egg rises when this mineral is given to hens in large doses (*Vecchi, 1933*). By increasing the manganese in the ration to 0.1 per cent, from a deficiency level of 0.0013 per cent, the amount of manganese in the yolk has been increased eightfold, from 0.004 mg. to 0.033 mg. (*Gallup and Norris, 1939*).

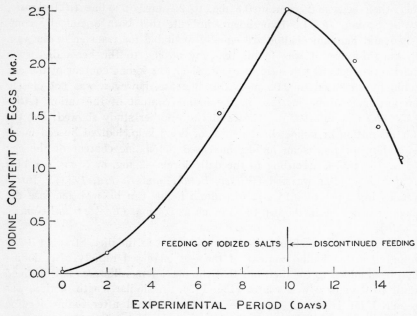

FIG. 195. Increase in the iodine content of the chicken egg during a period when 3 gm. of iodized salt (Rukotafutter's) were fed to the hen each day; subsequent decrease in the egg's iodine content after the administration of the salt was discontinued. (After Straub, 1932.)

Under normal conditions, only traces of selenium are present in the egg. However, when the hen is fed on grains grown in seleniferous soil (like that found in South Dakota, for example), selenium is transferred to the egg (*Moxon and Poley, 1938*). The amount of selenium in the egg is roughly proportional to the hen's intake of this element, as shown

SELENIUM IN RATION (parts per million)	SELENIUM IN ALBUMEN (milligram)	SELENIUM IN YOLK (milligram)
2.5	0.06	0.03
5.0	0.10	0.05
10.0	0.21	0.07

in the tabulation. It is evident from these data that selenium is preferentially deposited in the albumen. Selenium is usually found in the egg in higher concentration than in milk. Below is a comparison of the selenium content of several foodstuffs from seleniferous areas (*Smith, 1941*).

	SELENIUM
FOODSTUFF	(micrograms/100 gm.)
Cereals	87–1880
Vegetables	12–1780
Eggs	25– 914
Meats	117– 800
Milk	16– 127

It is interesting to note that little, if any, increase in the copper content of the yolk or albumen was observed after feeding copper sulfate ($CuSO_4$) (*Elvehjem, Kemmerer, Hart, and Halpin, 1929; Lesné, Zizine, and Briskas, 1938b*). It is possible that the deposition of copper in the egg depends to a greater extent upon the bird's age than upon the amount of copper fed. In the eggs of older hens (aged 3 to 4 years), 0.27 mg. of copper per yolk has been found, as compared with 0.43 mg. in the eggs of younger hens (1 to 2 years old) (*Lesné, Zizine, and Briskas, 1938a*).

There are wide differences in the concentration of trace elements in the drinking water of various localities (*Kehoe, Cholak, and Largent, 1944*). We may therefore assume that the amounts of some of these elements in the egg may also vary geographically, since a laying hen drinks an average of 150 gm. of water daily. The intake of water high in aluminum and fluorine, for example, might cause the deposition of somewhat greater than usual amounts of these minerals in the egg.

DEPOSITION OF RADIOACTIVE ISOTOPES

Recent studies have shown that radioactive isotopes of certain elements promise to be valuable for tracing the passage of these elements into the egg. Radioactive phosphorus is especially useful, not only because it retains its radioactivity for a relatively long period of time, but also because, in the laying hen, there is an exceptionally high turnover of phosphorus, the most abundant mineral of the egg.

The first study of the deposition of radioactive phosphorus in the egg was made by Hevesy and Hahn (*1938*), who investigated the synthesis of phospholipids. A more detailed study by Chargaff (*1942b*) revealed that intramuscular injection of radioactive sodium phosphate results in the formation of highly radioactive phospholipids (ovolecithin and ovocephalin) and phosphoproteins (ovovitellin) in the egg yolk. Later, Lorenz, Perlman, and Chaikoff (*1943*) showed that radioactive phosphorus can be deposited in all parts of the egg. The largest amount (about 1.2 per cent of the amount injected) was found in the egg yolk, and smaller quantities were found in the albumen and eggshell (less than 0.2 and 0.6

per cent, respectively, of the total injected). The deposition of phosphorus in the yolk was more rapid in heavily laying birds than in those producing eggs at a slow rate (see Chapter 4, "Formation").

The radioactive isotope of strontium is deposited largely in the eggshell, to some extent in the albumen, and practically not at all in the yolk. Radioactive iron, on the other hand, is transferred almost exclusively to the yolk (*Posin, 1942*).

BIOLOGICAL SIGNIFICANCE OF MINERALS IN THE EGG

It is now well known that many minerals are indispensable to life, although most of them are required in minute quantity only. The mineral composition of the egg, which may be influenced by such factors as the bird's nutrition and heredity, is important in its effect on embryonic development. The hatchability of eggs may be greatly impaired by a deficiency of some essential mineral element, or by an excess of an injurious or toxic mineral transmitted to the egg from the hen's diet.

Hatchability is particularly dependent upon the egg's complement of calcium and phosphorus. These minerals, to be normally assimilated by the embryo, not only require the activity of vitamin D but also must be present in the egg in adequate amounts and in the proper ratio.

For hatchability (*Gallup and Norris, 1939*) and normal formation of the chick (*Lyons and Insko, 1937a, 1937b*), the egg must contain sufficient manganese, which is considered an essential element in oxidation reactions in living cells. The presence of selenium in the egg, on the other hand, causes abnormal embryonic development (*Franke and Tully, 1935*) and thus considerably reduces hatchability.

The iron in the egg is used by the embryo for the production of hemoglobin, the oxygen carrier of the blood. Copper, as well as iron, is involved in the building of hemoglobin. Larger than normal amounts of iodine in the egg have an indirect effect on reproduction, by reducing the size of the egg (*Schmidt, 1932*) and consequently the size of the chick.

It seems evident, therefore, that trace elements, in spite of their presence in small amounts in the egg, may play an important role in the normal biological processes of the embryo. In fact, the various trace elements have been classified on the basis of their nutritive value and toxicity (*Calvery, 1942*).

PHYSICOCHEMICAL CONSTANTS

ALBUMEN

BOUND WATER (%)	25
COAGULATING TEMPERATURE (°C.)	61
DENSITY (GM./CM.³)	1.035
DIELECTRIC CONSTANT (ϵ)	68 (?)
ELECTRICAL COND.(MHO-CM.$^{-1}\times10^{-3}$)	8.68
FREEZING POINT (°C.)	-0.424
HEAT OF COMBUSTION (CAL./GM.)	5690
HYDROGEN-ION CONCENTRATION (pH)	7.6
REFRACTIVE INDEX (n_D^{25})	1.3562
SOLUBILITY COEFFICIENT FOR CO_2	0.71
SPECIFIC HEAT (CAL./GM.)	0.85
SPECIFIC RESISTANCE (OHM-CM.)	120
SURFACE TENSION (DYNES/CM.)	53
TITRATIVE BASICITY(CC.N/100HCL)	393
TRANSMISSION OF LIGHT(I/I_0):	
OUTER LIQUID LAYER	0.915
MIDDLE DENSE LAYER	0.482
INNER LIQUID LAYER	0.808
VAPOR PRESSURE(IN % OF NaCl)	0.756
VISCOSITY (POISES AT 0°C.)	25

INTACT EGG

BREAKING STRENGTH (KG.)	4.51
SPECIFIC GRAVITY	1.095
THERMAL CAPACITY (CAL./GM./°C.)	0.792

CALCAREOUS SHELL

CRUSHING STRENGTH (KG./CM.²)	3.4
DENSITY (GM./CM.³)	2.3
PERMEABILITY TO AIR (CC./CM.²/MIN./DIFF. 20CM.Hg)	19.5

SHELL MEMBRANE

DENSITY (GM./CM.³)	1.005

YOLK

BOUND WATER (%)	15
COAGULATING TEMPERATURE (°C.)	65
DENSITY (GM./CM.³)	1.035
DIELECTRIC CONSTANT (ϵ)	60 (?)
ELECTRICAL COND.(MHO-CM.$^{-1}\times10^{-3}$)	3.10
FREEZING POINT (°C.)	-0.587
HEAT OF COMBUSTION (CAL./GM.)	8124
HYDROGEN-ION CONCENTRATION (pH)	6.0
REFRACTIVE INDEX (n_D^{25})	1.4185
SOLUBILITY COEFFICIENT FOR CO_2	1.25
SPECIFIC RESISTANCE (OHM-CM.)	320
SURFACE TENSION (DYNES/CM.)	35
TITRATIVE ACIDITY(CC.N/100 NaOH)	184
VAPOR PRESSURE (IN % OF NaCl)	0.971
VISCOSITY (POISES AT 0°C.)	200

VITELLINE MEMBRANE

BURSTING STRENGTH (DYNES/CM.²)	4500

Approximate values for physicochemical constants of the biologically fresh hen's egg.

Chapter Seven

PHYSICOCHEMICAL PROPERTIES

The avian egg is a highly organized physicochemical system.

The avian egg is a well-organized system of considerable complexity. It is composed of several parts which differ distinctly in physical and chemical properties, and which are maintained in certain specific relationships by the operation of a number of measurable forces.

This physicochemical system, the egg, is regulated externally by the hard, yet porous, shell. Internally, the system consists of two phases, the yolk and the albumen, separated by the semipermeable vitelline membrane. The greatly unequal concentration of the organic and inorganic compounds on either side of the membrane creates a condition of imbalance, which tends slowly to approach equilibrium. However, the colloidal state of much of the egg's substance counteracts this tendency, by virtue of the large size of the suspended particles and their slight diffusibility. Amid constant change, a certain degree of stability is thus provided.

The intact egg, its separated component parts, and its individual chemical constituents behave in a characteristic manner under the application of various tests. It is the purpose of this chapter to review the quantitative data that have been accumulated as the result of such tests. The physicochemical properties of the egg, as thus revealed, are important; for, like all other properties, they are associated with the egg's biological function. The interpretation of their significance offers an interesting challenge to science.

PROPERTIES OF THE INTACT EGG

Certain distinct physical properties identify the avian egg while it is still intact. The following discussion considers those of special interest.

Specific Gravity

The specific gravity of the entire egg is calculated from its weight, and the weight of an equal volume of water.

$$\text{Specific gravity} = \frac{\text{Weight of egg}}{\text{Weight of equal volume of water}}$$

The average specific gravity of the strictly fresh chicken egg of normal shape is about 1.095. The specific gravity of off-shaped eggs, either long, elliptical, conical, or round, is slightly lower—1.088 to 1.090 (*Romanoff, unpublished*). Occasionally, the average specific gravity of a large number of eggs is quite low. For example, the average value for 100 eggs examined by Fronda and Clemente (*1934*) was 1.056. When consistently low values are found, it is probable that the eggs examined are not newly laid. After oviposition, all eggs rapidly lose weight through the evaporation of water and therefore decrease in specific gravity.

The specific gravity of the eggshell is nearly twice that of the egg contents. The entire egg's specific gravity is therefore largely influenced by the proportional amount, or thickness, of the shell (*Olsson, 1934*), as shown below by data on turkey eggs (*Phillips and Williams, 1944*).

Average Specific Gravity	Range in Thickness of Shell (millimeter)
1.070	0.28–0.30
1.080	0.33–0.36
1.090	0.38–0.41

The average specific gravity of turkey eggs is 1.085 (*Phillips and Williams, 1944*), and of duck eggs, 1.083 (*Kato and Ko, 1938*). Because of their relatively thick, heavy shells, the eggs of nearly all precocial birds are of comparable specific gravity. On the other hand, the eggs of altricial birds, like the pigeon and the canary, are of relatively low specific gravity, because the shells are proportionately thin and light in weight (see Chapter 3, "Structure").

Since the thickness of the shell is extremely variable among the eggs of each species (see Chapter 3, "Structure"), so also is the specific gravity of the egg. Figure 196 shows the frequency distribution, according to specific gravity, of large numbers of chicken, turkey, and duck eggs. It may be observed that the specific gravity is as low as 1.056 or as high as 1.116 in a few eggs, but that it falls within the range of 1.080 to 1.090 in the largest number of eggs.

The factors that determine the thickness of the eggshell—pathology, nutrition, and heredity (*Taylor and Martin, 1928*)—are also largely responsible for the specific gravity of the chicken egg. There is no doubt that eggs laid by the same bird possess a somewhat similar specific gravity. In addition, seasonal changes in the bird's feeding habits and rate of laying, as well as in the environmental temperature, are manifested by a variation in the egg's specific gravity from 1.087 in winter to 1.078 in summer (in the temperate zone). A similar trend has been observed in the Philippine Islands (*Fronda, Clemente, and Basio, 1935*), where the egg's specific gravity ranges from about 1.072 in January to 1.031 in

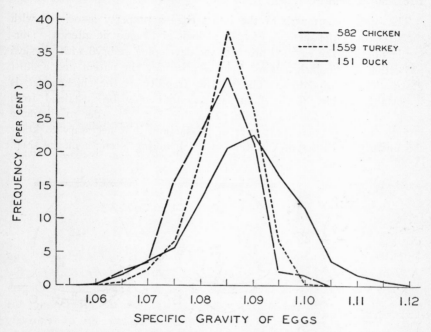

Fig. 196. Frequency distribution of eggs according to specific gravity. (After the date of Romanoff, unpublished, on chicken eggs; Phillips and Williams, 1944, on turkey eggs; and Kato and Ko, 1938, on duck eggs.)

March. However, some workers have recognized the possibility that low values found in summer in the temperate zone or during the dry season in the tropics may be due to loss of weight caused by evaporation of the egg contents.

Many attempts have been made to correlate the specific gravity of fertile eggs with their hatching power (*Mussehl and Halbersleben, 1923; Hays and Sumbardo, 1927; Munro, 1940; Phillips and Williams, 1944*) but apparently without significant results. The shells of thin-shelled eggs of low specific gravity apparently contain enough calcium to satisfy the needs of the developing embryo, provided that the requirements for vitamin D are also met.

Buoyancy. Early investigators determined the buoyancy of the egg by testing it in water (*Benjamin, 1914*). More recently, individual differences in the buoyancy of eggs have been measured by the Jolly balance (*Carr, 1939*).

Breaking Strength

The breaking strength of the intact egg—a property associated with the strength of the shell—is of both biologic and economic interest. During incubation, the eggshell provides the developing embryo with physical protection, particularly against being crushed by the parent and against destruction by predators. The relative fragility of the chicken egg is important because of the financial loss that may be sustained from excessive breakage during handling and shipping. In order to eliminate unduly fragile eggs at the source of production, it is necessary to know the underlying causes of variation in the breaking strength of the shell.

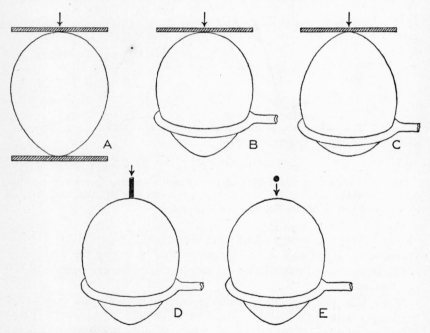

Fig. 197. Diagrams showing different methods of applying pressure or force in the quantitative measurement of the breaking strength of the eggshell.

A, equal pressure applied to both ends (after Romanoff, 1929a). *B* and *C,* pressure applied to one end of the egg (either *B,* the blunt, or *C,* the sharp end) (after Lund, Heiman, and Wilhelm, 1938). *D,* pressure applied by a pin 1 mm. in diameter (after Baskett, Dryden, and Hale, 1937). *E,* force applied by means of a falling ball (after Swenson and James, 1932).

Breaking strength has been most commonly determined by applying weight to one end, or both ends, of the egg, and by gradually increasing the weight until the shell is crushed (Fig. 197-*A*, *B*, *C*, and *D*). The breaking strength of the small end of the egg is generally greater than that of the large end. According to Lund, Heiman, and Wilhelm (*1938*), the average weight required to crush the small end of the chicken egg is 5.57 kg.; the large end, 4.73 kg. The curvature of the shell at the ends of the egg apparently has little influence on breaking strength (*Stewart, 1936*). Presumably the larger amount of calcareous material deposited at the small end of the eggshell accounts for the greater strength in this region. Since the small end of the shell is not always the thicker (see Chapter 3, "Structure"), it may be assumed that the large end of the egg is sometimes the stronger.

It has been observed that the average weight required to break the shell is from 1 to 2 kg. less when pressure is applied along the short axis of the egg than when it is applied along the long axis (*Romanoff, 1929a*).

It is extremely interesting to observe that the breaking strength (in terms of the weight required to break the shell) of the eggs of different species of birds is correlated with the size of the egg (*Romanoff, unpublished*).

SPECIES	EGG WEIGHT (grams)	BREAKING STRENGTH (kilograms)
Finch (*Taenis pygia castanotis*)	1	0.1
Quail (*Colinus virginianus*)	9	1.3
Pheasant (*Phasianus torquatus*)	30	3.5
Chicken (*Gallus gallus*)	60	4.1
Turkey (*Meleagris gallopavo*)	85	6.0
Peafowl (*Pavus cristatis*)	95	10.0
Swan (*Cygnus cygnus*)	285	12.0
Ostrich (*Struthio camelus*)	1400	55.0

The increase in breaking strength from the egg of the smallest bird (finch) to that of the largest (ostrich) is obviously not proportional to the increase in egg weight. It can be seen, therefore, that breaking strength is not entirely determined by the egg's external dimensions; rather, it is largely influenced by the thickness of the shell. Large eggs, as a rule, have thicker shells than small eggs, and consequently, the largest eggs are the strongest.

On the other hand, no relationship has ever been found between breaking strength and size among the eggs of the same species. When the average breaking strength varies from flock to flock, the variation is due mainly to hereditary, nutritional, and other factors that influence the proportional weight and thickness of the shell.

The direct relationship between breaking strength and shell thickness, especially at the egg's pointed end, is linear in the chicken egg (*Romanoff, 1929a; Stewart, 1936*), as shown in Fig. 198; the correlation is statistically high and significant (*Morgan, 1932; Stewart, 1936; Lund, Heiman, and Wilhelm, 1938*).

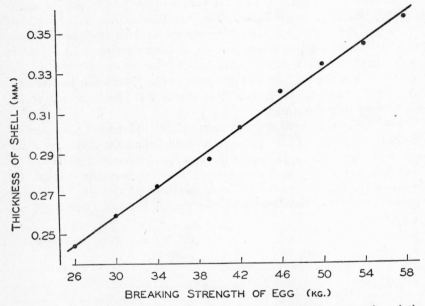

FIG. 198. Relationship between the chicken eggshell's breaking strength and the thickness of the shell at the pointed end; based on 3998 observations. (After Romanoff, 1929a.)

Since the thickness of the eggshell varies among the eggs of different hens (see Chapter 3, "Structure"), the breaking strength of the egg also varies. The accompanying table shows the distribution of 3998 eggs (laid by 91 hens) according to breaking strength (*Romanoff, 1929a*).

BREAKING STRENGTH (kilograms)	NUMBER OF EGGS
2.01–2.80	39
2.81–3.60	341
3.61–4.40	1177
4.41–5.20	1955
5.21–6.00	486

Variation in the breaking strength of the egg has been attributed largely to the amounts of various minerals, especially calcium (*Kennard, 1925*)—

in the presence of vitamin D—supplied to the hen during the laying period. A low intake of manganese, as well as of calcium, can decrease the breaking strength of the eggshell (*Gutowska and Parkhurst, 1942a*). To a great measure, the breaking strength of the shell is influenced by the same factors that are responsible for variation in shell thickness.

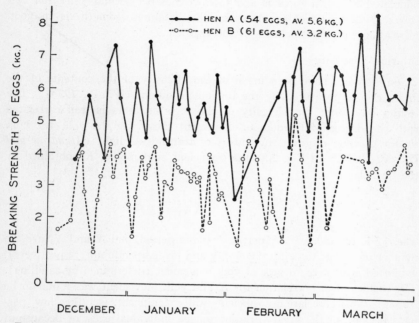

FIG. 199. Breaking strength of eggs laid by two hens under identical management. During the 16-week period, the eggs of hen *A* averaged 5.62 kg. in breaking strength; those of hen *B*, only 3.15 kg. (After Romanoff, 1929a.)

According to Heuser and Norris (*1946*), the breaking strength is lower during July and August than during the winter.

Eggs laid by the same individual are of more or less constant breaking strength. Figure 199 shows that two hens, producing at approximately the same rate for 16 weeks, each laid eggs of which the breaking strength never deviated greatly from a characteristic mean value.

The strength of the shell tends to be least variable during periods of high egg production (*Romanoff, 1929a*), when the secretory glands of the oviduct are apparently functioning in the most normal manner. In practice, therefore, these periods would be the most advantageous times to select hens on the basis of the characteristic strength of their shells, as revealed by tests of a few eggs.

Resistance to Cracking. Less force is required to crack, or "check," the eggshell than to crush it. Since tests for cracking should provide a true index of the egg's ability to withstand the abuses of being handled and shipped, attempts have been made to determine the actual force required to crack the egg by striking it (cf. Fig. 197-*E*). It has been calculated that this force is about 600 dynes per second (*Swenson and James, 1932*). Cracking strength, like breaking strength, is in good correlation with the thickness of the eggshell.

Interior Viscosity

Some studies have been made of the viscosity of the contents of the intact egg, as measured by the torsion pendulum (*Wilcke, 1936*). The results show that the viscosity of the egg's contents is well correlated with the degree of freedom of movement shown by the yolk (or yolk shadow), when observed by means of transmitted light (*Atanasoff and Wilcke, 1937*). In Fig. 200, the coefficient of damping, K (obtained by the formula,

$$K = \frac{2I(0.329)}{P}\left(\frac{1}{N_1} - \frac{1}{N}\right),$$

where I is the moment of inertia, P is the period, and N and N_1 are the number of swings to damp the empty and loaded pendulum, respectively), is plotted against the degree of yolk movement (determined by candling). The coefficient of correlation for K and yolk movement was found to be 0.2388. This correlation coefficient, while significant, is rather low, perhaps because of the great variation among individual eggs, or because of subjective errors in candling (see Chapter 9, "Food Value").

Optical Properties

Transparency. It is possible, by means of transmitted light, to see the contents of the egg through the shell, although not with great clarity. If the egg is white shelled, a fairly dim light is sufficient to show the position of the air cell, the color intensity of the yolk, and the yolk's freedom of movement, which is related to the density of the albumen. In eggs with pigmented shells, vision through the shell is very limited. In darkly pigmented and thick-shelled eggs (for instance, those of the emu), practically nothing can be observed through the shell.

Gane (*1937*) showed that the amount of light transmitted by the intact, white-shelled chicken egg is controlled to the greatest extent by the shell, and to a much lesser extent by the yolk and by the middle dense layer of albumen. Using a light of constant intensity (a 36-watt auto-mobile bulb with a reflector) and a Weston Photonic cell, he found that the ratio of the intensity of the transmitted light to that of the incident light, over an area of 1 sq. cm., varied from 0.007 to 0.057 at the egg's pointed end, and from 0.004 to 0.024 in the region of the yolk shadow.

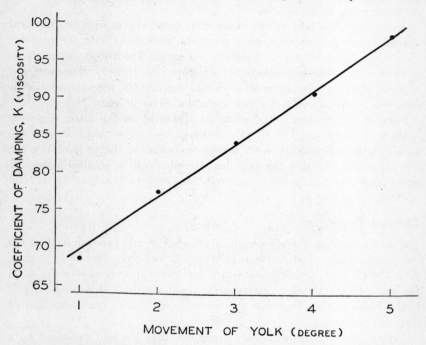

FIG. 200. Relationship between the egg's viscosity (coefficient of damping, K) and the yolk's freedom of movement (expressed in grades and determined by candling); based on observations of 3890 eggs. (After Atanasoff and Wilcke, 1937.)

Fluorescence. It is well known that a bird's egg fluoresces when exposed to ultraviolet light, probably because of the presence of porphyrin in the shell. In white-shelled chicken eggs, the fluorescent color may vary from dark violet-red, when the egg is fresh, to dark bluish red, after the egg has aged somewhat or has been exposed to light. Eggs with pigmented shells give a variety of colors. For example, a dark green emu egg emits a dark purple color upon excitation with ultraviolet light.

Electrical Studies

High-Frequency Conductivity and Dielectric Effect. Certain physico-chemical properties of the egg are revealed by the use of the short-wave vacuum tube oscillator (*Romanoff and Cottrell, 1939; Romanoff and Frank, 1941*). When the egg is introduced into the coil, reductions in current and in resonant frequency are observed. The maximum reduction in the current depends upon the conductivity of the egg; the reduction in resonant frequency—the result of increased coil capacity—depends upon the egg's dielectric effect.

An attempt has been made to correlate these effects with the physical characteristics and morphological structure, as well as with the fertility, of whole fresh eggs. It is possible to segregate fresh eggs on the basis of their radio-frequency conductivity (*Romanoff, 1944b; Romanoff and Hall, 1944*). At frequencies of about 14 megacycles, a wide variation is observed in the conductivity and dielectric effect of eggs.

Induced Polarization. The electrical properties of the intact egg were studied by Northrup (*1913*). He inserted two copper wires through the shell. After applying 24 volts to the electrodes in the ends of the egg for 1 minute and then joining these terminals to a millivoltmeter, he noted that the instrument registered from 0.1 to 0.2 mv. for several minutes.

Thermal Properties

Heat Production. Theoretically, fresh eggs should produce heat as the result of chemical or, in fertile eggs, metabolic activity. At the same time, they absorb heat, which is necessary for the evaporation of water from the egg. Under ordinary atmospheric conditions, the absorption of this latent heat may even exceed the production of heat from chemical or metabolic activities.

Langworthy and Barott (*1921*) observed that the heat produced by a fresh, infertile hen's egg amounts to 0.072 gram-calorie per day. The production of heat is much greater at high than at low temperatures, undoubtedly because of the accelerated rate of chemical activity.

Experimental findings have indicated that fertile eggs, on the other hand, absorb rather than give off heat during the first few hours of incubation (*Bohr and Hasselbalch, 1903*).

Thermal Capacity. Measurements made on several chicken eggs indicate that the specific heat of the entire egg is 0.772 normal calorie per gram per degree (*Romanoff, unpublished*). The thermal capacity of the egg is therefore much less than that of water (0.999) and falls in line with the specific heat of many dilute solutions.

Freezing Temperature. It is possible to cool the intact egg, without freezing it, to temperatures below the respective freezing points of the yolk or the albumen (which will be discussed later). Babin (*1938*) observed that the egg remained unfrozen at $-1.5°$ C. and $-2.0°$ C.; and he successfully prevented it from freezing at $-3.2°$ C. by continuously shaking it. Also, an egg occasionally may remain unfrozen for a week at $-11°$ C., although at this temperature the egg is usually frozen hard within 24 hours (*Moran, 1925*).

When the intact egg is frozen, its shell cracks. Cracking may be prevented by preliminary partial drying of the egg (to the extent of 1 to 4 per cent of its weight).

Effect of Boiling. Experiments have shown that, when the intact egg is boiled, a gradation in temperature exists for some time from the outermost portion of the albumen to the center of the yolk. According to Sieke (*1943*), the temperature of the center of the duck egg is only $65°$ C. to $74°$ C. after 8 minutes of boiling, and only $78°$ C. to $80°$ C. after 10 minutes.

The albumen and yolk do not change materially in weight when the egg is boiled (*Buckner, Insko, and Harms, 1943*). According to Carpiaux (*1903b*), however, boiling the egg may cause a loss in weight of 0.03 gm. to 0.1 gm. In the boiled egg, the percentages of such constituents as fat, nitrogen, phosphorus, and water are approximately the same as in the raw egg.

Gaseous Exchange

Many organic substances, as well as living organisms, absorb oxygen and eliminate carbon dioxide, and sometimes other gases. Gaseous exchange, of course, occurs in the fertile egg; it is less certain whether or not it takes place in the infertile egg.

Oxygen Consumption. If oxygen is absorbed by the fresh, infertile egg, the amount is so small that it cannot be measured (*M. Smith, 1931*). If "respiration" occurred, and the respiratory quotient (ratio of the volume of carbon dioxide eliminated to the volume of oxygen consumed) were unity, the daily amount of carbon dioxide lost by the egg would indicate an intake of 0.15 mg. of oxygen per day—an amount that could easily be detected (*Needham, Stephenson, and Needham, 1931*). The fresh, fertile egg, on the other hand, has been found to consume about 1.7 mg. (or 1.2 cc.) of oxygen per day, at $37.5°$ C. (*Romanoff, 1941*).

Oxygen consumption (if any) by the infertile egg would be due to oxidation within the egg, occurring in the course of chemical decomposition. The oxygen used by fertile eggs is largely metabolic. Although the embryo is dormant before incubation, the egg contents are already in a state of chemical and enzymatic activity, and hence some oxygen is required.

Carbon Dioxide Production. The output of carbon dioxide by both fertile and infertile eggs occurs at a relatively high rate (*Atwood and Weakley, 1924*). It has been demonstrated (*A. Smith, 1931*) that the initial liberation of carbon dioxide from the fresh, infertile hen's egg may be as high as 3.5 mg. (or 1.8 cc.) per day.

No doubt the carbon dioxide eliminated is in part a by-product of chemical decomposition in infertile eggs, and of slow metabolism in fertile eggs. However, the rather large output of carbon dioxide from the fresh egg, either fertile or infertile, is due chiefly to the de-solution of free carbon dioxide from the albumen—the *p*H of which increases as the carbon dioxide is lost (*Brooks and Pace, 1938a, 1938b*)—and to the slow generation of carbon dioxide in the egg, perhaps by the action of an acidic substance on the calcium carbonate of the shell (*M. Smith, 1931*).

Water Exchange

Loss and Gain of Water. Immediately after being laid, the egg, if exposed to air, begins to lose water by evaporation. Theoretically, the rate of evaporation, at a constant relative humidity and any constant temperature, should be proportional to the vapor pressure of the shell's total exposed pore area of about 140 sq. mm. (*Romanoff, 1943b*); that is, it should follow Dalton's law regarding the evaporation of a free water surface. Calculations from the experimental data of A. Smith (*1930*) indicate that there is agreement with this law; the slight deviations are almost within experimental error. However, in a biological evaporating surface, like that of the bird's egg, subsidiary effects of temperature on the diffusion of liquid water or of water vapor through the material and its coverings—eggshell and shell membranes—might have some modifying influence.

The eggshell is the most obvious source of resistance to the passage of water vapor. Its effect is well illustrated by the tabulated experimental

Condition of Egg	Daily Weight Loss (grams)
Intact	0.015
Filled with water, after removal of contents	0.017
Filled with water, after removal of contents and membranes	0.019
After removal of shell	0.548

data, obtained when holding the egg at 10° C. and 80 per cent relative humidity (*A. Smith, 1931*). From these data, it is also apparent that the shell membranes have a limiting effect on the rate of evaporation, although this effect is very slight compared with that of the shell.

By maintaining a partial vacuum around an egg held over concentrated sulfuric acid, it is possible to increase the rate of evaporation from the egg twenty-five to forty times, as shown in the tabulation (*A. Smith,*

Condition	Daily Weight Loss (grams)
In air	0.09
At 5 mm. negative pressure	2.38
At highest vacuum obtainable	3.65

1931). Under the highest vacuum, a total of 25 gm. of water was lost in 7 days by an egg in which the initial water content was about 38 gm.

At normal atmospheric pressure, the rate of evaporation from the egg is greatly influenced by the relative humidity and temperature of the environment. At any constant temperature, a decrease in relative humidity results in a linear increase in the rate at which the egg loses weight (Fig. 201-*A*). When the relative humidity remains unchanged and the temperature rises, weight is lost at an ever-accelerating rate (cf. Fig. 201-*B*).

Very little work has been done on determining the effect of air movement on the rate of evaporation of water from the egg contents at low temperatures; some investigators believe that air movement is not an important factor. However, at incubation temperatures (37.5° C.), an acceleration of 2.6 feet per second in air movement is sufficient to increase the daily weight loss by 0.01 gm. (*Romanoff, 1940*). It is logical to suppose that at any temperature, air circulation favors evaporation from the egg, as from other bodies.

Aggazzotti (*1913*) found that altitude influences the loss of weight by the egg; with reduced air pressure at high altitudes, the rate of evaporation is naturally greater.

In addition to environmental conditions affecting evaporation from the egg, there are features of the egg itself that influence the loss of weight. The size of the egg is perhaps the most important factor to be considered. Large eggs, although they lose more actual weight per unit of time, lose proportionally less of their original weight than small eggs. This fact is partially explained by the difference in the amount of surface exposed to evaporation by small and by large eggs. Since the surface areas of solids of similar shape do not vary directly with the weights of the solids, but rather with the two-thirds powers of their weights, large eggs have proportionally less surface than small eggs (*Dunn and Schneider, 1923*).

The permeability of the eggshell also affects the rate of evaporation, as has been shown when eggs are held at incubating temperature (*Romanoff, 1940*).

Effects of Immersion in Liquids. Baudrimont and St.-Ange (*1847*) were perhaps the first to note that gas bubbles arise from some points on the surface of the egg if the egg is subjected to a vacuum while immersed in a liquid. This observation led to the realization that the bird's eggshell is porous. The effect of immersing the intact egg in various fluids is of interest, especially to the food-preservation industry.

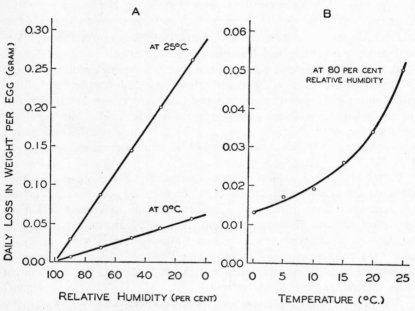

Fig. 201. The relationship between the rate at which water evaporates from a 58-gm. egg (as shown by daily loss in weight) and the temperature and relative humidity of the environment.

A, the daily loss in weight at two constant temperatures and various relative humidities (after Smith, 1933). *B,* the daily loss in weight at 80 per cent relative humidity and various temperatures (after A. Smith, 1930).

The egg, when held in distilled water, gains in weight because water diffuses into the egg. The rate of penetration varies with the surrounding temperature. At 21° C., fresh eggs are filled to maximum capacity within 1 week; at 13° C., within 2 weeks; and at 4° C., within 3 weeks (*Hall and Romanoff, 1943*).

Osborne (*1931*) found that the egg shows little change when immersed in ethyl alcohol for as long as 3 to 6 months. However, when the egg is kept in ether, the shell eventually cracks because ether slowly penetrates into the egg and causes the internal pressure to rise. Penetration by chloroform takes place rapidly, usually within a few hours. On the other hand, when immersed in glycerol or saturated ammonium sulfate, the egg slowly loses water.

PROPERTIES OF EGG COMPONENTS

When considering the egg as an organized system, it is necessary also to examine the physicochemical properties of its individual structural components. These properties are of both theoretical and practical interest. They bear a fundamental relationship, not yet fully understood, to the processes of embryonic development. In addition, the egg's importance in human economy may be augmented as the result of scientific studies that reveal means of improving incubation procedures and methods of storage and preservation, or that point out new uses of the egg in foods and industrial products.

SHELL AND MEMBRANES OF THE EGG

Studies of the eggshell's physicochemical characteristics have been very limited in scope and number. Furthermore, the shell has almost always been tested together with its external cuticular layer, and frequently with its inner linings, the shell membranes. The presence of the cuticle or the membranes, however, may considerably influence the results of experiments on the shell.

Cuticle

It has been frequently assumed, but never experimentally proved, that the shell's external cuticular covering greatly affects the permeability of the shell. If this is true, the cuticle assumes economic importance because of its relationship to the evaporation of water and to the penetration of bacteria through the eggshell. The fluorescent properties of the cuticle, and its role in gaseous exchange through the pores of the shell, are, in addition, of considerable biological interest.

Calcareous Shell

The calcareous portion of the eggshell, or the shell proper, may be characterized by such properties as its density, strength, permeability to liquids and gases, and ability to transmit light and other rays.

DENSITY

The density of the shell is nearly twice that of the contents of the bird's egg. The density of the chicken eggshell is between 2.14 and 2.47, according to Kelly (*1901*), Mussehl and Halbersleben (*1923*), Olsson (*1936*), Asmundson and Baker (*1940*), and Romanoff (*unpublished*). Among the eggs of different species, the density of the shell apparently is not a function of the size of the egg; however, in certain altricial birds' eggs, it is comparatively low. The findings of a few investigators are given in the tabulation.

	DENSITY OF SHELL		
	KELLY	ROMANOFF	ASMUNDSON, BAKER, AND
SPECIES	(*1901*)	(*unpublished*)	EMLEN (*1943*)
Precocial:			
Ostrich	2.55	2.52	....
Emu	2.50	....	....
Goose	2.54	....	....
Turkey	....	....	2.17
Chicken	2.47	2.30	2.26
Ringnecked pheasant	....	....	2.29
Grouse	....	2.24	....
Altricial:			
Mockingbird	....	....	1.59
Tricolored redwing	....	....	1.51
Barn swallow	....	....	1.48

There are some differences in the density of the shell in various parts of the egg, as shown by the following data on duck eggs (*Kato and Ko, 1938*):

REGION OF SHELL	DENSITY
Pointed end	2.30
Equatorial region	2.51
Blunt end	2.48

RESISTANCE TO PUNCTURE

According to Baskett, Dryden, and Hale (*1937*), the fowl's eggshell is strong enough to withstand a weight of 3.4 kg. per square centimeter (or 48.3 lb. per square inch). A hen's eggshells are of fairly constant strength, but, in average strength, they may differ considerably from those of other hens.

Shell strength and shell thickness appear to be intimately correlated. According to Lund, Heiman, and Wilhelm (*1938*), the coefficient of correlation between the shell's thickness and resistance to puncture is 0.835 ± 0.011.

PERMEABILITY

The permeability of the avian eggshell to water vapor and respiratory gases—oxygen and carbon dioxide—is of considerable interest, as it has a bearing both upon embryonic development and upon the preservation of eggs for human consumption. There is no doubt that excessively porous shells are responsible for the deaths of some embryos and for the rapid deterioration of many eggs during storage.

In early studies of shell permeability, attempts were made to measure the rate at which water flows through a small portion of eggshell (*Thunberg, 1902*). Later, however, it was found that the rate at which air or pure gases pass through the shell provides a more reliable criterion of shell permeability than the rate of water flow. In addition, it was established that more accurate results could be obtained by using large pieces of the shell (*Romanoff, 1943b*) than small (*Hüfner, 1892; Camus, 1904; Ferdinandoff, 1931; Penionzhkevich, 1936*). Air moves through the eggshell with equal velocity in either direction, inward or outward (*Moran and Haines, 1939*); water, on the other hand, passes into the egg nearly 50 per cent more slowly than out of it.

Average shell permeability varies among the eggs of different species. At any constant differential pressure, the rate of air flow per unit area is usually highest in the largest eggs and progressively lower in eggs of increasingly small size. The relationship between the size of the egg and the permeability of the shell is shown in the tabulation, where per-

SPECIES	PERMEABILITY TO AIR (cc. per sq. cm. per min. per 20 cm. Hg)	AREA (sq. cm.) WEIGHT (gm.)
Ostrich	60.0	0.38
Emu	50.0	0.51
Embden goose	35.0	0.78
Bourbon Red turkey	21.0	1.12
Runner duck	20.0	1.13
Leghorn chicken	19.5	1.18
Guinea fowl	19.0	1.33
Ringnecked pheasant	10.0	1.44
Ruffed grouse	5.0	1.78
Bobwhite quail	3.0	2.19

meability is expressed as the rate of air flow (in cubic centimeters per square centimeter per minute) from the inside of the dried shell, at a pressure difference of 20 cm. of mercury (*Romanoff, 1943b, and unpublished*). Throughout the course of embryonic development, both the proper rate of evaporation and the equilibrium between the water and the solid substance of the egg must be maintained. Examination of the data reveals that shell permeability is relatively high in large eggs and thus compensates for the small ratio of surface area to weight.

Of equal biological significance is the fact that the permeability of the fresh eggshell is much lower at the egg's sharp end than at its blunt end, where the air cell is usually situated. This difference in permeability is noticeable in eggs of all species of birds, as shown by the three examples listed (*Romanoff, 1943b; and unpublished*). Presumably, the difference

	SHELL PERMEABILITY (cc./sq. cm./min.)	
SPECIES	BLUNT END	SHARP END
Ostrich	73.0	46.3
Embden goose	54.1	8.2
Chicken	14.5	10.5

in the permeability of the blunt and the sharp end is more pronounced in the eggs of water fowl, such as geese, than in those of land birds.

Although shell permeability is fairly constant in the eggs laid by any one bird, it varies considerably in eggs laid by different birds of the same species. How great the variation may be is indicated in Fig. 202. The

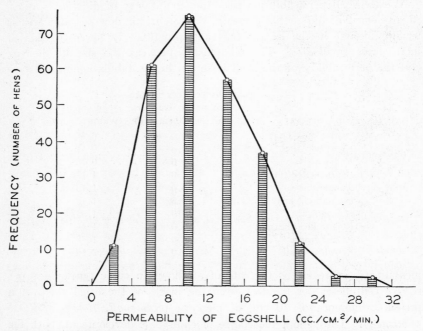

FIG. 202. Frequency distribution of 259 White Leghorn hens according to the average permeability of the shells of their eggs. (After Romanoff, unpublished.)

different individuals of a flock of 259 White Leghorn hens produced shells of which the average permeability ranged from 1.1 cc. up to 30.5 cc. per square centimeter per minute, although the average for the flock was 11.8 cc. The distribution curve would be almost symmetrical were it not for the very highly permeable shells of a few individuals' eggs. These eggs may be regarded as structurally abnormal; their high permeability is probably the result of some pathological condition of the shell-secreting portion of the oviduct.

Many other variables have been found to affect the chicken eggshell's permeability to air (*Romanoff, 1943b*). When the inner shell membrane is removed, shell permeability increases. The blunt end of the shell is usually more permeable than the sharp end, but, in shells of very low permeability, the reverse may be true. As the hen's age advances, she lays eggs of lower shell permeability. In each successive egg in laying cycles of 2 or 3 eggs, shell permeability is higher at the egg's blunt end and lower at its sharp end. However, there is apparently no correlation between permeability and the breaking strength of the entire egg or the thickness of the shell.

It has been observed that the eggshell varies in its permeability to different gases (*Hüfner, 1892; Romanoff, 1943b*). The comparative ability of air and of certain gases to penetrate the eggshell is shown in the table (*Romanoff, 1943b*).

Gas	Relative Penetration (per cent)
Hydrogen	140.98
Carbon dioxide	103.19
Nitrogen	100.86
Air	100.00
Oxygen	92.85

The more rapid passage of carbon dioxide, a heavy gas, than of oxygen, a light gas, is of extraordinary interest. The eggshell, like many animal membranes, violates the physical law (Graham's law) of the rate of gaseous diffusion.

Influence of Climate and Soil. It has been suggested that certain climatic conditions in different regions may have an effect on shell permeability. Coles (*1936*) states that eggshells tend to be of comparatively low permeability in areas where the relative humidity is low, and somewhat more porous in areas where the relative humidity is high.

SPECTRAL PROPERTIES

Transmission of Light. The ability of the shell to transmit light is important because of its effect on the candled appearance of the egg

contents (see Chapter 9, "Food Value"). The relative surface brightness of a white eggshell, as determined by means of a Bunsen photometer,
is about 86 per cent when a piece of opaque, white glass is used as a
standard of comparison (*Givens, Almquist, and Stokstad, 1935*).

It is surprising to find that the shell's thickness appears to have little
influence on the transmission of light through the shell. The most im-

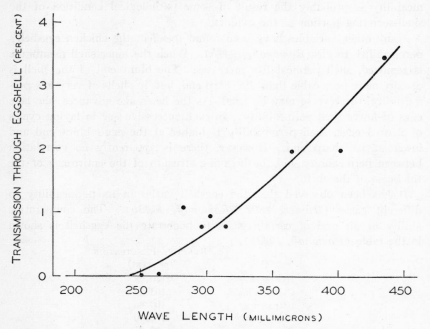

Fig. 203. Transmission of ultraviolet rays through a chicken eggshell of light color,
0.30 mm. thick. (After Benford and Howe, 1931.)

portant factor is the shell's water content. Givens, Almquist, and Stokstad (*1935*), using a piece of white glass as a standard, found that a
white eggshell, before being dried, transmitted 14 per cent more light
than the standard, and only 57 per cent as much afterward. Water,
which is found in the spaces between the calcite crystals in the shell
(particularly in the mammillary layer), is a medium of higher refractive
index than air. Some of the light reflected from the crystal surfaces may
therefore be refracted in a direction perpendicular to the surface of the
shell.

The factor next in importance is the shell's content of protein, especially
that comprising the surface cuticle and the matrix of the outer layer.

Protein is relatively opaque and hinders the passage of light. The membranes apparently have very little influence on light transmission; they probably reduce the brightness of transmitted light by not more than 10 per cent.

Transmission of Ultraviolet Rays. It has been shown (*Sheard and Higgins, 1930*) that the eggshell (together with the shell membranes) can transmit ultraviolet light of wave lengths as short as 300 mμ. However, only 1 to 2 per cent of energy, of wave lengths between 280 and 400 mμ, passes through a chicken eggshell of light color (*Benford and Howe, 1931*). The longer the wave length, the greater is the amount of energy transmitted, as shown in Fig. 203.

ADSORPTION OF CARBON DIOXIDE

In spite of the porous structure of the eggshell, no significant adsorption of carbon dioxide has been detected. Adsorption, when measured at 0° C. and an atmospheric pressure of 760 mm. of mercury, apparently does not exceed 0.01 cc. of carbon dioxide per gram of eggshell (*Brooks and Pace, 1938a, 1938b*).

Shell Membranes

The two membranes that line the shell display some specific physical properties. Among those studied are density, permeability, membrane potential, infrared absorption spectra, and transmission of ultraviolet rays.

DENSITY

The observations made by Olsson (*1934, 1936*) indicate that the density of the shell membranes is 1.005.

PERMEABILITY

The shell membranes are permeable not only to water but also to salt solutions and gases. Ample evidence of the permeability of the membranes to water is shown by the evaporation of water from eggs exposed to dry air and the flow of water into eggs submerged in distilled water.

The permeability of the membranes to a salt solution (0.9 per cent sodium chloride) was studied by Salvatori (*1936a*). He found that, at the start of his experiment, the outer and the inner membrane were both permeable to chloride ions, although the outer membrane was about 6 per cent more permeable than the inner membrane. After 8 hours of continuous diffusion, both membranes became saturated with chloride, so that the passage of chloride ions was completely suspended. He also found that the outer membrane permitted the passage of certain colloids, such as Congo Red, carboxyhemoglobin, and ovalbumin, whereas the inner membrane was impermeable to these substances.

The permeability of the membranes to gases is very high and presumably increases with dehydration (*Penionzhkevich, 1936*). The permeability of the outer membrane appears to be about 9 per cent less than that of the calcareous shell (*Romanoff, 1943b*), and therefore, when calculated, equals about 18 cc. per square centimeter per minute at a pressure difference of 20 cm. of mercury (from the inside of the egg).

MEMBRANE POTENTIAL

The electrical potential difference across the inner membrane (with the system from inside to outside: Pt/HCl 0.1 N/quinhydrone-membrane-distilled water/quinhydrone/Pt) averages about 0.352 volt in the fresh chicken egg, and 0.378 volt in the pigeon egg (*Salvatori, 1936b*).

INFRARED ABSORPTION SPECTRA

Stair and Coblentz (*1935*) examined the shell membranes for infrared absorption spectra, but, because of the highly diffusing character of the membranes, they could identify no absorption bands at wave lengths shorter than about 6 μ. At wave lengths longer than 6 μ, the absorption bands of the membranes, shown in Fig. 204, are similar to those of certain proteinous substances, including dried egg albumen.

TRANSMISSION OF ULTRAVIOLET RAYS

The ability of the shell membrane to transmit ultraviolet light, of wave lengths as short as 265 to 270 mμ, was demonstrated by Sheard and Higgins (*1930*). However, the percentage of transmission is very small.

Vitelline Membrane

The maintenance of the egg as a physicochemical system depends largely upon the presence of the vitelline membrane, which separates the yolk and the albumen. The fact that the vitelline membrane separates two solutions of very different osmotic pressure has been the subject of much discussion and has led to a certain amount of inquiry into the membrane's osmotic properties and permeability. In addition, its density, bursting strength, and ultraviolet absorption capacity have also been studied.

OSMOTIC PROPERTIES

It has been found that egg yolk and albumen, when separated by a thin collodion membrane, rapidly attain osmotic equilibrium (*Straub*

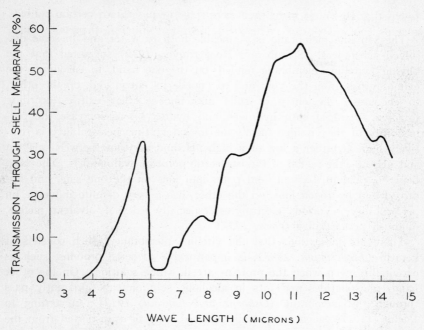

FIG. 204. Transmission of infrared rays through shell membranes. (After Stair and Coblentz, 1935.)

and Hoogerduyn, 1929; Needham, 1931b); yet, when separated in the apparatus by the much thinner vitelline membrane, they require a considerably longer time for equilibration (*Needham, 1931b*). On the other hand, the isolated vitelline membrane behaves like a dead collodion membrane when used in a dialyzing apparatus between solutions of simpler composition than the egg contents; it opposes no resistance to the passage of water and ions, even when it is placed between two salt solutions equivalent in concentration to the yolk and the albumen, respectively (*Needham, 1931b*).

These observations have led to the conclusion that neither the vitelline membrane, nor the yolk, nor the albumen is alone responsible for the maintenance of the steady state that exists in the intact egg; the phenomenon probably is attributable to some collaboration of the three components. Presumably, the osmotic gradient across the vitelline membrane in the intact egg is partially due to the fact that the attainment of equilibrium is hindered by the physical structure of the yolk, and, to a lesser extent, by that of the white (*Moran and Hale, 1936*). (See Chapter 3, "Structure".)

PERMEABILITY

The vitelline membrane is permeable both to water and to some dissolved substances. Straub and Hoogerduyn (*1929*) suggested that the "living" vitelline membrane tended, on the one hand, to encourage the exit of water from the yolk and to impede its entry, and, on the other, to encourage the entry of salts and impede their exit. According to Orru (*1940a*), both hydrogen and hydroxyl ions increase the permeability of the vitelline membrane to water; thus permeability is relatively high at either a low or a high pH, and very low between pH 4.6 and $pH \cdot 5.0$ (the region of the isoelectric point of ovalbumin). The ratio between calcium, barium, and potassium ions on the one side, and hydroxyl and hydrogen ions, on the other, has a very definite influence on permeability. Various electrolytes, especially salts of bivalent metals, diminish permeability (*Orru, 1939*).

There are indications that the vitelline membrane, which consists of keratin (*Liebermann, 1888*), is impermeable to some proteins, such as ovovitellin, even when the proteins are in saline solution. On the other hand, phosphatides and fats in alcoholic solution can apparently pass through the membrane (*Osborne and Kincaid, 1914*). According to Pucher (*1927*), sugars, especially glucose, may be transported from the albumen to the yolk. Orru (*1933d*) found that the vitelline membrane is permeable not only to glucose but also to sucrose and raffinose, to a degree independent of the concentrations of these substances in the external liquids.

The permeability of the vitelline membrane increases as the temperature rises and as the egg grows older (*Orru, 1940a, 1940b*).

BURSTING STRENGTH

The strength of the vitelline membrane is an important factor in the commercial preservation of eggs. In view of practical considerations, the intact membrane is usually used in studies of bursting strength, in preference to pieces of the membrane (*Haugh, 1933; Munro and Robertson, 1935; Moran, 1936c*).

In the new-laid egg, the bursting strength of the vitelline membrane (as determined by a hydrostatic method while the yolk is immersed in an isotonic sucrose solution) is approximately 4500 dynes per centimeter (*Moran, 1936c*). The addition of 0.25 per cent of trypsin to the sugar solution rapidly weakens the membrane. There is some evidence that the vitelline membrane is stronger in summer than in winter. However, the strength of the membrane is undoubtedly influenced by physiological factors, such as the hen's diet; and it is also probably determined to a large extent by heredity, since it is fairly uniform in the eggs of each hen (*Halnan and Moran, 1937*).

ULTRAVIOLET ABSORPTION

Uber, Hayashi, and Ells (*1941*) demonstrated that the vitelline membrane transmits the longer wave lengths of ultraviolet light to a greater

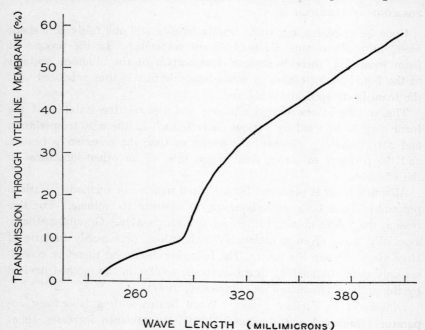

FIG. 205. Transmission of ultraviolet light through the vitelline membrane of the chicken egg. (After Uber, Hayashi, and Ells, 1941.)

degree than the shorter wave lengths (Fig. 205). Selective absorption occurs in the region around 280 mμ. Scattering of the rays presumably accounts for the relatively low transmission of ultraviolet light.

YOLK AND ALBUMEN

The physicochemical properties of the yolk and the albumen are of general scientific interest; information concerning these properties is also of obvious practical value. The following discussion is an attempt to bring together from scattered sources the most important data on the behavior of the egg's chief components under various treatments.

Colloidal Properties

Both egg albumen and egg yolk possess many of the characteristics of natural colloids. Egg albumen, upon being beaten, forms a foam; egg yolk, when mixed with other liquids, becomes a typical emulsoid. Both yolk and albumen, therefore, are lyophilic colloid systems, capable of producing gels that contain high percentages of the dispersion medium.

FOAMING OF ALBUMEN

Upon being beaten, egg white readily forms a stiff and fairly permanent foam, in which minute air bubbles are suspended. In the process of foam formation, there is surface denaturation of the albumen proteins at the liquid-air interface. A water-insoluble film is thus produced, and the foam is stiffened and stabilized.

The volume of the beaten albumen and the relative stability of the foam may be affected by various factors, such as the age, temperature, and pH of the egg albumen, the length of time the albumen is beaten, and the presence of water, yolk, sugar, salt, oil, or other substances in the albumen.

Albumen foam is produced by a limited number of methods. Various procedures have been used, however, to measure its volume. For this reason, the values obtained often are not comparable. Gravity methods, especially, have given unreasonably high values, presumably because of large air bubbles in the foam. The following discussion therefore considers only data obtained by direct methods, similar to the method devised by the authors (*Romanoff and Romanoff, 1944*).

Differences in Various Layers. When beaten with a laboratory apparatus (*Romanoff and Romanoff, 1944*), the albumen increases about three and one-half times in volume.[1] (The specific gravity of this foam is about 0.296.)

When beaten separately, the various layers of the albumen increase in volume to somewhat different extents, as shown below (*Romanoff and Romanoff, 1944*).

LAYER OF ALBUMEN	INCREASE IN VOLUME (per cent)
Outer liquid	372
Middle dense	359
Inner liquid	320

[1] With commercial beaters, one can presumably obtain a volume of foam nearly ten times that obtained with laboratory beaters (*Bennion, Hawthorne, and Bate-Smith, 1942*).

If the two liquid layers are mixed, as they have been by some investigators (*St. John and Flor, 1931*), the dense and the liquid albumen will show approximately the same percentage increase in volume, when beaten.

Effect of Age of Eggs. Although beating increases the volume of albumen from fresh eggs by about 350 per cent, it produces a 425 per

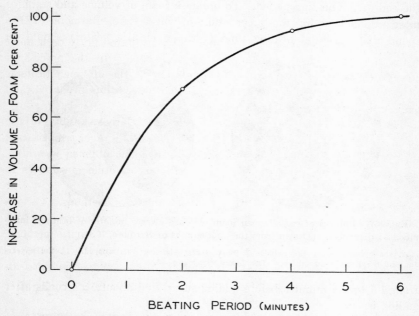

FIG. 206. Relationship between the length of the beating period and the increase in the volume of egg albumen foam. (After Henry and Barbour, 1933.)

cent increase in the volume of albumen from eggs held 2 weeks at room temperature (*Romanoff and Romanoff, 1944*). This difference may be attributed to the increase that occurs in the *p*H of the albumen as time passes. The data obtained by Barmore (*1934*) show that the stability of the foam decreases as the age of the egg increases.

Effect of Length of Beating Period. The length of time for which egg albumen is beaten has a pronounced effect both on the volume and the stability of the resulting foam. As shown in Fig. 206, the volume of the foam increases, although not linearly, as the length of the beating period is extended (*Henry and Barbour, 1933*). Presumably, under ordinary conditions, there is no increase in volume after 6 minutes of beating. The longer the albumen is beaten, the smaller are the air bubbles in the foam (*Barmore, 1934*). The most stable foam is obtained after only 2 minutes of beating, as shown by the penetrometer method. It is evident, therefore, that, if a stable foam is desired, egg albumen should not be beaten to its maximum volume. It may be seen, in Fig. 207, that a foam obtained after 5 minutes of beating is quite unstable, as shown by the increase in the size of the bubbles within 10 minutes after the end of the beating period. To obtain a foam of volume and stability most satisfactory for the preparation of angel cake, Barmore (*1936*) found that the total beating time (with an electric beater) should be

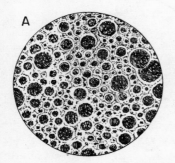 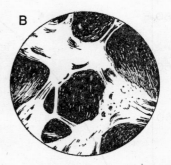

FIG. 207. Bubbles of egg albumen foam obtained after 5 minutes of beating; magnified 25 diameters. (Drawn from the photographs of Barmore, 1934.)

A, the appearance of the bubbles 2 minutes after beating was stopped. *B*, the appearance of the bubbles 10 minutes after beating was stopped.

from 1.5 to 2.5 minutes before adding sugar, and about 30 seconds after adding it.

Effect of p*H*. Experimental results obtained by Henry and Barbour (*1933*) and by Bailey (*1935*) indicate that the volume of beaten albumen tends to become greater as the *p*H of the albumen increases. Further-

	INCREASE IN VOLUME
*p*H	(per cent)
5.5	688
7.0	877
9.0	880
11.0	982

more, if the *p*H is below 8.0, considerable time is required for beating and for the production of a stable foam. It has been pointed out, however, that more stable foams, which show less leakage, are obtained from albumen of relatively low *p*H than from albumen of high *p*H (*Bailey, 1935*).

Effect of Temperature. Tests at the temperatures of 10° C., 15° C., 20° C., and 25° C. showed that the properties of albumen foams are not appreciably affected by changes of temperature within this range (*Henry and Barbour, 1933*). In addition, Barmore (*1934*) came to the conclusion that heating egg albumen to temperatures as high as 50° C. for 30 minutes has no effect on the quality of the foam produced, but that treatment at higher temperatures (60° C. and 65° C.) causes a decided decrease in the stability of the foam. The volume of albumen foam, also, is about 30 per cent smaller than average if the egg is immersed for 10 minutes, prior to beating, in water heated to 60° C. (*Romanoff and Romanoff, 1944*).

Low temperatures apparently do not adversely affect the capacity of egg albumen to form foam. Albumen that has been frozen solid at −3° C. and subsequently thawed yields a foam comparable in volume to that obtained from fresh albumen (*Henry and Barbour, 1933*).

Effect of Adding Water. It has long been a popular belief that the addition of water to egg white has the effect of improving the latter's foaming capacity. Experiments have borne out this supposition (*Henry and Barbour, 1933; Romanoff, unpublished*). There is a direct linear relationship between the volume of foam and the amount of water added (Fig. 208). If the proportion of water in the mixture does not exceed 40 per cent by volume, the stability of the foam is nearly as great as that obtained from the undiluted albumen, although the structure of the foams from water-containing samples appears, in general, to be more porous. Foams from samples containing 60 and 80 per cent of water, by volume, show considerable leakage on standing, before they dry; and there is a decided tendency for the structure of the foam to break down upon continued beating.

Effect of Adding Yolk. It is well known that the foaming of albumen is diminished by the addition of yolk (*St. John and Flor, 1931*). In demonstration of this fact, Bailey (*1935*) presented the following data:

YOLK ADDED (per cent)	RELATIVE FOAMING (per cent)
0.0	100
0.5	85
1.0	70
2.0	56
3.0	44

Effect of Adding Sugar or Salt. The addition of 50 per cent of sugar to albumen increases the stability of the foam, although the beating period must be more than doubled. On the other hand, the foam is rendered less stable when 2.5 per cent of sodium chloride (or common salt) is added (*Hanning, 1945*).

Effect of Adding Milk. It is generally known that egg albumen cannot be beaten to a foam if a small amount of milk is added. The assumption has been made that the milk fats interfere with the formation of

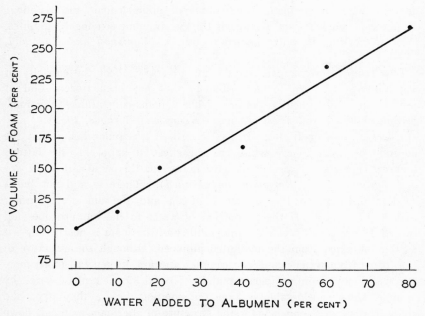

FIG. 208. Increase in volume of foam obtained upon adding various percentages of water to albumen before beating. (After Henry and Barbour, 1933.)

foam. This assumption was confirmed experimentally when it was found that stiff foam could not be obtained if four drops of whole milk, or three drops of 20 per cent cream, were added to egg albumen, but that as many as 446 drops of fat-free milk did not prevent foam formation (*Dizmang and Sunderlin, 1933*).

Effect of Adding Oil. It has long been thought that the presence of traces of oil has a detrimental effect on the foaming of egg albumen. This belief has been tested experimentally by adding cottonseed oil (*Henry and Barbour, 1933*) and olive oil (*Bailey, 1935*) to albumen, in amounts up to 1 per cent. The results are shown in the tabulation.

AMOUNT OF OIL	VOLUME OF FOAM (per cent)	
ADDED (per cent)	WITH COTTONSEED OIL	WITH OLIVE OIL
0.0	100	100
0.1	...	59
0.2	85	30
0.3	...	22
0.5	62	
1.0	28	

It is evident that the addition of oil causes a decrease in the volume of foam, to an extent that becomes more marked as increasingly large amounts of oil are added. The stability of the beaten product is apparently not affected by additions of small amounts of oil; but, if more than 0.5 per cent of oil is added, the foam breaks down after it stands.

Effect of Adding Chemicals. The addition of alkalies and alkaline salts —calcium hydroxide, sodium hydroxide, or sodium sulfite—to the albumen has no apparent effect on the formation of foam. On the other hand, acids and acid salts increase the stability of the foam considerably. Barmore (*1934*) tested several acid compounds and found that potassium acid tartrate had the most desirable effect. His data are given in the table.

ACIDS ADDED TO 53 CC. ALBUMEN	STABILITY OF FOAM (Drainage, gram/minute)
None	0.62
Hydrochloric (5 cc. 0.5 N)	0.50
Sulfuric (2.2 cc. 1.09 N)	0.35
Acetic (5 cc. 0.32 N)	0.35
Citric (0.15 gm.)	0.33
Potassium acid tartrate (0.25 gm.)	0.31

Acids apparently produce an irreversible change in the protein concentrated at the liquid-air interface, because the addition of enough base to neutralize an acid completely does not cause the foam to break down. However, the presence of acid (potassium acid tartrate) does not prevent small amounts of egg yolk from reducing the stability of the foam.

Physicochemical Characteristics of Albumen Foam. The bubbles in albumen foam are not spheres, but polyhedrons (cf. Fig. 207). It has been estimated that, in foam of specific gravity 0.137, the average diameter of the bubbles is 0.02 cm. (*Barmore, 1934*). As beating is prolonged, the specific gravity of the foam and the diameter of the bubbles decrease. After 6 minutes of beating, the specific gravity falls to about 0.088, and the diameter of the bubbles to about 0.01 cm. The space occupied by

air enlarges as the bubbles increase in number and surface area.   These
relationships are shown below (*Barmore, 1934*).

PROPERTIES	TOTAL VOLUME OF FOAM		
	402 cc.	459 cc.	552 cc.
Specific gravity	0.137	0.120	0.100
Average bubble diameter (cm.)	0.020	0.015	0.010
Total surface of bubbles (sq. cm.)	106,300	168,500	304,000
Space occupied by air (per cent)	86.9	88.0	90.0

Foam of low specific gravity, composed of very fine bubbles, drains
rapidly if it is allowed to stand.   Because of their large surface area,
extremely small bubbles are very thin-walled and collapse easily, in spite
of the decreased flow of liquid at their liquid-air interfaces.   Foam of
the highest stability is obtained after a relatively short beating period;
its specific gravity is between 0.150 and 0.170 (*Barmore, 1936*).

EMULSIFYING PROPERTIES OF YOLK

Egg yolk is a typical emulsion, a system of oil droplets suspended in
an aqueous medium (Fig. 209).   In addition, it contains protein, which
is a lyophilic colloid and which has a stabilizing effect.   Consequently,
egg yolk not only possesses the property of emulsifying other solutions
(*Ermolenko and Guterman, 1938*), but it also is a good stabilizer (as
when used in mayonnaise).

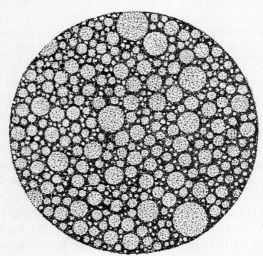

FIG. 209.   Microscopic view of egg yolk, showing suspended fat droplets.   (After
the photograph made by Kilgore, 1933.)

It has been generally assumed that lecithin is the constituent that makes egg yolk an effective emulsifier (*Corran and Lewis, 1924*). There is evidence, however, that egg yolk owes its emulsifying action to an unstable complex containing both lecithin and protein (*Snell, Olsen, and Kremers, 1935*), which has been called a "lecithoprotein."

Stability of Emulsions. Deterioration of emulsions of oils in egg yolk is brought about by the gradual coalescence and growth of the dispersed oil droplets, without the separation of free oil. In yolk-stabilized emulsions of kerosene and olive oil, for example, there is a decrease in specific interfacial area with time (*King and Mukherjee, 1940*).

Time (days)	Specific Interfacial Area (sq. cm. $\times 10^2$) With Kerosene Oil	With Olive Oil
0	150.0	29.9
7	90.2	28.9
14	34.9	21.9
30	25.8	17.1
60	14.6	

It may be concluded that finely dispersed emulsions of oils in egg yolk are no more stable than the coarser systems. The emulsion of kerosene oil was initially about five times as fine as that of the olive oil; but, after a few weeks, the two emulsions had dispersions of the same order.

King and Mukherjee (*1940*) also observed that emulsions stabilized with egg yolk immediately suffered phase inversion on the addition of calcium ions. Acids, however, brought about no phase inversion, although they caused initially coarser systems to be produced.

COAGULATION OF ALBUMEN AND YOLK

Knowledge of the conditions under which egg albumen coagulates is of special value in the proper preparation of hens' eggs as food for human consumption.

Both albumen and yolk can be coagulated not only by heat, but also by other agents, chemical, physical, or even mechanical. The mechanism involved in coagulation is discussed in the section on the coagulation of crystalline ovalbumin.

Coagulating Temperature. Some coagulation of egg albumen can be noticed at 57° C., but only after a long period of time. At 58° C., the albumen becomes faintly opalescent (*Haycraft and Duggan, 1890*). Visible coagulation begins at temperatures over 60° C. The three chief layers of albumen in the chicken egg exhibit the first indication of turbidity at slightly different temperatures, as shown below (*Romanoff*, *1943c*).

LAYER OF ALBUMEN	COAGULATION POINT (°C.)
Outer liquid	61.5
Middle dense	61.0
Inner liquid	60.3

Egg albumen begins to thicken into a gel as the temperature approaches 62° C., and, at 65° C., it will not flow. At 70° C., the mass, although fairly firm, is still jellylike and tender; at higher temperatures, it becomes very firm.

The yolk coagulates at a somewhat higher temperature than the albumen. It begins to thicken at 65° C. and stops flowing at about 70° C.

Species Differences. Early studies by Davy (*1863*) and Tarchanoff (*1884*) revealed that, when the eggs of altricial birds (such as the pigeon, the robin, and the starling) are boiled, the albumen either remains liquid and transparent or sets to a watery, translucent jelly. "Coagulation," when it occurs, takes place between 80° C. and 90° C.

Factors Influencing Heat Coagulation. After egg albumen has been concentrated somewhat by the evaporation of water, coagulation occurs at a lower temperature than usual. Conversely, dilution with water increases the coagulating temperature of both the albumen and the yolk (*Haycraft and Duggan, 1890*), as the tabulated data show.

WATER ADDED (volumes)	COAGULATING TEMPERATURE (°C.) ALBUMEN	YOLK
0	64.0	85.0
1	65.5	85.5
2	69.0	86.5
3	75.5	87.0
4	....	88.0

Some salts, when added to albumen, hasten coagulation; others retard it, as also do acids and alkalies. The addition of sugar inhibits coagulation (see discussion of crystalline ovalbumin). When an egg albumen solution is extracted by ether before being heated, coagulation does not occur even at 100° C.

Coagulation by Other Means. Bancroft and Rutzler (*1931*) showed that egg albumen can be coagulated by potassium iodide, potassium thiocyanate, urea, ammonium thiocyanate, sodium carbonate, and formaldehyde. Sodium hydroxide has only a slight effect.

Among other coagulating agents are mechanical agitation (*Wu and*

Ling, 1927); pressures of 5000 atmospheres or more, without increase in temperature (*Bridgman, 1914; Dow, 1940; Grant, Dow, and Franks, 1941*); and high-frequency sound waves (*Schmitt, Olson, and Johnson, 1928*).

SURFACE TENSION

All protein solutions are surface active. For this reason, the surface tensions of egg albumen and yolk are measurable. According to Harvey and Danielli (*1936*), the surface tension of albumen is about 53 dynes per centimeter; Peter and Bell (*1930*) found that the surface tension of a 12.5 per cent solution of dried albumen, at pH 7.8, was 49.9 dynes per centimeter. (The surface tension of water at 25° C. is about 72 dynes per centimeter.) A fall in surface tension is associated with the coagulation of albumen (*Amar, 1924*).

The surface tension of the yolk is only two-thirds that of the albumen (*Amar, 1924*).

GOLD NUMBER OF ALBUMEN

The protective efficiency of egg albumen as a lyophilic colloid may be expressed in terms of its gold number (*Zsigmondy, 1901*), which is 0.1–0.2 (*Schulz and Zsigmondy, 1903*). (The gold number is the amount of a protein, in milligrams, that is sufficient to protect 10 cc. of a standard colloidal gold solution from precipitation by electrolytes.) The gold number, however, is of less value for characterizing the whole egg albumen than it is for characterizing the pure albumen proteins.

BOUND WATER

Water found in biological materials may exist in a free or in a colloidally bound state. Bound water, which does not freeze at a temperature as low as −20° C., is present in both egg yolk and egg albumen; it is of possible importance in the osmotic equilibria with which the vitelline membrane has to deal (*Needham and Smith, 1931*).

By the calorimetric technique, St. John (*1931*) determined that in egg albumen there are about 1.97 gm. of bound water per gram of dry material. Figure 210, drawn from his data, indicates that, at a temperature of about −12.5° C., all the free water freezes; the remaining, or bound, water (which does not freeze even at −35° C.) constitutes about 26 per cent of the total water. Somewhat lower values for bound water were obtained by Jones and Gortner (*1932*), who employed the dilatometric method; they found the bound water to be approximately 1.55 gm. per gram of dry substance. Similar methods, as well as the cryoscopic method, have yielded still lower values (*Smith, 1933; Moran, 1935b*). It is probable, however, that about 25 per cent of the total water in the egg is bound water, as was found by the calorimetric method (*St. John and Caster, 1944*) and by the heat fusion method (*Caster and St. John, 1944*). The proportional amounts of bound

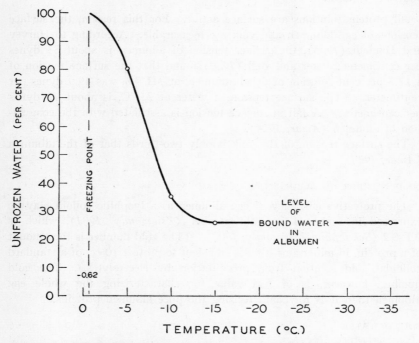

FIG. 210. Relationship between the freezing temperature and the percentage of unfrozen water in egg albumen. At temperatures colder than −15° C., only the bound water remains unfrozen. (After St. John, 1931.)

water in the thick and thin layers of the albumen are essentially the same (*St. John, 1931; Smith, 1934; Moran, 1935a*).

In a single experiment, Hill (*1930a*) determined that there is about 1.0 gm. of water per gram of dry yolk; that is, about 15 per cent of the total water of the yolk is bound water. The relative amount of bound water on either side of the vitelline membrane is therefore of roughly the same order.

Viscosity

The viscosity (internal friction; i.e., resistance to shear, agitation, or flow) of the yolk and the albumen may be expressed relatively, or in absolute units—poises.[2] Since both yolk and albumen are colloids, the internal friction between the molecules is much greater than that between the molecules in a solution, or in water.

According to Bellini (*1907*), the relative viscosity of the yolk of the fresh chicken egg is more than eight times as great as that of the albumen. If one uses the figures for the yolk's relative viscosity given by A. J. M. Smith (*1935a*), one may then compare the viscosities of yolk and albumen with that of water, as tabulated. According to Gane (*quoted by Smith,*

RELATIVE VISCOSITY

Water	1
Albumen (mixed layers)	53
Yolk	440

1934), the viscosity of egg yolk at high rates of shear is about 7000 times that of water. Gane (*1934*) gave the viscosity of the yolk, in absolute units, as about 200 poises at 0° C. (at which temperature the viscosity of water is 1.8 centipoises). The viscosity of yolk with a water content of about 49 per cent has also been given as 8 poises, at 25° C. (*Payawal, Lowe, and Stewart, 1946*).

Although the viscosity of the liquid albumen is only four times as great as that of water, the viscosity of the dense layer is over fifty times as great. The high viscosity of the dense albumen layer is attributable to its peculiar structure (see Chapter 3, "Structure"). When the relative viscosity of thick albumen is measured by a capillary tube method, it can be demonstrated that successive shearings produce a progressive change in viscosity in the direction of greater fluidity, until the viscosity is stabilized at a value approximating that of thin albumen. It is therefore apparent that the decrease in viscosity is due to the fact that the structure of the albumen is destroyed.

Changes in Albumen. When the whole albumen is dried to a constant weight over phosphorus pentoxide and then reconstituted by adding to the dry material an amount of water equal to that lost, the viscosity of the albumen, after reconstitution, is much lower than the original viscosity. Similarly, if the albumen is frozen at −5° C. and then thawed, the relative viscosity is again lower than that of undisturbed albumen (*A. Smith, 1935a*). These viscosity changes are shown below.

CONDITION OF ALBUMEN	RELATIVE VISCOSITY (per cent)
Undisturbed	100
Dehydrated and reconstituted	38
Frozen and thawed	54

[2] The reciprocal of absolute viscosity is fluidity.

Ostwald (*1913*) showed that, when egg albumen is heated, the viscosity continually decreases, except for a sudden increase between 57.5° C. and 60° C. The smoothed curve shown in Fig. 211-*A* indicates the changes in the viscosity of albumen over the temperature range of 60° C. to 70° C. (*Payawal, Lowe, and Stewart, 1946*). Maxima in viscosity occur at 63° C. and at 66° C., and a minimum at 64° C.

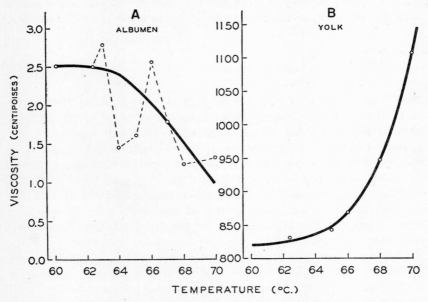

Fɪɢ. 211.   Effect of heating on the viscosity of *A,* albumen, and *B,* yolk. (After Payawal, Lowe, and Stewart, 1946.)

Changes in Yolk. It has been observed (*A. Smith, 1935a*) that the viscosity of the yolk is a function of the yolk's water content (Fig. 212). The addition of one part of water to every ten parts of yolk reduces the yolk's viscosity to one-tenth of its original value; the addition of two parts of water to ten of yolk reduces it to one-thirtieth. Correspondingly, when diluted yolk is increased in concentration, its viscosity rises (*Zawadzki, 1935*).

The addition of chloride salts of sodium, potassium, and calcium increases the viscosity of diluted yolk (*Zawadzki, 1933*). The maximum increase is observed when sodium chloride and potassium chloride are used.

High pressures have a definite effect on egg yolk viscosity, as shown below (*Ebbecke and Haubrich, 1937*).

Pressure (atmospheres)	Relative Viscosity (per cent)
Normal	100
400	118
800	159

Low temperatures have no specific effect on the viscosity of the yolk (*Moran, 1936a*). Heating the yolk above 60° C. increases its viscosity (cf. Fig. 211-*B*) until, at 70° C., coagulation occurs.

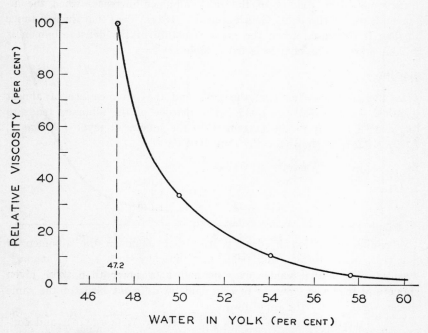

FIG. 212. Relationship between the viscosity and the water content of the yolk. The initial viscosity of yolk is taken as 100 per cent. The values were obtained by measuring the time required for 30 gm. of yolk to pass through a capillary tube 30 cm. long, under a pressure head of 45 cm. of mercury. (After A. Smith, 1935a.)

ELASTICITY OF ALBUMEN

The elastic properties of egg albumen have been studied by Freundlich and Seifrez (*1923*) by a method based on the withdrawal, by a magnet, of a nickel particle (18 μ in diameter) suspended in solutions of albumen of various concentrations. It was concluded that egg albumen has a very high elastic limit.

PLASTICITY OF ALBUMEN

St. John and Green (*1930*), after separating the liquid portion of the albumen from the dense, found different plasticity values for the two fractions. These values are in close agreement with the apparent fluidity.

PORTION OF ALBUMEN	PLASTICITY VALUE
Liquid	14.77 –20.7
Dense	0.167– 0.305

Furthermore, the fluidity of the thick albumen increases when the albumen is passed through a Gooch crucible. It is evident that the resistance to flow is decreased when the gross structure of the dense albumen is broken down. The change is irreversible.

Density

In the fresh chicken egg, the yolk and the albumen are of almost identical average density, 1.035. The density of the albumen tends to increase slightly from the outermost to the innermost layer (*Romanoff, 1940, 1943c*), as shown by the tabulated data.

LAYERS OF ALBUMEN	DENSITY (d_{25}^{25})
Outer liquid	1.032
Middle dense	1.036
Inner liquid	1.040
Chalaziferous	1.045

A similar gradation in density has been found in the albumen of pheasant, quail, turkey, and pekin and muscovy duck eggs (*Romanoff, 1943c*). The actual values may be higher or lower than those given above and may vary greatly in eggs laid by different birds of the same species, or even by the same bird.

Very interesting data have been presented by Witz (*1876*), and confirmed by Rakuzin and Flieher (*1923*), on the relationship between the concentration and the density of egg albumen, as shown in Table 37.

The animal and vegetal poles of the yolk differ noticeably in density. Baudrimont and St.-Ange (*1847*) showed (although presumably not on strictly fresh eggs) that the density of yolk taken from under the cicatricula (i.e., from the latebra) is 1.027, whereas that of yolk taken from the opposite pole is 1.032. The whole yolk, well mixed, has a density of 1.029.

Solubility

Egg albumen and egg yolk differ greatly in solubility. Albumen is water-soluble, since it consists chiefly of soluble ovalbumin. On the other

TABLE 37

CHANGES IN DENSITY OF EGG ALBUMEN WITH CHANGES IN CONCENTRATION
(After *Witz, 1876*)

Concentration of Albumen (per cent)	Baumé Scale (degrees)	Density
1	0.37	1.0026
2	0.77	1.0054
3	1.12	1.0078
5	1.85	1.0130
10	3.66	1.0261
15	5.32	1.0384
20	7.06	1.0515
25	8.72	1.0644
30	10.42	1.0780
35	12.12	1.0919
40	13.78	1.1058
45	15.48	1.1204
50	17.16	1.1352
55	18.90	1.1511

hand, yolk, which is composed largely of phosphoprotein, is insoluble in water and soluble in dilute salt solutions and in dilute alkalies.

Petit (*1871*) observed that animal charcoal has the property of removing albumen from its solution in liquids, whether acid, neutral, or alkaline.

Dehydration

Egg albumen can be dried (in vacuum) to light yellow flakes which have a crystalline appearance. This material, however, is different from the crystalline form of the albumen protein, ovalbumin.

The desiccation of the yolk leaves a granular, waxlike residue (*Amar, 1924*). When the exposed surface of the yolk is allowed to evaporate slowly, the loss of water proceeds steadily and apparently uniformly. At higher rates of evaporation, a gradient in water content is set up between the underlying portions of the liquid and the superficial layer, which eventually forms a solid crust (*Smith, 1932*).

RECONSTITUTION OF DEHYDRATED ALBUMEN AND YOLK

When egg albumen is dried (over phosphorus pentoxide) to a constant weight and then reconstituted by the adding of an amount of water equal to that lost, the thick and the thin parts of the albumen reappear in approximately their initial proportions (*A. J. M. Smith, 1935a*). The reconstituted albumen therefore resembles the albumen of the fresh egg.

On the other hand, yolk, which consists chiefly of lipids, cannot be returned to its original state after it has been dehydrated (*Moran, 1925*).

EFFECT OF DEHYDRATION

A moderate degree of heat turns egg albumen to an amber brown color (*Knecht, 1920*), which does not change further with the passage of time. Prolonged heating of the albumen at high temperature (about 70° C.) causes decomposition (*Tinkler and Soar, 1920*) and the evolution of some hydrogen sulfide.

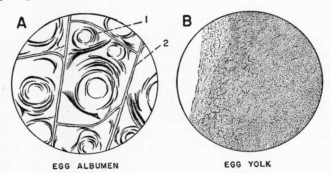

EGG ALBUMEN   EGG YOLK

FIG. 213. The appearance of dried films of albumen and yolk; magnified 30 times. (Drawn with the aid of camera lucida.)

A, egg albumen: 1, circular internal fissures; 2, linear cracks forming rectangles. *B*, egg yolk.

Egg albumen, when dried over sulfuric acid, forms transparent lamellae and spiral-like elements, covered by material of lesser density (*Amar, 1924*). If deposited upon a glass slide and permitted to dry, albumen takes on definite surface characteristics (*Glabau and Kepes, 1935; Romanoff, unpublished*), as shown in Fig. 213-A.

Egg yolk darkens upon being heated. It has also been observed (*Tinkler and Soar, 1920*) that the greenish black discoloration on the surface of the cooked (hard-boiled) yolk is due to the presence of finely divided iron sulfide, formed from the iron and sulfur in the yolk.

The desiccation of the yolk over sulfuric acid leaves a granular, waxy residue (*Amar, 1924*). The appearance of a film of dried yolk is shown in Fig. 213-B.

EFFECT OF FREEZING

Frozen egg yolk and albumen, like many other colloids, do not, as a rule, recover their original state upon thawing. Freezing causes more striking changes in the yolk than in the albumen.

According to Moran (*1925*), the yolk, if frozen at temperatures above −6° C., regains its normal fluidity on thawing. If the temperature, however, is carried below −6° C. and held below that point for some time, the yolk, when thawed, is changed permanently into a stiff, putty-like paste. The irreversible change is apparently caused by the separation of water from the lecithovitellin complex, which is then precipitated. Lipoid substances are incapable of reforming their original state once they are thrown out of solution from systems such as egg yolk.

In the presence of sucrose (10 per cent), egg yolk suffers practically no change in fluidity or appearance after being frozen at −11° C. and then thawed (*Moran, 1925*). The protective action of sugars is probably due to their ability to depress the freezing point of water (see Chapter 10, "Preservation").

It may be noted that pasty yolks which have been frozen quickly in liquid air are finer in texture, after thawing, than yolks which have been frozen slowly.

When albumen is frozen at any temperature, and then thawed, there is an increase in the liquid portion and a corresponding decrease in the dense part. Freezing apparently produces a separation of water from the dense portion, and this water is not reabsorbed when the albumen is thawed. The extent of water separation is controlled by the temperature reached during the freezing process, as shown in the tabulation (*Moran, 1925*). It is the temperature, and not the freezing or thawing rate,

TEMPERATURE (°C.)	LIQUID PORTION OF ALBUMEN (per cent)
(Control)	42
−3	50
−16	67

that determines the change. In eggs frozen at −3° C. and then transferred to −16° C., 65 per cent of the albumen may be liquid.

Precipitation, or coagulation, may be observed when the albumen has been stored in the frozen state for long periods of time. After 4 months of storage at −3° C., the dense portion of the albumen may reveal distinct signs of coagulation, as shown by the presence of white fibers.

In studies of the effect of freezing on a mixture of yolk and albumen (so-called whole egg magma), it was found that maximum gelation is reached in 60 to 120 days of storage at −21° C. to −18° C. (*Thomas and Bailey, 1933*). However, the addition of sodium chloride, together with a sugar (sucrose or dextrose), considerably lowers the degree of gelation (see Chapter 10, "Preservation").

Ionic Properties

TITRATABLE ACIDITY

Davy (*1863*) reported that the albumen of avian eggs is always alkaline, and the yolk always acid. Tarchanoff (*1884*) contended that the titratable acidity of the albumen in the eggs of precocial birds (chickens) is much greater than that in the eggs of altricial birds (pigeons, ravens). Recently, Shklyer (*1937*) noted that neutralization of the entire albumen required 393 cc. of 0.01 *N* hydrochloric acid, and neutralization of the entire yolk, 184 cc. of 0.01 *N* sodium hydroxide; acid thus exceeded base in the amount of 209 cc. Shklyer's results are in good agreement with those of Forbes, Beegle, and Mensching (*1913*).

	AMOUNT (cubic centimeters)
Total acid	322
Total base	106
Excess of acid	216

In general, it is impossible, especially by means of titration with the usual indicators, to determine with accuracy the exact alkalinity of the albumen or the exact acidity of the yolk. The effect of proteins and other constituents of the egg on indicators is still obscure. Lepper and Martin (*1927*) have shown that pure ovalbumin (in concentrations up to 7 per cent) does not cause any serious colorimetric error with either Neutral Red or Phenol Red, whereas egg white (in concentrations up to 20 per cent) decreases the pH reading by 0.1.

ANION-CATION RATIO

The concentrations of positive and negative ions, calculated as millimoles and milliequivalents, are unevenly divided between yolk and albumen. There is an excess of anions over cations in the yolk, and more cations than anions in the albumen (*Carpiaux, 1908*). The average anion-cation ratio for the whole egg is 2.3, for the yolk 2.8, and for the albumen 0.54 (*Needham, 1931a, p. 302*).

The average concentrations of ions in the chicken egg, in milliequivalents, is shown in the tabulation on the next page.

In the eggs of altricial birds, like the pigeon, the albumen is relatively richer in alkali than it is in the eggs of precocial birds, such as the chicken (*Needham, 1931a, p. 274*).

It is not surprising to find, in the literature, a very wide variation in the values given for ionic strength, because of the different amounts of

Ions	Albumen	Yolk	Entire Egg Contents
Cations			
K	4.0	3.4	3.7
Na	4.3	3.6	4.1
Mg	1.0	1.9	1.4
Ca	0.7	8.1	6.3
Fe	0.2	0.7	....
Anions			
SO$_4$	0.3	....	0.1
PO$_4$	0.9	35.0	30.2
Cl	4.2	3.7	4.0

minerals observed in the egg (*Poleck, 1850; Rose, 1850; Voit, Hermann, Forster, Feder, and Stumpf, 1877; Carpiaux, 1908; Iljin, 1917; G. Clark, 1925; Bialaszewicz, 1926; Isaki, 1930; Straub and Donck, 1934; and others*).

HYDROGEN-ION ACTIVITY (*p*H)

In the new-laid chicken egg, the *p*H of the albumen approaches that of the hen's blood; it is about 7.6. On the other hand, the *p*H of the yolk is about 6.0. Approximately the same values have been observed in the eggs of other precocial birds (*Romanoff, 1943c; Romanoff, 1944c*).

Slight differences have been noted in the *p*H values of the various layers of egg albumen. The layer closest to the yolk usually has a lower *p*H than the outermost layer (*Romanoff and Romanoff, 1929; Romanoff, 1943c*).

Role of Carbon Dioxide. The presence of bicarbonates in egg albumen was reported by School (*1893*). Much later, Healy and Peter (*1925*) suggested that the albumen of fresh eggs contains both bicarbonates and carbon dioxide. It is now established that the albumen contains carbon dioxide, both combined and free.

There is a relationship between the concentration of free carbon dioxide in the egg and the *p*H of the egg contents. The loss of dissolved carbon dioxide results in a rapid increase in the *p*H of the albumen. The observations of Romanoff and Romanoff (*1933*) and of Brooks and Pace (*1938b*) clearly show that the *p*H of the albumen depends on the partial pressure of carbon dioxide. The range in the *p*H of the albumen may be from 9.7 in air to 6.5 in an atmosphere of 100 per cent carbon dioxide (see Chapter 10, "Preservation"). It is of interest to note, however, that changing the carbon dioxide pressure has little effect on the *p*H of the yolk.

OXIDATION-REDUCTION POTENTIAL

Attempts have been made to determine the oxidation-reduction potential of the hen's egg. It was found (*Pavlov and Isakova-Keo, 1929*) that, in the fresh egg, the potential showed a shift to the positive side, and that the potential of the yolk was more positive than that of the albumen. The values obtained for the oxidation-reduction potential (as *Eh*) of the yolk and albumen of the fresh egg were 292 mv. and 205 mv., respectively.

When determined by means of a Beckman potentiometer, with a platinum electrode, the *Eh* of albumen (pH 7.78) was found to be 414 mv., at a temperature of 25° C.; that of yolk (pH 6.11), 333 mv. (*Romanoff and Yushok, unpublished*). The deviations in these values from those given above, and the fact that the oxidation-reduction potential of the albumen is higher than that of the yolk, are difficult to explain.

ELECTRICAL CONDUCTIVITY

Because of the presence of ions in the yolk and albumen, conductance of electric current is possible. The degree of conductivity indicates the relative concentration of electrolytes. Electrical conductivity is usually expressed in reciprocal ohms (mhos) of 1 gram-equivalent of electrolyte, measured in a solution between electrodes 1 cm. apart.

Among the earliest studies of the conductivity of the hen's egg are those made by Bellini (*1907*), Wladimiroff (*1926*), and Perov and Dolinov (*1932*). Romanoff and Grover (*1936*) extended their investigation to the eggs of other species of birds. Their data for the specific conductivity of the yolk and albumen of fresh eggs, at audio frequency (1000 cycles) are reproduced here.

	CONDUCTIVITY [(mho-cm^{-1}) $\times 10^{-3}$]	
SPECIES	ALBUMEN	YOLK
Chicken	8.68	3.10
Pheasant	8.14	3.16
Quail	8.00	3.09
Turkey	7.55	2.90
Duck	7.73	2.39

It has been suggested (*Rudolph, 1938*) that the specific conductivity of egg albumen is influenced by the albumen's total phosphoric acid content. However, the concentration of inorganic phosphorus alone has no decisive effect.

Studies made by Orru (*1936*) revealed that there is a linear increase in the conductivity of egg albumen as the temperature rises from a very low point to 60° C. The conductivity of the yolk (*Orru, 1935*) also increases with temperature; but, at 70° C. to 75° C., the conductivity diminishes. It increases again at higher temperatures. It has been suggested that the decrease in conductivity is coincident with the coagulation of the lecithovitellin and livetin of the yolk.

In the albumen, high-frequency conductivity is highest in the middle dense layer, lowest in the inner liquid layer, and of intermediate value in the outer liquid layer (*Romanoff and Frank, 1941*).

Electrical Resistance. As already stated, electrical conductivity is reciprocal to electrical resistance. Bellini (*1907*) first measured the electrical resistance of fresh egg yolk and albumen. The data of Romanoff and Grover (*1936*) for electrical conductivity, if converted, would give values of 0.32 ohm-cm. and 0.12 ohm-cm. for the yolk and the albumen, respectively, of the chicken egg.

DIELECTRICS

The specific property designated as the dielectric constant, ϵ, may be of some interest in the study of avian eggs, because it bears a certain relationship to the moisture content of colloidal systems.

According to Fürth (*1923*), the dielectric constants (measured as specific inductive capacity) of yolk and albumen are, respectively, 60 and 68, when determined by the method of Drude. These values, compared with those of many pure proteins, are rather low.

Optical Properties

REFRACTIVITY

The refractive index (a quantitative measurement of the refraction of light, which is usually referred to the D-line of sodium, wave length 5893 angstrom units) has been determined for albumen, and recently for yolk also. The refractometric method is readily applicable to egg albumen, because the albumen gives a sharp and easily read line of demarcation between the illuminated and dark segments of the field in the refractometer. The yolk gives a less distinct line of demarcation.

Studies of fresh eggs laid by chickens of various breeds revealed that the average refractive index of the albumen is 1.3562, and of the yolk, 1.4185 (*Rice and Young, 1928*). In the eggs of other species of birds— namely, the pheasant, the quail, the turkey, and the pekin and muscovy ducks—nearly identical values have been observed (*Romanoff, 1943c*).

The refractive indices of the various layers of the albumen are noticeably dissimilar, and distinct differences in chemical composition are therefore indicated. In the chicken egg, the refractive indices of the albumen layers are as tabulated. Similar differences in the refractive

ALBUMEN LAYER	REFRACTIVE INDEX	
	MORAN AND HALE *(1936)*	ROMANOFF AND SULLIVAN *(1937)*
Outer liquid	1.3492	1.3529
Middle dense	1.3566	1.3552
Inner liquid	1.3582	1.3582
Chalaziferous	......	1.3606

indices of the various layers of the albumen have been observed in the eggs of other species of birds (*Romanoff and Sullivan, 1937; Romanoff, 1943c*).

Moran and Hale (*1936*) observed that the refractive index of yellow yolk is higher than that of white yolk. The values were found to be 1.4200 and 1.4140 for yellow and white yolk, respectively.

Relation to the Content of Total Solids. As early as 1910, Herlitzka presented evidence that the refractive index of egg albumen varies with

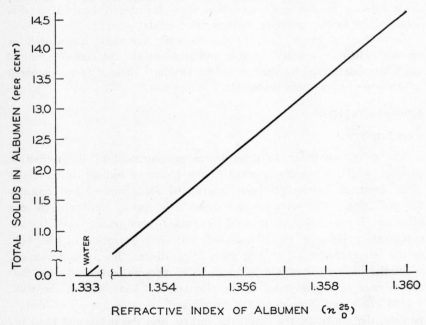

Fig. 214. Relationship between the index of refraction and the total solids content of egg albumen. (After Almquist, Lorenz, and Burmester, 1932.)

the concentration of the albumen solution. The practical application of
this principle to the estimation of the egg's content of solids is more
recent (*Holst and Almquist, 1931a; Almquist, Lorenz, and Burmester,
1932; Urbain, Wood, and Simmons, 1942*). The direct linear relationship that exists in egg albumen between the refractive index and the
percentage of total solids is shown in Fig. 214. As may be seen in Fig.
215, there is a similar relationship between the refractive index and the
content of total solids in egg yolk (*Bailey, 1936*).

Influence of Temperature. It has been shown (*Janke and Jirak, 1934;
Bailey, 1936; Romanoff, unpublished*) that the refractive index of a
substance varies with fluctuations in temperature, because of changes
in the density of the material. There is a linear decrease in refractive
index as the temperature increases. Observations made at temperatures
other than 25° C. therefore require correction by the addition or sub-
traction of about 0.00015 for each degree (centigrade) of change in
temperature. The temperature coefficient of yolk is 0.0001, and of albumen, 0.0002.

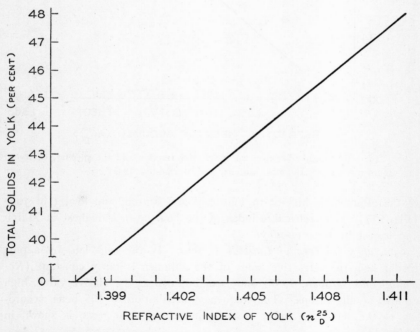

FIG. 215. Relationship between the index of refraction and the total solids content
of egg yolk. (After Urbain, Wood, and Simmons, 1942.)

Relation to Specific Gravity. In the albumen, there is a direct linear relationship between index of refraction and specific gravity (*Almquist, Lorenz, and Burmester, 1932*). This relationship is presented graphically in Fig. 216.

Seasonal Variation. Observations made by Romanoff and Sullivan (*1937*) reveal that, in the chicken egg, the refractive index of each layer

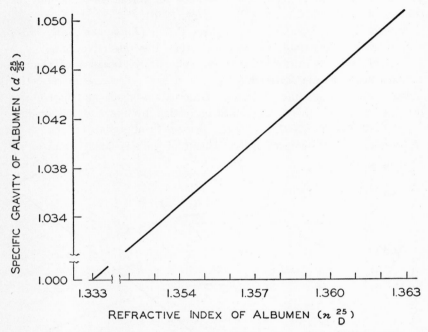

Fig. 216. Relationship between the index of refraction and the specific gravity of egg albumen. (After Almquist, Lorenz, and Burmester, 1932.)

of the albumen is highest in February and March, and lowest in June (Fig. 217). The refractive index of the yolk, also, is highest in March and lowest in June (*Bailey, 1936*).

Constancy in Eggs of Individual Hens. In the eggs laid by a particular hen, the refractive index of the albumen is fairly constant (*Romanoff and Sullivan, 1937*) and is maintained at its relative value, high or low, at all seasons. The frequency distribution of 105 hens, according to the refractive index of the albumen of their eggs, is shown in Table 38.

TABLE 38

FREQUENCY DISTRIBUTION OF HENS ACCORDING TO THE AVERAGE REFRACTIVE INDEX
OF THE ALBUMEN OF THEIR EGGS *

Number of Hens	Number of Eggs Examined	Refractive Index of Albumen Layers (n_D^{25})			
		Outer Liquid	Middle Dense	Inner Liquid	Chalaziferous
4	17	1.3497	1.3516	1.3536	1.3570
6	27	1.3508	1.3532	1.3548	1.3597
22	96	1.3515	1.3535	1.3552	1.3600
31	141	1.3525	1.3542	1.3557	1.3603
27	122	1.3535	1.3550	1.3569	1.3606
15	66	1.3542	1.3569	1.3581	1.3619

* After Romanoff and Sullivan (*1937*).

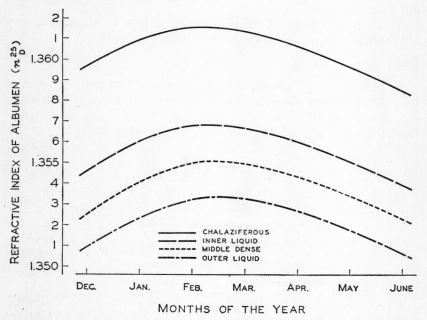

FIG. 217. Seasonal changes in the refractive index of the various layers of chicken egg albumen. (After Romanoff and Sullivan, 1937.)

Spectral Properties

TRANSMISSION OF LIGHT

An outstanding property of egg albumen is the ability to transmit light. The transmission of light by the different layers of the albumen is not related to the percentage of total solids in these layers (*Almquist, Givens, and Klose, 1934*). Light transmission, or the ratio of transmitted light to incident light (I/I_o), in the three major albumen layers of the fresh chicken egg is as follows:

ALBUMEN LAYER	TRANSMISSION OF LIGHT (I/I_o)
Outer liquid	0.915
Middle dense	0.482
Inner liquid	0.808

It is evident that the transmission of light is lowest in the middle dense layer. The transmission of light by the albumen of fresh eggs is in close negative correlation with the percentage of mucin in the albumen and varies with the temperature and pH, which affect the physical condition of the mucin.

It may be assumed that the transmission of light by the yolk is very low, although there are no quantitative data to this effect.

LIGHT ABSORPTION SPECTRA

An early study (*Lewin, Miethe, and Stenger, 1908*) indicates that, in the light absorption spectrum of egg yolk, there are pronounced bands at the wave lengths of about 450 and 480 mμ. There is also some absorption at wave lengths of 380, 400, and 430 mμ (Fig. 218).

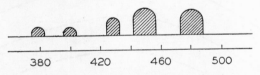

WAVE LENGTH (MILLIMICRONS)

FIG. 218. Absorption spectrum of egg yolk in ether solution. (After Lewin, Miethe, and Stenger, 1908.)

INFRARED ABSORPTION SPECTRA

The spectrum of a dry film of egg albumen shows strong absorption bands in the infrared region, but none in the region either of visible light or of light of wave lengths shorter than 2.5 μ (*Stair and Coblentz, 1935*). A band of maximum absorption extends from 2.95 to 3.5 μ, and a similar band from 5.9 to 8 μ (Fig. 219). Beyond 8 μ, there are wide bands of selective absorption.

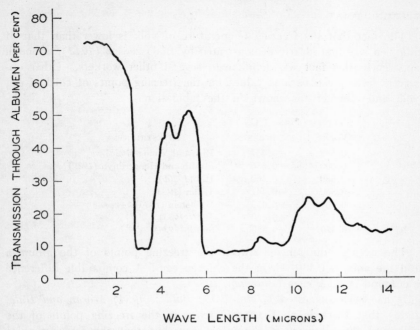

Fig. 219. Transmission of infrared rays through egg albumen. (After Stair and Coblentz, 1935.)

FLUORESCENCE

Both the albumen (*Dingemans, 1932*) and the yolk (*Gouzon, 1934*) have been observed to fluoresce to some extent. Fluorescence is associated largely with the amount of riboflavin present in the fresh egg.

Thermal Properties

HEAT OF COMBUSTION

According to Berthelot and André (*1890*), the average heat of combustion of chicken egg yolk is 8124 calories per gram, and that of the albumen, 5690 calories per gram.

SPECIFIC HEAT

The average specific heat of egg albumen has been determined as 0.85 normal calorie per gram (*St. John, 1931*).

FREEZING POINT

The fact that the freezing temperature of yolk is lower than that of albumen was first clearly demonstrated by Bialaszewicz (*1912*), although apparently this fact was familiar to several other workers. Other observers have given various values for the freezing points of chicken egg yolk and albumen, as shown in the tabulation.

FREEZING POINT (°C.)

YOLK	ALBUMEN	INVESTIGATOR
−0.587	−0.437	Rice and Young (*1928*)
−0.600	−0.456	Straub and Hoogerduyn (*1929*)
−0.580	−0.460	Hill (*1930b*)
−0.560	−0.447	Johlin (*1933*)
−0.601	−0.453	Weinstein (*1933*)
−0.570	−0.420	Hale (*1933*)
−0.570	−0.450	Smith (*1934*)

The average difference between the freezing points of the albumen and the yolk is 0.139° C. in the chicken egg. A comparable difference was found in duck eggs (*Bateman, 1932*).

It has been suggested (*Hale, 1933; Baldes, 1934; Moran and Hale, 1936*) that there is no sharp difference in the freezing points of the yolk and the albumen in the intact egg, but that the freezing point gradually rises from the center of the yolk outward to the albumen (*Moran and Hale, 1936*). It has also been observed (*Smith, 1932*) that

REGION IN THE EGG	FREEZING POINT (°C.)
Center of yolk	−0.587
Yolk between center and periphery	−0.568
Yolk adjacent to vitelline membrane	−0.454
Albumen	−0.424

the freezing-point depression of the dense albumen is about 0.005° C. less than that of the outer liquid albumen. These differences in freezing point within the egg indicate that there are gradients in osmotic pressure, both in the yolk and in the albumen.

Effect of Concentration. The freezing point of either yolk or albumen is raised by dilution of the material and lowered by concentration. This fact has been very clearly demonstrated by Smith (*1934*), who determined the freezing points of a series of samples of albumen and yolk that had been diluted or concentrated to various extents. The direct linear relationship between freezing point and water content is shown in Fig. 220.

The experimental immersion of the intact yolk in water results in a rapid elevation of the freezing point, not only because the yolk takes up water, but also because it simultaneously loses traces of salts. In about 70 hours, the freezing point rises to approximately −0.2° C. (*Smith and Shepherd, 1931*).

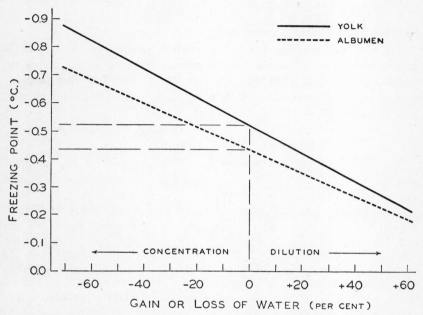

FIG. 220. Freezing points of yolk and albumen as affected by dilution and concentration of these components of the egg. (After Smith, 1934.)

Effect of Carbon Dioxide. The freezing points of the yolk and the albumen in the newly laid egg are considerably affected by the escape of carbon dioxide from the egg. The freezing points of both components become considerably elevated during the first 12 hours after the egg is laid, coincident with the rapid escape of carbon dioxide (*Smith, 1934*).

| | FREEZING POINT (°C.) | |
TIME AFTER LAYING	YOLK	ALBUMEN
10 min.	−0.628	−0.522
1 hr.	−0.599	−0.475
12 hr.	−0.580	−0.450

Within 12 hours after the egg is laid, the freezing points of both the yolk and albumen change rapidly. The subsequent changes are relatively slow (*Smith and Shepherd, 1931*), because they are caused not so much by loss of carbon dioxide as by the gradual passage of water across the vitelline membrane from albumen to yolk, and also by changes in concentration that occur in albumen and yolk.

Experimental saturation of the yolk and albumen with carbon dioxide may cause the freezing point of the yolk to be depressed to $-0.744°$ C., and that of the albumen to $-0.639°$ C. (*Smith, 1934*).

Induced Oxidation

Egg yolk and albumen can be oxidized by passing air through them at ordinary temperatures, in the presence of reducing agents and some metallic hydroxides (*Palit and Dhar, 1930a*). The following percentages of oxidation were observed to occur in 5 per cent solutions of yolk and albumen:

| | OXIDATION (per cent) | |
INDUCTORS	YOLK	ALBUMEN
Cerous hydroxide	69.4	83.3
Sodium hydroxide	28.5	22.5
Sodium bicarbonate	12.0	9.7

PHOTOCHEMICAL OXIDATION

Yolk and albumen may be oxidized by passing air through them in the presence of sunlight (*Palit and Dhar, 1930a, 1930b*). Oxidation was found to be more than twice as extensive in a 5 per cent solution of yolk as in a solution of albumen of the same concentration. (In 10 hours, an average of 69.6 per cent of the yolk and 27.1 per cent of the albumen was oxidized.)

Color Changes Produced by Various Reagents

It is interesting to note a few of the color changes that occur when egg albumen comes in contact with various reagents (*Reichard, 1910; Rakuzin, Brando, and Pekarskaja, 1915; Hunter, 1923*).

A partial list of color reactions is shown in Table 39. From this table, it is evident that many reagents produce color changes in albumen solutions or cause the formation of precipitates of various hues. It is not surprising that a foreign color may occasionally be observed when albumen has been beaten with copper or iron cooking utensils.

Free Carbon Dioxide in Albumen and Yolk

Content. According to Straub and Donck (*1934*), the albumen of the fresh hen's egg contains an average of 55 mg. of carbon dioxide, although there is none in the yolk. Mathieu and Urbain (*1908*) used a vacuum

TABLE 39

COLOR REACTION OF EGG ALBUMEN WITH VARIOUS REAGENTS *

Reagents Used	Color of Albumen Solution † or its Precipitate
Alizarin	Brilliant orange precipitate
a-Naphthol	Yellow solution (with precipitate formation)
Anthraquinone	Light gray solution
Aurin	Bright chrome solution (and similar precipitate)
Benzidine	Pale violet solution (similar precipitate floating)
Cupric sulfate with sodium hydroxide	Pale blue flakes
Cuprous sulfate	Pale blue precipitate
Diazoaminobenzene	Deep red solution (and precipitate)
Eosin	Red-brown bulky precipitate
Ferric chloride	Pale yellow precipitate
Resorcinol	Yellow-green solution (and precipitate)
Thiocarbanilide	Complete white precipitate

* After Hunter (*1923*).
† Albumen flakes diluted 40 times.

pump to extract the gases from the albumen of the fresh egg and obtained about 22 cc. of carbon dioxide. The equivalent weight of this amount is in close agreement with the above value.

Solubility. From measurements of the total uptake of carbon dioxide at 25° C. by the yolk and albumen of the fresh, infertile egg, it has been calculated that the coefficient of absorption, *a* (Bunsen solubility coefficient), is 1.25 for the yolk and 0.71 for the albumen (*Brooks and Pace, 1938a*). The solubility coefficient of water is 0.75. It has been suggested that the lipids in the yolk are responsible for the greater solubility of carbon dioxide in the yolk than in the albumen.

The retention of carbon dioxide by egg albumen is inversely proportional to the partial pressure of carbon dioxide in the air. Albumen apparently retains carbon dioxide approximately as well as does blood serum. At a pressure of 0.05 atmosphere, retention by the blood serum is about 0.065 Warburg unit, and, by egg albumen, 0.045 unit (*Brooks and Pace, 1939*).

Effect. In the presence of excessive amounts of carbon dioxide, the transparency of egg albumen disappears; that is, the albumen becomes cloudy. This effect is most pronounced in the dense layer of the albumen, which contains the highest proportion of ovomucin. The opacity usually disappears after the carbon dioxide is allowed to escape. According to Almquist (*1933*), the cloudiness is associated with the phenomenon of syneresis—the contraction of the ovomucin fibers, accompanied by the squeezing-out of the liquid albumen. Moran (*1936a*) suggests, however, that it is due to a coagulation of ovomucin.

RELATIONSHIP BETWEEN YOLK AND ALBUMEN

Osmotic Relationship

It has been pointed out that the concentration of each dissolved substance in the yolk is independent of its concentration in the albumen. For example, the amounts of sodium, potassium, and chlorine in the yolk exceed the amounts in the albumen (although not to the same extent). Consequently, the total osmotic pressure is different on either side of the vitelline membrane and is higher in the yolk than in the albumen. This difference is probably of great embryological importance, because of its relationship to the movement of water yolkward during the early incubation period.

In the new-laid chicken egg, the difference in the osmotic pressures of the yolk and the albumen equals about 1.8 atmospheres (*Straub and Hoogerduyn, 1929*). According to Needham and Smith (*1931*), this difference persists for at least 2 months at room temperature, diminishing slowly. At 0° C., however, equilibration takes an even longer time and is reached only after many months of storage (*Moran, 1925*).

To maintain the difference in osmotic pressure, it is probable that the work of preventing the dilution of the yolk must be done largely at the vitelline membrane. Furthermore, Hale (*1933*) suggested that there is no sharp change of pressure at either of the surfaces of the vitelline membrane. There is, rather, a gradient from the center of the yolk outward, of varying but always slight slope; this gradient ends at a surface the position of which depends upon the rate of evaporation from the shell. This hypothesis has been confirmed by Baldes (*1934*), and by Moran and Hale (*1936*).

VAPOR PRESSURE

Baldes (*1934*) observed that the vapor pressure (expressed in terms of the sodium chloride solution of the same vapor pressure) in different parts of the egg is as follows:

	VAPOR PRESSURE (per cent NaCl)
Interior of the yolk	0.971
Yolk close to vitelline membrane	0.844
Albumen	0.756

Osmotic Behavior of Yolk in Artificial Solutions

Very complicated osmotic phenomena occur when the intact yolk is suspended in artificial solutions. Osborne and Kincaid (*1914*) demonstrated that the yolk, like a red blood corpuscle, swells when placed in distilled water, is not visibly changed when placed in 0.9 per cent saline, and shrinks to a corrugated globe in pure glycerol; however, unlike a blood cell, it does not shrink in 10 per cent saline but swells.

Although there is considerable variation in the capacity of different yolks to absorb distilled water (*Orru, 1933a*), the rate at which water is absorbed increases with increasing temperature (*Orru, 1933c*). When the yolk is immersed in distilled water for 2 hours at 20° C., an increase in the *p*H of the water, from *p*H 5.7 to *p*H 6.3, indicates the passage of ions through the vitelline membrane. The electrical conductivity of the water also increases, sometimes by as much as ten times (*Orru, 1933b*).

Osmotic equilibrium is attained less rapidly in hypotonic solutions, when the yolk is gaining water, than in hypertonic solutions, when the yolk is not losing water but is gaining dissolved substances, as well as a certain amount of water. The rate of equilibration seems to be of the same order whether a nonelectrolyte (glycerol, glucose) or an electrolyte (Ringer's solution) is employed (*Smith and Shepherd, 1931*). The rate is almost twice as rapid at 15° C. as at 0° C. Equilibrium is much more rapidly approached in artificial solutions than in the intact egg.

According to Straub and Hoogerduyn (*1929*), the injection of small amounts (2 or 3 mg. per egg) of morphine, cocaine, or potassium cyanide interferes with the function of the vitelline membrane and leads to an unusually rapid equilibration.

Electrical Potential between Yolk and Albumen

As early as 1860, Davy described the results of his experiments on the electrical potential of the egg. "Using a delicate galvanometer and a suitable apparatus, on plunging one wire into the white and the other into the yolk, the needle was deflected to the extent of 5°; and on changing the wires, the course of the needle was reversed. . . . But when the white and yolk were well mixed, then no distinct effect was noticeable."

With modern, more sensitive equipment, the difference in the electric potential of yolk surface and albumen was found to be 0.2 mv. (*Romanoff and Bless, 1942*). In a few cases, however, an extraordinarily high potential difference (about 1.5 mv.) was observed, although it was limited to a small region on the yolk, usually remote from the blastoderm.

A much greater potential difference (up to 2 or 3 mv.) between the yolk and the albumen has been observed by Capraro and Fornaroli (*1939a*), presumably because of fertilization and early development in the egg that was used.

The difference in potential between albumen and yolk has been explained as due to the resistance opposed by the egg substances to the passage of diffusible products (*Capraro and Fornaroli, 1939a*), that is, as due to the existence of phase boundaries between the yolk and the albumen (*Romanoff, 1944a*).

PROPERTIES OF DIALYZED YOLK AND ALBUMEN

Both yolk and albumen have been purified of electrolytes by dialysis.[3] In this process, the electrolytes pass slowly through a collodion membrane which is fairly permeable to water. The removal of electrolytes may be hastened, and the colloidal material can be obtained in a much higher degree of purity, by dialyzing with the aid of an electric current. This technique, known as electrolysis, removes electrolytes that cannot be removed by ordinary dialysis, or even by ultrafiltration (suction under pressure).

It has been shown that the dialyzates of egg yolk and albumen against water have identical freezing points of $-0.45°$ C. (*Grollman, 1931*). However, the ultrafiltrates of yolk and albumen, obtained at pressures of 60 to 70 cm. of mercury, have freezing points slightly higher than those of the fresh material, as shown below (*Smith, 1932*).

	FREEZING POINT (°C.)	
	YOLK	ALBUMEN
Fresh material	-0.560	-0.445
Ultrafiltrate	-0.548	-0.433

There is a marked decrease in the osmotic pressure of dialyzed egg albumen, as shown below by data for 1 per cent solutions (*Moore and Parker, 1902*).

CONDITION OF ALBUMEN	OSMOTIC PRESSURE (millimeters of Hg)
Not dialyzed	13.9–16.0
Dialyzed	7.6– 9.1

---

[3] Dialysis is the process of fractional separation of colloids from crystalloids, by diffusion; it is not to be confused with *osmosis*, during which nothing but solvent passes through the semipermeable membrane.

Dialyzed egg albumen markedly raises the surface tension of water; however, globulin—contained in albumen—lowers it (*Iscovesco, 1910*). The addition of alkali (*Perov, 1935*) to dialyzed albumen is attended by a decrease in surface tension and an increase in refractive index.

Lepeschkin (*1922*) showed that the dialysis of albumen for several days makes it noncoagulable by heat, although, after 3 weeks of dialysis, it becomes coagulable again.

ISOLATED CHEMICAL COMPOUNDS

Many physicochemical methods have been used to determine the properties of the various purified organic substances of the egg. Physico-chemical techniques are especially useful in the study of proteins, both in the natural and the denatured state, as well as of lipids, carbohydrates, and pigments.

The space available here does not permit much interpretation of the data presented. An attempt to include a discussion of theoretical impli-cations would require an extensive consideration of many aspects of modern science. In addition, present-day conceptions in physical chem-istry, and especially in that branch of the science which deals with pro-teins, are subject to continual modification, as new evidence appears regarding the nature of matter. The following pages should be regarded as a relatively brief compilation, for reference purposes, of the most im-portant data gathered on the physicochemistry of the pure substances of the egg. For general knowledge, the reader should consult textbooks and monographs.

PROTEINS

The egg protein whose physicochemical properties have been most extensively investigated is ovalbumin, often termed egg albumin. Of the five chief proteins in egg white, ovalbumin constitutes the largest fraction, or about three-quarters of the total protein of the albumen.[4] The next most frequently studied protein is ovovitellin, which constitutes about four-fifths of the total amount of protein in the yolk. These two proteins, since they are the major ingredients of the egg, are the most important of egg proteins.

There are very few data concerning the physicochemical properties of the other proteins that are present in the egg in relatively small quan-tities. Therefore, the following review will deal largely with ovalbumin and ovovitellin.

---

[4] There is some evidence that ovalbumin does not always exist in the egg in a form identical with "purified" ovalbumin (*Svedberg, 1931*).

It is well to remember that the egg proteins not only possess properties specific to the protein molecule, but that they also behave in many respects as lyophilic colloid systems.

Crystalline Ovalbumin

Ovalbumin can be obtained in crystalline form. A review of the physicochemical properties of crystalline ovalbumin may be of special interest both to chemists and to biologists.

PREPARATION

Methods for preparing crystalline ovalbumin have long been known. One of the earliest is that of Hopkins and Pinkus (*1898*), who crystallized ovalbumin by adding a saturated ammonium sulfate solution to egg albumen and acidifying the filtrate from this mixture with acetic acid. This method was slightly modified by Sörensen and Höyrup (*1916*), who used sulfuric acid instead of acetic acid; it has subsequently been modified by LaRosa (*1927*), and others. A good method is given in detail by A. Cole (*1933*). In the opinion of Sörensen (*1925*), the crystallization of ovalbumin does not differ in nature from the crystallization process in general; it is simply the slow and difficult crystallization of a substance from a supersaturated solution.

It was noted by Dale and Hartley (*1916*) that ovalbumin from ducks' eggs is more difficult to crystallize than that from chickens' eggs, a larger addition of acid being required to initiate the first crystallization. When once obtained in clean crystals, duck ovalbumin may be recrystallized as readily as chicken ovalbumin.

After it has been recrystallized several times (up to four), ovalbumin becomes one of the few proteins that may be considered as a distinct chemical entity. Of the proteins, ovalbumin probably exhibits the greatest degree of constancy in composition, molecular weight, and solubility, under varying conditions.

The colorless crystals of ovalbumin are brittle, very refractive, rhombic needles (Fig. 221), the ends of which are formed by the intersection of two planes at an obtuse angle. The dimensions of the largest crystals are about 0.2 x 0.015 x 0.001 mm. Upon drying in air, ovalbumin crystals lose their shape (*Clark and Shenk, 1937a*). They are soluble in water over the entire titrative range of pH; their solubility is an advantage in quantitative studies.

The temperature at which eggs are kept influences the speed with which ovalbumin crystallizes. Bidault and Blaignan (*1927*) show that ovalbumin will crystallize within 3 weeks from fresh eggs kept at 18° C., but that as many as 11 months are required for the complete crystallization of ovalbumin from eggs kept at 0° C. Temperature similarly affects the equilibrium between the solution and the crystals. Sörensen and Höyrup (*1917*) demonstrated the fact that the velocity of crystallization of ovalbumin is nearly 100 per cent greater at 20° C. than at 0° C.; thus, an ovalbumin solution that is in equilibrium at 0° C. will produce additional crystals if the temperature is raised to 20° C. Conversely,

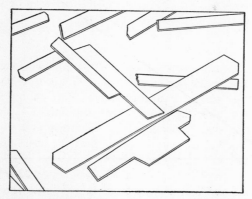

Fig. 221. The appearance of ovalbumin crystals. (After Sörensen and Höyrup, 1917.)

some of the crystals formed at 20° C. will dissolve at the temperature of 0° C.

It has also been pointed out (*Sörensen and Höyrup, 1917*) that the crystallization of ovalbumin will advance farthest at the optimal *p*H, which is about 4.58. An increase in *p*H causes a decided alteration in the shape of the crystals; at *p*H 5.3, they are not needles, but prisms.

Properties

DENSITY

The density of ovalbumin crystals, observed by means of flotation in media of different densities (at 20° C.), is probably between 1.202 (*Bull, 1941a*) and 1.268 (*Adair and Adair, 1936*). When calculated from the values for dried ovalbumin, the density of a 14.6 per cent solution of ovalbumin at 20° C. is 1.359 (*Chick and Martin, 1913*); however, the density of ovalbumin in solution varies with the concentration of the protein (Fig. 222). It has been found that the limit of saturation for ovalbumin at 17° C. is 15.36 per cent (*Rakuzin and Flieher, 1923*), and at 25.2° C., 28.15 per cent (*Chick and Lubrzynska, 1914; Rakuzin and Flieher, 1923*). It has therefore been concluded that natural egg albumen is a saturated solution.

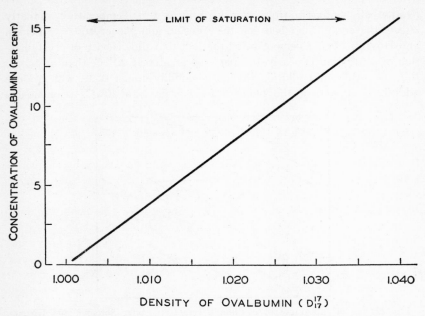

FIG. 222. Relationship between the concentration and the density of an aqueous solution of ovalbumin. (After Rakuzin and Flieher, 1923.)

SOLUBILITY

Ovalbumin is very soluble in water and also soluble in dilute salt solutions and dilute acids and alkalies. It forms molecularly dispersed solutions, systems which are in a state of physicochemical equilibrium; that is, which obey phase rule. Ovalbumin is precipitated by saturation of the solution with ammonium sulfate.

Ovovitellin is soluble in a 10 per cent solution of sodium chloride and is thrown out of solution upon the addition of water (*Moran, 1925*). When in contact with water, it slowly loses its phospholipid and becomes more and more insoluble in dilute salt solution (*Jukes and Kay, 1932a*). The table shows the known solvents for some of the egg proteins.

PROTEIN	SOLVENT
Ovalbumin	Water; dilute solutions of salts; dilute acids and alkalies
Ovoglobulin	Dilute salt solutions
Ovomucin	Dilute alkalies
Ovovitellin	Dilute acids and alkalies; 10 per cent sodium chloride solution
Ovolivetin	Water
Ovokeratin	Insoluble

Egg proteins, notably ovalbumin, are typical colloids with amphoteric properties; that is, in solution they behave as acids or bases, depending upon the hydrogen-ion concentration of the dispersion medium.

Dissociation Curve. The amphoteric nature of ovalbumin may be shown quantitatively by titration with acid and with alkali, in spite of the fact that ovalbumin, even after protracted dialysis, retains some of the anions and cations originally bound to the crystals. The reaction of the protein with a strong acid or base is described by the dissociation curve. This curve expresses the acid- and base-binding capacity of ovalbumin. The curve for ovalbumin represents reversible equilibria over the range of about pH 2 to pH 11 (*Cannan, Kibrick, and Palmer, 1941*).

The dissociation curve of ovalbumin has been most commonly obtained by titration with hydrochloric acid and sodium hydroxide. The titration curves obtained by various investigators (*Loeb, 1924; Hendrix and Wilson, 1928; Booth, 1930; Wu, Liu, and Chou, 1931; Prideaux and Woods, 1932; Loughlin, 1933; Kekwick and Cannan, 1936a; and others*), using different methods, are not without certain discrepancies, especially on the extreme acid or the extreme alkaline side.

Perhaps the most complete and typical dissociation curve is shown in Fig. 223. From this curve, it is evident that, at the isoelectric point of pH 4.8, the hydrogen-ion-combining capacity of ovalbumin is lowest. It reaches a maximum of 30 to 32 hydrogen-ion equivalents per gram molecule at a point slightly below pH 2. At the alkaline extreme of pH 11.5 to pH 12, there is no evidence of a stoichiometric end point indicative of a maximum dissociation capacity. The highest figure, which represents 26 equivalents per gram molecule, is therefore not the true end point. It has been suggested (*Pauli, 1928*) that the maximum binding capacity of 1 gm. of ovalbumin, in gram equivalents, is 110×10^{-5} positive groups and 134×10^{-5} negative groups.

There are no experimental data from which to construct the dissociation curves for the other egg proteins. These curves would presumably be much the same, because the curves of individual proteins usually differ only in the relative slope and span of specific regions. These differences are believed to reflect differences in the relative numbers of certain types of ionizing groups present in the proteins. Each protein is multivalent both as an acid and as a base, and, in the neighborhood of its isoelectric point, exists largely as dipolar ions.

Polar Groups. Ovalbumin has approximately twenty-eight cationic, and an equal number of free carboxylic, groups (*Cohn, 1939*). The nature and number of polar and nonpolar side chain groups are shown in Table 40. It has been calculated (*Cohn, 1936*) that, in ovalbumin, the maximum distance for a single dipole is 41 angstrom units. The presence of a large number of polar groups is responsible for the extreme solubility of ovalbumin in water.

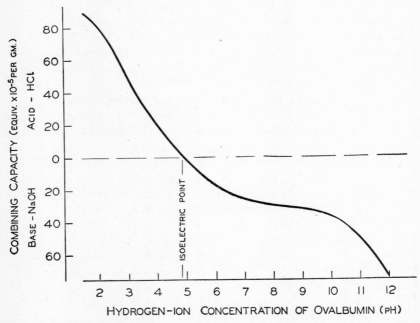

Fɪɢ. 223. Dissociation curve showing the acid-base-combining capacity of a dilute solution of ovalbumin. (After Kekwick and Cannan, 1936a.)

Isoelectric Point. During electrophoresis, proteins migrate to the anode in alkaline solutions, and to the cathode in acid solutions. At a certain pH, the protein ceases to migrate. This pH is known as the isoelectric point of the protein. It has been demonstrated (*Errera and Hirshberg, 1933; Myers and France, 1940*) that prolonged electrolysis of crystallized ovalbumin solutions always brings the pH to the isoelectric point of the protein.

The isoelectric point of egg proteins is upon the acid side of aqueous neutrality. The isoelectric point of ovalbumin has been determined by different methods and by a number of investigators; values varying from pH 4.55 (*Tiselius, 1930*) to pH 4.88 (*Sörensen, Linderström-Lang, and Lund, 1927*) have been obtained. There is also evidence that it is ap-

TABLE 40

Nature and Number of Polar and Nonpolar Side Chain Groups of Ovalbumin *

Chain Groups and Their Contributing Amino Acids	Number of Groups
Cationic groups:	
Guanidine (= arginine)	12
Ammonium (= lysine)	12
Imidazole (= histidine)	4
Anionic groups:	
Carboxyl (= aspartic + glutamic)	28
Phenolic-hydroxyl (= tyrosine)	8
Sulfhydryl or ½(S—S)(= 2 × cystine)	6
Nonionic polar groups:	
Sulfhydryl or ½(S—S)(= 2 × cystine)	6
Amide (= ammonia)	27
Hydroxyl (serine — threonine)	(?)
Methionyl (= methionine)	12
Indole (= tryptophane)	2
Nonpolar groups:	
Paraffin (= alanine + valine + leucine)	45
Benzene (= phenylalanine)	12
Pyrrolidine (= proline)	12
Glycine	(?)

* After Cohn (*1939*).

proximately at pH 4.8 (*Chou and Wu, 1934; Smith, 1936*). Kay and Marshall (*1928*) observed that the isoelectric point of ovolivetin is between pH 4.5 and pH 5.0. The isoelectric point of ovoglobulin is apparently in the neighborhood of pH 5.6 to pH 5.8 (*Longsworth, Cannan, and McInnes, 1940*), and that of ovomucoid at pH 4.5 (*Hesselvik, 1938*).

E. R. B. Smith (*1935*) presented evidence that the isoelectric point of ovalbumin becomes lower when the solution is concentrated. The decrease is about 0.03 pH unit for each 1 per cent increase in concentration.

It has also been demonstrated (*Smith, 1936*) that the apparent isoelectric point of ovalbumin is influenced in a definite manner by the nature and concentration of the ions in the buffer used. The relationship between the ionic strength of the medium and the isoelectric point of ovalbumin is always linear. The slope of the line varies with the different ions (Fig. 224), but it is usually positive for the multivalent and negative for the univalent cations.

The apparent isoelectric point of ovalbumin, however, is practically unaffected by the common crystallizing salts, sodium sulfate and ammonium sulfate, used in the preparation of crystalline ovalbumin.

At the isoelectric point, because of the absence of ionic movement, certain physical properties of proteins—such as electrical conductivity, viscosity, osmotic pressure, and solubility—are at their lowest values. Therefore, the measurement of these properties has been frequently used to determine the isoelectric point of proteins.

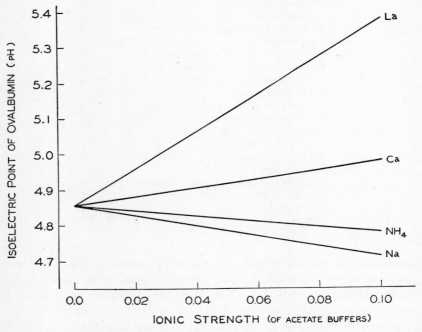

FIG. 224. The isoelectric point of ovalbumin as affected by the concentration of the cations of certain acetate buffers. (After Smith, 1936.)

SIZE OF PARTICLES, OR "MOLECULAR WEIGHT"

On the basis of sulfur and phosphorus content, the molecular weight of ovalbumin was estimated over a century ago as more than 50,000 (*Mulder, 1838*). Numerous observations made afterward (Table 41) lead to the conclusion that the weight of the molecule may be between 34,000 and 45,160.[5] Attempts have also been made to calculate the weight of the ovalbumin molecule from the total number of amino acid residues present (see Chapter 6, "Chemical Composition").

[5] Svedberg and Nichols (*1926*) have shown that crystalline ovalbumin contains a fraction, of molecular weight 170,000, that cannot be removed by repeated crystallization but is precipitated by electrolysis.

TABLE 41

MOLECULAR WEIGHT OF OVALBUMIN DETERMINED BY VARIOUS METHODS

Molecular Weight	Method	Investigator
36,800	Chemical analyses	Cohn, Henry, and Prentis (*1925*)
34,000	Chemical analyses	Calvery (*1932*)
37,500	Chemical analyses	Bergmann and Niemann (*1937*)
36,800	Chemical analyses	Bernhart (*1940*)
34,500	Ultracentrifuge	Svedberg and Nichols (*1926*)
40,500	Ultracentrifuge	Svedberg and Pederson (*1940*)
34,000	Osmotic pressure	Sörensen (*1917*)
43,000	Osmotic pressure	Marrack and Hewitt (*1929*)
46,000	Osmotic pressure	Taylor, Adair, and Adair (*1932*)
35,000	Osmotic pressure	Briggs (*1935*)
45,160	Osmotic pressure	Bull (*1941a*)

The ovalbumin molecule is of colloidal size and dialyzes slowly through a fairly porous collodion membrane. Sjögren and Svedberg (*1930*) showed that the ovalbumin molecule is spherical and has a radius of 21.7^{-8} cm. According to Lee and Wu (*1932*), the ovalbumin molecule occupies its maximum area at pH 4.8, the isoelectric point.

The molecular weights of ovovitellin and ovolivetin have been determined as 192,000 and 64,000, respectively.

Diffusion. Longsworth (*1941*), using electrophoresis, determined the diffusion coefficients of a 1.4 per cent ovalbumin solution, at 0° C., and various hydrogen-ion concentrations within the stability range. The accompanying tabulation gives the results he obtained after correcting

pH	DIFFUSION COEFFICIENT $(D_n \times 10^7)$
2.43	3.78
3.05	3.96
4.64	3.99
7.83	4.11
11.81	4.13

for *n*, the viscosity of the buffer solution used as solvent. As Groh (*1926*) also noted, the diffusion coefficient tends to become higher as the pH increases. This effect is probably due to a slight consolidation of the protein molecule, since about fifty-eight protons are lost in the transition from pH 3 to pH 12 (*Longsworth, 1941*). It has been reported that the diffusion coefficient exhibits a pronounced minimum in the neighborhood of the isoelectric point (*McBain, Dawson, and Barker, 1934*), where there is least mutual acceleration or retardation of ions and charged particles.

The diffusion coefficient of ovalbumin at temperatures higher than 0° C., although larger—at 20° C., it rises to 7.67 (*Tiselius and Gross, 1934*), and at 30° C., to 8.70 (*Polson, 1939a*)—is comparable to those given above, when corrected to 0° C., and to the same *p*H values.

It has been shown (*Holzapfel, 1938*) that the diffusion coefficient increases with the concentration of the protein solution. On the other

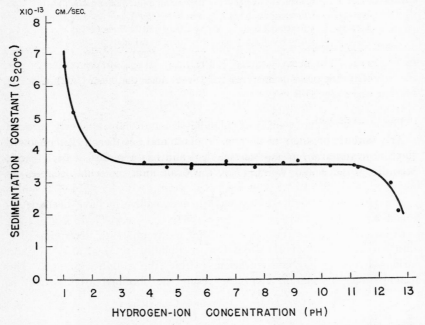

FIG. 225. Relationship between the sedimentation constant of ovalbumin and the *p*H of the solution. (After Sjögren and Svedberg, 1930.)

hand, it has also been found to be nearly independent of concentration (*Polson, 1939b*).

Ovalbumin, although it may dialyze through a membrane in appreciable amount (*McBain, Dawson, and Barker, 1934*), is retained by a sufficiently dense membrane.

The diffusion coefficient of ovomucoid has been found to be 3.4 at 7.5° C. (*Herzog, 1907*).

Sedimentation Equilibrium. According to Nichols (*1930*), the sedimentation constant of ovalbumin, in the presence of electrolytes, is probably stable from *p*H 3 to *p*H 7. The stability region of electrodialyzed ovalbumin is from *p*H 4 to *p*H 9 (*Sjögren and Svedberg, 1930*).

The mean value for the sedimentation constant within the stability range was found to be 3.55×10^{-13} cm. per second at 20° C. (*Svedberg, 1930; McBain and Leyda, 1938; Gralén and Svedberg, 1939*). At a higher temperature, i.e., 30° C., the value would correspond to about 4.06×10^{-13} cm. per second.

Below *p*H 3, the sedimentation constant increases, "indicating the formation of aggregates of denatured proteins"; at *p*H values higher than 9, the sedimentation constant decreases, "indicating the breaking-up of the whole material" (*Sjögren and Svedberg, 1930*). The changes in the sedimentation constant of ovalbumin between *p*H 1 and *p*H 13 are shown in Fig. 225. The sedimentation equilibrium of ovalbumin indicates that the stability region is somewhat narrower than the range within which normal molecules still exist.

OSMOTIC PRESSURE

The osmotic pressure of an ovalbumin solution is directly proportional to the concentration of the solution, as shown in Fig. 226. The osmotic pressure, as determined by direct measurement against a water manometer,

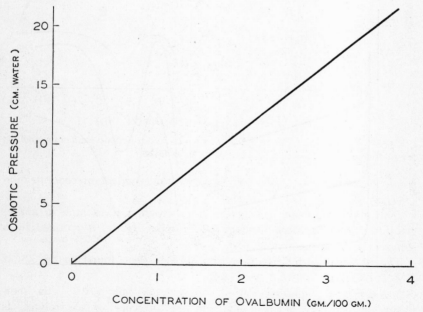

FIG. 226. Relationship between the concentration of an ovalbumin solution (in a 0.05 *M* sodium acetate buffer, at *p*H 4.7) and the osmotic pressure of the solution, as determined by the direct method against a water manometer. (After Bull, 1941a.)

at the concentration of 22.66 gm. of ovalbumin hydrate per 100 gm. of water, attains a maximum of 86 cm. (*Sörensen, 1917*). This value may be considered the upper limit for lyophilic colloid systems.

Presumably little, if any, effect is produced on the osmotic pressure of ovalbumin by the addition of acids and alkalies (*Lillie, 1907; Sörensen, 1917*). Sodium and potassium salts considerably lower the osmotic pressure (*Lillie, 1907*).

SURFACE PROPERTIES

Surface Tension. Molecular cohesion is measured by surface tension (γ). This attractive force operates over a short distance, not exceeding one or two molecular diameters. The surface tension of aqueous solutions of ovalbumin has been extensively studied, and various relationships determined.

The values found for the surface tension of ovalbumin solutions have ranged from 49.9 (*Peter and Bell, 1930*) to 70.3 dynes per centimeter (*Clark and Mann, 1922*). Until recently, it was not well understood that the surface tension is subject to the influence of such variables as the age

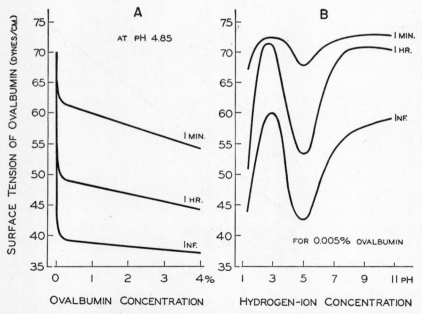

Fig. 227. The surface tension of chicken ovalbumin solutions of various ages, as affected by *A*, the concentration, and *B*, the *p*H, of the solution. (After Hauser and Swearingen, 1941.)

of the surface and the concentration and pH of the protein solution, as has been admirably demonstrated by Hauser and Swearingen (*1941*). The effects of these factors are shown in Fig. 227, which includes aging curves for infinite time, in order to indicate the limiting surface tension values approached with increasing age.

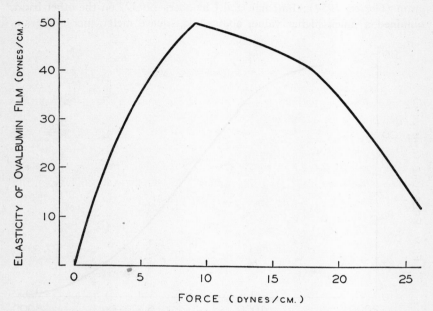

FIG. 228.   Force-elasticity curve for ovalbumin film on 0.01 M phosphate buffer, at pH 7.0.   (After Bateman and Chambers, 1939.)

Surface Elasticity.   The ability of a substance to recover from distortion caused by pressure or stress and to return to its original shape is known as elasticity.   An extensive study of the surface elasticity of ovalbumin solutions was made by Bateman and Chambers (*1939*), who found that there is a definite relationship between elasticity and applied force (Fig. 228).   At first elasticity increases linearly and, at a certain force, reaches a maximum value; it then begins to decline, causing irreversible change in the film.

Spreading in a Monomolecular Layer.   Ovalbumin, like many other substances, can be spread on the surface of water in a layer that is only one molecule thick.   To explain spreading, it has been suggested that one part of each molecule is soluble in water and another part insoluble (*Gorter, 1934*).

Bull (*1938b*) found that the area of spread of a 0.495 per cent ovalbumin solution is 1.04 square meters per milligram. This value also compares favorably with that of 1.06 found by Philippi (*1936*). Several other investigators obtained values much lower than the above, ranging from 0.66 (*Fosbinder and Lessig, 1933*) to 0.9 square meter per milligram (*Gorter, 1934*); Bateman and Chambers (*1939*) on the other hand, obtained a much higher value, about 1.45 square meters per milligram.

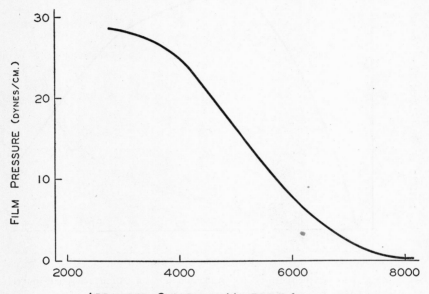

FIG. 229. Relationship between film pressure and the area of the ovalbumin molecule in a monomolecular layer. (After Bull, 1938b.)

Presumably, there is no difficulty in obtaining homogeneous films of ovalbumin on substrata varying in *p*H from 2.4 to 11.0 (*Fosbinder and Lessig, 1933*). However, there is general agreement that the maximum limiting area is attained in the neighborhood of the isoelectric point (*Gorter and Philippi, 1934; ter Horst, 1936; Seastone, 1938; Bateman and Chambers, 1939*).

The area of the film increases with temperature. The velocity of spreading is highest at 25° C. (*ter Horst, 1936*).

The mean limiting thickness of the film varies from 7.5 to 30.0 angstrom units (*Gorter and Grendel, 1926*). Presumably, films more than 10 angstrom units thick are incompletely spread.

Ovalbumin films give nearly linear surface pressure-area curves, above the pressure (3 dynes) at which they become solid (*Neurath, 1936*). The coefficient of compressibility of ovalbumin, calculated from the straight portion of this curve (Fig. 229), is 0.0157 dyne per centimeter.

Adsorption of Water. It has been shown (*Bull, 1944*) that, when the amount of water vapor adsorbed by ovalbumin is plotted against the

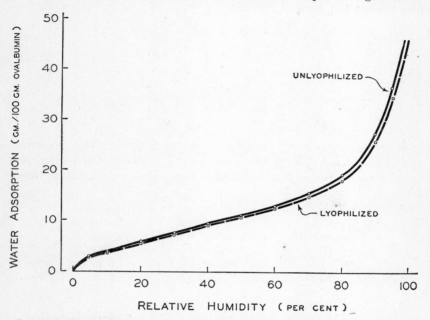

FIG. 230. Water adsorption of ovalbumin (both lyophilized and unlyophilized) at various relative humidities. (After Bull, 1944.)

relative humidity, a typical S-shaped curve is obtained (Fig. 230). There is a little greater adsorption with unlyophilized ovalbumin than with lyophilized (dehydrated under a vacuum while frozen).

In the low-pressure region, the average value for the free energy change per gram of water adsorbed at 25° C. was calculated as 50.6 calories (*Bull, 1944*).

Hydration. In studying the hydration of ovalbumin in systems of two components (protein and water), it is convenient to consider, first, the hydration of the protein in the form of crystals, and, second, the hydration of the molecules in solution.

The hydration of ovalbumin crystals in ammonium sulfate was studied by Sörensen and Höyrup (*1917*). The hydration value was determined as 0.22 gm. per gram of dry protein. This value is in close agreement with the value of 0.21 gm. given later by Adair and Robinson (*1931*).

The hydration of a solution of crystalline ovalbumin was determined by Moran (*1935b*). Solutions were placed in collodion sacs and surrounded by ice. As the temperature fell, water passed out from the sac. The following values were obtained, expressed in grams of water (free plus bound) per gram of protein:

TEMPERATURE (°C.)	AMOUNT OF WATER (gram)
−1	0.60
−2	0.54
−3	0.49
−5	0.44
−10	0.40
−20	0.37

In another experiment, the solutions were subjected to high pressures; and it was found that, at the same activity of water, the results were more or less similar to those obtained in the freezing experiment, as shown below.

PRESSURE (lb./sq. in. at 0° C.)	ACTIVITY OF WATER	HYDRATION (gm. water/gm. dry ovalbumin)
500	0.796	0.53
1,000	0.952	0.46
2,000	0.904	0.38
13,200	0.513	0.25
32,000	0.195	0.23
38,200	0.146	0.22

It is evident that the limiting hydration for both ovalbumin crystals and ovalbumin in pure water solution is about 0.22 gm., which is comparable to the firmly bound water observed in many other proteins.

The hydration of ovalbumin in a saturated solution of sodium and potassium chloride was found, however, to be only about 0.13 gm. (*Adair and Moran, 1935*).

Freezing. Studies of the effect of freezing a solution of ovalbumin (*Bull, 1932*) indicate that freezing produces an increase in the total volume of the suspended particles. It is assumed that this change is due to an increase in the amount of bound water consequent to the disaggregation of the particles. Although the state of aggregation is decreased in solutions of low concentration, it is evidently increased in solutions of high concentration (*Lange and Nord, 1935*).

Gold Number. Egg albumen proteins, like the whole albumen, can be characterized by their relative efficiency as protective colloids—their ability, that is, to protect suspensoids from precipitation by electrolytes. The accompanying table gives the gold numbers for several albumen proteins, as determined by Schulz and Zsigmondy (*1903*).

PROTEIN	GOLD NUMBER
Ovalbumin	2.0 –8.0
Ovomucoid	0.04–0.08
Ovoconalbumin	0.03–0.06
Ovoglobulin	0.02–0.05

By the use of the gold number technique, the presence of very small quantities of other egg proteins can be detected in ovalbumin.

VISCOSITY

The viscosity (which is defined as the internal friction of a liquid) of an ovalbumin solution is influenced by many variables and is not characterized by the linear changes observed in truly viscous systems. Among

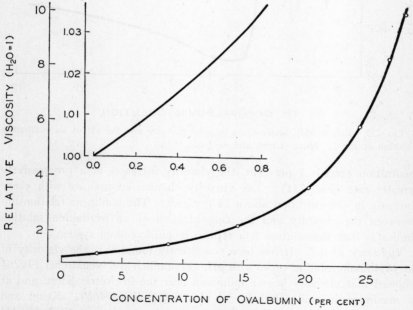

FIG. 231. Influence of the concentration of ovalbumin on the viscosity of the solution at room temperature. (After Chick and Lubrzynska, 1914, for higher concentrations, and Bull, 1940a, for lower protein concentrations at pH 8.)

the most important of the many factors that determine the viscosity of
ovalbumin solutions are concentration, *p*H, and temperature.

Influence of Protein Concentration. Of all the factors affecting viscosity, the concentration of the protein solution perhaps has the greatest
influence. At any concentration below 1 per cent, an increase in concentration leads to an almost linear increase in viscosity; but, as the con-

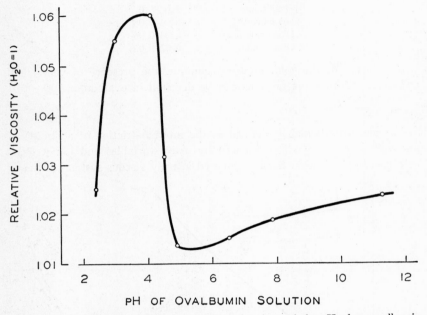

Fig. 232. Relationship between the relative viscosity and the *p*H of an ovalbumin
solution at 20° C. (After Kruyt and de Jong, 1934.)

centration exceeds 1 per cent, the viscosity increases at a progressively
greater rate (Fig. 231). The viscosity changes enormously with every
increase in concentration above 25 per cent. The nonlinear relationship
between the viscosity and the concentration of an ovalbumin solution
indicates that the solution is a typical lyophilic colloid system.

Influence of pH. It has been consistently found that the viscosity of
an ovalbumin solution varies with *p*H. Freundlich and Neukircher (*1926*)
showed that viscosity is at a minimum near the isoelectric point, and at
a maximum in acid solutions; so also did Bull (*1940b*). Kruyt and
de Jong (*1934*) presented detailed data on changes in viscosity with *p*H
(Fig. 232).

Bull (*1940b*) also demonstrated that the electrostatic charge on ovalbumin is important in increasing the viscosity both on the acid and on the basic side of the isoelectric point. His measurements on so-called electroviscous effects (the specific viscosity, divided by the volume of the protein in 1 cc. of solution, as a function of pH) were of significant magnitude, although much lower than the values observed with some other proteins.

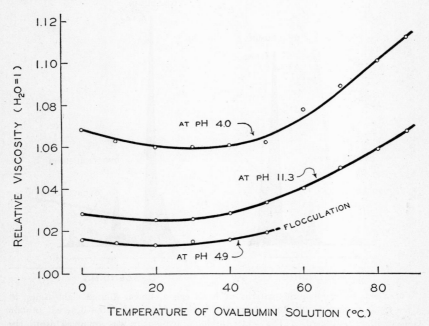

FIG. 233. Relationship between the relative viscosity and the temperature of an ovalbumin solution (2 per cent). Flocculation occurs in ovalbumin solutions between pH 4.5 and pH 7.8, at temperatures above 50° C. (After Kruyt and de Jong, 1934.)

Influence of Temperature. Kruyt and de Jong (*1934*) demonstrated that there is a very slight decrease in the viscosity of ovalbumin as the temperature rises from 0° C. to 20° C. According to Akagi (*1930*), the viscosity of a 0.2 per cent solution of ovalbumin changes very little between 20° C. and 60° C. At temperatures above 60° C., an increase in viscosity is observable at every pH, except near the isoelectric point (Fig. 233).

ELECTRICAL PROPERTIES

Electrophoresis. Electrophoresis (the measurement of the migration of negative colloidal particles to the positive pole, and positive colloidal particles to the negative pole, in an electric field) has been extensively used in the study of egg proteins. This technique affords both a method for the accurate determination of the isoelectric point and means for the separation and purification of different proteins, since proteins are of unequal ionic mobility. Egg albumen, because of its availability, has

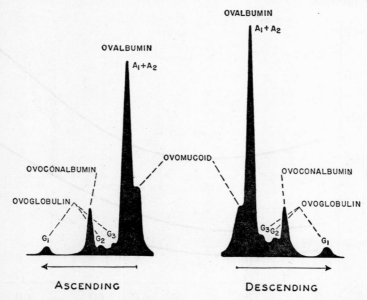

FIG. 234. Electrophoretic patterns of hen's egg albumen, diluted eight times, at *p*H 4.45, in buffers of 0.1 ionic strength. Various fractions of the different protein molecules are indicated. (The water-insoluble ovomucin was removed from the solution.) (After Longsworth, Cannan, and MacInnes, 1940.)

been frequently used as a source of pure proteins for biological investigation.

Electrophoretic Patterns. A typical electrophoretic pattern for egg albumen, obtained by the moving boundary method (*Tiselius, 1937*), is shown in Fig. 234. In this pattern, the ordinates represent gradients in refractive index, and the abscissae indicate the position of each component in respect to electrophoretic mobility. Thus, the area under each peak is proportional to the concentration of the corresponding component. As may be observed, egg albumen shows peaks corresponding to all its constituents, with the exception of one of the two fractions of ovalbumin. Ovomucin, owing to its insolubility in the selected buffer (solutions of sodium salts), has not been studied electrophoretically.

The two fractions of ovalbumin, which have different electrophoretic mobilities, are apparent in the patterns obtained at intermediate, but not at high or low, pH values. In Fig. 235, two separate peaks are therefore distinct only in the patterns obtained at pH 5.33 and pH 9.54. It has been shown (*MacPherson, Moore, and Longsworth, 1944*) that there is some change in the electrophoretic pattern of an ovalbumin solution upon aging. The largest of the two fractions decreases and completely disappears in less than a year (Fig. 236).

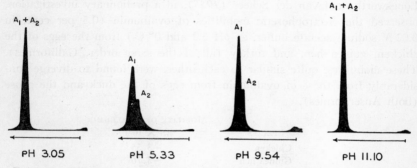

FIG. 235. Different electrophoretic mobility patterns of ovalbumin solutions obtained at various hydrogen-ion concentrations. (After Longsworth, Cannan, and MacInnes, 1940.)

Electrophoretic Mobilities. The accompanying data show the mobilities [6] of various albumen proteins from the chicken egg (at 0° C. and pH 5.33, in sodium acetate buffer of ionic strength of 0.1), as computed from measurements on the electrophoretic patterns (*Longsworth, Cannan, and MacInnes, 1940*). The differences in electrophoretic mobility are striking enough to distinguish each protein fraction contained in egg

PROTEIN	MOBILITY ($u \times 10^5$)
Ovalbumin:	
Fraction 1	−2.51
Fraction 2	−2.21
Ovomucoid	−1.83
Ovoconalbumin	1.19
Ovoglobulin:	
Fraction 1	7.09
Fraction 2	0.44
Fraction 3	−0.37

[6] The mobility, u, is defined as the distance of migration per second per volt, expressed in centimeters.

albumen, although the purity of some of the fractions is still doubtful. Seven fractions are listed here, but ten have been observed in electrophoretic patterns obtained when the three principal layers of the albumen are examined separately. These ten fractions are not distributed among the different layers in the same proportions (*Frampton and Romanoff, 1947*).

It has been suggested that species differences in proteins might be studied by the electrophoretic method (*Tiselius, 1937*). Landsteiner, Longsworth, and van der Scheer (*1938*), in a preliminary investigation, observed the electrophoretic mobilities of ovalbumin (0.5 per cent, in 0.02 N sodium acetate buffer, at pH 5.2 and 0° C.) from the eggs of the chicken, guinea hen, and turkey (all of the same order, Galliformes). These mobilities, quite similar to each other, were found to diverge considerably from those of ovalbumin from eggs of the duck and the goose (both Anseriformes).

	MOBILITY OF OVALBUMIN
SPECIES	$(u \times 10^5)$
Chicken	-3.55
Guinea hen	-3.76
Turkey	-3.53
Duck	-4.27
Goose	-4.60

Electrophoretic mobility is a function of pH and of ionic strength and depends upon the isoelectric point of the protein. Since the isoelectric points of all egg albumen proteins are different, the mobilities vary at any selected pH, and there is asymmetry in the ascending and descending boundaries of the electrophoretic patterns. The pH of any protein, when removed from its isoelectric point, determines not only the direction, but also the rate, of electrophoretic mobility. This fact is indicated in Fig. 237, which shows the relationship between pH and the mobilities of four egg proteins. The ionic strength of the buffer is also of the first order of importance. It can shift the isoelectric point (cf. Fig. 224) and thus produce a change in ionic mobility, as has been observed on proteins other than those of the egg (*Davis and Cohn, 1939*).

Streaming Potential. Potential differences produced when an ovalbumin solution was forced through a diaphragm were studied by Briggs (*1928*). He observed that the streaming potential of an ovalbumin solution is influenced by the hydrogen- and hydroxyl-ion concentration, through the range of pH 3.5 to pH 7.0 (Fig. 238).

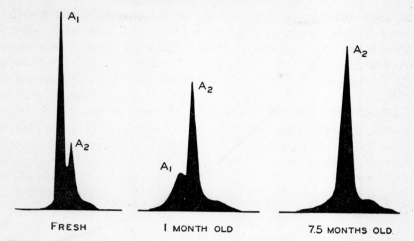

FIG. 236. Changes in the relative amounts of the two electrophoretic components, A_1 and A_2, of hen's ovalbumin during storage of an isoelectric salt-free solution. Similar changes occur in powders, but at a much slower rate. (After MacPherson, Moore, and Longsworth, 1944.)

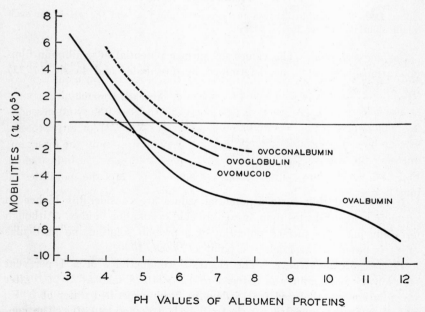

FIG. 237. The electrophoretic mobilities of various albumen proteins (at 0° C., in buffer solutions of a constant ionic strength of 0.1) as influenced by the pH of the solution. (After Longsworth, Cannan, and MacInnes, 1940; Longsworth, 1941.)

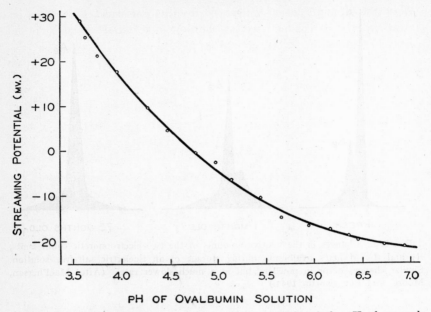

Fig. 238. Relationship between the streaming potential and the *p*H of an ovalbumin solution. (After Briggs, 1928.)

Surface Potential. The values for surface potential of ovalbumin films
on various substrates, as determined by Fosbinder and Lessig (*1933*)
are as follows:

Substrate	*p*H	Average Surface Potential (millivolts)
Hydrochloric acid	2.0	331
Citrate-Na₂HPO₄	2.4	344
NaOH-citrate	6.4	254
Phosphate	8.0	247

It is evident that surface potential values are considerably higher on
strongly acid substrates than on weakly acid or alkaline buffers. Although
*p*H influences the surface potential of a protein solution, no particular
effect is noted at the isoelectric point (*Pchelin, 1940*).

Electrical Conductance. The electrical conductivity of a 1 per cent
solution of ovalbumin was reduced to 1.3×10^{-5} mho at 18° C., after
the solution was dialyzed for 15 days in flowing distilled water at 0° C.,
and then further purified by electrodialysis (*Nichols, 1930*). The conductivity of a 2 per cent solution reached its minimum value of 6.5×10^{-6}
mho after 7 days of electrolysis (*Neurath, 1936*).

When the protein solution is removed from its isoelectric point, there is a great increase in conductivity, as shown by the tabulated data for a 3 per cent solution at 25° C. (*Loughlin, 1932*).

*p*H of Ovalbumin	Specific Conductivity (mho $\times 10^{-5}$)
2.3	10.5
3.0	6.6
3.9	1.0
4.3	0.4
5.3	1.2

In 1895, Sjöqvist showed that the addition of increasing amounts of ovalbumin to a constant amount of 0.025 N hydrochloric acid reduced the molar conductivity of the acid solution, until a constant value was reached.

Dielectric Measurements. The dielectric constant (ϵ) of solid ovalbumin was given by Fürth (*1923*) as 5.6. The dielectric constant of a 2 per cent ovalbumin solution is about 83 at the isoelectric point and becomes increasingly high as the *p*H is removed in either direction (*Garreau and Marinesco, 1929*). There are minima at *p*H 4.4 and *p*H 5.25 (*Shutt, 1934*).

Other investigations have dealt largely with a factor more sensitive to changes, that is, the increment in dielectric constant, determined by both bridge and calorimetric methods. The value for ovalbumin has been found to be 0.09 (*Cohn, 1938*), and, in another instance, 0.17 (*Oncley, 1942*). This increment is small, compared with that of many other proteins of higher molecular weight than ovalbumin. Furthermore, owing to the relatively high conductivity of ovalbumin solutions, the dispersion data are much more difficult to resolve. Figure 239 shows the results with a fairly concentrated solution (about 9 per cent), over a characteristic range of frequencies. The critical, or optimal, frequency for such measurements is approximately at the midpoint of the dispersion curve, about 3 megacycles.

Perhaps of greater interest and usefulness in dielectric studies is the dipole moment (μ)—the product of the charge and the distance between the two charges in the solution, of the order of 10^{-18} electrostatic unit. It provides information about the shape or asymmetry of a molecule and is a quantitative measure of the extent to which a molecule is polar. Among proteins, ovalbumin, owing to its low molecular weight, has the lowest dipole moment, both per gram and per mole. Shutt (*1934*) and

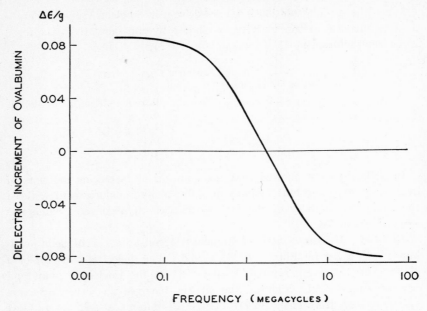

Fig. 239. Dielectric dispersion curve of ovalbumin solution (about 9 per cent). (After Oncley, Ferry, and Shack, 1940.)

Cohn (*1938*) gave the value as 180, and Oncley (*1942*) as 250 per mole. These values are about one-sixth that of a high-molecular-weight protein, edestin (*Oncley, 1942*).

OPTICAL PROPERTIES

Optical Rotation. Aqueous solutions of ovalbumin are levorotatory. The specific rotatory power (in sodium vapor light, D-line) of ovalbumin from the eggs of various species of birds is as follows:

Species	Specific Rotation ($[a]_D$)	Investigator
Chicken	$-37.5°$	Young (*1922a*), Holden and Freeman (*1930*), Almquist and Greenberg (*1934*)
Turkey	$-34.9°$	Worms (*1906*)
Duck	$-37.1°$	Panormoff (*1905b*)
Dove	$-36.3°$	Panormoff (*1905b*)

In light of different wave lengths, the specific rotation of chicken ovalbumin varies, as illustrated by the data of Hewitt (*1927*).

Wave Length (angstrom units)	Specific Rotation ($[a]_\lambda^{20}$)
4359	$-83.9°$
5461	$-44.5°$
5780	$-38.3°$
6660	$-27.5°$

There is a broad zone of minimum rotation which extends through the isoelectric region, from approximately pH 4.5 to pH 10.0 (*Almquist and*

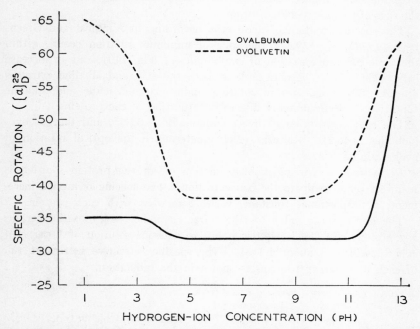

Fig. 240. Relationship between pH and the rotatory power of ovalbumin and ovolivetin. (After Almquist and Greenberg, 1934.)

Greenberg, 1934). However, beyond these limits, the rotatory power rapidly increases upon the addition of acids or alkalies (Fig. 240) and reaches the highest values, about $-60.0°$, on the alkaline side.

The concentration of protein in the solution has hardly any influence on the specific rotation of ovalbumin (*Jessen-Hansen, 1927*). In the iso-electric region, heating nearly to the coagulation temperature does not seem to alter the optical rotation (*Pauli and Kölbl, 1935*).

The specific rotation of ovomucoid has been determined as $-61.6°$ (*Mazza, 1930; Onoe, 1936*). As the tabulated data show, Mörner (*1912*)

OPTICAL ROTATION OF OVOMUCOID

SPECIES	($[a]_D$)
Chicken	$-70.9°$
Eider duck	$-72.7°$
Garrot	$-88.2°$
Grebe	$-67.1°$
Guinea fowl	$-74.0°$
Pelican	$-74.3°$

gave somewhat higher values for the optical rotation of ovomucoid from the eggs of various species of birds.

The specific rotation of ovolivetin, according to Almquist and Greenberg (*1934*), is $-39.1°$; the zone of minimum rotation occurs within the same pH range as that of ovalbumin (cf. Fig. 240), and an increase in specific rotation is noticeable at both the acid and alkaline extremes of pH. These changes in rotation under extreme acidic and alkaline conditions are undoubtedly due to denaturation of the proteins.

The specific rotation of ovoconalbumin is $-37.5°$, and that of ovoglobulin, $-46.2°$. Rakuzin (*1916*) determined the optical rotation of ovokeratin as $-27.8°$.

Refractivity. The index of refraction of an egg protein solution is in direct relationship to the concentration. The specific refractive increment, a proportional constant, varies somewhat with each protein.

The approximate value for the refractive index (n_D^{20}) of the various egg proteins (at the isoelectric point, in aqueous solution of 1 per cent concentration) is about 1.3335. The specific refractive increment (a) of each of four egg proteins is shown in the tabulation.

PROTEIN	SPECIFIC REFRACTIVE INCREMENT (a)	INVESTIGATOR
Ovalbumin	0.00177	Haas (*1918*), Hand (*1935a*)
Ovalbumin	0.00185	Barker (*1934*)
Ovalbumin	0.00192	Kay and Marshall (*1928*)
Ovomucoid	0.00160	Robertson (*1909–1910*)
Ovovitellin	0.00130	Robertson (*1909–1910*)
Ovolivetin	0.00190	Kay and Marshall (*1928*)

Although the specific refractive increment varies greatly from solvent to solvent, the extrapolated figure for the refractivity of a pure protein in solution is practically independent of the solvent (*Hand, 1935a*).

It is well known that a change in pH produces a change in refractive index, the value of which reaches its maximum at the isoelectric point (*Kondo and Iwamae, 1936*). Temperature also has a slight effect (*Herlitzka, 1910*). The refractive index of an aqueous solution of ovalbumin also varies with the viscosity (*Hand, 1935b*). With good control of the above variables, measurement of the refractive index offers an easy, quick, and accurate method for determining the concentration of the protein in a solution (*Bull, 1938a*).

Scattering of Light. The light-scattering power (i_o/I, the ratio of the intensity of scattered light, i_o, to the intensity of the beam, I) of several proteins, including ovalbumin, was measured by Putzeys and Brosteaux (*1935*). It was found that in light-scattering power, the various proteins stood in exactly the same relationship as in molecular weight. The average divergence for ovalbumin, for the light of the yellow doublet, was $+3.2$ per cent. The divergence was of the same order for the green wave length; but for total light it was more pronounced.

SPECTROSCOPIC PROPERTIES

Ultraviolet Absorption Spectrum. Egg proteins absorb ultraviolet light of wave lengths between 2500 and 3000 angstrom units.

The absorption curve for ovalbumin is of a characteristic configuration (*Svedberg and Tiselius, 1926; Sjögren and Svedberg, 1930; Hicks and Holden, 1934; Arnow, 1935; Crammer and Neuberger, 1943*). At *p*H 4.9 there is a point of maximum absorption at about 2800 angstrom units (Fig. 241).

The measurement of the absorption of ultraviolet light has been used to identify the amino acid content of proteins. Studies of the ultraviolet absorption spectra of amino acids indicate that the increased absorption of wave lengths above 2500 angstrom units can be attributed to the presence in ovalbumin of the aromatic amino acids, tyrosine, tryptophane, and phenylalanine.

Hicks and Holden (*1934*) pointed out that the hydrogen-ion concentration of ovalbumin causes little variation in the ultraviolet absorption curve. Although the position of the maximum and minimum is the same at *p*H 2.2, *p*H 5.5, and *p*H 11.2, the peak of the curve is nevertheless much lower at *p*H 5.5 than at the two other hydrogen-ion concentrations (*Sjögren and Svedberg, 1930*). The greater light absorption close to the extremes of acidity and alkalinity is probably due to the presence of decomposition products, as will be seen later.

The ultraviolet absorption spectrum of ovomucoid has a more narrow band than that of ovalbumin, between 2590 and 2900 angstrom units at *p*H 7 (*Mazza, 1930*). This band is shifted toward the near ultraviolet in an alkaline solution, but not in an acid medium. There are indications that this shifting is due principally to the presence in ovomucoid of a large quantity of phenylalanine, and the almost complete absence of tryptophane.

According to Marchlewski and Wierzuchowska (*1928*), the absorption band for ovoglobulin (in water) is between the wave lengths of 2407 and 2975 angstrom units, and for ovovitellin (in 0.04 N sodium hydroxide), between 2645 and 3441 angstrom units.

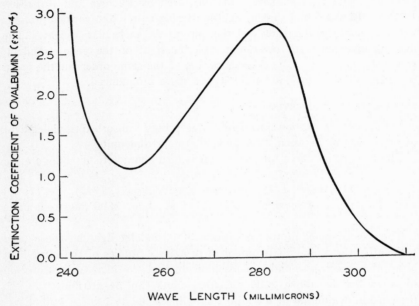

FIG. 241. Absorption, by ovalbumin, of ultraviolet light of various wave lengths, as shown by molecular extinction coefficient. (After Crammer and Neuberger, 1943.)

Infrared Spectra. Infrared spectra, to 6.2 μ, have been determined for several proteins, including ovalbumin (*Vlès and Heintz, 1935*). The spectra are similar, and each shows a real band corresponding to the iso-electric point.

X-Ray Diffraction. During recent years, attempts have been made to study the molecular structure of crystalline ovalbumin by means of X-ray diffraction patterns. The patterns show two principal spacings, 4.6 and 10.0 angstrom units. These two spacings appear to represent the two principal linkages between the polypeptide chains (*Clark and Shenk, 1937a*).

PROPERTIES OF DENATURED PROTEINS

One of the most striking characteristics of proteins is the instability of their physical state when they are subjected to certain treatments. The change from the native to the so-called denatured condition is of considerable interest. Investigation into the specific properties of a denatured protein may shed some light on the structure of the native protein molecule.

Denaturing Agents

Denaturation of egg proteins, specifically ovalbumin, may be brought about by numerous agents. Among these are heat (*Sörensen and Sörensen, 1925*); strong acids (*Cubin, 1929*) and alkalies (*Booth, 1930*); urea (*Hopkins, 1930; Clark, 1943*); guanidine hydrochloride (*Greenstein, 1938*); detergents (*Anson, 1939; Lundgren, Elam, and O'Connell, 1943*); salicylates (*Pauli and Weissbrod, 1935*); organic solvents, such as alcohol (*Wu, 1927*) and acetone; ultraviolet light (*Clark, 1935; Bernhart, 1939*); X-rays (*Rajewsky, 1930*); rays of radium salts (*Fernau and Pauli, 1915*); sound waves (*Chambers and Flosdorf, 1936*); high pressures (*Bridgman, 1914; Grant, Dow, and Franks, 1941*); and agitation (*Bull and Neurath, 1937*). In the opinion of Bull (*1941b*), it is much easier to find chemicals that will cause protein denaturation than it is to find chemicals that will not.

Character of Denaturation. According to one theory (*Wu, 1931; Mirsky and Pauling, 1936*), denaturation may be considered as a change in the protein molecule from a highly specific structure to a more random arrangement. The change is analogous to the change that occurs when a substance passes from a crystalline to an amorphous state. Recently, it has been suggested that denaturation consists of an unfolding of the protein molecule.

Stages of Denaturation. From extensive studies of denaturation by various agents, Wu (*1929*) proposed, and Clark (*1935*) and Bull (*1938c*) reaffirmed, the theory that the denaturation of proteins consists of three reactions: (1) intramolecular rearrangement, or the denaturation proper; (2) flocculation of the denatured molecules, preparatory to coagulation; and (3) the formation of an insoluble coagulum.

The denaturation of a perfectly clear ovalbumin solution is revealed by a considerable increase in viscosity. The flocculum, which is formed at the isoelectric point, is completely soluble in acids and bases; the flocculation reaction is temporarily reversible. The coagulation reaction renders the flocculum insoluble.

All three stages have been observed in heat, urea, and surface denaturation (*Bull, 1938c*), as well as upon denaturation with ultraviolet light (*Clark, 1935*).

Energy of Activation. The temperature coefficient of heat denaturation is very large. Lewis (*1926*) estimated the critical increment to be as large as 130,000 calories. The temperature coefficient for both urea and surface denaturation is somewhat less than that for heat denaturation (*Lenti, 1945*). Some attempts have been made to calculate, from the temperature coefficients, the so-called energies of activation of proteins in denaturation. Unfortunately, the energy of activation of ovalbumin has been estimated for the extreme acid region only. The value calculated by Bull (*1938c*), 35,000 calories, compares favorably with the value of 36,000 to 48,000 calories previously obtained by Cubin (*1929*).

Velocity of Reaction. The velocity of the reaction during the denaturation of ovalbumin is influenced chiefly by temperature and pH, and also by the concentration of the solution, and by the presence of salts and other substances.

Lewis (*1926*) showed that the velocity of heat denaturation passes through a minimum at pH 6.76, close to the point where the concentration of hydrogen ions is equal to that of hydroxyl ions. At 70.2° C., the unimolecular velocity constant, K (obtained by plotting the concentration of ovalbumin against time, in seconds), varies as shown in the tabulation.

pH OF OVALBUMIN SOLUTION	VELOCITY OF REACTION $(K \times 10^5)$
4.98	510.0
5.50	134.0
6.10	30.7
6.35	14.8
6.89	5.5
7.29	19.2
7.72	38.4

If precautions are taken to keep the pH of the solution constant, a steady, unimolecular process can be obtained (*Chick and Martin, 1912*).

The other variables will be discussed below, when the specific influence of various denaturants is considered.

Specific Influence of Various Denaturants

Heat Denaturation. Of the different kinds of denaturation, that produced by heat has been most thoroughly investigated. As early as 1888, Corin and Berard observed that the coagulation of egg albumen occurred fractionally, at five different temperatures ranging from 57.5° C. to 82° C.

The quantitative studies of Chick and Martin (*1910*) indicate that, the higher the temperature, the more rapid the heat coagulation of oval-bumin. Furthermore, at every temperature, the rate of coagulation continually diminishes as coagulation proceeds (Fig. 242).

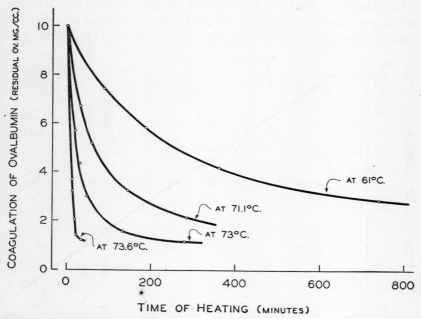

FIG. 242. Rate of coagulation of ovalbumin in solution (about 1 per cent) at various temperatures. (After Chick and Martin, 1910.)

Subsequent investigations revealed that the temperature at which denaturation occurs is highly variable and depends upon a number of factors.

It has long been known that proteins are much less readily denatured by heat when dry than when moist. Lewith (*1890*) used powdered salt- and globulin-free albumen, dried in a vacuum to various degrees of dehydration. After sealing small samples in thin-walled glass tubes, he determined the coagulating temperature of each sample. His results, reproduced below, led to the hypothesis (*Chick and Martin, 1910*) that denaturation is essentially a reaction between protein and water. This interpretation has also been used to explain the relatively greater steriliz-ing effect of moist heat, as compared to dry (*Rahn, 1932, p. 326*).

MOISTURE IN ALBUMEN (per cent)	COAGULATING TEMPERATURE (° C.)
0	170
6	145
18	85
25	77
50	56

In addition, it has been demonstrated (*Barker, 1933b*) that the temperature of denaturation of dry, crystallizable ovalbumin increases when there is a decrease in the relative humidity of the air (Fig. 243). This

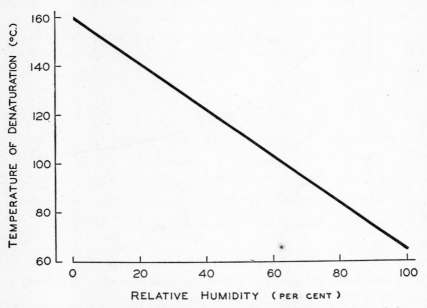

FIG. 243. Relationship between relative humidity and the temperature of denaturation of dry, crystallizable ovalbumin. (After Barker, 1933b.)

fact constitutes good evidence that low relative humidity affects the freedom of water molecules. At low humidity, the restriction in the movement of the water molecules might be more important in lowering the denaturation velocity than the actual reduction in the water concentration.

It has also been shown that the *p*H of an ovalbumin solution may have some effect on the temperature of denaturation (*Arnow, 1935*).

*p*H OF OVALBUMIN	COAGULATING TEMPERATURE (° C.)
1.65	32
4.81	74
6.63	74

The temperature of coagulation is raised about 2° C. at very low salt concentration $(10^{-3} N)$, but it is lowered 5° to 6° C. by increased concentration of such salts as magnesium or potassium sulfate (*Pauli and Weissbrod, 1935*). Salicylates favor coagulation at low concentrations but have an inhibiting effect at high concentrations. Urea retards coagulation (*Pauli and Weissbrod, 1935*), and sucrose and glycerol inhibit it (*Beilinsson, 1929*), as does dextrose, when present in large enough amounts (*Bancroft and Rutzler, 1931*).

Urea Denaturation. The extent to which ovalbumin is denatured by urea depends upon the concentration of the protein, the concentration of urea, and the pH and temperature of the solution.

Hopkins (*1930*) demonstrated that the denaturation of an ovalbumin solution by urea is made more extensive by increases in the concentration of the protein. At 22° C., with 30 per cent urea, the results were as shown.

Concentration of Ovalbumin (per cent)	Immediate Condition
2	No visible change
5	Precipitate
7	Heavy cloud

Clark (*1943*) found that ovalbumin solutions (0.9 per cent) at room temperature are not denatured by 20 per cent urea but are denatured slowly by 25 per cent urea and very rapidly by 35 per cent urea. When the concentration of urea is in excess of 50 per cent, there is possibly some dissociation of the ovalbumin molecule (*Williams and Watson, 1937*).

At the isoelectric point of ovalbumin, the denaturing effect of urea is greatest (*Clark, 1945*).

The effect of temperature on the denaturation of ovalbumin by urea has been demonstrated by measuring the opalescence of Tyndall beams, in apparent foot-candles (*Clark, 1943*). From the accompanying figures,

Temperature of Ovalbumin (° C.)	Tyndall Beam (foot-candles)
0	0.7
10	0.9
20	1.3
30	2.0
40	3.7

it is evident that the rate of denaturation (by 35 per cent urea) is accelerated as the temperature increases, especially above 20° C.

Denaturation by Shaking. Shaking a protein solution produces surface denaturation. It has been noted that the denaturation occurs at the interface between the ovalbumin solution and the air. Surface energy brings about denaturation similar to that produced by heat energy.

The rate of surface denaturation is influenced by the concentration of the ovalbumin (*Bull, 1938a; Wu and Wang, 1938*). The higher the protein concentration, the lower is the rate of denaturation (Fig. 244).

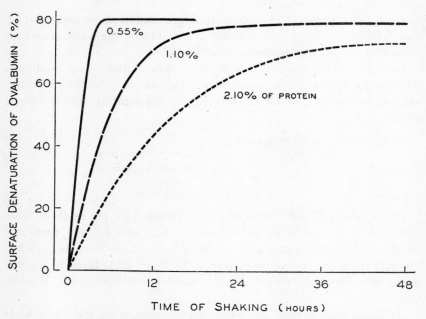

Fig. 244. Rate of surface denaturation of ovalbumin solutions of various concentrations, at *p*H 4.88. (After Bull and Neurath, 1937.)

The rate of surface denaturation is also largely dependent upon the *p*H of the protein solution (*Bull and Neurath, 1937; Wang and Wu, 1940*). Unlike heat denaturation, surface denaturation occurs at a maximum rate at or near the isoelectric point of ovalbumin (Fig. 245).

Beating an ovalbumin solution to a foam, or bubbling air through the solution, results in surface denaturation similar to that caused by shaking, and the entire protein can be collected in insoluble form.

Denaturation by Radiant Energy. There have been numerous studies of the effect of ultraviolet radiation on egg proteins, especially ovalbumin (*Young, 1922b; J. Clark, 1925; Stedman and Mendel, 1926; Spiegel-*

Adolf and Krumpel, 1929; Rajewski, 1929, 1930; Clark, 1935; Arnow, 1935; Bernhart, 1939; Bernhart and Arnow, 1939; and others).

Dreyer and Hanssen (*1907*) showed that, of the egg proteins, ovovitellin is most easily coagulated by ultraviolet light. Next in sensitivity is ovoglobulin; ovalbumin is least sensitive. It has been pointed out (*Rajewski, 1930*) that active absorption by ovalbumin takes place between

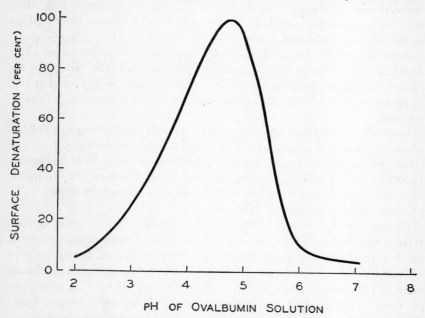

Fig. 245. Effect of the *p*H of ovalbumin solution (1.05 per cent) on the degree of surface denaturation obtained after agitation for 12 hours (96 movements per minute, with a 7-cm. stroke). (After Bull and Neurath, 1937.)

300 and 254 mμ, and that the reaction of the protein to ultraviolet radiation is independent of temperature. Bernhart (*1939*) observed, however, that secondary protein derivatives may be formed by ultraviolet radiation at a temperature of 25° C.

Studies of the effect of X-rays on ovalbumin are less numerous (*Rajewski, 1929, 1930*). The reaction of the protein to X-rays was found to be essentially identical to the reaction to ultraviolet radiation, with the exception that the effect of X-rays is greatly influenced by temperature (*Rajewski, 1930*).

There have been only a few studies on the effect of the rays emanating from radium salts (*Fernau and Pauli, 1915; Spiegel-Adolf and Krumpel,*

1929). It was found that the influence of radium radiation is much greater than that of heat activation (*Spiegel-Adolf and Krumpel, 1929*).

Denaturation by High Pressure. Studies by Bridgman (*1914*) showed that egg albumen can be coagulated by high hydrostatic pressure (122,000 kg. per square centimeter). Grant, Dow, and Franks (*1941*) recently investigated the effect of high pressures on an ovalbumin solution. At every pressure between 1000 and 7500 kg. per square centimeter, they observed a noticeable coagulum, the amount of which became more copious as the pressure increased.

Denaturation by Sound Waves. By the use of ultrasonic waves, Wu and Liu (*1931*) were able to coagulate ovalbumin. The reaction proceeded actively in the presence of air, carbon dioxide, or oxygen, but not in an atmosphere of nitrogen or hydrogen, or in a gas-free solution under vacuum.

Similarly, a coagulum was observed by Flosdorf and Chambers (*1933*) when ovalbumin, in aqueous solution and at the isoelectric point, was subjected to the action of intense sound waves of audible frequency (1000 to 15,000 cycles). Later, Chambers and Flosdorf (*1936*) found that the denaturation occurred only at acoustic intensities sufficient to promote vigorous cavitation of the solution.

Coagulation by Alcohol. The action of alcohol on ovalbumin is considered identical with that of heat (*Wu, 1927*). Denaturation occurs at or near the isoelectric point. However, the protein can merely be precipitated if the concentration of the alcohol is below a certain minimum value. This reaction, like the precipitation of neutral salts, is reversible. As the concentration of the alcohol is increased beyond a certain limit, the rate of denaturation begins to accelerate. Finally, denaturation becomes irreversible.

Changes Caused by Denaturation

The denaturation reaction produces numerous changes in native ovalbumin. Some of these changes are quite evident, whereas the nature of others has not been established. The change in the structure of the protein molecule is still under dispute. It is possible, however, to separate the characteristic changes caused by denaturation from other changes produced, for example, by hydrolysis, or by the combination of proteins with various salts.

PHYSICAL PROPERTIES

Increase in Viscosity. It has been repeatedly observed (*Loughlin, 1932; Liu, 1933; Hand, 1935b; Bull, 1940a*) that the viscosity of denatured ovalbumin is higher than that of native ovalbumin. However, viscosity is greatly modified by the concentration and the pH of the protein solution.

Under any constant conditions, an increase in concentration produces an almost linear increase in viscosity (Fig. 246). The effect of concentration, observable in native ovalbumin solutions, is intensified after

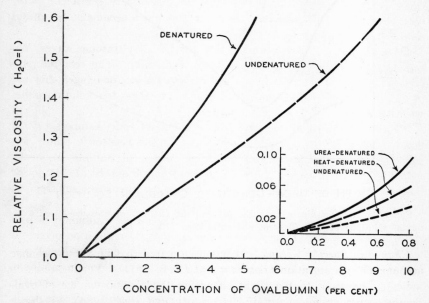

Fig. 246. Relationship between the concentration and the viscosity of denatured and undenatured ovalbumin solutions. (Curves for higher concentrations after Hand, 1935b; for lower concentrations, shown in insert, after Bull, 1940a.)

denaturation of the protein. The viscosity of a denatured ovalbumin solution is also dependent to a certain extent upon the method of denaturation; for example, urea-denatured ovalbumin has a higher viscosity than heat-denatured ovalbumin.

From Fig. 247, it is evident that the pH of the solution also affects the viscosity of denatured ovalbumin. The viscosity of urea-denatured ovalbumin is least at pH 5.7; it is much greater at either the acid or the alkaline side of this point and is greatest at the extreme of alkalinity.

It is reasonable to assume that the increase in viscosity that accompanies denaturation is associated with the loss of the ovalbumin molecule's symmetry, as will be seen later.

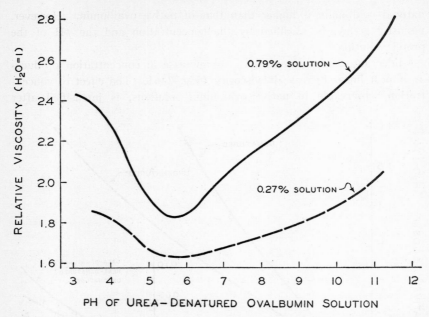

Fig. 247. Relationship between the *p*H and the viscosity of urea-denatured oval-bumin solutions of two different concentrations. (After Liu, 1933.)

Change in Specific Volume. It has been stated that no volume change accompanies denaturation (*Loughlin and Lewis, 1932*). Further research indicates, however, that a slight change in the specific volume of oval-bumin occurs upon denaturation (*Neurath and Bull, 1936*), as shown below.

OVALBUMIN	SPECIFIC VOLUME AT 25° C.
Undenatured	0.79020
Surface-denatured	0.76829
Heat-denatured	0.77280

Change in Density. When dry ovalbumin comes into contact with water, the contraction in the volume of the ovalbumin is very small. Consequently, there is no appreciable difference in the density of native and denatured protein in aqueous solution (*Nichols, 1930; Loughlin and Lewis, 1932*). Similarly, the densities do not differ significantly in hydrogen (*Bull, 1938c*).

When xylene is used as the displacing liquid, an increase in the density of ovalbumin is noticeable after denaturation (*Neurath and Bull, 1936*), as shown in the tabulation. The fact that, in xylene, denatured ovalbumin

OVALBUMIN   DENSITY IN XYLENE AT 25° C.

Undenatured	1.2655
Surface-denatured	1.3016
Heat-denatured	1.2946

has a higher density than undenatured ovalbumin is possibly due to the inability of xylene to penetrate into the free spaces in the molecule.

Loss of Ability to Recrystallize. Denatured ovalbumin has not been crystallized out of solution. A change from the crystalline to the amorphous state is evidently associated with the structural changes in the protein molecule that occur upon denaturation.

AMPHOTERIC PROPERTIES

Insolubility in Pure Water at the Isoelectric Point. One of the most characteristic changes that denaturation causes in ovalbumin is the loss of solubility in pure water, or in dilute salt solutions, at the isoelectric point (pH 4.8). This insolubility has been rightly used as a definition of denaturation, and the rate or extent of denaturation is measured by the formation of insoluble protein.

However, heat-denatured ovalbumin, which is dissolved by acid and alkali, can be dissolved at the isoelectric point by a number of substances. Among these are urea, guanidine hydrochloride, detergents, and salicylates.

Changes in Acid-Base-Combining Power. The extent to which the acid-base-combining power of ovalbumin is changed after denaturation is still a matter of dispute. Hendrix and Wilson (*1928*) and Barker (*1933a*) observed a marked change accompanying heat denaturation. Booth (*1930*), however, found the combining power to be unaltered, if the denatured protein is titrated immediately after treatment.

Changes in pH. The pH of ovalbumin changes during denaturation. There is a slight but definite shift, the direction and degree of which is determined by the initial pH of the ovalbumin solution.

When ovalbumin is denatured at the isoelectric point (pH 4.8), the shift in pH proceeds most rapidly. During surface denaturation, the pH value may increase by 0.6 (*Bull and Neurath, 1937*), and, during heat denaturation, by 1.0 (*Pauli and Kölbl, 1935*). When ovalbumin is denatured on the alkaline side, the pH value may decrease by 0.5; when denatured on the acid side, the change in pH may range from a decrease of 1.0 to an increase of as much as 3.0.

Similar changes are observed during urea denaturation. When urea, in concentrations of 10 to 50 per cent, is added to isoelectric ovalbumin solutions, there is a rise in pH value of 0.4 to 1.0, depending upon the concentration of urea (*Clark, 1943*).

SURFACE PROPERTIES

Decreased Affinity for Water. The attraction of a protein for water decreases during denaturation. Sörensen (*1917*) found that the water of hydration in undenatured ovalbumin is 0.22 gm., whereas in heat-denatured ovalbumin there is 0.187 gm. of water bound per gram of protein. Somewhat similar results were obtained by Neurath and Bull (*1936*).

OVALBUMIN	BOUND WATER (gram/gram of protein)
Undenatured	0.36
Surface-denatured	0.20
Heat-denatured	0.16

The data show that the affinity of ovalbumin for water may vary with the type of denaturation; heat-denatured ovalbumin obviously has less affinity for water than the surface-denatured protein.

The relationship between the water content of ovalbumin and the volume contraction per gram of protein is shown in Fig. 248. When the hydration of the molecule is complete, there is no further contraction.

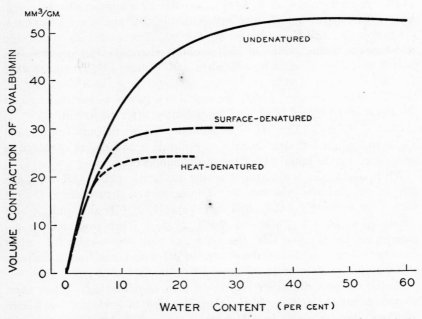

FIG. 248. Relationship between volume contraction and water content of denatured and undenatured ovalbumin. (After Neurath and Bull, 1936.)

Change in Water Adsorption. Experiments indicate that heat-denatured ovalbumin takes up approximately 80 per cent as much water, at every relative humidity, as does undenatured ovalbumin (Fig. 249). These findings disclose a very interesting, and apparently anomalous, situation. The denaturation process itself probably involves a reaction of the pro-

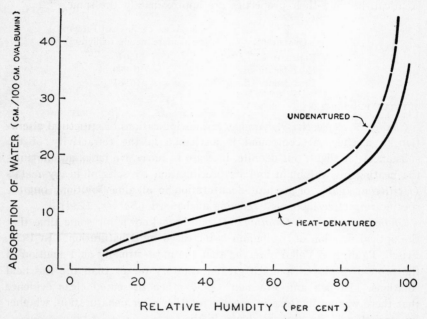

FIG. 249. Adsorption of water by native and heat-denatured ovalbumin, at various relative humidities. (After Barker, 1933b.)

tein with water. At the same time, it is evident that the denatured protein is less heavily hydrated than the material from which it is derived (*Sörensen and Sörensen, 1925; Adair and Robinson, 1931; Barker, 1933b*).

Decrease in Surface Tension. The surface tension of ovalbumin decreases slightly upon exposure of the protein to ultraviolet light, as shown below (*Young, 1922b*).

Exposure to Ultra-violet Light (hours)	Surface Tension (γ)
0	69.74
1.5	69.54
3.0	68.70
4.5	68.52

Change in Spread Area of Monomolecular Film. The maximum area of a monomolecular layer of denatured ovalbumin is only slightly less than that of a similar film of undenatured ovalbumin. The values given by Bull (*1938b*) for 0.495 per cent solutions, at a *p*H of about 3, are shown here. He concludes that the three maximum spread films are probably identical, because their properties are approximately the same.

OVALBUMIN	AREA OF SPREAD FILMS (square meters/milligram)
Undenatured	1.04
Heat-denatured	0.990
Urea-denatured	0.987

OPTICAL PROPERTIES

Change in Refractivity. Among various indications of structural alteration in a chemical compound is a change in the refractivity of the substance. A slight, but definite, increase in refractive index accompanies the heating of ovalbumin and its precipitation by salts of heavy metals (*Herlitzka, 1907*), or its heat denaturation in alkaline solution; but the specific refractive increment remains unchanged (*Barker, 1934*).

Changes in Optical Rotation. It has been known for some time that the optical rotation of ovalbumin is increased by denaturation. In 1925, Frisch, Pauli, and Valkó observed that the most strongly acid solution of the protein displayed a much greater rotatory power than the less acid solutions. Holden and Freeman (*1930*) then presented clear evidence that there was an increase in specific rotation after denaturation, whether by alcohol, acid, or heat, as shown in the table.

DENATURING AGENT	SPECIFIC ROTATION ($[a]_{5461}$)
None	$-37.3°$
Alcohol	$-64.8°$
Hydrochloric acid (0.5 N)	$-94.0°$
Heat, to 100° C.	$-99.0°$

Later, these observations were supplemented by evidence (*Clark, 1943*) indicating that the optical rotation of ovalbumin (0.6 per cent) increases with the degree of denaturation, as, for example, when ultraviolet radiation is the denaturing agent. Similarly, during urea denaturation, there

ULTRAVIOLET RADIATION (minutes)	DENATURATION (per cent)	OPTICAL ROTATION ($[a]_D^{22}$)
20	35	$-32°$
40	50	$-42°$

is an increase in the optical rotation of ovalbumin, of greater magnitude as the concentration of urea is increased.

CONCENTRATION OF UREA (per cent)	OPTICAL ROTATION OF OVALBUMIN ($[a]_D^{22}$)
25	$-27.5°$
45	$-28.0°$
60	$-36.0°$

Observations made by Holden and Freeman (*1930*) also showed that there is a slow change in optical rotation in acid solution during a period of 12 days. There is considerable support for the view that denatured proteins cannot be successfully characterized by their specific optical rotation. Barker (*1933a*) first demonstrated the fact that the rotatory power of heat-denatured ovalbumin varies in an orderly manner with the time and temperature of heating, although it usually approaches a certain limiting value. As Fig. 250 indicates, optical rotation increases steadily at 70° C. but very rapidly attains a maximum value at 80° C. The limiting value is primarily a function of the *p*H and the concentration

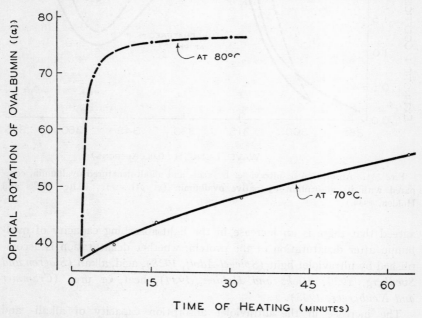

FIG. 250. Changes in the optical rotation of heat-denatured ovalbumin during 1 hour's heating at two different temperatures. The observations were made with a 1 per cent solution at *p*H 7.4, using a mercury arc as a source of light (5461 angstrom units). (After Barker, 1933a.)

of the protein solution during the actual heating. Lowering the *p*H (at least within the range of *p*H 6.6 to *p*H 9) and increasing the concentration of the ovalbumin solution both increase the specific rotation.

SPECTRAL PROPERTIES

Changes in Absorption Spectrum. A few observations have been made on the absorption of light by denatured ovalbumin. It is generally

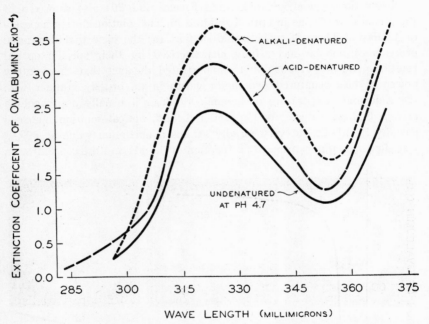

Fig. 251. Absorption of ultraviolet by acid- and alkali-denatured ovalbumin, compared with its absorption by native ovalbumin (at *p*H 4.7). (After Hicks and Holden, 1934.)

agreed that there is an increase in the light-absorbing capacity of ovalbumin after denaturation of the protein, whether denaturation is accomplished by ultraviolet light (*Spiegel-Adolf, 1928*), acid, alkali (*Sjögren and Svedberg, 1930; Hicks and Holden, 1934*), heat, or urea (*Crammer and Neuberger, 1943*).

The increase in the ultraviolet absorption capacity of alkali- and acid-denatured ovalbumin is shown in Fig. 251. The positions of the maximum and minimum are nearly the same on all three curves. The

Denaturant	Extinction Coefficient $(E \times 10^{-4})$
None	1.49
Heat	2.60
Urea	2.78
Alkali	2.75
Acid	2.70

tabulated data indicate the change in the magnitude of the extinction coefficient observed at 295 mμ, with various denaturants, and at pH 12 (*Crammer and Neuberger, 1943*). These changes in the absorption spectrum after denaturation are interpreted as indicating that the phenolic groups of the denatured ovalbumin molecule are able to ionize at a pH lower than 12.

Changes in X-Ray Diffraction. The X-ray diffraction pattern of denatured ovalbumin is similar to that of undenatured ovalbumin (*Miller, Chesley, Anderson, and Theis, 1932; Clark and Shenk, 1937a*). The pattern consists of two rings, of roughly constant dimensions, but varying in sharpness (Fig. 252).

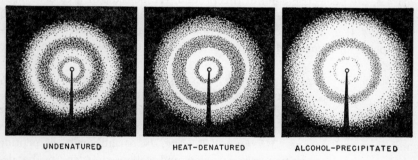

UNDENATURED   HEAT-DENATURED   ALCOHOL-PRECIPITATED

X-RAY DIFFRACTION OF OVALBUMIN

Fig. 252. X-ray diffraction pattern of native ovalbumin, compared with that of heat-denatured and alcohol-precipitated ovalbumin. (After Clark and Shenk, 1937a.)

ELECTRICAL PROPERTIES

Alteration in Electrophoretic Mobility. Several times, it has been concluded that there is no measurable alteration in the electrophoretic mobility of ovalbumin after denaturation (*Abramson, 1932; Longsworth, 1941; MacPherson, Heidelberger, and Moore, 1945*). It has, in fact, been asserted that ovalbumin, completely denatured by heat or acid, shows a mobility only slightly lower than that of the undenatured substance. On the other hand, Lundgren, Elam, and O'Connell (*1943*) found the mobility of heat-denatured ovalbumin to be -6.7×10^{-5}, rather than -5.6×10^{-5}, the mobility of the undenatured protein. They also showed that the presence, in ovalbumin, of a small amount (0.3 per cent) of a detergent (alkylbenzene sulfonate) changes the electrophoretic mobility to that of the completely denatured ovalbumin.

PARTICLE SIZE

Many attempts have been made to identify the changes that occur in the protein molecule during denaturation. Interest has been centered chiefly on alterations in the weight and shape of the molecule.

Change in Molecular Weight. It is becoming fairly certain that denaturation produces no perceptible change in the molecular weight of ovalbumin (*Burk and Greenberg, 1930; Huang and Wu, 1930; Burk, 1937*), although, in the opinion of Bull (*1938c*), several terminal amino acid residues could be broken off without causing a detectable change in molecular weight.

A molecule of undenatured ovalbumin occupies its maximum area at pH 4.8, the isoelectric point of this protein. A molecule of urea-denatured ovalbumin, on the other hand, occupies its maximum area at pH 7.0 (*Lee and Wu, 1932*). It can, therefore, be concluded that the isoelectric point of denatured ovalbumin is higher than that of the native protein.

Change in Shape of Molecule. According to one hypothesis (*Wu, 1931*), the molecular structure of a native protein is orderly and compact, whereas that of a denatured protein is disorderly and diffuse. Other studies indicate that a molecule of ovalbumin, when denatured, unfolds and changes its shape (*Mirsky and Pauling, 1936*), although the peptide chain probably remains greatly folded and collapsed, even in the denatured molecule. At present, it is impossible to describe the exact structural changes that take place in the molecule; but the fact that changes occur is evident from the manner in which the physical and chemical properties of a denatured protein differ from those of the native substance.

Increase in Asymmetry of the Molecule. Using the formula of Polson (*1939b*), who assumed that the protein molecule is a prolate ellipsoid, Bull (*1940a*) substituted specific viscosity values obtained experimentally and thus calculated the ratio of the long to the short axis of the native and denatured ovalbumin molecule. His results indicate that the molecule, whatever its shape, becomes more asymmetrical after denaturation.

Condition of Ovalbumin	$\dfrac{\text{Major Axis}}{\text{Minor Axis}}$ of Molecule
Undenatured	3.9:1
Heat-denatured	7.4:1
Urea-denatured	9.2:1

Presumably, molecules of denatured ovalbumin are not only asymmetric but also polar. They tend to orient themselves in respect to each other, so that their nonpolar heads are together (*Bull, 1938c*).

CHEMICAL CHANGES

There has been a great deal of controversy as to the degree of molecular dissociation brought about in ovalbumin by various denaturants. Sörensen and Sörensen (*1925*) concluded that heat coagulation is not accompanied by decomposition. Nevertheless, extensive studies show a noticeable liberation of sulfhydryl, and other, groups. A few experimental observations also indicate that a small amount of nitrogen could be split off and some sulfur liberated during the denaturation of various proteins, including ovalbumin.

Liberation of Sulfhydryl Groups. It has been known for some time that denatured, but not native, ovalbumin gives the nitroprusside color reaction for sulfhydryl (—SH) groups (*Heffter, 1907; Arnold, 1911*). A strong nitroprusside reaction is given by denatured ovalbumin in guanidine hydrochloride solution, and a weaker reaction in urea solution (*Anson, 1941*).

The first quantitative estimation of the —SH groups in denatured ovalbumin was made by measuring the amount of cystine reduced to cysteine by a suspension of the precipitated protein (*Mirsky and Anson, 1935*). Porphyrindin, a strong oxidizing agent, has also been used (*Kuhn and Desnuelle, 1938*). The presence of —SH groups in denatured ovalbumin may be shown by the failure of ferricyanide to be reduced by denatured ovalbumin treated with iodine (*Anson, 1940a*), mercuric chloride, iodoacetate (*Mirsky, 1941*), or mercuric benzoate (*Anson, 1941*); and also by the liberation of iodine when iodosobenzoate is used as an oxidizing agent for the —SH groups of denatured ovalbumin (*Hellerman, Chinard, and Ramsdell, 1941*).

It is still not entirely clear whether or not all the —SH groups are liberated during the denaturation of ovalbumin by various agents. There is considerable evidence (Table 42) that heat liberates only one-half, and urea about 80 per cent, of the —SH groups liberated by guanidine hydrochloride.

Some sugars, especially the simple ones, inhibit sulfhydryl formation

TABLE 42

<small>LIBERATION OF SULFHYDRYL GROUPS (AS CYSTEINE) FROM DENATURED OVALBUMIN BY
VARIOUS AGENTS</small>

Denaturing Agent	—SH as Cysteine (per cent)	Investigator
Heat	0.56	Mirsky and Anson (*1935*)
	0.62	Mirsky and Anson (*1936a*)
	0.63	Todrick and Walker (*1937*)
	0.50	Greenstein (*1938*)
	0.55	Rosner (*1940*)
Average	0.57	
Urea	1.00	Greenstein (*1938*)
	0.87	Rosner (*1940*)
	0.95	Mirsky (*1941*)
Average	0.94	
Duponol *	1.24	Anson (*1940b*)
	0.95	Mirsky (*1941*)
Average	1.10	
Guanidine hydrochloride	1.28	Greenstein (*1938*)
	1.24	Anson (*1940b*)
	1.32	Hellerman, Chinard, and Ramsdell (*1941*)
	0.95	Mirsky (*1941*)
	1.24	Hellerman, Chinard, and Deitz (*1943*)
Average	1.21	

* Trade name for a group of fatty alcohol sulfates.

in heat-denatured ovalbumin. Ball, Hardt, and Duddles (*1943*) demonstrated that the amount of cysteine, as calculated from the sulfhydryl content, is decreased by 14 to 17 per cent in the presence of *d*-glucose, *d*-fructose, *d*-mannose, *d*-arabinose, and *d*-xylose, although by only 1 per cent in the presence of *d*-mannitol.

Detection of Disulfide, Tyrosine, and Tryptophane Groups. When ovalbumin is denatured, various groups, other than the sulfhydryl groups, appear in detectable forms. Among these are disulfide (S—S) groups (*Walker, 1925*), tyrosine groups (*Mirsky and Anson, 1936b; Herriott, 1936*), and tryptophane groups (*Mirsky and Anson, 1936b*).

The number of S—S groups may be estimated by reducing them to —SH groups with thioglycollic acid (*Mirsky and Anson, 1935*); the increase in —SH groups equals the number of S—S groups originally present.

Splitting-Off of Nitrogen. Hendrix and Dennis (*1938*) have made numerous chemical analyses for the nitrogen content of both native ovalbumin and ovalbumin denatured by acid, alkali, and agitation. They have invariably found about 0.5 per cent less nitrogen in denatured

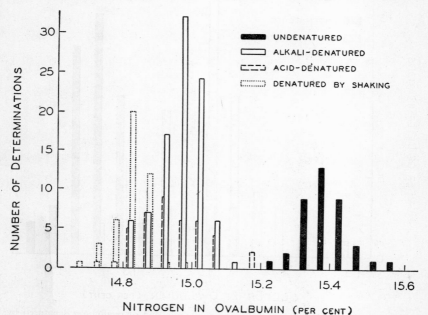

Fig. 253. The percentage of nitrogen in native ovalbumin and in ovalbumin denatured by alkali, acid, and shaking. (After Hendrix and Dennis, 1938.)

ovalbumin than in native (Fig. 253). These observations have been confirmed by MacPherson and Heidelberger (*1945a*), who noted, however, that the nitrogen split from acid- and heat-denatured ovalbumin was less than 0.1 per cent of the total nitrogen denatured. Nitrogen loss during alkali denaturation varied from less than 0.1 per cent to more than 1 per cent.

Liberation of Sulfur. Long ago, Hopkins (*1900*) observed that lead acetate paper was blackened during the denaturation of ovalbumin by shaking, and he therefore concluded that ovalbumin lost sulfur, either as hydrogen sulfide or some other volatile sulfide. His findings could not be corroborated by Wu and Ling (*1927*); but Bull (*1938c*) noticed that a peculiar odor, resembling that of wet wool, is produced during both heat and surface denaturation. Recently, Hendrix and Dennis (*1943*) have demonstrated that there is a measurable loss of sulfur from ovalbumin during denaturation by acid or alkali (Fig. 254).

It has been suggested (*Wu and Wu, 1925*), but not confirmed (*Mastin and Rees, 1926*), that tyrosine is split off when ovalbumin is coagulated

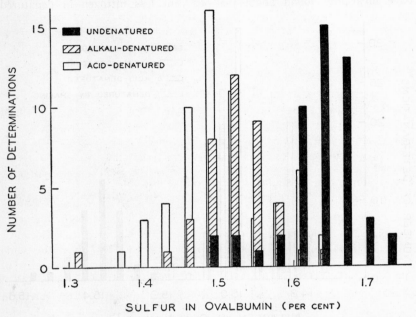

FIG. 254. The sulfur content of native ovalbumin and of ovalbumin denatured by alkali and acid. (After Hendrix and Dennis, 1943.)

by heat. Possibly tyrosine may have been found in samples of ovalbumin that were not completely purified.

SPECIFIC BIOLOGICAL PROPERTIES

After denaturation, some specific biological properties of ovalbumin are lost. For example, digestibility by certain enzymes is increased. On the other hand, the species specificity of ovalbumin and other egg proteins (as shown especially by immunological properties) may become less marked (see Chapter 8, "Biological Properties").

Increase in Digestibility. It is a generally accepted fact that denatured egg proteins are more easily digested by enzymes than are the native substances. This change has been demonstrated *in vitro* (see Chapter 8, "Biological Properties") and by experiments on animals and human beings (see Chapter 9, "Food Value").

Effect of Formaldehyde

It has been frequently reported that ovalbumin, upon treatment with dilute formaldehyde, becomes resistant to heat coagulation (*Blum, 1897; Abiko, 1935; Kekwick and Cannan, 1936b*) and to acid coagulation (*Cubin, 1929*). Formaldehyde changes the properties of proteins. Its greatest effect is to harden them, but the nature of the reaction remains largely unexplained. Possibly the polypeptide chains become more regularly aligned (*Clark and Shenk, 1937b*).

Phosphorylation of Ovalbumin

It has been demonstrated (*Heidelberger, Davis, and Treffers, 1941*) that as many as twenty to thirty phosphoryl groups may be introduced into crystalline ovalbumin. Phosphorylated ovalbumin acquires certain new chemical and physical properties, the most pronounced of which are incoagulability by heat and increased viscosity.

Conversion from Globular to Oriented Fibrous Form

Ovalbumin can be converted from a globular to an oriented fibrous form (see Chapter 11, "Industrial Uses"). After the protein filaments are soaked in various reagents (acetic anhydride, glacial acetic acid, etc.) and then in water, the converted protein molecules are given the same spatial arrangement as beta-keratin. The X-ray diffraction patterns of ovalbumin, after such treatment, are characterized by a reflection of 9.8 angstrom units on the equator and 4.7 angstrom units on the meridian (*Astbury, Dickinson, and Bailey, 1935*). The above values agree with the side chain spacing and the "backbone" spacing, respectively.

As many as five diffraction rings have been obtained from ovalbumin treated with aqueous methanol, formamide, or urethane (*Nutting, Senti, and Copley, 1944*). The rings from the formamide-treated ovalbumin occurred at 10.7, 4.7, 3.7, 2.2, and 2.0 angstrom units. The nine forms found in the fiber pattern of beta-keratin and heated ovalbumin are also found in chemically treated ovalbumin.

From the sharpness and the length of the diffraction arcs, it has been concluded that the qualities of the converted ovalbumin preparations are comparable with those of such natural fibers as silk and wool.

Hydrolysis of Egg Proteins

The dissociation of a protein molecule into smaller units, with the splitting-off of ammonia, is known as hydrolysis. This process, quite clearly, is not denaturation and should be classed separately.

The ammonia freed during the hydrolysis of ovalbumin by strong

acids or alkalies constitutes an important fraction of the nitrogen of the
protein. Ammonia is liberated at essentially the same rate by both acid
and alkali. The curve of ammonia production rises rapidly at first; the
reaction then proceeds at a fairly uniform rate and continues indefinitely.
As shown in Fig. 255, ammonia is liberated, during alkaline hydrolysis,
more rapidly as the temperature and the concentration of the alkali are

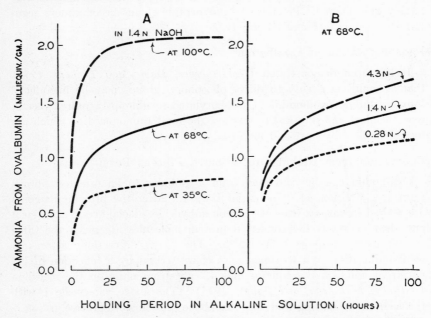

FIG. 255. Rate of formation of ammonia from ovalbumin in alkaline solutions.
(After Warner and Cannan, 1942.)

A, at various temperatures. *B,* in the presence of different concentrations of NaOH.

increased (*Warner and Cannan, 1942*). It is generally assumed that
it is chiefly the primary acid amide group that is hydrolyzed during the
more rapid phase of the reaction. The secondary evolution of ammonia
has been attributed to deamination (*Shore, Wilson, and Stueck, 1935*).
It has also been suggested that, in alkaline hydrolysis, part of the ammonia originates in some alkali-labile groups existing in the protein itself
(*Warner and Cannan, 1942*).

The rate at which free amino groups are liberated, during hydrolysis,
is most rapid initially and later levels off (Fig. 256). Among the free
amino acids formed are cystine, in acid hydrolysis, and tyrosine and
tryptophane, in alkaline hydrolysis (Fig. 257). According to Warner

(*1942*), the free amino acids formed during alkaline hydrolysis do not rise to the value expected from amino determinations.

Perhaps it is of interest to note that the racemization of the amino acid residues of egg proteins, by the action of dilute alkali, reveals a difference in the molecular structure of ovalbumin from chicken and duck eggs (*Dakin and Dale, 1919*). There is some difference in the degree

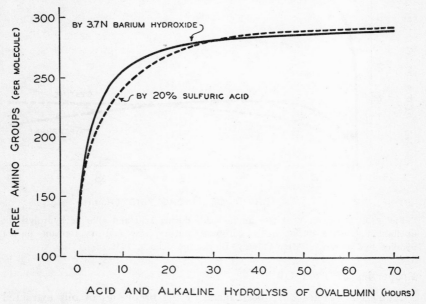

Fig. 256. Rate of hydrolysis of ovalbumin by 3.7 N barium hydroxide and by 20 per cent sulfuric acid, at 100° C. (After Warner, 1942.)

of racemization of histidine, leucine, and aspartic acid. The proteins of each species are also distinguishable by their serological reactions (see Chapter 8, "Biological Properties").

LIPIDS

Lipids from egg yolk, as well as from various plant and animal sources, possess rather constant characteristics which serve as a means for their identification. One general property that distinguishes lipids from other compounds is their total insolubility in water, and their solubility in such substances as chloroform and ether.

The specific physicochemical properties of lipids from egg yolk (there are but negligible amounts of lipids in the other egg components) are of scientific as well as of commercial interest.

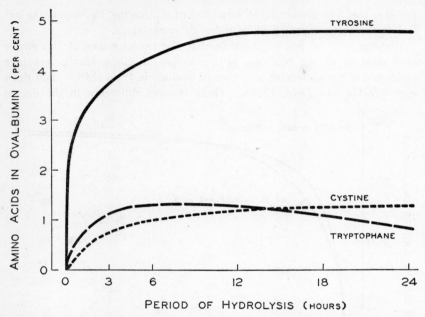

Fig. 257. Appearance of free amino acids during acid and alkaline hydrolysis of ovalbumin (cystine during acid hydrolysis, and tyrosine and tryptophane during alkaline hydrolysis). (After Calvery, Block, and Schock, 1936.)

Properties of Yolk Oil

Numerous studies have been made of the properties of oils extracted from egg yolk with suitable solvents. Certain physical and chemical properties of yolk oils from various birds' eggs are expressed quantitatively in Table 43.

Properties of Fats and Compound Lipids

Since nearly two-thirds of egg yolk oil consist of neutral fats and only about one-third of compound lipids, the physicochemical properties of the neutral yolk fats differ very little from those of crude yolk oil (see Table 43). On the other hand, the phospholipids of the yolk possess many physicochemical properties distinct from those of fats, although they are quite similar to phospholipids from other sources. Consequently, for many details, the reader can be referred to general treatises on phospholipids.

The phospholipids of the yolk are optically active. They are dextro-

TABLE 43

PHYSICOCHEMICAL PROPERTIES OF EGG YOLK OIL *

	Species			
Property	Chicken	Turkey	Duck	Goose
Physical:				
Specific gravity (at 15° C.)	0.914	....	....	....
Melting point (°C.)	23.0	....	....	....
Solidification point (°C.)	9.0	....	....	....
Refractive index (n_D) at 40° C.	1.4650	1.4638	1.4674	1.4651
Chemical:				
Acid number	5.23	7.7	5.3	4.7
Iodine absorption number	75.1	65.8	73.2	63.3
Saponification value	187.0	184.7	193.0	199.2
Reichert-Meissl number	0.53	....	0.27	....
Hehner number	84.3	86.0	87.5	87.1
Polenske number	0.28	....	0.25	....
Acetyl number	3.8–11.9	....	....	....
Ester value	182.7	177.0	200.3	194.5
Free (soluble) acids:	2.25	2.24	2.77	2.57
Melting point (°C.)	37.0	....	....	....
Iodine absorption number	80.1	....	....	....
Unsaponifiable matter (per cent)	4.37	....	....	....

* Ether or chloroform extract.

References

CHICKEN: Parke (*1867*); Amthor and Zink (*1897*); Kitt (*1897*); Pennington (*1909–10*); Serono and Palozzi (*1911*); Thomson and Sorley (*1924*); Romanoff (*1932*); Vita and Bracaloni (*1933*); Teixeira e Silva (*1946*).
TURKEY: Hepburn and Miraglia (*1937*).
DUCK: Thomson and Sorley (*1924*); Hepburn and Katz (*1927*).
GOOSE: Hepburn and Katz (*1927*).

ortatory, as shown in the accompanying tabulation. In general, the degree of rotation is influenced to some extent by the concentration, temperature, and nature of the solvent. (The cerebrosides are also optically active; ovophrenosin is dextrorotatory, and ovokerasin, levorotatory.)

YOLK PHOSPHATIDES	OPTICAL ACTIVITY ($[a]_D^{40}$)	INVESTIGATOR
Ovolecithin	+11.34°	Ulpiani (*1901*)
Ovocephalin	+6.00°	Levene and West (*1918*)
Ovosphingomyelin	+7.54°	Levene (*1916*)

Phosphatides, like proteins, are amphoteric. The electrophoretic velocity of ovolecithin, under a potential gradient of 1 volt per centimeter, has been found to be 13.0×10^{-5} cm. per second, at 20° C. The changes in the electrophoretic velocity of ovolecithin with pH are shown in Fig. 258.

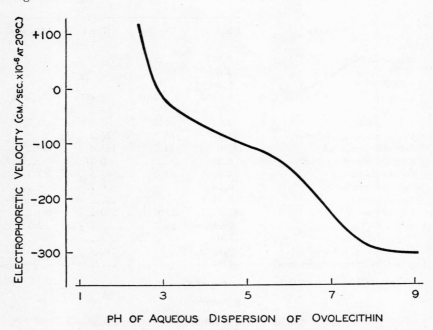

Fig. 258. Relationship between electrophoretic velocity and pH of ovolecithin (dispersed in water). (After Price and Lewis, 1933.)

The isoelectric points of neutral fats and of phospholipids are not well established. The isoelectric point of ovolecithin, for example, has been variously reported, as shown in the table.

ISOELECTRIC POINT (pH)	INVESTIGATOR
2.7	Fujii (1924)
4.7	Sueyoshi and Kawai (1932)
2.6	Price and Lewis (1929)
6.4	Bull and Frampton (1936)

There is no doubt that these inconsistencies are due to the presence of various impurities in ovolecithin. It has been shown that the isoelectric point becomes higher when albumin is added (*Sueyoshi and Kawai, 1932*) and when amino nitrogen is present (*Bull and Frampton, 1936*), although the addition of cholesterol produces no change (*Price, 1933*).

In addition, the isoelectric point of ovolecithin apparently decreases as time passes (*Sueyoshi and Kawai, 1932; Bull and Frampton, 1936*); how rapidly may be seen in Fig. 259.

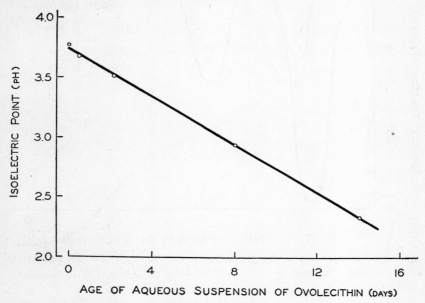

Fig. 259. Relationship between isoelectric point and age of aqueous suspension of ovolecithin. (After Bull and Frampton, 1936.)

The iodine number of ovolecithin is somewhat lower than that of neutral fats. Henriques and Hansen (*1903*) suggested that the iodine number of the egg's neutral fat is more variable, and more easily influenced by the fats in the hen's diet, than is the iodine number of lecithin. However, the iodine number, both of the fat and of lecithin, decreases when hens are placed on a nearly lipid-free diet (*McCollum, Halpin, and Drescher, 1912*).

The surface tension of ovolecithin reaches its maximum value at *p*H 2.6 (*Price and Lewis, 1929*) and falls to a very marked minimum on either side of this point (Fig. 260). In the presence of cholesterol, surface tension is at its maximum at *p*H 4.0 rather than *p*H 2.6. The addition of salts, such as sodium chloride and calcium chloride, has unexpectedly complex effects.

The molecular weight of pure ovolecithin is 797, and of commercial lecithin, 766 (*Price and Lewis, 1929*). These values are in close agreement with that of 771 observed in 1907 by Stern and Thierfelder.

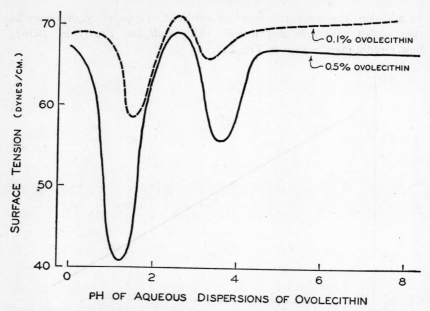

Fɪɢ. 260. Relationship between *p*H and surface tension of 0.1 per cent and 0.5 per cent ovolecithin (dispersed in water). (After Price and Lewis, 1929.)

CARBOHYDRATES

There is little to be added to the discussion of the carbohydrates of the egg given in Chapter 6, "Chemical Composition."

It has been shown that one polysaccharide unit isolated from ovalbumin has a molecular weight of 1200 (*Neuberger, 1938*).

PIGMENTS

As noted in Chapter 6, "Chemical Composition," the major pigments of the egg are the carotenoids of the yolk, ovoflavin of the yolk and the albumen, and oöporphyrin of the shell membranes, the shell, and the cuticle. Among the carotenoids are the carotenes (alpha- and beta-) and the xanthophylls (cryptoxanthin, lutein, and zeaxanthin).

The color of the various carotenoids differs (*Kuhn and Brockmann, 1932*); the xanthophylls, especially lutein and zeaxanthin, have a value (compared to a solution of 14.5 mg. of azobenzene in 100 cc. of 96 per cent alcohol) higher than that of the carotenes, as shown below.

CAROTENOID	COLOR VALUE (mg./cc. petroleum ether)
Carotenes:	
Alpha-carotene	0.00235
Beta-carotene	0.00235
Xanthophylls:	
Cryptoxanthin	0.00242
Lutein	0.00252
Zeaxanthin	0.00252

The melting points of some of the egg pigments have been determined as follows:

PIGMENT	MELTING POINT (°C.)	INVESTIGATOR
Carotene	166–168	Peterson, Hughes, and Payne (*1939*)
Lutein	193	Willstäter and Escher (*1912*)
Zeaxanthin	207	Kuhn, Winterstein, and Lederer (*1931*)
Ovoflavin	240–284	Kuhn and Wagner-Jauregg (*1933*); Karrer and Schöpp (*1934*)

In optical activity, these pigments differ greatly from each other. After excluding some variation caused by the concentration of the solutions, the average values are as follows:

PIGMENT	SPECIFIC ROTATION ($[a]_D$)	INVESTIGATOR
Lutein	+145 (in acetic acid)	Kuhn, Winterstein, and Lederer (*1931*)
Zeaxanthin	−55 (in acetic acid)	Kuhn, Winterstein, and Lederer (*1931*)
Ovoflavin	+23.5 (in water)	Kuhn and Wagner-Jauregg (*1933*)

The characteristic spectroscopic bands of egg pigments have been used extensively for the identification of these substances, and for determining their degree of purity. However, the observed points of maximum absorption (extinction or absorption maxima), which are independent of concentration or stratum thickness, are very similar in the spectra of all the carotenoid pigments of the egg. According to Kuhn and Brockmann (*1932*), the absorption maxima in ethyl alcohol are as follows:

CAROTENOID PIGMENT	ABSORPTION MAXIMA (millimicrons)	
Alpha-carotene	477	509
Beta-carotene	492	530
Cryptoxanthin	483	519
Lutein	475	508
Zeaxanthin	483	519

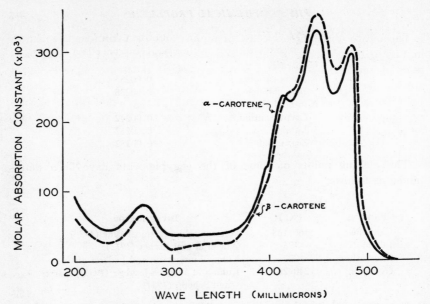

Fig. 261. Absorption spectra, in the region of ultraviolet and visible light, of alpha- and beta-carotene derived from egg yolk and dissolved in hexane. (After Smakula, 1934.)

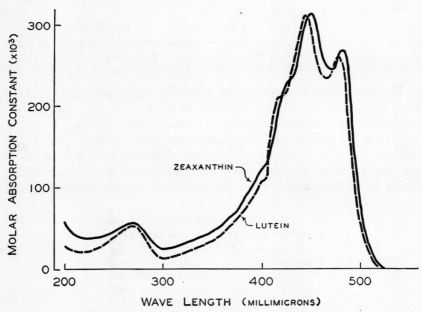

Fig. 262. Absorption spectra, in the region of ultraviolet and visible light, of zeaxanthin and lutein derived from egg yolk and dissolved in alcohol. (After Smakula, 1934.)

The entire absorption spectra of alpha- and beta-carotene are shown in Fig. 261, and those of lutein and zeaxanthin in Fig. 262. The positions of the maxima are apparently somewhat different, but the curves are all very similar.

In the absorption spectrum of ovoflavin, the maximum is at 267 mμ, as shown below by optical density (*Kuhn and Wagner-Jauregg, 1933*).

WAVE LENGTH (millimicrons)	OPTICAL DENSITY ($\log I_o/I$)
220	0.95
267	1.25
358	0.38
374	0.39
445	0.48

The maximum spectral absorption of oöporphyrin is presumably at about 395 mμ (*Theorell, 1938*).

It has been determined that the molecular weight of xanthophyll is about 640 (*Mattikow, 1932*), and that of ovoflavin, 500 (*Kuhn and Wagner-Jauregg, 1933*).

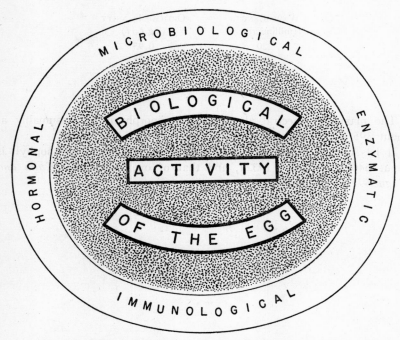

Types of biological activity shown by the bird's egg.

Chapter Eight

BIOLOGICAL PROPERTIES

*The avian egg is charged with potential biological
activity.*

The egg is composed of substances, widely found in living matter, that possess intrinsic biological properties. Although these substances enter into complex recombinations during the development of the avian embryo, it is not necessary to incubate the egg in order to demonstrate the active nature of its contents. This is amply revealed by reactions both *in vivo* and *in vitro*.

Because of its biological potentialities, the egg is valuable material for the investigation of fundamental physiological processes, including the body's responses to foreign substances, the production of immune antibodies, and hydrolysis by enzymes. In addition, a number of interesting and largely unexplained effects may be produced by some of the egg's constituents.

The egg, primarily a store of food for the avian embryo, sometimes becomes the passive carrier of microorganisms, which may avail themselves of its nutrients. It then acquires biological activity of a secondary nature.

MICROBIOLOGY OF THE EGG

The interior of the new-laid egg is usually free of microorganisms, chiefly because of the natural protection provided by the egg's physical structure and by the chemical constitution of the albumen. Contamination of the egg contents, however, occasionally occurs, either before the egg is laid or shortly thereafter. As a result, the egg may eventually decompose, or it may be responsible for the dissemination of disease among poultry or human beings.

INCIDENCE OF MICROFLORAL CONTAMINATION

It is remarkable that the interior of the egg rarely contains microorganisms, for bacteria and molds are usually present in profusion on the surface of the eggshell. The number of mold spores on the shell of

491

the fresh egg averages 200 to 500 (*Tomkins, 1937a*). The bacterial count per shell may be about 35,000 when the culture is incubated at 37° C., or 130,000 at 20° C.; the maximum count at the latter temperature is sometimes as high as 8 million (*Haines, 1938*). Eggshell microorganisms are derived from the environment. Stuart and McNally (*1943*) have shown that the shell is usually sterile before the egg is laid and seldom is contaminated when the egg passes through the cloaca, unless the hen suffers from diarrhea (*Haines, 1939*).

Very diversified types of microorganisms may be present on the eggshell. The approximate composition of the flora found in one investigation (*Haines, 1938*) is shown in the table. The species of bacteria or

Type of Microorganism	Distribution (per cent)
Nonsporing rods	38
Sporing rods	30
Cocci	25
Yeasts	4
Actinomyces	3

fungi on the exterior of the egg vary according to circumstances, and a detailed discussion of the numerous types that have been found is therefore of little avail. The majority are air, soil, and water saprophytes.

In spite of the numerous microorganisms on the surface of the eggshell, bacteria are not usually found in the yolk in more than 10 per cent of all fresh eggs, or in the albumen in more than 3 per cent (*Rettger, 1913; Hadley and Caldwell, 1916; Haines, 1938*). Although higher percentages of contamination have been reported, such as 18 per cent (*Maurer, 1911*), 54 per cent (*Poppe, 1910*), and 87 per cent (*Stiles and Bates, 1912*), it is quite likely that these figures were obtained because of the technical difficulties encountered in working with material of such unusual consistency and physical structure.

The types of microorganisms found in the fresh egg are extremely varied, but, like those on the exterior of the shell, they are usually common saprophytes. Most of them grow more readily at 20° C. than at 37° C. When large lots of eggs are examined, as many as thirty-six (*Pennington, 1909–10*) or forty (*Hadley and Caldwell, 1916*) different species may be isolated. *Bacilli* and *Micrococci* are apparently most frequently encountered. In one investigation, over 64 per cent of the bacterial flora consisted of very short rods, most of which were Gram-negative and nonmotile; these were not identified (*Hadley and Caldwell, 1916*). Molds, as well as bacteria, have been demonstrated in the interior of fresh eggs (*Baumgarten, 1884; Pennington, 1909–10; Rettger, 1913;*

Horowitz-Vlasova, Balin, and Novotelnov, 1932; Zagaevsky and Lutikova, 1937), although they may not constitute more than 4 per cent of the total flora (*Rettger, 1913*). Among the types that have been found are various species of *Aspergillus, Penicillium,* and *Mucor.*

The incidence of microfloral contamination in fresh eggs cannot be correlated with any one particular factor. The laying of eggs that contain

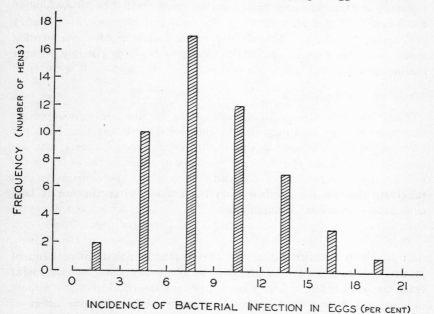

Fig. 263. Frequency distribution of 52 hens according to the percentage of bacterially contaminated eggs they laid during 1 year. (After Hadley and Caldwell, 1916.)

bacteria is apparently not a consistent trait of the individual bird (except when certain pathological conditions exist, as will be explained later). Studies conducted over long periods of time indicate that no hen lays bacteriologically sterile eggs continuously (*Bushnell and Maurer, 1914; Hadley and Caldwell, 1916*). The proportion of contaminated eggs laid annually by the individuals of a group of fifty-two fowl, for example, varied from 2.8 to 18.6 per cent; the largest number of birds laid eggs of which 6.0 to 8.9 per cent contained bacteria (Fig. 263).

Similarly, there appears to be no relationship between the incidence of contamination in eggs and the hen's rate of production (*Bushnell and Maurer, 1914; Hadley and Caldwell, 1916*), or her age (*Maurer, 1911;*

Hadley and Caldwell, 1916). A slight increase in the proportion of eggs containing bacteria is said to occur with the advent of warm weather (*Lamson, 1909; Maurer, 1911*). It is quite likely, however, that a rise in the egg's bacterial content during the summer is due to the effect of high temperatures and other environmental conditions that quickly increase the egg's susceptibility to bacterial invasion.

Fertile and infertile eggs ordinarily are contaminated to approximately equal extent (*Pennington, 1909–10; Bushnell and Maurer, 1914; Hadley and Caldwell, 1916*). The slightly lower incidence of contamination found in fertile eggs in one study (*Rettger, 1913*) is possibly without significance.

Mode of Contamination

There are only two possible explanations for the presence of microorganisms in eggs that may still be considered fresh and that have not been subjected to adverse environmental conditions. Bacteria that are incorporated into the egg during its formation, either in the ovary or the oviduct, may be considered congenital in origin. Microorganisms that penetrate through the eggshell from the outside, after the egg is laid, constitute extragenital contamination.

CONGENITAL

It has been repeatedly observed that contamination is more frequent in the yolk than in the albumen. The possibility that bacteria might enter the yolk before it leaves the ovary must be considered.

It is well known that *Salmonella pullorum*, the causative agent of pullorum disease (bacillary white diarrhea), may enter the yolk in the ovary. The bird becomes infected by way of the digestive tract (*Rettger and Stoneburn, 1909*), and the organism is carried to the ovary by the blood stream (*May, 1924*). Experiments have not shown conclusively that ovarian infection follows the feeding or inoculation of certain bacteria (*May, 1924; Miles and Halnan, 1937*), but many organisms apparently attain the ovary by the same avenues traveled by *Salmonella pullorum*. Among those found both in the ovary and the egg are *Salmonella anatum*, *Salmonella typhimurium*, *Salmonella enteritidis* (*Bunyea, 1942b*), and *Mycobacterium avium* (*Koch and Rabinowitsch, 1907*). The transitory presence of harmless bacteria in the ovary is also likely, for these tend to occur in small groups of eggs laid successively by an individual bird (*Maurer, 1911*).

It is also possible that the egg sometimes acquires bacteria as it passes through the oviduct. This organ opens into the body cavity at one end and into the cloaca at the other and is thus exposed to infection from two potential sources. Lamson (*1909*) stated that bacteria were often present in the oviduct. Normally, however, it is sterile, except in the cloacal region (*Rettger, 1913*). Horowitz (*1903*) showed that organisms artificially introduced into the oviduct die in 48 to 72 hours and therefore suggested that the oviduct maintains its sterility by means of phagocytosis, the secretion of bactericidal substances, and peristaltic action directed toward the cloaca.

The inclusion within the egg of such foreign material as feces, feathers, worms, and pebbles, as well as the formation of double eggs (see Chapter 5, "Anomalies"), indicates that the oviduct is nevertheless capable of antiperistalsis. Perhaps antiperistalsis can impel bacteria upward from the cloaca so rapidly that they are included in an egg before the oviduct's antibacterial action can take effect. The resistance of *Salmonella pullorum* to the bactericidal activity of the oviduct has been clearly shown. The organism was injected into the oviduct and subsequently was found in the ovary (*Rettger, Kirkpatrick, and Card, 1919*). Whether or not other microorganisms are similarly able to survive has not been demonstrated.

The theory that the egg may acquire bacteria at the time of fertilization is not borne out by comparisons of the incidence of infection in fertile and in infertile eggs.

EXTRAGENITAL

Potentially, the pores of the shell are avenues for the entrance of bacteria and fungi into the egg's interior. Although there is wide variation in the size of the pores, some are inevitably large enough to allow microfloral penetration.

As early as 1851, von Wittich demonstrated the ability of molds to enter the egg through the shell; and, in 1895, Wilm found the cholera organism similarly capable of invasion. By the experimental smearing of the eggshell surface with cultures, or by the immersion of the entire egg into liquid suspensions of bacteria, it has repeatedly been shown that the shell is pervious to such organisms as *Escherichia coli* (*Wilm, 1895; Zagaevsky and Lutikova, 1944c*), *Eberthella typhi* (*Piorkowski, 1895; Lange, 1907*), *Salmonella paratyphi* (*Lange, 1907; Poppe, 1910*), *Staphylococcus aureus*, "Bacillus No. 9" (*Pernot, 1909*), *Proteus vulgaris* (*Zagaevsky and Lutikova, 1944c*), fluorescent organisms (*Rievel, 1939*), *Serratia marcescens*, and *Pseudomonas aeruginosa* (*Zagaevsky and Lutikova, 1944c*). In addition, suspensions of *Saccharomyces ellipsoideus* and a species of *Pseudomonas* have been experimentally drawn through the shell by suction, although from interior to exterior (*Haines and Moran, 1940*).

Under the experimental conditions mentioned above, the eggshell is wet. The presence of moisture on the exterior of the egg is conducive to bacterial invasion. In fact, instantaneous penetration of the wet shell may be effected by certain bacteria, as shown by a study in which eggs

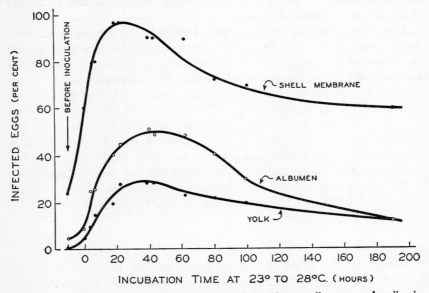

FIG. 264. Incidence of contamination in shell membrane, albumen, and yolk observed during 200 hours' incubation of eggs, the shells of which had been smeared with a culture of *Pseudomonas aeruginosa*. (After Stuart and McNally, 1943.)

were swabbed externally with a liquid culture of *Pseudomonas aeruginosa* (*Stuart and McNally, 1943*). The organism was recovered immediately from the shell membrane in 60 per cent of the treated eggs; 18 hours later, it was found in 95 per cent (Fig. 264). Rapid changes in environmental temperature favor the microfloral invasion of the egg, because they often result in the condensation of a thin film of water on the shell surface. When an egg becomes soiled, it is usually moistened at the same time. The eventual entrance of bacteria through the shell is then very probable, especially as the presence of dirt adds immeasurably to the numbers of contaminating organisms. Washing the egg merely increases its vulnerability. If the washing water is colder than the egg, the egg contents contract, and water and bacteria may be drawn through the shell by suction (*Haines and Moran, 1940*). Dirty eggs, washed eggs, and eggs that have been allowed to "sweat" all show a higher incidence of rotting in storage than clean, untreated eggs (see Chapter 10, "Preservation").

Experimental evidence suggests that enlargement of the pores by the hyphae of molds may facilitate the entrance of bacteria (*Zagaevsky and Lutikova, 1944b, 1944c*). Two lots of eggs were smeared externally with cultures of *Serratia marcescens, Pseudomonas aeruginosa, Proteus vulgaris,* or *Escherichia coli* and were then kept for 4 months, the first lot at a temperature of 0.5° C., the second at 18° C. Although the first lot remained internally sterile, the second became contaminated to the extent shown in the tabulation (*Zagaevsky and Lutikova, 1944c*). When

ORGANISM	INCIDENCE OF CONTAMINATION (per cent)
Serratia marcescens	70
Pseudomonas aeruginosa	50
Proteus vulgaris	30
Escherichia coli	20

the inoculum consisted of the same organisms mixed with *Aspergillus niger,* the percentage of internal contamination at both temperatures was as follows:

ORGANISM	INCIDENCE OF CONTAMINATION (per cent)	
	AT 0.5° C.	AT 18° C.
Serratia marcescens	4	100
Pseudomonas aeruginosa	10	100
Proteus vulgaris	0	70
Escherichia coli	0	50

The data given above also indicate how environmental temperature influences the egg's susceptibility to microfloral invasion. At refrigeration temperatures, most bacteria are almost completely inhibited; but molds are capable of fairly vigorous growth, if they are provided with sufficient moisture (see Chapter 10, "Preservation").

The motility of the invading organism may be another factor that determines the ease with which penetration of the shell is effected (*Lange, 1907*). Poppe (*1910*) concluded that motile rods and spirilla were able to pass through the shell, but that nonmotile rods and cocci were not. Nonmotile types, often found in the egg contents, perhaps represent a secondary invasion.

As the new-laid egg cools, bacteria are possibly drawn in with the air that forms the air cell. This would be especially likely to occur if a film of moisture were present on the surface of the shell (*Haines and Moran, 1940*).

Certain types of bacteria, having penetrated the shell, may make their way into the interior of the egg with great rapidity. This fact, also, was demonstrated experimentally by smearing eggs with cultures of *Pseudomonas aeruginosa*. In 10 per cent of the eggs so treated, the organism was found immediately in the albumen, and, in 5 per cent, it attained the yolk without delay (cf. Fig. 264). Within 24 hours, it had reached the yolk in 30 per cent of the eggs; and, in 42 hours, the albumen of 55 per cent.

DEFENSES OF THE EGG AGAINST MICROORGANISMS

There are a number of reasons for the infrequent internal contamination both of fresh eggs and of eggs that have been held for long periods of time. The physical organization and the chemical constitution of the egg together form an efficient defensive system for combating microorganisms, whether of congenital or extragenital origin.

Physical Defenses

The shell obviously provides the egg with its greatest natural protection against the microorganisms of the environment. Normally the pores in the shell are impervious to microbial penetration. They are filled with an organic substance (presumably mucin) that, when dry, does not permit bacteria and fungi to enter. On the other hand, if this organic material is dissolved or partially removed by abrasion, the pores are opened, and microfloral invasion immediately becomes possible (see Chapter 10, "Preservation").

The shell membranes, formed of interlacing fibers, may perhaps act as a filter for the removal of many of the microorganisms that succeed in penetrating through the pores of the shell. This possibility is indicated by the much lower incidence of infection in the albumen than on the shell membranes of eggs experimentally subjected to external contamination (cf. Fig. 264). It has also been shown that the shell membranes are capable of resisting penetration by such bacteria as *Pseudomonas aeruginosa*, *Serratia marcescens*, and *Sarcina flava*, when broth cultures of these organisms are inoculated directly on the membranes over a shell-free area (*Zagaevsky and Lutikova, 1944b*). Some of the protective effect of the shell membranes may also be due to chemical action.

The albumen is the third successive barrier to microflora. Microorganisms of extragenital origin which pass through the shell and membranes are often killed in the albumen before they can reach the yolk and its abundance of utilizable food materials. The arrangement of the egg's internal components therefore, of itself, constitutes protection. Bacteria will grow luxuriantly in the egg, if the yolk has been broken and mixed with the albumen (*Zagaevsky and Lutikova, 1944b*).

Chemical Defenses

The egg's chemical defenses against bacteria are contained in the albumen. The ability of the albumen to protect the yolk from con-

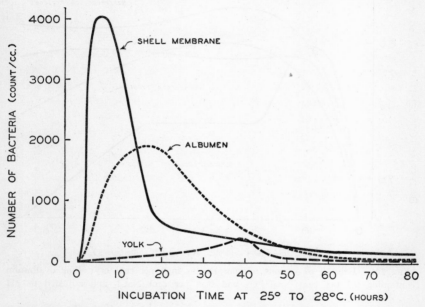

Fig. 265. The extent of contamination in shell membrane, albumen, and yolk observed during 80 hours' incubation of eggs smeared externally with a culture of *Pseudomonas aeruginosa*. (After Stuart and McNally, 1943.)

tamination by the microorganisms of the external environment is clearly shown in Fig. 265. Bacteria, after penetrating through the shell and membranes, multiply in the albumen for a short time and then rapidly decrease in number. Contamination of the yolk is delayed and comparatively light.

As early as 1890, Wurtz noted that various organisms, such as typhoid bacilli and pyogenic cocci, could not survive in egg albumen. Albumen, when left in an open, sterile dish, may remain free of bacterial growth for as long as 2 months (*Laschtschenko, 1909*).

The antibacterial action of egg albumen is partially due to the inability of many bacteria to utilize native proteins. Organisms such as *Eberthella typhosa, Escherichia coli,* and *Salmonella enteritidis* decrease in number when incubated in solutions of pure proteins, including ovalbumin, and produce no demonstrable proteolysis (*Bainbridge, 1911*). The actively proteolytic *Clostridium histolyticus* utilizes denatured ovalbumin, but it

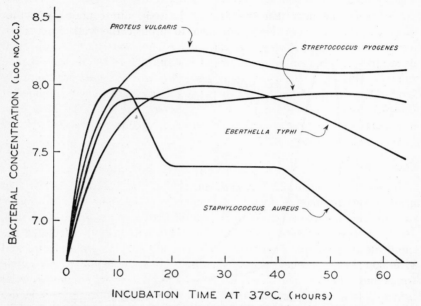

FIG. 266. Growth of various microorganisms in 1 per cent crystalline ovalbumin (containing 0.2 per cent Na_2HPO_4 and 0.36 per cent NaCl, and adjusted to pH 7.4). (After Grob, 1943.)

does not attack the native substance (*Haines, 1937*). Before they can elaborate their proteolytic enzymes, many bacteria must have simple nitrogenous compounds upon which to subsist (*Haines, 1934*); some also require the presence of the calcium ion (*Haines, 1932*). Even in stale eggs, only slight chemical changes in the proteins are demonstrable (see Chapter 10, "Preservation"). However, very minute amounts of protein degradation products are often sufficient to sustain proteolytic organisms.

The effectiveness of the protection that the albumen provides for the yolk therefore depends to some extent upon the nature of the invading organisms. The difference in the ability of several species of bacteria to multiply in ovalbumin may be seen in Fig. 266. It is interesting to note that these species all show better growth when trypsin is added to the ovalbumin (*Grob, 1943*).

The frequently high alkalinity of the albumen probably contributes to the inhibition of bacterial growth. Bacteria can survive only between pH 4 and pH 9 (approximately). Each species has its optimum pH; proteolytic bacteria prefer a neutral or very slightly alkaline medium. The pH of the albumen may eventually reach 9.5, the time it requires to do so depending upon the environmental temperature (see Chapter 10, "Preservation"). Such species of bacteria as *Pseudomonas fluorescens, Serratia marcescens, Proteus vulgaris,* and *Escherichia coli* are suppressed in samples of albumen adjusted to pH 9.5 and pH 10.7. In less alkaline samples (pH 8.5) *Bacillus cereus* and *Bacillus mycoides* are similarly inhibited. On the other hand, all these species grow well in albumen if the pH value is between 5.3 and 8.3. Bactericidal properties have even been imparted to egg yolk by adjusting it to approximately pH 9.0 (*Sharp and Whitaker, 1927*).

GERMICIDAL SUBSTANCES

It has been recognized for many years that the failure of bacteria to grow in egg albumen is due chiefly to the presence there of a substance, or substances, possessing definitely germicidal activity (*Laschtschenko, 1909; Rettger and Sperry, 1912*). Laschtschenko (*1909*) attributed the rapid destruction of *Bacillus subtilis* to an enzyme, but Sharp and Whitaker (*1927*) believed it was caused by a thermolabile, dialyzable compound. Schade and Caroline (*1944*) also detected an antibacterial principle that was thermolabile and dialyzable. Friedberger and Hoder (*1932*) claimed that albumen flocculated many bacteria, and that this phenomenon was caused by an adsorbable substance inactivated by boiling. The most interesting discovery was made by Fleming (*1922*). He found that albumen, in common with tears and various other secretions, completely lyses many organisms.

Lysozyme. The lytic principle in egg albumen is an enzyme, lysozyme. This substance has been crystallized both as the salt of several acids, and in the pure form. One type of lysozyme crystal is obtained at reactions below pH 7.0, and another between pH 7.0 and pH 11.0 (Fig. 267). As much as 88 per cent of the total lysozyme of egg albumen can be obtained by crystallization at 4° C. and pH 9.5, after adding 5 per cent sodium chloride and a few pure lysozyme crystals (*Alderton and Fevold, 1946*).

Lysozyme is present not only in the egg but also in many body fluids and secretions, including tears, saliva, blood serum, sputum, and nasal mucus; in such tissues as cartilage, liver, tonsil, intestine, and stomach; and in certain plants, notably the turnip. Although it was originally thought that lysozyme from all sources was the same substance, varying in concentration in different tissues and fluids, it is now known that the form contained in plants, at least, is chemically different from that found in egg albumen (*Meyer, Hahnel, and Steinberg, 1946*).

Certain properties of lysozyme can be enumerated, as the result of many studies (*Fleming, 1922; Fleming and Allison, 1922a, 1922b, 1924,*

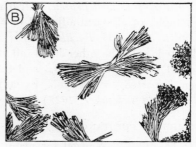

BELOW PH 7   BETWEEN PH 7 AND PH 11

LYSOZYME CRYSTALS

FIG. 267. Forms of lysozyme crystals obtained, *A*, below *p*H 7.0, and *B*, between *p*H 7.0 and *p*H 11.0. (After Alderton and Fevold, 1946.)

1927; Abraham, 1939; Alderton, Ward, and Fevold, 1945; Meyer and Hahnel, 1946). It is probably a pure protein, possibly a globulin, with isoelectric point at about *p*H 11.0, and molecular weight of nearly 18,000. It is soluble in water and normal saline solution, but insoluble in alcohol, ether, acetone, chloroform, xylol, and toluol. It is nondialyzable, but it is adsorbable upon porcelain, charcoal, cellulose, and certain types of clay. Although its rapidity of action increases with temperature up to 60° C., all lytic power disappears upon heating it to temperatures higher than 70° C. The exact degree of heat necessary to destroy lysozyme and the time required to do so vary with the reaction of the medium and are not the same for lysozyme from all sources. Pure lysozyme loses 60 per cent of its activity in 80 minutes of heating at 96° C. and *p*H 2.8, and it is even more thermolabile in alkaline solution. The reaction at which the lysozyme of egg albumen has its maximum effect is *p*H 5.3, but the optimum *p*H varies according to the source of the lysozyme. Salt in concentrations of more than 2 per cent is inhibitory, but 0.2 to 0.5 per cent salt somewhat increases activity. Enzymes, such as trypsin, papain, and bacterial and mold proteinases, do not digest lysozyme; but pepsin is capable of destroying 56 per cent of its activity in 24 hours, at 30° C.

Lysozyme acts on both living and dead bacteria. After it has dissolved a suspension of organisms, its lytic power increases in proportion to the number of bacteria lysed; but the increase in activity is specific to the organism dissolved and does not extend to others. These facts indicate that bacteria themselves contain lysozyme (*Fleming and Allison, 1922a, 1922b*). The occurrence of lysozyme both in microbes and in

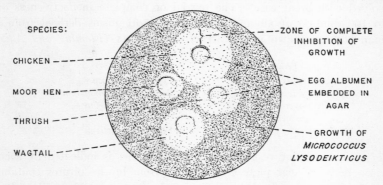

Fig. 268. Comparison of the ability of albumen from the eggs of several species of birds to inhibit the growth of *Micrococcus lysodeikticus* on an agar plate. (After Fleming and Allison, 1924.)

tissues suggests that its bactericidal action is incidental, and that its function is primarily metabolic. There is evidence that it is concerned with the depolymerization or hydrolysis of mucopolysaccharides (*Meyer, Thompson, Palmer, and Khorazo, 1934; Meyer and Hahnel, 1946*).

Egg albumen is one of the most potent sources of lysozyme. Visible lysis occurs when a suspension of *Micrococcus lysodeikticus* (an organism that is unusually susceptible to the effects of lysozyme) is incubated for 24 hours in egg albumen diluted 1 to 50,000,000 (*Fleming, 1922*). One per cent albumen, in salt solution, will produce complete lysis of *Micrococcus lysodeikticus* in 30 seconds at 50° C. (*Fleming and Allison, 1924*). However, on the basis of total protein content, the lysozyme activity of tears is somewhat higher than that of albumen (*Meyer and Hahnel, 1946*).

When a small amount of albumen is embedded in an agar plate, the growth of *Micrococcus lysodeikticus* is inhibited in a well-defined area surrounding the albumen. By this cultural method, it has been shown that the albumen in the eggs of several species of birds is bacteriolytic to varying degrees. The antibacterial power of the hen's albumen is greater than that of the thrush's, wagtail's, and moor hen's (Fig. 268).

It is comparable to that exhibited by the duck's albumen, although the latter perhaps has the greater bactericidal action in dilutions higher than 1 to 100. There is some variation in the lysozyme activity of hens' eggs, but this variation is less than the interspecies difference (*Fleming and Allison, 1924*).

Not all bacteria are equally susceptible to the lytic power of egg albumen. Airborne species are in general less resistant than those that may be isolated from the human body, although *Streptococcus fecalis* is dissolved with great ease. The partial or complete ineffectiveness of a 1 to 100 dilution of albumen upon certain organisms derived from the human body is shown in the accompanying lists (*Fleming and Allison,*

Partial Lysis	No Lysis
Bacillus anthracis	*Eberthella typhosa*
Pseudomonas aeruginosa	*Salmonella typhimurium*
Salmonella paratyphi	*Shigella dysenteriae*
Salmonella schottmuelleri	*Vibrio comma*
Shigella paradysenteriae	*Vibrio paracholerae*

1924). The varying resistance of several species is indicated by a comparison between agar plate counts made from broth cultures containing diluted albumen, and plate counts made from control broth cultures (*Fleming and Allison, 1924*).

	Bacterial Colonies	
Species	Albumen, 1:2	Control
Streptococcus fecalis	1	317
Eberthella typhosa	16	51
Salmonella schottmuelleri	31	89
Vibrio comma	32	57
Salmonella paratyphi	33	113
Shigella paradysenteriae	62	72

The resistance of some of the above organisms to the bacteriolytic action of egg albumen may perhaps be explained by the fact that bacteria become less susceptible to lysozyme when they are grown in proximity to body tissues containing this substance. They also become resistant when they are first grown in very high dilutions of egg albumen and later transferred to successively lower dilutions (*Fleming and Allison, 1927*). Figure 269 illustrates the difference between a resistant and a nonresistant strain, as shown by the agar plate counts made after incubating the organisms for 1 hour in various dilutions of albumen. The fact that strains that are resistant to the lysozyme of albumen are also much less susceptible to intracellular digestion by human leucocytes

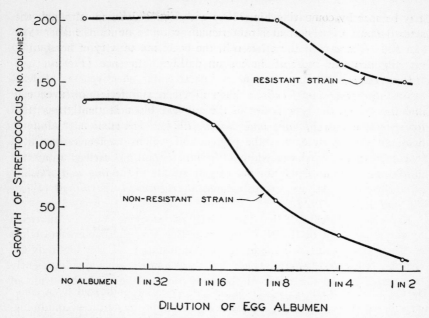

FIG. 269. Comparison of the ability of a resistant and a nonresistant strain of *Streptococcus* to withstand the bactericidal effect of egg albumen. (After Fleming and Allison, 1927.)

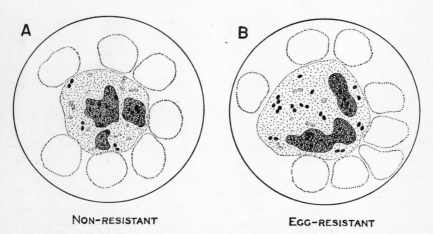

FIG. 270. Comparison of the susceptibility of *A*, nonresistant and *B*, resistant strains of *Streptococcus* to intracellular digestion by human leucocytes. Note the better staining properties of the cocci in *B*, and the large number of shadows in *A*. (After Fleming and Allison, 1927.)

may be seen by comparing Fig. 270-*A* with Fig. 270-*B*. Only a few well-stained cocci of a normal strain remain within one of the leucocytes pictured; whereas, in the other cell, the cocci are of a type resistant to egg albumen, and most of them stain darkly.

Ingested egg albumen retains its bactericidal effectiveness, at least against *Streptococcus fecalis*. When albumen is injected intravenously into the rabbit, the lytic power of the animal's blood is greatly increased for several hours (*Fleming and Allison, 1924*). Desiccation of albumen does not destroy lysozyme, the presence of which can be demonstrated in commercially dried egg white (*Fleming, 1922*). However, admixture of albumen with an equal amount of yolk results in the loss of practically all antibacterial activity, except against *Micrococcus lysodeikticus* (*Fleming and Allison, 1924*).

Experimental evidence that the activity of lysozyme could be increased by the addition of biotin led to the suggestion that the bactericidal effect of the former substance might be a manifestation of the biotin-avidin complex (*Meyer, 1944*). These findings, however, have subsequently been explained as the result of the autolytic tendencies of the strain of *Micrococcus lysodeikticus* that was used (*Meyer and Hahnel, 1946*). It has also been reported that lysozyme and avidin are interchangeable in activity (*Laurence, 1944*); and, again, that avidin has lytic properties, but that lysozyme produces none of the effects of avidin (*Meyer and Hahnel, 1946*). Resemblances between lysozyme and avidin may be due to the presence of both substances in relatively impure concentrates of either; for highly purified preparations of lysozyme and avidin do not possess any common activity. The identity of avidin and lysozyme is extremely doubtful, in view of their different isoelectric points and molecular weights (*Alderton, Lewis, and Fevold, 1945*). However, it is possible that they may be related, and that one may be derived from the other (*Meyer and Hahnel, 1946*).

Avidin, nevertheless, probably aids in inhibiting bacterial growth in egg albumen by depriving microorganisms of biotin (*Eakin, Snell, and Williams, 1940*). Bacteria requiring the presence of biotin in the culture medium are completely suppressed by avidin, whereas those that can synthesize biotin are not affected. It may be noted that among the latter are such organisms as *Salmonella enteritidis*, *Serratia marcescens*, *Aerobacter aerogenes*, and *Proteus vulgaris* (*Landy, Dicken, Bicking, and Mitchell, 1942*), all of which have been found in the egg.

That the shell membranes of the egg may have some germicidal effect is indicated by the rapid decrease in the numbers of bacteria found within a few hours after their penetration through the shell (cf. Fig. 265).

It is not known whether or not there is any specific substance in the shell membranes that might cause this phenomenon.

At pH 6.4 and above, egg albumen can completely inhibit various species of bacteria by making iron unavailable. This inhibitory, iron-binding constituent of albumen has been electrophoretically identified as ovoconalbumin (*Alderton, Ward, and Fevold, 1946*).

THE EGG IN BACTERIAL CULTURE MEDIA

Under certain circumstances, the contents of the egg are well able to support the growth of microorganisms. The egg, especially the yolk, furnishes an abundance of excellent nutriment. Certain species of bacteria cannot grow on simple culture media but thrive on media in which the egg is incorporated. For this reason, the egg is used in laboratories for enrichment purposes, much as are blood, body fluids, and various vitamins. It would probably be in extensive general use in bacteriological work were it not for the difficulty with which it is opened aseptically, and the inconvenience that may be caused by its coagulation during the sterilization of media.

As long ago as 1888, Hueppe reported that many microorganisms could be grown in the egg; and, in 1896, Capaldi used the yolk alone as a bacterial nutrient. In 1902, Dorset employed media containing the entire egg contents for the cultivation of the tubercle bacillus and noted development of colonies in 7 to 8 days after inoculation. Besredka and Jupille (*1914*) later observed not only that such organisms as *Diplococcus pneumoniae, Eberthella typhosa* and various *Streptococci* grew well in a mixture of meat broth and egg, but also that the human and bovine strains of the tubercle bacillus could easily be differentiated in this medium. Further investigation resulted in the discovery that tubercle bacilli could be cultivated in a medium consisting only of yolk, diluted with water and treated with a soda solution (*Besredka, 1921*).

It is now common practice to incorporate egg yolk, or the entire egg contents, into culture media for the tuberculosis organism. In such media, the egg has been used in combination with a great many different substances. There is convincing evidence, however, that it provides sufficient nourishment alone, without the addition of broth, agar, potato, cream, or other materials, except water (*Corper and Cohn, 1933*). Apparently the egg's growth-promoting factor for *Mycobacterium tuberculosis* is contained in the yolk lipids. When the alcohol-ether extract of egg yolk is added to a synthetic agar medium, in proportions equal to the concentration naturally present in the yolk, tubercle bacilli grow as well as on yolk media. The unique growth-promoting substance is entirely lacking in the albumen. Albumen, when coagulated, is nevertheless a fairly good nutrient for the tubercle bacillus (*Boissevain and Schultz, 1938*). In fact, a medium containing albumen and asparagin has been found valuable in detecting the presence of the tuberculosis organism in cerebrospinal fluids that appear negative on direct examination (*Saenz and Costil, 1932*). It is also said that *Mycobacterium tuberculosis* grows well throughout the entire contents of the intact, infertile egg, when inoculated through the air cell (*Gelarie, 1944*).

Egg yolk media have also been used successfully for the culture of *Neisseria gonorrhoeae* (*Kawamoto, 1936*). A mixture of one part of a suspension of unheated egg yolk and five parts of a base medium has been found valuable in studying *Achromobacter larvae* (*White, 1919*). A medium upon which *Hemophilus gallinarum* grows well may be made by incorporating one egg yolk in 100 cc. of nutrient agar, containing 0.8 per cent sodium chloride (*Cunningham and Stuart, 1944*). In addition, certain fungi which apparently require simple organic nitrogenous compounds can utilize egg albumen in the presence of growth-promoting substances, or "bios" (*Farries and Bell, 1930*).

Because of its avidin content (see Chapter 9, "Food Value"), egg albumen is useful in differentiating microorganisms on the basis of their biotin requirements (*Landy, Dicken, Bicking, and Mitchell, 1942*).

TRANSMISSION OF DISEASES

The egg may occasionally harbor and transmit bacteria that cause disease. It is guilty chiefly of disseminating various afflictions of poultry and other birds, domesticated and wild. It has also been implicated, to a certain extent, in the spread of contagion among human beings, and it therefore assumes some significance from the viewpoint of public health.

Avian Diseases

Several infections of birds may be transmitted by way of the egg. Pullorum disease is the most important. It is also possible for avian tuberculosis to be egg-borne, but contagion by contact is far more usual. There are various other diseases, of bacterial and viral origin, whose dissemination by means of the egg is known or suspected.

PULLORUM DISEASE

Pullorum disease (bacillary white diarrhea) has been of widespread occurrence throughout the world. It affects not only chickens but also turkeys, pigeons, geese, guinea fowl, peafowl, canaries, turtledoves, and various wild birds (*Bunyea, 1942a*). It became especially prevalent among poultry with the advent of large-scale artificial incubation. Although it is now fairly well under control, it has caused enormous losses to the poultry industry because of its high mortality rate, which is sometimes 90 per cent or more.

The disease usually manifests itself in chicks within a month after hatching, often sooner. Its symptoms include listlessness, loss of appetite, subnormal temperature, progressive weakness, and diarrhea of light color and sticky consistency. Death may occur suddenly or after a prolonged period of increasing debility. Large numbers of "dead-in-shell" embryos may indicate the presence of pullorum infection.

Pullorum disease may be transmitted by contact with infected birds (*Rettger, Kirkpatrick, and Card, 1919*) and by the ingestion of food contaminated with the causative organism, *Salmonella pullorum* (*Rettger and Stoneburn, 1909*). The egg, however, is largely responsible for transmitting the disease from one generation to the next and thus perpetuating a cycle of infection. Chicks that have survived an attack of the disease may grow to maturity and harbor *Salmonella pullorum* in their tissues and organs, including the ovary (*Rettger and Stoneburn, 1911; Rettger, 1914*). The bacteria thus find their way into the egg, where they multiply during incubation. The embryo becomes infected and either dies before hatching or hatches out as an infected chick. The disease spreads rapidly by contact even within the incubator, where down from the chick aids in disseminating it (*Hinshaw, Upp, and Moore, 1926*). An incubator that has contained pullorum-infected eggs may spread the contagion to the next hatch (*Bunyea and Hall, 1930*).

Fortunately, hens that are carriers of *Salmonella pullorum* are usually poor producers (*Rettger and Stoneburn, 1911*). Furthermore, the organism is present in only a small proportion of their eggs, probably 5 to 7 per cent (*Kaupp and Dearstyne, 1927; Cunningham, 1944*). Occasionally, however, an individual bird may lay eggs in which the incidence of infection is as high as 35 per cent (*Tittsler, 1927*) or 50 per cent (*Runnells and Van Roekel, 1927*).

Detection of carriers may easily be accomplished by means of the agglutination test. The prompt elimination of all birds that react positively should eventually result in greatly decreasing the incidence of pullorum disease.

AVIAN TUBERCULOSIS

Tuberculosis is fairly prevalent among domesticated birds, including chickens, and among wild birds in captivity. It is caused by *Myco-*

bacterium avium, which is morphologically very similar to *Mycobacterium tuberculosis hominis* but different in cultural characteristics, optimum growth temperature, and pathogenicity for certain laboratory animals. The organism's usual mode of entrance is by way of the bird's alimentary canal. Avian tuberculosis is rarely of the pulmonary type and is ordinarily manifested by lesions and tuberculous nodules in various internal organs, including the ovary (*Liverani, 1934*). The occasional presence of tubercle bacilli in the egg is thus explained. However, tuberculous hens do not lay well (*Fitch, Lubbehusen, and Dikmans, 1924*). In addition, the tuberculosis organism seldom makes its way into more than a very small proportion of their eggs. Examinations of large numbers of eggs from infected fowl have sometimes failed entirely to reveal the presence of the organism (*Bonnet and Leblois, 1939*); *Hülphers, 1939*). In other instances, the incidence of infection has been variously found to be 0.3 to 0.4 per cent (*Beller and Henninger, 1930; Lichtenstein, 1932*), 5.7 per cent (*Klimmer, 1932*), 10 per cent (*Klimmer, 1931*), and 13.9 per cent (*Stafseth, Biggar, Thompson, and Neu, 1934*); it has even been reported to be as high as 48.4 per cent (*Raebiger, 1929*).

The egg probably plays but a small part in the dissemination of tuberculosis among birds. If infected eggs are incubated, hatchability is lower than normal; and the chicks that hatch, although tuberculous (*Sibley, 1890*), usually do not live long enough to spread the disease (*Fitch and Lubbehusen, 1928*).

OTHER DISEASES

In addition to *Salmonella pullorum,* several other members of the *Salmonella* group of bacteria cause diseases in birds. One of these infections is fowl typhoid, which occurs among chickens, turkeys, pigeons (*Bunyea, 1942b*), and guinea fowl (*Johnson and Anderson, 1933*) and which is produced by *Salmonella gallinarum.* This organism has been found in the ovary of the hen and also in the unabsorbed yolk sac of the chick (*Beaudette, 1925; Beach and Davis, 1927*). It therefore seems possible that the egg may be an agent of transmission, although not the principal one (*Doyle, 1926*).

Other *Salmonella* diseases are the so-called paratyphoid infections. One of the most serious and widespread of these is "keel disease," which affects young ducklings and is often fatal. This is primarily an intestinal disturbance, with severe thirst as a prominent symptom. After drinking, ducklings sometimes draw themselves to full height, stagger, fall to the ground, and die. The infective agent is *Salmonella anatis,* which may be transmitted through the egg, much as is *Salmonella pullorum* (*Rettger and Scoville, 1920*).

Salmonella typhimurium (or *Salmonella aertrycke*), also a paratyphoid organism, may infect almost all species of domesticated birds. Its symptoms are chiefly gastrointestinal. Ducklings are especially prone to epidemics of *Salmonella typhimurium* origin. This organism may be harbored in the duck's ovary and can pass through the pores of the egg-shell (*Wesselmann, 1936*). It is often found in ducks' eggs, which, when infected, usually fail to hatch (*Dalling and Warrack, 1932; Scott, 1932*). It has been demonstrated in dead-in-shell turkey embryos and in the ovary of turkey hens (*Lee, Holm, and Murray, 1936; Cherrington, Gildow, and Moore, 1937*), as well as in the unabsorbed yolk of a squab (*Beaudette, 1926*). It has been found both on the shells and in the contents of eggs laid by turkeys to which it was fed experimentally (*Gauger and Greaves, 1946*). If it is present in fecal contamination on eggshells, it can apparently infect chicks in the incubator, after hatching (*Schalm, 1937*). *Salmonella enteritidis*, a third paratyphoid organism, often found in ducks' eggs, also can penetrate into the egg from the outside of the shell (*Warrack and Dalling, 1933*). In view of the above evidence, the egg may be considered a possible agent for the dissemination of both *Salmonella typhimurium* and *Salmonella enteritidis* among birds.

Lymphomatosis, or leukosis (fowl paralysis), is a virus disease that manifests itself as a complex syndrome. Paralysis and iritis are among its symptoms. There is considerable evidence indicating that infection is spread not only by contact but also through the agency of the egg, for the disease may appear among birds that are hatched and reared in complete isolation. In one instance, approximately 23 per cent of an isolated flock died from lymphomatosis before attaining the age of 2 years (*Waters, 1945b*). Iritis appeared in nearly 20 per cent of another large group of chickens that had no contact with diseased birds up to 15 months of age; furthermore, it occurred in over 40 per cent of those individuals whose parents were infected, but in only 26.7 per cent of the progeny of leukosis-free parents (*McClary and Upp, 1939*).

It has been suggested that another virus disease, avian encephalomyelitis, may be egg-borne. The inoculation of chicks with material from the unabsorbed yolk sacs of infected birds does not produce typical clinical symptoms, but it nevertheless results in brain lesions histologically resembling the characteristic lesions of encephalomyelitis (*Jungherr and Minard, 1942*). The virus of Newcastle disease has also been found in the ovary (*Farinas, 1930*) and in the egg yolk (*Picard, 1928*); but it is not known whether or not the disease is spread by the incubation of infected eggs.

Relationship to Public Health

Foodstuffs are well-known disseminators of bacterial and parasitic diseases. The egg, like various fruits, nuts, and vegetables that are similarly enclosed in a protective covering, is much less guilty of transmitting pathogenic organisms than the majority of foods commonly consumed by man; in general, it is probably one of the safest articles in his diet. However, it has been responsible for many individual cases and mass outbreaks of "food poisoning."

BACTERIAL FOOD POISONING

Many members of the *Salmonella* group of bacteria cause "food poisoning." There are approximately 100 different species of *Salmonellae*, closely resembling each other and distinguishable only if serological as well as cultural methods are employed. They are distinctly dangerous organisms. The list of *Salmonellae* responsible for serious and often fatal infections is continually growing. Although gastroenteritis and enteric fever are the most frequent manifestations of salmonellosis, various species of *Salmonellae* are known to have caused septicemia, endocarditis, osteomyelitis, pleurisy, peritonitis, meningitis (*Bornstein, 1942*), cholecystitis, appendicitis, and salpingitis (*Seligmann and Hertz, 1944*). *Salmonellae* are particularly insidious because their presence in food is usually undetectable; they produce little or no change in odor or appearance.

Ducks' eggs are implicated in food poisoning more often than eggs of other domestic birds, because they more frequently contain *Salmonella* organisms. Ducks are prone to lay their eggs in wet places near ponds and stagnant water, which may be contaminated. The shells are usually moistened and dirtied before the eggs are gathered, and infection of the contents is thus favored. Ducks' eggs are not eaten to any great extent in the United States. It is chiefly in European countries that their consumption has produced *Salmonella* poisoning, which has its highest incidence in the summer (Fig. 271). Foods such as custards, ice cream, mayonnaise, soups, and pies, that contained ducks' eggs, have caused severe outbreaks in Great Britain (*Scott, 1930, 1932; Savage, 1932*) and on the continent (*Fromme, 1933, 1934a, 1934b; Fürth and Klein, 1933; Clarenburg and Pot, 1935*) as well as in the United States (*Brown, Combs, and Wright, 1940*). In these cases, the *Salmonellae* involved were *Salmonella enteritidis* and *Salmonella typhimurium*, the latter of which is the species most frequently encountered in human salmonellosis

(*Bornstein and Schwarz, 1942*). *Salmonella anatis,* not generally thought to be pathogenic to man, is also on record as having caused food poisoning at least once (*Kauffmann and Silberstein, 1934*).

Hens' eggs are, in general, much more free of *Salmonella* organisms than those of ducks. Chickens prefer a dry environment; they are less

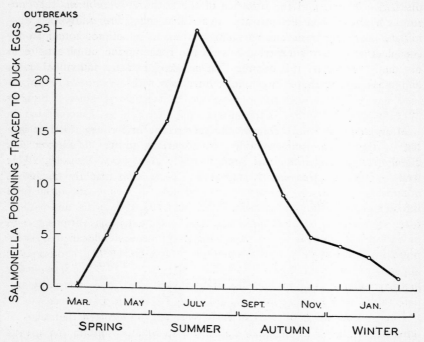

Fig. 271. Seasonal distribution of 115 outbreaks of poisoning caused by foods containing ducks' eggs, and from which living organisms of the *Salmonella* group (chiefly *Salmonella typhimurium*) were isolated. (After Savage, 1932, from records of the British Isles, 1919–1931.)

subject to *Salmonella* infections of the paratyphoid group than are ducks; and their eggs are laid in surroundings where shell contamination with *Salmonellae* is not particularly likely to occur. An examination of 1600 hens' eggs revealed no species of *Salmonella* other than *Salmonella pullorum* and *Salmonella gallinarum* (*Simeone, 1944*). Probably neither of these types is pathogenic to man, although *Salmonella pullorum* produces severe gastrointestinal disturbances in kittens and rabbits (*Rettger, Hull, and Sturges, 1916; Rettger, Slanetz, and McAlpine, 1926*). Upon one occasion, *Salmonella gallinarum* was recovered from the feces of an individual suffering from gastroenteric fever (*Müller, 1933*). The pathogenicity of *Salmonella gallinarum* var. *duisburg* has also been demonstrated (*Kauffmann, 1934*). In another instance, food poisoning in twenty-eight individuals was traced to mayonnaise made with hens' eggs that were infected with *Salmonella montevideo* (*Watt, 1945*). *Salmonella typhimurium* has been isolated from the hen's egg in pure culture (*Beller and Zeki, 1934*), and the presence of the pathogenic *Salmonella columbensis* has also been demonstrated in chicken eggs (*Braga, 1939*).

Salmonella typhimurium var. *copenhagen* is sometimes found in pigeons' eggs. Gastroenteritis following the consumption of pigeons' eggs is usually caused by this organism (*Clarenburg and Dornickx, 1932*).

The possibility that many human infections with *Salmonella* organisms may have been derived from eggs must be considered, since a number of *Salmonellae* that are pathogenic to man have been demonstrated in fowl and may therefore conceivably be present in the egg. Below is a list of these organisms, compiled from various sources (*Jungherr and Clancy, 1939; Bornstein and Saphra, 1942; Darby and Stafseth, 1942; Monteverde and Simeone, 1944a, 1944b; Seligmann and Hertz, 1944*).

S. aberdeen	S. kentucky	S. oregon
S. amersfoort	S. london	S. panama
S. bareilly	S. maleagridis	S. paratyphi
S. bredeney	S. manhattan	S. saint paul
S. chester	S. minnesota	S. san diego
S. cholerae suis	S. montevideo	S. schottmuelleri
S. derby	S. muenchen	S. senftenberg
S. eastbourne	S. newington	S. urbana
S. give	S. newport	S. wichita
S. hvittingfoss	S. oranienburg	S. worthington

Although there is no definite evidence that the egg has acted as the vehicle for transmitting any of these species to man, contamination of the eggshell with infected feces may easily occur and be followed by penetration of the bacteria into the egg contents. In fact, it has been experimentally demonstrated that hens, when fed *Salmonella bareilly*, soon start to excrete the organism, which may then be present on the shells of their eggs (*Gibbons and Moore, 1946*).

Food poisoning may also be caused by certain *Staphylococci*. The egg has never been involved directly in cases of staphylococcal poisoning, but it is an ingredient of many dishes, such as custards, to which this type of infection has been traced. Refrigeration is not particularly effective in suppressing the growth of *Staphylococci* in a food that provides a favorable environment. Reheating the food in an oven for 25 minutes at 190° C., or for 20 minutes at 232° C., usually kills the bacteria (*Stritar, Dack, and Jungewaelter, 1936*).

TUBERCULOSIS

The susceptibility of man to avian tuberculosis is somewhat doubtful; in general, he appears to be quite resistant. However, human infection with *Mycobacterium avium* has been reported a number of times (*Klimmer, 1930*). It is possible that, in certain individuals, this organism can adapt itself to the temperature of the human body (which is lower than that of birds) and retain its virulence. If the alleged instances of avian tuberculosis in human beings were indeed such, it cannot be stated whether or not the disease was contracted after eating infected eggs. The usual slow course of tuberculosis necessarily causes difficulty in tracing the source of infection.

Hülphers (*1939*) found that hens contracted tuberculosis when inoculated with material from infected eggs boiled 3 minutes, whereas 4 minutes of boiling apparently rendered the eggs sterile. According to Löwenstein (*1925*), the avian tubercle bacillus remains viable in the albumen after 3 minutes of cooking, and in the yolk after 5 to 10 minutes. If an individual were susceptible to *Mycobacterium avium,* possibly the ingestion of insufficiently cooked eggs could lead to infection.

The likelihood of this occurrence is greatly reduced by the fact that a very low percentage of market eggs contains the organism of avian tuberculosis. As previously stated, the tubercle bacilli are deposited in only a few of the eggs laid by tuberculous hens, which are poor producers. However, upon occasion, the organism retains its virulence in the egg for lengthy periods of time, perhaps from 4 to 5 months (*Verge, 1934*). After 5 or 6 months, it is apparently no longer viable (*Eber, 1932*).

Hogs are quite susceptible to infection with the avian type of the tubercle bacillus (*Klimmer, 1930*). If consumed raw, eggs from tuberculous hens are potentially more dangerous to swine than to human beings.

HORMONES AND ENZYMES OF THE EGG

A full complement of hormones and enzymes, substances that regulate and activate physiological reactions, is not necessarily present in the new-laid egg. For embryonic development to occur, certain enzymes must exist in the fertile egg, until such a time as the embryo itself can elaborate them. A number of enzymes and hormones may be found in the active state in eggs that have not been incubated, and in which there presumably has been no development. Additional enzymes, and hormones as well, appear upon incubation; others possibly disappear.

HORMONAL ACTIVITY OF THE YOLK

A number of investigations have indicated the presence of estrogenic hormone in the yolk of the chicken egg. According to Marlow and Richert (*1940*), there are 5 rat units of estrogenic material in 1 kg. of yolk. Subcutaneous injection of yolk lipids into rabbits appears to exert the same positive influence on uterine growth as lipoids derived from

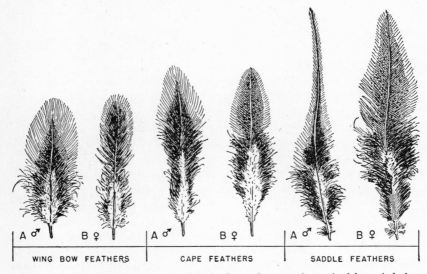

A ♂	B ♀	A ♂	B ♀	A ♂	B ♀
WING BOW FEATHERS		CAPE FEATHERS		SADDLE FEATHERS	

FIG. 272. Wing bow, cape, and saddle feathers of an ovariectomized hen, *A*, before, and *B*, after yolk injections. All types of plumage assume female characteristics after the injections. (After Kopeć and Greenwood, 1929.)

mammalian reproductive organs (*Fellner, 1925*). In one instance, injected egg yolk produced reversion to the female type of plumage in an ovariectomized Brown Leghorn hen, many of whose feathers were typically male in shape and color. About 5 weeks after the last of five injections, female plumage characteristics began to replace the male and became increasingly apparent during the following 5 weeks. Altogether, 6 or 7 months passed before the effects of the treatment disappeared. In Fig. 272 may be seen the changes which occurred in the shape of wing bow, cape, and saddle feathers (*Kopeć and Greenwood, 1929*).

Operative removal of yolks from the ovary significantly lowers the high blood calcium level associated with the laying condition in the hen. The intraperitoneal injection of 50 to 100 cc. of egg yolk, like the injection of progynon-B or theelin, induces a pronounced increase in the blood calcium concentration both of immature pullets and of capons (*Altmann and Hutt, 1938*).

Egg yolk, administered orally, sometimes produces an effect similar to that of prolan. Growth and function of testes and ovaries were stimulated in month-old rats fed a quarter of a yolk daily for 10 days (*Iwata, Susuki, and Iwata, 1941*).

It has also been suggested that egg yolk contains a substance that stimulates the thyroid. The weight and activity of this gland may increase when yolk is fed (*Tsuji, 1922*).

In addition, the oral or parenteral administration of an extract of chicken egg yolk may produce hypoglucemia and the disappearance of glycogen from the liver. This effect resembles that of insulin, but it is slower to appear and endures for a longer time (*Shikinami, 1928*). By feeding ten egg yolks in the morning, it has been found possible to decrease the blood sugar of diabetics by 30 mg. per cent in 4 to 5 hours (*Holland, Hinsberg, Kohls, and Nickel, 1934*). The hypoglucemic principle of the yolk is destroyed by heating at 110° C. for 1 hour (*Shikinami, 1928*). Ovolecithin alone may have hypoglucemic effects (see "Biological Effects of Cholesterol and Lecithin," p. 567), although this fact does not necessarily preclude the presence, in the yolk, of a hormone resembling insulin.

ENZYMES OF THE EGG

It is well known that a great many enzymes make their appearance in the developing egg. A smaller number of these substances has been demonstrated in the unincubated egg, and with less certainty. Some reports of investigations on egg enzymes have not indicated whether fertile or infertile eggs were used, and therefore the possible effects of embryonic development during the egg's passage through the oviduct may not have been taken into consideration.

Koga (*1923*) states that there is diastase in both the albumen and the yolk, and that the yolk contains approximately thirty-two times as much of the enzyme as the albumen. He also demonstrated a lecithinase and a monobutyrase in the yolk, and a small concentration of the latter enzyme in the albumen. On the other hand, he showed a tributyrase to be active in the albumen to a greater extent than in the yolk. A lipase that attacks ethyl butyrate was found by Pennington and Robertson (*1912*) in samples of the mixed contents of infertile eggs, although in negligible concentration until after the onset of deteriorative processes. Phosphatase activity is said to occur in the yolk, but not in the albumen, of infertile eggs (*Thompson, 1944*).

The albumen apparently contains both polypeptidases and dipeptidases, whose activity is increased after holding the albumen for several days at *p*H 8 (*Schultze, 1943*). According to van Manen and Rimington (*1935*), two erepsins are present in the albumen, one with a *p*H optimum of 5.5, the other with a *p*H optimum of 7.0 to 8.0. Koga (*1923*) believes erepsin to be present in the yolk and possibly in the albumen, and he also states that the albumen has very weak fibrinolytic activity. No true proteinase has been found in the yolk (*Koga, 1923*), whole egg contents (*Pennington and Robertson, 1912*), or the thick or thin portion of the albumen (*van Manen and Rimington, 1935*), although Thompson (*1944*) suggests that albumen may contain trypsin because of its ability to accomplish partial digestion of collagen fibers. It is well established, on the other hand, that egg albumen, like blood serum (*Purjesz and Weiss, 1925*) is able to resist tryptic digestion (*Vernon, 1904; Sugimoto, 1913*), but there is no evidence that the inhibitors in albumen and serum are identical (*Balls and Swenson, 1934b*). The inhibitor in albumen is thought to be an antienzyme that prevents activation of trypsin by combining either with kinase (*Delezenne and Pozerski, 1903*), or with both kinase and inactive trypsin (*Balls and Swenson, 1934b*). It is destroyed by continued boiling (*Balls and Swenson, 1934b*), but not by dehydration (*Cohn and White, 1935*). The three layers of the albumen possibly differ in their content of the trypsin-inhibiting principle, which appears to be relatively lacking in the dense albumen (*Balls and Swenson, 1934a*) and chiefly concentrated in the inner thin portion (*Hughes, Scott, and Antelyes, 1936*). However, a phosphate buffer extract of the dense albumen inhibits the action of trypsin upon casein and gelatin (*Pace, 1937*). Albumen also prevents the liquefaction of gelatin by culture filtrates of such bacteria as *Clostridium sporogenes* and *Clostridium bifermentans* (*Pozerski and Guélin, 1938*).

Various other enzymes may be present in the egg. Catalase has been found in the albumen, and in lower concentration in the yolk (*Pennington and Robertson, 1912; Rullman, 1916*). The yolk apparently contains a salicylase and a histozyme, and the albumen an oxidase that forms a brown pigment from pyrocatechin, adrenalin, and dioxyphenylalanine (*Koga, 1923*). The presence of lysozyme in the albumen has already been discussed (see "Defenses of the Egg against Microorganisms," p. 498).

HYDROLYSIS BY ENZYMES

In discussing the biological activity of the egg, it is important to consider the effect of various enzymes upon the egg's constituents. The nutritional value of the egg, as of any food, is influenced by the extent to which its components can be broken down and made ready for absorption; and the role of enzymes in this physiological process is well known. Similarly, some of the more or less toxic and pathological effects of the egg may be explained, at least partially, by the egg's intrinsic capacity to resist enzymatic action under certain conditions.

In their natural environment, enzymes act under the influence of a multitude of factors, and the reactions that they catalyze are never isolated phenomena. By contrast, laboratory experiments with enzymes are highly artificial. However, the enzymatic hydrolysis of the egg or its individual constituents in the test tube is at least similar to, if not identical with, the course of digestion in the body. The complex nature of natural enzymatic processes may therefore be studied by observing reactions *in vitro*. In addition, hydrolysis by enzymes, like that by acids or alkalies, provides a means for investigating the composition of the egg.

Since only one enzyme was obtained in the pure crystalline form prior to 1930, there has been little time for intensive investigation of the nature and action of individual enzymes. Furthermore, methods for the isolation of the chemical constituents of the egg are far from perfect. For these reasons, some of the results and conclusions reported in the following pages may appear to be somewhat uncertain and, occasionally, even conflicting.

PEPSIN

Hydrolysis of Albumen

The peptic digestion of raw egg albumen *in vitro*, when prolonged, is not strictly comparable to its digestion *in vivo*. Raw albumen characteristically remains in the stomach only a short time, and the action of pepsin upon it is therefore considerably restricted. In the test tube, native albumen (0.5 per cent) may appear to be unaffected by pepsin (0.1 per cent) after 90 minutes of incubation, if digestion is measured by titration values. A change is evidently produced, however, for preliminary peptic digestion of albumen greatly facilitates the hydrolysis of albumen by trypsin (*Cohn and White, 1935*). Under different experimental conditions, raw albumen may be digested by pepsin with ease (*Calvery, Block, and Schock, 1936*). Determination of the increase in nitrogen uncoagulable by heat and acid has shown that appreciable digestion occurs *in vitro* during incubation for only 1 hour. By the same method, it has been demonstrated that pepsin attacks coagulated albumen far more readily than the uncoagulated substance, as revealed by the tabulated data (*Harte, 1945*). The course of the peptic hydrolysis of

DIGESTION TIME (hours)	INCREASE IN UNCOAGULABLE N (per cent)	
	RAW ALBUMEN	COAGULATED ALBUMEN
1	61	98
2	88	127
4	138	195

coagulated albumen during 160 hours is shown in Fig. 273, where it may be seen that the rate of digestion is most rapid during the early stages

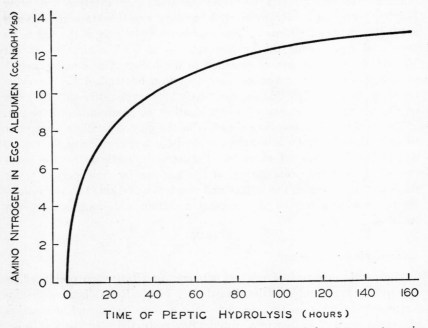

Fig. 273. Hydrolysis of coagulated albumen (5 per cent) by 1 per cent pepsin during 160 hours' incubation at pH 2.0. (After Young and Macdonald, 1927.)

of the reaction. When hydrolysis of 5 per cent albumen by 1 per cent pepsin is continued for 10 days, about 25 per cent of the total nitrogen is finally converted into free amino nitrogen (*Young and Macdonald, 1927*).

Acidification of the digestive mixture with different acids affects the peptic digestibility of coagulated egg albumen. The use of hydrochloric acid results in the most extensive hydrolysis. Other acids, in concentrations equinormal to 0.2 per cent hydrochloric acid, have been listed as follows in the order of decreasing effectiveness: nitric, oxalic, phosphoric, sulfuric, tartaric, lactic, citric, acetic, and boric (*Berg, 1909*). When digestion occurs in the presence of hydrochloric acid, the concentration of chloride ions remains almost unchanged, but the concentration of hydrogen ions decreases. Apparently the hydrolytic products have a greater combining capacity for hydrogen ions than does the original protein (*Rohonyi, 1912*).

HYDROLYSIS OF OVALBUMIN

The peptic digestion of crystalline ovalbumin has been studied with greater quantitative exactitude than that of crude albumen. The speed

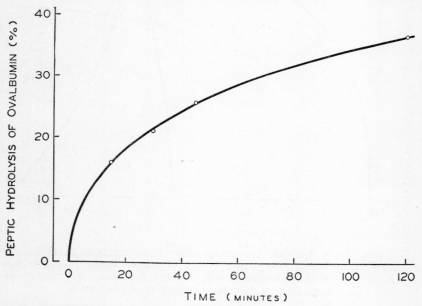

FIG. 274. Hydrolysis of 3.2 per cent native ovalbumin during 2 hours' incubation with 0.2 per cent pepsin at 37.7° C. and pH 1.6. (After McFarlane, Dunbar, Borsook, and Wasteneys, 1927.)

and completeness of the process is known to depend upon time, temperature, enzyme and substrate concentration, and the pH of the solution. These are among the usual factors that influence all enzyme reactions.

Proteolysis by pepsin is characteristically most rapid initially, whatever the substrate. Figure 274 shows that the rate of digestion of a 3.2 per cent solution of ovalbumin by pepsin (0.2 per cent) is greatest during the first 20 minutes and is subsequently retarded. During the first 2 hours, over 35 per cent of the protein is hydrolyzed; and, within 4 hours from the start, the digestion of two-thirds of the substrate is accomplished. The additional ovalbumin digested in the next 8 hours is little more than 20 per cent of the amount originally present. Hydrolysis of the remaining 10 per cent of the protein occurs at such a slow rate that over 2 weeks are required to complete the digestive process. Under various experimental conditions, the extent of hydrolysis and the

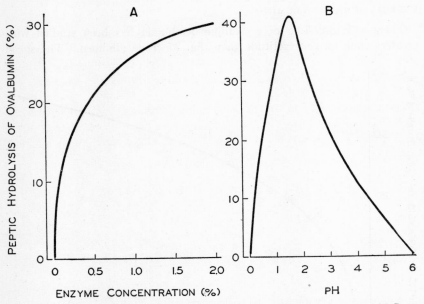

FIG. 275. Peptic hydrolysis of native ovalbumin (3.2 per cent) at 37.7° C., as influenced, A, by enzyme concentration (during 45 minutes' incubation at pH 1.6); and B, by the pH of the solution (during 1 hour's incubation with 0.2 per cent pepsin). (After McFarlane, Dunbar, Borsook, and Wasteneys, 1927.)

time necessary to reach equilibrium may differ, but the speed of the reaction is always greatest at the start.

Pepsin is most active from pH 1.5 to pH 2.0. The maximum effect of this enzyme upon ovalbumin is observed approximately at pH 1.6, as shown in Fig. 275-B.

As Fig. 275-A indicates, the higher the concentration of pepsin, up to 2.0 per cent, the more extensive is the hydrolysis of a 3.2 per cent solution of ovalbumin during 45 minutes of incubation. However, the relationship is not a direct proportion, and the effect of increasing pepsin concentration is limited. This fact is made clear by the tabulated data, obtained when pepsin in various concentrations was allowed to act on a 4.5 per cent ovalbumin solution for 1 hour (*Morrell, Borsook, and*

Pepsin Concentration (per cent)	Amount of Hydrolysis (milligrams of N/100 cc.)
2.1	194
4.2	220
6.0	220
7.0	210

Wasteneys, 1927). If enzyme concentration is sufficiently high, hydrolysis may be retarded.

Increasing the concentration of ovalbumin may also inhibit hydrolysis. When the effect of varying the enzyme concentration is eliminated by using an excess of enzyme with all concentrations of substrate, the amount of hydrolysis, during 1 hour's incubation at 35° C., demonstrably decreases as the strength of the ovalbumin solution is increased (*Morrell, Borsook, and Wasteneys, 1927*). The influence of substrate concentration is even more evident when the ovalbumin has been coagulated by heat.

Ovalbumin Concentration (per cent)	Amount of Hydrolysis (per cent)
3	42.0
5	34.2
7	26.6
12	19.9

As shown in Fig. 276, the optimum temperature for the peptic digestion of coagulated ovalbumin, regardless of concentration, is about 60° C. Above this point, the inhibiting effect of the inactivation of pepsin overcomes the accelerating effect of increasing temperature. By comparing the curve for the digestion of 1.2 per cent ovalbumin with that for the digestion of 2.5 per cent ovalbumin, it may be seen that there is a relatively much smaller increase in the percentage hydrolysis of the more concentrated material as the temperature rises from 20° C. to 60° C.

PEPTIC SYNTHESIS OF OVALBUMIN

Most hydrolytic reactions catalyzed by enzymes are reversible. Some of the effects described in the preceding paragraphs are possibly explained by the ability of pepsin, under appropriate circumstances, to synthesize protein from the products of peptic hydrolysis. Whether or not the "plastein" formed is a true protein is not as yet entirely certain; nevertheless, hydrolysis would tend to be retarded if there were any shift in equilibrium toward the protein side of the reversible reaction.

The amount of plastein that pepsin synthesizes from a peptic digest at pH 4.0—the optimum pH for synthesis (*Wasteneys and Borsook, 1924*)—is in direct linear proportion to the concentration of hydrolytic

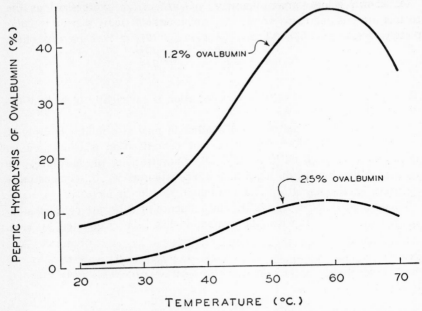

FIG. 276. Effect of temperature on the rate at which coagulated ovalbumin (1.2 per cent and 2.5 per cent) is hydrolyzed by 0.1 per cent pepsin during 1 hour at *p*H 1.6. (After Morrell, Borsook, and Wasteneys, 1927.)

products in the solution. In solutions corresponding to less than 8 per cent protein, however, synthesis does not occur at all; and, in very high concentrations, there is a falling-off from the linear relationship (*Borsook and Wasteneys, 1925*). It has been demonstrated that the addition of hydrolytic products to a solution of ovalbumin, before pepsin is allowed to act, inhibits the digestion of ovalbumin to an extent directly proportional to the amount added. The inhibition of hydrolysis by high enzyme concentrations may be due in part to the fact that proteoses and peptones are usually present in commercial pepsin (*Morrell, Borsook, and Wasteneys, 1927*).

In addition, high enzyme concentration perhaps retards digestion because of the more extensive synthesis that occurs when the strength of the enzyme solution is increased (*Wasteneys and Borsook, 1924*).

Pepsin Concentration (gram/10 cc.)	Protein N (per cent of total N of digest)
0.01	13.1
0.10	23.0
0.80	33.2

As shown below, larger quantities of protein are synthesized as the temperature is increased to 72° C., above which point pepsin is completely destroyed (*Borsook and Wasteneys, 1924*). Perhaps this effect

Temperature (°C.)	Protein N (per cent of total N)
13	1.5
21	14.5
37	25.3
72	30.0
80	0

of temperature upon synthesis is partly responsible for the inhibition of digestion observed at temperatures higher than 60° C., especially when substrate concentration is high.

Lengthy exposure of a peptic digest of ovalbumin to the action of pepsin, after all the original protein has disappeared, results in a diminution of the amount of protein that can later be synthesized. Apparently a slow secondary hydrolysis decomposes to a certain extent the complex necessary for synthesis (*Borsook, MacFadyen, and Wasteneys, 1930*).

PRODUCTS OF HYDROLYSIS

According to Tiselius and Eriksson-Quensel (*1939*), molecules of ovalbumin attacked by pepsin are broken down very rapidly to the end products of the reaction, with the formation of few, if any, intermediary split products. The remaining ovalbumin molecules are unaffected. Consequently, at any time during the peptic hydrolysis of ovalbumin, the digestive mixture consists of two fractions. One fraction is of high molecular weight and has the electrophoretic mobility and sedimentation velocity of unchanged ovalbumin. The second is of low molecular weight and is composed of the products of hydrolysis. Although the latter fraction increases in amount as hydrolysis proceeds, it does not change in nature, for its diffusion constant remains essentially the same throughout the entire course of digestion. Ultracentrifugal sedimentation studies show, however, that it is not entirely homogeneous, and that it probably consists of peptides of varying complexity.

Chemical analysis similarly indicates that at least 85 per cent of the end products of the peptic digestion of ovalbumin are derived directly from the ovalbumin molecule itself (*McFarlane, Dunbar, Borsook, and Wasteneys, 1927*). Proteoses, peptones, and subpeptones have been identified. During hydrolysis, the proteose fraction may decrease slowly from its maximum value and therefore probably undergoes a certain amount of secondary hydrolysis. A very slight eventual gain in the sub-peptone fraction perhaps indicates that peptone also is broken down to some extent (Fig. 277). According to Rudd (*1924a*), the proteoses may be separated into 4 fractions.

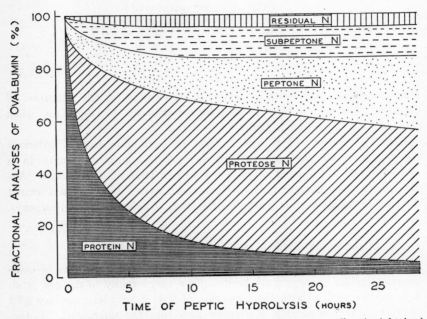

FIG. 277. Fractional analysis of a peptic digest of 5 per cent ovalbumin (obtained by incubation with 0.2 per cent pepsin at 37° C. and *p*H 1.6). (After McFarlane, Dunbar, Borsook, and Wasteneys, 1927.)

Dialysis through membranes of graded permeability has also permitted separation of a peptic digest of ovalbumin into fractions of varying composition, as shown by the tabulated data (*Desai, 1930*).

FRACTION	TOTAL N (per cent)	AMINO N (per cent)	PEPTIDE N (per cent)	NONAMINO N (per cent)
1	8.7	1.7	3.1	3.9
2	26.5	6.2	17.6	2.7
3	40.4	12.4	27.0	1.0
4	24.4	2.7	15.5	6.2

At the end point of the peptic digestion of ovalbumin (either native or heat-coagulated), 24 or 25 per cent of the total nitrogen of ovalbumin may be liberated as amino nitrogen, as determined by the Van Slyke method (*Desai, 1930; Calvery, 1933b*). Subsequent treatment by trypsin results in no further hydrolysis. Papain hydrocyanic acid liberates from the peptic digest 20 per cent of the total nitrogen as amino nitrogen; polypeptidases, 25 per cent. Since calculations have indicated that approximately one-third of the peptide linkages may be split by pepsin, it has been considered probable that the chief products formed are tripeptides (*Calvery, 1933b*). It is also thought that they may be peptides

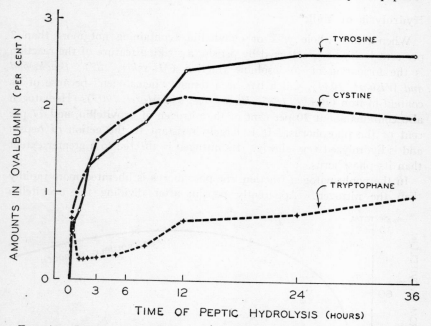

FIG. 278. Increase in percentages of tyrosine, cystine, and tryptophane, during peptic hydrolysis of ovalbumin. (After Calvery, Block, and Schock, 1936.)

containing 8 amino acids (*Tiselius and Eriksson-Quensel, 1939*). The presence of dipeptides in the peptic digest is indicated by the liberation of amino nitrogen by dipeptidase (*Calvery, 1935*). During exhaustive peptic hydrolysis, diketopiperazine nitrogen decreases from 37.5 per cent to 19.9 per cent of total nitrogen in a digest of native ovalbumin, but it increases from 11.2 per cent to 27 per cent in a digest of heat-denatured ovalbumin (*Blanchetière, 1927*). Figure 278 shows the changes in the amounts of cystine, tyrosine, and tryptophane during the peptic digestion of undenatured ovalbumin. The values (expressed as percentages of ovalbumin) increase most rapidly during the first 12 hours. At the end of 36 hours, the percentage of cystine is larger than that obtained by acid hydrolysis, but the percentages of tyrosine and tryptophane are lower than those produced by alkaline hydrolysis (*Calvery, Block, and Schock, 1936*). Peptic hydrolysis results in a volume contraction of 17.5 to 18.0 cu. mm. per millimole of amino nitrogen liberated; in alkaline hydrolysis, there is an increase in volume of about 14.5 cu. mm. per millimole. Different cleavage products, as well as different ionization, are apparently indicated (*Rona and Fischgold, 1933*).

Hydrolysis of Yolk

When either whole yolk or ovovitellin (containing not more than 2 per cent lecithin) is attacked by pepsin, a striking feature of the reaction is the formation of an insoluble fraction (*Miescher, 1871; Blackwood and Wishart, 1934*), which has been termed "hematogen" because of its content of iron (*Bunge, 1885; Hugounenq and Morel, 1905*). Hematogen also contains about 20 per cent of the nitrogen of ovovitellin, and 77 per cent of the phosphorus. It is largely resistant to the action of pepsin and is hydrolyzed very slowly. Its nitrogen is affected to a greater extent than its phosphorus.

In the nonhematogen fraction, the phosphorus is liberated more rapidly than the nitrogen. Apparently pepsin, after dividing the ovovitellin

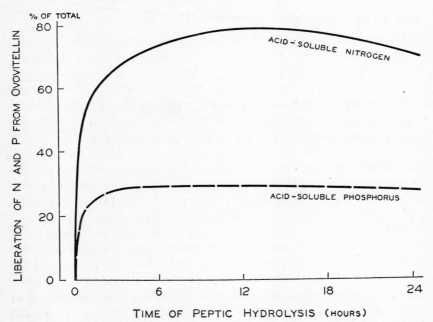

FIG. 279. Liberation of acid-soluble nitrogen and phosphorus from ovovitellin (0.75 per cent) during hydrolysis by pepsin. (After Blackwood and Wishart, 1934.)

molecule into hematogen and a predominantly nitrogenous fraction, very speedily frees all the phosphorus from the latter fraction by splitting peptide linkages near a phosphorus group. The nitrogenous residue is at first acid insoluble, but it is soon rendered soluble (*Blackwood and Wishart, 1934*). The final result of the peptic hydrolysis of ovovitellin is the liberation, in acid-soluble form, of 70 to 80 per cent of the total nitrogen, but of not more than 30 per cent of the phosphorus (Fig. 279).

RENNIN

According to Herzog and Margolis (*1909*), commercial rennin is capable of hydrolyzing ovalbumin to the extent of about 30 per cent. The course of the reaction resembles that of the digestion of ovalbumin by pepsin, although it proceeds much more slowly. Like pepsin, rennin acts with greater velocity as its concentration is increased up to a certain point. Beyond this point, the effect of a further increase in concentration is imperceptible. It is quite possible, however, that the preparation used by these workers contained pepsin as well as rennin.

TRYPSIN

Hydrolysis of Albumen

Trypsin does not digest raw albumen readily in the test tube, although a certain degree of hydrolysis is demonstrable after a time, coincidental, probably, to the disappearance of antitrypsin (*Dauphinee and Hunter, 1930*). When the criterion for the extent of digestion is the increase in the amount of nitrogen soluble in trichloroacetic acid, the action of trypsin is almost imperceptible during 24 hours' incubation at pH 7.4, with a substrate-enzyme nitrogen ratio of 60:1. Titration with 0.05 N sodium hydroxide, on the other hand, indicates that some hydrolysis may occur within the first 5 hours, when the substrate-enzyme ratio is 5:1 (*Cohn and White, 1935*). However, the course of digestion becomes

INCUBATION TIME (minutes)	HYDROLYSIS (cc. 0.05 N NaOH)
60	0.05
120	0.11
240	0.15
300	0.17

greatly accelerated when incubation with trypsin is preceded by a short period of incubation with pepsin. In fact, hydrolysis of the pepsin-treated albumen, at the end of 20 minutes, is more extensive than that of the untreated albumen at the end of 5 hours. A comparison of the following data with those given above makes clear the effect of pepsin (*Cohn and White, 1935*):

INCUBATION TIME (minutes)	HYDROLYSIS (cc. 0.05 N NaOH)
20	0.20
60	0.55
120	0.77
240	1.06
300	1.06

The tryptic digestibility of coagulated albumen is far superior to that of the native substance, as one may see by comparing the two curves shown in Fig. 280; and, in fact, digestibility is improved even by incomplete coagulation of albumen. Heating albumen at 100° C. for only 1.5 minutes, prior to incubating it with trypsin, results in a fourfold increase in the extent of hydrolysis occurring in 5 hours. As the preliminary heating period is extended to 30 minutes, the magnitude of hydrolysis becomes correspondingly greater. These facts are indicated by the

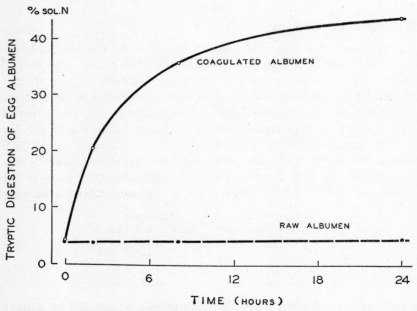

FIG. 280. Comparison of tryptic digestibility (*in vitro*) of raw and coagulated egg albumen, as shown by the percentage of total nitrogen converted to nitrogen soluble in 10 per cent trichloroacetic acid. (After Harte, 1945.)

tabulated titration values obtained at the end of 5 hours (*Cohn and White, 1935*). Exposure to a temperature of 100° C. for a time longer

HEATING PERIOD (minutes)	HYDROLYSIS (cc. 0.05 N NaOH)
0	0.17
1.5	0.68
5.0	1.15
30.0	1.22

than 30 minutes does not further the hydrolytic process. Apparently, boiling gradually destroys the tryptic inhibitor present in raw albumen and completes the inactivation of this antienzyme in half an hour (*Cohn and White, 1935*).

HYDROLYSIS OF OVALBUMIN

Purified ovalbumin, when subjected to the attack of trypsin, behaves much as does albumen; that is, native ovalbumin resists hydrolysis initially but eventually is broken down, whereas the denatured substance is readily digested from the beginning. These facts are illustrated in Fig.

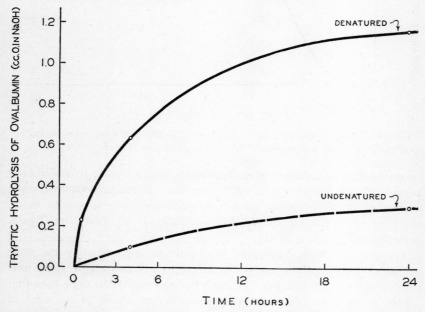

FIG. 281. Comparison of rates at which native and heat-denatured ovalbumin (2.5 per cent) are hydrolyzed by 0.1 per cent trypsin. (After Haurowitz, Tunca, Schwerin, and Göksu, 1945.)

281. Coagulated ovalbumin may be completely digested in 24 hours, but the hydrolysis of native ovalbumin may require 14 days under the same experimental conditions (*Calvery, 1933b*). Denaturation improves the ability of trypsin to hydrolyze not only ovalbumin, but also serum globulins; fibrous proteins, as opposed to globular proteins, are as readily hydrolyzed before denaturation as afterward. It has therefore been

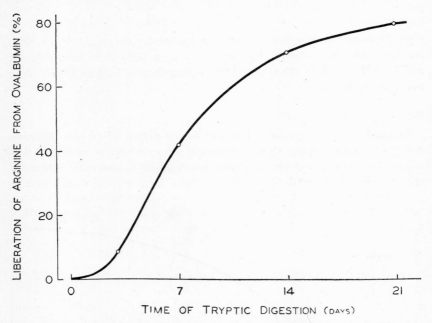

Fig. 282. Rate of liberation of arginine from 10 per cent ovalbumin during hydrolysis by 0.15 per cent trypsin. (After Dauphinee and Hunter, 1930.)

suggested that coagulation of ovalbumin facilitates tryptic hydrolysis not because of the destruction of an antitrypsin, but rather because the denaturation of this protein, as of any globular protein, results in an unfolding of peptide chains (*Linderström-Lang, 1939; Haurowitz, Tunca, Schwerin, and Göksu, 1945*).

Trypsin hydrolyzes coagulated ovalbumin to approximately the same extent as does pepsin; that is, according to calculations, it splits one-third of the peptide linkages (although not necessarily the same ones that pepsin attacks) and liberates as amino nitrogen about 25 per cent of the total nitrogen of the ash- and moisture-free substance. The carboxyl groups liberated equal the amino groups in number. There is no further action by pepsin on the trypsin split products of ovalbumin. Papain hydrocyanic acid, protaminase, polypeptidases, and erepsin respectively liberate an additional 25 per cent, 6 per cent, 35 per cent, and 48 per cent of total nitrogen as amino nitrogen. When acting successively, trypsin, protaminase, and carboxypolypeptidase probably split a total of 60 per cent of the peptide linkages (*Calvery, 1933b*). Crude preparations of pancreatic juice have been found to hydrolyze native ovalbumin to the extent of 70 (*Herzog and Margolis, 1909*) or 78 per cent (*Dauphinee and Hunter, 1930*) and may liberate leucine, isoleucine, and tryptophane after prolonged incubation (*Levene and Beatty, 1907a*); tyrosine, also, has been identified during the early stages of hydrolysis (*Brown and Millar, 1906*). Although approximately 80 per cent of the arginide links may finally be hydrolyzed by similar preparations of trypsin (*Dauphinee and Hunter, 1930*), the liberation of arginine is very slow initially (Fig. 282).

Trypsin, like pepsin, can resynthesize plastein from a concentrated enzymatic hydrolyzate (either peptic or tryptic) of ovalbumin. The products of tryptic and peptic synthesis are similar. The maximum yield of synthesized material is obtained at pH 5.8; when the solution is more alkaline, some hydrolysis also occurs (*Wasteneys and Borsook, 1925*). The rate of synthesis can be doubled if an emulsion of egg yolk lipoids is added at the start of the reaction (*Marston, 1926*).

Hydrolysis of Yolk

In 36 days, trypsin may convert into a soluble form about half of the phosphorus of ovovitellin containing lecithin (*Plimmer and Bayliss, 1905*). The tryptic hydrolysis of ovovitellin (with a maximum lecithin content of 2 per cent) may liberate as much as 70 to 80 per cent of the total nitrogen in acid-soluble form, and 70 per cent of the phosphorus. Whether there is preferential liberation of nitrogen or phosphorus depends upon the concentration of the enzyme. In the higher concentrations of trypsin, the liberation of nitrogen is favored, as indicated by the tabulated figures obtained at the end of 2 hours' incubation (*Blackwood and Wishart, 1934*). When 0.75 per cent trypsin is employed, phosphorus is liberated

TRYPSIN CONCENTRATION (per cent)	ACID-SOLUBLE N (per cent of total N)	ACID-SOLUBLE P (per cent of total P)
0.1	15.1	19.3
0.25	27.2	34.5
0.5	40.7	38.2
1.0	56.7	40.6
2.5	74.5	58.2
5.0	82.5	61.8

more rapidly than nitrogen, until about half of the total nitrogen and phosphorus have been rendered soluble in trichloroacetic acid. Thereafter, liberation of phosphorus may be slower than that of nitrogen (Fig. 283).

The insoluble hematogen formed upon peptic digestion of egg yolk is easily attacked by trypsin. Three polypeptides, the nitrogen-phosphorus

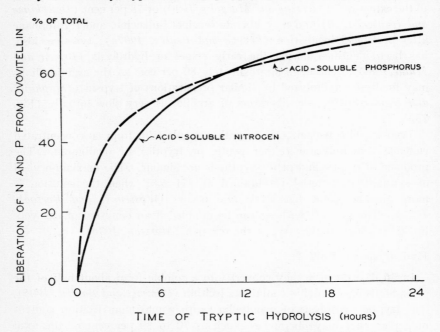

Fig. 283. Rate of liberation of acid-soluble nitrogen and phosphorus from ovovitellin (1 per cent) during hydrolysis by 0.75 per cent trypsin. (After Blackwood and Wishart, 1934.)

ratios of which are respectively 1.75, 2, and 3, have been isolated from a tryptic digest of hematogen (*Posternak and Posternak, 1927*).

CHYMOTRYPSIN

Chymotrypsin, at pH 8.5, hydrolyzes approximately the same number of peptide linkages in the native ovalbumin molecule as do pepsin and trypsin, if fresh enzyme is added to insure complete digestion (Fig. 284). According to Ross and Tracy (*1942*), the splitting of eighty-nine peptide linkages in a molecule containing only eight tyrosine and twelve phenylalanine residues indicates that this enzyme acts upon groups other than these, although it is not generally thought to do so. These investigators observed that malonyl ovalbumin (ovalbumin denatured by treatment with carbon suboxide) is more readily digested initially than native ovalbumin; yet it is eventually less completely hydrolyzed. Apparently the final extent of hydrolysis is less because of the presence of malonyl residues on lysine, or in some other combination more stable than the easily hydrolyzed malonyl-tyrosine linkage.

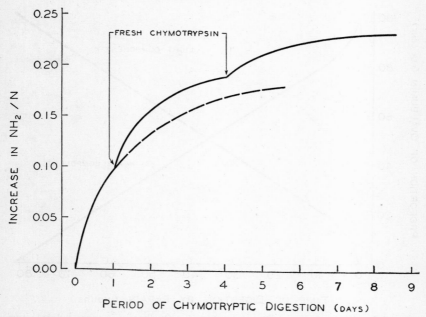

FIG. 284. Exhaustive chymotryptic digestion of ovalbumin (in concentration of 3.5 mg. of nitrogen per mole). The broken curve shows the extent of digestion when the initial enzyme concentration of 0.19 mg. of nitrogen per mole is not increased. (After Ross and Tracy, 1942.)

PAPAIN

Papain, activated by hydrocyanic acid, digests heat-denatured ovalbumin (at pH 5.0) more rapidly than native ovalbumin (*Haurowitz, Tunca, and Yurd, 1943*). Its action on the coagulated substance is approximately twice as extensive as that of pepsin or trypsin, neither of which effects further hydrolysis following hydrolysis by papain (*Calvery, 1933b*).

A papain digest of native ovalbumin can be separated by dialysis into two portions, a heavy fraction and a light fraction. The light fraction increases at the expense of the heavy (Fig. 285). The change in the proportional amounts of the two fractions occurs as a linear function of time, except during an initial, more rapid period (about 5 hours long, when the concentrations of ovalbumin and papain are 4.0 per cent and 0.3 per cent, respectively). Neither of these fractions is homogeneous. The results of cataphoresis and diffusion measurements have indicated that the light fraction probably consists of amino acids and the lower polypeptides (*Annetts, 1936*). Polypeptidase and dipeptidase liberate

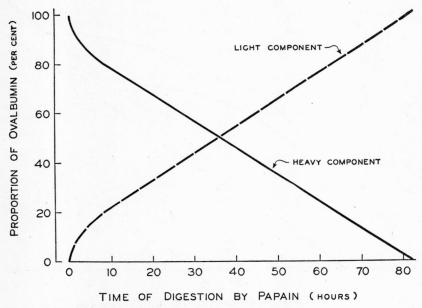

FIG. 285. Digestion of ovalbumin by papain. Ovalbumin is separated into two fractions, the lighter of which increases at the expense of the heavier. (After Annetts, 1936.)

from a papain digest an additional 12 per cent and 24 per cent, respectively, of the total nitrogen as amino nitrogen (*Calvery, 1933b*). It has been suggested that during hydrolysis by papain there is first a loosening of bonds in all the ovalbumin molecules, and then a gradual splitting-off of very small portions of the molecules (*Svedberg and Eriksson, 1933; Annetts, 1936*).

LECITHINASES, PHOSPHATASES

Kidney and small intestinal mucosa contain a lecithinase that hydrolyzes egg yolk lecithin, with the liberation of inorganic phosphate (Fig. 286). The reaction takes place with maximum velocity at body temperature. The optimum pH is 7.5, as shown in Fig. 287 by the amounts of inorganic phosphorus liberated between pH 5.5 and pH 9.5. The rate of hydrolysis diminishes after the initial period, apparently because of slow destruction of the enzyme, rather than because of the accumulation of hydrolytic products. Probably either glyceride or choline phosphate

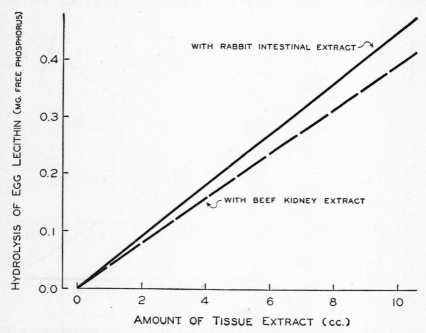

FIG. 286. Hydrolysis of egg yolk lecithin by the phosphate-splitting enzyme of rabbit intestines and beef kidney, as shown by the liberation of free inorganic phosphorus by various amounts of the tissue extracts during 48 hours. (After King, 1931.)

is first liberated, and a later cleavage of either, or both, of these yields free phosphate (*King, 1931*). It has been shown that choline is a product of the hydrolysis of egg yolk lecithin by the lecithinase of rat intestine (*Kahane and Lévy, 1944*). The phosphatase of takadiastase also liberates choline from ovolecithin, as shown in Fig. 288. The optimum pH for this reaction is 4.0.

Lecithinase-A, contained in the pancreas as well as in snake venom, converts egg lecithin into the hemolytic substance, lysolecithin, by splitting off a fatty acid. Lysolecithin in turn can be hydrolyzed by the lecithinase of rabbit intestinal mucosa. The reaction is about twice as rapid

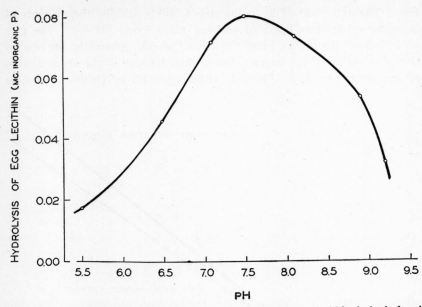

FIG. 287. Influence of *p*H upon the extent to which egg lecithin is hydrolyzed by extract of small intestine during 12 hours of incubation.   (After King, 1934.)

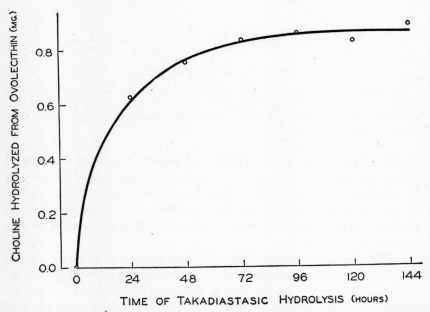

FIG. 288. Rate of takadiastasic hydrolysis of ovolecithin, as shown by the liberation of choline.   (After Yosinaga, 1936.)

as the hydrolysis of lecithin, and its optimum is somewhat higher, pH 7.8 (*King, 1934*).

The action of bone phosphatase upon egg lecithin is more feeble than that of lecithinase. The comparative rates of hydrolysis during 7 hours of incubation are given in the tabulation (*King, 1934*).

| TIME | HYDROLYSIS (per cent) | |
(hours)	LECITHINASE	PHOSPHATASE
1	1.2	0.4
2	2.4	0.8
7	4.8 .	1.5

GLUTAMINASE

It has been observed that kidney glutaminase hydrolyzes the glutamine in dialyzates of egg albumen and yolk. This enzyme may therefore be used in estimating the glutamine content of the egg (*Archibald, 1944*).

IMMUNOLOGICAL PROPERTIES

The egg possesses antigenic properties; it is, in fact, an aggregation of antigens. Many of the proteins and other individual chemical entities contained in it are distinct and separate antigens and are capable of stimulating the production of antibodies singly as well as together.

The purified proteins of the egg have been used extensively in research in the field of immunology. The actual nature of antibodies and antigens, and the mechanism of the various reactions between them, have been greatly clarified through studies in which the egg served as the source of immunizing proteins. Ovalbumin in particular has been employed, not only because it is readily available and may be isolated with relative ease, but also because its physical and chemical properties are well known.

Egg antigen-antibody reactions are manifested by such commonly observed phenomena as precipitation, complement fixation, anaphylactic shock, and allergic symptoms. Most antigen-antibody reactions vary qualitatively or quantitatively, not only according to the species of animal or even the individual that elaborates the antibody, but also according to the substance that acts as the antigen. Immunological properties therefore serve as additional means of characterizing various chemical components of the egg. The great delicacy of immunological tests also reveals many relationships and distinctions not detectable by other methods.

ALBUMEN

The antigenic activity of egg albumen was discovered many years ago. In 1900, Uhlenhuth injected the albumen of 2 or 3 eggs at one time into the peritoneal cavity of the rabbit and later was able to demonstrate the presence, in the animal's blood, of antibodies that caused precipitation in egg albumen diluted with 100,000 volumes of water. Similar effects were observed after the subcutaneous injection of diluted albumen (*Hollande and Gaté, 1918*). In addition, fatal anaphylactic shock was produced in guinea pigs by giving them an injection of whole egg white, and repeating the injection 10 or 12 days later (*Vaughan and Wheeler, 1907*). Species differences in the sensitivity of animals to the whole albumen were revealed when dogs failed to exhibit any symptoms of anaphylaxis, even after a shocking dose of 5.0 cc. of 50 per cent albumen per kilogram of body weight (*Manwaring, Marino, Mc-Cleave, and Boone, 1927*).

In the course of early immunological investigations, it was also recognized that the whole albumen was biologically related to the bird's blood (*Uhlenhuth, 1901*). The serum of an animal that had received injections of hens' egg albumen was found to give the precipitation reaction when mixed with chicken blood serum (*Levene, 1901*). Correspondingly, it was observed that an antiserum against defibrinated fowl blood could precipitate not only its homologous antigen, but also egg albumen (*Gengou, 1902*).

Other experiments showed that, in antigenic activity, the egg albumen of one species resembled that of other species, but was nevertheless different. Antiserum against hens' egg albumen was seen to react with pigeon albumen, but more weakly than with chicken albumen (*Uhlenhuth, 1900*).

Egg albumen contains several different antigens, which are immunologically unrelated, in spite of their common biological origin. The development of methods for the separation of albumen into its constituent proteins made it possible to study the individual behavior of these proteins.

In 1911, Wells first showed that there are at least three antigens in crude albumen. He found that guinea pigs sensitized to the entire albumen would react anaphylactically to any of the three protein fractions with which he worked. He also noted that the same animals, after recovering from shock induced by a particular albumen protein, were refractory to further injections of the same protein; yet they would still react to either one of the remaining two, or to the entire albumen.

He made it clear that ovalbumin and ovomucoid were distinct when he demonstrated the inability of ovalbumin to produce anaphylactic shock in guinea pigs sensitized to ovomucoid, and vice versa.

Later, four of the protein constituents of the albumen were clearly differentiated by means of the precipitin reaction in undiluted rabbit antisera. The tabulated data show that the highly diluted individual proteins react only with their homologous antisera, except for a cross reaction between ovomucin and ovoglobulin, presumably caused by the impurity of the latter substance (*Hektoen and Cole, 1928*).

ANTIGEN	PRECIPITATION REACTION (titer) IN ANTISERA FOR			
	OVALBUMIN	OVOGLOBULIN	OVOMUCIN	OVOMUCOID
Ovalbumin	1,000,000	0	0	0
Ovoglobulin	0	1,000,000	0	0
Ovomucin	0	100,000	100,000	0
Ovomucoid	0	0	0	1,000,000

Relationship to the Blood of the Bird

The immunological similarity between whole albumen and the blood plasma of the bird is due to the identity of ovoconalbumin with blood albumin. The lack of resemblance between egg albumen and either blood euglobulin or blood fibrinogen is indicated by the failure of these blood constituents to form precipitates with antiserum against crude albumen. Blood albumin, however, reacts with albumen antiserum. Similarly, of the five major albumen proteins, the noncrystallizable ovoconalbumin alone precipitates antiserum against blood albumin (*Hektoen and Cole, 1928*).

Species Specificity

The albumens of all birds' eggs possess common immunological characteristics. On the other hand, the albumen of each species is identical with that of no other species. Antisera against the albumen of many birds' eggs may give cross reactions not only with the albumen of closely related species but also with the albumen from avian species that are farther removed zoologically (*Sievers, 1939*). In Fig. 289, it may be seen that antisera for the albumen of such birds as the curlew and the greenshank give cross reactions (complement fixation) with the albumen from nearly all the birds represented. However, when species are as closely related to each other as the goose and the duck, or the chicken and the pheasant, the reaction is much stronger than when the relationship is distant. The albumen from the eggs of certain species—the sandmartin and the black-headed gull, for example—apparently always has low antigenic activity, since the antisera give weak reactions even with the homologous antigens.

The individual albumen proteins resemble the whole albumen in specificity. In the eggs of each species, ovalbumin is immunologically related to ovalbumin in the eggs of other species, yet it is also demon-

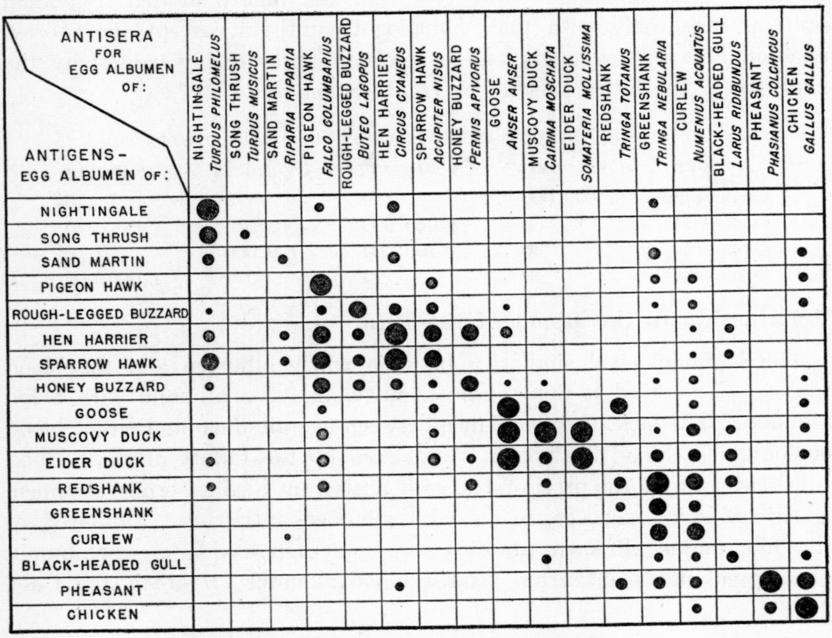

FIG. 289. Immunological relationships between the albumen of various birds' eggs, as shown by the complement fixation reaction between albumen and homologous and heterologous antisera. The size of the solid black circle indicates the relative strength of the reaction, the largest circle representing the strongest reaction. (After Sievers, 1939.)

strably different. The ovomucoids and ovoconalbumins of all species are likewise similar, yet distinguishable immunologically. The constituents of egg albumen, as well as those of blood, are therefore useful for taxonomic classification.

OVOMUCOID

By comparison with the other albumen proteins, ovomucoid is a rather weak antigen (*Wells, 1911; Hektoen and Cole, 1928*), possibly because the methods by which it is prepared reduce its activity (*Lewis and Wells, 1927*). It does not cause sufficiently pronounced anaphylactic reactions in guinea pigs to demonstrate clearly the relationships between the ovomucoids of different species. However, when rabbits are immunized with ovomucoids, complement fixation tests clearly indicate the close relationship between the duck and the goose, and between the hen, guinea fowl, and turkey, as shown by the tabulated data (*Lewis and Wells, 1927*).

OVOMUCOID ANTISERA

	GOOSE (milligram of antigen)	HEN (milligram of antigen)	
OVOMUCOIDS OF	0.5	0.5	5.0
Goose	++++	0	0
Duck	++++	0	0
Hen	0	++++	++++
Guinea fowl	0	+	+++
Turkey	0	0	+
Ostrich	0	0	0

When immunization with hen ovomucoid is carried further, species differences tend to become less apparent in the complement fixation test. Hen ovomucoid antiserum, for example, then reacts strongly in the more concentrated dilutions of all the above ovomucoids. In the higher dilutions, however, the greater similarity of the hen ovomucoid to the guinea hen ovomucoid, and its distinct difference from the others, may still be seen. If antiserum for an avian ovomucoid, such as that of the turkey, is sufficiently potent, it may give complement fixation with ovomucoid from turtle eggs; but the reaction still indicates that turtle ovomucoid is less closely related to all the avian ovomucoids than these are to each other (*Lewis and Wells, 1927*).

By the use of the uterine-strip anaphylaxis reaction, the relationships between the above-mentioned avian ovomucoids can be corroborated (*Lewis and Wells, 1927*). A strip of muscle from the uterus of a guinea pig that has been sensitized to one of the ovomucoids will contract strongly when placed in a solution of any of the other ovomucoids. If desensitized to one of the heterologous antigens, however, the uterine muscle still retains its sensitivity to the antigen to which it was originally sensitized. This behavior demonstrates the fact that the ovomucoids are very similar, yet not identical (although goose and duck ovomucoids are so alike that desensitization to one may also desensitize to the other).

OVOCONALBUMIN

The ovoconalbumin of certain birds' eggs is immunologically related to that of the chicken egg (*Hektoen and Cole, 1937*). Rabbit antiserum against chicken blood albumin (identical to chicken ovoconalbumin in antigenic activity) produces precipitation in 1:10,000 dilutions of the mother liquors remaining after the crystallization of the ovalbumins of the pearl guinea fowl, the amherst pheasant, and the golden pheasant (*A. Cole, 1938*). Since the mother liquors contain the noncrystallizable ovoconalbumins, the relationship of the ovoconalbumins of these birds to that of the chicken may be inferred.

OVALBUMIN

Like other albumen proteins, ovalbumin possesses certain properties which it exhibits whatever its species origin; yet, in each species, its immunological activity is nevertheless distinct. Ovalbumin has been studied by means of precipitin, absorption, and uterine-strip anaphylaxis tests.

The fact that hen and duck ovalbumin share common avian characteristics is revealed by the reaction of the guinea pig uterus to either protein, when sensitized to only one of them (*Dale and Hartley, 1916*). If a guinea pig has been injected with hen ovalbumin, one horn of the excised uterus will contract, 3 weeks later, to a dose of 0.05 mg. of hen ovalbumin, and the other horn to the same quantity of duck ovalbumin (Fig. 290). However, if the uterus is desensitized to the heterologous protein, it still retains its sensitivity to the homologous antigen (*Dakin and Dale, 1919*). The species specificity of hen and duck ovalbumins is thus indicated.

Precipitin reactions reveal relationships between the ovalbumins of the chicken, pearl guinea fowl, and amherst and golden pheasants. The table shows the highest dilutions of the various antigens that have been found to give precipitates with antisera for these ovalbumins (*A. Cole, 1938*).

	ANTISERUM AGAINST OVALBUMIN OF			
OVALBUMIN OF	HEN	PEARL GUINEA FOWL	AMHERST PHEASANT	GOLDEN PHEASANT
Hen	100,000	10,000	10,000	10,000
Pearl guinea fowl	1,000,000	100,000	100,000	100,000
Amherst pheasant	0	100,000	100,000	100,000
Golden pheasant	0	100,000	100,000	100,000

The chicken is thus seen to be more closely related to the guinea fowl than to either of the pheasants. Absorption tests further emphasize this fact. When rabbit antiserum against amherst pheasant ovalbumin is absorbed with chicken ovalbumin, the subsequent addition of the various antigens indicates that the chicken ovalbumin removes the precipitins for both the chicken and the guinea fowl ovalbumins, but not those for the two pheasants. Similarly, when antiserum against pearl guinea fowl ovalbumin is absorbed with the ovalbumin of either of the pheasants, the latter do not remove the precipitins for the chicken and the guinea fowl ovalbumins (*A. Cole, 1938*).

Although the ovalbumins of the chicken, turkey, guinea hen, duck, and goose all give precipitates with rabbit anti-chicken ovalbumin serum, the

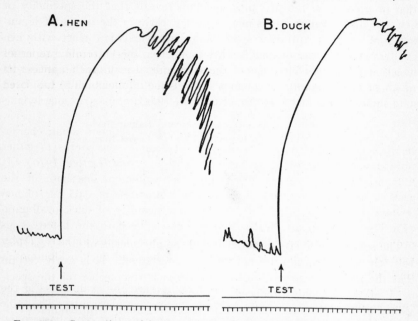

FIG. 290. Contractions of horns of uterus from guinea pig sensitized with ovalbumin from the chicken egg. (After Dale and Hartley, 1916.)

A, first horn: effect of hen ovalbumin. *B,* second horn: effect of duck ovalbumin.

strength of the reaction with each of these ovalbumins decreases in the order given. This fact is shown by the diminishing volume of the precipitate obtained upon complete exhaustion of chicken ovalbumin antiserum with the respective ovalbumins. If the reaction with the chicken ovalbumin is taken as 100, the relative values for the reactions are as shown (*Landsteiner and van der Scheer, 1940*). In absorption tests,

		GUINEA		
CHICKEN	TURKEY	HEN	DUCK	GOOSE
100	50	42	25	19

each ovalbumin removes from the chicken ovalbumin antiserum the antibodies for itself, and for the more weakly reacting proteins. Thus, after absorption with turkey ovalbumin, precipitation occurs only with chicken ovalbumin, but absorption with goose ovalbumin does not affect the precipitins for any of the other antigens.

Much as the complement fixation test shows that the species specificity of chicken and guinea hen ovomucoids becomes less distinct as immunization progresses, so the precipitin reaction reveals that the specificity of hen and duck ovalbumins is most apparent early in the course of immunization. Rabbit anti-duck ovalbumin serum may not react with hen ovalbumin when the animal is bled soon after the start of a series of injections; but later bleedings of the same animal may give an antiserum which precipitates both ovalbumins in the same dilutions. The tabulated data indicate this fact (*Hooker and Boyd, 1934*).

ANTIGEN DILUTION (Power of 10)	ANTI-DUCK OVALBUMIN RABBIT SERUM			
	EARLY BLEEDING		LATER BLEEDING	
	HEN	DUCK	HEN	DUCK
1	0	+	+	+
2	0	+	+	+
3	0	+	+	+
4	0	0	±	±
5	0	0	0	0

The species specificity of antigens and the phenomena of cross reactions between related species are not entirely understood. It is fairly certain that the molecular structure of each antigen is responsible for the specificity of the antibody produced. Cross reactions appear to indicate the ability of a single pure substance to engender multiple antibodies. Possibly there are several distinct determinant groupings in each antigen; or perhaps a single determinant structure may call forth diverse antibodies. It is probable that, in the case of egg proteins, cross reactions are due to a general similarity in the structure of the individual antigens, rather than to the presence of certain identical molecular groupings in the related antigens (*Landsteiner and van der Scheer, 1940*).

Tuberculin-Type Reactions

When a tuberculous guinea pig is injected with tuberculin, the animal's response is, according to the technique of the test, either a characteristic skin reaction, or a protracted type of anaphylactic shock which usually culminates in death. For reasons not entirely understood, the injection of egg albumen into a tuberculous animal, previously sensitized to albumen, results not in typical anaphylactic shock but in a response that is almost identical with the protracted tuberculin shock (*Dienes and Schoenheit,*

1927; Dienes, 1930a). The tuberculin type of skin sensitiveness is strongest and develops earliest if the albumen is injected directly into a tuberculous lesion (*Dienes, 1929*). The antigenic activity of albumen is greatly increased in tuberculous animals; in inducing hypersensitiveness and precipitin formation, 0.15 mg. in repeated doses has been found to have the expected effect of 15.0 mg. (*Dienes, 1927*).

The protracted type of shock may be produced by ovalbumin, ovoglobulin, and ovomucoid, as well as by whole albumen (*Dienes, 1931a*). The tuberculin type of skin reaction is induced as easily with ovoglobulin as with whole albumen, but ovalbumin gives a typical anaphylactic skin reaction before the tuberculin type appears (*Dienes and Schoenheit, 1927; Dienes, 1931b*). The skin sensitivity is specific to the albumen protein used (*Dienes, 1930b*).

Reactions of Ovalbumin

The immunological characteristics of ovalbumin have received more attention and have been studied in greater detail than those of the other albumen proteins. As a result, considerable information is available concerning its antigenic properties, and the nature of its antigen-antibody reactions.

PRECIPITIN REACTION

Extremely small amounts of ovalbumin, injected into the rabbit, stimulate the production of precipitins. The responses of different rabbits may vary, however. Hektoen and Cole (*1932*) detected the presence of precipitins in the serum of one animal that had received, altogether, a total of 30 cc. of a 1:100,000 dilution of ovalbumin, or the equivalent of 0.000,29 gm. of ovalbumin. This serum gave precipitation with a 1:100,000 dilution of antigen. Another rabbit required 115 cc. of antigen, or 0.0017 gm. of ovalbumin, before precipitins could be found in its serum, which reacted with no dilution of antigen over 1:10,000. Two other animals failed entirely to produce antibodies.

The precipitin test for ovalbumin, under ordinary circumstances, is sensitive to an antigen dilution of 1:200,000 (*Heidelberger and Kendall, 1935*). In this test, one reagent, either antigen or antibody, is kept constant in amount, and various dilutions of the other are added to it. After precipitation has occurred, analysis of the supernatant fluid shows that, in most of the mixtures, an excess either of antigen or of antibody remains uncombined. If antigen is greatly in excess, no precipitation takes place, since the compounds formed are soluble; they may be precipitated, however, by the addition of antibody (*Heidelberger and Kendall,*

1935). Whether or not soluble compounds are formed when antibody is in excess depends upon the species of animal whose serum is used. Rabbit antisera, for example, never form soluble compounds in the antibody excess region. At a certain point (the "equivalence point") in the series of dilutions, little or no antibody or antigen can be detected

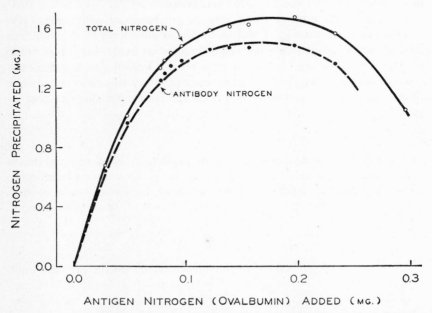

Fig. 291. Amounts of total nitrogen and antibody nitrogen precipitated in 1.0 cc. of ovalbumin antiserum (from third bleeding of rabbit) by various concentrations of ovalbumin. (After Heidelberger and Kendall, 1935.)

in the supernatant; having been present in equivalent amounts, the two have combined completely and precipitated.

The quantity of the precipitate formed is least in those tubes where either antigen or antibody is in excess. In Fig. 291, it may be seen that the smallest amounts of precipitate (measured as total nitrogen precipitated) are found at either end of the series of antigen dilutions. Maximum precipitation does not necessarily coincide with the equivalence point in antigen-antibody systems in general, but it usually does so (at least approximately) in the ovalbumin-anti-ovalbumin reaction (*Culbertson, 1932*).

The *p*H value of the ovalbumin and rabbit antisera mixtures has no effect upon the amount of precipitate formed at the equivalence point, at least between pH 5.2 and pH 9.5. However, at either the antigen excess or antibody excess end of the reaction, more precipitate is formed at pH 7.0 than toward the acid and alkaline limits (pH 4.5 and pH 10.0, respectively) within which precipitation occurs. When ovalbumin is present in excess, the precipitate formed tends to remain in suspension, unless the solution is neutral. Neutralization of an acid or alkaline solution results in the formation of additional precipitate (*Culbertson, 1934*).

The exact proportion of antibody to ovalbumin in the precipitate depends upon the ratio in which these reagents are mixed, and upon which of the two is present in constant amount. However, the precipitate always contains more antibody than antigen. The fact that the amount of antigen nitrogen in the precipitate increases as ovalbumin is added in larger quantities is indicated in Fig. 291 by the divergence of the two curves, since the antigen nitrogen is represented by the difference between total nitrogen and antibody nitrogen.

The ratio of antibody nitrogen to ovalbumin nitrogen in the precipitate from rabbit serum may vary throughout the range of reaction (*Heidelberger and Kendall, 1935*) as shown in the table. On the basis of very

| | PRECIPITATE ANTIBODY N |
SUPERNATANT	ANTIGEN N
Antibody in excess	16.2
No antibody or antigen	9.7
Antigen in excess	7.4

similar findings, the corresponding molecular compositions of the precipitates have been calculated and are probably much as shown, where ovalbumin is represented as "Ea" and antibody as "A" (*Heidelberger, 1938*).

EXTREME ANTIBODY EXCESS	ANTIBODY EXCESS	ANTIGEN EXCESS
EaA_5	EaA_3	Ea_2A_5

Rabbit antisera against ovalbumin may differ in their content of precipitins. Considerable variation is indicated by the fact that the equivalence point may be reached with dilutions of ovalbumin containing, for example, from 0.043 mg. to 0.096 mg. of nitrogen per cubic centimeter (*Culbertson, 1932*). However, there is a constant ratio of antibody nitrogen to antigen nitrogen in the precipitate obtained at the equivalence point, no matter what is the dilution of antigen at this point (as may be seen in Fig. 292).

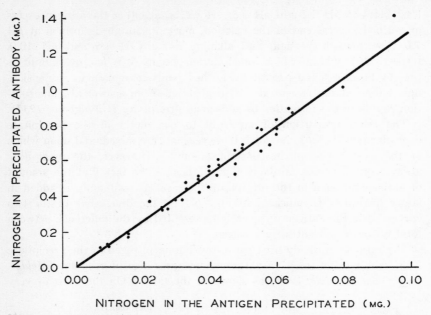

FIG. 292. Constancy of the ratio of antibody nitrogen to antigen nitrogen in the precipitate formed in 1.0 cc. of rabbit anti-ovalbumin serum, when antigen and antibody are completely precipitated. Each point represents a reaction with a different antiserum. (After Culbertson, 1932.)

As immunization proceeds, rabbit antiserum against ovalbumin becomes capable of forming larger amounts of precipitate with the same quantity of antigen. Figure 293 indicates that, with each successive bleeding of an animal immunized over a period of 3 months, there is an increase in the total nitrogen precipitated from 1.0 cc. of antiserum by all dilutions of ovalbumin. A broadening of the equivalence zone with each bleeding may also be seen, as well as an increase in the amount of ovalbumin nitrogen in the precipitate at the equivalence point. Probably, in the later stages of immunization, the antibodies elaborated can react with a larger number of distinct groupings on the ovalbumin molecule. The decrease in the species specificity of the precipitin reaction, during the course of immunization, might be explained on this basis (*Heidelberger and Kendall, 1935*).

The horse, as compared with the rabbit, is very slow in producing precipitating antibodies against ovalbumin. As may be seen in Fig. 294, the precipitating zone with horse anti-ovalbumin serum is much narrower than with rabbit anti-ovalbumin serum; and, where antibody is

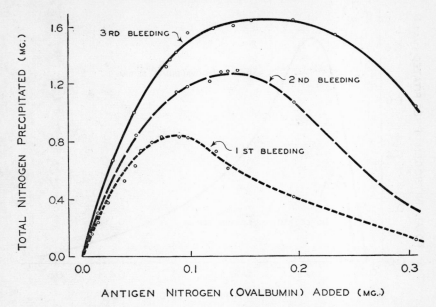

FIG. 293. Total nitrogen in the precipitates formed, by various amounts of oval-bumin, in 1.0 cc. of anti-ovalbumin sera from three successive bleedings of the same rabbit. (After Heidelberger and Kendall, 1935.)

greatly in excess of antigen, no precipitation occurs. The precipitating antibodies against ovalbumin are contained in the highly soluble pseudo-globulin fraction of the horse's serum, and not, as in the rabbit's serum, in the euglobulin fraction. The reaction between ovalbumin and horse anti-ovalbumin serum resembles other horse protein-anti-protein reactions (*Pappenheimer, 1940*).

ANAPHYLAXIS

Minute quantities of ovalbumin are as effective in producing hyper-sensitivity as in stimulating the production of precipitins. If sensitized with as little as 0.000,000,5 gm. of ovalbumin, guinea pigs may react slightly to a second injection several days later, although fatal anaphylaxis does not usually result unless the sensitizing dose is at least 0.000,001 gm., and the shocking dose 0.0005 gm. (*Wells, 1911*). Doses of 0.04 gm. to 0.05 gm. have been found most satisfactory for the production of hyper-sensitivity (*Armit, 1910*).

To sensitize an animal to ovalbumin, it is not always necessary that the protein be injected into the circulation. Wells (*1911*) found that

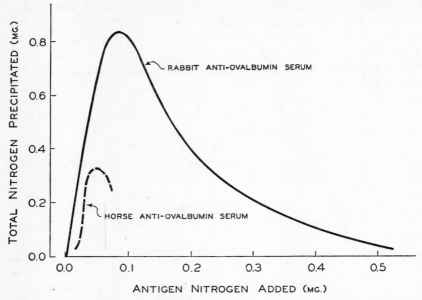

Fig. 294. Total nitrogen precipitated, by various amounts of ovalbumin, in anti-ovalbumin serum of the horse and of the rabbit. (Curve for reaction with horse antiserum after Pappenheimer, 1940; curve for reaction with rabbit antiserum after Heidelberger and Kendall, 1935.)

if guinea pigs were raised from the time of weaning on dried ovalbumin they gave strong anaphylactic reactions to this protein at the age of 30 to 60 days. When 100 days old, however, they reacted only slightly. Injection of ovalbumin failed to increase the degree of their sensitivity, since they had apparently been desensitized by the long-continued feeding of the protein. Rectal administration of ovalbumin may also produce hypersensitivity in both young and adult guinea pigs, although it is not so effective as oral administration (*Maie, 1922*). These findings indicate that the unaltered protein is capable of penetrating the wall of the gastrointestinal tract.

Although one injection of ovalbumin is normally sufficient to sensitize a guinea pig so that it will react anaphylactically 1 to 3 weeks later, the rabbit usually requires several injections, repeated at intervals of a few days. Anaphylaxis has been produced in sensitized guinea pigs by the injection of serum from rabbits into which ovalbumin had been injected 8 to 16 days previously; the antigen was therefore still present in the rabbits' blood (*Wollman and Bardach, 1935*). The symptoms of shock are less pronounced in the rabbit than in the guinea pig. When anaphylaxis is fatal, the autopsy findings are somewhat different in the two species. In the guinea pig, which exhibits much respiratory difficulty during shock, the lungs are markedly expanded, whereas the right side of the rabbit's heart is greatly dilated. These changes, and most of the symptoms of shock, are attributable to smooth muscle contractions. The usefulness of the isolated guinea pig uterus in making immunological tests depends upon the response of sensitized smooth muscle during anaphylaxis.

Apparently histamine is intimately involved in the mechanism of anaphylaxis. In the guinea pig, anaphylaxis is similar to the type of shock induced by the injection of histamine; in fact, it has been shown that histamine is liberated during the anaphylactic reaction. Anaphylaxis in guinea pigs sensitized to whole egg albumen can be prevented by the intrajugular injection of histaminase 15 minutes prior to the shocking dose of albumen. The enzyme presumably destroys histamine as soon as the latter is set free (*Karady and Browne, 1939*). In the rabbit, on the other hand, there is a decrease in the histamine of the blood and the plasma during anaphylactic shock caused by ovalbumin or other antigens. Anaphylactic shock differs in various respects from histamine shock. Histamine, therefore, does not appear to be primarily responsible for the production of anaphylactic symptoms in this species, although histamine metabolism is disturbed during shock (*Rose, 1941*).

EFFECTS OF DENATURATION

When denatured by heat, acid, alkali, or other means, ovalbumin is markedly altered in immunological behavior. Its original specificity, although not always entirely lost, is obviously changed, for denatured ovalbumin reacts weakly with antiserum for native ovalbumin. In addition, denaturation often decreases the antigenic activity of ovalbumin (*Wu, Tenbroeck, and Li, 1927*).

Crystalline ovalbumin, when heated at 100° C. for 5 minutes, no longer produces precipitation in antiserum for native ovalbumin. On the other hand, ovalbumin may be exposed to a temperature of 70° C. for 30 minutes without losing all its original specificity. It is sufficiently altered to permit a reaction with the antiserum for ovalbumin that has been heated at 100° C. (*Fürth, 1925*). Similar results have been obtained upon heating whole albumen to approximately the same temperatures (*Sharma, 1932*). Albumen apparently withstands heating at 90° C. for 5 minutes without losing its capacity to sensitize guinea pigs to the raw substance (*Pinoy and Fabiani, 1937*).

The precipitation reaction between acid-denatured ovalbumin and its homologous antiserum is of the same general type as that between native ovalbumin and anti-native ovalbumin serum (*MacPherson and Heidelberger, 1940*), as indicated by the shape of curve *A* in Fig. 295. Cross reactions, however, reveal that extensive change in specificity occurs during acid denaturation. Curves *B* and *C*, respectively, indicate that

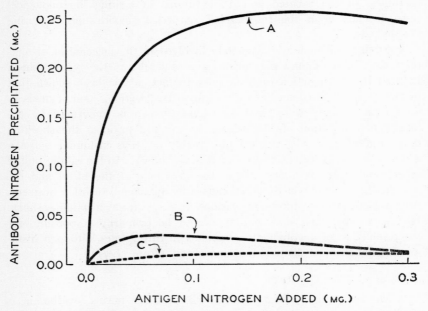

FIG. 295. Antibody nitrogen precipitated in reactions between *A*, acid-denatured ovalbumin and its homologous antiserum; *B*, acid-denatured ovalbumin and anti-native ovalbumin serum; *C*, native ovalbumin and anti-acid-denatured ovalbumin serum. (After MacPherson and Heidelberger, 1940.)

only a small quantity of antibody nitrogen is precipitated from native ovalbumin antiserum by acid-denatured ovalbumin, and from acid-denatured ovalbumin antiserum by native ovalbumin. After acid denaturation, a larger amount of antigen is required to inhibit precipitation in the homologous antiserum in the region of antigen excess (Fig. 296). The ratio of antibody nitrogen to antigen nitrogen in the precipitate is lower at every point in the reaction. Maximum precipitation in the acid-denatured ovalbumin system occurs in the antigen-excess zone.

After ovalbumin is denatured by alkali, the precipitin reaction is again somewhat different (cf. Fig. 296). In the region of antibody excess, the reaction resembles that between the acid-denatured protein and its antiserum; but, in the antigen-excess region, inhibition of precipitation is not quite so gradual and the precipitate shows a similar antibody-antigen ratio. Maximum precipitation occurs near the equivalence point (*MacPherson and Heidelberger, 1945b*).

Alkali-denatured ovalbumin, like acid-denatured ovalbumin, reacts only weakly with antiserum against native ovalbumin; it does not react at all

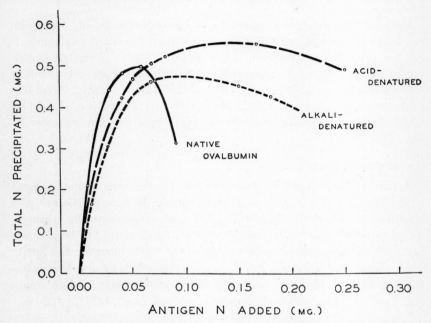

FIG. 296. The precipitin reactions between native ovalbumin, acid-denatured ovalbumin, and alkali-denatured ovalbumin and their respective homologous (rabbit) antisera, as shown by total nitrogen precipitated by various amounts of antigen. (After MacPherson and Heidelberger, 1945b.)

if degraded. In cross reactions with acid-denatured ovalbumin antiserum, alkali-denatured ovalbumin precipitates less nitrogen in the region of antigen excess than does the homologous antigen, and there is a higher ratio of antibody nitrogen to antigen nitrogen in the precipitate. In alkali-denatured ovalbumin antiserum, acid-denatured ovalbumin again precipitates more nitrogen in the antigen-excess zone than does alkali-denatured ovalbumin; but inhibition of precipitation begins more quickly, as larger amounts of antigen are added. In the antibody-excess region, the cross reactions of both antigens with heterologous antisera resemble the reactions with homologous antisera (*MacPherson and Heidelberger, 1945b*).

Dialyzed crystalline ovalbumin, coagulated by sound waves of audible frequency (1000 to 15,000 cycles per second), exhibits considerable decrease in antigenic activity, and also an altered specificity (*Flosdorf and Chambers, 1935*). It becomes somewhat similar to ovalbumin denatured by heat in acid and alkaline solutions, although some native specificity may be retained (probably because of incomplete separation of coagulum from supernatant).

Various other treatments alter the antigenic properties of ovalbumin. By anaphylactic tests, it has been shown that racemized ovalbumin is distinct from native ovalbumin (*Ten Broeck, 1914*). Putrefying ovalbumin loses its ability to react with ovalbumin antiserum, even while it still contains coagulable protein (*Hektoen and Cole, 1929*). According to Arloing and Langeron (*1925*), digestion by trypsin reduces the ability of ovalbumin to produce anaphylactic hypersensitivity but increases its shocking capacity.

Nonprotein Antigens

Apparently it is possible, upon occasion, for the carbohydrate present in egg albumen to act as an incomplete antigen, or hapten. Guinea pigs sensitized to the whole albumen may respond anaphylactically when injected with the protein-free carbohydrate of the albumen. In large doses, this carbohydrate may sensitize weakly to the whole albumen, but not to itself (*Sevag and Seastone, 1934*). Neither anaphylaxis nor desensitization results when the carbohydrate of egg albumen is injected into guinea pigs passively sensitized to albumen (*Ferry and Levy, 1934*).

Allergy to Egg Albumen

An additional manifestation of the antigenic properties of egg albumen is the ability to cause allergy. The term "allergy," as it is usually employed, refers specifically to the occurrence of various clinical symptoms in man, as the result of antigen-antibody reactions within the body. Allergy is thus akin to anaphylaxis because it produces observable phenomena in a living organism. It differs from anaphylaxis in several important respects, among which are the facts that it occurs naturally, rather than artificially; that it may not only be acquired through contact with an allergen (allergy-producing antigen) but may also be inherited; and that its symptoms vary from one individual to another. In allergy, antibodies are not precipitins, and often they can be demonstrated only indirectly.

ATOPIC INFANTILE DERMATITIS

Skin tests indicate that hypersensitivity to egg albumen is one of the most frequent causes of atopic infantile dermatitis, or eczema (*György, Moro, and Witebsky, 1930; Woringer, 1932a*). Although the skin test is a useful method of diagnosis, a negative result does not necessarily indicate absence of sensitivity (*Woringer, 1932c*). Another diagnostic procedure is the Prausnitz-Küstner test. This is essentially the passive transfer of local sensitivity to an area of the skin of a normal individual. In this test, serum from an eczematous infant is injected intracutaneously, and, 1 to 3 days later, egg albumen is injected in the same site. An inflammatory reaction results, if egg albumen is the allergen causing the infant's dermatitis (*György, Moro, and Witebsky, 1930*). This technique can be used for the quantitative determination of antibodies in the serum. Normal skin is injected with mixtures of allergic serum and egg albumen in dilutions varying from 1:500 to 1:1,500,000. The antibody content of the serum is indicated by the limits within which a reaction occurs when a subsequent injection of albumen is made (*Woringer, 1933*).

In atopic dermatitis, albumen antibodies are complement fixing. Quantitative titrations for serum antibodies may give values from 1:1000 to 1:5,000,000, with an average of about 1:50,000. The titer of the serum and the severity of the dermatitis are not always directly correlated (*Woringer, 1932a*). An excess of antigen inhibits complement fixation (*Woringer, 1932d; Sharma, 1932*). In normal infants, the antibodies formed against egg albumen give complement fixation only with relatively high concentrations of antigen. For this reason, it has been concluded that albumen antibodies in congenitally hypersensitive individuals differ qualitatively from those that may be elaborated by normal children (*Woringer, 1932b*). This theory is supported by the fact that the skin reaction differs in allergic children and in normal ones artificially sensitized to albumen (*Tezner, 1935*).

In rodents, antigens and antibodies can be transmitted from mother to fetus by way of the placenta. Possibly these substances can also be transmitted through the human placenta, which bears certain similarities to the rodent placenta (*Ratner, Jackson, and Gruehl, 1927a, 1927b, 1927c; Holford, 1930; Ratner and Gruehl, 1931*). Some types of congenital allergy to egg albumen are perhaps thus explained. Unaltered egg albumen may also pass into the milk—specifically, the milk whey (*Donnally, 1930*)—when eggs are included in the nursing mother's diet. It is doubtful that human milk contains antigenic protein in sufficient quantity to sensitize infants (*Smyth and Bain, 1931*), but, in certain instances, egg albumen in breast milk is possibly a factor in allergy (*Ratner, 1928*). By removing eggs from the diet of the mother, eczema may be abated in allergic infants (*Shannon, 1922*). It has repeatedly been found that the wall of the alimentary tract, including the colon (*Smyth and Stallings, 1931*), is permeable to egg albumen in infants less than 10 days old. By the precipitin reaction, it can be shown that albumen may appear in the urine (*Lawatschek, 1914; Grulee and Bonar, 1921*) and stools (*Grulee, 1920*) following the ingestion of egg white. Positive reactions have been seen in passively sensitized areas of the skin within 45 minutes after the feeding of diluted egg albumen to the newborn (*Wilson and Walzer, 1935*).

Ovalbumin is the protein chiefly responsible for atopic allergy to egg albumen (*Bosch, György, and Witebsky, 1932*). Ovomucin, however, is probably also involved, although its complement fixation reaction with the serum of sensitive children is much weaker than that of ovalbumin. Ovoglobulin may also fix complement, but possibly only because of contamination with traces of ovalbumin (*Bosch, György, and Witebsky, 1931*).

OTHER MANIFESTATIONS

Allergy to egg albumen in adults and in older children may be diagnosed by skin tests (*Bloch and Prieto, 1929*). It is shown in a number of ways, including such common manifestations as urticaria, asthma, and various gastrointestinal disturbances (*Bronfenbrenner, Andrews, and Scott, 1915; Pagniez, Vallery-Radot, and Haguenau, 1921; Iwakura, 1939*). Seizures resembling anaphylaxis are rare, but they have been reported (*Pagniez, Vallery-Radot, and Haguenau, 1921*). A response similar to shock has been produced experimentally in a few human subjects, but only with difficulty, after the feeding of very large quantities of albumen over a period of several months. The reaction was characterized by collapse and a sudden drop in blood pressure, sometimes accompanied by urticaria (*Bastei, 1926*).

Positive skin tests often may be seen within 1 hour after the ingestion of albumen by passively sensitized individuals. It is apparent, therefore, that unaltered egg albumen can pass through the wall of the gastrointestinal tract in adults, as in infants. In fact, absorption of the unchanged protein possibly occurs in 85 per cent of adults (*Sussman, Davidson, and Walzer, 1928*) and probably takes place in at least 74 per cent of children between the ages of 1 and 10 years (*Wilson and Walzer, 1935*). Experiments on dogs indicate that the normal path of absorption is through the lymph channels (*Alexander, Shirley, and Allen,*

1936). Ulcers and other gross lesions of the gastrointestinal mucosa, especially in the upper portion of the tract, are readily permeable to albumen (*Marks, 1936*). In view of these facts, it may easily be understood how allergic conditions are produced.

Allergy to an unidentified constituent of the egg, neither protein nor lipoid, has also been noted (*Jadassohn, 1926*).

YOLK

The immunological properties of the yolk's constituents have not been investigated so thoroughly as have those of the albumen, although some attention has been given to ovolivetin. The yolk's high content of fats places it in a different category from the albumen, since lipoids do not ordinarily engender antibodies unless injected in combination with antigenic substances. The ability of the fraction of the yolk extractable by alcohol to stimulate the production of lipoid antibodies is indicated by the flocculation which may occur (*Guggenheim, 1929*) when antiserum for yolk extract is mixed with lecithin or beef heart extract (lipoid-containing antigens). The alcohol and ether extracts of yolk behave as antigens distinct from the whole yolk (*Seng, 1913*). Antiserum for the whole yolk reacts not only with the whole yolk but also with the extract; whereas the yolk extract antiserum reacts only with its homologous antigen (*Guggenheim, 1929*). The whole yolk, less specific an antigen than the alcohol extract, is also less species specific.

Species Specificity

When used as an antigen, the whole yolk does not differentiate avian species as well as does the whole albumen. Rabbit antiserum against the yolk of the hen's egg gives precipitation with the egg yolks of many birds, including members of the various families to which belong the woodpecker, finch, lark, warbler, thrush, titmouse, starling, crow, vulture, falcon, pigeon, grouse, pheasant, plover, swan, goose, duck, and gull. As may be seen in Fig. 297, chicken yolk antiserum reacts more strongly with the yolks of the swan, goose, and duck than it does with those of the chicken's closest relatives. The yolks of the canary and the sparrow, members of the same family, give greatly different reactions. It can therefore be inferred that the whole yolk possesses immunological characteristics that are more or less common to all species of birds, and that its antigenic activity exhibits but little regularity consistent with taxonomic classification (*Seng, 1913*).

Species differences in whole yolk can be demonstrated by cross-testing with the complement fixation reaction. Thus complement is fixed when the antiserum against the yolk of the pheasant combines with its homologous antigen diluted 1:81,920; but complement fixation does not occur in mixtures of pheasant yolk antiserum and sparrow hawk yolk that is diluted more highly than 1:5120 (*Sievers, 1939*). It has also been

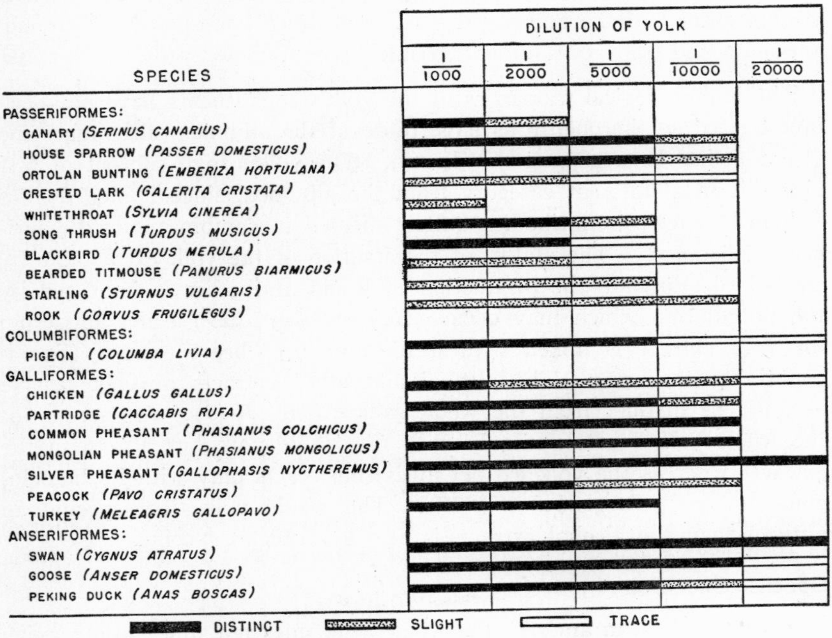

SPECIES	DILUTION OF YOLK				
	$\frac{1}{1000}$	$\frac{1}{2000}$	$\frac{1}{5000}$	$\frac{1}{10000}$	$\frac{1}{20000}$
PASSERIFORMES:					
CANARY (SERINUS CANARIUS)					
HOUSE SPARROW (PASSER DOMESTICUS)					
ORTOLAN BUNTING (EMBERIZA HORTULANA)					
CRESTED LARK (GALERITA CRISTATA)					
WHITETHROAT (SYLVIA CINEREA)					
SONG THRUSH (TURDUS MUSICUS)					
BLACKBIRD (TURDUS MERULA)					
BEARDED TITMOUSE (PANURUS BIARMICUS)					
STARLING (STURNUS VULGARIS)					
ROOK (CORVUS FRUGILEGUS)					
COLUMBIFORMES:					
PIGEON (COLUMBA LIVIA)					
GALLIFORMES:					
CHICKEN (GALLUS GALLUS)					
PARTRIDGE (CACCABIS RUFA)					
COMMON PHEASANT (PHASIANUS COLCHICUS)					
MONGOLIAN PHEASANT (PHASIANUS MONGOLICUS)					
SILVER PHEASANT (GALLOPHASIS NYCTHEREMUS)					
PEACOCK (PAVO CRISTATUS)					
TURKEY (MELEAGRIS GALLOPAVO)					
ANSERIFORMES:					
SWAN (CYGNUS ATRATUS)					
GOOSE (ANSER DOMESTICUS)					
PEKING DUCK (ANAS BOSCAS)					

███ DISTINCT   ▒▒▒ SLIGHT   ▭ TRACE

FIG. 297. Relative strength of precipitin reactions between yolk of various birds' eggs and rabbit antiserum against chicken egg yolk. (After Seng, 1913.)

concluded that the lipoid antibodies in the antisera against alcoholic extract of yolk are species specific, at least for the chicken and the goose. Complement fixation occurs in mixtures of alcoholic extract of goose yolk and its homologous antiserum, but not in mixtures of alcoholic extract of chicken yolk and antiserum against goose yolk extract (*Guggenheim, 1929*).

THE FORSSMAN ANTIGEN

The yolk of certain birds' eggs contains the Forssman antigen, more properly called the Forssman hapten because it cannot engender antibodies unless injected with another substance. The Forssman antigen (the antibodies of which are characterized by their ability to hemolyze sheep blood cells) occurs in the cells of a great many animals. Its distribution is not, however, strictly according to taxonomy. Although it is not found in the eggs of all species of birds, it is usually either present or absent in the eggs of all the species constituting a single family.

The Forssman antigen, which is alcohol soluble, can be demonstrated in chicken egg yolk by the flocculation reaction that occurs between the alcohol extract of the guinea pig's kidney (which contains the Forssman antigen) and the antiserum for alcohol extract of yolk. A similar reaction may occur when antiserum against a saline emulsion of whole yolk is used. The antiserum for either of the yolk preparations is capable of producing hemolysis of sheep cells. However, complement fixation does not necessarily take place when whole yolk is mixed with antiserum against guinea pig kidney extract. The availability of the Forssman antigen in the yolk of the hen's egg may therefore be somewhat variable (*Guggenheim, 1929*). The Forssman antigen is entirely lacking in the albumen (*Matsuda, 1926*).

Immunization with alcohol extract of yolk can result in the production of lipoid antibodies. For this reason, a positive complement fixation reaction between guinea pig kidney extract and antiserum for yolk extract is significant only if there is a negative reaction between yolk extract antiserum and lecithin or beef heart extract. Both reactions have not always been observed. Below, a few families of birds are listed according to the probable presence or absence of the Forssman antigen in their eggs (*Sievers, 1938*). The goose may also be numbered among

FORSSMAN ANTIGEN

PRESENT	ABSENT	
Shrikes	Crows	Cuckoos
Grebes	Finches	Owls
Gulls, terns	Wagtails	Falcons
Cranes	Flycatchers	Ducks
Grouse	Swallows	Pigeons
Pheasants, chickens	Woodpeckers	Plovers

those species from whose egg yolk the Forssman antigen is absent (*Guggenheim, 1929*). The eggs of the hen harrier do not contain it, but it is found in those of other Accipitres (*Sievers, 1938*).

Yolk Proteins

The proteins of the yolk are distinct antigens. Their antigenic activity is unrelated to that of the albumen proteins. Ovovitellin, for example, does not sensitize guinea pigs anaphylactically to ovalbumin or ovomucoid (*Wells, 1911*).

OVOLIVETIN

In certain respects, the ovolivetin of the hen's egg, unlike ovovitellin, resembles the globulin of the fowl's blood serum. Its relationship to this protein fraction is indicated by the fact that rabbit anti-chicken pseudoglobulin serum gives as pronounced a precipitin reaction with ovolivetin as it does with pseudoglobulin. However, a relationship almost as close is revealed by the reaction obtained between ovolivetin and rabbit anti-chicken euglobulin serum. Guinea pigs sensitized with either pseudoglobulin or euglobulin react anaphylactically to ovolivetin, although the reaction is somewhat stronger in the pseudoglobulin-sensitive animals. On the other hand, the complement fixation test distinguishes ovolivetin, pseudoglobulin, and euglobulin as three distinct antigens. Apparently ovolivetin is very similar to both globulins of the fowl's blood, without being entirely identical, chemically or structurally, with either (*Jukes and Kay, 1932c*).

Antibodies are contained in the globulin fraction of the blood. The close relationship between serum globulin and the ovolivetin of the yolk suggests the possibility that antibodies circulating in the bird's blood stream might be transferred to this protein in the egg.

In 1901, Dzierjgowski found that diphtheria antitoxin, when injected into hens, passes into the yolks of their eggs and becomes concentrated in the water-soluble fraction (livetin). It has also been demonstrated that the active immunization of hens by injection with diphtheria toxoid results in the appearance of antitoxin in the ovolivetin fraction of the yolk. With repeated injections, the antitoxin titers of the blood serum and of the ovolivetin show a parallel increase, as indicated by the tabulated data (*Jukes, Fraser, and Orr, 1934*). In fact, at the end of a

| | TITER (unit of antitoxin/cubic centimeter) | |
	SERUM	OVOLIVETIN
After first 7 injections	0.1–0.5	0.02–0.1
After last injection	1	0.1 –0.5

course of immunization, the titer of the ovolivetin may be nearly as high as that of the blood serum, as has been shown in studies on ducks (*Fraser, Jukes, Branion, and Halpern, 1934*). No antitoxin is associated either with the lecithovitellin of the yolk, or with the albumen.

Ovolivetin has also been indicated as the carrier of immunity from one generation to the next. Diphtheria antitoxin, if present in ovolivetin before incubation, may be detected in the serum of the newly hatched chick (*Dzierjgowski, 1901; Ozawa, 1936*), and also in the serum of ducklings hatched from the eggs of diphtheria-immune birds. Apparently no constant relationship exists between the yolk antitoxin and the antitoxin of the duckling's serum. The yolk antitoxin titer may be higher in the eggs of one bird than in those of another, yet the eggs with the lower titer may hatch ducklings with the higher serum antitoxin titer. Diphtheria immunity, transmitted through the egg, is only transitory; antitoxin cannot be demonstrated in the serum of ducklings 3 weeks after hatching (*Fraser, Jukes, Branion, and Halpern, 1934*).

Tetanus antitoxin is similarly transmitted to the chick (*Ramon, 1928*). So, too, are antibodies against fowlpox (*Hallauer, 1936*) and Newcastle disease (*Brandly, Moses, and Jungherr, 1946*). Immunity to the last two diseases disappears within the first month of life.

OTHER BIOLOGICAL EFFECTS

Many aspects of the egg's potential biological activity have been investigated, in addition to those already discussed. Some of the findings are very interesting in their revelation of the profound effect that certain constituents of the egg may have upon the animal organism.

Physiological Effects of Injected Albumen

A single intravenous injection of a large amount of egg albumen is sometimes fatal to animals; but a small amount, so administered, quickly disappears from the circulation (*Kenton, 1938*) and can be detected within a few hours in the bile (*Matsuda, 1931*), brain, liver, kidney, and spleen (*Vaughan, Cumming, and McGlumphy, 1911*). As observed first by Berselius, in the early nineteenth century, and subsequently by Lehman (*1864*), injected egg white is excreted by the kidney. In man, however, small subcutaneous doses are not necessarily followed by albuminuria. The protein appears in the urine of the rabbit within 10 to 30 minutes after its intravenous administration, and somewhat later when it is given subcutaneously (*Chiray, 1907*). Crystalline ovalbumin may be eliminated by the dog within 5 minutes after its injection; it is often still demonstrable in the urine 74 hours later (*Briggs, 1938*). The rabbit excretes both crude albumen and purified ovalbumin for at least 24 hours (*Opie, 1924*).

After its parenteral administration, egg albumen is eliminated largely in an unchanged form. The urine voided after an injection of albumen may, when in turn injected, sensitize other animals to albumen (*Van Alstyne and Grant, 1911; Vaughan, Cumming, and Wright, 1911; Pearce, 1912*).

The excretion of injected egg albumen protein is accompanied by the excretion of the animal's blood proteins. The urine of the rabbit, for example, thus acquires the ability to sensitize guinea pigs to rabbit blood serum (*Vaughan, Cumming, and Wright, 1911*). The serum proteins may be diminished by as much as 20 per cent of their total amount within 24 hours (*Vaughan, Cumming, and McGlumphy, 1911*). After crystalline ovalbumin is injected into the dog, serum pseudoglobulin and serum albumin may be detected in the animal's urine, usually within 10 to 60 minutes (*Briggs, 1938*).

When the injection of albumen or ovalbumin is repeated over a period of time, various responses of the organism to the continued administration of foreign protein become evident. The animal usually becomes immunized, and certain antigen-antibody reactions occur.

When the protein is given intravenously to the dog, in repeated doses of increasing size, it is excreted after each injection, but in amounts that do not increase to a corresponding extent. Some of it is retained in the animal's body and is eliminated intermittently for several weeks after the final injection. Apparently some of it is never eliminated (*Briggs, 1938*). When given subcutaneously, from 32 to 96 per cent of it may fail to be excreted (*Brull, 1933*). Blood proteins may still appear in the urine in very small amounts for months, or even years, after the egg protein is no longer found there. The concentration of pseudoglobulin in the urine often equals or exceeds that of serum albumin (*Briggs, 1938*). If massive injections are continued for 2 months, the hemoglobin content of the dog's blood may eventually decrease, and the erythrocytes and leukocytes may be somewhat reduced in number. Autopsy usually shows an enlarged spleen, congestion of the liver, kidneys, and gastrointestinal mucosa, and degenerative and proliferative sclerosing lesions of the large arteries. Kidney damage may be extensive, with narrowing of the lumina of the arterioles, thickening of the glomerular capsules, and distension of the tubules, which may contain hyaline casts and much stringy albuminous material. These findings in the kidney indicate clogging of the glomerular filtration membrane, perhaps by albumen protein or a complex containing it, and suggest a possible reason for the delayed elimination of albumen (*Hueper, 1942*). It is also thought that part of the albumen may be metabolized by the body, or eliminated in combination with other substances (*Briggs, 1938*).

When the rabbit is given repeated small subcutaneous injections of ovalbumin in the same skin area, the albumen is possibly held at the site of injection for as long as 5 days, instead of being excreted. The skin becomes inflamed locally, even necrotic (Arthus phenomenon), due to an antigen-antibody reaction which apparently destroys or fixes the antigen (*Opie, 1924*).

In sensitized guinea pigs and rabbits, repeated intravenous injections of ovalbumin (given slowly enough to avoid anaphylaxis) lead to massive areas of coagulative necrosis of the liver parenchyma, with congestion of the portal veins and sinusoids by leukocytes. These phenomena are also

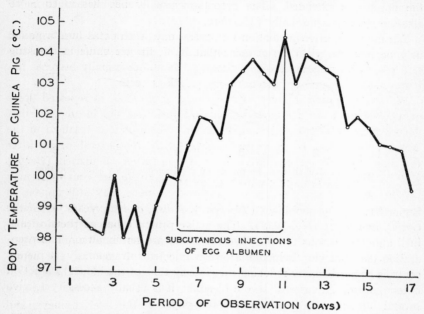

Fig. 298. Effect of subcutaneous injections of egg albumen (2 cc., five times daily) on the body temperature of a guinea pig. (After Cohen, 1920.)

associated with an antigen-antibody reaction (*Hartley and Lushbaugh, 1942*).

Fever is another effect produced by the parenteral introduction of albumen. When the protein is given subcutaneously or intravenously, at short intervals and in small doses, the temperature soon becomes elevated in rabbits (*Vaughan, Cumming, and Wright, 1911*) and in guinea pigs (Fig. 298), but not in dogs (*Cohen, 1920*).

Hyperglucemia has been produced in the rabbit by the intraperitoneal injection of albumen. The effect was delayed, but of long duration (*Chahovith and Arnovlevitch, 1927*).

By micromanipulation, crystalline ovalbumin has been injected into the amoeba, in amounts varying from two to seven times the volume of the nucleus. Subsequent changes in the appearance of the animal are shown in Fig. 299. The plasmalemma becomes raised in the form of hyaline "blisters," which rapidly spread around the entire periphery; and rotary streaming of the cytoplasm occurs. If repeated injections are given, the granular and hyaline areas remain distinct, and the hyaline portion is eventually swept to the posterior end of the amoeba. Sometimes the nucleus is extruded. The cytoplasm usually becomes much more alkaline than it is normally (*Marshak, 1944*).

Egg albumen, after incubation for several days with potassium bromide, may be made to yield a substance that is of distinct value in checking

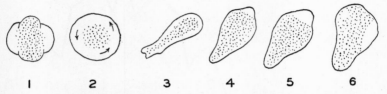

FIG. 299. Successive changes in the form of *Amoeba dubia* after injection of 2 per cent ovalbumin in distilled water. (After Marshak, 1944.)

hemorrhages in hemophiliacs (*Timperley, Naish, and Clark, 1936; Clark, Gaunt, and Timperley, 1938*). The active principle has not been identified, and the reasons for the observed effect are not known. The preparation does not cure hemophilia; when injected intravenously, it merely controls hemorrhage by reducing the clotting time of the blood. Its action is temporary. In cases of severe bleeding, it is usually necessary to give several injections, at intervals of a few days.

Protective Effect of Ovalbumin against Selenium Poisoning

Selenium, by accumulating in the tissues of farm animals, may produce a pathological condition known as "alkali disease." A diet of high protein and low carbohydrate content reduces the toxicity of selenium (*Smith, 1939*). Ovalbumin is among the proteins that afford a high degree of protection against the ill effects of this mineral. When it constitutes 20 per cent of a diet containing 10 parts per million of naturally occurring wheat selenium, ovalbumin may entirely prevent the development of pathological lesions in rats (*Smith and Stohlman, 1940*).

There have been no known cases of chronic selenium poisoning in man, in spite of the fact that human absorption of selenium is almost universal in regions where the soil is seleniferous (*Smith, 1941*). It is doubtful that the egg would contribute significantly to selenium poisoning (if such exists among human beings), in view of the protective effect of ovalbumin.

Effect of Albumen on Permeability of Intestinal Wall

Raw egg albumen increases the permeability of the intestinal wall to bacteria (*Arnold, 1929*). When bacteria are suspended in fresh egg white, and injected directly into the duodenum of the dog, they may be detected in the femoral vein (*Nedzel and Arnold, 1931a*) and in the gall bladder (*Nedzel and Arnold, 1931c*) within half an hour, in numbers approximately three times as great as when they are injected in a saline medium. Correspondingly, the presence of egg albumen in the duodenum increases the number of bacteria that may be eliminated into this portion of the intestine from the circulation (*Nedzel and Arnold, 1931b*).

Biological Effects of Cholesterol and Lecithin

In view of the relatively high cholesterol content of egg yolk, it is not surprising that blood, bile, and liver cholesterol values should be elevated by the ingestion of yolk in unusually large quantities. This effect has been demonstrated both in human subjects and in animals. In one instance, the blood cholesterol levels (total, free, and esterified) in several normal women were determined during a month on a low cholesterol regime. These levels were then compared with those found during a subsequent month on the same diet to which had been added 4 egg yolks daily and a certain amount of liver. The average blood cholesterol values of these women during the two periods are given below (*Okey and Stewart, 1933*).

| | BLOOD CHOLESTEROL | | |
	TOTAL (mg./100 cc.)	FREE (mg./100 cc.)	ESTERIFIED (mg./100 cc.)
Low cholesterol diet	154.0	93.5	60.6
High cholesterol diet	167.5	97.7	70.2

When dogs are fed from 4 to 6 egg yolks daily, or the equivalent of approximately 1.5 gm. of cholesterol, the resulting increase in the output of cholesterol in the bile is comparable to that which occurs when 1 gm. of bile salt is given. The stimulating action of 3 gm. of bile salt is not augmented by egg yolk. These facts are indicated by the data in the table (*Wright and Whipple, 1934*). When a dog's daily secretion of

DAILY FEEDING	INCREASE IN BILE CHOLESTEROL (per cent)
Bile salt (1 gm.)	60
Egg yolks (4–6)	40
Bile salt (3 gm.)	124
Bile salt (3 gm.) and 3 egg yolks	111

bile is collected through a fistula and refed, egg yolk adds to the effect of bile in increasing the daily output of bile cholesterol; but bile salt secretion diminishes. Bile cholesterol reaches its maximum when a combination of bile, bile salt, and egg yolk is given. However, the actual amount of cholesterol that appears in the bile after the ingestion of egg yolk is but a small fraction of the amount in the egg (*Wright and Whipple, 1934*).

Animals that are fed diets rich in cholesterol or egg yolk tend to develop fatty livers (*Abelin, 1935*). In rabbits, cholesterol atheroma frequently appears (*Bailey, 1914*). In rats fed enough egg yolk to supply 1 per cent cholesterol, the liver cholesterol ester values, after 2 months, may be ten times as great as when the diet contains yolk protein, vegetable fat, and no cholesterol (*Okey and Yokela, 1936*).

The lecithin contained in the yolk modifies the yolk's tendency to produce fatty livers. At the end of 60 days, the liver fatty acid and cholesterol ester values in rats fed whole egg yolk may be only one-half those found in rats fed lecithin-free diets containing yolk protein, vegetable fat, and pure cholesterol. Egg yolk-fed animals do not show the enlargement of the liver which may appear in cholesterol-fed animals. Upon continuing the egg yolk diet for 120 days, however, the effect of lecithin tends to disappear in female rats. Their liver fatty acid and cholesterol ester values may exceed those of the rats of both sexes on the lecithin-free diet. These sex differences in the rat have been considered analogous to the well-known sex differences in the incidence of arteriosclerosis and hepatic disorders among human beings (*Okey and Yokela, 1936*). Lecithin, like pancreatic tissue, prevents the development of fatty livers when it is fed to insulin-treated, depancreatized dogs (*Hershey, 1930; Hershey and Soskin, 1931*). Egg yolk does not contain enough lecithin, however, to counteract the effects of cholesterol in rats fed egg yolk exclusively for very long periods (*Rosenkrantz and Bruger, 1946*).

Choline is now known to be the active constituent of lecithin that prevents fatty livers (*Best and Huntsman, 1932*). When given in sufficient quantity, choline may even be curative (*Best, Ferguson, and Hershey, 1933*). It removes neutral fat and cholesterol esters from the liver (*Best, Channon, and Ridout, 1934*) by stimulating the production of phospholipids (*Welch, 1936; Perlman and Chaikoff, 1939*). High egg-yolk diets result in the elevation of the dog's plasma choline phospholipid and total phospholipid values, but rarely in an increase in plasma lipids (*Glomset and Bollman, 1943*). In cases of parenchymatous liver disease, eggs have been well tolerated and have improved absorption of fat from the intestinal tract; the only contraindications are caused by nonhepatic intolerance (*Sainz and Rodrigues, 1943*).

Lecithin also influences carbohydrate metabolism in diabetics and, to a lesser extent, in normal individuals. It is said to increase the effect of insulin in producing hypoglucemia and diminishing glucosuria (*Contini, 1934; Vacirca and Bertola, 1937; Sugita, 1940*), although a temporary hyperglucemia may be the first result of its injection (*DaRin, 1936*).

The preparation of lecithin (including egg lecithin) in the pure form is difficult, and the compound is not stable. It is probable that some of the effects of lecithin, both *in vitro* and *in vivo,* are due to its contamination with various other substances, or with oxidation compounds, and such products of decomposition as free fatty acids. For example, when lecithin is administered intravenously, its toxicity varies inversely with its purity. Rabbits may die within 4 minutes to 12 hours after the injection of 1.75 gm. of commercial egg lecithin per kilogram of body weight; yet lecithin freshly extracted from eggs and purified in so far as possible may be given in quantities up to 7 gm. per kilogram before being fatal. Within 14 days of storage at 55° C., such a purified preparation of lecithin may develop toxic properties, which become pronounced by the end of 53 days (*Ashby, 1946*).

Lecithin, when added to red blood cells *in vitro,* may be hemolytic. This effect is usually observed neither in high nor in low concentrations of lecithin, but only in lecithin solutions of intermediate strength (*Levin, 1935*). Lecithin also accelerates the hemolytic action of a variety of substances and imparts hemolytic properties to other substances that do not ordinarily possess them. Paradoxically, it may sometimes retard hemolysis; and it reinforces the inhibiting influence of cholesterol on hemolytic agents (*Lee and Tsai, 1942*). Investigation of these antagonistic effects of lecithin has yielded many conflicting results, chiefly because the lecithin preparations used have not been uniform.

It is well known, however, that the addition of lecithin to snake venom greatly increases the hemolytic activity of the latter (*Kellaway and Williams, 1933; De, 1940; Roy and Chopra, 1941*), and that this effect is due to the conversion of lecithin to the hemolytic substance, lysolecithin, by the lecithinase which venom contains. Lysolecithin is capable of destroying red blood cells and producing hemoglobinuria in rabbits, when injected intravenously in amounts as small as 50 to 100 mg. per kilogram of body weight (*Gronchi, 1932*). Lysolecithin is also one of the products of the spontaneous decomposition of lecithin (*Fiori, 1930*). Its presence, in traces, is probably responsible in part for the hemolytic properties of lecithin, which become more marked as lecithin preparations increase in age (*Levin, 1935*).

There is also rather general agreement that lecithin inhibits hemolysis by saponin, and that its antihemolytic action varies not only according to its own concentration, but also according to the concentration of saponin (*Grönberg and Lundberg, 1928; Levin, 1936a, 1936b; Lee and Tsai, 1942*). It is said that lecithin similarly retards hemolysis by crude bile acid, digitonin, serum hemolysin (*Lee and Tsai, 1942*), *Bothriocephalus* (tapeworm) extract, hypotonic salt solution (*Grönberg and Lundberg, 1928, 1929*), and streptococcal hemolysin (*Roy and Chopra, 1941*). The hemolytic action of quinine salts, optochin, and *Vibrio* hemolysin remains unchanged in the presence of lecithin (*Roy and Chopra, 1941*), whereas that of oleic acid, sodium taurocholate, and sodium glycocholate is accelerated (*Lee and Tsai, 1942*). It is not entirely clear why lecithin should affect the action of any of these substances; its influence upon some of them has been attributed to the presence of impurities (*Roy and Chopra, 1941*).

On the other hand, it has been noted that purified ovolecithin, containing no lysolecithin or other hemolytic contaminants, may still have an adverse effect upon the blood, when given parenterally to animals. Repeated intravenous injections of lecithin, administered to the rabbit throughout a period of 74 days, may decrease the erythrocyte count by 1 million cells and the hemoglobin content of the blood by 10 per cent. The fragility of the red corpuscles increases. Marked reticulocytosis eventually occurs, and nucleated red cells appear in the circulation (*Tompkins, 1943*). The number of white blood cells, especially the lymphocytes and monocytes, is elevated (*Stassano and Billon, 1902; Bain, 1912; Tompkins, 1936a, 1943*). The bone marrow becomes hyperplastic; the spleen enlarges, and shows various changes which would partially account for the destruction of the erythrocytes. There is a generalized infiltration of macrophages, containing lipoids and fragments of red blood cells. These findings resemble conditions observed both in lipoid storage diseases and in hemolytic anemias (*Tompkins, 1943*). However, it has also been observed that the red blood cells of rabbits may increase in number following the administration of lecithin intravenously (*Stassano and Billon, 1902*), subcutaneously, or by mouth (*Bain, 1912*).

Lecithin appears to have a beneficial effect on the condition of tuberculous rabbits. When inoculation with a single massive dose of tubercle bacilli is followed by a series of intravenous injections of egg lecithin, given over a period of several months, the course of tuberculosis may be demonstrably checked. Comparison of lecithin-treated animals with control animals may reveal considerable difference in the degree of pulmonary involvement (*Tompkins, 1936b*). Also, the death rate

	TYPE OF PULMONARY TUBERCULOSIS	
	HEALED OR HEALING (per cent)	ACTIVE OR SEVERE (per cent)
Lecithin-treated	54	46
Untreated	17	83

among untreated tuberculous rabbits may be more than three times that of animals receiving lecithin. It has been suggested that the inhibitory influence of lecithin upon the course of tuberculous infection may be due not only to the stimulated production of macrophages (*Tompkins, 1936b*), but also to the fact that the macrophages appearing after the injection of lecithin (unlike those formed after the introduction of tubercle bacilli and other foreign bodies) do not undergo degenerative changes into epithelioid and giant cells. On the contrary, they become hyperactive physiologically (*Tompkins, 1936a*).

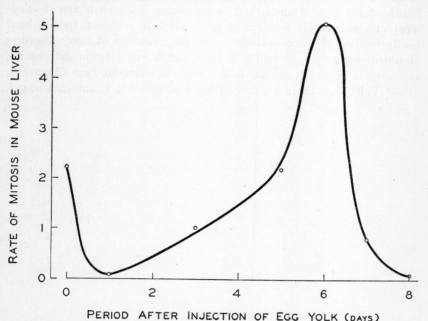

FIG. 300. Effect of the intraperitoneal injection of boiled egg yolk (in Ringer's fluid) on the mitotic rate in the liver of the mouse, aged 23 days. (After Wilson and Leduc, 1947.)

The total egg yolk lipoids, administered parenterally to female goats in the amount of 5 to 20 cc. daily, may have an effect similar to that of mixtures of pure lecithin and cholesterol given by the same route; that is, body weight and milk secretion may increase, and the percentage of total phosphorus, inorganic phosphorus, casein phosphorus, total lipoids, and cholesterol contained in the milk may rise. Excessively large doses of egg lipoids, however, may cause diminution in body weight and quantity of milk secreted, especially if the animal is undernourished (*Torrisi, 1935a, 1935b*).

In conclusion, it is interesting to note that hard-boiled egg yolk, suspended in Ringer's solution and injected into young mice, at first inhibits and later stimulates the mitotic rate in the liver cells (Fig. 300). The injection of Trypan Blue, or of suspensions of liver or kidney, produces the same effect, which cannot as yet be explained.

The Egg in Dietotherapy

Its great variety of nutritive substances makes the egg a highly concentrated food. When used in the diet, the egg aids in maintaining the health of the normal adult, and in promoting the growth and development of children. It is often of great value in restoring the depleted bodily reserves of the convalescent. In the treatment of many disorders of nutritional origin, including a large number of vitamin deficiencies, and various types of anemia, it is almost an essential (see Chapter 9, "Food Value"). There are few diseases in which it is contraindicated.

Part III

BIO-ECONOMIC IMPORTANCE

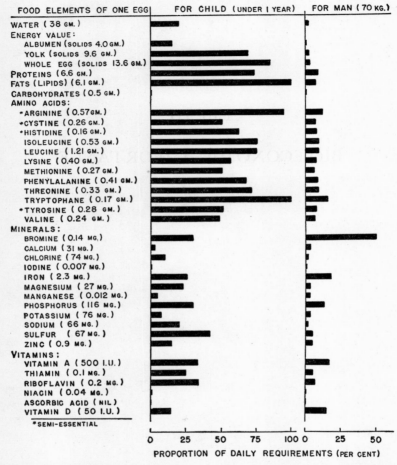

The extent to which 1 egg satisfies some of the daily dietary requirements of a man and of a child less than 1 year old.

Chapter Nine

FOOD VALUE

The avian egg is a highly nutritious and protective food.

Birds' eggs have been used as food by human beings since early antiquity. Before the establishment of stable agricultural communities, wild birds' eggs were included in the diet of primitive peoples.

In historic times, the eggs of the more prolific and easily domesticated birds, such as the duck, goose, guinea fowl, turkey, and especially the chicken, have almost completely supplanted all others in the human diet. Occasionally, where a wild species of large bird has an extensive nesting site, its eggs may be used locally. In England and Germany, plovers' eggs are considered a great delicacy and are sold in the markets.

Compared with the hen's egg, no other single food of animal origin is eaten and relished by so many peoples the world over; none is served in such a variety of ways. Its popularity is justified not only because it is so easily procured and has so many uses in cookery, but also because it is almost unsurpassed in nutritive excellence.

WORLD CONSUMPTION AND SUPPLY OF EGGS

For many years, the trend has been toward an ever larger consumption of eggs throughout the world. The average number of eggs eaten annually by the individual varies greatly in different nations. Figure 301 shows that, in 1935, this number approximated 300 in some countries, yet it was less than 50 in others. In that year, the per capita consumption of eggs in the United States was unusually low, but it increased considerably as economic conditions improved during the following decade. The number of eggs consumed annually by the American fluctuates greatly from year to year (Fig. 302) and is, in general, fairly well correlated with the periodic rise and fall in prosperity. Nevertheless, the egg, pound for pound, ranked in 1935 as the second most abundantly used food in the United States and was surpassed in popularity only by milk (Fig. 303).

The consumption of eggs is distributed geographically more or less

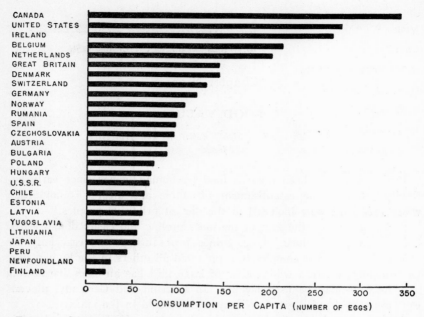

Fig. 301. The average per capita consumption of eggs in various countries of the world in 1935. (Compiled from various sources.)

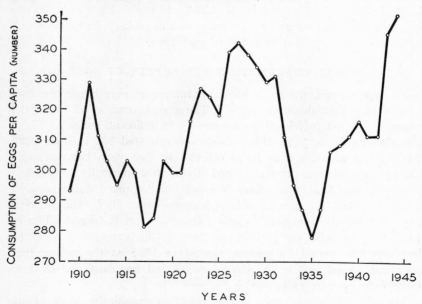

Fig. 302. Changes in the per capita consumption of eggs in the United States between 1909 and 1945. (After Agricultural Statistics of the United States.)

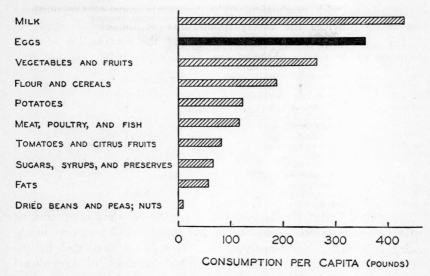

Fig. 303. Per capita consumption of eggs in the United States in 1935–36, compared with that of other foods. (After Stiebeling, 1942.)

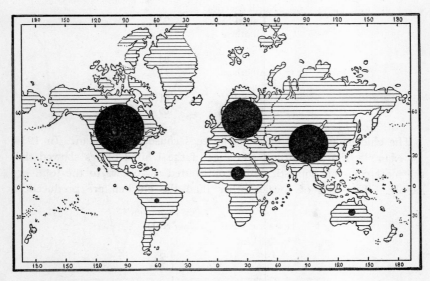

Fig. 304. Distribution of the world's production of chicken eggs in 1935. (From *International Yearbook of Agricultural Statistics*, 1940.)

in accordance with production. The temperate zone of the northern hemisphere provides the vast preponderance of the global supply of chicken eggs (Fig. 304). The table shows the contribution of each continent to the total number of eggs produced in 1935.

CONTINENTS	EGGS PRODUCED (billions)
North and Central America	36.1
Europe	32.0
Asia	31.7
Africa	2.8
Australia	0.6
South America	0.2
Total	103.4

The United States, Russia, and China are the principal egg-producing nations on their respective continents, but their relative lead over neighboring countries is not equally great. This fact is indicated by the tabulated data, taken from the International Yearbook of Agricultural Statistics for 1940.

NATION	EGG SUPPLY (per cent)
North America:	
United States	93
All others	7
Asia:	
China	83
All others	17
Europe:	
U.S.S.R.	31
All others	69

The chicken produces most of the eggs consumed by man. In 1934, for example, the agricultural statistics of the United States showed that the most common domestic birds contributed as follows to the total egg supply. The eggs of species other than the chicken are produced in

DOMESTIC BIRD	EGGS PRODUCED (per cent)
Chicken	92.6
Turkey	4.5
Goose	1.6
Duck	1.3

relatively larger proportion in some countries, but not in numbers greatly exceeding those required for flock replacement and for the hatching of meat birds. In this discussion of the egg's nutritive value, the hen's egg is meant, unless otherwise stated.

NUTRITIVE VALUE

From the nutritional standpoint, the widespread use of the egg in the human diet is well deserved. The hatching of the chick or nestling attests to the normal ability of the egg to supply sufficient nutriment for the avian embryo. It does not necessarily follow that the egg is equally adequate for the nourishment of species other than birds, or that its nutrients are readily available for the support of post-embryonic life processes. There is much evidence, nevertheless, that the addition of the egg greatly improves the diet of man and various animals. The egg not only is excellent for body maintenance, but also promotes growth, lactation, and reproduction, which make rigorous demands and are very likely to uncover deficiencies in a food.

Of the three most important dietary essentials—proteins, fats, and carbohydrates—the egg is composed largely of the first two. Its proteins are remarkably complete. It also supplies various minerals, some in significant amounts. It contains nearly all the known vitamins and is an excellent source of at least two. However, there is considerable variability in its content of a number of food substances. In addition, its nutritive value depends to a certain extent upon what proportion of the diet it constitutes, whether it is eaten raw or cooked, and whether the whole egg, or yolk or albumen alone, is consumed. In general, it is very well digested and assimilated, although not equally so under all circumstances.

WHOLE EGG

The small amount of unabsorbed residue that remains after large quantities of whole egg are consumed indicates that the egg is extremely digestible (*Lebbin, 1902*). This conclusion is borne out by the relatively few cases of digestive disturbance attributed to eggs, as compared with other foods (Fig. 305). Large numbers of eggs, both raw and cooked, have been eaten at one time without ill effects (*Lebbin, 1902; Aufrecht and Simon, 1908*).

Most of the digestive processes that prepare the egg for absorption occur in the intestine in the presence of the pancreatic and intestinal juices. How rapidly the egg passes from the stomach therefore has little relationship to its digestion in the remainder of the alimentary tract (*Maile and Scott, 1935*). However, digestion is likely to be less complete when the passage of any food is rapid than when it is slow, provided the response of the digestive glands is comparable in both cases.

The manner in which the egg is prepared has been found to affect the time spent by the egg in the stomach, where raw eggs apparently cause somewhat less acid secretion than cooked eggs. Soft-boiled, poached, and shirred eggs leave the stomach in the shortest time, usually about 2

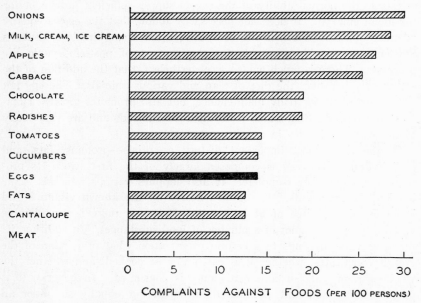

FIG. 305. Number of complaints made against various foods, per 100 patients. Among symptoms of distress were: 1, vomiting and diarrhea; 2, gas, belching; 3, heartburn; 4, regurgitation; and 5, headache. (After Alvarez and Hinshaw, 1935.)

hours; raw, fried (with little fat), hard-boiled, fried (with excess fat), deviled, and scrambled eggs, and omelettes remain for successively longer periods, up to 3.5 hours (*Aufrecht and Simon, 1908; Miller, Fowler, Bergeim, Rehfuss, and Hawk, 1919*). *In vitro* experiments, using 20 cc. of standard pepsin solution and 5 gm. of whole egg prepared in various ways, indicate that different cooking methods influence the time required for complete digestion, as shown in the tabulation (*Andross, 1940*). In the dog, hard-boiled eggs are eventually absorbed two or three times as well as soft-boiled eggs (*Hosoi, Alvarez, and Mann, 1928*). One human subject was observed to digest the raw, whole egg slightly more completely than the soft-boiled egg (*Aufrecht and Simon, 1908*),

	Digestion Time *in vitro*
Type of Cooked Egg	(hours)
Poached (plain water, or vinegar added)	40
Fried at 137° C.	40
Soft-boiled	50
Scrambled	50
Poached (salt water)	50
Fried at 235° C.	Over 50
Hard-boiled	Over 50

but it is not likely that the raw egg is universally the more digestible of the two.

When the whole egg is eaten raw to the exclusion of other foods, certain toxic effects of the albumen usually become apparent, and the capacity of the egg to sustain life is not realized to its fullest extent. There is considerable evidence that raw eggs alone do not support experimental animals in good condition indefinitely. Only for a limited time do young rats grow well on a diet of raw whole eggs (*Friedberger and Seidenberg, 1927; Stenqvist, 1928*), or mature rats maintain normal health. On the other hand, when the whole egg is cooked sufficiently, it sustains rats excellently, since heating largely destroys the harmful properties of the albumen. The difference in the ability of raw and of hard-boiled eggs to support growth in rats is illustrated in Fig. 306. For approximately 20 days, the growth rate is almost the same, when the egg is fed in either form. During the next 100 days, growth is greatly retarded in the rats that receive raw egg. These animals lose weight during the next 60 days and die, whereas those fed cooked egg continue to grow (*Stenqvist, 1928*).

Young rats fed raw whole egg sometimes grow more rapidly at first than those fed the same amount of cooked egg (*Friedberger, 1926; Friedberger and Seidenberg, 1927*). This fact indicates that some of the egg's nutritive value may be destroyed by heat, the extent of the loss depending upon the degree of heat and the time of exposure to it (*Friedberger and Seidenberg, 1927*). However, when the usual methods of preparation are employed, this loss is not great (*Scheunert and Wagner, 1928, 1929*); and, in fact, the capacity of whole eggs to sustain the growth of young rats for extended periods increases with the time and temperature of ordinary cooking. For example, eggs boiled for 10 minutes are superior in growth-promoting value to eggs boiled for 3.5 minutes, or cooked for 10 minutes at 70° C. (*Friedberger and Abraham, 1930*).

Many of the home methods of preparing the egg are wasteful, because some of the egg contents adheres to the eggshell or to the cooking

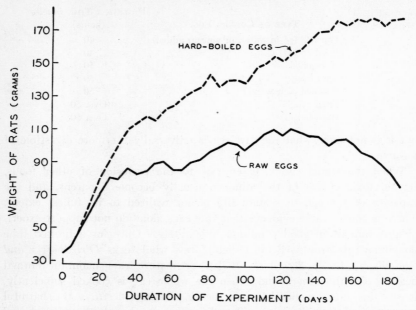

Fig. 306. Relative value of exclusive diets of raw and hard-boiled eggs, as shown by growth rates of rats. (After Stenqvist, 1928.)

utensils. Much of the albumen of a poached egg flakes off into the water. Analyses of eggs cooked in different ways indicate that the loss during preparation may be considerable, as shown by the reduction in the protein content of the egg (*Andross, 1940*). The boiled egg is

Type of Cooked Egg	Loss of Protein (per cent)
Boiled	0
Fried at low temperature	1.5
Omelettes	3.0
Poached	7.5
Fried at high temperature	8.9
Scrambled	13.5

obviously the most economical. Hard-boiled eggs are of the best consistency when started in cold water and boiled 15 minutes, a procedure that permits complete coagulation of the yolk as well as of the albumen and allows the two components to be separated easily. If the hard-boiled egg is quickly cooled in cold water, albumen does not adhere to the shell. In addition, the diffusion of hydrogen sulfide to the surface of the egg

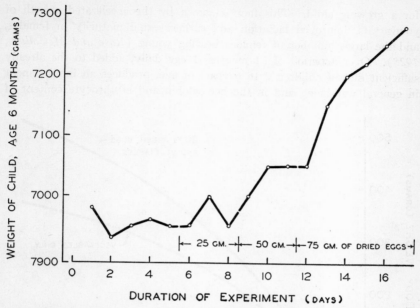

FIG. 307. Changes in the growth of a 6-months-old child after the addition of various amounts of dried egg to the basal diet. (After Rantasalo, 1929.)

is accelerated, because of reduced pressure, and discoloration of the yolk by ferrous sulfide is thus minimized.

Dehydration of the egg apparently causes little decrease in food value, at least in that of its most important constituents (*Lhamon, 1919*). Clayton (*1930*) found that dried egg, in sufficient quantity to meet the protein requirements of growing rats, was superior to all other sources of protein tested, except glandular meats. The addition of dried egg to a basal diet accelerates the growth of an infant 6 months old, in proportion to the amount of egg fed (Fig. 307). The fact that both young and adult rats digest dehydrated egg to the extent of almost 95 per cent (*Sumner, 1938*) is additional evidence that any deterioration due to desiccation is insignificant.

Fertilized eggs are apparently a better source of nutriment for growing turkeys than unfertilized eggs. Spring eggs are likewise superior to those laid in the fall (*Tallarico, 1933*).

The whole egg, when used as a supplement to other foods, assumes great importance. This is true even when the egg is added to a diet that is already entirely adequate. A good diet for rats is measurably improved when whole egg is added in amounts equivalent to 1 egg a day for a growing child. This fact is shown by the accelerated growth of young rats, improved lactation and earlier sexual maturity in females, and the larger number of females bearing young (*Rose and McCollum, 1928*). Over a period of 21 months, 1 egg daily, added to the already sufficient diet of children 2 to 6 years of age, produces an improvement in general well-being and in the hemoglobin and erythrocyte content of

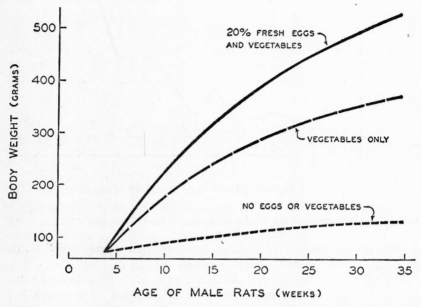

FIG. 308. The growth rate of rats fed a basal diet, compared with their growth rates on the same diet supplemented with leafy vegetables, and with vegetables combined with eggs. (After Wan, Wu, and Chang, 1935.)

the blood (*Rose, 1926; Rose and Borgeson, 1935*). As a supplement to an adequate ration, the egg is superior to various vegetables (*Rose and MacLeod, 1928b*). Rats gain weight at an improved rate when leafy vegetables are incorporated into their diet, but growth is still better upon the further addition of egg (Fig. 308). A ration of 70 per cent bread and 30 per cent whole egg maintains health and permits normal reproduction and lactation in rats. This ration is superior to bread combined with carrots, spinach, almonds, or lean beef and is inferior only to bread and milk (*Rose and MacLeod, 1928a*). One egg compares well with 8 ounces of milk in caloric value, and in its content of protein, fat, vitamin A, thiamin, and riboflavin; it contributes much more iron and vitamin D. Milk surpasses the egg, however, in its complement of calcium, phosphorus, carbohydrates, and vitamin C (*Kon, 1940*). For dogs, eggs are better than meats (*Daggs, 1931*), although they are not so good as liver.

It can readily be seen that the nutritive excellence of the egg enhances the value of any food in which it is incorporated. Its widespread use in cookery for purposes of leavening, thickening, binding, and emulsifying (see Chapter 11, "Industrial Uses") considerably improves the human diet.

ALBUMEN

The white of the egg may be considered as essentially a concentrated solution of proteins. Cooked albumen is well utilized; but raw albumen is not entirely digestible, and its utilization is therefore limited.

Raw Albumen

In the human stomach, raw egg albumen excites very little glandular secretion. It rapidly passes onward, usually in about 1.5 hours (*Miller, Fowler, Bergeim, Rehfuss, and Hawk, 1919; Farrell, 1928*); nor is it retained long in the stomach of the cat or dog (*Cannon, 1904; London and Sulima, 1905*). Furthermore, there is an antienzyme in raw egg white that interferes with digestion by trypsin (*Vernon, 1904; Abderhalden and Pettibone, 1912; Balls and Swenson, 1934b*). It has been suggested that this deterrent to digestion is present in both the dense and liquid portions of the albumen, but presumably it is in an active form only in the liquid portion (*Balls and Swenson, 1934b*). On the other hand, the two portions of the albumen are equally nutritious (*Guerrant and Rudy, 1939*), and therefore the antienzyme apparently has little effect on general food value.

When given in large amounts to man and to such animals as rabbits and dogs, a considerable proportion of raw albumen remains undigested, as shown in the tabulation. Raw albumen is apparently somewhat more

SPECIES	AMOUNT FED (cubic centimeters)	DIGESTIBILITY (per cent)	INVESTIGATOR
Rabbit	120 (dried)	61–70	von Knieriem (*1885*)
Dog	120–160	51–62	Bateman (*1916*)
Man	1000 (one dose)	50	Wolf (*1912*)
Man	350–420, daily	80	Rose and MacLeod (*1922*)

digestible when it is well beaten (*Rose and MacLeod, 1923*).

RAW ALBUMEN	COEFFICIENT OF DIGESTIBILITY (per cent)
Unbeaten	82.3
Beaten	86.1

The poor utilization of raw albumen is revealed by studies of nitrogen elimination. Feeding of raw egg white is followed by tardy excretion of nitrogen in small amount, as compared with meat (Fig. 309-*A*).

A small and variable amount of albumen is unchanged by the digestive processes and passes through the wall of the intestine (*Cramer, 1908; Walzer, 1927*) in many human adults (*Sussman, Davidson, and Walzer,*

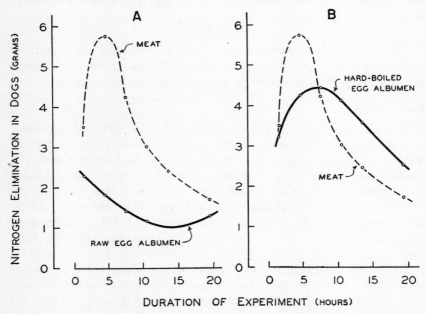

Fig. 309. Nutritive value of *A,* raw egg albumen, and *B,* hard-boiled egg albumen, compared with that of meat (as shown by nitrogen balance in female dogs). (After Mendel and Lewis, 1913.)

1928) and particularly in babies less than 10 days old (*Lawatschek, 1914; Wilson and Walzer, 1935*). In some individuals, the intestinal wall is apparently highly permeable to the relatively small molecules of ovalbumin and perhaps of other albumen proteins. The toxic effects of the albumen, which are more severe in young than mature animals (*Parsons, 1931a*), may be attributable in part to this unaltered portion of the albumen.

Animals, large or small, cannot subsist long on a diet composed entirely of egg albumen. Small animals, such as rats and mice, succumb in a very few days, young animals sooner than mature ones. Full-grown rats may survive a week on nothing but raw albumen, but they lose weight very rapidly (*Friedberger and Abraham, 1930*). Dried albumen is equally as injurious as fresh albumen (*Boas, 1927*).

When rats are fed diets containing various percentages of raw albumen, the resulting ill effects become more pronounced as the proportion of egg white is increased (*Bateman, 1916*). Although amounts up to 20 per cent are well tolerated, coagulable protein appears in the softened feces when 40 per cent is given, and 60 per cent causes diarrhea and loss of weight.

Cooked Albumen

The digestibility of albumen is improved by cooking. Coagulated egg white excites more gastric secretion than the raw substance, and it remains in the stomach for a longer time, although not so long as other proteins (*Cannon, 1904; London and Sulima, 1905*). The peptic digestibility of coagulated albumen, at least *in vitro,* varies according to the method of coagulation; it is best when the albumen is heated at 70° C. (*Frank, 1911*) and becomes poorer as coagulation is accomplished at higher temperatures (*Bizarro, 1913*). Digestion by pepsin usually, but not always (*Young and Macdonald, 1927*), is made more difficult by an increase in the time of exposure to heat, whether the albumen is brought to 100° C. from temperatures as low as 40° C. (*Frank, 1911*) or heated at constant temperatures as high as 140° C. (*Bizarro, 1913*). On the other hand, the digestion of albumen by trypsin is markedly improved after albumen has been heated for short periods of time at 100° C. or 140° C. (*Bizarro, 1913*). Heat destroys the antitryptic enzyme (*Vernon, 1904*), and therefore the beneficial effect of cooking is doubtless due largely to the fact that the digestion of albumen in the small intestine is facilitated.

Human digestion of cooked egg albumen, even in large amounts, is approximately 86 per cent complete (*Rose and MacLeod, 1923*). The utilization of cooked albumen by the dog, as measured by nitrogen excretion, is almost 90 per cent, or more than one-third better than raw albumen (*Bateman, 1916*). In rapidity of utilization, cooked egg white is similar to meats (cf. Fig. 309-*B*), and far superior to uncooked albumen (cf. Fig. 309-*A*).

Cooked egg albumen is an excellent restorer of serum albumin and total plasma proteins in the dog, and presumably in man. In this respect, it is almost as effective as beef serum and surpasses beef muscle, beef liver, and casein (*Pommerenke, Slavin, Kariher, and Whipple, 1935; Weech and Goettsch, 1939; Weech, 1942*). Such regenerative activity presupposes good digestion and absorption.

Cooked albumen is almost completely digested when it is fed in small quantities as the sole source of protein in the diet; as, for example, when it constitutes an 8 per cent supplement for rats (*Henry, Kon, and Thompson, 1940*). However, to support normal growth in young rats, an 18 to 20 per cent supplement is necessary (*Bond, 1922; Mitchell, 1925*). On the other hand, when as much as 66 per cent is fed, growth is soon

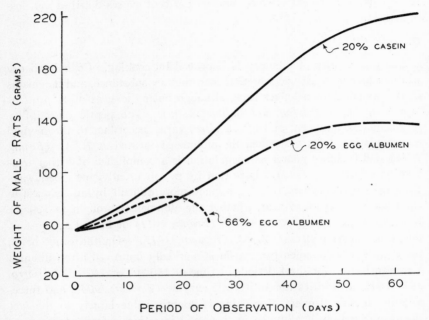

FIG. 310. The growth rate of rats fed 20 per cent casein compared with the growth rates of rats fed 20 per cent and 66 per cent coagulated egg albumen. (After Bell, 1933.)

retarded (Fig. 310), and the animals may develop acute nephritis and die (*Bell, 1933*). Albumen is somewhat less able to support growth than casein (*Mitchell, 1925*), as may also be seen in Fig. 310.

Heating reduces the injuriousness of egg albumen. Coagulation at 70° C. destroys its capacity to cause diarrhea in dogs (*Bateman, 1916*). However, cooked egg white is so much more palatable to animals than the raw substance that, when fed *ad libitum*, it is consumed in much larger quantities. Consequently, rats and mice fed an exclusive diet of cooked egg white do not live appreciably longer than those fed raw albumen alone, as shown in the table (*Friedberger, Abraham, and Seidenberg, 1928*).

TREATMENT OF ALBUMEN	AMOUNT CONSUMED (grams)	SURVIVAL (days)
None (raw)	5.0	1 –2
Cooked 10 min. at 70° C.	15.1	1 –2
Cooked 3 min. at 100° C.	17.0	1 –3
Cooked 20 min. at 100° C.	27.0	1.5–3

YOLK

Egg yolk is a concentrated food containing a wide variety of nutrients in forms which are readily available, not only to the developing chick, but also to other animals, young and mature. The yolk is approximately half solids. Almost two-thirds of these are lipids of high energy value, and one-third is protein of the phosphorus-containing type, found elsewhere in large proportion only in milk.

Yolk in the native state is finely divided and highly emulsified, in readiness for digestion by the stomach, which retains it (whether raw, soft-boiled, or hard-cooked) for approximately the same length of time as it retains whole egg. Egg yolk stimulates good secretion of gastric juices (*Miller, Fowler, Bergeim, Rehfuss, and Hawk, 1919*). In fact, egg yolk is so easily digested that it is an excellent source of additional nutriment for infants (*de Sanctis, 1922*), for whom it is recommended at as early an age as 2 months. Yolk has even been suggested as a substitute for milk during the first 2 days of life (*Midelton, 1907*). Infants quickly attain a positive balance for many of the egg constituents soon after the egg is included in the diet (*Ruotsalainen, 1928*). Egg yolk is also an excellent means of promoting weight increases in hospital patients in poor nutritive condition (*Steiner, 1941*).

Animals fed exclusively on egg yolk have been known to survive indefinitely in good health. Rats maintained on an egg-yolk diet for 126 days grew without interruption and were able to reproduce (*McCollum, 1909*). Friedberger and Abraham (*1929, 1930*) observed no signs of dietary deficiency in rats fed nothing but egg yolk for 9 months or a year. Yolk in the natural state, or with the fats extracted, is well tolerated by dogs (*Bateman, 1916*). It is due to the protective qualities of the yolk that animals fed whole egg are able to survive the detrimental effects of the albumen.

Ordinary methods of cooking, such as are employed in the household, probably do not noticeably diminish the nutritive value of egg yolk, in so far as this is indicated by the growth of young rats for a limited period (*Scheunert and Wagner, 1929*). However, the yolk suffers a nutritional loss when it is heated for extended periods of time at excessive temperatures. This fact is shown by the retarded growth rate of young rats fed yolk cooked for 30 minutes at 110° C. (Fig. 311). Alteration in food value might be due either to reduced digestibility or to diminution in the vitamin content of the yolk. The latter interpretation is favored by Friedberger and Abraham (*1930*), although their animals fed on cooked egg yolk showed no symptoms of deficiency.

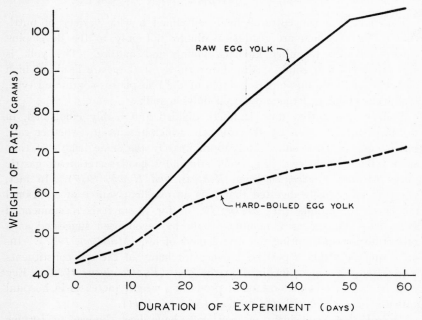

FIG. 311. Comparison of the nutritive value of raw and of hard-boiled egg yolk as the sole food for young rats, as shown by growth rates. (After Friedberger and Abraham, 1929.)

PURE SUBSTANCES OF THE EGG

Proteins

A protein is utilized only in so far as it fulfills the minimal requirements for each of the essential amino acids. If one of these is absent entirely and is not furnished by supplementary foods, there is an immediate loss of nitrogen from the body; and this, of course, cannot be made up by a larger intake of the same deficient protein. On the other hand, if all the essential amino acids are present in a protein, but not in ideal proportions, the organism's minimal requirements for the amino acids that are relatively lacking can be met if consumption of the protein is sufficiently large. Some of the amino acids are then ingested in excessive quantities, and the products of their catabolism appear in the excretions. A protein may be tested for amino-acid balance and general excellence by determining the smallest amount of it that supports life processes.

There is no deficiency or lack of balance in egg proteins, whose amino acid content is excellent. Of the nine amino acids known to be essential in human nutrition, and the four that are necessary under certain conditions (cf. Table 44, column 1), all are found in the egg in amounts comparable with, or exceeding, the amounts present in milk, muscle meats, and plant proteins (*Block and Bolling, 1944*). Each of the two major egg proteins, ovovitellin of the yolk and ovalbumin of the egg white, is

TABLE 44

THE NUTRITIVE VALUE OF EGG PROTEINS IN TERMS OF THE AMINO ACID REQUIREMENTS
OF MAN

Amino Acids	Amounts in the Egg * (grams/egg)	Daily Maintenance Requirements of 70-Kg. Man † (grams/day)	Fraction Furnished by 1 Egg (per cent)
Essential for maintenance of N-balance: ‡			
Isoleucine	0.36	3.7	9.7
Leucine	1.21	12.6	9.6
Lysine	0.38	5.2	7.3
Methionine	0.27⎤	4.1	12.9
(and cystine)	0.26⎦		
Phenylalanine	0.41	4.7	8.7
Threonine	0.33	3.6	9.2
Tryptophane	0.17	1.1	15.5
Valine	0.24	3.9	6.2
Probably essential or semi-essential:			
Arginine	0.57	4.7	12.1
Cystine (see above)	0.26		
Histidine	0.16	2.0	8.0
Tyrosine	0.28	3.9	7.2

* Authors' data.

† Block (*1943*).

‡ Albanese, Holt, Brumback, Hayes, Kadji, and Wangerin (*1941*), Holt, Albanese, Shettles, Kadji, and Wangerin (*1942*), Rose, Haines, and Johnson (*1942*), Albanese, Holt, Brumback, Kadji, Frankston, and Wangerin (*1943*), Albanese, Holt, Frankston, Kadji, Brumback, and Wangerin (*1943*), Albanese, Randall, and Holt (*1943*), Block (*1943*), Holt, Albanese, Brumback, Kadji, and Wangerin (*1941*), Rose, Haines, Johnson, and Warner (*1943*).

able to meet the protein needs of growing rats (*Osborne and Mendel, 1915*). When the whole egg is used, the individual proteins supplement each other in so satisfactory a manner that only an insignificant fraction remains unused.

In Table 44 are shown the daily requirements of a 70-kg. man for the essential amino acids, the amounts of these amino acids contained in

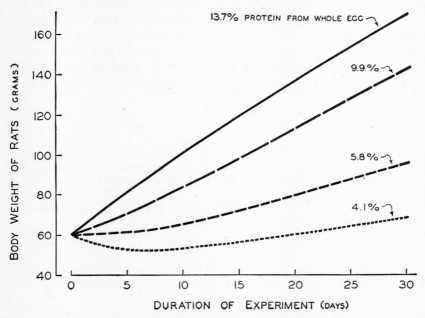

FIG. 312. The growth rate of female rats, as influenced by various dietary levels of whole egg protein. (After Barnes, Maack, Knights, and Burr, 1945.)

the egg, and the extent to which 1 egg can meet the daily requirement for each acid. Of the truly essential amino acids, only methionine (including cystine) and tryptophane seem to be present in excessively large amount in the egg, in terms of human needs. On the other hand, these two amino acids, as well as the semiessential arginine, are apt to be scarce or lacking in the human diet.

In supporting the most exacting of life processes—reproduction, lactation, and growth—the proteins of the egg, when constituting 20 or 30 per cent of the diet, are better than those of milk and muscle meats (*Clayton, 1930; Daggs, 1931*), both of which are considered good sources of complete protein. When eggs are fed to rats as 10 or 15 per cent of the diet, the growth-promoting value of the whole egg is superior to that of wheat germ, corn germ, soybean flour, cottonseed flour, skim-milk powder, and casein (*Jones and Widness, 1946*). The influence of various dietary levels of egg protein (ether-extracted dried whole eggs) on the growth of female rats may be seen in Fig. 312. It is interesting to note that growth is not improved by feeding more than 13.7 per cent of egg protein, at least under the conditions of the experiment that provided the data

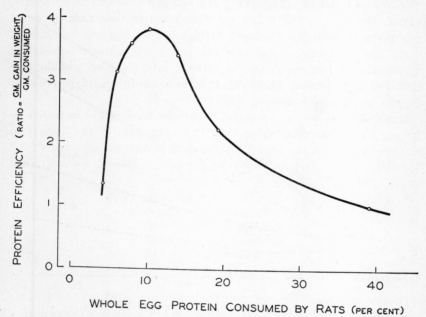

FIG. 313. Efficiency of whole egg protein at various dietary levels, as shown by ratio of weight gain in rats to amount of protein consumed. (After Barnes, Maack, Knights, and Burr, 1945.)

for Fig. 312 (*Barnes, Maack, Knights, and Burr, 1945*). The efficiency of the rat's utilization of egg protein (as measured by the ratio of gain in body weight to amount of protein consumed) increases sharply up to a maximum at the 10 per cent level and thereafter declines almost as rapidly (Fig. 313).

Egg proteins are digested by young rats to the extent of 93 to 95 per cent (*Mitchell and Carman, 1924; Sumner, 1938*). In man, their digestibility is somewhat greater, sometimes as high as 98 per cent (*Lebbin, 1902*), and compares favorably with that of the most easily digested proteins from other sources, as follows (*Murlin, Edwards, and Hawley, 1944*):

	DIGESTIBILITY
PROTEIN	(per cent)
Beefsteak	97
Egg protein	96
Soybean protein	89

Egg proteins are unquestionably superior to those of various cereals (*Osborne and Mendel, 1913, 1915; Mitchell and Carman, 1924; Murlin, Nasset, and Marsh, 1938*) and are more effective than milk or wheat endosperm in maintaining nitrogen balance in man (*Sumner and Murlin, 1938; Sumner, Pierce, and Murlin, 1938*). In rats, nitrogen equilibrium is maintained when egg protein constitutes only 8 per cent of the diet (*Mitchell and Carman, 1924, 1926*), or even as little as 5 per cent, if the rats are mature (*Sumner, 1938*).

Man's utilization of egg proteins is likewise excellent, as indicated by the egg's high biological value (100 times the ratio of the amount of nitrogen retained in the body to the amount of nitrogen absorbed during digestion). This fact is shown in the table (*Murlin, Edwards, and Hawley, 1944*).

	BIOLOGICAL VALUE
PROTEIN	(per cent)
Whole egg	97
Beefsteak	84
Soybean protein	81

OVOVITELLIN

Of the individual proteins contained in the egg, ovovitellin has the best amino acid balance (see Chapter 6, "Chemical Composition"). Furthermore, ovovitellin is a phosphoprotein, a type of protein found in quantity only in milk and eggs. As shown in Fig. 314-*A*, rats maintain better than normal growth for more than 100 days on a diet containing 18 per cent of ovovitellin (*Osborne and Mendel, 1913*). As little as 9 per cent of ovovitellin permits uninterrupted growth in rats for 150 days, at a rate which, although subnormal, is nevertheless better than that observed when an equivalent amount of casein is fed (*Osborne and Mendel, 1915*). Bateman (*1916*) found ovovitellin to be utilized excellently by dogs; to the extent of 86 per cent, according to Falta (*1906a*).

As to utilization, ovovitellin compares favorably with proteins from other sources. When it is fed to dogs, the products of its metabolism appear in the excretions as rapidly as those from meat, as shown in Fig. 315-*A*.

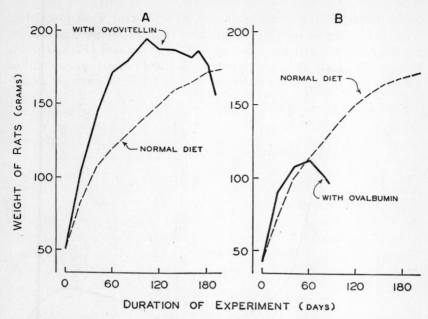

Fig. 314. Value of *A,* ovovitellin, and *B,* ovalbumin as protein supplements, as shown by growth of female rats. (After Osborne and Mendel, 1913.)

OVOLIVETIN

Ovolivetin constitutes somewhat less than one-fourth of the protein of the yolk. It appears as well balanced in essential amino acids as ovovitellin (see Chapter 6, "Chemical Composition"), but its effectiveness in animal nutrition has not been determined.

OVALBUMIN

In amino acid composition, ovalbumin appears to be a complete protein. As the sole source of protein in the diet, ovalbumin maintains rats for a longer time than casein or blood protein (*Falta and Noeggerath, 1906*). Normal growth in rats continues for about 6 weeks when all the protein is derived from ovalbumin fed as 18 per cent of the diet (cf. Fig. 314-*B*). In fact, when fed as only 9 per cent of the diet, ovalbumin has supported continuous growth in rats for 140 days (*Osborne and Mendel, 1913*); and growth, although less rapid than normal, is somewhat better than it is when the diet contains the same percentage of casein.

Raw ovalbumin in large amounts is not well tolerated by dogs (*Bateman, 1916*) or man (*Falta, 1906a*). In man, ovalbumin is utilized only to the extent of 70 per cent when raw, 87.5 per cent when coagulated. When it is added to the normal diet in reasonable quantities, such as 80 gm. daily, its assimilation is increased to 90 or 95 per cent (*Falta, 1906b*). In dogs, it is slowly and incompletely utilized, and causes profuse diarrhea (*Mendel and Lewis, 1913; Bateman, 1916*). Nitrogen excretion follow-

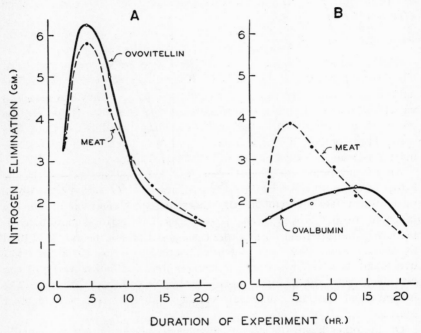

FIG. 315. Rate of nitrogen elimination in female dogs when fed meat and when fed the purified egg proteins, *A*, ovovitellin, and *B*, ovalbumin. (After Mendel and Lewis, 1913.)

ing the ingestion of ovalbumin is delayed and small in amount, compared with that following the ingestion of meat (cf. Fig. 315-*B*). Ovalbumin is thus apparently responsible for the poor digestibility of raw egg albumen, in which it is the most abundant protein.

OVOMUCOID

Ovomucoid seems to be fairly well balanced in amino acid content (see Chapter 6, "Chemical Composition"). Its nutritive value is not known, but it is tolerated well by the digestive system of dogs (*Bateman, 1916*).

OVOMUCIN, OVOCONALBUMIN, AND OVOGLOBULIN

Ovomucin, ovoconalbumin, and ovoglobulin constitute only a small share of the egg's total protein content. Neither their amino acid composition nor their effectiveness as foods has been investigated to any great extent, although experiments with dogs indicate that ovoconalbumin is difficult to digest, and apparently ovoglobulin is not very well utilized (*Bateman, 1916*).

Fats

Of the nutrients present in egg yolk, the fats are the most abundant. They constitute almost one-third of the total weight of the yolk in the natural state, and two-thirds of dehydrated yolk. They consist of a variety of fats and fatlike substances of high energy value. True fats, the glycerides of various fatty acids, make up almost two-thirds of the yolk's entire lipid content, the phospholipids and other compound lipids forming the remainder (see Chapter 6, "Chemical Composition").

All the egg lipids are so well emulsified that they are digested to the extent of 30 to 40 per cent in the human stomach (*Levites, 1909*), where unemulsified fats are only slightly digested. The digestion of the egg fats that are not broken down in the stomach is carried almost to completion in the intestine. As a result, 96 per cent of the fat is utilized by man (*Lebbin, 1902*). A somewhat poorer utilization (93.8 per cent) was found, however, when human subjects were fed half a pound of egg fat along with other foods (*Langworthy and Holmes, 1917*). The utilization of egg fats compares favorably with that of fats from other sources.

On the other hand, the various components of egg fat are not equally well utilized by man. The true fats (glycerides of fatty acids) are digested and absorbed so completely that only 2 per cent of the amount ingested is found in the feces; a somewhat greater proportion (9 per cent) of the most important phospholipid, lecithin, remains unabsorbed (*Lebbin, 1902*).

As the table shows, young male rats fed corn or coconut oil with 0.25 per cent egg lecithin gain somewhat more weight in 3 weeks than

FAT IN DIET	GAIN IN WEIGHT (grams)
Corn oil	52
Corn oil and egg lecithin	55
Coconut oil	52
Coconut oil and egg lecithin	57
Butter fat	68

animals receiving the oils without lecithin, although they gain less than rats fed butter fat (*Schantz, Boutwell, Elvehjem, and Hart, 1940*).

On the other hand, it has been observed that the growth of mice may be slightly retarded either by feeding 80 mg. of egg lecithin per day (*Robertson, 1916*), or by feeding 100 mg. of egg lecithin per day to their mothers, if the mice are suckling (*Robertson and Cutler, 1916*). It is possible that impurities were present in the lecithin used in these experiments.

Carbohydrates

The egg is an insignificant source of carbohydrates, which constitute only about 1.0 per cent of the total egg contents. Of this amount, at least one-half is combined with proteins or lipids (see Chapter 6, "Chemical Composition").

ENERGY VALUE

Eggs, although almost indispensable in the human diet, are not eaten primarily for their energy value. As a source of animal heat, they are too expensive. Although their cost is 5 per cent of the total sum spent annually for food in the United States (*Stiebeling, 1936, 1939*), they contribute only 1.7 per cent to the caloric intake of the population. However, they supplement the fuel value of many foods in which they are incorporated as ingredients.

In animal metabolism, fats yield 9.5, and proteins and carbohydrates each 4.0, Cal. per gram. (The Calorie is the average amount of heat necessary to raise the temperature of 1 kg. of water 1° C., between the freezing and boiling temperatures of water.) Since the egg contains only a negligible amount of carbohydrates, we need consider only fats and proteins as contributing to its fuel value.

Because of the variability in the size of hens' eggs, their energy value is not a constant quantity. A 32.3-gm. egg, with its shell, yields only 50 Cal. (*Tso, 1925a*); a 62.6-gm. egg, 97.2 Cal. (*Tangl, 1903*); and the standard 58-gm. egg, 92.4 Cal. (*Herring, quoted by Brody, 1945, p. 877*). The average energy value of 74 to 78 Cal. is used for the egg contents, in dietary calculations (*Chatfield and Adams, 1940; Berryman and Chatfield, 1943*).

The energy values of the eggs of various species of birds are dissimilar, not only because the eggs are unequal in size, but also because they are different in composition. The data in the table give the comparative calorific values for 100-gm. portions of the mixed contents of the eggs of several species of domesticated and game birds.

Species	Calories/100 Gm.	Investigator
Duck	202	Plimmer (*1921*)
	190	Hepburn and Katz (*1927*)
Goose	173	Hepburn and Katz (*1927*)
Turkey	169	Hepburn and Miraglia (*1937*)
Guinea fowl	162	Langworthy (*1917*)
Hen	162	Plimmer (*1921*)
	158	Chatfield and Adams (*1940*)
Plover	148	Langworthy (*1917*)
Sparrow	116	Tangl (*1903*)

The relatively high energy value of duck and goose eggs seems to be associated with their large fat content (*Hepburn and Katz, 1927; Hepburn and Miraglia, 1937*). In contrast, the eggs of altricial birds are low in calorific value. The fat content of avian eggs may bear some relationship to the length of the incubation period in different species; the longer this period, the greater is the need for stored energy.

The eggshell, rarely if ever used for human food, has little energy value. The shell of the chicken egg yields about 1.6 Cal. (*Herring, quoted by Brody, 1945, p. 877*). This heat is derived from the membranes and shell matrix, which are predominantly keratin (see Chapter 6, "Chemical Composition") and are not digestible. The eggshell and membranes contribute about 10 per cent to the weight of the eggs of precocial birds, such as the chicken and turkey (see Chapter 3, "Structure"). Since only a small part of these structures is combustible, the calorific value of the egg, per unit weight, is less when the shell and shell membranes are included.

Only one-fourth of the egg contents is combustible, the greatest part being water (see Chapter 6, "Chemical Composition"). Most of the water is contained in the albumen, which yields only one-ninth of its weight in dry matter. The yolk, however, is about one-half solids. Furthermore, fat is abundant in the yolk and lacking in the albumen. Protein is found in both components, but the yolk contains the larger amount. The greater part of the energy supplied by the egg is consequently derived from the yolk, as indicated in the following data.

Since the yolk is the main source of energy in the egg, the size of the yolk in proportion to the total weight of the egg contents becomes a significant factor. In the smaller eggs of each species, the proportion of yolk is greater (see Chapter 3, "Structure"). Consequently, the calorific

ENERGY VALUE (Calories/100 gm.)

SPECIES	YOLK	ALBUMEN	INVESTIGATOR
Duck	402	43	Hepburn and Katz (*1927*)
	409	49	Plimmer (*1921*)
Goose	409	47	Hepburn and Katz (*1927*)
Turkey	374	49	Hepburn and Miraglia (*1937*)
Guinea fowl	355	46	Langworthy (*1917*)
Hen	381	51	Plimmer (*1921*)
	315	52	Sherman and Wang (*1929*)
	353	46	Chatfield and Adams (*1940*)

value of small eggs is higher per unit weight than that of large eggs. The data in the table indicate the relative superiority of small chicken eggs as a source of energy (*Tso, 1925a*).

WEIGHT OF ENTIRE EGG (grams)	ALBUMEN (Calories)	YOLK (Calories)	EGG CONTENTS (Calories/100 gm.)
32	8	42	177
46	14	54	168
55	17	63	165
67	21	72	155

The calculated energy value of the completely dried egg contents is 615 Cal. per 100 gm. Several samples of commercially dried eggs, containing 6.6 per cent water, have been found to yield 675 Cal. per 100 gm. Still higher values, ranging from 699 to 708 Cal. per 100 gm. of dry substance have been found (*Tangl, 1903; Herring, quoted by Brody, 1945*). Pidan, or Chinese preserved eggs, from which about one-third of the water is removed, yield only 282 Cal. per 100 gm. (*Hepburn, Fegley, Sohn, and Cox, 1933*).

In Table 45 may be seen the proportion of the day's energy requirements which 1 whole egg and its individual components can give human beings of various ages. The data indicate that the egg is an excellent source of energy for infants, but an unimportant one for a full-grown man.

THE MINERALS OF THE EGG IN NUTRITION

The egg is an admirable source not only of such basic nutrients as proteins and fats, but also of indispensable minerals (free, or bound in organic molecules), some of which are not found in abundance in other foods. Among these are phosphorus and iron. The egg's content of the latter seems to be insignificant, but it is not so when compared with the daily human requirement. Some essential minerals, such as sulfur,

TABLE 45

ENERGY VALUE OF CHICKEN EGGS IN HUMAN NUTRITION
(In terms of daily requirements)

	Daily Requirements * (Calories)	Percentage Furnished by a 58-Gm. Egg		
		Albumen (15 Cal.)	Yolk (69 Cal.)	Whole (84 Cal.)
Child:				
Under 1 year	100	15.0	69.0	84.0
5 years	1500	1.0	4.6	5.6
10 years	2000	0.8	3.4	4.2
Girl, 15 years	2800	0.5	2.5	3.0
Boy, 15 years	3200	0.5	2.1	2.6
Woman (50 kg.):				
Inactive	2100	0.7	3.3	4.0
Very active	3000	0.5	2.3	2.8
Man (70 kg.):				
Inactive	2500	0.6	2.8	3.4
Very active	4500	0.3	1.5	1.8

* Recommended allowances (*J. Am. Diet. Assoc. 17:565, 1941*).

magnesium, potassium, sodium, and chlorine, are plentiful in the egg but are also liberally supplied by the ordinary mixed diet. Calcium is the only mineral for which the need is great and which the egg supplies very inadequately. The egg contains traces of other minerals that man requires only in small quantities. These, in order of abundance in the egg, are zinc, copper, bromine, manganese, and iodine.

MAJOR MINERALS

Phosphorus

The egg is one of the two foods (the other being milk) that furnish large quantities of readily available, organically combined phosphorus. Almost two-thirds of the egg's phosphorus is contained in the phospholipids, and more than one-fourth in the phosphoproteins; a small amount is present in the inorganic form (see Chapter 6, "Chemical Composition"). More than 99 per cent of the egg's total phosphorus is in the yolk.

In higher animals, lipid phosphorus is utilized equally as well as the inorganic form of the mineral; to rats, however, the phosphorus of the yolk proteins is somewhat less available than inorganic phosphorus (*Bunkfeldt and Steenbock, 1943*). Sufficient vitamin D in the diet makes phosphorus of all types equally available.

Phosphorus retention, in both man and the lower animals, depends on how closely the supply approaches the minimum requirement, as well as on several other factors, including the phosphorus-calcium ratio and the vitamin D content of the diet. Human requirements vary with the age of the individual, since infants and children must have phosphorus for growth as well as for maintenance. Very young, bottle-fed infants require the highest daily intake per unit of body weight, probably at least 45 to 50 mg. per kilogram, and possibly three times as much (*Jeans, Stearns, Goff, McKinley, and Oelke, 1933*). For safety, Shohl (*1939*) suggests twice the minimum requirement, or 95 mg. per kilogram. Children 8 to 11 years of age can maintain phosphorus balance on 34 mg. per kilogram of body weight, daily (*Wang, Kern, and Kaucher, 1930*). The requirement for the human adult is more moderate, between 0.88 and 1.3 gm. per day, or 12 to 13 mg. per kilogram of body weight (*Owen, 1939; Sherman, 1941*).

An infant would require, each day, half an egg per kilogram of body weight for bare maintenance of phosphorus balance, or 1 whole egg per kilogram for safety, if his sole source of phosphorus were eggs, each containing 116 mg. of the mineral. Since the infant's diet invariably includes milk, the phosphorus of the egg need be used only as a supplement. The egg then adds significantly to the phosphorus intake, without greatly increasing bulk.

To obtain the required amount of phosphorus from eggs alone, it would be necessary for an adult to consume, each day, 8 eggs; an 8-year-old, 20-kg. child, 6 eggs. The ordinary mixed diet, however, includes foods of plant origin, containing phytin, the source of a considerable amount of phosphorus. Although the phosphorus of phytin is not so available as that of the egg (*Krieger, Bunkfeldt, Thompson, and Steenbock, 1941*), it is ingested in such large quantities that the egg need be considered only as a valuable supplement.

Calcium

The entire egg contains more than 2.0 gm. of calcium; but the egg is nevertheless a poor source of this mineral, since 99 per cent of its calcium is discarded with the shell. The daily calcium requirement for an infant is 120 mg. per kilogram (*Shohl, 1939, p. 351*); the minimum for a man is from 0.45 gm. per day (*Sherman, 1941*) to 0.52 gm., and for a woman, 0.55 gm. (*Leitch, 1937*). Sherman (*1941*) suggests a safety factor of 50 per cent, such that the daily intake be 0.7 gm. Assuming the entire amount of calcium in the egg contents to be utilizable, about 25 eggs would be necessary to supply the quantity needed by an adult. Nevertheless, egg yolk, like cod-liver oil (see "Vitamin D," p. 616), may be an effective supplement to a calcium-deficient diet (*Tso, 1926b*).

Magnesium

This mineral occurs in the egg in the amount of 27 mg., of which 90 per cent is found in the yolk (see Chapter 6, "Chemical Composition"). Although magnesium is essential for growth, its exact function and minimum requirements have not been determined. It has been suggested (*von Wendt, 1925*) that 0.01 gm. per kilogram of body weight, or about 0.7 gm. per day, should be sufficient for an adult, although the ordinary diet furnishes only 0.34 gm. daily (*Shohl, 1939*). Infants and children require a proportionately greater amount than adults. Considerable retention of magnesium was observed in a bottle-fed infant who received 14.0 mg. per kilogram daily (*Shohl and Sato, 1923*). A magnesium intake of 13.0 mg. per kilogram is considered safe for children of pre-school age (*Daniels and Everson, 1936*).

Fifteen to thirty per cent of the magnesium requirement of an infant can be supplied by 1 egg per day; but 15 to 30 eggs daily would be needed by an adult, if there were no intake of magnesium from other sources. Magnesium, however, is not a critical mineral in nutrition and occurs in many foods.

Potassium, Chlorine, and Sodium

The egg contains substantial quantities of potassium, chlorine, and sodium—82 mg., 76 mg., and 66 mg., respectively. These are extremely necessary in the body economy; but they are found in such large amounts and are so widely distributed in foods that the contribution of the egg is not of particular significance.

Sulfur

This mineral is found in the egg contents in the amount of 67 mg. It is present in all the cells of the body and is necessary for growth and maintenance. Human requirements are not known, but a normal day's intake for an adult is about 1.5 gm., of which an egg can furnish only 4 to 5 per cent. Sulfur is widely distributed in the ordinary foods of the diet and is readily assimilated, so that there is no need to be dependent upon the egg for a supply of this mineral.

All the sulfur of the egg contents is bound in organic compounds, mainly, if not entirely, in the amino acids, cystine and methionine (*Marlow and King, 1936*). Because of the ease with which their sulf-hydryl groups are oxidized and reduced, these amino acids play a significant part in cellular respiration. Cystine and methionine are provided by the egg in such proportions as to be completely utilizable.

Iron

The egg contains about 2 mg. of iron. Iron is almost entirely confined to the yolk (*Lindow, Elvehjem, and Peterson, 1929*); very little is found in the albumen (*McHargue, 1925*), where it is bound to protein (*Lenti,*

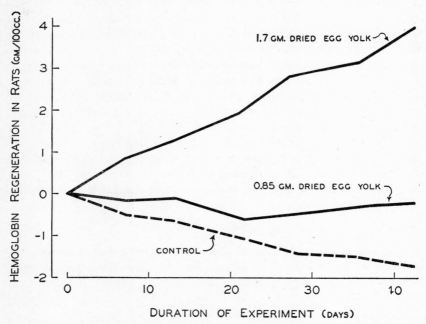

Fig. 316. Hemoglobin regeneration in the rat, as influenced by the feeding of dried egg yolk at two different levels. (After Rose, Vahlteich, and MacLeod, 1934.)

1939). In the yolk, iron is said to be in free ionic form, possibly colloidal ferric hydroxide (*Hill, 1930*), although there is evidence that it is largely present in organic combination with ovovitellin (*Saha and Mazumder, 1945*). At least 97 per cent of the iron in the yolk is in a readily available form (*Sherman, Elvehjem, and Hart, 1934*).

The iron of the egg, however, is efficiently used in hemoglobin formation only when the diet also furnishes adequate amounts of copper. In the metabolism of iron, copper is necessary (*Hart, Steenbock, Waddell, and Elvehjem, 1928*) to convert into hemoglobin the iron stored in the liver and spleen (*Cunningham, 1931; Elvehjem and Sherman, 1932*).

Egg yolk, without supplemental copper, contributes toward hemoglobin regeneration to a certain extent (Fig. 316); but, in general, the iron of the egg is less easily utilized than that of muscle or glandular meats (*Miller, Forbes, and Smythe, 1929; Whipple and Robscheit-Robbins, 1930*) or bran (*Rose, Vahlteich, and MacLeod, 1934*). The sulfur of the egg renders copper inactive in the body by combining with it to form

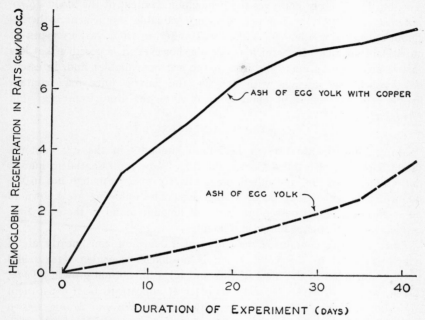

Fig. 317. Hemoglobin regeneration in rats fed ash of egg yolk with and without copper sulfate supplements. (After Rose, Vahlteich, and MacLeod, 1934.)

copper sulfide, which is insoluble in dilute hydrochloric acid (*Sherman, Elvehjem, and Hart, 1934*). For this reason, copper supplements must be large enough to insure an excess of copper uncombined with sulfur. Figure 317 shows that the effectiveness of dried egg yolk in regenerating hemoglobin in the rat is greatly improved when a very small amount of copper is added to the diet.

Human requirements for iron differ, depending on age and sex. Infants need 1.0 mg. per kilogram of body weight (*Josephs, 1939*); small children, about 0.6 mg. per kilogram (*Leichsenring and Flor, 1932; Daniels and Wright, 1934; Ascham, 1935; Porter, 1941*); adult males, a minimum of 5.0 mg. daily (*Farrar and Goldhamer, 1935*), or, more safely, 8.0 mg. (*Sherman, 1941*). Women require slightly more iron than men, particularly during pregnancy. The egg is therefore a relatively good source of iron, provided that the diet contains sufficient copper, as it usually does. One egg supplies 40 per cent, 15 per cent, and 25 to 40 per cent, respectively, of the daily iron requirement of a 5-kg. infant, a 20-kg. child, and an adult. The addition of an egg a day to a diet that is already adequate for children increases iron intake by only 10 per cent, but it definitely improves the hemoglobin content of the blood (*Rose and Borgeson, 1935*). The egg is a very valuable supplement in hemoglobin-regenerating diets for individuals suffering from various types of nutritional anemia. In pernicious anemia, however, it is possible that low gastric acidity may make the iron of the egg unavailable; and, in essential hypochromic anemia with achylia, the gastric juice may contain neither enzyme nor acid enough to liberate the iron from egg yolk (*Lederer, 1939*).

TRACE MINERALS

Many minerals are present in minute amounts in the egg contents (see Chapter 6, "Chemical Composition"). Some are essential in human and animal nutrition, and others are positively toxic, although not in the quantities in which they occur in the egg. The following are known to be essential, listed in descending order of amount found in the egg: zinc, copper, bromine, manganese, and iodine.

Zinc. An egg contains a maximum of 1.0 mg. of zinc, nearly all of which is found in the yolk. Zinc is no longer considered toxic and must be included in the diets of rats and mice for normal growth (*Todd, Elvehjem, and Hart, 1934; Day, 1942*). It appears to be a constituent of insulin (*Scott, 1934*), carbonic anhydrase (*Keilin and Mann, 1939*), and uricase (*Holmberg, 1939*). The daily requirement for young children, suggested by Scoular (*1939*), is 0.3 mg. per kilogram. One egg per day therefore can supply 10 to 15 per cent of the amount needed by a child.

Copper. This mineral, indispensable in human and animal nutrition, occurs in the amount of 0.3 mg., at most, in the contents of the egg. This is but a small portion of the quantity suggested by Daniels and Wright (*1934*) as the minimum requirement for a child (0.1 mg. per kilogram). Furthermore, the sulfur in the egg makes its copper, which is already insufficient, partly or entirely unavailable. The poor availability of the copper in eggs is not a serious problem, however, since the foods in general use contain copper (*Lindow, Elvehjem, and Peterson, 1929*) in such amounts that a mixed diet probably far exceeds the minimum requirement.

The function of copper in the formation of hemoglobin has been discussed previously, in the section on iron.

Bromine. This mineral is present in the egg in quantities no greater than 0.2 mg., practically all of which is contained in the yolk. Bromine is necessary for the growth of rats, in the extremely small amount of 0.5 mg. per kilogram of body weight; but a larger amount is required if the young are to survive for any appreciable length of time (*Winnek and Smith, 1937*).

Little is known concerning human requirements. One egg per day can probably furnish sufficient bromine for maintenance, but more is undoubtedly necessary for viability in infants.

Manganese. The egg contains no more than 0.02 mg. of this mineral. It appears necessary for fertility in male rats, and for lactation and maternal instinct in females (*Orent and McCollum, 1931; Daniels and Everson, 1935*). The human requirement is not known, but Everson and Daniels (*1934*) suggest 0.2 to 0.3 mg. per kilogram of body weight daily. The egg is therefore not a good source of manganese for man; but manganese is not a critical mineral, since it occurs in green plants and cereals.

Iodine. Normally this mineral is found in the egg in very small amounts, 0.004 to 0.01 mg., of which the greater part is in the yolk. The iodine content of the egg depends upon that of the hen's diet. In the spring and early summer, when plants grow so rapidly that their iodine absorption may lag, ingestion of the mineral and its deposition in the egg may be lowest. In those parts of the world where iodine is lacking (goitre belts), iodine may be present in the egg in especially small quantities (*Hercus and Roberts, 1927*). By feeding the hen sufficiently large amounts of iodine, it is possible to increase the iodine concentration in the egg seventy-five- to one hundred-fold (*Wilder, Bethke, and Record, 1933*). Intravenous injection of the hen with sodium iodate also results in the production of eggs with an iodine content much higher than normal (*Berkesy and Gönczi, 1933a*). "Iodized" eggs, obtained through either method of treatment, have been used in various European countries for therapeutic purposes (*Rosenberger, 1932; Leone, 1936; Vezzani, DeValle, Meynier, and Simonetta-Cuizza, 1936*).

The minimum human requirement for iodine, 0.1 mg. per day (*Cole and Curtis, 1935*), can be satisfied only to a very small extent by 1 egg, particularly if the egg is laid by a hen whose diet is not unusually rich in iodine.

THE VITAMIN CONTENT OF THE EGG

However excellent the organic and inorganic nutrients of the diet, they fail to sustain the higher animals in good health for extended periods of time unless small amounts of a variety of organic substances, the vitamins, are also consumed. A deficiency in any particular vitamin produces specific symptoms and may be fatal if sufficiently severe and prolonged. The vitamins originally were thought to be single substances. It is now known that the activity of almost every vitamin is shared by several chemically related compounds, which sometimes occur together and which often differ in potency. Although designated by letters of the alphabet, the vitamins may be more conveniently classified, according to their solubilities, as fat soluble and water soluble.

Since the egg is used so widely as human food, it is of importance to know which vitamins it contributes to the diet, and in what amounts. It is obvious that the egg's vitamin content at the start of incubation is sufficient to satisfy the immediate requirements of the avian embryo. This fact does not necessarily indicate that the egg contains all the vitamins essential to man, since the needs of various species differ. Furthermore, certain vitamins are known to be formed during incubation (*Snell and Quarles, 1941*), and to be present only to a limited extent in the unincubated egg.

The distribution of the various vitamins in the egg is determined by their solubilities. The fat-soluble vitamins are limited to the yolk, which contains all the fat of the egg contents; the water-soluble vitamins occur throughout the yolk and the albumen, which are approximately one-half and nine-tenths water, respectively (see Chapter 6, "Chemical Composition").

FAT-SOLUBLE VITAMINS

Included in this group are four vitamins, A, D, E, and K, also called antixerophthalmic, antirachitic, antisterility, and antihemorrhagic, respectively, according to the most obvious deficiency symptom that each cures or prevents. These vitamins are contained in the unsaponifiable fraction of the fats of the yolk.

Vitamin A

Vitamin A is a product of animal metabolism, manufactured in the liver and usually stored in that organ until used. It is derived from certain carotenoid pigments found in plants. Of the thirty or more such pigments, yellow, orange, or red in color, only four are vitamin-A precursors. These are alpha-, beta-, and gamma-carotene, and cryptoxanthin. Their vitamin activity depends upon the presence of the beta-ionone ring in the molecule (*Kuhn, 1933*). Beta-carotene[1] possesses two of these rings and is twice as potent a vitamin-A precursor as any of the others, each of which possesses only one such ring.

Vitamin A is essential for all animals. A severe deficiency is characterized chiefly by xerophthalmia and keratinization of the epithelium. In man, an early symptom of insufficiency is night blindness.

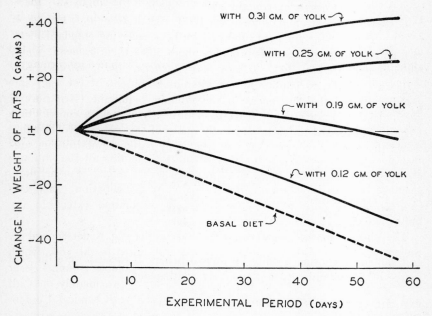

FIG. 318. Changes in the weight of rats on a basal diet deficient in vitamin A, and on the same diet supplemented with various amounts of raw egg yolk per week. (After Bisbey, Appleby, Weis, and Cover, 1934.)

The vitamin-A activity of the egg is easily demonstrated. The egg, when added to a vitamin A-deficient diet, cures xerophthalmia in rats (*Murphy and Jones, 1924*); it also improves growth. Figure 318 compares the weight of rats on a vitamin A-deficient diet with the weight of rats receiving various yolk supplements. The egg, however, is an unusual animal product in that its vitamin-A potency is due not only to the vitamin itself but also to the presence of precursors.

[1] Pure beta-carotene has been accepted as the standard of reference for vitamin-A activity, 0.6 μg (gamma) of this substance being defined as the International Unit (I.U.), which is equal to one U. S. Pharmacopoeia Unit, or 0.7 Sherman-Munsell unit.

The yolk of the egg is a better source of vitamin A than any other natural food of animal origin except liver, and butter fat from cows on rich pasturage. Since the albumen has no vitamin-A activity, the potency of the mixed egg contents per unit weight is about one-third that of the yolk alone.

No single average value can be given for the amount of vitamin A in the egg; the quantity is variable and dependent upon a number of factors. Market eggs may contain from 200 to 800 I.U. (*Boucher, 1941; Cruickshank, 1941*), and sometimes 1000 I.U. (*Rose and Vahlteich, 1938*). The average for eggs produced by hens receiving standard diets, with no vitamin supplement, is probably about 500 I.U. (*Baumann, Semb, Holmes, and Halpin, 1939; Hauge and Zscheile, 1942; Bird, 1943*).

Because of the variability in the vitamin-A content of the egg, and also because of the differences in the vitamin requirements of normal individuals, it is not possible to state definitely how great a contribution 1 egg makes toward the necessary daily quota of this vitamin. The table shows the number of eggs that the League of Nations publications sug-

	DAILY REQUIREMENT OF VITAMIN A (I.U.)	NUMBER OF EGGS REQUIRED DAILY
Growing child	6000–8000	6–12
Man	3000	3– 5
Woman	3000	3– 5
Pregnant or lactating woman	5000	5– 8

gest should be consumed daily by human beings, if the egg were the sole source of the vitamin (a most unlikely contingency).

VARIABILITY IN EGGS OF INDIVIDUAL HENS

Individual hens differ greatly in their capacity to absorb and transform the vitamin A-active substances of their feed and transfer them to the yolk (*Sherwood and Fraps, 1935*). Successive eggs from the same hen may also vary in vitamin-A potency, probably because of daily variation in the intake of the vitamin or of its precursors (*Cruickshank, 1941*). The amount of vitamin A deposited in eggs is fairly irregular even when feed and environmental conditions are rigidly controlled (*Hughes and Payne, 1937; Baumann, Semb, Holmes, and Halpin, 1939*).

Of the amount of vitamin A ingested by the hen, only the quantity in excess of that necessary for the maintenance of life processes is available for transfer to the yolk. The liver's store of this excess is drawn upon for deposition in the egg, as indicated by the fact that the vitamin-A content of the hen's liver varies inversely with the vitamin-A content of the yolk (*Russell and Taylor, 1935a*). For this reason, division of the available surplus vitamin among the eggs of an intense layer results in eggs with less vitamin activity than that of the eggs from a slow producer (*Koenig, Kramer, and Payne, 1935*).

INFLUENCE OF THE HEN'S DIET

The vitamin-A activity of the egg is proportional to the amount of the vitamin ingested by the hen. However, when the percentage of the

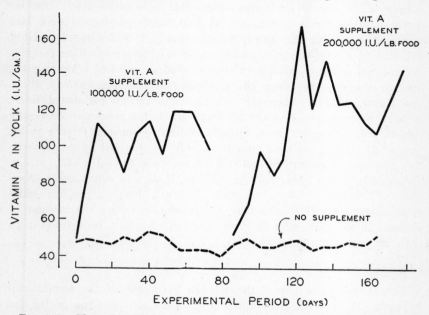

FIG. 319. The relationship between the vitamin A of the hen's diet and the vitamin-A content of the egg yolk, as shown by the effect of feeding shark-liver oil at two levels. (After Deuel, Hrubetz, Mattson, Morehouse, and Richardson, 1943.)

vitamin in the ration is increased, the vitamin content of the egg is not augmented to an equal extent. As shown in Fig. 319, doubling the dietary supplement of the vitamin enhances the yolk's vitamin-A potency but falls far short of doubling it. The efficiency with which the hen transfers vitamin A from her feed to the egg, when an ordinary ration is given, falls within the range of 8 to 39 per cent (*Sherwood and Fraps, 1935; Russell and Taylor, 1935a*). When large supplements of the vitamin are administered, the efficiency of transfer not only becomes highly variable (cf. Fig. 319) but also greatly decreases (*Thomas and Quackenbush, 1933*), and it may even fall to 0.2 per cent (*Cruickshank and Moore, 1937*). The amount of vitamin D in the feed apparently has no effect on vitamin-A transfer (*DeVaney, Titus, and Nestler, 1935*).

Grasses, alfalfa, and yellow corn, often fed to laying hens, provide large amounts of carotenoid pigments (*Sjollema and Donath, 1940*), including a certain proportion of the vitamin-A precursors. Without further supplement, a diet rich in these feeds will lead to the production of eggs containing considerable quantities of the actual vitamin and its precursors, as well as the unaltered, inactive carotenoid pigments. The natural occurrence of the active carotenoids in association with the inactive has given rise to the concept that highly pigmented yolks are also high in vitamin A. The belief is generally true, but not invariably so (*Bisbey, Cover, Appleby, and Weis, 1933*). Hens fed diets containing uniform amounts of pigment tend to produce yolks of uniform color, but not always of the same vitamin-A content (*Bisbey, Appleby, Weis, and Cover, 1934*). When the hen consumes certain pigmented feeds, such as peppers (see Chapter 3, "Structure"), the yolk may be deeply colored without having great vitamin activity. Conversely, pale yolks, rich in the almost colorless vitamin A, may be obtained by supplementing carotenoid-poor rations with shark- or cod-liver oil (*Ellis, Miller, Titus, and Byerly, 1933; Bisbey, Appleby, Weis, and Cover, 1934; Russell and Taylor, 1935b; Cruickshank, 1941*). Feeding the hen white corn instead of yellow corn may also markedly reduce the amount of pigment in the yolk, without decreasing the vitamin-A content by more than 20 per cent (*Russell, 1934*).

SEASONAL VARIATION

The annual cycle affects the vitamin-A content of the egg in two respects. If the flock is permitted free access to vegetation during the growing seasons and is fed an unsupplemented diet during the remainder of the year, the level of vitamin A in the ration varies greatly. Under such conditions, winter eggs have a vitamin-A content of only 311 I.U., in contrast to that of 730 I.U. in summer eggs (*Booher and Marsh, 1941*). In addition, the amount of vitamin A in the egg varies seasonally, merely because it is in close inverse relationship to the number of eggs that the hen lays (see Chapter 1, "Mode of Laying").

The vitamin-A content of the egg is also affected by the length of time that the hen has been in production. The first eggs of the laying season are richer in the vitamin than those laid at the close (*Sherwood and Fraps, 1932*), whether or not the ration is supplemented with vitamin A. As the hen's stores of the vitamin are depleted, less can be transferred to the yolk.

When the vitamin-A intake is uniform throughout the year, the influence of the rate of laying upon the vitamin content of the egg is clearly seen. Figure 320 shows the effect of feeding the vitamin at two levels, the lower of which, 224 I.U., is near the level of bare maintenance. On a diet containing this quantity of the vitamin, hens do not lay at the maximum rate. As the rate of production increases in the spring, less of

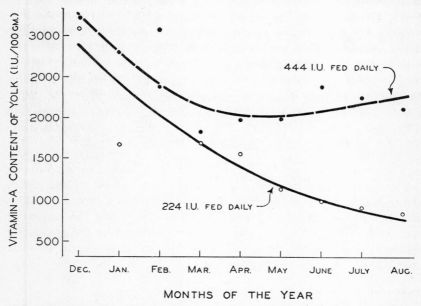

FIG. 320. Changes in vitamin-A content of egg yolk during the year, as influenced by the level of vitamin A fed to the hen. (After Sherwood and Fraps, 1934.)

the vitamin can be transferred to the yolk. Toward midsummer, laying has depleted the body reserves of the vitamin. These cannot be restored by a ration supplying only a minimum of vitamin A; and thus, as productivity declines, the amount of the vitamin in the egg still continues to diminish. When the ration furnishes 444 I.U. of vitamin A daily, or twice as much as is necessary for maintenance, the vitamin-A content of the egg lessens appreciably during the early spring; but in midsummer, when the peak of production is past, the egg tends to become somewhat richer in the vitamin as the number of eggs laid decreases. The vitamin ingested is sufficient to restore the depleted body reserves to some extent.

As yet, breed differences in the vitamin-A content of eggs have not been demonstrated. It is possible, however, that eggs laid by the slow-producing meat breeds may have a higher average vitamin-A content than those of the heavily-laying breeds, provided the hens' rations are comparable.

EFFECT OF COOKING, STORAGE, AND DESICCATION

Vitamin A is fairly thermostable, at least at the temperature of boiling water, and is more readily destroyed by oxidation and sunlight than by heat. However, there are some indications that a partial destruction of vitamin A may take place when the egg is cooked. Rats fed cooked yolk apparently grow more slowly than those fed raw yolk (cf. Fig. 311), possibly because cooking reduces the egg's vitamin-A content (*Friedberger and Abraham, 1930*). A value of 540 I.U. of vitamin A, found in cooked yolk from June eggs (*Coward and Morgan, 1935*), is significantly less than the figure of 730 I.U. for the uncooked summer egg (*Booher and Marsh, 1941*). Cooking for 7 to 10 minutes perhaps destroys as much as 37.5 per cent of the vitamin (*Caro and Locatelli, 1936*), but exposure to moderate heat for 3 to 5 minutes, the more usual practice, probably causes a much smaller loss.

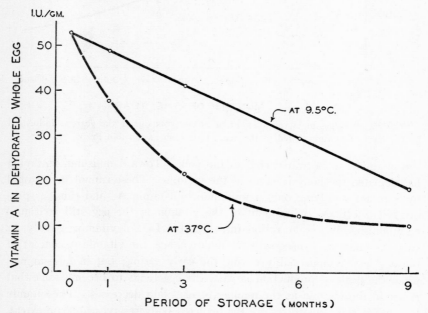

Fig. 321. Changes in the vitamin-A content of dehydrated whole egg powder during periods of storage at two different temperatures. (After Klose, Jones, and Fevold, 1943.)

When held for prolonged periods, the egg sometimes loses much of 'its vitamin-A potency. Its content of this vitamin may be reduced 75 per cent during a year in cold storage, and 50 per cent during 18 months in water-glass solution (*Manville, 1926*). However, eggs held in the frozen state for 9 years are almost equal to fresh eggs in their ability to cure the symptoms of vitamin-A deficiency in rats (*Jones, Murphy, and Moeller, 1925*).

The spray drying of eggs apparently causes little or no destruction of vitamin A (*Hauge and Zscheile, 1942; Klose, Jones, and Fevold, 1943; Denton, Cabell, Bastron, and Davis, 1944; Schrenk, Chapin, and Conrad, 1944; Cruickshank, Kodicek, and Wang, 1945*). However, the particulate form of dehydrated eggs permits an oxidative loss of vitamin A during storage. The extent of the loss depends upon the length of the storage period and the holding temperature. At refrigeration temperatures, the vitamin-A content of dried eggs is reduced more than 60 per cent in 9 months (Fig. 321). When the temperature is 37° C., a similar reduction occurs in 3 months, and an 80 per cent reduction in 9 months. Loss of

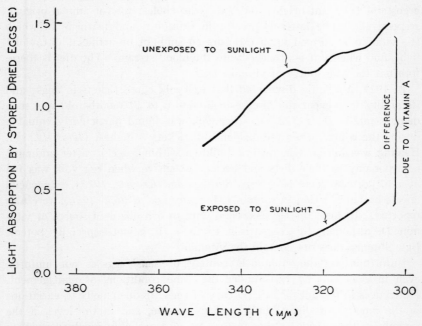

FIG. 322. Reduction in the vitamin-A content of dried eggs exposed to sunlight, as shown by changes in the spectrophotometric absorption curve. (After Denton, Cabell, Bastron, and Davis, 1944.)

vitamin-A potency in dried eggs may be prevented, or greatly retarded, by storage at $-18°$ C. (*Hauge and Zscheile, 1942*), or at $15°$ C. in an atmosphere of nitrogen (*Cruickshank, Kodicek, and Wang, 1945*).

Spectrophotometric measurements of dried egg powder reveal that its content of vitamin A may be destroyed by sunlight. In Fig. 322, the loss in vitamin A is indicated by the difference in the two curves, the upper of which represents the absorption before exposure to sunlight, and the lower, the residual absorption afterward.

In China, the preservation of eggs as pidan does not markedly reduce the vitamin-A content (*Tso, 1925b, 1926a*).

Vitamin D

Of all vitamins, antirachitic D is present in the fewest natural foods and is generally least available in the diet, despite the fact that it is essential in the proper metabolism of calcium and phosphorus. Although usually spoken of as a single substance, the vitamin is actually a group of several, of which only two are of importance. These, calciferol and demethyl-dihydro-calciferol, result from the activation of the provitamins, ergosterol (of plant origin) and 7-dehydro-cholesterol (of animal origin), respectively. The latter is present in skin, fur, and feathers, in which the vitamin is formed upon exposure to sunlight or artificial ultraviolet light, and whence it is absorbed into the blood stream. The diet is therefore not the sole source of vitamin D.

In 1921, Mellanby discovered that egg yolk cured rickets in dogs; and its ability to prevent the disease in infants 6 to 14 months old was soon demonstrated (*Hess, 1923a*). Egg yolk was found particularly valuable during the winter, when the incidence of rickets is highest (*Hess, 1923b*). Healing was initiated in rachitic children within 3 weeks after adding 2 eggs per day to their diets and became extensive when egg yolk was fed at a 10 per cent level (*Casparis, Shipley, and Kramer, 1923*). Beneficial effects on calcification in rabbits (*Mellanby and Killick, 1926*) were also reported. In fact, the egg is second only to fish oils as a source of vitamin D, although, by comparison to these, it is not especially potent. Egg albumen does not contain the vitamin.

Unfortunately, the vitamin-D content of hens' eggs is not uniform, even in eggs laid by hens under the same conditions of management. Differences in the antirachitic potency of eggs are due chiefly to variations in the amount of direct sunshine hens receive, and in the level of the vitamin in their diets. Other factors are also probably involved, such as the rate of laying, and the dissimilar physiological requirements of individual hens for the vitamin.

EFFECT OF ULTRAVIOLET IRRADIATION

The vitamin D of the egg is derived from the surplus which the hen does not need for the maintenance of life processes. Hens that receive no direct sunlight or supplementary antirachitic vitamin in their diet do not lay well (see Chapter 1, "Mode of Laying"). Their eggs may contain from 5.0 I.U.[2] of vitamin D per yolk (*Bethke, Record, Wilder, and Kick, 1936*), to about 11.0 or 12.0 I.U. per yolk (*Branion, Drake, and Tisdall, 1934a, 1935*). When the diet is low in the vitamin, the vitamin-D content of the egg is determined largely by the amount of ultraviolet light to which the hen is exposed (*Hughes, Payne, Titus, and Moore, 1925*). Hens on limited range, and therefore receiving a certain amount of direct sunlight, may lay eggs in which the vitamin D of the yolk ranges from 25 to 70 I.U., depending on the season (*DeVaney, Munsell, and Titus, 1936*). Sunlight, however, is not reliable as the sole source of vitamin D, since it is variable in amount from season to season, and even from day to day. Exposing the hen to artificial ultraviolet light is a fairly effective means of enhancing the vitamin-D potency of the egg. Irradiation of the hen for only 10 minutes a day sometimes results in a tenfold increase in the antirachitic activity of her eggs (*Hart, Steenbock, Lepkovsky, Kletzien, Halpin, and Johnson, 1925*). In one instance, however, 20 minutes of irradiation daily raised the average vitamin-D content of the yolk by only 25 per cent (*Branion, Drake, and Tisdall, 1934b*).

When the total amount of vitamin D available to the hen each day remains the same throughout the year, the quantity of the vitamin in each egg diminishes somewhat when laying increases in intensity. If, on the other hand, the hen's diet is unsupplemented in winter, and she is allowed open range during the warmer months, there is inevitable seasonal variation in the vitamin-D content of the egg. Egg-laying and the antirachitic potency of the egg then increase simultaneously in the spring. Eggs laid in June have been found 50 per cent richer in vitamin D than eggs laid in April, when there was only half as much sunlight as in June (*Bisbey, Appleby, Weis, and Cover, 1934*). Summer eggs may be more than four times as potent as winter eggs (*Steenbock and Scott, 1932*). The table shows how the egg's approximate vitamin-D content may increase steadily during the spring (*DeVaney, Munsell, and Titus, 1933*). Autumn eggs, although a fairly good source of the vitamin, decrease in antirachitic potency as the season progresses (*Maughan and Maughan, 1933*).

[2] One International Unit of vitamin D is equivalent to the activity of 0.025 μg. of the crystalline vitamin.

VITAMIN-D CONTENT OF EGG

MONTH	(International Units)
February	25
April	49
May	58
June	70

By direct irradiation of the yolk, the vitamin-D activity of the egg can be multiplied 15 or 20 times (*Steenbock and Black, 1925*). By the same means, egg oil, containing 3.3 I.U. of vitamin D per gram, may be given a potency of 33.3 I.U. per gram; and the vitamin-D content of the unsaponifiable matter of egg oil can be increased from 16.6 I.U. per gram to 831 I.U. (*Reder, 1938*). This is interesting, in view of the discovery that egg oil is more rapidly absorbed from the gastrointestinal tract (of the rat) than is cod-liver oil (*Reder, 1942*).

EFFECT OF VITAMIN-D SUPPLEMENTS

Supplementing the hen's ration with vitamin-D concentrates is the most dependable method of obtaining eggs of good antirachitic potency throughout the year. A fivefold increase in the vitamin-D content of eggs laid by confined hens may result when a 2 per cent supplement of cod-liver oil is added to the diet (*Bethke, Kennard, and Sassaman, 1927*). Larger supplements cause further increases in the amount of the vitamin in the yolk (Fig. 323). However, there are limitations to the use of such substances as cod-liver oil, sardine oil, and irradiated ergosterol (viosterol). Massive doses are not well tolerated, and they are usually followed by a decline in egg production, and sometimes by almost complete cessation of laying (*Branion and Smith, 1932; Titus and Nestler, 1935; Bethke, Record, Kick, and Kennard, 1936*). Furthermore, when the diet contains excessively high levels of vitamin D, there is a great reduction in the efficiency with which the ingested vitamin is transferred to the yolk, although the eggs produced are true vitamin-D concentrates. These facts are indicated by the tabulated data based on the observations of Branion, Drake, and Tisdall (*1934a*).

VITAMIN D IN RATION	VITAMIN D IN YOLK (I.U.)	INCREASE IN POTENCY (times)	RELATIVE EFFICIENCY OF TRANSFER
None	13	. .	. . . .
Cod-liver oil (1 per cent)	99	8	1.00
100× viosterol (1 per cent)	2,699	207	0.27
10,000× viosterol (1 per cent)	59,400	4,569	0.06

For practical purposes, a cod-liver oil supplement in the amount of 1 or 2 per cent of the confined bird's diet is optimum, both for egg production (*Titus and Nestler, 1935*) and for efficient transfer of vitamin D to the yolk (*DeVaney, Munsell, and Titus, 1933*). Cod-liver oil in excess of 2 per cent of the diet is apt to impart a fishy taste to the egg (*Cruickshank, 1939*). Although viosterol may sometimes be equally as effective as cod-liver oil at the 1 per cent level (*Branion, Drake, and*

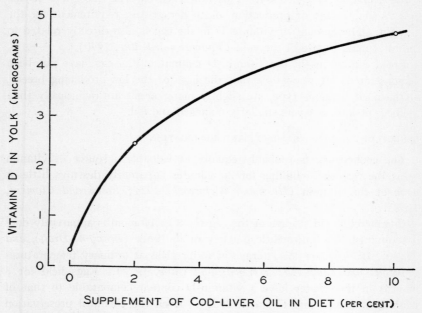

Fig. 323. Relationship between the vitamin-D content of the egg yolk and the amount of cod-liver oil in the diet of the hen. (After Branion, Drake, and Tisdall, 1934a.)

Tisdall, 1934a), it is usually less so; and its inferiority is especially noticeable when it is fed at the higher levels (*DeVaney, Munsell, and Titus, 1933; Branion, Drake, and Tisdall, 1934a; Guerrant, Kohler, Hunter, and Murphy, 1935*). The data below show the comparative

Amount/100 Gm. Feed (International Units)	Vitamin D in Yolk (International Units)
Cod-liver oil	
540	90
5400	850
Viosterol	
540	50
5400	550

effects of equal units of the two preparations upon the vitamin-D content of the egg (*Bethke, Record, Wilder, and Kick, 1936*).

Pullets possibly are not able to transfer vitamin D to their eggs as well as older hens, since the vitamin-D content of pullets' eggs is apparently less than that of eggs laid by mature birds (*Guerrant, Kohler, Hunter, and Murphy, 1935*). This difference may be due either to the pullet's higher rate of production, or to her greater requirement for the vitamin. The amount of vitamin D in the egg is also directly related to the total mass or size of the yolk (*Thomas and Wilke, 1940*).

Breed differences in the vitamin-D content of the egg have not been demonstrated. It is possible that the eggs of the low-producing breeds, of the meat or game type, are slightly more potent antirachitically than the eggs of intense layers similarly managed and fed.

EFFECT OF COOKING, STORAGE, AND DESICCATION

The cooked egg is probably equally as valuable a source of vitamin D as the raw egg. Boiling for 20 minutes apparently destroys little or none of the vitamin (*Hess and Weinstock, 1924; Lesné and Clement, 1932*).

Commercial cold storage of the egg for 8 to 10 months apparently does not diminish the antirachitic potency of the yolk (*Branion, Drake, and Tisdall, 1934c*), nor does storage in water glass at ordinary temperatures (*Marlatt and Clow, 1927; Woods and Clow, 1927*). Eggs held for 8 months in lime water have a vitamin-D content comparable to that of fresh eggs (*Lesné and Clement, 1932*). Even the chemical preservation of eggs as pidan, in China, does not affect their antirachitic activity to any appreciable extent (*Tso, 1926b*).

Furthermore, the vitamin D of the egg is not measurably reduced during the process of spray drying (*Klose, Jones, and Fevold, 1943; Denton, Cabell, Bastron, and Davis, 1944; Cruickshank, Kodicek, and Wang, 1945*). Antirachitic potency has been found undiminished in dehydrated eggs stored at 9.5° C. for 9 months in normal atmosphere, and at 15° C. for 5 months in a nitrogen gas pack (*Klose, Jones, and Fevold, 1943; Cruickshank, Kodicek, and Wang, 1945*). A loss of 29 per cent of the vitamin-D content of an egg powder held 9 months at 37° C. has been reported (*Klose, Jones, and Fevold, 1943*), although another sample was apparently unaffected during storage at the same temperature for 4 months (*Denton, Cabell, Bastron, and Davis, 1944*).

THE EGG AS A SOURCE OF VITAMIN D IN THE HUMAN DIET

The human requirement for vitamin D, taken orally, varies greatly according to the individual's age and condition, and the amount of direct sunlight he receives. During pregnancy and lactation the demand is probably greater than at any other time (*Macy, Hunscher, McCosh, and Nims, 1930; Hunscher, 1930*), about 750 or 800 I.U. per day. Premature infants must have approximately 600 to 800 I.U. daily; normal full-term infants, from 300 to 600 I.U. (*Stearns, Jeans, and Vandecar, 1936*). Growing children and adolescents need perhaps 350 I.U. daily to insure proper bone and tooth development (*Boyd, Drain, and Nelson, 1929*). The vitamin-D needs of the normal adult are obscure; 300 to 400 I.U. are assumed necessary, but less will probably suffice in spring and summer if the individual spends much of his time in direct sunlight.

The egg is difficult to evaluate as a source of vitamin D, in view of the variability in its content of the vitamin. However, if 50 I.U. is taken as the vitamin-D potency of the average egg (*Bird, 1943*), the percentage of the daily requirement for this vitamin that 1 egg contributes to the diet of various classes of individuals will be as tabulated.

	DAILY REQUIREMENT (International Units)	CONTRIBUTION OF 1 EGG (per cent)
Pregnant or lactating women	750–800	6– 7
Premature infants	600–800	6– 8
Full-term infants	300–400	13–17
Young children, adolescents	350	14
Normal adults	300–400	13–17

Although these data apparently indicate that the addition of 1 egg to the diet does not provide a substantial increase in vitamin D, the limited distribution of this vitamin in the foods normally consumed makes any source important.

Vitamin E

Vitamin E is known to be essential for normal reproduction in various animals and is assumed to be required by man, although it is difficult to obtain conclusive evidence to this effect. The substance that possesses the highest vitamin-E activity has been identified as alpha-tocopherol. Its presence in the egg was established by experiments which showed that rats on a vitamin-E deficient diet became able to reproduce when one-third of a whole egg was added daily to their ration (*Evans and Burr, 1927*). Alpha-tocopherol is contained in the nonlecithin fraction of the lipids extracted from the yolk (*Serono and Montezemolo, 1943*). The relative vitamin-E potency of the egg is influenced by the amount of the vitamin in the hen's feed (*Barnum, 1935*).

Vitamin K

The antihemorrhagic vitamin K, or menadione, is essential for the proper formation of prothrombin, and therefore for normal blood-clotting. Man's daily requirement for this vitamin has not yet been established, but from 1 to 2 mg. per day are apparently sufficient to correct conditions of prothrombin deficiency.

In the egg, vitamin K is confined to the yolk. Hard-boiled yolk gives protection to chicks fed vitamin K-deficient diets; the albumen does not (*Almquist and Stokstad, 1936*).

The vitamin-K potency of the egg is variable, depending upon the level of the vitamin in the hen's diet. When chicks are fed vitamin K-deficient rations, their resistance to hemorrhagic disease is in direct relationship to the amount of vitamin K fed to the hens from whose eggs the chicks hatched (*Almquist and Stokstad, 1936*). Differences in the degree of prothrombin deficiency in newly hatched chicks kept on vitamin K-free diets from the start indicate that there is also seasonal variation in the egg's content of vitamin K. Eggs laid during the late winter and early spring appear to contain less of the vitamin than eggs laid at other times of the year (*Thayer, McKee, Binkley, MacCorquodale, and Doisy, 1939*).

WATER-SOLUBLE VITAMINS

The water-soluble vitamins consist chiefly of the vitamin-B complex, a group of substances of diverse chemical composition. Vitamin C, or ascorbic acid, is also water soluble.

Vitamin-B Complex

Of the twelve or more water-soluble vitamins which probably constitute the vitamin-B complex, eight of those already identified have been found in the egg. These are thiamin, riboflavin, niacin, pantothenic acid, inositol, pyridoxine, biotin, and folic acid. The number becomes nine if choline is considered a vitamin.

In general, B-complex vitamins are widely distributed in foods, but in concentrations that differ greatly from one food to another. The egg is a good source of some of these substances, and a poor source of others. In addition, the egg's content of certain water-soluble vitamins is demonstrably variable, depending chiefly upon the composition of the hen's diet.

THIAMIN

Thiamin functions as an essential part of cocarboxylase, an enzyme of carbohydrate metabolism preventing the accumulation of pyruvic acid in the tissues. Although a severe deficiency of thiamin causes polyneuritis, a variety of symptoms may result when the insufficiency is moderate or slight.

Apparently the chicken egg's entire content of thiamin is confined to the yolk. A small amount has been reported in the albumen of the duck's egg, the yolk of which is also a better source of this vitamin than that of the hen's egg (*Leong, 1940*). The growth of young rats, if retarded by a thiamin-deficient diet, is not restored to normal by the addition of chicken egg albumen (coagulated) to the ration (*Chick and Roscoe, 1929*). Also, rats die of polyneuritis when a thiamin-free diet is supplemented by chicken egg albumen alone. On the other hand, egg yolk, as the sole supplement of such a diet, permits almost normal growth, when fed in the amount of 1.0 gm. per rat per day; if the daily quantity given is only 0.5 gm. per rat, polyneuritic symptoms are prevented (*Bethke, Record, and Wilder, 1936*). The antineuritic properties of the hen's egg, however, are not sufficient to protect the pigeon from polyneuritis and eventual death, even when as much as 30 per cent of the diet consists of whole egg (*Hoagland and Lee, 1924*).

The amount of thiamin present in the hen's egg has been variously reported. Estimates of the concentration of thiamin in the yolk range from 3.5 μg. per gram (*Booher and Hartzler, 1939*) to 4.8 μg. per gram (*Pyke, 1937*). Analyses of the whole egg have yielded values from 1.2 μg. per gram (*Cheldelin and Williams, 1942*) to 2.6 μg. per gram (*Pyke, 1939*). Differences probably are due not only to methods of determination but also to the fact that the thiamin content of the egg varies greatly according to the concentration of the vitamin in the diet of the hen. Eggs produced by birds on high-thiamin diets are two or three times richer in the vitamin than the eggs of birds on low-thiamin diets (*Ellis, Miller, Titus, and Byerly, 1933*). It is also possible that there are breed differences, or at least familial differences, in the hen's ability to transfer thiamin from her feed to her eggs. It has been noted that certain strains of White Leghorns lay eggs containing considerably more thiamin than those of Rhode Island Reds and Barred Plymouth Rocks fed comparable rations (*Scrimshaw, Hutt, Scrimshaw, and Sullivan, 1945*). Furthermore, the eggs of White Leghorns, when fed to pigeons, appear to have a slightly greater antineuritic value than those of the other two breeds (*Hoagland and Lee, 1924*).

Thiamin is somewhat thermolabile, except in acid solution. Cooking may slightly reduce the egg's content of this vitamin (*Hoagland and Lee, 1924*). During storage, the thiamin of the egg probably does not decrease, since it has been found undiminished in eggs held for 3 weeks at incubation temperature, at room temperature, and at 5° C. (*Scrimshaw, Thomas, McKibben, Sullivan, and Wells, 1944*). Chinese preserved ducks' eggs (pidan), however, suffer an almost complete loss of thiamin (*Tso, 1926a*). Spray drying may destroy either little or no thiamin (*Herraiz and de Alvarez Herrero, 1943*), or as much as 30 per cent of the amount present in the liquid egg, probably depending upon conditions during dehydration. When stored under relatively high temperatures, such as 20° C. to 37° C., dried eggs may lose from 50 to 100 per cent of their thiamin in 9 months, but the concentration of the vitamin is not appreciably lessened during the same storage period at 9° C. (*Klose, Jones, and Fevold, 1943; Cruickshank, Kodicek, and Wang, 1945*).

An average egg supplies approximately 5 per cent of the daily amount of thiamin needed by an adult man.

RIBOFLAVIN

Riboflavin may also occur as riboflavin phosphate or as part of a flavoprotein complex. It is present in enzymes concerned with tissue respiration. It gives a greenish yellow fluorescence in neutral aqueous solution. In different species of animals, deficiency of riboflavin is manifested in a variety of ways, among which are retarded growth, dermatitis, cataracts, paralysis, and nerve degeneration. In man, dermatitis, vascularizing keratitis, ocular symptoms, and various abnormalities of the tongue and mucous tissues of the mouth indicate insufficient riboflavin intake.

Riboflavin is present in both yolk and albumen and is responsible for the greenish yellow tinge of the latter. Its amount per unit of weight is greater in the yolk than in the albumen. The estimated concentration of riboflavin in the yolk ranges from 3.9 μg. per gram (*Bauernfeind and Norris, 1939; Norris and Bauernfeind, 1940*) to 7.6 μg. per gram (*Snell and Strong, 1939*); in the albumen, from 1.35 μg. per gram (*Engel, Phillips, and Halpin, 1940*) to 4.96 μg. per gram (*Murthy, 1937*). Although the yolk, in the natural state, contains from one and one-half to two and one-half times as much riboflavin per gram as the albumen, this ratio is reversed when the egg is dried, because of the proportionally greater loss of water by the albumen.

The riboflavin content of the egg, especially that of the albumen, is extremely variable. Several factors are responsible for this variability. Although a hen may lay successive eggs in which the amount of ribo-flavin remains at a fairly constant level, individual birds differ greatly in their ability to transfer riboflavin to their eggs and in their metabolic requirements for this vitamin. For example, in the yolks of the eggs

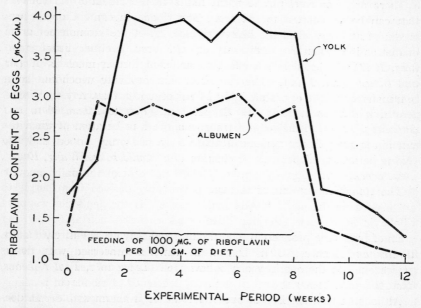

Fig. 324. Increase in the riboflavin content of the yolk and albumen when 1000 μg. of riboflavin (in the form of liver extract) are added to each 100 gm. of the hen's diet. (After Norris and Bauernfeind, 1940.)

laid by six birds fed the same ration, the average riboflavin concentration was found to range from 2.96 μg. to 4.69 μg. per gram (*Peterson, Dearstyne, Comstock, and Weldon, 1945*).

In addition, the quantity of riboflavin deposited in the egg is dependent largely upon the amount in the diet of the hen. The improvement in the growth-promoting value of the yolk and the albumen, after the hen's ration is supplemented with riboflavin, may be seen in the table (*Lepkovsky, Taylor, Jukes, and Almquist, 1938*). Figure 324 shows

RIBOFLAVIN SUPPLEMENT IN HEN'S DIET	WEIGHT GAINED BY RATS (grams/day)	
	FED ON YOLK	FED ON ALBUMEN
None	1.03	1.39
2.5 mg./lb. feed	1.95	2.85

the effect of increasing the amount of riboflavin in the hen's diet from 100 μg. per 100 gm. to 1000 μg. Within 2 weeks, the average concentration of riboflavin in the albumen, 1.29 μg. per gram, increases to 2.89 μg.; in the yolk, it increases from 1.58 μg. per gram to 3.98 μg. After the original diet is resumed, the riboflavin content of the egg gradually decreases, until, about 4 weeks later, it has returned to the lower level.

There are, however, physiological limits to the amount of riboflavin that can be transferred to the egg. The riboflavin content of the egg is almost as great when the hen receives 5 μg. of the vitamin per gram of diet as it is when she receives 10 μg. (*Petersen, Lampman, and Stamberg, 1947*); 20 μg. does not effect a significant further increase (*Norris and Bauernfeind, 1940*). The maximum efficiency with which riboflavin is transferred to the egg is probably 24 per cent (*Jackson, Drake, Slinger, Evans, and Pocock, 1946*) to 30 per cent (*Petersen, Lampman, and Stamberg, 1947*). Efficiency of transfer appears to be greatest when the vitamin is given in the amount of about 3 μg. per gram of feed, and also during periods of highest egg production (*Jackson, Drake, Slinger, Evans, and Pocock, 1946*).

The riboflavin content of the egg is probably derived from surpluses not needed by the hen for body maintenance. If the hen's diet supplies riboflavin in amounts affording little or no surplus—such as 0.8 μg. per gram of feed—egg production soon declines and may cease entirely (*Hunt, Winter, and Bethke, 1939*). During the winter months, egg production is proportional to the quantity of riboflavin fed (*Lepkovsky, Taylor, Jukes, and Almquist, 1938*).

Although riboflavin is destroyed by light, it is thermostable. It does not disappear during spray drying of the egg, nor during subsequent storage of the dehydrated product (*Klose, Jones, and Fevold, 1943; Denton, Cabell, Bastron, and Davis, 1944; Cruickshank, Kodicek, and Wang, 1945*). The activity of this vitamin is retained in cooked eggs (*Aykroyd and Roscoe, 1929*).

From the viewpoint of human nutrition, the egg is counted among the good sources of riboflavin, because 1 egg supplies nearly 7 per cent of an adult man's daily requirement of riboflavin.

NIACIN

Niacin (nicotinic acid) is a component of enzymes that function in glycolysis and respiration, and its deficiency in man's diet is the chief causative factor in pellagra. Various methods of estimating the amount of this vitamin in the egg have given different results, but the concentration in the entire egg contents is probably very low, between 0.7 and 0.9 μg.

per gram (*Snell and Quarles, 1941; Cheldelin and Williams, 1942; Tepley, Strong, and Elvehjem, 1942*). Some investigators have found a higher concentration of niacin in the yolk than in the albumen, whereas others have observed the reverse.

Niacin is an extremely stable compound and is not destroyed either during spray drying of the egg or subsequent storage (*Klose, Jones, and Fevold, 1943; Cruickshank, Kodicek, and Wang, 1945*).

One egg supplies only about 0.2 per cent of the adult male's daily requirement of niacin. The egg's negligible contribution of niacin to the human diet has not encouraged inquiry into the factors controlling the concentration of this vitamin in the egg. Apparently the chick embryo does not require niacin already formed but is capable of synthesizing it during incubation (*Snell and Quarles, 1941*).

PANTOTHENIC ACID

Pantothenic acid is presumably essential in human nutrition, although its exact function is not as yet completely understood. Its deficiency in man either is not recognized or else it does not occur because of the vitamin's wide distribution in foods. However, certain species of animals are known to require pantothenic acid, an inadequacy of which is associated with various nervous disorders, and sometimes with changes in fur pigmentation.

The concentration of pantothenic acid in the egg is high, exceeded only by that of inositol, choline, and possibly riboflavin. The whole egg probably contains between 11 and 14 μg. per gram (*Snell and Quarles, 1941; Cheldelin and Williams, 1942*). Most of the pantothenic acid is in the yolk. By various methods of assay, its concentration has been found to be 0.76 to 27 μg. per gram in the albumen, and 46.8 to 65 μg. per gram in the yolk (*Snell, Aline, Couch, and Pearson, 1941; Jukes, 1941*).

The pantothenic acid concentration in eggs from different hens fed the same ration may vary by 100 per cent; but, in general, the amount of the vitamin in the bird's diet regulates the amount deposited in the egg. Supplementing a pantothenic acid-deficient diet with 8.4 mg. of *dl*-sodium pantothenate per 100 gm. of feed causes the pantothenic acid content of the whole egg to rise from 254 μg. to 3400 μg. per 100 gm., as shown in Fig. 325. Relatively, the yolk gains slightly more than the albumen (*Snell, Aline, Couch, and Pearson, 1941*). It has not yet been demonstrated whether or not there are limits to the extent to which dietary pantothenic acid may be transferred to the egg.

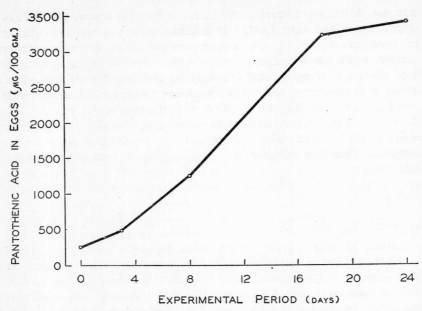

F<small>IG</small>. 325. Increase in the amount of pantothenic acid in the egg contents after the addition of 8.4 mg. of *dl*-sodium pantothenate to each 100 gm. of the hen's ration. (After Snell, Aline, Couch, and Pearson, 1941.)

Pantothenic acid does not disappear from the egg during spray drying, probably because it is less readily destroyed by moist than dry heat; nor is it lost when dried eggs are stored (*Klose, Jones, and Fevold, 1943*).

INOSITOL

Inositol, a hexa-hydroxy-cyclohexane isomer of *d*-glucose, occurs in several forms in tissues. Like pantothenic acid, it is known to be essential for the prevention of various symptoms in lower animals, but its specific function in human nutrition is undetermined. It is usually present in food in larger absolute amounts than any other vitamin, and in the egg its concentration (as total inositol) is high, from 220 to 330 μg. per gram of whole egg (*Woolley, 1942; Cheldelin and Williams, 1942*). In the free form, its concentration is much less, from 46 μg. (*Woolley, 1942*) to 73 μg. per gram (*Snell and Quarles, 1941*).

PYRIDOXINE

Pyridoxine functions in the body in an unknown manner. It possibly aids in synthesizing fat from protein (*McHenry and Gavin, 1941*), and its deficiency perhaps contributes to the occurrence of pellagra (*Spies, Bean, and Ashe, 1939*). It is known to be essential to a number of lower animals, which may develop acrodynia and anemia when they receive insufficient pyridoxine. Rat-curative methods of assay indicate that the egg is a fairly good source of pyridoxine (*Schneider, Ascham, Platz, and Steenbock, 1939*); but yeast-growth methods, perhaps more specific, show that the concentration of pyridoxine in the egg is about 0.22 μg. per gram and therefore low when compared with that in many other foods (*Cheldelin and Williams, 1942*). Pyridoxine appears to be contained chiefly in the yolk and to be largely or wholly absent from the albumen.

BIOTIN

Biotin is thought to be essential for most animals. It is present in foods in extremely minute quantities, but it is very potent. In the hen's egg, there is about 0.37 μg. of physiologically active biotin per gram of yolk (*Snell, Eakin, and Williams, 1940*). In the albumen, where the concentration of biotin is approximately 0.05 μg. per gram, or possibly two or three times this amount (*György and Rose, 1942*), this vitamin exists in a somewhat peculiar state. It is very stably combined with another substance, avidin, and is thereby rendered entirely inactive. When the egg is cooked, biotin is released from the avidin-biotin complex. By comparison with other foods, the egg is then a rich source of biotin. Raw albumen, however, contains avidin in sufficient excess to inactivate the biotin of the entire uncooked egg.

The discovery of avidin and biotin, and of the capacity of the one to bind the other, explained the rather puzzling phenomenon of "egg-white injury" as a biotin deficiency. This condition was first observed in rats that were fed undenatured, dried egg albumen as the sole source of protein in the diet (*Boas, 1924*). Egg-white injury is manifested as an eczematous dermatitis, accompanied by denudation, roughening of the fur, loss of weight, and such evidences of nervous disorder as spasticity of gait and abnormal, "kangaroo" posture. The outcome is death, which occurs only a few weeks after the onset of symptoms.

At first, it was thought that desiccation changed the nutritive value of albumen in some respect, either by destruction of an essential factor (*Findlay and Stern, 1929*), or by the formation of a toxic substance (*Fixsen, 1931*). It was soon noted, however, that raw fresh albumen was able to cause egg-white injury (*Parsons, 1931b*), even when constituting as little as 18 per cent of the diet (*Salmon and Goodman, 1934*). Albumen coagulated at 65° C. was found to be less toxic than raw egg white (*Parsons, 1931b*). Furthermore, the toxic properties disappeared entirely when albumen was heated at a temperature of 80° C. for 5 minutes, steamed for 1.5 hours, autoclaved for 6 hours at 15 pounds' pressure, digested by pepsin, or incubated for 3 days with hydrochloric acid (*Parsons and Kelly, 1933a, 1933b*). Alcohol denaturation had no effect (*Parsons and Kelly, 1933b*), and digestion by papain concentrated

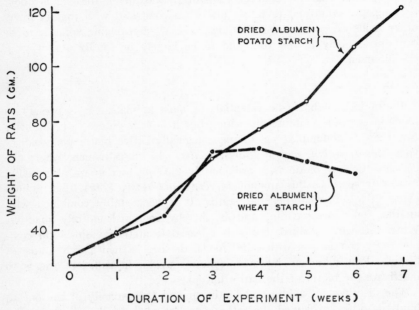

Fig. 326. Growth of rats receiving dried egg albumen and wheat starch compared with the growth of rats fed dried egg albumen and potato starch (which contains biotin). (After Boas, 1927.)

the toxicity in the filtrable digest (*Parsons, Janssen, and Schoenleber, 1934*). The toxicity remained in the protein fraction after complete saturation with ammonium sulfate but was absent from purified oval-bumin (*Parsons and Kelly, 1933b*) and ovoglobulin (*Fixsen, 1931*).

In addition, it was shown that whole raw egg cured the syndrome in rats, if twice as much yolk as albumen was fed (*Parsons and Lease, 1934*). Chicks recovered only when they received five to ten times as much yolk as albumen (*Lease, 1937*). Protective properties were found in various other foods; in fact, it was concluded (*Boas, 1927*) that the first recorded instance of egg-white injury had occurred because wheat starch was included in the diet instead of potato starch (Fig. 326). A protective substance, first called "vitamin H," was eventually isolated from liver and kidney (*Lease, 1936*), and recognized (*György, Melville, Burk, and duVigneaud, 1940*) as identical with the yeast-growth factor, biotin, discovered in egg yolk by Kögl and Tönnis (*1936*). Egg-white injury in chicks was shown to be associated with a deficiency of biotin in the tissues (*Eakin, McKinley, and Williams, 1940*).

Avidin was finally isolated from egg albumen, and its ability to inactivate biotin *in vitro* was demonstrated (*Eakin, Snell, and Williams, 1940*).

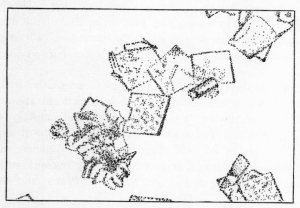

Fig. 327. Crystals of avidin, magnified 100 times. (After Pennington, Snell, and Eakin, 1942.)

The typical symptoms of egg-white injury were produced in rats fed cooked egg albumen and avidin (*György, Rose, Eakin, Snell, and Williams, 1941*). Avidin is apparently a basic protein containing a high carbohydrate moiety (*Pennington, Snell, and Eakin, 1942*). Its isoelectric point is at pH 10 (*Woolley and Longsworth, 1942*), and its molecular weight is probably between 70,000 (*Woolley and Longsworth, 1942*) and 87,000 (*Eakin, Snell, and Williams, 1941*). It has been obtained in crystalline form (Fig. 327).

When rats are fed avidin, negligible quantities of biotin are found in their feces until the feces are steamed. Biotin is then released from the avidin-biotin complex (*György and Rose, 1941*). The fundamental cause of egg-white injury is the inability of biotin to be absorbed from the intestines because of its unavailable form. The avidin-biotin complex withstands acidification to pH 1.8 (*György, Rose, and Tomarelli, 1942*) and is not attacked by proteolytic enzymes (*György and Rose, 1943*).

Avidin, when administered parenterally, is without toxic effect; in fact, it is sometimes curative. Apparently small amounts of biotin, bound to avidin and contained in avidin preparations, are liberated under parenteral conditions (*György and Rose, 1941*). Since biotin is released from the complex by oxidation with hydrogen peroxide, the metabolism of avidin-biotin is possibly linked with oxidation-reduction systems (*György and Rose, 1943*).

Biotin deficiency has been experimentally induced in four human subjects, who developed pruritic dermatitis, muscle pains, hyperesthesia, localized paresthesia, anorexia, and precordial distress. The hemoglobin content of the blood decreased, and the serum cholesterol was elevated. All these symptoms disappeared after the parenteral administration of 150 to 300 μg. of biotin daily for 3 to 5 days (*Sydenstricker, Singal, Briggs, DeVaughn, and Isbell, 1942*).

When it was observed that several types of neoplasm, found in man and lower animals, contained unusually high concentrations of biotin (*West and Woglom, 1941*), it was hoped that malignant growths could be retarded by the administration of excessive amounts of avidin. Unfortunately, the feeding of large quantities of egg albumen, in combination with avidin concentrates, has not given any encouragement to this theory (*Kensler, Wadsworth, Sugiura, Rhoads, Dittmer, and duVigneaud, 1943; Rhoads and Abels, 1943*).

FOLIC ACID

Folic acid (pteroylglutamic acid) is an essential antianemic vitamin for the chick and a valuable therapeutic agent in the treatment of various types of anemia in man (*Spies, 1946*). Although it occurs in very small amounts in foods, its concentration in the whole egg, 0.86 μg. per gram (*Cheldelin and Williams, 1942*), is insufficient to place the egg among important sources.

CHOLINE

Choline is tentatively included among the vitamins because it is essential at least to certain lower animals, apparently because of the methyl groups that it supplies. Its chief deficiency symptom is a fatty liver. It is contained in lecithin, which constitutes approximately 8.6 per cent of egg yolk (see Chapter 6, "Chemical Composition"). There are about 17.13 mg. of choline per gram of yolk (*Engel, 1943*), and the egg is therefore one of the richest known sources of this substance.

Vitamin C (Ascorbic Acid)

Attempts to demonstrate the presence of ascorbic acid in the egg have failed. Apparently vitamin C is absent from both the yolk and the albumen, even when hens are fed an antiscorbutic ration (*Hauge and Carrick, 1925; Dougherty, 1926; Suomalainen, 1939*). However, it has been claimed that egg yolk may contain about 0.0003 per cent of ascorbic acid (*Machida and Sasaki, 1937*). It is also interesting to note that ascorbic acid is said to be formed when fresh egg yolk is incubated with minced rat liver, but not when it is incubated with guinea pig liver, or with rat liver heated to 100° C. (*Nowinski and Ferrando, 1942*).

EVALUATION OF EGG QUALITY

The internal quality of the intact egg cannot readily be judged. The egg is one of the very few foods that do not offer easily detectable clues regarding their suitability for consumption. The prospective retail buyer is able to observe only the egg's size, shape, shell color, intactness, and cleanliness, which give him little or no information about the condition of the egg's contents. His sole criterion is the grade in which the egg has been placed before it is offered for sale. At present, no method of evaluating the egg's quality is entirely accurate and also easily adaptable to commercial use.

The term "egg quality" is itself difficult to define. Its meaning depends to some extent upon how the egg is to be used. Many eggs that are excellent for general cooking would not be enjoyed at the breakfast table. In addition, consumers differ in their conception of egg quality and frequently judge the egg by quite fallacious criteria. For example, the belief, prevalent in certain geographical regions, that white eggs or brown eggs have greater food value or "richer" yolks may usually be traced to the larger local production of eggs of one color or the other. The practice of sorting and selling eggs by sizes originated because many persons are willing to pay a higher price for the larger eggs.

Freshness, however, is universally desired in the egg. When the egg gives certain indications that it can no longer be considered new laid, it is said to be losing, or to have lost, quality. Sometimes deterioration occurs very rapidly, if the new-laid egg is kept under improper environmental conditions (see Chapter 10, "Preservation"). A decline in apparent quality does not necessarily diminish the nutritive value of the egg or impair its usefulness in cookery. In addition, eggs vary considerably in internal condition at the time they are laid, and therefore some eggs are inevitably judged as being of poor quality while they are still fresh.

THE INTACT EGG

About 20 per cent of all eggs sold in retail stores in the United States have been inspected by candling. This procedure is at present the sole commercial method of judging the quality of the intact egg.

Candling

Candling is made possible because of the translucency of the eggshell and the differences in the ability of the other elements of the egg to

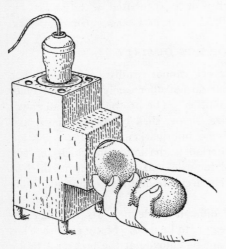

transmit light. In candling, the egg is held before a light, in a room darkened so that no light, or very little, falls upon the surface of the eggshell. The candling light is enclosed completely, except for a small circular aperture into which the end of the egg fits. The interior of the egg is fairly well illuminated, because the light rays are scattered as they are transmitted.

Some candling lamps may have reflectors, lenses for concentrating the light, colored filters for counteracting the effects of the natural color of the egg contents and shell, and features of construction that eliminate glare and possible strain on the operator's eyes.

FIG. 328. A typical electric lamp for candling chicken eggs and the manner in which eggs are held during the candling operation.

eyes. However, simple homemade candling equipment is adequate for occasional use.

In candling, the operator takes 2 (sometimes 3) eggs in each hand. With his right hand, he inserts the large end of an egg into the aperture of the candling lamp at a slight angle (Fig. 328) and twists the egg first to one side and then to the other. He then repeats the entire procedure with an egg from his left hand, meantime shifting the eggs in his right hand so that the uncandled egg is brought into candling position. These manipulations require dexterity, but a skilled operator can work very rapidly.

Before the candle, the egg appears yellow to rosy red (depending upon its shell color), with a darker, redder area in the yolk region. The condition of the eggshell is revealed, the air cell is visible, and the amount of light transmitted by the albumen may be judged. The size, distinctness, color, and mobility of the yolk may be observed. Physical defects, foreign bodies, embryonic development, and spoilage are usually apparent. The egg is graded on the basis of what is known concerning the candled appearance of the average new-laid egg and the changes that occur with loss of "quality."

The Eggshell. The ease with which the chicken egg may be judged in candling is dependent largely upon the tint of the eggshell. Brown egg-

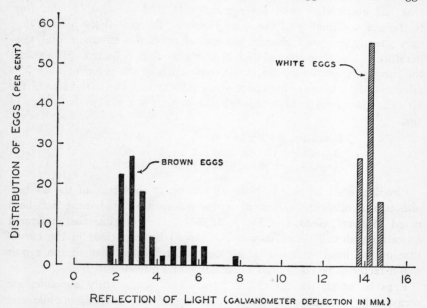

FIG. 329. Reflectance of light by brown-shelled and white-shelled chicken eggs, as measured by a photoelectric cell. (After Doan, 1934.)

shells greatly reduce the transmission of light. In addition, the shell pigment appears red by transmitted light and thus introduces the complicating factor of color. Standards of judgment for brown eggs are difficult to formulate, because the range of pigmentation is so much wider than in white eggs. This fact can be well demonstrated by measuring the light reflected from white eggs and from brown eggs (Fig. 329).

Candling instantly reveals any mottling of the eggshell. Mottled areas are much more translucent than the remainder of the shell. They may vary in size from that of a pinpoint to fairly large patches (see Chapter 10, "Preservation"). There is no relationship between shell mottling and the internal quality of the fresh egg, although the type of mottled shell known as "glassy" may be thinner than average, and therefore more subject to breakage (*Almquist, 1933*).

Candling is valuable for the detection of fine cracks in the shell. Many such cracks are entirely invisible to the naked eye but are very apparent before the candle.

Air Cell. The air cell of the candled egg appears as a circular area darker than the remainder of the egg. Eggs are graded chiefly upon the depth of the air cell. Air-cell size is often related to the amount of water that has evaporated from the egg contents, and presumably, therefore, to the age and quality of the egg. However, the size of the air cell may vary greatly in new-laid eggs, depending upon the environmental temperature at the time the egg is laid and after it is collected. The colder the surroundings, the greater is the contraction of the egg contents. The tabulated data indicate the effect of temperature on the initial size of the air cell (*Almquist, 1933*). Air-cell size is thus not always indicative of age.

Dimension	At 15.6° C.	At 1.1° C.
Diameter	1.49 cm. (0.59 in.)	1.92 cm. (0.76 in.)
Depth	0.28 cm. (0.11 in.)	0.36 cm. (0.14 in.)

Candlers must be able to estimate accurately the depth of the air cell, although a remarkable lack of agreement in their judgment has been noted (*Stewart, Gans, and Sharp, 1932b*). It is possible that judgment concerning air-cell depth may be influenced by variations in the curvature of the inner wall of the air cell. The relative size of the egg is also a factor to be considered when estimating air-cell size.

Eggs in which the air cell is "tremulous," or slightly movable, are occasionally encountered. Sometimes the air cell is freely mobile and travels beneath the shell as the egg is turned; or it may consist of small air bubbles, in most instances also movable. These conditions are usually the result of excessively rough handling of the egg, sufficient to cause a separation between the inner and outer shell membranes. Air-cell faults disqualify the egg for storage, but they are unrelated to fitness for immediate consumption. On the other hand, "bubbly" albumen, resembling a "broken" air cell, has been observed in eggs laid by birds convalescing from pneumoencephalitis (*Lorenz and Newlon, 1944*).

Yolk Shadow. The yolk shadow is another feature upon which the candler bases his judgment of the egg. Eggs are graded down as the yolk increases in darkness, visibility (or distinctness of outline), and mobility. These characteristics of the yolk shadow have been chosen as criteria because of certain known facts regarding the changes that occur, in time, in the interior of the egg. The color of the yolk may deepen; possibly for this reason, there is much consumer prejudice against yolks that are "too dark." The yolk enlarges, and consequently the yolk shadow becomes somewhat more visible. Because of physical changes in the mucin fibers, the percentage and firmness of the dense layer of the albumen decrease (*Almquist, Givens, and Klose, 1934*). These changes in the mucin fibers (see Chapter 10, "Preservation") also improve the transmission of light through the albumen, and the definition of the yolk shadow is therefore sharpened. As the albumen liquefies, the yolk meets less resistance and spins more rapidly. If the yolk turns freely, it may swing away from its central position and closer to the shell, and its visibility then increases. In addition, the yolk can sometimes shift its position along the egg's longitudinal axis.

Unfortunately, there are various factors that complicate interpretation of the yolk's appearance in the candled egg. The yolk shadow is not an infallible indicator of yolk color. It is possible for distinct shadows to be cast by a considerable proportion of yolks of a pale or medium shade; and it is also possible for some dark yolks to be ill defined. These facts are indicated in the table (*Almquist, 1933*). Furthermore, the color

Yolk Color	Distinct Yolk Shadow (per cent)
Pale	24
Medium	47
Dark	77

of the yolk is characteristic of the individual bird's eggs and is greatly affected by the hen's diet (see Chapter 3, "Structure"). Yolk color changes only slightly as the egg grows older (see Chapter 10, "Preservation") and therefore is not especially indicative of the age of the egg.

The distinctness of the yolk shadow also depends upon the size and shape of the egg. In small eggs, or narrow eggs, the yolk is close to the shell, and therefore its shadow is readily seen because of the short distance the light must traverse. In addition, the yolk is proportionately larger in small eggs than in large eggs.

Normally, there is probably great variability in the yolk's freedom to move within the inner liquid layer of albumen. The mobility of the yolk and the darkness of its shadow are not necessarily correlated with the condition of the albumen. The percentage of dense albumen in eggs with extremely dark yolk shadows may be very similar to that found in eggs with extremely light yolk shadows (*Almquist, 1933*). There is

	· DENSE ALBUMEN
YOLK SHADOW	(per cent)
Light	59.7
Dark	58.6

also experimental evidence indicating that rapid rotation of the yolk is sometimes more closely associated with a high percentage of dense albumen than with a low percentage (*Halnan and Day, 1935*). It is probable that the yolk shadow and the manner in which the yolk spins are influenced to a considerable extent by the condition of the chalazae. If these support the yolk loosely, the egg may be misjudged before the candle as one in which the percentage of dense albumen is very low, and therefore as an egg of inferior quality.

DEFECTS DETECTABLE BY CANDLING

Candling usually reveals various defects, some of which render the egg inedible. Others merely cause the egg to be graded down, or classed as unmarketable.

Blood spots on the yolk and "meat spots" in the albumen (see Chapter 5, "Anomalies") can ordinarily be detected, although the chalazae may sometimes be mistaken for blood spots. The edibility of the egg is not affected by these abnormalities, but their presence in the egg is very distasteful to many persons and usually causes the egg to be graded low. Other edible but unmarketable eggs are those in which the yolk or albumen is discolored because of various ingredients of the hen's diet (see Chapter 3, "Structure"). Green or brown yolks change the color of the yolk shadow. Pink albumen is noticeable in white-shelled eggs, but is not easily seen in brown-shelled eggs.

"Heat spots," caused by early embryonic development, are often visible before the candle and are permissible. On the other hand, a blood ring, indicating the presence of a dead embryo of more advanced stage, disqualifies the egg. The egg is also discarded if the albumen is bloody, or if it is yellow because of seepage of the yolk. If the egg contains foreign bodies or a "stuck" yolk (a yolk that adheres so firmly to the shell membranes that it cannot be dislodged), it is classed as inedible.

Candling eliminates the majority of eggs that are moldy or rotten, or that contain large embryos, because such eggs are opaque in varying degree. Candling does not detect sour or musty eggs, or eggs with green albumen (*Pennington, Jenkins, and Betts, 1918*).

LIMITATIONS OF CANDLING

Candling is a far from satisfactory method of judging the internal condition of the egg, as the foregoing discussion has indicated. It is used only because of the lack of a better method. The possibilities of error are great, not only because of the variables of the egg itself, but also because subjective opinions are involved. The speed with which commercial candlers must work and the element of fatigue inevitably reduce accuracy. In addition, there is a tendency to place too great an emphasis upon the size of the air cell (*Stewart, Gans, and Sharp, 1933b*).

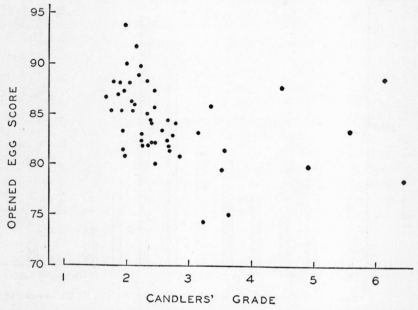

FIG. 330. The relationship between candlers' average scores and the opened egg scores for 51 dozen fresh chicken eggs. (After Stewart, Gans, and Sharp, 1933b.)

There may be wide discrepancies between candlers' grades and the actual condition of the egg, as revealed when the egg is opened. This fact is indicated in Fig. 330, although it is evident that there is a rough general agreement between the scores obtained by the two methods.

Candling assures the elimination of practically all inedible eggs, but it is probable that edible eggs are graded correctly only to the extent of 75 per cent, at the most. However, for practical purposes, this degree of accuracy is adequate, since studies indicate that the judgment of candlers is fairly trustworthy in regard to eggs of high and low quality. Disagreement occurs chiefly in classifying the intermediate grades (*Stewart, Gans, and Sharp, 1932a, 1932b*), which ordinarily, of course, are wholesome and nutritious.

Many attempts have been made to find means of grading eggs ac-

curately by mechanical devices not subject to errors of human judgment. Unfortunately, none of these has as yet been entirely successful,
although the application of electronic principles (*Romanoff, 1944b*) holds
promise (see Chapter 10, "Preservation"). This method is completely
mechanical; it is also sufficiently sensitive to distinguish, in intact, new-
laid eggs, differences in electrical conductivity which are well correlated

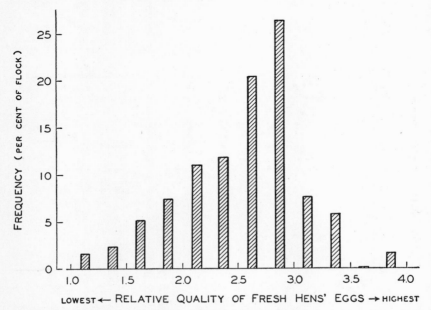

Fig. 331. Frequency distribution of a flock of 137 White Leghorn hens according to the quality of their new-laid eggs, as determined by measurements of radio-
frequency conductivity. (From the data of Romanoff, unpublished.)

with variations in internal quality. Figure 331 shows the wide range in
the quality, evaluated electronically, of 2083 fresh eggs from 137 hens.
In addition, the electronic method clearly indicates how constant is the
quality of the eggs laid by each hen. By testing a few eggs from every
hen in the flock, desirable breeders can be selected. The future average
quality of eggs may thus be improved.

THE OPENED EGG

There is a certain amount of common household knowledge regarding
the relationship between the appearance and the quality of the opened
egg. The housewife's discrimination against an egg with a flat, easily

ruptured yolk and liquefied albumen is based upon observations that these internal conditions are likely to develop as the age of the egg advances and are often associated with stale flavor. An egg with these characteristics, however, is not necessarily old, or unpleasant to the taste. There is considerable variation in the opened appearance of new-laid eggs. Furthermore, "watery" albumen is not undesirable unless the egg

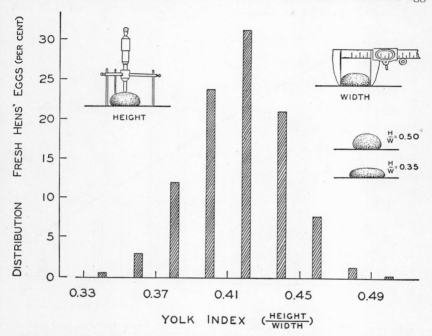

FIG. 332. Frequency distribution of yolk index (height of the yolk divided by its width) in 595 fresh hens' eggs. (After unpublished data of G. O. Hall.)

is to be poached or fried, or used in a whole-egg cake (*Pyke and Johnson, 1941*). In fact, it has been claimed that the liquid portion of the albumen, when beaten, gives a larger volume of foam than does the dense portion; and also that "watery" eggs make better angel-food cake than eggs with albumen of good consistency (*St. John and Flor, 1931; St. John, 1936*).

Various attempts have been made to express the general internal condition of the egg by means of mathematical formulae based on measurable characteristics. These efforts have resulted in numerous methods for evaluating the egg quantitatively.

Yolk. The yolk index is the ratio of the height of the broken-out yolk to its width (*Sharp, 1929b*). After the egg is opened on a flat surface, the albumen is carefully removed. Since the yolk tends to become flatter as it stands and as it grows warmer, measurements are made after a standard interval has elapsed, and at a certain temperature. Height is determined with a depth micrometer, and width with outside vernier calipers.

In new-laid eggs, there are wide variations in the yolk index, which may range from 0.30 to 0.50, although it is usually between 0.39 and

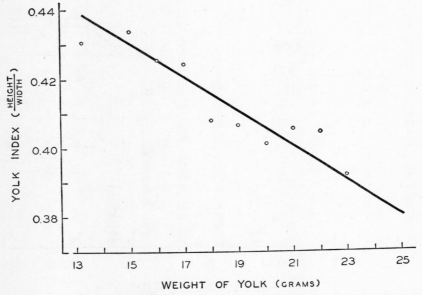

FIG. 333. Relationship between yolk weight and yolk index. (After Stewart, Gans, and Sharp, 1933a.)

0.45 (Fig. 332). Since a low yolk index is associated with stale eggs (see Chapter 10, "Preservation"), the new-laid egg is considered of poor quality if its yolk index is low.

The size of the yolk in the new-laid egg is a factor upon which the yolk index depends. The larger the yolk, the lower is its index (Fig. 333). The value of the yolk index is fairly constant in the eggs produced by a particular individual, but it may vary considerably from the eggs of one bird to those of another (Fig. 334). In view of this fact, it is possible that the hen may inherit a tendency to lay eggs of a certain characteristic yolk index, although the evidence in favor of this hypothesis is not clear-cut (*Van Wagenen and Hall, 1936*). In addition, the yolk index apparently varies seasonally and becomes lower, in the eggs of both pullets and hens, from early spring to midsummer (*Hunter, Van Wagenen, and Hall, 1936*), as indicated in Fig. 335.

It has not been possible to show that there is any relationship between the hen's diet and the yolk index. The inclusion of rye in the ration perhaps has an adverse effect (*North, 1935*), but other green feeds and various proteins and vegetable fats are apparently without influence.

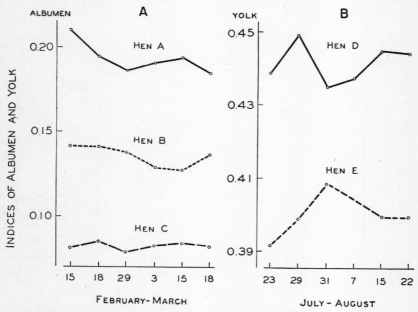

FIG. 334. Constancy of *A*, albumen index, and *B*, yolk index, in the eggs of individual hens. (After the data of Wilhelm and Heiman, 1936; and G. O. Hall, unpublished.)

However, when hens receive refined cottonseed oil as 4 per cent of the diet, they may lay eggs in which the yolk ruptures so easily that the yolk index cannot be measured (*Heywang and Titus, 1941*).

The color of the yolk may vary greatly. Consumer preference is sometimes very strongly in favor of yolks of light, dark, or medium color, depending chiefly upon nationality or local prejudice. Devices have been made to measure yolk color quantitatively (see Chapter 3, "Structure").

Albumen. The larger the percentage of dense albumen and the firmer its consistency, the higher is the quality of the egg generally considered to be. It is possible to estimate the condition of the egg by separating the different albumen layers (either by straining the albumen through a sieve, or by pipetting off the liquid portion), and then measuring the volumes or weights of the different layers (*Romanoff, 1943c*). However, the egg's quality, or observed grade, is probably more closely correlated with the consistency of the dense albumen than with its quantity (*Heiman and Wilhelm, 1937a*). An albuminous sac of poor consistency

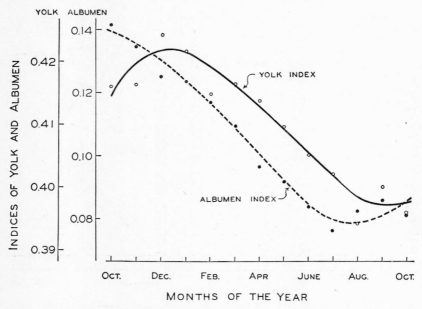

Fig. 335. Seasonal changes in the quality of eggs, as shown by indices of yolk and albumen. (After Hunter, Van Wagenen, and Hall, 1936; and Wilhelm and Heiman, 1938.)

spreads over a wide area when the egg is opened and does not stand up around the yolk. For this reason, measurements of the height and area of the albuminous sac are often used for evaluating the condition of the egg. The height alone gives a fairly good indication of quality.

The albumen index is the ratio obtained by dividing the height of the apparent dense albumen by the average of its long and short diameters (*Heiman and Carver, 1936a, 1936b*). In new-laid eggs, there is considerable variation in this value, which may fall between 0.050 and 0.174, although it is usually between 0.090 and 0.120 (Fig. 336). The albumen index, like the yolk index, is much the same in all eggs laid by a single hen (cf. Fig. 334).

The albumen index of chicken eggs varies throughout the year (cf. Fig. 335). The condition of the dense albumen becomes steadily poorer from April to August (*Hunter, Van Wagenen, and Hall, 1936*). It cannot be conclusively stated whether a decline in quality during the warm months is truly seasonal, or whether it is due to the effect of high

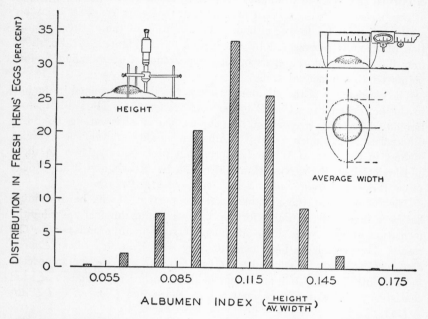

FIG. 336. Frequency distribution of albumen index (height of albumen divided by its average width) in 1139 fresh Leghorn eggs. (After Heiman and Carver, 1936a.)

environmental temperature (*Lorenz and Almquist, 1936*). In pullets' eggs, the albumen index decreases continuously throughout the entire laying year (usually from October to July). This trend probably bears little relationship to temperature (*Wilhelm and Heiman, 1938*). It may be a function of the increasing age of the bird, because it is observable regardless of the season in which the birds start to lay (*Jeffrey, 1941*).

The ratio between the area of the albuminous sac and its weight constitutes the albumen area index (*Parsons and Mink, 1937*). Some use has also been made of a score that varies logarithmically with dense albumen area, after correction for weight by a mathematical formula

(*Hoover, 1938*). The measurement of area may be facilitated by the use
of a slightly concave plate marked with concentric circles (Fig. 337).

In addition, the albumen may be scored on a numerical scale from 1
(highest quality) to 5 (lowest quality), with intervals of 0.5 (*Sharp,
1934; Van Wagenen and Wilgus, 1935*). A series of nine photographs
of opened eggs, viewed from above and from the side, is used as a standard
of comparison. In Fig. 338, it may be seen that eggs less than 1 day old
may score from 1.5 to 4.5, although the condition of the greatest proportion of them is represented by a score of 1.76. Figure 338 also shows
four illustrations of the egg's opened appearance, corresponding to four
points on the numerical scale.

The consistency of the albuminous sac is possibly determined by factors
in the hen's heredity (*Van Wagenen and Hall, 1936*). For this reason,
familial differences may affect the significance of observations on breed
differences in albumen quality. Albumen quality, nevertheless, seems
highest in the eggs of White Leghorns and progressively lower in those
of Rhode Island Reds, New Hampshires, Barred Rocks, and White Rocks
(*Van Wagenen, Hall, and Wilgus, 1937; Scott, 1941*).

The histology of the albumen-secreting portion of the hen's oviduct
has been investigated in an effort to discover the anatomical basis for the
formation of eggs with dense albumen of good quality and poor. The
results, however, have been contradictory (*R. K. Cole, 1938; Conrad
and Scott, 1942a*). Chemically, the two types of egg are similar, except
for a slightly greater sodium content in those with dense albumen of good
consistency. The difference may not be significant (*Conrad and Scott,
1942a*).

Flavor. Flavors must be judged subjectively, through the senses of
human tasters. Flavor evaluation involves not only individual differences in preference and discrimination, but also the difficulty of describing sensation verbally. It is possible, however, to achieve a fair
degree of accuracy in classifying egg flavors, if the test conditions are
carefully controlled, and if sampling is done by a panel of judges whose
reactions do not differ too widely.

The flavor of the table egg is an important criterion of quality. When
eaten, the new-laid egg may easily be distinguished from a stale one.
Even when the egg is combined with other ingredients, its taste, fresh or
stale, often remains detectable. In addition to the flavor changes which
normally accompany the aging of the egg, foreign taints may be acquired
during storage (see Chapter 10, "Preservation").

Occasionally, the flavor of the new-laid egg is unnatural. A "fishy"
off-flavor is sometimes encountered. It may be caused by feeding exces-

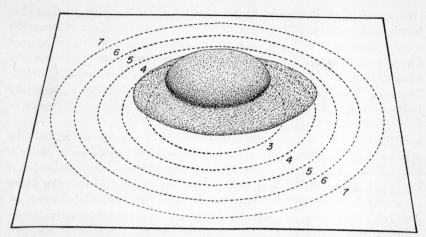

FIG. 337. A device for measuring the surface area of the albuminous sac. (After Hoover, 1938.)

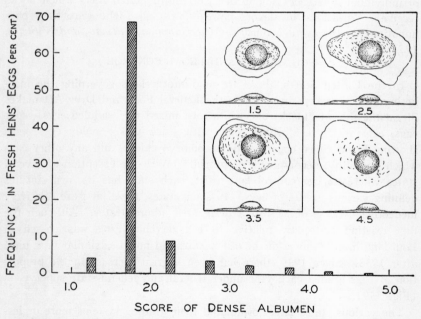

FIG. 338. Frequency distribution of dense albumen score in 4796 fresh eggs. (After Van Wagenen and Wilgus, 1935.)

sive amounts of low-grade fishmeal (*Cruickshank, 1939*), and it has been noted in the eggs of hens that received two capsules of cod-liver oil daily (*Thompson, Albright, Schnetzler, and Heller, 1932; Albright and Thompson, 1935*). Frequently, fishy flavor cannot be traced to any dietary factor; and in this case, it is apparently a peculiarity of the individual bird. The production of "fishy" eggs is possibly a hereditary trait, since it sometimes occurs in closely related strains or families of fowl (*Vondell, 1933*).

A distinct taste of garlic may be imparted to the egg by feeding the hen 4 gm. of garlic pods every day. The flavor is detectable in the yolk about 5 days after the first feeding, but sometimes it may not be found in the albumen until 30 days have passed. It is doubtful that the flavor of onions is similarly transferred to the egg (*Albright and Thompson, 1935*). The feeding of large amounts of cabbage and ordinary household garbage appears to have no effect on egg flavor (*Vondell and Putnam, 1945*).

The normal flavor of the egg may vary somewhat, probably because of individual differences among hens, or the dissimilar diets which flocks receive. Variations in flavor, however, are so slight that they pass unnoticed by the average consumer (*McCammon, Pittman, and Wilhelm, 1934*).

EGG GRADES AND LEGAL CONTROL

In the United States, there are no uniform laws governing the marketing of eggs, except in so far as the Federal Food and Drug Act makes illegal the sale of "black rots, white rots, mixed rots (addled eggs), sour eggs, eggs with green whites, eggs with stuck yolks, moldy eggs, eggs showing blood rings, eggs containing embryo chicks, and any other eggs which are filthy, decomposed, or putrid." With the exception of several states where no egg control laws exist, each state has its own statutes defining various grades of eggs. In some states, these laws are relatively ineffective because of lack of a designated administrator. Although the first existing state law relative to egg merchandising was passed in Louisiana in 1921, most of the other states did not pass similar laws until after 1933. Since 1939, there has been a great increase in the proportion of eggs graded each year under federal and federal-state supervision (Fig. 339).

The various states have formulated their grading systems more or less according to definitions of several grades of eggs promulgated in 1923 by the Federal Bureau of Agricultural Economics. In 1946, the four principal wholesale egg grades were U. S. Special, U. S. Extra, U. S. Standard, and U. S. Trade, all of which must have clean, sound shells. Stained eggs that correspond in internal quality to the last three of the foregoing grades may be classed in those grades, but must be designated "Stained." Eggs with checked or cracked shells are graded as U. S.

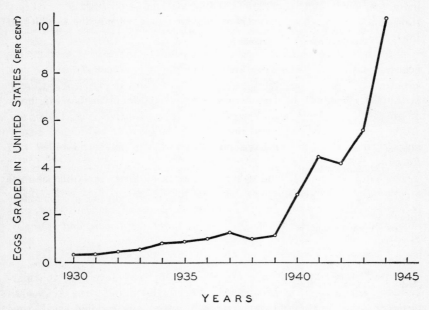

Fig. 339. Quantities of shell eggs graded in the United States by federal and federal-state graders each year from 1930 to 1944 inclusive (shown as percentages of total number of eggs consumed annually). (After Agricultural Statistics of the United States.)

Check. The air cell in eggs of Special, Extra, and Standard grades must not be more than ⅛, ⅔, and ⅜ inch in depth, respectively. It must be practically immobile in U. S. Special eggs but some degree of mobility is permitted in the other two grades. Germ development may be slightly visible in eggs of Standard grade and plainly visible in eggs classed as "Trades," but it is not permissible in eggs of the two highest grades.

The official United States standards of quality for eggs sold in the retail market are shown in Table 46.

In many regions, eggs are classified not only according to candled appearance, cleanliness, and intactness, but also according to weight. Such classification is usually controlled by law. The tentative United

TABLE 46

UNITED STATES STANDARDS FOR QUALITY OF INDIVIDUAL SHELL EGGS

Quality	Shell	Air Cell	White	Yolk
AA Quality	Clean Unbroken Practically normal	⅛ in. or less in depth Practically regular	Clear Firm	Well centered Outline slightly defined Free from defects
A Quality	Clean Unbroken Practically normal	⅖ in. or less in depth Practically regular	Clear May be reasonably firm	May be fairly well centered Outline may be fairly well defined Practically free from defects
B Quality	Clean Unbroken May be slightly abnormal	⅜ in. or less in depth May show movement not over ⅜ in. May be free if not over ⅜ in.	Clear May be slightly weak	May be off center Outline may be well defined May be slightly enlarged and flattened Definite but not serious defects
C Quality	Clean Unbroken May be abnormal	May be over ⅜ in. in depth May be free or bubbly	Clear May be weak and watery Small blood clots or spots allowed	May be off center Outline may be plainly visible May be enlarged and flattened May show clearly visible germ development but no blood May show other serious defects
Stained	Unbroken May be stained or soiled			
Dirty	Unbroken May be dirty			
Check	Checked or cracked Not leaking			
Leaker	Broken Contents leaking			

States weight classes for consumer grades of shell eggs are shown in the tabulation. For purposes of sales appeal, eggs are often sorted by color also.

	MINIMUM WEIGHT	
	PER EGG (grams)	PER DOZEN (ounces)
Jumbo	64	27
Extra large	59	25
Large	54	23
Medium	46	20
Small	35	15

Canadian egg-grading practices are very similar to those of the United States. On the continent of Europe, candling is done chiefly to eliminate inedible eggs, and to remove those of best quality for export. Exported eggs are sold almost entirely by weight. In England, eggs are graded chiefly by weight, but are also judged by air-cell size and candled appearance. It must be pointed out that the danger of deterioration of the egg during the summer is less in European countries than in the United States, because the climate is more moderate. This fact is possibly the reason why less emphasis is placed on candling.

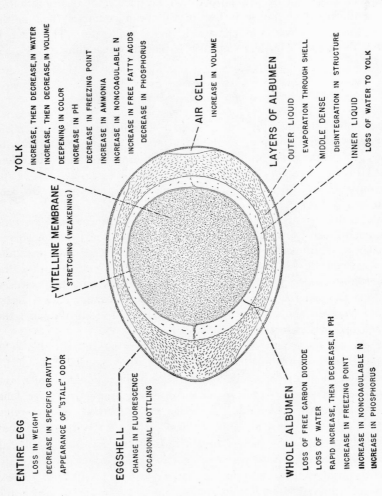

ENTIRE EGG
LOSS IN WEIGHT
DECREASE IN SPECIFIC GRAVITY
APPEARANCE OF "STALE" ODOR

EGGSHELL
CHANGE IN FLUORESCENCE
OCCASIONAL MOTTLING

WHOLE ALBUMEN
LOSS OF FREE CARBON DIOXIDE
LOSS OF WATER
RAPID INCREASE, THEN DECREASE, IN PH
INCREASE IN FREEZING POINT
INCREASE IN NONCOAGULABLE N
INCREASE IN PHOSPHORUS

VITELLINE MEMBRANE
STRETCHING (WEAKENING)

YOLK
INCREASE, THEN DECREASE, IN WATER
INCREASE, THEN DECREASE, IN VOLUME
DEEPENING IN COLOR
INCREASE IN PH
DECREASE IN FREEZING POINT
INCREASE IN AMMONIA
INCREASE IN NONCOAGULABLE N
INCREASE IN FREE FATTY ACIDS
DECREASE IN PHOSPHORUS

AIR CELL
INCREASE IN VOLUME

LAYERS OF ALBUMEN
OUTER LIQUID
EVAPORATION THROUGH SHELL
MIDDLE DENSE
DISINTEGRATION IN STRUCTURE
INNER LIQUID
LOSS OF WATER TO YOLK

The most important changes that time produces in the intact avian egg.

Chapter Ten

PRESERVATION

The avian egg is an unstable physicochemical system.

Its peculiar structure and the unequal distribution of its chemical constituents make the egg an unstable system. In addition, the interior of the egg is imperfectly protected from the environment and is therefore exposed to the forces of another system. For these reasons, the egg is in a continual state of readjustment, which takes place at a rate controlled largely by external factors.

Physical disintegration within the egg is normal. By contrast, the type of decomposition that results from microfloral growth may be considered pathological. Ordinarily, the egg is adequately protected from invading microorganisms by its shell and the shell membranes. In addition, the albumen possesses bactericidal properties, and any bacteria that succeed in passing through the shell usually die or at least cannot grow. Occasionally, however, the egg's protective mechanism is destroyed, and the egg then becomes vulnerable to microbial attack.

The egg has its maximum value as an article of food at the time it is laid. The start of readjustment within it marks the beginning of its deterioration as a dietary commodity. The difficulty with which the egg may be replaced, and its somewhat seasonal production, have long made its preservation desirable. Although many methods of preserving the intact egg have been attempted, none has as yet been devised that can entirely prevent any changes from taking place. Essentially, all that preservation can accomplish is to avert microfloral invasion and to retard physicochemical deterioration for a reasonable length of time. These results are achieved either by controlling the environment in which the egg is placed, or by treating the egg so that it is less easily affected by external conditions.

When the egg is removed from its shell, it is deprived of its strongest natural defense. The loss of its original physical form leaves the egg exposed to inevitable contamination by microorganisms and accentuates the effect of deleterious environmental conditions. Because the opened egg is so perishable, methods for its preservation are adaptations of those ordinarily used to prevent the decomposition of other foodstuffs of a similar nature.

CHANGES IN EGGS WITH ADVANCING AGE

The passage of time may greatly alter the egg in many ways. Certain gross changes can be detected while the egg is yet intact. Other effects of age may be obvious in the appearance of the opened egg; many more are revealed only by extremely delicate tests. If microorganisms gain entrance to the egg and succeed in growing within it, the result may be putrefactive decomposition and complete destruction of nutritive value.

The rate at which events take place within the egg is dependent fundamentally upon variables of the environment, especially temperature and humidity.

THE "FRESH" EGG

The biologically fresh egg and the fresh egg of commerce are not the same. Theoretically, only a new-laid egg is fresh. When marketed as an article of food, however, the egg is considered fresh if candling reveals that it still has a relatively small air cell, and that no obvious internal changes have occurred. When properly handled and held under suitable environmental conditions, the egg may still be commercially acceptable as fresh 2 or 3 weeks after having been laid (Fig. 340).

The age of the egg is therefore a function of the environment; it is more or less relative, the result of the interaction between the two variables, original constitution and holding conditions.

GROSS CHANGES

The intact egg may show its age by loss in weight, by a decrease in specific gravity, by enlargement of the air cell (as revealed in candling), and sometimes by the development of a perceptible odor.

Loss in Weight

Loss in weight is one of the most obvious changes in the aging egg. It is caused chiefly by the evaporation of moisture, at first principally from the albumen. To a small extent, it may also be due to the escape of gases, such as carbon dioxide, ammonia, possibly nitrogen, and sometimes hydrogen sulfide, most of which are products of the chemical breakdown of the egg's organic constituents.

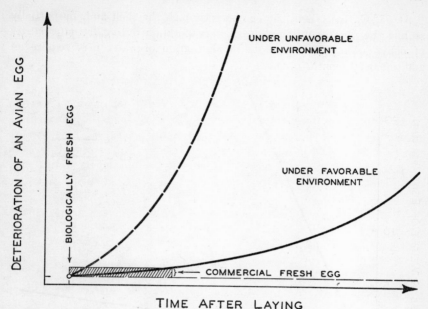

FIG. 340. Diagram showing the deterioration of an avian egg as the function of time and environment, and the distinction between the biologically fresh egg and the egg commercially acceptable as fresh.

Evaporation of water from the egg is a continuous process. It begins the moment the egg is laid, and it does not cease until the egg is completely dehydrated.

Loss in weight occurs in almost linear relationship with time, under all constant environmental conditions (Fig. 341). However, in terms of the egg's original weight, the daily loss diminishes throughout the holding period (*A. Smith, 1931; Romanoff, 1940; Romanoff, 1943b*). The rate of weight loss is accelerated at higher temperatures (cf. Fig. 341), and retarded at higher relative humidities.

The table gives the approximate time that is required, at different temperatures and two relative humidities, for the complete dehydration of a hen's egg of average shell permeability.

TEMPERATURE (°C.)	TIME (months)	
	AT R.H. 50 PER CENT	AT R.H. 80 PER CENT
0	60	96
12.5	26	42
25.0	12	20
37.5	5	8

At 55° C., the oils of the egg ooze through the shell and, by partially sealing the pores, considerably delay evaporation during the later stages of dehydration. To evaporate the last gram of water may require an additional 10 months or more.

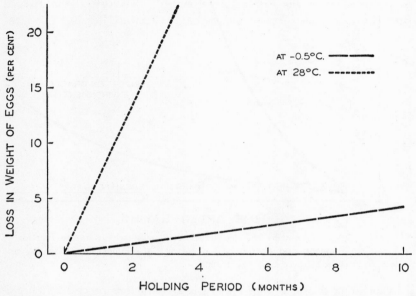

Fig. 341. Loss of weight in hens' eggs when held at about 82 per cent relative humidity and at two different temperatures. (After data of Moran and Piqué, 1926, for weight loss at −0.5° C.; and data of Fronda and Clemente, 1936a, for weight loss at 28° C.)

Experiments (*Romanoff, 1940*) show that there is a very slight falling-off in the rate of weight loss as complete dehydration is approached, and that this falling-off is greater, the higher the holding temperature. In eggs held 10 months at 15° C., the final rate of loss is about 0.88 per cent of the initial rate. It is 0.93 per cent of the initial rate in eggs held for the same period at 0.4° C. (*A. Smith, 1931*).

The rate at which the egg loses weight by evaporation tends to be greater if the permeability of the shell is high (*Romanoff, 1940*), if the egg is comparatively small, or if the air surrounding the egg is in rapid movement (*A. Smith, 1930; Romanoff, 1940*). (See Chapter 7, "Physicochemical Properties.")

INCREASE IN THE SIZE OF THE AIR CELL

Almost immediately after the egg is laid, the air cell is produced by contraction of the egg contents as they cool from the hen's body temperature of about 41° C. (*Lee, Robinson, Yeates, and Scott, 1945*) to the temperature of the external environment. Thereafter, the air cell grows larger as the egg contents evaporate. Commercially, loss in weight

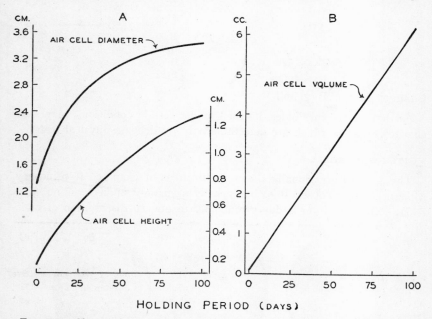

Fig. 342. Changes in *A*, dimensions, and *B*, volume of air cell in the hen's egg held at approximately 28° C. and 82 per cent relative humidity. (Dimensions after Fronda and Clemente, 1936b; volume calculated from same data.)

and enlargement of the air cell together are known as "shrinkage." The size of the air cell, as revealed by the candling lamp, therefore has a certain significance when the condition of the egg is evaluated.

The diameter, height, and volume of the air cell are functions of time, when the temperature and relative humidity of the environment are constant. At first, the diameter and the height grow rapidly larger, but the rate of increase soon diminishes and eventually becomes very slow in older eggs (Fig. 342-*A*). The volume of the air cell, on the other hand, increases at a constant rate as time passes (cf. Fig. 342-*B*) and is thus proportional to the loss of water from the egg.

Decrease in Specific Gravity

As the egg contents lose weight, the egg's specific gravity decreases, since the volume of the entire egg, of course, remains the same. In eggs held at 28° C. and 82 per cent relative humidity, for example, the specific gravity may change as follows (*Fronda and Clemente, 1936a*):

AGE OF EGGS (days)	AVERAGE SPECIFIC GRAVITY
0	1.063
15	1.027
30	0.992
60	0.938
99	0.825

Changes in the Eggshell

As the egg becomes older, its shell, under certain conditions, may show changes, some of which are detectable only by delicate physical tests.

MOTTLING OF SHELL

The eggshell occasionally shows spotty areas of increased translucency, sometimes easily observable, but usually seen only upon candling (Fig. 343). This condition is due presumably to uneven distribution of mois-

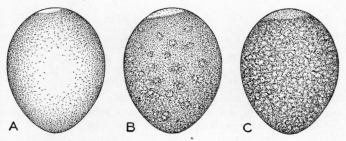

FIG. 343. Various degrees of mottling of the eggshell, as seen by transmitted light. *A*, no mottling (fresh egg); *B*, slight mottling; *C*, severe mottling.

ture in the shell. Contradictory evidence has been presented to the effect that the incidence of mottling becomes greater both as the humidity increases (*Holst, Almquist, and Lorenz, 1932a*) and as it decreases (*Jeffrey and Darago, 1940*). The effect of the holding temperature is apparently less than that of the relative humidity.

Mottling of the shell is not necessarily an indication of age. Although it appears most frequently in held eggs, it may be seen in eggs less than 24 hours old.

FLUORESCENCE OF SHELL

In time, the fluorescent properties of the eggshell change. The reddish fluorescence of white-shelled chicken eggs, for instance, becomes reddish violet and, later, blue (*Haitinger, 1928; Wehner, 1930; Gaggermeier, 1932; Straub and Kabos, 1938*). The rate at which fluorescence changes is retarded by a decrease in the holding temperature (*Zäch, 1929*) and by exclusion from direct daylight (*Grini, 1939*).

Changes in Odor

Fresh eggs have no odor, other than a slight smell of lime. Upon aging, however, they may acquire a characteristic stale odor which can be detected through the shell. Other, more unpleasant, odors are consequent to the chemical breakdown of the egg's contents, caused either by the action of microorganisms, or by adverse environmental conditions, especially high temperatures.

Eggs may quickly absorb foreign odors from the environment, especially if they are kept in proximity to disinfectants, mold growths, or decaying

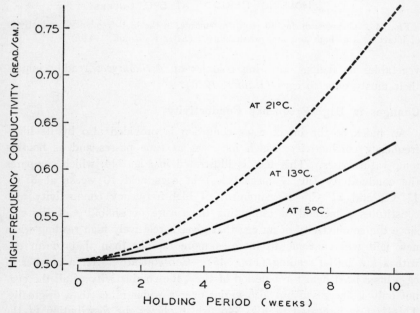

FIG. 344. Changes in the high-frequency conductivity (at about 27 megacycles) of intact eggs during 10 weeks of storage at various temperatures. (After Romanoff and Hall, 1944.)

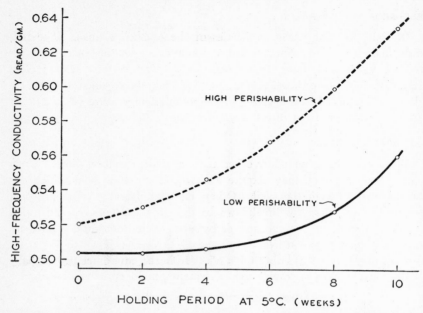

Fig. 345. Changes, during 10 weeks of storage, in the high-frequency conductivity of intact eggs of high and low perishability. (After Romanoff, 1944b.)

vegetables or fruits. Growing colonies of *Actinomyces* readily impart their musty odor to eggs (*Haines, 1930*).

Changes in High-Frequency Conductivity

An index of the intact egg's condition is provided also by its high-frequency conductivity, which increases as time passes and as holding temperature rises. This fact is illustrated in Fig. 344, which compares the conductivity (at 27 megacycles) of eggs held 10 weeks at 5° C., 13° C., and 21° C. Measurement of high-frequency conductivity also constitutes a method of estimating the future perishability of the egg, since the conductivity of an egg that gives a relatively high reading when new laid will continue to increase more rapidly than that of an egg with a low initial reading (Fig. 345). Consequently, after a number of eggs have been held for a period of time, the conductivity of all the eggs not only is increased but also varies over a wider range than originally. This fact is indicated in Fig. 346, which shows the distribution of the high-frequency conductivity readings on 95 eggs when fresh, and after having been held for 12 weeks at 21° C.

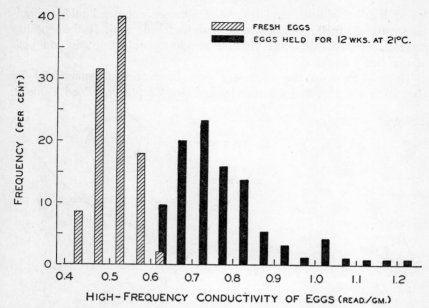

FIG. 346. Frequency distribution of intact eggs according to high-frequency conductivity when fresh and after having been held for 12 weeks at 21° C. (After Romanoff, unpublished.)

INTERNAL CHANGES

Evidence of increasing age is apparent in the interior of the egg. The egg's components change in both absolute and relative amount and in color, and sometimes histologically. These modifications, like all others, may occur rapidly or slowly, depending upon environmental conditions.

General Appearance of the Egg Contents

When the aging egg is hard-boiled and sectioned, the yolk is no longer well centered in the albumen (Fig. 347). The eccentricity of the yolk is due to an increase in the density of the albumen, and a decrease in the density of the yolk. The yolks in old eggs may also be much enlarged and wrinkled.

In the broken-out new-laid egg, the middle dense layer of albumen (the albuminous sac) stands up in oval shape around the yolk and has a thick, gelatinous consistency. As time passes, the albuminous sac decreases in height, at first rapidly, and later more slowly (Fig. 348); it also loses its firmness, so that it spreads out thinly over a wide area when the egg is opened (cf. Fig. 347). Eventually, the dense albumen can no longer be distinguished, and the entire albumen appears liquid. Simultaneously, the yolk becomes large and flattened; and, if the egg has deteriorated sufficiently, the vitelline membrane may rupture, and yolk and albumen mingle.

These changes in the aging egg may be expressed numerically by changes in the yolk and albumen indices (see Chapter 9, "Food Value").

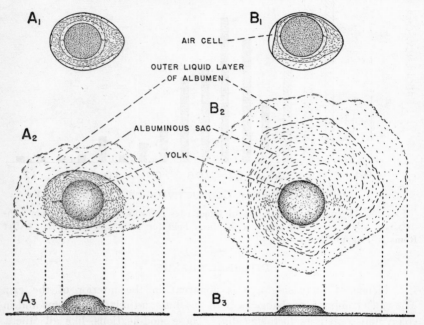

FIG. 347. Comparison between structure of *A*, the fresh egg, and *B*, the aging egg.

A_1 and B_1, longitudinal section through hard-boiled eggs; A_2 and B_2, opened eggs viewed from above; A_3 and B_3, opened eggs viewed from the side.

The yolk index decreases as the egg grows older, at a rate determined by the holding temperature. When this is as low as 2° C., only very slight flattening of the yolk may occur in more than 3 months; but, at 25° C., the index may decrease greatly in less than 3 weeks (Fig. 349). The decrease in the albumen index, as in the yolk index, is retarded by low temperatures. If the egg is held at −1.0° C., the albumen index, at the end of 6 months, may still be within the range found in fresh eggs; but if the holding temperature is 32° C., the albumen index may decrease 40 per cent in 20 hours (Fig. 350).

Since the decrease in the height of the albumen proceeds logarithmically (cf. Fig. 348), it has been noted that the condition of the egg

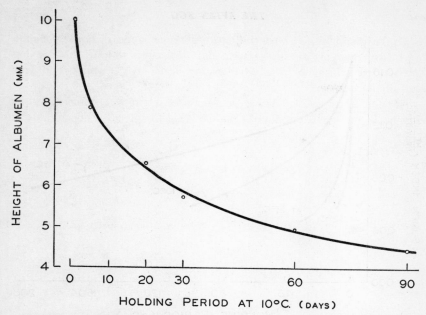

Fig. 348. Changes in the height of the albumen when the egg is held 90 days at 10° C. (After Haugh, 1939.)

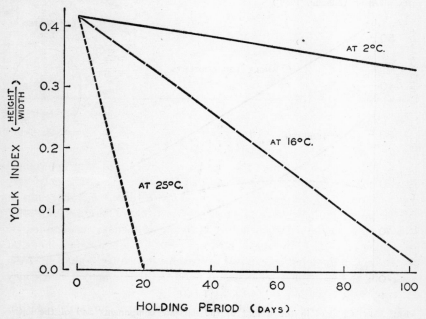

Fig. 349. Changes in yolk index when the egg is held at various temperatures. (After Sharp and Powell, 1930.)

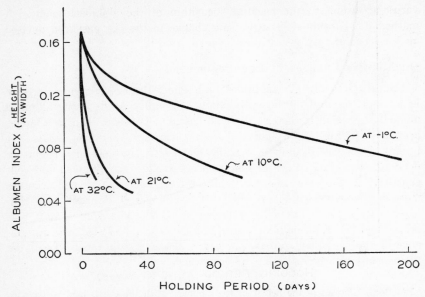

FIG. 350. Changes in the albumen index when the egg is held at various temperatures. (After Wilhelm, 1939.)

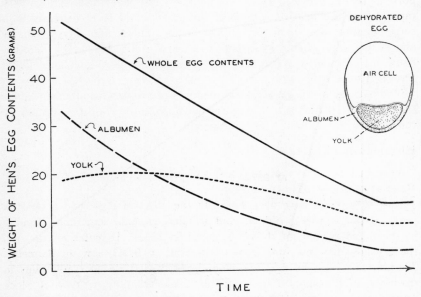

FIG. 351. Changes in the weight of the separate components and of the entire contents of a hen's egg of average size from the time the egg is laid until it becomes completely dehydrated. (See Chapter 3, "Structure," and Chapter 6, "Chemical Composition.") (Reconstructed from the authors' data.)

varies according to the negative logarithm of the albumen height. A mathematical formula for expressing change in the egg has been devised on the basis of this observation (*Haugh, 1937*).

Quantitative Changes in Egg Components

Under practically all conditions of holding, the egg contents decrease in amount, because of the evaporation of water. At first, the albumen alone loses weight, whereas the yolk gains somewhat. These changes are shown in Fig. 351 by curves based on various experimental data. It is only when dehydration is well advanced that the yolk also begins to decrease in weight. When dehydration is complete, the albumen and yolk constitute a relatively small portion of the egg, and the air cell occupies most of the space within the shell. The albumen has shrunk to a thin layer. It is hard, brittle, and amber colored. The yolk is soft and greasy and has a strong odor resembling slightly rancid oil. The appearance of the egg's dehydrated contents is characteristic, regardless of the temperature at which the egg has been held.

PROPORTIONAL CHANGES IN LAYERS OF ALBUMEN

As the albumen decreases in total amount, changes occur in the proportions of its various layers. The middle dense layer disintegrates very rapidly, and the relative volume of the outer liquid layer increases at a corresponding rate. The percentage of the inner liquid layer remains almost the same, when the egg is held at 25° C. (Fig. 352). On the other hand, in 5.5 months of storage at 0° C., the volume of the inner liquid layer may decrease by about 45 per cent (*Moran, 1937a*).

The changes in the proportional amounts of the various albumen layers are hastened by higher temperatures but apparently are little affected by the relative humidity of the environment.

Histological Changes

Very little is known about the histological changes that occur in the albumen with the passage of time.

Rosette crystals have been found in the yolks of eggs held at least a year in cold storage. Similar, but smaller, crystals sometimes develop in the liquid portion of the albumen (Fig. 353). In yolks a year old, the crystals may measure from 0.018 mm. to 0.063 mm. in diameter. The crystals may be more numerous and larger, up to 0.109 mm. in diameter, in the yolks of eggs stored 20 months (*Wiley, 1908*). Since crystals are sometimes observed in the dense layer of the albumen in fresh eggs, it has been suggested that crystals in the yolk result from

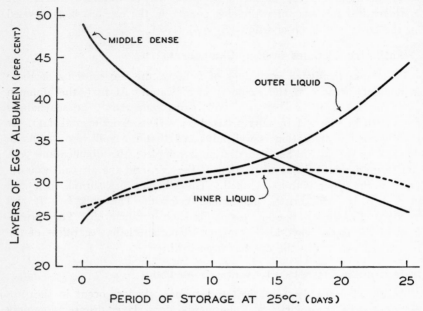

FIG. 352. Changes in the proportional amounts of the various layers of the albumen during storage of the egg at 25° C. and 80 per cent relative humidity. (After Moran, 1937b.)

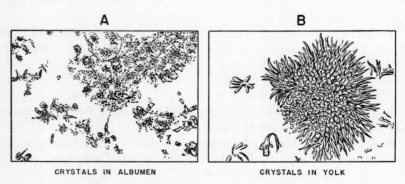

FIG. 353. Crystals found in the albumen and yolk of a hen's egg stored 2 years; magnified about 80 times. (After Schaible and Bandemer, 1947.)

the infiltration of albumen. Chemically, the crystals appear to be a complex protein calcium phosphate (*Schaible and Bandemer, 1947*).

Color Changes

Color changes (other than those of bacterial origin) may occur in the contents of the aging egg. The yolk may grow darker, especially if the holding conditions are unfavorable (*Parker, 1927b*). The yolk shows little color change in eggs from which rapid evaporation is prevented, as, for example, by oil treatment of the shell (*Almquist, 1933*). The darkening therefore is due possibly to a concentration of the yolk's pigment by dehydration when storage is prolonged, or when the egg is held at low relative humidity.

Mottling of the yolk is seen in held eggs more frequently than in fresh eggs. It may be due to inability of the vitelline membrane to prevent penetration into the yolk of proteins from the albumen (*Almquist, 1933*).

As the egg deteriorates, the very slight greenish or yellowish tinge of the albumen, observed in the fresh egg, may grow more yellow (*Haines, 1940*), but not noticeably so until after at least 25 days at 25° C. (*Moran, 1937a*). The color of the inner liquid layer is slightly deeper than that of the outer liquid layer.

INFLUENCE OF THE DIET OF THE HEN

Certain color changes which sometimes occur in held eggs may be the result of the ingestion of malvaceous plants by the hen (*Schaible, Moore, and Moore, 1933; Almquist and Lorenz, 1933b; Lorenz, 1939*). Pink albumen, and yolks that are olive, brownish, pinkish, or even salmon red, may develop after 1 to 11 months at 0° C. These colors usually disappear when the egg is cooked.

Olive yolks are found in eggs from hens fed raw cottonseed, which contains gossypol, a toxic yellow dye. The olive color is produced by a reaction between gossypol and iron liberated from yolk proteins during storage (*Swensen, Fieger, and Upp, 1942*). Olive yolks develop more quickly than the pink or red discoloration. The incidence of olive yolks in storage eggs is directly proportional to the percentage of cottonseed meal in the hen's diet, as shown in the table (*Upp, 1934*).

COTTONSEED MEAL (per cent)	OLIVE YOLKS (per cent)
0	0.3
10	5.1
20	18.8
30	31.5

In eggs that develop pink albumen during storage, the protein-fat ratio of the yolk may be somewhat higher than normal, as shown below (*Lorenz and Almquist, 1934*). A similar increase in the protein-fat ratio

	NORMAL EGGS	"PINK WHITE" EGGS
Total solids (per cent)	47.8	39.7
Crude fat (per cent, dry basis)	68.4	63.3
Protein (per cent, dry basis)	31.9	36.5
Protein-fat ratio	0.47	0.58

occurs in salmon-colored yolks, in which the albumen proteins, ovalbumin and ovoconalbumin, may be detected. It is therefore apparent that egg albumen passes through the vitelline membrane in "cottonseed" eggs with salmon yolks (*Schaible, Bandemer, and Davidson, 1946*). The reaction with iron appears to be confined to the ovoconalbumin fraction. The pink discoloration of albumen may be attributed to the diffusion of iron through the vitelline membrane from the yolk (*Schaible and Bandemer, 1946b*). The development of pink or red colors in stored eggs is also thought to be the result of a reaction involving the Halphen substance, or a substance closely related to it, which is found in malvaceous plants (*Lorenz and Almquist, 1934; Lorenz, 1939*).

Color changes in the stored egg may also result from feeding the hen peanut, soybean, or linseed meal as the only protein concentrate of the diet, in the amount of 30 per cent of the mash (*Thompson and Albright, 1934*). In only 14 days, yolk color may darken to shades ranging from deep orange through red. In extreme cases, the yolk may become brown, reddish brown, or slate gray. Occasionally, the ingestion of peanut, soybean, or corn gluten meal may be responsible for a pinkish discoloration in the albumen.

Changes in Flavor

Since our experience of flavors, as distinguished from true tastes, is largely olfactory, the odor and flavor of the aging egg are closely related. However, an objectionable flavor in the egg is usually less pronounced than the odor of the broken-out contents, probably because volatile odoriferous substances escape rapidly (*McCammon, Pittman, and Wilhelm, 1934*).

As the egg ages, it develops a characteristic "storage flavor," first in the yolk, and later in the albumen. Storage flavor is always more marked in the yolk (Fig. 354). The exact chemical substances responsible for this flavor are not known, but they are probably the products of certain changes in the yolk fats.

As might be expected, the absorption of foreign odors, even to a very slight extent, affects the flavor of the egg. In commercial storage, the packing material may impart its odor to the egg and thus change the flavor noticeably. Strawboard fillers are especially likely to do so (*Sharp, Stewart, and Huttar, 1936*).

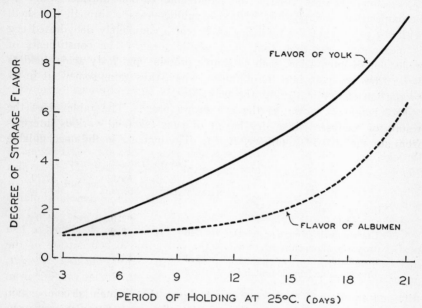

FIG. 354. Changes in the flavor of yolk and albumen in eggs stored at 25° C. and 80 per cent relative humidity. (After Davies, 1937.)

During holding, the egg may lose certain odors originally present in it, such as those caused by meat and bone scraps in the hen's diet (*McCammon, Pittman, and Wilhelm, 1934*). On the other hand, the fishy flavor which sometimes results from the feeding of excessive amounts of cod-liver oil has been known to persist for as long as 6 months (*Albright and Thompson, 1935*).

PHYSICOCHEMICAL CHANGES

The aging of the egg is accompanied by pronounced changes in its physical properties. Fundamentally, these changes are caused by the movement of water and by the escape of carbon dioxide. A certain degree of chemical decomposition may also occur, but the disintegration of the egg is more accurately measured by delicate physical tests than by ordinary chemical analysis.

Changes in Chemical Constituents

The chemical changes that take place in the held egg are few; they are most readily detectable when the egg has been stored under adverse conditions, especially high temperature. Some products of chemical disintegration—such as volatile oils, ammonia, carbon dioxide, water of decomposition, and possibly hydrogen sulfide—escape through the shell, with the result that the total dry matter of a completely dehydrated egg is measurably less than that of a new-laid egg. The constituents of molecular disintegration, such as simple proteins and fatty acids, tend to move from the yolk into the albumen, which thus gains somewhat in its proportion of dry matter, as the yolk loses.

Proteins break down as the egg grows older. The table gives the results of analyses (on a dry basis) of eggs taken at various intervals from storage at 0.5° C. (*Wiley, 1908*). The increase in the uncoagulable

	TOTAL NITROGEN (per cent)		
	INITIAL VALUE	AT 12 MONTHS	AT 20 MONTHS
Nitrogen in yolk:			
Coagulable	92.4	92.0	34.0
Uncoagulable (simple proteins)	7.6	8.0	66.0
Nitrogen in albumen:			
Coagulable	83.2	84.7	76.0
Uncoagulable (simple proteins)	16.8	15.3	24.0

nitrogen fraction of each component shows that protein decomposition is more rapid in the yolk than in the albumen. So, also, does the relatively greater increase in ammonia nitrogen in the yolk than in the albumen, after 10 months at a temperature slightly above freezing (*Hendrickson and Swan, 1918*). In eggs of somewhat inferior quality,

	AMMONIA NITROGEN (milligrams/egg)	
	INITIAL VALUE	AT 10 MONTHS
Albumen	0.10	0.11
Yolk	0.52	1.14
Whole egg	0.62	1.25

Jenkins and Pennington (*1919*) found that the ammonia content rose, during 10 months of storage, from 1.1 mg. per egg to 1.6 mg. per egg. After the infertile egg has been held at the incubation temperature of 37.5° C. for 3 weeks, the value may be as high as 6.3 mg. for the entire egg contents (*Romanoff, 1940*). In discolored stored eggs from hens fed cottonseed products, the ammonia nitrogen in the albumen may be ten times as high as it is in normal eggs stored for comparable periods.

On the other hand, the ammonia nitrogen in the yolk is only two-thirds as high (*Bandemer, Schaible, and Davidson, 1946*).

The decomposition of fats proceeds very slowly. Thomson and Sorley (*1924*) estimated the free fatty acids in the fat of fresh egg yolk to be 1.72 per cent and noted but very little increase after 3 months' storage (at an unspecified holding temperature). After a year, the free fatty acids increased to 3.12 per cent, and, after 5 years, to 5.15 per cent. When the egg is held at 37.5° C., the yolk loses fat (as shown by ether extract) and the albumen gains, because fat derivatives diffuse through the vitelline membrane from yolk to albumen (*Romanoff, 1940*).

Glucose also may pass through the vitelline membrane, but it proceeds from albumen to yolk. During 20 days at incubation temperature, the glucose content of the albumen in infertile eggs may fall from 460 mg. to 430 mg. per 100 gm., and that of the yolk may rise from 245 mg. to 275 mg. per 100 gm. (*Pucher, 1927*).

In held eggs, the phosphorus content of the albumen may rise, as shown below (*Janke and Jirak, 1934*).

	PHOSPHORUS IN ALBUMEN (milligrams/100 gm.)
Fresh eggs	64
Eggs held 8.5 months at 0° C.	1087
Eggs held 8 months at 14° C.	4362

The total phosphorus content of the yolk decreases during a year's storage at refrigeration temperatures, although the ratio of acid-soluble phosphorus to total phosphorus increases. This fact may indicate that phosphatides are converted to some other phosphorus-containing substance that is acid soluble, such as inorganic phosphorus (*Wiley, 1908; Erikson, Boyden, Insko, and Martin, 1938*).

The actual amount of calcium in the yolk has been found both to increase by as much as 15 per cent during storage for a year at refrigeration temperatures (*Erikson, Boyden, Martin, and Insko, 1938*), and to decrease by 1.08 per cent during 21 days at incubation temperatures (*Romanoff, 1940*). A rise in calcium in the egg contents supposedly is due to some chemical action upon the shell.

Iron, confined chiefly to the yolk of the fresh egg, gradually passes into the albumen as the egg grows older. As the table shows, the iron con-

STORAGE PERIOD (months)	IRON IN ALBUMEN (micrograms/100 gm.)
1.0	74
6.0	110
8.5	118

tent of the albumen may change in eggs held in cold storage for various lengths of time (*Schaible, Davidson, and Bandemer, 1944*). The increase in the iron concentration of the albumen is greater than normal in eggs showing the type of discoloration associated with the presence of cottonseed feeds in the hen's diet (*Schaible and Bandemer, 1946a*).

Movement of Water in the Egg

The peculiar structure of the egg and its high moisture content make it inevitable that the movement of water within it should be complex and conspicuous. While water is evaporating from the egg, it is also diffusing through the vitelline membrane, at first in one direction, and later in the other.

LOSS OF WATER FROM THE ALBUMEN

During the early holding period, the albumen loses water not only by evaporation through the shell, but also by diffusion to the yolk. In the later stages of dehydration, water passes from the yolk to the albumen and thence to the outside, until it is entirely dissipated into the external atmosphere (Fig. 355).

Superficial examination of the opened, deteriorated egg shows an apparent "liquefaction" of the albumen; actually, the albumen continually increases in density, because of its loss of water. This fact is indicated by a constant increase in the refractive indices of all layers of the albumen, although the refractive indices of the chalaziferous and inner liquid layers decrease somewhat at first (Fig. 356).

EXCHANGE OF WATER BETWEEN YOLK AND ALBUMEN

When the egg is laid, a difference of about 1.8 atmospheres in osmotic pressure exists across the vitelline membrane, in spite of the fact that the membrane is permeable both to water and to dissolved substances (see Chapter 7, "Physicochemical Properties"). The movement of water between the egg components is the result of this osmotic gradient.

The direction of diffusion is initially from albumen to yolk, because the osmotic pressure of the yolk is the greater. It is probable that this equilibrative process begins even before the egg is laid (*Bialaszewicz, 1912*); if environmental conditions permit, it continues until equilibrium is attained. The movement of water reverses itself when the albumen, by dehydration, has become the more concentrated of the two egg components.

The noticeable enlargement of the yolk in the aging egg is due to its increased content of water. In eggs held at 30° C., the percentage of

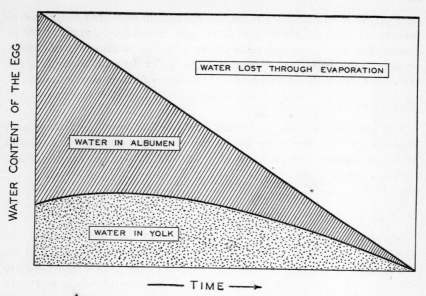

FIG. 355. Diagrammatic representation of the movement of free water in the egg from the time the egg is laid until it is completely dehydrated.

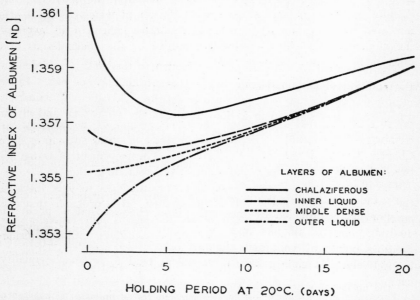

FIG. 356. Changes in the refractive index of the various layers of albumen when the egg is held at 20° C. and 60 per cent relative humidity. (After Romanoff and Sullivan, 1937.)

water in the yolk may rise, in only 10 days, from 48.02 per cent to 54.33 per cent (*Holst and Almquist, 1931b*). As the yolk gains water, its total content of solid matter decreases proportionately. In 3 weeks of storage at 37° C., the yolk's solids content may decline from approximately 52 per cent to as little as 44 per cent.[1] If the yolk contains less than 46 per cent solids, it usually does not remain intact when the egg is opened (*Sharp and Powell, 1930*).

The viscosity of the yolk naturally decreases as the water content rises. A very small dilution lowers the viscosity considerably, as shown below (*A. Smith, 1935a*).

WATER CONTENT (per cent)	RELATIVE VISCOSITY (water = 1)
47.2	100.0
50.0	32.7
54.0	10.5
57.6	3.5

The rate at which water diffuses into the yolk is influenced by the holding temperature. In 60 days of storage in saturated air at 0° C., the yolk's initial water content of about 48 per cent may increase to 49 per cent; but, at 20° C., it may rise to 52 per cent (Fig. 357). The increase is thus four times as much at 20° C. as at 0° C. (*A. Smith, 1932*). At all temperatures, the rate of gain is most rapid initially and subsequently levels off, although this change is almost imperceptible at 0° C. (cf. Fig. 357).

The movement of water from the albumen to the yolk is retarded by the addition of carbon dioxide to the air surrounding the egg. In an atmosphere of 100 per cent carbon dioxide, the yolk gains approximately 25 per cent less water than it does in air, in the same time and at the same temperature (*A. Smith, 1932*). It is possible that slower equilibration in the presence of carbon dioxide is due to the formation, around the yolk, of a gelatinous layer of albumen, which interferes with diffusion (*Moran, 1936b*).

An index to the disappearance of the osmotic gradient between the egg components is provided by the concomitant changes that occur in the freezing points of albumen and yolk. In the new-laid egg, the average freezing points of yolk and albumen are −0.60° C. and −0.45° C., respectively. As time passes, these values tend to approach equality.

[1] The uptake of water by the yolk is more accurately measured by dry-weight determination than by direct weighing, which involves an uncertain assumption as to the yolk's original weight.

Initially, the rate of change is fairly rapid; but, under usual storage conditions, it ultimately becomes extremely slow. At 0° C., many months would probably elapse before the freezing points would become the same. At higher temperatures, equalization is accelerated and, at 25° C., may be completed in 75 days (Fig. 358). The tabulation shows the freezing points of yolk and albumen (originally −0.622° C. and

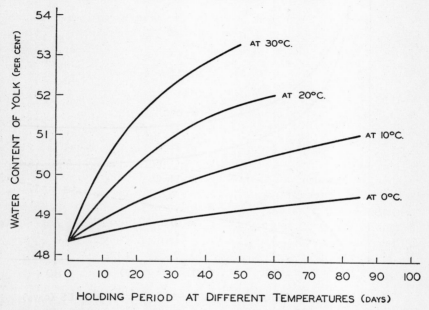

Fig. 357. The rate at which the percentage of water increases in the yolk when the egg is held at various temperatures. (After A. Smith, 1932.)

−0.452° C., respectively) at the end of 76 days of storage at different temperatures (*Smith and Shepherd, 1931*). These figures also show that,

HOLDING TEMPERATURE (°C.)	FREEZING POINTS (°C.) YOLK	ALBUMEN
0	−0.538	−0.492
10	−0.519	−0.495
18	−0.502	−0.503
25	−0.507	−0.514

at the two higher temperatures, the freezing point of the albumen eventually becomes lower than that of the yolk. A reversal in the original osmotic relationship between the two egg components is thus indicated.

The addition of carbon dioxide to the storage atmosphere does not alter the process of change in freezing points, beyond hastening the initial rate somewhat. This effect may be due merely to the higher solubility of carbon dioxide in albumen than in yolk (*A. Smith, 1932*). It is therefore probable that the equalization of freezing points is

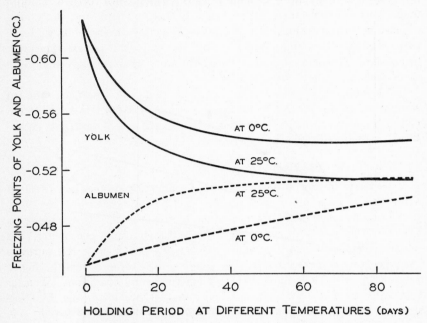

FIG. 358. The equilibration of the freezing points of yolk and albumen in the hen's egg held at 0° C. and at 25° C. (After Smith and Shepherd, 1931.)

affected by the egg's carbon dioxide content only during the early holding period, when carbon dioxide is escaping rapidly from the egg.

The fact that the tendency to approach osmotic equilibrium is not a simple process in the egg is indicated by the manner in which the actual curves for the changes in freezing points deviate from the curves that can be calculated from the known water exchange. Furthermore, the movement of ions across the vitelline membrane is not necessarily excluded as a factor in the attainment of osmotic equilibrium. Nevertheless, even when ionic movement is taken into consideration, the actual changes still do not correspond with the calculated (*A. Smith, 1932; Smith and Shepherd, 1931*).

Experimental evidence indicates that there is considerable resistance to water movement both between and within the components of the intact egg. When the albumen is rapidly concentrated by evaporation, the rate of change in the freezing point of the yolk is little affected. When separated from the egg and used in artificial osmotic systems, neither albumen, nor yolk, nor vitelline membrane behaves as it does in the intact egg; on the contrary, equilibrium is attained with relative rapidity, under such circumstances (*Needham, 1931b*). There are, no doubt, osmotic gradients within both albumen and yolk. In the yolk, at least, their existence is plainly indicated by the fact that an exterior crust forms over a separated yolk's liquid interior, when rapid evaporation takes place (*A. Smith, 1932*).

In general, however, it can be said that the approach toward equilibrium is due to the process of diffusion, as conditioned by certain factors inherent in the constitution of the egg, and not as yet understood in their entirety.

WEAKENING OF THE VITELLINE MEMBRANE

A decrease in the strength of the vitelline membrane is correlated with the diffusion of water into the yolk. Enlargement of the yolk stretches the membrane and consequently weakens it. If its breaking strength has fallen to approximately 2500 dynes per centimeter, the vitelline membrane ruptures when the egg is opened on a flat surface.

In general, deterioration in the vitelline membrane occurs in linear relationship with time, under constant environmental conditions. During 6 months at 0° C. and 90 per cent relative humidity, the bursting strength may decrease by 10 to 20 per cent. However, there is evidence (*Gane, 1935*) that the strength of the membrane may actually increase during the first few days of storage, as illustrated in Fig. 359. (The values shown in Fig. 359 are higher than those usually found, probably because of the methods of determination.)

Since the diffusion of water into the yolk is accelerated as the temperature rises, the vitelline membrane weakens more rapidly at higher temperatures. Vitelline membranes in infertile eggs held 15 days at 39° C. may possess an average bursting strength of only 1000 dynes per centimeter (*Moran, 1936c*). The addition of carbon dioxide to the atmosphere retards deterioration. After the egg has been stored in 100 per cent carbon dioxide for 8 months at 0° C., the membrane is only slightly weaker than it is in the fresh egg (*Gane, 1935*).

Although the vitelline membrane loses its strength as time passes, it is possible that its elasticity increases, within certain limits (*Moran, 1936c*).

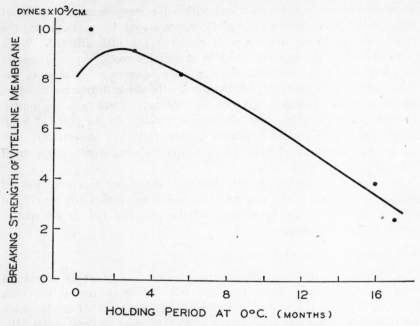

Fɪɢ. 359. Changes in the breaking strength of the vitelline membrane when the egg is held at 0° C. and about 80 per cent relative humidity. (After Gane, 1935.)

Liberation of Carbon Dioxide

The escape of carbon dioxide begins the moment the egg is laid. Since fresh egg yolk contains no free carbon dioxide (see Chapter 7, "Physicochemical Properties"), the albumen must of necessity be the source at this time.

From an initial maximum, the output of carbon dioxide falls off rapidly at first, then more slowly, until it becomes fairly well stabilized at a slow and constant rate (Fig. 360). A very gradual decline toward an eventual zero point probably takes place but cannot be easily demonstrated.

Depending upon the environmental temperature, the production of carbon dioxide may begin at a rate as high as 9.0 mg. per egg per day and eventually fall to approximately 0.1 mg. to 0.2 mg. per egg per day, at which level it may continue for 100 days or more (*M. Smith, 1931*). The higher the temperature, the more rapidly is carbon dioxide lost (cf. Fig. 360). The season of the year in which the egg is laid appears to be of little influence on any aspect of the liberation of carbon dioxide (*M. Smith, 1931*).

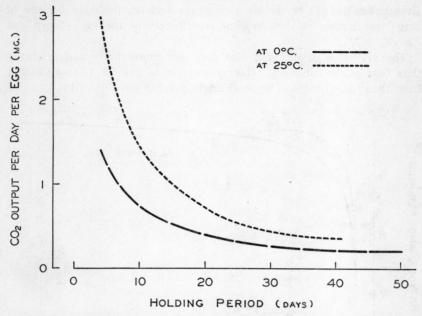

Fig. 360. The rate at which carbon dioxide escapes from the hen's egg at 0° C. and at 25° C. (After M. Smith, 1931.)

Apparently the new-laid egg contains a store of carbon dioxide, which is given off until the albumen and the surrounding air are in carbon dioxide equilibrium. This tendency to approach equilibrium is indicated by the fact that carbon dioxide, if introduced into the atmosphere, is absorbed by the egg (*Sharp and Powell, 1927*). However, simple de-solution does not explain the long-continued production of carbon dioxide by the egg after equilibrium presumably is attained. Probably carbon dioxide is slowly generated within the egg, as the result of various chemical changes in both albumen and yolk, or of chemical action upon the shell (*M. Smith, 1931*).

CHANGES IN *p*H VALUE OF ALBUMEN AND YOLK

Immediately after the egg is laid, the albumen starts to grow more alkaline. Under certain conditions, the albumen's original *p*H value of 7.6 (which approaches that of the hen's blood) may rise in 1 week to as high as *p*H 9.0 or *p*H 9.7, where it usually remains fairly constant for some time (Fig. 361). At *p*H 9.7, egg albumen is one of the most alkaline natural biological fluids known (*Sharp and Powell, 1927*).

Eventually, the *p*H of the albumen again declines, probably because of actual breakdown of the chemical constituents of the egg (*Sharp and Powell, 1931*).

The *p*H value of the yolk also rises, but more slowly and gradually than that of the albumen. It may increase to *p*H 6.8 (*Sharp, 1929a*) from the original value of approximately *p*H 6.0 (cf. Fig. 361).

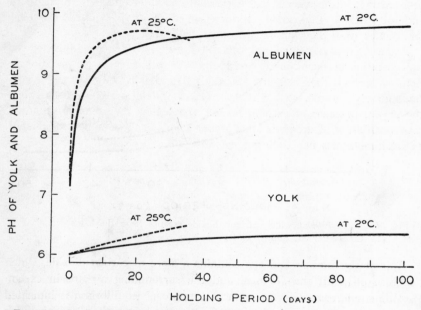

Fig. 361. Changes in *p*H of yolk and albumen when the egg is held at 2° C. and at 25° C. (After Sharp and Powell, 1931.)

The increase in *p*H value of the albumen is thus marked and abrupt, with an early leveling-off at a constant value, and an eventual decrease. By contrast, the increase in the *p*H value of the yolk is linear and steady, and relatively small (*Baird and Prentice, 1930*).

In both albumen and yolk, the *p*H value increases more rapidly at higher temperatures (cf. Fig. 361). The time required for the *p*H of the albumen to reach the value of 9.25 varies at different holding temperatures, as the table shows (*Sharp and Powell, 1931*). At 37° C., the *p*H of the albumen begins to decline again in about a week.

TEMPERATURE (°C.)	TIME (days)
3	11
15	5
37	2

The fundamental cause of the rise in pH value, or increase in alkalinity, is the loss of carbon dioxide from the egg. In addition to dissolved carbon dioxide, the albumen contains bicarbonates of sodium and potassium; these substances constitute the buffer system of the egg. As carbon dioxide is lost through the shell, the concentration of bicarbonate ions in the albumen also decreases, and the buffer system of the egg becomes progressively disorganized. It is possible, however, that the increase in alkalinity of the albumen is retarded somewhat by the diffusion of acid phosphates from the yolk (*E. C. Smith, 1935*).

Under increased partial atmospheric pressures of carbon dioxide, the concentration of bicarbonate ions in the albumen rises as carbon dioxide diffuses through the shell into the egg. The pH of the albumen is correspondingly lowered. At 25° C., the following relationships exist between the partial pressure of carbon dioxide, the pH of the albumen, and the concentration of bicarbonate and carbonate ions in the albumen, calculated as gram-molecules per 1000 gm. (*Brooks and Pace, 1938b*).

CO_2 Pressure (atmosphere)	pH	ALBUMEN HCO_3^-	$CO_3^=$
0.0003	9.61	0.0231	0.0117
0.01	8.43	0.0504	0.0017
0.03	7.99	0.0551	0.0007
0.05	7.88	0.0568	0.0004
0.10	7.50	0.0594	0.0002
0.97	6.55	0.0652	......

The original pH of the albumen may be maintained by the addition of the proper concentration of carbon dioxide to the air (in which the partial pressure of carbon dioxide normally is 0.0003 atmosphere). The pH of the yolk, protected as it is by the albumen, is but little affected by atmospheric carbon dioxide.

DISINTEGRATION OF THE GEL STRUCTURE OF THE ALBUMEN

The tendency for the structure of the albumen to lose its heterogeneity and to approach homogeneity is one of the earliest indications of the aging process in the egg. The albuminous sac gradually disappears, and the entire albumen becomes liquid.

The deterioration in the dense albumen is due probably to physical changes in ovomucin, which is present in the form of fibers enmeshing liquid material. In the fresh egg, the albuminous sac contains the largest portion of the ovomucin. As time passes, however, there is a measurable tendency for the mucin content of the dense layer to decrease and for that of the liquid portions to increase, as shown in the table (*Moran, 1937a*). These figures suggest that mucin does not disappear as a

MUCIN CONTENT (grams/100 gm., dry weight)

LAYER	AT 24 HOURS	AT 11 DAYS
Outer liquid	0.22	0.56
Middle dense	2.44	2.05
Inner liquid	?	0.33

chemical entity, although other analyses indicate that there may be a slight decrease in the mucin content of the entire egg albumen during storage (*Conrad and Scott, 1939*).

Histological examination of albumen after storage for 8 days at 35° C. has indicated that the fibers of ovomucin may maintain their integrity, at least under certain conditions. It is possible, however, that their elasticity is altered (*Conrad and Scott, 1939*).

The hypothesis has been advanced that proteolysis, due to enzymic activity, causes the breakdown of the dense albumen; but this theory is now regarded as doubtful (*Sharp, 1937; Balls and Hoover, 1940*). No sulfhydryl or disulfide groups—products of protein hydrolysis—have been detected in the reactive form in native egg albumen (*Anson, 1940a*).

The disintegration of the albuminous sac is more or less coincidental to the rise in pH value and probably is correlated to a large extent with increased alkalinity. Since a gel of mucin will dissolve so completely under prolonged treatment with dilute alkali that neutralization will not bring about precipitation, ovomucin might well be affected as the albumen becomes more alkaline (*Bate-Smith, 1936*). In the presence of insufficient carbon dioxide, the network of ovomucin fibers apparently loses its structure and breaks up. Too high a concentration of carbon dioxide, on the other hand, causes syneresis, or contraction of the fibers, with the result that the enmeshed liquid is squeezed out (*Almquist and Lorenz, 1932a; Sharp, 1937*). For these reasons, the condition of the albumen is best between pH 7.5 and pH 8.5. The more rapid deterioration of the albumen at higher temperatures is due possibly to the accelerated loss of carbon dioxide or to other causes as yet undetermined.

Eggs that originally contain a high percentage of dense albumen show relatively less deterioration of the albuminous sac during storage than those in which there is a small proportion of dense albumen (*Almquist and Lorenz, 1935*).

MICROBIOLOGICAL CHANGES

In the majority of new-laid eggs from healthy hens, the egg contents are bacteriologically sterile (*Rettger, 1913; Zagaevsky and Lutikova, 1937; Trossarelli and Massano, 1938*). If environmental conditions are favorable, the egg may retain its freedom from bacteria for lengthy periods of time. As the table shows, the percentage of uncontaminated eggs in a lot held at 0° C. was not greatly altered at the end of 6 months (*Haines, 1938*).

	PROBABLY STERILE (per cent)
Fresh eggs:	
Albumen	98
Yolk	93
Eggs held 6 months:	
Albumen	92
Yolk	90

The exterior of the eggshell also is usually sterile when the egg is laid (*Stuart and McNally, 1943*), but it immediately becomes contaminated in the nest. Subsequently, the shell acquires profuse and heterogeneous microflora from innumerable sources. The egg normally is equipped with means of preventing the invasion of microorganisms and of combating them should they gain entrance, but, if these defenses are weakened or destroyed, internal contamination may occur. The probability of eventual decomposition is then increased.

The structure of the egg ordinarily offers effective physical barriers to bacteria, which cannot penetrate through the intact shell as long as the outer openings of the pores are filled with the mucinous material of the cuticle. In addition, the yolk, with its excellent supply of nutrient substances, occupies an interior position in the egg and is surrounded by the albumen, a much less favorable medium for microbial growth. Microorganisms that succeed in penetrating into the egg often find it difficult to maintain themselves against the chemical defenses of the albumen (see Chapter 8, "Biological Properties"), especially against its potent bactericidal principle (*Fleming and Allison, 1924*). Furthermore, the proteins of native egg albumen are not in a form that can be utilized by many types of bacteria; and the alkalinity of the albumen in the aging egg is unfavorable to bacterial growth (*Sharp and Whitaker, 1927*).

The egg, however, is instantly exposed to microfloral invasion if the shell is cracked or broken. In addition, bacterial and fungal penetration may readily occur if the pores are opened by the removal of the cuticle, whose protection is destroyed by abrasion, for example, or by the presence of a mere trace of water on the shell. The ease with which a stalked type of organism may gain entrance through the pores is illustrated in Fig. 362. Certain bacteria, once they have entered the egg, are more resistant than others to the egg's chemical defenses. If even slight protein decomposition has occurred within the egg, such organisms are able to support themselves until they can elaborate their own proteolytic enzymes. The products of their enzymatic activity will permit the growth of other bacteria which ordinarily cannot survive in the egg's interior. The presence of blood spots possibly increases susceptibility of the egg to bacterial spoilage by supplying more easily assimilable nutrient material for bacteria.

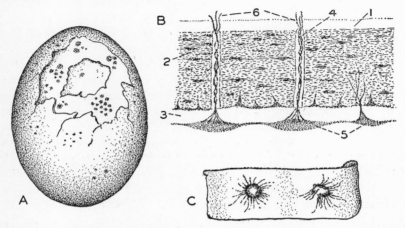

Fig. 362. Illustration of internal growth of "stalked" bacteria (genus *Leptothrix*) in the chicken egg. (After Cattaneo, 1877.)

A, an egg with shell removed from a large area and membrane also removed from a smaller area to show penetration of bacteria through the eggshell. *B*, semidiagrammatic illustration of cross section of eggshell, magnified 300 times; the numbers indicate: 1, cuticle; 2, calcareous shell; 3, shell membranes; 4, pore canal; 5, accumulation of bacterial growth under the shell membranes; 6, external protrusion of threads of bacteria. *C*, portion of inner shell membrane, showing penetration by threads of bacteria; magnified 300 times.

The growth of molds on the exterior of the eggshell depends to a large extent upon the humidity of the environment. If the moisture content of the atmosphere is sufficiently great, molds are able to avail themselves of the nutriment present on the shell surface, until this material is exhausted. At high humidities, it is possible that additional food is brought to the molds through the pores of the shell. Dirt on the eggshell furnishes food in such large amounts that molds can support themselves at unusually low humidities (*Sharp and Stewart, 1936*). Molds, like bacteria, are capable of penetrating into the egg (Fig. 363), presumably in much the same way. Little is known, however, of the chemical defenses of the albumen against molds, or of the manner in which these organisms utilize the proteins of the egg (*Haines, 1939*).

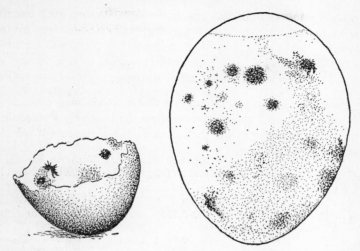

FIG. 363. Mold growth inside the shell of an egg stored at high humidity. On the right, appearance of the intact egg before the candling lamp; on the left, view of a portion of the shell, showing moldy areas on the inner surface. (After Moran and Piqué, 1926.)

Chemical Changes Due to Bacteria

Chemical evaluation of bacterial decomposition in the egg is difficult because of the fact that spoilage is so often associated with a complex infection by many types of microorganisms, each of which has its own specific action upon the chemical constituents of the egg. In spoiled eggs, there may usually be found a chaotic mixture of the end products and by-products of bacterial metabolism. For this reason, few attempts have been made to correlate putrefactive chemical changes with the bacterial species responsible for them.

By hydrolysis, bacteria may convert the proteins of the egg into amino acids, which thereafter may be decomposed to yield bases and acids, often simultaneously, with the elimination of nitrogen in the former case and carbon dioxide in the latter. The lipids and lecithin of the yolk also may be attacked by bacteria. Hydrolysis or oxidation may liberate fatty acids or form aldehydes and ketones, and also probably other intermediary compounds (*Mitchell, 1940*).

Decomposition may be accompanied by a great increase in the ammonia content over the normal level of about 0.6 mg. per egg. Different quantities of ammonia are found in different types of rots, as shown in the table (*Houghton and Weber, 1914*).

AMMONIA
(milligrams per egg)

Spots	6.3
Light rots	8.0
Rots	11.8
Black rots	76.7

The extent of chemical disintegration that bacteria may cause is indicated in the following data, which also show the relatively greater decomposition that may occur in the yolk than in the albumen (*Mitchell and Horwitz, 1941*).

	VOLATILE ACIDS (cc. of 0.01 N/100 gm.)	VOLATILE BASES (cc. of 0.01 N/100 gm.)
Yolk:		
Fresh	0	80–100
Decomposed	180–6,030	760–11,210
Albumen:		
Fresh	0	3–7
Decomposed	10–200	100–3,500

Below is given a comparison between certain constituents of normal yolks and of yolks in which two types of putrefaction—sulfidic and rancid—have occurred (*Mitchell, 1940*).

	NORMAL YOLK	PUTREFIED YOLK SULFIDIC	RANCID
Formic acid (mg./100 gm.)	(?)	2	9
Volatile acids (cc. of 0.01 N/100 gm.)	0	327	1,699
Bases (cc. of 0.01 N/100 gm.)	1,012	3,178	12,314
Lipids:			
Acidity (cc. of 0.05 N/100 gm.)	3	12	46
Volatile acids (cc. of 0.05 N/100 gm.)	0	373	1,628

Rots

In certain types of rot, one organism is definitely predominant. Sometimes the chief characteristics of a rot of mixed microbial content can be reproduced in a fresh egg by one specific organism isolated from the rot. The classification of a few rots according to the causative agent is therefore possible; but, in general, classification is very difficult and necessarily somewhat arbitrary.

ROTS DUE TO BACTERIA

Certain bacteria are known to be chiefly responsible for green and black rots, and sometimes for red rots.

Green Rots. Green rots are usually caused by members of the *Pseudomonas* group of organisms, which are commonly found on the surface of the eggshell. Certain species of *Pseudomonas* multiply in the albumen, from which they synthesize a characteristic fluorescent green pigment (*Haines, 1938*). The yolk may not be affected, although the vitelline membrane may become thick, white, and opaque, or occasionally black. This type of rot is encountered fairly frequently among commercial eggs, because in its early stages little change occurs except for the production of green pigment not easily detected by candling.

Red Rots. Red or pink rot is also caused by bacteria of the *Pseudomonas* group. Before the candling lamp, it appears as a pronounced reddish tinge in the yolk. When the egg is opened, the albumen may be found either liquefied, or still viscous; it has a grayish cast and contains colored patches varying from golden brown, through rusty red, to deep blood-red. The vitelline membrane is often opaque and very thick and white in places, although it is occasionally pink and brittle.

Black Rots. One type of black rot is caused by *Proteus melanovogenes,* first isolated by Miles and Halnan (*1937*). This rot is characterized by the presence of gas; and, since the shell membranes apparently become impermeable, the pressure created within the egg is often sufficient to burst the shell and scatter the egg's contents. The yolk becomes black, hard, and solid, and the albumen is entirely liquefied, turbid, granular, and murky brown or greenish brown. The egg has a pronounced fecal odor. Other strains of the *Proteus* group produce somewhat similar black rots, in which the appearance of the albumen and of the yolk is slightly different. The egg, when candled, appears dark and opaque.

Another type of black rot is caused by a certain species of *Pseudomonas.* Before the candling lamp, the yolk appears as an opaque, liquid mass, floating freely in the albumen. When the egg is opened, the albumen is seen to be liquefied, fluorescent green or greenish brown, sometimes granular and viscous. The yolk is a soft greenish black mass.

Miscellaneous Rots. A number of other types of rotting have been described. Some of these have been attributed to various bacteria.

A pink color sometimes develops in a coagulated deposit on the egg membranes, the albumen becomes a dirty grayish white, and the egg acquires a characteristically putrid odor. A Gram-negative coccobacillus has been isolated from rots of this type and provisionally placed in the genus *Serratia* (*Bennetts, 1931*).

In another type of spoilage, the yolk liquefies, and the albumen becomes turbid and watery and contains fluorescent green pigment. *Pseudomonas fluorescens, Micrococcus roseus,* and *Staphylococcus aureus* have been isolated from eggs containing rots of this description (*Pavarino, 1929*).

It also has been observed (*Wüdik, 1937*) that *Pseudomonas fluorescens* is responsible for the production of gases, and possibly for the rupture of the egg, when incubated at 37° C.

"Sour" eggs, so called because of the peculiarly pungent odor which they emit when opened, are not readily detected by candling. The albumen may be turbid, and the yolk and albumen may be somewhat mingled. These eggs contain extremely large numbers of bacteria, among which *Escherichia coli* is usually present (*Pennington, Jenkins, St. John, and Hicks, 1914*).

Motile, Gram-positive, sporulating rods have been isolated from eggs containing "floating and cloudy" yolks (*Anderson and Platt, 1936*). These eggs appear normal except for the chalazae, which are swollen, opaque, and partially digested. Sometimes, however, there may be incomplete coagulation of the yolk proteins.

Among the bacteria said to cause other miscellaneous types of spoilage is an aerobic, Gram-negative, motile rod which liberates sulfur from proteins and which has been called *Bacillus thioaminophiles* (*Tissier, 1926; Lagrange, 1935*). Another organism, termed *Bacillus sinicus,* has been isolated from Chinese eggs with coagulated yolks (*Tissier and Lagrange, 1925*).

"Mixed rots," "addled eggs," and "white" or "light" rots (decomposed eggs which appear light before the candle) are other terms frequently used to describe various forms of spoilage in eggs. Few attempts have been made to determine the specific organisms responsible for these rots, or to classify them systematically. They may be regarded as types of general bacterial decomposition, probably due to heterogeneous flora.

Superficial Fungal Growth

A white, fragile, soft growth of various species of *Penicillium, Thamnidium,* and *Mucor* appears on the surface of the eggshell during cold storage, when the humidity is sufficiently high. The higher the humidity, the more luxuriant and dense is the growth, which is known commercially as "whiskers," and which may be from 1 to 10 mm. in length (*Sharp and Stewart, 1936; Moran and Piqué, 1926*). When eggs are held in a saturated atmosphere, the time required for molds to appear on the eggshell depends upon the temperature. This fact is indicated by the following data (*Tomkins, 1937b*):

TEMPERATURE (°C.)	APPEARANCE OF MOLDS (weeks)
1	12
10	6
20	3

The various species of *Mucor* confine themselves to the shell surface and are apparently incapable of penetrating into the interior of the egg (*Mallmann and Michael, 1940*). Their presence does not appear to affect the physical quality of the egg.

FUNGAL ROTTING

Aside from the genus *Mucor,* most of the fungi found on the surface of the shell are capable of penetrating through it and multiplying within the egg. Their growth may be confined at first to the inner surface of the shell and the outer surface of the shell membrane, and to the membranes in the air cell (*Mallmann and Michael, 1940*). "Pin-spots" of different colors are produced by various molds: yellow or blue spots by *Penicillia,* dark green or black by species of *Cladosporium,* and red or pink by types of *Sporotrichum* (*Weston and Halnan, 1927; Tomkins, 1937a*). Most fungi find a favorable environment within the egg. As they multiply, they may penetrate into the albumen, where they often cause coagulation of the areas immediately surrounding their growth. Sometimes molds spread along the chalazae; species of *Sporotrichum,* in so doing, produce pink pigment and thus a type of "red rot" (*Haines, 1939*). If molds reach the yolk, the vitelline membrane may rupture, and the molds then flourish luxuriantly. Eventually, the egg contents may become semisolid.

Influence of Microorganisms on Flavor

Microfloral decomposition of the egg is usually accompanied by pronounced odors. However, certain microorganisms, growing either on the exterior or interior of the egg, may impart their specific odor and flavor to the egg contents when there is no noticeable spoilage.

As previously noted, eggs may absorb the musty odor given off by *Actinomyces* growing elsewhere. A second source of this odor is surface moldiness of the eggshell. Certain species of *Achromobacter* and *Pseudomonas,* when present in the egg contents, are also said to be responsible for mustiness (*Turner, 1927; Levine and Anderson, 1930*).

Atypical coliform organisms have been found to cause a "fishy" flavor in eggs without producing definite rotting (*Haines, 1938*).

PRESERVATION OF EGGS

The purpose of preserving eggs is to delay physicochemical deterioration and to prevent microbial spoilage. Preservation makes possible a better distribution of eggs throughout the year, so that their price is stabilized, and the dietary elements which they supply are continuously available.

Spring is the most advantageous season for withdrawing eggs from the usual commercial channels and setting them aside for later use. Prices are lowest in March, April, May, and June, when egg production is highest in the northern hemisphere. These are the months, therefore, when the largest numbers of eggs are placed in cold storage warehouses and the output of frozen and dried eggs is greatest.

PRELIMINARY HANDLING

The history of the egg destined for preservation is of primary importance in determining the condition of the intact egg at the end of the holding period, or the quality of the product manufactured from the opened egg. Intact eggs intended for prolonged storage should be of greatest excellence and must be properly handled to prevent their deterioration before they reach the warehouse. The prevention of microbiological contamination is especially important, since the most ideal methods of preservation fail if decomposition is incipient when the egg is placed in storage. Sanitation in handling eggs should therefore be observed at all times.

Deterioration of eggs prior to their preservation may be materially reduced if they are laid in clean surroundings, gathered as soon as possible after being laid, candled to remove those with physical defects, and placed immediately in a humidified room at a temperature close to 0° C. Eggs should never be washed with plain water before being stored.

Dirty Eggs

Microorganisms are, of course, present in larger numbers on the shells of dirty eggs than on those of clean eggs. Dirty eggs are therefore much more likely to decompose, especially if they are exposed to summer temperatures. The chances of spoilage are particularly great if the egg is soiled with chicken feces, which not only contain many bacteria, but which, by moistening the shell, make it more pervious to microorganisms.

Experience has shown that dirty eggs, whether washed or unwashed, must be rejected for storage, as there is much more spoilage among them than among normally clean eggs. In a study of eggs stored at −1.5° C. to 0° C. for 11 months, the incidence of decomposition was found to be lowest in clean, untreated eggs, greater in dirty eggs, and highest in dirty eggs that had been washed (Fig. 364). If dirty eggs are fresh and sound, they may be washed and then frozen or dried.

Washing the egg removes the cuticle, an important element of the egg's defense against microorganisms. Either wet washing, or cleaning by means of various abrasives, opens the pores of the shell. If the initial temperature of the egg is higher than that of the washing fluid, the latter is readily drawn in through the pores by suction as the egg cools, and bacteria may enter with it (*Haines and Moran, 1940*). After the egg is

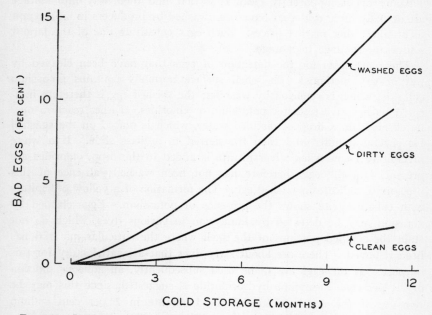

FIG. 364. Incidence of rotting in clean, dirty, and washed eggs held in cold storage. (After Jenkins, Hepburn, Swan, and Sherwood, 1920.)

washed, microorganisms do not necessarily penetrate the shell immediately, but their subsequent invasion is greatly facilitated (*Haines, 1939*).

Losses in storage as a result of washing dirty eggs can be reduced, however, if the washing water contains certain chemical agents. The data in the table show the proportions of inedible eggs found, after 24 weeks' storage, among dirty eggs washed with various solutions (*Funk, 1938*).

WASHING SOLUTION	INEDIBLE EGGS (per cent)
Tap water only	25.3
Chlorine solution (0.15 per cent)	6.6
Ethyl alcohol (70 per cent)	5.2
Sodium hydroxide (0.35 per cent)	2.2

When there is a sudden change in the temperature at which the egg is held, moisture may condense on the shell, which is then said to "sweat." A very thin film of water is sufficient to promote the growth of dormant microorganisms on the shell surface and to give them an opportunity to penetrate through the pores (*Moran and Piqué, 1926; Rosser, White, Woodcock, and Fletcher, 1942*).

Many of the apparently clean eggs that find their way into storage are actually dirty eggs that have been washed by producers in an attempt to obtain higher market prices. Such eggs constitute one of the largest sources of spoilage in storage.

There are tests for the detection of eggs that have been cleaned by wet or dry methods. The shell surface normally contains potassium chloride, which is removed by washing; the washed egg is therefore indicated by negative tests for potassium or chlorides. Either reaction may be observed in a drop of distilled water, which is placed on the eggshell for several minutes and then transferred to a glass slide. If a white precipitate forms when silver nitrate is added to the drop, chlorides are present, and the egg therefore has not been washed; absence of this precipitate indicates a washed egg. The formation of a yellow precipitate with cobalt nitrite shows the presence of potassium. Eggs cleaned by abrasion may be detected by immersion in various dyes, which do not stain the scraped portions of the shell where the cuticular material has been removed. There are about sixty dyes that are satisfactory for this purpose. If the eggs are to be used subsequently, an aqueous solution of 0.1 per cent Rosaniline hydrochloride is suggested, since this may be removed if the eggs are soaked for 1 minute in 2 per cent sodium bisulfite and then washed with water (*Sharp, 1932*).

Eggs which have been scoured also show dark areas in ultraviolet light. The removal of stamped grading marks (used chiefly in Europe) can be detected by uneven fluorescence (*Vilter and Schmidt, 1933*).

PREVENTION

Usually at least 10 per cent of all eggs become dirty before they are collected, unless preventive measures are taken (*Pennington and Pierce, 1910*). If birds are not allowed to walk in mud or in their droppings, their feet and ventral feathers remain clean, and dirt is not carried to the nests and transferred to the eggs. Keeping nests clean by frequently changing the nesting material greatly reduces the percentage of dirty eggs. The incidence of soiled eggs may also be lowered considerably by making nests of shavings, oat hulls, sawdust, or excelsior, rather than of straw; and it may be decreased by 50 per cent simply by gathering the eggs four times a day (*Funk, 1937*).

Cooling of Eggs

The cooling of the egg at the place of production is an important step in preliminary handling. If the egg is not cooled as soon as possible

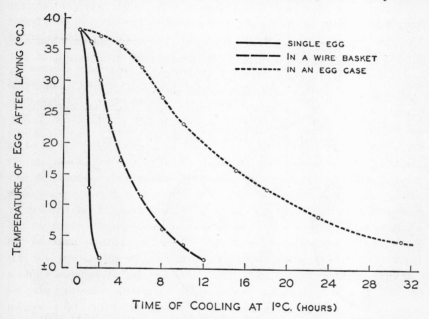

Fig. 365. Rate at which an egg cools at 1° C. when alone and when surrounded by many others in a wire basket and in an egg case. (After Funk, 1935.)

after it is laid, the processes of deterioration may soon begin, especially if the egg is exposed to high temperatures in summer.

The rate at which eggs lose heat after they are laid depends primarily, of course, upon the environmental temperature. In the same type of container, they cool almost twice as quickly in a refrigerator at 1° C. as in a storage room at 10° C. (*Funk, 1935*). High humidity and circulating air also hasten cooling. During the first half hour, an egg cools as rapidly at 13° C. in moist, moving air as it does at 4.5° C. in still, dry air (*Nicholas, 1936*). However, too rapid air movement may hasten the evaporation of the egg contents and cause undue enlargement of the air cell.

Eggs cool at various rates in containers of different types. In a wire basket full of eggs, the eggs in the center cool almost three times as rapidly as those in the center of a full egg case (Fig. 365). The rate of heat loss from a single egg on a wire tray is, in turn, nearly six times that of an egg in the center of a full wire basket. Therefore, eggs should not be placed in cases while they are still warm, and they should be cooled on trays if possible.

Rough Handling

Rough handling, especially in transportation, may not only damage the eggshell but may also disturb the architectural arrangement of the egg contents and thus accelerate physical disintegration.

Jarring may produce separation of the shell membrane from the egg membrane, with the consequent formation of a somewhat movable or "tremulous" air cell. Sometimes a "loose" air cell, consisting of several air bubbles, is formed when the inner membrane ruptures (Fig. 366-*B*).

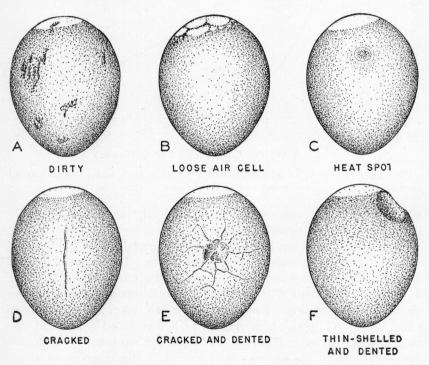

FIG. 366. Eggs which are edible when fresh, but which are usually rejected by the market and are unsuitable for storage. The conditions in *B*, *C*, and *D* are detectable only by candling.

Eggs with air cells of this type are not suitable for storage. Prolonged shaking, if sufficiently violent, may also result in relaxation of the albuminous structures, including the chalazae, which hold the yolk in its central position (*Almquist, Nelson, and Lorenz, 1934*).

Results of Improper Handling

It may be said in conclusion that many eggs are rendered unsuitable for preservation, or become entirely inedible, merely because of improper handling, or poor control of environmental conditions. Among the eggs which may be considered fit for immediate use, but which must be rejected for storage, are dirty eggs (cf. Fig. 366-*A*), and fertile eggs that have been held at sufficiently high temperatures to develop "heat spots" (cf. Fig. 366-*C*). In the same class are eggs with cracks (cf. Fig. 366-*D*), dents (cf. Fig. 366-*E* and *F*), and loose air cells, all the result of rough treatment. Eggs must be rejected for human consumption if candling them shows that they contain blood rings (Fig. 367-*A*), adherent

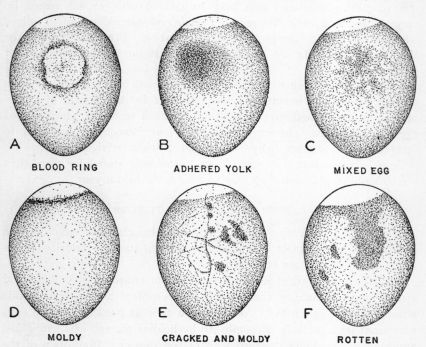

A BLOOD RING	**B** ADHERED YOLK	**C** MIXED EGG
D MOLDY	**E** CRACKED AND MOLDY	**F** ROTTEN

FIG. 367. Eggs which are rejected by the market, both for immediate consumption and for storage (upper row), or which are definitely inedible (lower row). None of the conditions shown can be detected without candling the egg.

yolks (cf. Fig. 367-*B*), mixed contents (cf. Fig. 367-*C*), or bacterial or fungal growths (cf. Fig. 367-*D, E,* and *F*).

PRESERVATIVE METHODS

Many of the methods in general use for the prevention of decomposition in foodstuffs are employed for the preservation of eggs, both intact and opened. These methods depend mainly upon low temperatures, chemical agents of various types, and dehydration. Sometimes two or more methods are combined.

Low temperatures greatly retard the progress of physical and chemical disintegration in eggs, as in all organic matter, and in addition may destroy many microorganisms and inhibit the growth of others. For these reasons, refrigeration is almost essential to the storage of the intact egg, and actual freezing is successfully used in the preservation of the liquid egg.

Gases that retard deterioration in the intact egg may be added to the storage atmosphere; and eggs may be treated with, or submerged in, various preservative compounds. Certain substances are often mixed with liquid eggs for the purpose of stabilization. Precautions are taken that chemical agents are nontoxic and impart no undesirable odors or flavors to the egg.

Dehydration at high temperatures removes most of the free water from the egg and thus eliminates a vital substance for the growth of microorganisms, and a necessary vehicle for chemical reactions. Only opened eggs can be dried. Intact eggs, however, are sometimes subjected to certain heat treatments before being placed under refrigeration.

Eggs, whether intact or opened, are not usually stored for more than 6 to 9 months, although they may sometimes be held a year.

PRESERVATION OF INTACT EGGS

Methods of preserving intact eggs may be classified according to whether the environment is modified or the egg itself is treated. Sometimes both methods are combined.

Fundamentally, all methods depend upon retarding microbiological growth and maintaining the original water content and carbon dioxide tension of the egg's interior for as long a time as possible. These ends may be at least partially accomplished by holding the egg in a humidified atmosphere, sometimes enriched with carbon dioxide. Prevention of the escape of carbon dioxide and water vapor has also been achieved, with varying success, by packing the egg in a number of dry substances, immersing it in certain liquids, treating the shell with sealing agents, or coagulating, by means of heat, a thin layer of albumen adjacent to the shell. The checking of microfloral growth is incidental to some of these methods, and antibiotic agents are occasionally used in conjunction with others.

Some methods of egg preservation are suitable for commercial use on a large scale, and the application of others is limited to homes. A number of methods have been found impractical and are no longer extensively employed, and a few are still in the experimental stage.

Dry Packing

Preservation of intact eggs by dry packing has been frequently attempted in the past and is occasionally still practiced today. The following substances are among those investigated as possible packing material:

SUBSTANCE	EARLY INVESTIGATOR
Bran	Wheeler (*1890*)
Calcium iodate	Vautin and Whiffen (*1926*)
Chaff	Vosseler (*1908*)
Excelsior saturated with formaldehyde	Prenzlau (*1910*)
Oats	Jarvis (*1899*)
Peat dust	Anonymous (*1897*)
Salt	Wheeler (*1890*)
Sand	Vosseler (*1908*)
Sand and burned gypsum or pulverized lime	Falck-Loendahl (*1924*)
Sawdust	Vinson (*1908*)
Shale dust mixed with gypsum, ashes, sulfur, borax, saltpeter, and salicylic acid	Slukes (*1902*)
Silicates of iron, lime, and aluminum	Roberts (*1891*)
Soda lime	Utescher (*1898*)
Straw	Vosseler (*1908*)
Wood ashes	Svoboda (*1902*)

If the packing material is loose, it does not prevent evaporation or decomposition. If it is compact, it can only retard, and not prevent, the growth of microorganisms, although some antibiotic packing substances have been suggested. Dry packing is not feasible commercially, because of the necessity of transporting the excess weight of the packing material with the eggs.

CHINESE METHODS

For many years, the Chinese have preserved eggs by methods somewhat similar to dry packing. However, the egg is not retained in its original state but rather converted into an entirely different article of food

(*Wang, 1929*), probably by bacterial action. For example, a product known as "hulidan" results when eggs are coated individually with a mixture of salt and wet clay or ashes and stored for a month. This process darkens and partially solidifies the yolks and gives the eggs a salty taste. More pronounced changes are found in "dsaudan," or eggs that are kept for at least 6 months, packed in cooked rice and salt. The shells soften, the membranes thicken, and the egg contents coagulate and take on a characteristic winey flavor. A great delicacy, "pidan," is made by storing eggs 5 months or more, after covering them with lime, salt, and wood ashes, mixed with a tea infusion. Eggs treated in this manner bear little resemblance to fresh eggs. The yolks are greenish gray, and the albumen is a coffee-brown jelly, containing tyrosinelike crystals (*Wang, 1929*). After the eggs have been kept for several years, their contents become dark brown throughout. The taste has been described as "limey," and the odor as "ammoniacal" (*Tso, 1926a*).

An analysis of the fat of preserved Chinese egg yolks 2 years old is shown in the table, in comparison with an analysis of fresh yolk fat (*Thomson and Sorley, 1924*). It is obvious that pronounced chemical

	FRESH YOLK	CHINESE YOLK
Unsaponifiable matter (per cent)	0.36	3.64
Iodine value	74.73	84.20
Acid value	4.47	51.88
Free fatty (oleic) acid (per cent)	2.25	26.08

changes occur in the lipids of the yolk, when the egg is preserved by the Chinese method.

Immersion in Liquids

Immersion in liquids is a fairly old method of preserving eggs. It primarily prevents the evaporation of moisture from the egg. Depending upon the liquid used, it may also inhibit bacterial decomposition by chemical action or by physical means, such as the occlusion of air. Low temperatures are essential to the successful preservation of eggs by immersion in liquids.

Many liquids have been used experimentally, with varying success, but only a few have attained much popularity. Among those investigated as egg preservatives are the following:

LIQUID USED	EARLY INVESTIGATOR
Borax and salt solution	Anonymous (*1899*)
Brine	Kronmann (*1899*)
Chlorinated lime solution	Calvert (*1873*)
Chlorine water	Calvert (*1873*)

Fluosilicic acid	Teisler (*1902*)
Lime water	(See text)
Lime water, boric acid, and glycerol	Barral and Trésorier (*1907*)
Magnesium oxide solution	Utescher (*1921*)
Molasses	de Villèle (*1903*)
Phenol	Calvert (*1873*)
Potassium hydroxide solution	Chin (*1933*)
Sodium hydroxide solution	Chin (*1933*)
Sodium silicate	Thieriot (*1897*)
Sulfite lye solution	Wise (*1901*)

Of all the various liquids tested, only two, lime water and water glass, have been used extensively for keeping eggs over long periods of time.

LIME WATER

Lime water has been known as an egg preservative since the eighteenth century (*Swingle and Poole, 1923*), but its use was not widespread until the last quarter of the nineteenth century. It has been employed chiefly in homes, although in fairly recent times some commercial establishments, especially in Holland, stored large numbers of eggs in enormous vats of lime water (*Spamer, 1931*).

Lime water is easily prepared by slaking quicklime (calcium oxide), and then adding water. The proportions used may be 1 lb. of lime to 1 gal. (*Swingle and Poole, 1923*), 2 gal. (*Miller, 1927*), or 5 gal. (*Shutt, 1927*) of water, although a saturated solution of lime contains only 12.6 lb. of calcium oxide to 1000 gal. of water (*Miller, 1927*). When the excess lime has settled out from the mixture, the supernatant liquid is poured off and used.

The carbon dioxide in the atmosphere reacts with the lime water and precipitates calcium carbonate over the surface of the solution. The lime water, therefore, is eventually weakened if the container is kept uncovered (*Shutt, 1927*), or if the layer of calcium carbonate is broken (*Miller, 1927*). Salt is sometimes added to the solution, but experiments have shown that the surface layer of calcium carbonate will not remain intact if salt is present in a concentration of 1 lb., or more, to 10 gal. of water (*Miller, 1927*).

The effectiveness of lime water as an egg preservative is probably due in part to its alkalinity, which discourages the growth of microorganisms, although not with complete success in all cases. Lime water deposits a thin film of calcium carbonate on the eggshell and thus partially seals the pores. Apparently it also has some chemical action which renders the shell membranes relatively impermeable (*Moran and Piqué, 1926*). Lime water is quite likely, however, to impart its own taste to the egg (*King-horne, 1920*), especially if salt has been added to the solution (*Shutt, 1927*). This preservative is therefore less desirable than water glass, although it is somewhat cheaper.

WATER GLASS

Since the beginning of the twentieth century, water glass has been used with considerable success for the domestic preservation of eggs, and it still is popular. It is a viscid, syrupy, nonvolatile solution of sodium silicate. It is prepared by diluting one part of the commercial concentrated solution with ten parts of water, although higher dilutions occasionally are suggested (*Swingle and Poole, 1923*). If the solution is not diluted sufficiently, it becomes a gel, which makes the handling of the eggs extremely difficult. Almost any type of container may be used, but glass jars with tightly fitting tops are perhaps preferable, because they prevent evaporation and gelling (*Hall, 1945a*). The surface of the solution should at all times be at least 2 inches above the uppermost layer of eggs. Eggs and solution may be added at any time until the container is full.

Water glass does not penetrate through the eggshell. It is a poor solvent of gases (*Jones and Dubois, 1920*) and deposits a precipitate of silica on the eggshell (*Moran and Piqué, 1926*). An anaerobic condition, inhibitory to many bacteria, is thus created within the egg. There are also indications that water glass possesses intrinsic antiseptic powers. Various dilutions of water glass have been found to kill *Escherichia coli* within very short periods of time, as shown in the table (*Swingle and Poole, 1923*):

DILUTION	TIME (minutes)
1:20	5
1:40	10
1:80	15

Water glass imparts no odor or taste to eggs, although it causes them to lose their fresh flavor and become noticeably "flat." Because of the precipitate deposited on the egg, the blunt end of the shell must be pricked if the egg is to be boiled, to prevent the shell from bursting when heat expands the air within the air cell (*Pierce, 1909*).

One of the most important factors in the successful use of water glass is the storage temperature, which should be as close to refrigeration as possible and preferably not over 4.5° C. However, water-glass eggs kept for several months at temperatures as high as 13° C. to 16° C. (55° F. to 60° F.) are often satisfactory for cooking purposes.

Various experiments have clearly shown that water glass is not a sufficiently good preservative to overcome the deleterious effects of relatively high storage temperatures. Hall (*1945a*) stored three lots of eggs in water glass at average temperatures of 2° C., 14° C., and 26° C., respectively. At the end of 6 months, the eggs kept at the lowest temperature compared favorably with fresh eggs, although their odor and flavor were slightly stale. Displaced yolks were found in about 72 per cent of the eggs held at 14° C., but the vitelline membranes did not rupture when the eggs were opened. By contrast, the presence of stuck yolks caused 76 per cent of the eggs held at room temperature to be graded as inedible when the eggs were candled. Little or no dense albumen was observed in opened eggs of this group. The result of additional physicochemical tests on all three lots of eggs (*Romanoff, 1948*), summarized in Table 47, also indicate the progressively greater deterioration which occurs at the higher storage temperatures.

TABLE 47

INFLUENCE OF HOLDING TEMPERATURE ON PHYSICOCHEMICAL CHANGES IN EGGS
PRESERVED IN WATER GLASS FOR 6 MONTHS *

		Water-Glass Eggs		
Observation	Fresh Eggs	At 2° C.	At 14° C.	At 26° C.
Fluorescence of eggshell:				
Red (per cent)	100	46	43	0
Orange (per cent)	0	54	51	60
Blue (per cent)	0	0	6	40
Hydrogen-ion concentration:				
Albumen (pH)	7.60	7.92	8.31	8.47
Yolk (pH)	6.00	5.94	6.28	6.11
Refractive index:				
Albumen (n_D)	1.350	1.353	1.359	1.359
Yolk (n_D)	1.422	1.417	1.409	1.410
Coagulation point:				
Albumen (°C.)	61.2	61.4	61.0	58.8
Foaming properties:				
Albumen (% increase)	453	379	337	48
Yolk index † $\dfrac{\text{height}}{\text{width}}$	0.403	0.396	0.315	0.201

* From the data of Romanoff (*1948*).
† From the data of Hall (*1945a*).

Cold Storage

In the United States, the commercial cold storage of eggs was first undertaken approximately in 1890 and was facilitated greatly by the invention of mechanical refrigeration. The peak holdings of each year,

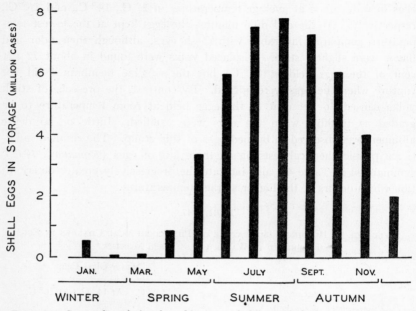

Fig. 368. Seasonal variation in cold storage holdings of shell eggs in the United States in 1940.

always in the month of August, increased rapidly until 1930, but they have subsequently declined.

YEAR	THOUSANDS OF CASES
1920	6,872
1930	11,198
1940	7,784

By withdrawing surplus eggs from the market during the season of highest production and selling them later during times of relative scarcity, extreme fluctuations in price can be avoided. The largest number of eggs go into the warehouses in the spring, and their purchase for cold storage continues at a somewhat reduced rate during the summer (Fig. 368). Withdrawals are made chiefly during the autumn and winter. Holdings are usually lowest in February or March.

Lengthening the laying season has had its effect upon the cold storage of eggs. Eggs are now placed in the warehouses earlier in the year than formerly and remain there for shorter periods of time, so that they do not accumulate in such large numbers and are of better quality. As the production of eggs becomes less seasonal, and the facilities for the transportation of fresh eggs continue to improve, the commercial cold storage of eggs may decrease in economic importance.

COLD STORAGE IN NORMAL ATMOSPHERE

In cold storage, intact eggs are held at the lowest possible temperature that will avoid their freezing and the bursting of their shells. They may, however, be held at temperatures lower than the freezing points of albumen and yolk (see Chapter 7, "Physicochemical Properties"). It has been found experimentally that intact eggs do not freeze at temperatures between $-1.5°$ C. and $-2.0°$ C. The range of temperature in the large, specially designed warehouses may therefore be from $-0.5°$ C. to $-2.2°$ C., although it is sometimes from $0°$ C. to $-1.5°$ C. Attempts are made to keep the temperature as uniform as possible, but fluctuations are sometimes difficult to prevent.

The usual range of relative humidity in the cold-storage room is between 80 and 90 per cent, since experience has shown that 85 per cent is the most practical level (*Moran, 1936b*). Maintenance of the proper relative humidity depends upon the constancy of the temperature. A sudden fall of $0.33°$ C. may produce a rise of 2.2 per cent in relative humidity (*Sharp and Stewart, 1936*). In addition, the humidity is not uniform within the egg case; it increases from exterior to interior. The lower relative humidity at the periphery of the case is due to the hygroscopic properties of the packing material, which can act as a water-absorbing unit (Fig. 369).

The relative humidity at which eggs are stored represents a compromise between the level which best retards loss of water from the egg and that which inhibits the growth of molds. Low temperatures usually check the multiplication of bacteria, but mold spores germinate during refrigeration whenever sufficient moisture is available. Figure 370 shows the relationship between relative humidity and the time required for mold growth on eggs to become visible at $5°$ C. and at $0°$ C. The appearance of molds is much more rapid at 95 per cent relative humidity than at 90 per cent. If the relative humidity in cold storage is higher than 90 per cent, the eggshell, sooner or later, is covered with a growth of white, fragile hyphae ("whiskers") 1 to 2 mm. long. These are followed by dark molds, if the relative humidity is over 96 per cent. At relative humidities between 85 and 90 per cent, molds eventually appear, but they are less conspicuous (*Sharp and Stewart, 1936*).   A relative humidity of 85 per cent is therefore obviously the upper limit of safety if the egg is not to become moldy.   This level, however, is not high enough to prevent evaporation from the egg contents, and loss of moisture proceeds at the rate of about 0.12 per cent per week (*Moran, 1937d*).

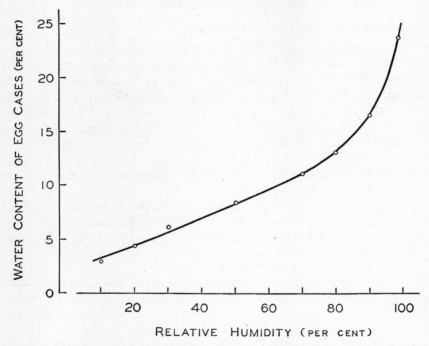

Fig. 369. Hygroscopic properties of ordinary wooden egg cases as shown by the water content at 15° C. and various relative humidities. (After A. Smith, 1935b.)

The molds that grow on eggs in cold storage are common species.   The molds in the storage room, especially on egg cases and packing material, are the same as those found on the eggshell (Fig. 371).   In many warehouses, it is the practice to disinfect the walls and floors before each new consignment of eggs arrives.   Disinfection, however, does not affect the interior of the egg cases.   A more effective method of suppressing molds is to introduce mycostatic agents into the flats and fillers.   However, if the mycostats used are the type that kills mold spores and mycelia upon contact only, mold growth is merely prevented from spreading; the germination of spores is not inhibited.   Some species of molds are markedly resistant to such contact mycostats as sodium borate, cupric sulfate, and cupric acetate, unless these substances are used in concentrations far too strong to be practical. On the other hand, molds rarely appear when the flats and fillers are treated with compounds that slowly give off mycostatic vapors. Among these substances are various derivatives of phenol, especially sodium pentachlorophenate, sodium trichlorophenate,

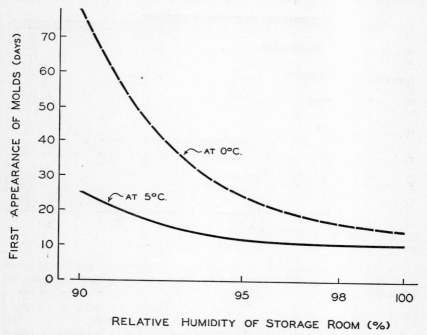

Fig. 370. Relationship between relative humidity in the storage room and the time required for molds to appear on eggs held at 0° C. and 5° C. (After Haines, 1939.)

and sodium tetrachlorophenate, of which the first is apparently the best, under commercial conditions (*Mallmann and Michael, 1940*). These mycostats, used in concentrations of 0.81 to 0.92 per cent, have prevented the development of molds on eggs during 8 months of storage at 0° C. and 92 to 95 per cent relative humidity. No detectable foreign flavor is imparted to the eggs. It is said that *o*-phenyl phenol is also effective (*Tomkins, 1937b*). Mycostatic agents may be applied to flats and fillers at the time of manufacture; egg cases should also be treated, unless they are made of Sitka spruce, which is naturally resistant to molds (*Mallmann and Michael, 1940*).

Eggs must be defrosted slowly when they are removed from cold storage, so that moisture will not condense on the shells and permit microbial growth and spoilage before the eggs reach the consumer. A defrosting period of 18 to 24 hours is advisable (*Moran and Piqué, 1926*).

Cold storage is not an ideal method for the preservation of eggs. Air conditioning still presents a practical problem, since most of the moisture

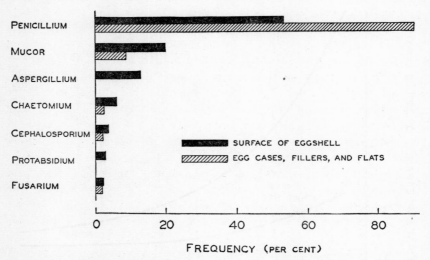

FIG. 371. The incidence of various molds on eggs compared with the frequency with which the same molds are found on egg containers. (After Mallmann and Michael, 1940.)

of the atmosphere is deposited as ice on the refrigerating coils. Deterioration is inevitable in cold storage eggs. The characteristic flavor of the storage egg usually develops, and many of the disintegrative processes previously described may eventually occur. For example, the ammonia content of the egg increases steadily during storage, as the table shows (*Hendrickson and Swan, 1918*).

COLD STORAGE (months)	AMMONIA CONTENT OF EGGS (milligrams/100 gm.)
0	1.3
3	2.1
6	3.2
9	4.0

The practice of adding carbon dioxide or ozone to the storage atmosphere has resulted from attempts to develop better methods for storing eggs.

COLD STORAGE IN ATMOSPHERE ENRICHED WITH CARBON DIOXIDE

Early in the nineteenth century, Berard (*1821*) suggested the preservation of immature fruits in an atmosphere free of oxygen. Since then, various experiments have been made with the use of inert gases, such as carbon dioxide, to prevent spoilage of foodstuffs. Practical application of gas storage was first made on British ships importing perishable foods to England.

Carbon dioxide lends itself well to the preservation of the egg, whose condition is intimately related to its content of this gas. As early as 1873, Calvert was able to keep eggs well for 3 months in an atmosphere of nitrogen, hydrogen, and carbon dioxide. Later, Lescardé (*1908*) evacuated the air from eggs, replaced it with carbon dioxide, and stored the eggs at refrigeration temperatures in an atmosphere of 94 per cent carbon dioxide and 6 per cent nitrogen. During the last 20 years, the commercial use of carbon dioxide in the preservation of eggs has become fairly extensive, especially in England.

The addition of carbon dioxide to the storage atmosphere retards the deteriorative physical changes which normally occur within the egg with the passage of time. The beneficial effects of carbon dioxide may be seen even when its concentration is as low as 0.55 per cent (*Sharp and Stewart, 1931*).

Under increased carbon dioxide tension, the egg absorbs the gas until equilibrium is reached. As carbonic acid is formed within the egg, the alkalinity of the egg contents is neutralized. Thus, by the addition of the proper amount of carbon dioxide to the atmosphere of the storage room, the original hydrogen-ion concentration of the egg can be retained. As the environmental temperature decreases, lower concentrations of carbon dioxide are sufficient to hold the pH of the egg at 7.6. At 20° C., at least 10 per cent of carbon dioxide is required, whereas 3 per cent is all that is needed at 0° C. (*Sharp, 1929a*).

Eggs stored under carbon dioxide show less liquefaction of the dense albumen, but the albuminous sac decreases in amount. After storage for 5.5 months at 0° C., the percentage of each albumen layer varies according to the atmospheric concentration of the gas, as the following table shows (*Moran, 1937a*). The higher the concentration of carbon dioxide, the smaller is the volume of the dense layer (*Moran, 1937d*). However, the mucin content of the dense albumen is not reduced, and the viscosity of the outer liquid albumen is lower than when the egg is stored in air (*Bate-Smith, 1937*). These two facts indicate that there

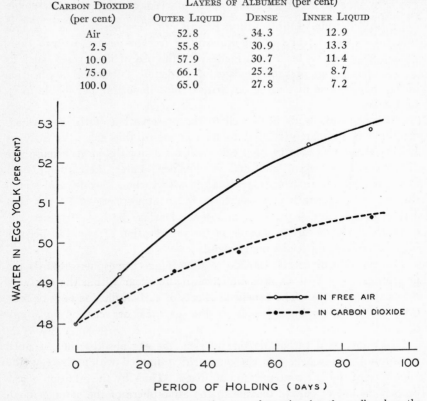

Carbon Dioxide (per cent)	Layers of Albumen (per cent)		
	Outer Liquid	Dense	Inner Liquid
Air	52.8	34.3	12.9
2.5	55.8	30.9	13.3
10.0	57.9	30.7	11.4
75.0	66.1	25.2	8.7
100.0	65.0	27.8	7.2

Fig. 372. Comparison between rate of water absorption by the yolk when the egg is stored at 15° C. in air and when it is stored in carbon dioxide in concentrations greater than 2.5 per cent. (After Moran, 1937d.)

is little or no dissolution of mucin. At *p*H 6.5, which is reached when the concentration of carbon dioxide is approximately 60 per cent, the albumen becomes cloudy, perhaps due to the coagulation of ovomucin (*Moran, 1936a*).

The diffusion of water yolkward is effectively restrained by the presence of as little as 2.5 per cent of carbon dioxide in the atmosphere (Fig. 372). The permeability of the vitelline membrane apparently diminishes, for the freezing points of the yolk and the albumen are only slightly equalized (*Moran, 1936a*). The relative benefits of higher concentrations of carbon dioxide, such as 10 to 100 per cent, are negligible.

Neither the odor nor the flavor of the egg deteriorates as much in the presence of carbon dioxide as it does when the egg is stored in normal air (Fig. 373). The taste may be somewhat flat, but it does not resemble the usual storage taste. A concentration of 1 per cent carbon dioxide is as effective as higher concentrations in preserving good flavor.

Carbon dioxide inhibits, rather than destroys, the microflora of the egg. Furthermore, its inhibitory effect is not pronounced, unless the gas is present in a concentration of more than 60 per cent, and the temperature

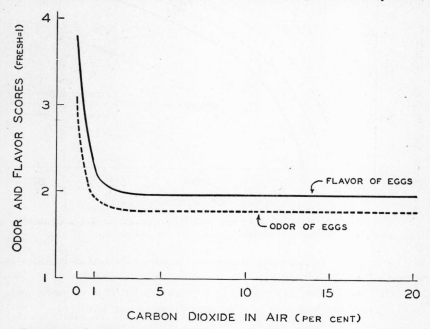

Fig. 373. The flavor and odor scores of eggs after storage for 14 months at −1.0° C. in air containing various percentages of carbon dioxide. (After Sharp and Stewart, 1931.)

is close to freezing. The appearance of molds may then be retarded for as long as 9 months, in saturated air. At 10° C., however, various types of *Penicillium* and *Sporotrichum* grow and can penetrate into the interior of the egg when the carbon dioxide concentration is as high as 60 to 80 per cent (Fig. 374). Of the bacteria that are the most troublesome causes of spoilage in eggs, various members of the genus *Proteus* are fairly resistant to carbon dioxide; but, at refrigeration temperatures, these species are usually somewhat restrained even in normal air. *Achromobacter* is comparatively susceptible to carbon dioxide. Some strains of *Pseudomonas* that grow well in air at 0° C. do not multiply in carbon dioxide. They nevertheless survive, even in an 80 per cent concentration

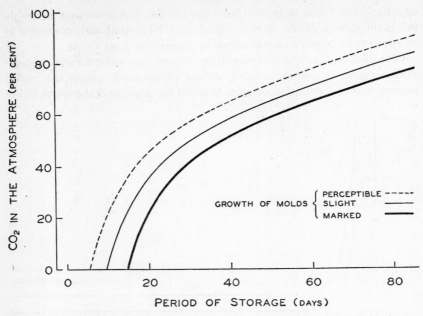

Fig. 374. The growth of molds on eggs stored (over water) at 10° C. in air containing various percentages of carbon dioxide. (After Moran, 1937c.)

of the gas, and will reassert themselves whenever favorable conditions are provided (*Haines, 1939*).

Carbon dioxide replaces oxygen to a greater extent than nitrogen, and in high concentrations may be dangerous to warehouse workers.

COLD STORAGE IN ATMOSPHERE ENRICHED WITH OZONE

Ozone possesses the unique property of arresting fungal growth on biological surfaces. In the proper concentrations, and at certain temperatures, it greatly delays the appearance of molds on the eggshell, even when the relative humidity of the storage room is as high as 90 per cent (*Ewell, 1936*). At this humidity, it does not restrain molds that have already penetrated through the eggshell (*Moran, 1937d*).

Ozone must be used at refrigerating temperatures if it is to be effective. To prevent mold growth on clean eggs, at 90 per cent relative humidity, ozone must be maintained in a concentration of at least 0.6 p.p.m.[2] by weight. A concentration of 1.5 p.p.m. is recommended, however, in order to destroy fungi whose development may have begun,

[2] The abbreviation p.p.m. = parts per million.

and to insure the continual presence of ozone in quantity sufficient to provide protection. Ozone in low concentrations is of questionable effectiveness against bacteria (*Haines, 1935*).

Ozone is especially well adapted to use in egg storage rooms. Eggs absorb it less readily than do other foodstuffs, and maintenance of its concentration is thus less difficult than in the presence of meat, fruit, or vegetables. The relative absorption of ozone (in a concentration of 1 p.p.m.) by several foods in half an hour is shown in the table (*Ewell, 1930*).

	PART PER MILLION
Fresh beef	0.38
Fresh apples	0.33
Cauliflower	0.21
Eggs	0.03

The effectiveness of ozone is due to the fact that it decomposes spontaneously and releases nascent oxygen, which is a powerful oxidizing agent in the presence of moisture. Although ozone has a characteristic pungent odor, it is an excellent deodorant. Eggs held in low concentrations of ozone do not acquire any trace of mustiness and are said to be indistinguishable in taste from fresh eggs, even after 8 months of storage (*Pennington and Horne, 1924*). When present in concentrations as high as 10 p.p.m., ozone imparts a strong, metallic flavor to eggs (*Haines, 1935*).

There are certain limitations to the use of ozone. Its odor is objectionable, and it may be very corrosive, especially to rubber fittings of ducts, doors, etc.; in higher concentrations, it is irritating to the mucous membranes (*Ewell, 1936, 1938*). Due to its instability, it must be made where it is to be used, and it must be continuously supplied. Various engineering problems are thus encountered. Because it is an oxidizing agent, it may produce rancidity of fats, and its use in the preservation of dried eggs is therefore precluded (*Ewell, 1936*).

Little is known, as yet, concerning the possible effectiveness of ozone in checking the physical deterioration of the egg.

Shell-Sealing Treatments

It is obvious that many of the problems of conserving eggs could be overcome if the shell of the egg could be sealed so that water vapor and carbon dioxide could not escape, and microorganisms could not penetrate. With this objective in mind, numerous investigators have experimented with a great variety of materials. A partial list follows:

Material Used	Early Investigator
Agar-agar	Willis (*1927*)
Alum	Jacobsen (*1910*)
Cactus juice	Clairemont (*1914*)
Celluloid	Doyle and Doyle (*1935*)
Collodion	de la Mota (*1917*)
Cottonseed oil	Subirana (*1917*)
Cottonseed oil and guttapercha	Coleman (*1914*)
Gelatin	Mills (*1891*)
Irish moss	Willis (*1927*)
Lanolin	Fowler and Edser (*1926*)
Lard	Campanini (*1908*)
Linseed oil	Fryklind (*1901*)
Magnesium chloride	Christensen (*1916*)
Mineral oil	Almy, Macomber, and Hepburn (*1922*)
Nitrocellulose and camphor	Jerne (*1911*)
Paraffin, beeswax, and boric acid	Aston and Stevens (*1924*)
Paraffin and formaldehyde	Jacoby (*1910*)
Paraffin, tallow, and boric acid	Greensmith (*1934*)
Peanut oil and beeswax	French (*1935b*)
Plaster, alum-treated	Garrigon (*1907*)
Rubber	Rollman (*1934*)
Soap	Urner (*1921*)
Sodium silicate (in vacuum)	Stead (*1882*)
Sulfuric acid	Reinhardt (*1899*)
Tung oil	Davis and Metz (*1916*)
Varnish	Barlow (*1902*)
Vaseline	Thieriot (*1897*)
Wax and shellac	Lipcscy (*1930*)

Coating the egg with many of the suggested substances involves smearing each egg individually. Such a procedure is adaptable to preservation on a small scale only and cannot be used commercially. Another disadvantage of certain sealing agents is the fact that they tend to crack as they cool and dry. At present, light oils are the materials chiefly employed.

Other processes for retarding deterioration in eggs make use of heat, either to destroy bacteria, or to produce a sealing effect within the egg.

COATING WITH OIL

Oiling the eggshell is a fairly successful method of rendering it less permeable. This process is not new. At least as early as 1807, Dutch farmers immersed eggs in linseed oil for several hours before storing them for future use (*Spamer, 1931*).

Oil treatment of the shell may be accomplished simply by dipping eggs, held in wire containers, into a bath of tasteless, odorless, colorless, edible oil. The eggs are immersed for only a moment and are then removed, and the excess oil is allowed to drain.

If oil treatment is to be effective, it should be done preferably at the point of production, the day after the egg is laid. A lapse of 24 hours allows part of the egg's carbon dioxide to escape and therefore prevents the albumen from becoming turbid, and the albuminous sac from shrinking unduly, during subsequent storage (*Evans and Carver, 1942*). Washed or "sweating" eggs may be oiled, if they are dried prior to treatment. Oiling is not a substitute for refrigeration. Oiled eggs must be held at low temperatures to prevent physical deterioration and microfloral growth, especially the growth of external molds (*Gibbons, Fulton, and Hopkins, 1942*).

Vegetable oils, such as cottonseed, linseed, and peanut oils, are good sealing agents; but mineral oils are preferable, since they are less subject to oxidative change during storage. Heavy mineral oils appear to be more effective than those of low viscosity (*Almy, Macomber, and Hepburn, 1922; Evans, 1942; Rosser, White, Woodcock, and Fletcher, 1942*). However, it is more convenient to use light oils which flow well without being warmed. In addition, treatment with such oils, at ordinary temperatures, does not expose the egg to the danger of being changed internally by heat. The greatest protection is afforded by oils whose pourpoint is a few degrees above the storage temperature (*Swenson, Slocum, and James, 1936*).

Oil treatment is a fairly efficient method of retarding evaporation from the egg. After 7 months of storage at $-0.5°$ C., the weight loss in oiled eggs may be much less than in untreated eggs, as shown below (*Mallmann and Davidson, 1944*).

	WEIGHT LOSS (per cent)
Unoiled eggs	6.66
Oiled eggs	0.61

The rate at which carbon dioxide escapes from oil-treated eggs is considerably reduced, especially during the early holding period, as the tabulated data indicate (*Swenson, 1939*). Correspondingly, the pH value of the albumen rises less rapidly in an oiled egg than in an

	CARBON DIOXIDE LOST (mg./24 hr.)	
	48–96 HR.	96–1000 HR.
Unoiled egg	10.0	5.0
Oiled egg	5.5	3.0

untreated egg. Oiling does not prevent a perceptible increase in alkalinity as the time of storage lengthens.

After the egg has been oiled, the changes in the albumen during storage resemble those observed when the egg is held under increased partial pressure of carbon dioxide. The outer liquid portion of the albumen increases in amount; and, although the albuminous sac shrinks, it retains much of its consistency (*Evans and Carver, 1942*).

Eggs that have been oiled are not likely to absorb foreign odors. If the proper type of odorless oil is used, no flavor is imparted to the egg (*Swenson, 1939*).

Partial sealing of the eggshell with oil reduces the possibility of penetration by microorganisms and inhibits the growth of molds, although only to a slight extent at high temperatures (*Reedman and Hopkins, 1942*). Oil treatment should not be applied to eggs that have been exposed to any major contamination. It is possible to add to the oil such antibiotic agents as pentachlorophenol, which has given good results when used in this manner in a concentration of 0.25 per cent (*Mallmann and Davidson, 1944*).

An oiled egg may be detected by dipping one end into ethyl ether for a second or two. As the ether evaporates, it leaves an oily ring, even if the egg has been sanded (*Sharp, 1932*). An oil-treated egg is also indicated by the turbidity produced when a piece of the shell is boiled with alcoholic sodium hydroxide and the solution then diluted with water (*Hoover, 1939*); but this method, of course, requires that the egg be broken.

SEALING UNDER VACUUM

The shell-coating method has been modified so that it actually impregnates the shell with the sealing agent. The egg is first immersed in a sealing solution and then subjected to reduced atmospheric pressure. When normal pressure is restored, the tendency of the air to enter the pores of the shell causes the solution also to be drawn in. One of the first sealing agents used experimentally in this manner was aluminum soap dissolved in gasoline (*Jones and Dubois, 1920*); but the gasoline, although it quickly evaporated, imparted its flavor to the egg.

Oil may be successfully used in the vacuum-impregnation method. During the treatment, compensation may also be made for the egg's normal loss of carbon dioxide by substituting carbon dioxide for the air within the egg. The eggs are placed in oil in an airtight container, from which the air is exhausted. The eggs are then lifted above the level of the oil, and carbon dioxide is introduced into the chamber. Oil and carbon dioxide are drawn into the pores of the eggshell as the vacuum is released (*Swenson, 1939*).

The oil does not penetrate through the egg membranes. The fact that it is actually incorporated into the shell is shown by the tabulated analyses of the shells and membranes of untreated eggs, oil-dipped eggs, and eggs oil-treated by the vacuum method (*Swenson, 1939*).

	OIL (per cent)
Unoiled eggs	0.065
Oil-dipped eggs	0.58
Vacuum oil-dipped eggs	2.36

Eggs oiled under vacuum lose weight more slowly than ordinary oil-dipped eggs, when stored for 11 months (*Swenson, 1939*).

	WEIGHT LOSS (per cent)
Oil-dipped eggs	1.60
Vacuum oil-dipped eggs	0.10

Increase in pH value is also retarded to a greater extent than after plain oil dipping, as shown below (*Swenson, 1939*).

	pH VALUE	
	AT 8 MONTHS	AT 11 MONTHS
Oil-dipped eggs	8.30	8.63
Vacuum oil-dipped eggs	7.80	8.20

Vacuum carbon dioxide treatment of eggs, however, is more expensive than plain oil dipping.

"THERMOSTABILIZATION" IN OIL

Rotation of the egg for 10 minutes in an oil bath at the temperature of 60° C. provides the protection which oiling gives and also devitalizes the embryo in fertile eggs, so that development will not occur if the egg is subsequently held at high storage temperatures (*Funk, 1943a*). In addition, there are probably some pasteurization effects against various bacteria (*Funk, 1943b*), although heating the egg to 65° C. for 20 minutes has failed to sterilize it completely (*Wedemann, 1940*). Partial coagulation of the albumen occurs during this treatment, but to such a slight extent that only a faint opalescence is visible when the egg is broken out onto a black background. However, the foaming properties of the albumen are greatly reduced, and the egg is therefore rendered unsuitable for various culinary purposes, especially cake baking. Otherwise, the preservative effects of this treatment are better than those of many others.

"FLASH" HEAT TREATMENT

Immersion of the egg for a short time in boiling water, or other liquid, gives a result different from that of longer treatment at a lower temperature. This method of prolonging the keeping qualities of eggs has been employed by housewives for many years. It is the basic principle of various processes for which patents have been issued during the last century (*Atwell and Crawford, 1867; Gray, 1887; Clairemont, 1914*). Some of these older processes also involve the use of solutions that coat the surface of the eggshell.

The object of the "flash" heat treatment is to coagulate a thin film of albumen immediately beneath the egg membrane and thus to seal the shell internally. The crucial factor upon which success depends is the time that the egg remains in the hot liquid. Too short an immersion fails to coagulate enough albumen, and too prolonged treatment results in a layer of coagulated albumen so thick that it is visible when the egg is opened. Experimentation indicates that 5 seconds is probably the optimum period of exposure to boiling water (*Romanoff and Romanoff, 1944*). When the egg cools, the thin layer of coagulated albumen adheres to the egg membrane and cannot be seen in the opened egg.

Examination of treated eggs indicates that their contents are not changed by brief dipping in boiling water. The foaming properties of the albumen, important in baking, are not decreased. This fact is shown by the tabulated observations of the volume increase produced by beating

	INCREASE IN VOLUME (per cent)
Untreated eggs	424
Treated eggs	448

the albumen of eggs held 2 weeks at 20° C. In addition, fertilized eggs, subjected to the same treatment, remain capable of supporting embryonic development (*Romanoff and Romanoff, 1944*). This fact indicates that the major portion of the egg's contents does not undergo chemical change when the egg is treated.

Another advantage of the flash heat treatment is the fact that it is a simple method requiring no thermostatic controls or stirring devices. It is also possible that some of the bacteria present on the surface of the egg may be washed away and destroyed.

If the egg has a large air cell, the effectiveness of the flash heat treatment is reduced. Too large an area of albumen remains uncoagulated beneath the air cell.

PRESERVATION OF OPENED EGGS

When removed from the shell, the egg contents can be rendered relatively nonperishable and stable by two methods of preservation, freezing

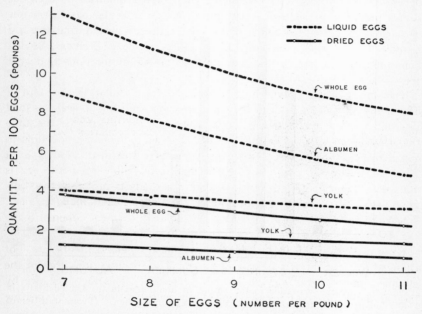

Fig. 375. Yield of frozen and dried egg products obtained from 100 eggs of different sizes. (After Masurovsky, 1942.)

and drying. By either method, the edibility of the egg can be safeguarded and its deterioration delayed for a reasonable length of time. In addition, processed eggs occupy less space than shell eggs and are more easily handled during transportation and in storage. Figure 375 indicates the quantities of liquid (or frozen) and dried eggs, in the form of whole egg, yolk, and albumen, that are obtained from 100 eggs of various sizes. Egg products are also more convenient and economical for large-scale use than shell eggs. The fact that yolk and albumen are available separately makes processed eggs popular among food manufacturers. Various problems are encountered, however, in the production, transportation, and storage of frozen and dehydrated eggs.

Processed eggs made their appearance approximately at the beginning of the twentieth century. Although the first egg products were usually edible, they were often much impaired in flavor and frequently showed signs of deterioration. They could be used with fair success, however, as ingredients of cooked foods. The practical application of technical

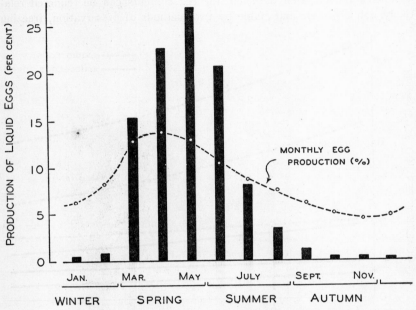

FIG. 376. Seasonal variation in the output of liquid eggs in the United States in 1940 compared with the seasonal variation in egg production. (From statistics of the United States Department of Agriculture for 1940.)

knowledge, and emphasis on sanitation in the packing plant, eventually effected a great improvement in the quality of frozen and dried eggs. Since 1915, the volume of all egg products has grown steadily larger. During the second world war, military needs enormously increased the output of dried eggs, which were required to meet exacting standards of excellence, in order to withstand the effects of every type of climate.

The manufacture of egg products is a seasonal industry. Over 80 per cent of the annual volume of frozen and dried eggs is manufactured in the spring and early summer, during the periods of highest egg production (Fig. 376).

Preparation of Liquid Egg

At any time during the preparation of liquid egg,[3] contamination by bacteria may occur in sufficient degree to initiate decomposition. In addition, desirable properties of the egg may be lost at certain critical stages of treatment. For these reasons, efforts are made to control every manipulation carefully. Plants for breaking shell eggs are especially equipped and designed to accomplish preliminary handling and actual processing with the greatest speed, and with the least impairment of the original quality of the egg contents.

The preparation of liquid egg is the same prior to either freezing or drying. Although every packing plant has its own practices, much the same general procedure is followed everywhere in the United States, in order to conform with legal requirements.

Eggs, as soon as they arrive at the plant, are carefully sorted. Those with damaged shells are removed, and the sound eggs are transferred to a holding room, where they are chilled to a temperature of 15.6° C. Pre-chilling makes it possible to recognize, during candling, certain defects not apparent in warm eggs (*Goresline and McFarlane, 1944*). It also facilitates separation of yolk and albumen. From the holding room, the eggs are brought to the candling room, where the inedible eggs are eliminated. Eggs that are suitable for use are then washed (sometimes with a sterilizing agent), drained, and taken to the breaking room.

The eggs are broken by being struck sharply against a knife mounted horizontally, edge uppermost. If yolk and albumen are not to be separated, the egg contents are dropped into a small cup, or onto a gently sloping table with metal traps or other devices. Separation of yolk and albumen is made by dropping the egg onto a small concave disk placed near the knife or attached to the cup. The albumen runs off, and the yolk remains on the disk. After every second or third egg, the contents of the cup are smelled and critically examined, and not until then are they added to the bulk container. All buckets of liquid egg are re-examined by another operator before being emptied into receiving tanks.

Liquid whole egg, because it is viscous and lumpy, must be blended before further processing, in order to secure a smooth and homogeneous mixture. Blending is accomplished in churns, which stir, but do not beat, the liquid egg (*Pennington, 1941*). Churns are of two types, those that merely mix, and those that are refrigerated in order to chill

[3] A variety of terms is used to describe the egg contents prior to processing. Among these are "egg meats," "egg pulp," "egg batter," "egg mélange," and "liquid egg."

the liquid egg nearly to freezing. The vitelline membranes, chalazae, and any shell particles are removed by passing the churned material through a fairly fine sieve. Liquid yolk is also churned and is sometimes strained. Liquid albumen is usually forced through a fine mesh screen or sieve in order to break up its structure. After it has been churned, liquid egg is stored in refrigerated tanks at 7.2° C. until the final processing, which takes place after the lapse of as little time as possible.

In the future, methods of preparing liquid egg may perhaps be improved by automatic grading and breaking machines, more effective sterilization of shell eggs prior to breaking, and the mechanical separation of yolk and albumen (*Loy, 1944*).

Handling of Liquid Egg

If the frozen or dried egg product is to be edible, if it is to meet accepted standards of taste and appearance, and if it is to be a successful substitute for the fresh egg, the original properties of the liquid egg must be retained in so far as possible. The chemical changes caused by bacteria must be prevented, and the physicochemical changes produced by the freezing or drying process itself must be kept to a minimum.

SANITATION IN HANDLING

The opened egg is extremely perishable. If liquid egg becomes sufficiently contaminated with microorganisms, it is subject to rapid spoilage before final processing. In addition, the bacterial content of liquid egg largely determines the quality of reconstituted frozen and dried eggs, and also of many foodstuffs. Egg products are used extensively, for example, in mayonnaise and confectionery, which receive little or no heat treatment during preparation. For these reasons, constant precautions are taken to keep the bacterial content of liquid egg as low as possible, and to guard against the multiplication of the organisms that are inevitably introduced, even under the best conditions.

Sanitation in an egg-processing plant is a serious problem, since there are numerous opportunities for microflora to infect the liquid egg. The sources of contamination are the egg itself and the processing environment.

The Egg as a Source of Contamination. Fresh, sound shell eggs are the essential raw material of good egg products; only eggs of greatest initial excellence are suitable for freezing or drying. Damaged eggs become contaminated with microorganisms in considerable numbers before any evidence of decomposition is noticeable; and such eggs, if used, contribute countless bacteria to liquid egg. Eggs that become cracked or dented in the plant, subsequent to candling, may be used if they are broken out and processed immediately; but, in the United States, the final product must be labeled as substandard (*McFarlane, Watson, and Goresline, 1945*).

Egg packing material that has been soiled by broken eggs is also a source of contamination to sound eggs, as it provides a favorable growth medium for microflora. Such packing material, and all decomposed and otherwise inedible eggs detected during candling, are removed from the plant at frequent intervals, to prevent spread of infection.

Some eggs, though unspoiled, contain large numbers of bacteria. In others, spoilage is incipient but detectable either with difficulty or not at all. Approximately 2 per cent of all eggs that withstand the test of candling still contain imperfections which are not likely to be noticed when the eggs are broken out, unless careful inspection is made. Sour eggs, and eggs in which the albumen is beginning to turn green, are not easily recognizable except by close scrutiny after they are opened. The type of "musty" eggs caused by *Achromobacter perolens* (*Spanswick, 1930*) are particularly insidious, since they are distinguishable by their odor alone. Eggs of all these types, like dirty eggs, are possible sources of contamination.

Contamination by bacteria from dirty eggs can be prevented if the eggs are washed and sterilized before the breaking operation. After they have been washed, however, dirty eggs are acceptable only if processed immediately, before microorganisms have the opportunity to penetrate through the shells. The shells of clean eggs can also be a source of infection during breaking manipulations, especially if shell particles accidentally fall into the liquid egg.

For these reasons, the washing and sterilization of all eggs are now established practices in many packing plants. These procedures reduce the bacterial count of the liquid egg material by at least a thousand times (*Zagaevsky and Lutikova, 1944a*). Washing methods depend largely upon the size and facilities of each plant. In some instances, eggs are washed by machine, and in others, by hand, with the aid of powders or other agents which loosen dirt. The water is warmer than the eggs, to prevent its being drawn through their shells. Wet or dry sandblasting is also used. After being cleaned, the eggs are rinsed in an odorless, nonirritating bactericidal agent. Among the germicides that have been suggested is a derivative of quaternary ammonium chloride, composed of approximately equal quantities of lauric and myristic fatty acid esters of colamino formyl methyl pyridinium chloride (*Epstein, Harris, and Katzman, 1943*). Rinsing washed dirty eggs for 5 minutes in this solution, in a concentration of 0.05 per cent, reduces the number of bacteria which still remain on the shells from as many as 6 million per egg to only a few thousand, or destroys them entirely (*Penniston and Hedrick, 1944*). In Russia, chlorinated lime is used as a disinfectant; in a concentration of 1.5 per cent, this kills even spore-forming bacteria in 5 minutes, not only on the shell surface, but also within the structure of the shell and on the shell membrane. Its odor, however, lingers for some time, unless neutralized by spraying the eggs with a 10 per cent solution of soda or sodium hyposulfite. Precautions must also be taken to protect workers from the ill effects of chlorinated lime (*Zagaevsky and Lutikova, 1944a*).

Recently, a method has been devised by which eggs are cleaned, and the microflora of the shells simultaneously reduced. The eggs are sprayed with water heated to 72° C., and are then irradiated with short-wave ultraviolet light. The surface contamination on eggs so treated has been found to be less than 50 per cent of that on naturally clean, unwashed eggs (*Gunderson and Gunderson, 1945*).

The Processing Environment as a Source of Contamination. The danger of contaminating liquid egg with bacteria originating in the environment is extremely great. In egg-processing plants, all equipment is made as clean and as nearly sterile as possible, and the personnel is required to conform to certain sanitary regulations, especially in the breaking room. Temperature, humidity, ventilation, and insect control are essential, and all apparatus is arranged to expedite operations, since speed is of great importance.

All doors and windows are screened, and ventilation systems are equipped with filters to aid in keeping the air dry, clean, and odorless. Egg-packing plants are preferably not constructed in neighborhoods where there is smoke, dust, or odors caused by industry or any other type of enterprise.

Attempts are made to suppress bacterial multiplication by performing all operations at as low a temperature as is feasible. In the holding room, the temperature is kept within the range of 0° C. to 4.5° C., for the preliminary cooling of the eggs. In the candling room, it is not higher than 12° C.; in the breaking room, not over 18° C. The temperature of the liquid egg is reduced to 4.5° C., or less, within 30 minutes, and is maintained at that point until the final step in processing. The bacterial count of liquid egg increases greatly within a very short time, the rate of increase depending upon the temperature of the egg material (Fig. 377). Cooling the liquid egg to 3.3° C. restrains the growth of bacteria for a few days only. Under normally sanitary conditions, 48 hours is the upper limit of safety at 7.2° C.; and at 11.1° C., liquid egg cannot be held even for 24 hours (*Gibbons and Fulton, 1943*).

It is important that extreme cleanliness be maintained and sanitary precautions observed throughout the entire plant, and especially in the breaking room. Here walls and floor are waterproof, so that they may

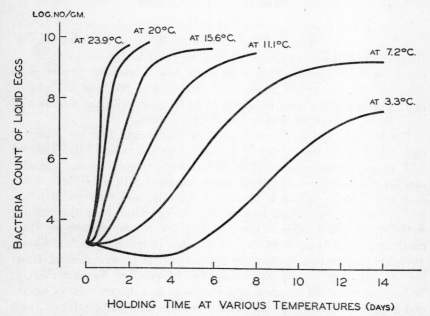

FIG. 377. Increase in number of bacteria in liquid eggs held at various temperatures. (After Gibbons and Fulton, 1943.)

be hosed with hot water and sprayed with bactericidal solutions. All pieces of large equipment—benches, tables, receiving shelves, and conveyors—are constructed and placed so that they are readily cleansed and sterilized. Smaller utensils, such as pails, drip trays, knives, and cups, are made from material that will withstand live steam sterilization. They are discarded if they become dented or rusty, and therefore difficult to clean. The containers for liquid egg are never placed on the floor or allowed to come in contact with the containers for shell eggs. The vats and tanks in the breaking room are kept covered.

Egg breakers are required to take a complete set of clean equipment several times a day. Continuous use of the same implements, even during one morning, greatly increases the bacterial content of the liquid egg, as the table indicates (*Zagaevsky and Lutikova, 1944a*). Utensils are changed, of course, as soon as a questionable egg is discovered.

Time	Bacteria (per cubic centimeter)
7:30 A.M.	485,000
9:00 A.M.	867,000
12:00 M.	1,000,000

Some of the bacterial content of liquid egg is undoubtedly derived from the operators' fingers, which are not sterile and are continually wet. Breakers keep their fingertips as dry as possible by constant use of paper towels or tissues. They must wash their hands with odorless soap whenever they enter the room, and always after breaking an inedible egg. Cosmetics, the scent of which might interfere with the detection of "off" odors, are not allowed in the breaking room.

The washing and sterilization of those implements that are routinely replaced several times a day is performed in rooms specially equipped with tanks, sterilizing apparatus, and racks for rapid drying. Smaller utensils are rinsed in clear, cool water, cleansed with a brush in a hot detergent solution, which is changed frequently, and then rinsed again. Many bacteria are carried over to the rinsing water from the washing water, which may contain as many as 5 million microorganisms per cubic centimeter after being used for 40 minutes (*Zagaevsky and Lutikova, 1944a*). Sterilization is therefore imperative. It is accomplished either by subjection to flowing live steam for 30 minutes, by immersion in boiling water for 30 minutes, or by soaking in a bactericide. Chlorine and hypochlorite solutions (containing not less than 100 p.p.m. of available chlorine in the United States, or 50 p.p.m. in Canada) are increasing in use for sterilization purposes in egg-packing plants. However, extremely thorough rinsing of equipment is necessary before treatment with these solutions. They cannot penetrate a film of egg material, and they are sometimes ineffective against bacteria in cracks or seams. In addition, care must be taken that they do not corrode metal implements.

It is also possible for contamination to occur in the large pieces of equipment, such as churns, mills, settling vats, fermentation vats, receiving tanks for liquid egg, and all pipe lines, pumps, surface coolers, and the like, through which the liquid egg passes. Egg material, if it becomes lodged in apparatus of this type, is difficult to remove and constitutes an excellent culture medium for bacteria. Special detergents and methods of cleansing are required. Receiving tanks and churns are rinsed with water, scrubbed with a washing compound, rinsed again, and flushed with a sterilizing agent. If possible, they are steamed. High pressure lines (with nozzles removed), homogenizers, and viscolizers are similarly treated, as are spray nozzles, orifices, cores, and whizzers.

In egg-freezing plants, the containers in which the eggs are frozen are sterilized before receiving the egg material. The freezing room is equipped to freeze the liquid egg within 60 to 72 hours after packaging, at the most; and the storage room is usually kept at a temperature of −12.0° C. Care is taken not to overload the storage room, and thus produce a rise in temperature. The packages are stacked so that air may circulate around them, and no packages are kept on the floor.

In egg-drying plants, spray driers are equipped with air intake filters, and air is drawn from sources free of odors and dust. Driers are constructed so that they may be easily cleaned. Handling of the dry egg powder by operators is not permitted. The storage room for dehydrated eggs is kept clean, dry, odorless, and at a temperature of less than 10° C.

The personnel facilities in egg-breaking establishments must conform with the legal requirements for dressing-rooms, lockers, and lavatories in any food-processing plant. Separate ventilation is provided for lavatories, and no personnel rooms open into the processing rooms. Workers are provided with clean, long-sleeved uniforms, must confine their hair, and may not wear jewelry.

BACTERIAL CONTENT OF LIQUID EGG

Liquid egg has a relatively low bacterial count if it is derived from shell eggs of good history and is produced under proper conditions. Official standards of quality based on microbiological content have been proposed but not fixed as yet in the United States (*Loy, 1944*). However, in commercial practice, a bacterial count of 10,000 to 50,000 viable organisms per gram, as shown by growth on agar at 20° C. after 5 days, is generally accepted as indicating a product of excellent quality, whereas a count of 1 million or more per gram signifies one that should be regarded with suspicion. The bacterial count of liquid egg may vary considerably from one plant to another, but it usually increases in summer (Fig. 378).

There are various opinions as to the permissible numbers of coliform bacteria. Suggestions for the proper upper limits for coliform bacteria range from 1000 per gram to 100,000 per gram (*Pennington, 1941*). The coliform count, which usually parallels the total bacterial count (*Schneiter, Bartram, and Lepper, 1943*), is used as an index of fecal pollution, and therefore of sanitary conditions at the packing plant. Since *Escherichia coli* is demonstrable in the intestinal tract of chickens, its presence in liquid egg may also be the result of recent fowl pollution (*Brownlee and James, 1939*). Contamination with chicken manure is indicated if uric acid can be detected in liquid egg (*Duggan, 1946*).

The bacterial content of liquid egg may be determined by the plate count method, but this requires so much time that its usefulness is limited in the processing plant. The direct microscopic test is more

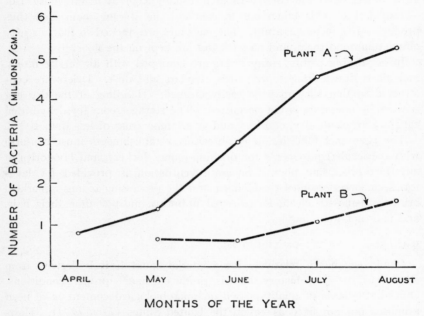

FIG. 378. Seasonal variation in bacterial count of liquid whole egg in two commercial plants. (After Pennington, Jenkins, and Stocking, 1916.)

rapid. A sample of liquid egg is collected on a sterile applicator and spread evenly on a glass slide in a strip 1.0 cm. wide. It is then dried, stained with Methylene Blue, rinsed, and examined by red light. The number of bacteria per square centimeter is determined by counting those in a certain number of microscope fields, and the bacteria per gram may then be calculated by formula. This method may also be used to estimate the bacterial contamination of equipment. The entire test may be completed in 30 minutes (*Mallmann and Churchill, 1942*).

The direct microscopic count does not, of course, replace the standard method of estimating viable bacteria by incubating a sample of known volume on a culture plate at 20° C. or 37° C., or both, and counting the colonies which appear after a certain length of time, usually 3 to

5 days. Certain bacteria often found in liquid egg do not grow so well as others on all media, however.

Occasionally a sample of liquid egg may be greatly contaminated and yet not show any signs of spoilage (*Scott and Gillespie, 1943*). Indirect chemical tests give the most reliable information concerning the bacterial content in samples of this type. Recently, the reductase test used in milk analysis has been adapted to the examination of liquid eggs (*Scott and Gillespie, 1943, 1944*). A more sensitive dye, Resazurin, replaces Methylene Blue, and the temperature of the bath is reduced to 30° C. These changes in technique are made because the most common contaminants of liquid egg, the *Pseudomonas* group, have a low optimum temperature and a weak reducing power. The dye reduces from a greenish hue, through mauve, to a bright pink, which is a convenient endpoint because it is not influenced by the color of the egg. At least three grades of liquid egg can be defined by this test. Samples that do not reduce the dye in 8 hours are of first quality; and two poorer qualities have been arbitrarily established to correspond to shorter periods of time in which the dye may be reduced. The chief disadvantage of this test is the fact that it requires 8 hours to complete and is therefore somewhat slow for use in an egg-packing plant.

PHYSICOCHEMICAL PROPERTIES OF LIQUID EGG

Processing may so alter the physicochemical nature of liquid egg that reconstitution of the product is difficult. In particular, the egg's original viscosity and colloidal constitution may be affected.

The greatest problem in either the freezing or drying of eggs is presented by the water content of the liquid egg. The formation of ice crystals disrupts the colloidal system of the egg, which thus suffers a fundamental change. When the egg is thawed, its consistency may be considerably altered. To dehydrate the egg, heat must be applied to remove water. Unless the drying process is carefully controlled, the dried product may reabsorb water with difficulty, and its texture, color, and flavor may be impaired.

In the United States, egg products are required to meet certain standards for content of total solids: 25.5 per cent for whole eggs, and 43 per cent for yolks. The proper composition may be achieved before processing by adjustment of the total solids of the liquid egg. Addition of the correct amounts of yolk or of albumen produces an increase or a decrease, respectively, in egg solids, the percentage of which is most easily and rapidly determined by the refractometric method. Direct readings for whole egg and yolk are difficult because of turbidity, but the material may be rendered optically homogeneous through digestion by such enzymes as trypsin and papain (*Urbain, Wood, and Simmons, 1942*).

The edibility and freshness of liquid egg is also defined by law in terms of the acid value of ether-extracted fat, and of the content of ammoniacal nitrogen and dextrose. Estimation of these substances gives an index of the extent of bacterial activity. Even slight bacterial decomposition of fresh liquid egg may be accompanied by an appreciable increase in acidity, due to the liberation of acid products of bacterial metabolism. The normal content of lactic acid is substantially increased, and formic and acetic acids may be formed. The correlation between the bacterial content and the development of these substances is indicated in the table, which compares a fresh sample of liquid egg with one that was allowed to stand at 29.4° C. for 18 hours (*Lepper, Bartram, and Hillig, 1944*). A significant decrease in sugar usually accompanies

| | BACTERIA (millions/gram) | ACIDS (milligrams/100 gm.) | | |
		LACTIC	FORMIC	ACETIC
Fresh material	0.5	24	0	Trace
Held material	1,180.0	200	285	188

bacterial growth in liquid egg, especially when large amounts of acid are formed. Volatile-base (ammoniacal) nitrogen increases, but to a varying extent, according to the type of decomposition.

Preservation by Freezing

The first eggs frozen in the United States appeared on the market in 1899 and were produced on a small scale in Iowa. Soon afterward, the industry obtained its start when a larger plant was established in St. Paul, Minnesota. During this period, China was producing large quantities of frozen eggs and was exporting much of her output to the United States. However, between 1921 and 1928, American purchases of the Chinese product fell from 18 million pounds annually to 8 million pounds and have since become negligible. This trend was caused by the phenomenal expansion of the industry in the United States, which is shown by the quantities of frozen eggs held in storage on August first of

YEAR	MILLIONS OF POUNDS
1916	6
1920	20
1925	43
1930	116
1935	116
1940	155
1943	351

various years since 1916. The storage holdings of frozen eggs, as of shell eggs, are usually at the peak on the first of August (Fig. 379). With the growth of the egg-freezing industry, there has been a steady decrease in the ratio of peak holdings of shell eggs in cold storage to peak hold-

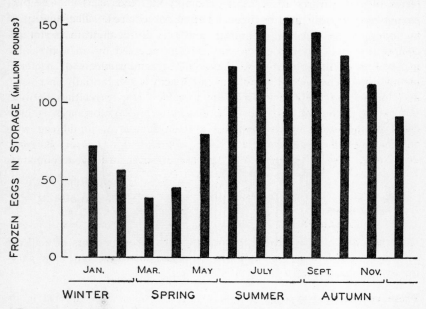

FIG. 379. Seasonal variation in storage holdings of frozen eggs in the United States in 1940. (From statistics of the United States Department of Agriculture for 1940.)

ings of frozen eggs (calculated as quantities equivalent to cases), as indicated in the table.

YEAR	RATIO $\dfrac{\text{SHELL EGGS}}{\text{FROZEN EGGS}}$
1916	97:3
1920	92:8
1925	89:11
1930	78:22
1935	72:28
1940	65:35
1943	48:52

Freezing the egg, and subsequently storing it at low temperatures, prevents its decomposition by inhibiting microorganisms, or actually destroying them. Certain changes are produced in the egg's physical properties, but these changes may be largely controlled and kept to a minimum. Consequently, the most valuable qualities, including flavor, are preserved sufficiently well to allow frozen eggs to be substituted for fresh eggs in many instances. In fact, bakers, and other manufacturers of foods, prefer frozen eggs. A large supply of frozen eggs may be purchased economically at one time and used as required. The quality of frozen eggs is dependable and uniform and does not deteriorate on storage.

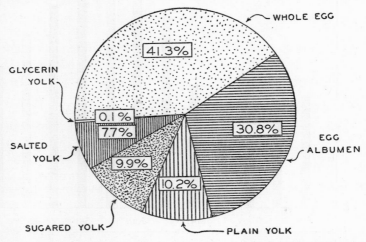

FIG. 380. Relative amounts of various forms of frozen eggs produced in the United States in 1940. (From statistics of the United States Department of Agriculture.)

The fact that frozen yolk and albumen are available, already separated, is also a distinct advantage, especially to candy manufacturers. Makers of ice cream and salad dressing find that frozen eggs give better consistency to their products than fresh eggs. Up to the present time, large industries have been the chief users of frozen eggs, which have been packed in standard cans of 10- to 30-lb. capacity. As home refrigeration facilities improve and the popularity of frozen foodstuffs increases, it is possible that eggs may be frozen in smaller containers, perhaps with individual detachable units each equivalent to 1 egg (*Schaible and Card, 1943*). The sole disadvantage of frozen eggs for home use is the fact that the loss of their physical structure prevents their being fried, poached, or boiled.

About 40 per cent of the total output of frozen eggs is prepared as whole egg, and the remainder as separated albumen and yolk, in approximately equal quantities. Frozen yolk is produced in three principal forms—plain, salted, and sugared. These are manufactured in more or less equal amounts. A very small percentage of frozen yolk products contains glycerin (Fig. 380). Usually frozen whole egg, too, contains salt and sugar.

FREEZING METHODS

Liquid egg is usually frozen in the cans in which the product is to be shipped and sold. Freezing may be accomplished in about 72 hours by subjecting the egg to temperatures of $-18.0°$ C. to $-21.0°$ C. The freezing time may be reduced by one-half if the containers are air-blasted (*Pennington, 1941*). Rapid freezing methods have now largely replaced the slower processes. The temperature may be within the range of $-23.3°$ C. to $-28.9°$ C., and sometimes as low as $-40°$ C. to $-45.6°$ C. Complete freezing before packaging may be achieved in 7 seconds by feeding the liquid egg in a thin film to a mechanically refrigerated roll.

PHYSICOCHEMICAL CHANGES IN FROZEN EGGS

Egg albumen suffers no significant chemical breakdown as a result of freezing, although after it is thawed it is more liquid than originally (*Moran, 1924*). Carbon dioxide escapes from the albumen during slow freezing and subsequent storage, much as it escapes from the intact egg.

In the yolk, however, freezing produces a great change, which becomes especially apparent when the yolk is thawed. Whole egg, because of its content of yolk, is also altered. During the freezing process, water separates from certain yolk solids, which then cohere in small lumps and become progressively harder as further separation of water takes place. The water collects in localized spots and forms ice crystals. When liquid egg is frozen in cans, the solids tend to migrate toward the center, where they form a core, visible as a "hump" in the middle of the top surface. After the frozen egg is defrosted, it does not return to its original consistency but is a thick, lumpy, tenacious mass; in other words, it is in a state of gelation. Gelation increases during the storage of the frozen egg and reaches a maximum in 60 to 120 days, usually in 90 days (*Thomas and Bailey, 1933*).

The precipitated material in thawed yolk has been identified as lecithin, or a lecithoprotein (*Urbain and Miller, 1930*). If sufficient water is lost from the lecithoprotein complex as the egg is frozen, moisture cannot be reabsorbed when thawing takes place, and the result is an irreversible change in the yolk's consistency.

The greater the mechanical disruption of the liquid egg, the less gelation occurs. The degree of gelation appears to be directly proportional to the viscosity of the liquid egg at the time of freezing; it is therefore inversely proportional to the extent to which the liquid egg is churned. Churning the egg material at approximately 1750 revolutions per minute, for varying lengths of time, has the following effect (*Thomas and Bailey, 1933*) on the viscosity (expressed as mobilometer number, which is higher, the greater the viscosity) of frozen whole eggs, thawed after 30 days of storage. If the egg is passed through a colloid mill before freezing, little or no gelation is observed in the thawed frozen egg.

TIME CHURNED	MOBILOMETER NUMBER
50 sec.	32
1 min.	20
12 min.	2

It is also possible that gelation is dependent to some extent upon the egg's content of total solids and fat. Gelation is less extensive in samples of frozen egg containing relatively small amounts of these constituents (*Thomas and Bailey, 1933*).

Consistency is improved if freezing is accomplished rapidly while the container is agitated (*Finnegan, 1941*). The egg is frozen too quickly to permit extensive separation of water from the yolk solids, and small rather than large ice crystals are formed. However, the rate of freezing is not entirely uniform throughout the can, and for this reason additional measures are usually taken. (It is said that no difficulties are encountered in freezing eggs by the refrigerated roll method.)

The addition of certain edible substances, such as sugars, sodium chloride, or glycerin, increases the osmotic pressure and depresses the freezing point of liquid egg. As a result, the egg solids retain sufficient water entirely to prevent their precipitation and the formation of a core. Since the colloidal nature of the egg is not destroyed, moisture is reabsorbed during the thawing of the egg, and the egg's approximate original consistency is restored. Lowering the freezing point by only a few hundredths of a degree has a pronounced effect. Figure 381 illustrates the much greater degree of gelation which occurs in untreated frozen whole egg than in egg products containing 0.1 per cent salt and 1.0 per cent sucrose, respectively. The corresponding freezing points of these three samples is given in the table (*Thomas and Bailey, 1933*). A mixture of sodium chloride and sucrose in the same concentrations as shown in the table is often added to commercial frozen whole egg and prevents gelation more effectively than either substance alone.

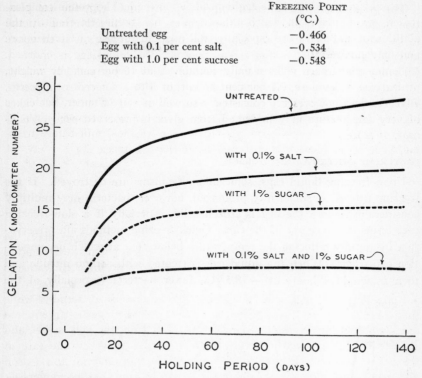

FREEZING POINT
(°C.)

Untreated egg	−0.466
Egg with 0.1 per cent salt	−0.534
Egg with 1.0 per cent sucrose	−0.548

FIG. 381. Influence of salt (sodium chloride) and sugar (sucrose) on the degree of gelation occurring in frozen whole egg during storage. (After Thomas and Bailey, 1933.)

Sucrose is less effective than an equal weight of either dextrose or levulose, since the osmotic pressure of each of the latter sugars is approximately 1.75 times that of sucrose. Addition of sucrose or dextrose in equimolecular quantities, of course, inhibits gelation to an identical extent; but it is not practical, commercially, to use more than the smallest possible amount of sugar. The relative merits of the same weights of sucrose, dextrose, and levulose in preventing the coagulation of lecithin during freezing are shown in the table (*Urbain and Miller, 1930*). Since

SUGAR	LECITHIN COAGULATED (per cent)
Sucrose	98.8
Dextrose	0.8
Levulose	0.6

the emulsifying action of the egg depends upon the lecithoprotein complex, frozen egg products in which this complex has been protected by the addition of stabilizers are especially well adapted for use in baked goods and mayonnaise.

Commercial frozen yolks usually contain about 10 per cent, by weight, of dextrose or levulose, 0.7 per cent of salt, or 3 to 5 per cent of glycerin. Glycerin does not preserve consistency so well as salt or sugar, but cakes of very fine texture may be baked from glycerin-preserved eggs (*Woodroof, 1942*).

BACTERIAL CONTENT AND STORAGE

Upon freezing liquid eggs, some microorganisms are destroyed. If the freezing process is sufficiently prolonged, however, bacteria may multiply considerably before the egg is frozen solid, especially if a slow-freezing core forms in the center of the can. Quick freezing methods aid preservation by rapidly reducing the temperature below that at which decomposition can occur. When the liquid egg is treated with carbon dioxide and then frozen in 2 hours at $-78.5°$ C., fewer bacteria (especially of the

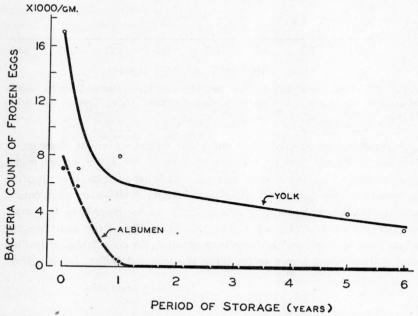

Fig. 382. Changes in bacterial content of frozen yolk and albumen of good quality during prolonged storage. (After Schneiter, Bartram, and Lepper, 1943.)

colon aerogenes type) survive than when the egg is frozen by ordinary methods (*Swenson and James, 1935*).

The stabilizers that prevent gelation also inhibit bacteria by increasing osmotic pressure. Glycerin, however, is not especially effective against molds (*Godston, 1934*).

During storage at subfreezing temperatures, the bacterial content of frozen eggs may be considerably reduced, especially in albumen, the bactericidal principle of which apparently remains active at low temperatures. It has been found, for example, that in 6 years the bacterial count may decrease from 17,000 to 2700 per gram in frozen yolk, and from 7000 to less than 10 per gram in frozen albumen (Fig. 382). A comparable reduction in bacterial content was observed when originally putrid frozen egg of very high initial contamination was stored experimentally.

During the first few weeks of the storage period, those microorganisms that grow best at 20° C. may multiply to some extent in frozen yolk (containing 10 per cent sugar). On the other hand, there is a steady decline in the number of coliform and other bacteria whose optimum growth temperature is 37° C. (Fig. 383). After frozen whole egg has been stored for 6 months, the proportions in which the various bacterial genera are present in the total flora may differ somewhat from the original proportions (Fig. 384).

The fact that freezing does not entirely destroy all the bacteria present in the liquid egg indicates the potential danger that decomposition may occur during or after defrosting. This possibility is increased if a slowly thawing core has been formed in the container. Obviously, the more quickly thawing is accomplished, the less opportunity there is for bacteria to multiply and produce spoilage. Reduction of the thawing time by heat, however, is not advisable. The outer portions of egg material in a large can may become warm enough to support bacterial growth before defrosting is completed. The ordinary method of thawing, by immersion of the container in cold running water at about 11.5° C., requires approximately 24 hours. During this time, the bacterial content of the egg may increase by nearly 700 per cent (Fig. 385). Constant agitation of the can has been found to reduce the defrosting period to 8 hours, and to limit bacterial increase to 100 per cent. The consistency of the thawed egg is not impaired by this procedure, in spite of the prevailing belief that only slow reversion to the liquid condition permits proper resorption of water by the egg solids (*Brownlee and James, 1939*).

The quality of frozen eggs may be judged by the same bacteriological and chemical tests as liquid egg. In addition, an official odor test for

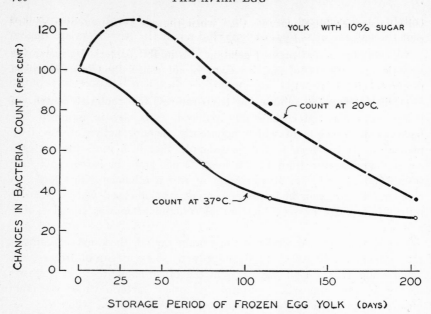

FIG. 383. Bacterial count of frozen yolk (containing 10 per cent sugar) during 200 days of storage. Count of organisms with optimum growth temperature of 20° C. is compared with that of organisms growing best at 37° C. (After Nielson and Garnatz, 1940.)

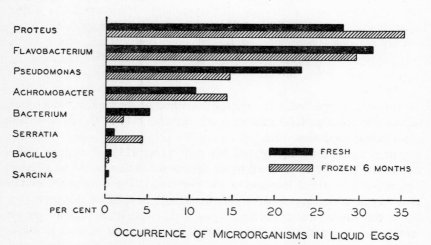

FIG. 384. Frequency of occurrence of various types of microorganisms in fresh eggs and in frozen eggs. (After Johns and Bérard, 1946.)

edibility is recognized in the United States. A 1-inch electric drill is sent through the mass of frozen egg and removed. Any odors are emphasized and easily detected, because driving the drill warms the egg material. A frozen-egg product of good quality is odorless, even after prolonged storage, but a decomposed product smells discernibly putrid or sour (*Pennington, 1941*). Long exposure of the liquid or frozen egg

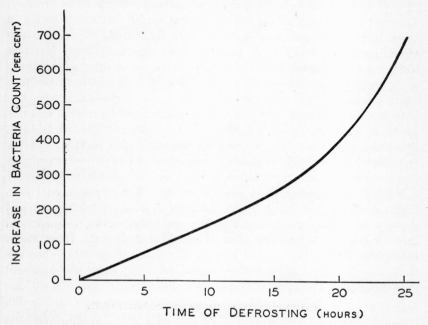

FIG. 385. Relationship between the time required to thaw frozen egg and the relative increase in the bacterial content of the egg. (After Brownlee and James, 1939.)

to the air, during either processing, storage, or defrosting, may result in oxidation of egg fats to rancidity. Foreign odors may also be absorbed, especially during processing.

The refrigeration of frozen eggs during transportation requires careful control. Freight cars are thoroughly insulated and precooled for 24 hours before being loaded, and trucks also are insulated and refrigerated, usually with dry ice (*Pennington, 1941*).

On the whole, frozen eggs of initial good quality, as assessed by organoleptic, bacteriological, and chemical tests, suffer little damage during prolonged storage at subfreezing temperatures. Under sanitary conditions, they are able to withstand at least two complete thawings and refreezings, without significant change in odor, appearance, or bacterial content (*Schneiter, Bartram, and Lepper, 1943*).

Preservation by Drying

By removal of the water from the egg, bacterial and fungal growth is made relatively impossible, and the dried product therefore may be shipped without refrigeration and stored at temperatures above freezing. The bulk and weight of the egg is reduced so far that much saving in space is achieved, and handling is greatly facilitated. There are also special needs that can be well filled by dried eggs, but not by intact or frozen eggs. The most serious objection to the use of dried egg products is the inconvenience caused by the necessity of reconstituting them to a liquid consistency by the addition of water.

In the United States, few dried eggs were produced previous to 1930. The first American egg-drying plant on record was established in 1878, although a patent for a dehydrating process was sought in 1865. During the nineteenth century, desiccated eggs were manufactured chiefly in China. Early in the twentieth century, a number of plants for drying eggs began operation in the United States, but their products were soon condemned by government health authorities as being adulterated and decomposed. As a result, the industry gradually became dormant and then practically nonexistent after World War I brought rising costs. Importations from China grew steadily in volume until 1922, when the tariff was raised on Chinese dried eggs. In 1931, there was a second tariff increase. The tariff wall, together with the outbreak of civil war in China in 1927, and a decline in domestic egg prices, established a favorable basis for competition. The industry was started anew and was given impetus for tremendous expansion when war was declared in Europe in 1939. The increase in the production of dried eggs between 1930 and 1943 is shown in the table.

Year	Thousands of Pounds
1930	489
1935	3,000
1940	7,487
1943	261,972

Normally, more than two-thirds of all dried egg products are prepared as yolk, and at least one-fourth as albumen (Fig. 386). During the second world war, however, more than 90 per cent of the total was manufactured as whole egg.

The general quality of dehydrated eggs was much improved as the result of wartime needs. Originally, dried eggs were employed chiefly in commercially manufactured foodstuffs and were considered adequate if they were wholesome and approximately fulfilled the function of shell eggs. During the war, the complete replacement of fresh eggs by dried eggs in the diets of armies and many civilian populations required superior products of great stability, good flavor, and versatility of use. Such products can now be prepared, as the result of intensive investigation of the factors controlling the properties of the egg in the dehydrated form.

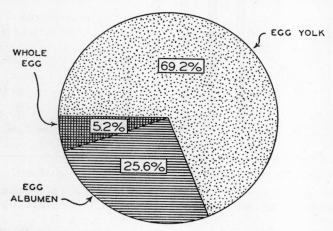

FIG. 386. Relative quantities of various types of dried eggs produced in the United States in 1940. (After statistics of the United States Department of Agriculture.)

The production of dried eggs, like that of frozen eggs, is normally seasonal and at its peak in the spring, when fresh eggs are most plentiful (Fig. 387). Eggs laid from March to June, inclusive, yield dried-egg products of somewhat better quality than eggs laid during the remainder of the year (*Pearce, Reid, Metcalfe, and Tessier, 1946*).

METHODS OF PREPARATION

In all processes for the desiccation of eggs, water is removed by evaporation in the presence of heat, either by spray, drum, or pan drying methods. The first of these yields a powdered product, and the latter two a flaked product. Albumen is usually pan dried, and whole egg and yolk spray dried.

Albumen. Preliminary fermentation of albumen is necessary if the dried product is to retain its color and solubility during storage. Unfermented dried albumen becomes a dark reddish brown as it ages and

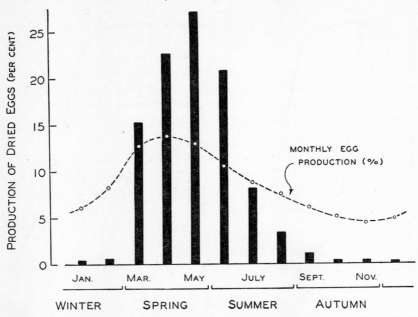

Fɪɢ. 387. Seasonal variation in the manufacture of dried eggs in the United States compared with the seasonal variation in egg production.

does not reconstitute well. Fermentation also preserves the foaming properties of the albumen and, by reducing viscosity, facilitates the handling of albumen during processing.

The value of fermentation was discovered in China, where it was the practice to hold the albumen in large wooden casks, at about 20° C., for 36 to 60 hours before drying (*Blomberg, 1932*). In modern commercial plants in the United States, the albumen is placed in large vats and sometimes slightly heated to temperatures of 23° C. to 29.4° C. Within the first 12 hours of fermentation, there is no great increase in bacterial content, but the pH rises from 7.45 to 9.1, probably because of the escape of carbon dioxide. Subsequently, there is a steady decrease in pH, from 9.1 to 6.25, at a rate that depends upon the number of bacteria initially present (Fig. 388-*A*). The albumen gradually separates into two components, a thin portion, containing a sediment which settles to the bottom of the vat; and a thick, foamy portion, which rises slowly to the top and finally forms a gelatinous scum (Fig. 389). This scum contains most of the ovomucin and a large proportion of the glycoproteins (*Stuart and Goresline, 1942a*). The liquid portion is at first very

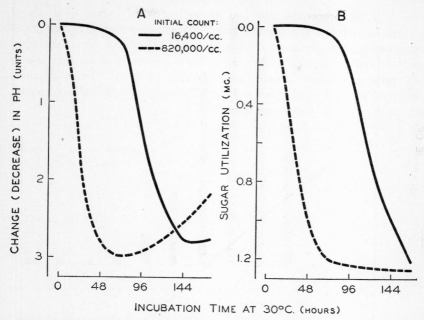

FIG. 388. Changes in *A*, *p*H value, and *B*, sugar utilization in fermenting albumen with high and low initial counts of *Aerobacter aerogenes*. (After Stuart and Goresline, 1942b.)

stringy, but it gradually becomes watery over a period of about 72 hours. At the end of this time, fermentation is usually stopped by adding ammonia, lactic acid, or tartaric acid (*Stewart and Kline, 1941*). During fermentation, the bacterial count may rise to several hundred million. In the first 72 hours, there is a marked decrease in the sugar content of the albumen, but no changes occur either in protein nitrogen or combined amino and amide nitrogen value. The speed with which the sugar disappears depends upon the initial concentration of bacteria (cf. Fig. 388-*B*). However, if fermentation is allowed to continue for 96 hours, there is no further reduction in sugar content. Instead, there is marked evidence of proteolysis, accompanied by objectionable odors and a pronounced increase in *p*H value (*Stuart and Goresline, 1942a*).

The best grades of commercially dried albumen result when fermentation is caused by bacteria of the genera *Aerobacter* or *Escherichia*, probably *Aerobacter aerogenes* and *Escherichia freundii* (*Stuart and Goresline, 1942a*). These are apparently natural shell contaminants and must be present in relatively large numbers if fermentation is to be rapid and regular. The process is slow and incomplete when caused by types of bacteria that infect the egg internally. Strains of *Proteus, Serratia,* and *Pseudomonas* produce a rapid increase in the amount of formol nitrogen, due to their proteolytic action, and only a relatively small decrease in *p*H value and sugar (*Stuart and Goresline, 1942b*). Dried albumen produced after fermentation by these organisms is dull and amorphous, rather than bright and "crystalline."

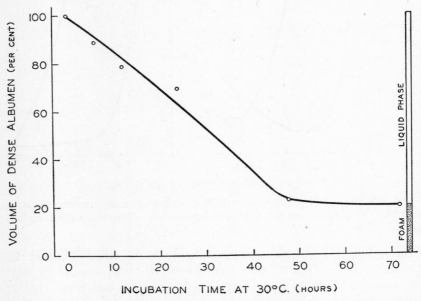

FIG. 389. Increase in the liquid portion of fermenting albumen and eventual separation of foam. (After Stuart and Goresline, 1942a.)

When fermentation has been completed, both the scum and the sediment are discarded, although the scum is sometimes used after it has been liquefied by the addition of citric acid and pepsin (*Swenson, 1938*). The liquid albumen is poured into shallow metal (usually aluminum) pans or trays, which have been thinly coated with mineral oil. These are placed in ventilated cabinets, which are heated to 45° C. or 50° C., and in which the trays remain for 6 to 16 hours (*Swenson, 1941*). The finished product has a final moisture content of about 8 per cent and is in flake form, although it is often pulverized before it is packaged (*Termohlen, Warren, and Warren, 1938*).

Ordinarily, albumen cannot be spray dried because it coagulates, and its solubility is thus destroyed. However, the spray-drying method has been modified so that it can be used for dehydrating albumen. The albumen may be hydrolyzed by acids (*Fischer, 1935; McCharles and Mulvaney, 1937*) or trypsin (*Balls and Swenson, 1936a*) to thin it and thus to facilitate spray drying. Hydrolysis also saves time and materially reduces the bacterial content. However, it retains the natural sugar of the albumen. In addition, enzymatically hydrolyzed albumen is not so stable in the dried form as fermented albumen (*Balls and Swenson, 1936b*).

Another problem in the spray drying of albumen is the extremely fine, dustlike texture of the product, which makes recovery difficult. Suction devices have been used successfully for the collection of the powder (*Termohlen, Warren, and Warren, 1938*).

Whole Egg and Yolk. In spray drying, powerful pumps force the liquid egg through nozzles, under pressures up to 3000 lb. per square inch, into a chamber where filtered air is circulated rapidly by blowers. The temperature of the air varies considerably in different plants and may be as high as 160° C. The egg material enters the chamber as a fine mist, and its water content is evaporated almost instantaneously. Drying is facilitated if the liquid egg is preheated before it enters the drying chamber. The dried powder falls down the sloping sides of the drier and is removed continuously from the bottom. It is quite hot as it emerges; cooling devices, such as shallow trays, are therefore usually provided to lower its temperature. Sometimes the conveyor to the packaging machine is also the cooling unit. Before the dried egg is placed in barrels, it is sifted to break up any lumps which may be present. Most of the powder that fails to collect in the drier escapes with the exhausted air and falls to the bottom of an auxiliary chamber (*Termohlen, Warren, and Warren, 1938*).

Depending upon drying conditions, the moisture content of spray-dried egg may vary considerably. It is advisably kept below 5 per cent; but too low a percentage of water may be associated with poor quality caused by excessive heat absorption during the drying process. Essentially, the water content depends upon the relative humidity within the drying chamber, as determined by the air temperature and the rate of feeding the liquid egg. If the flow rate of the egg is sufficiently high, evaporation may be accomplished without overheating, and the air within the chamber becomes considerably cooled. With an inlet air temperature of 121° C., the moisture content of the dried egg and the exhaust temperature may vary as the input of liquid changes (*Woodcock and Reid, 1943*). These relationships are shown in the table. A maxi-

Flow Rate (cubic centimeter/minute)	Exhaust Temperature (°C.)	Moisture Content (per cent)
35.1	76	1.18
64.4	62	3.71

mum production at a relatively (but not excessively) high inlet temperature usually results in dried eggs of the best quality compatible with low moisture content.

The particles of spray-dried egg are roughly spherical in shape, but of various sizes. Some are solid, and others contain voids. Dehydration

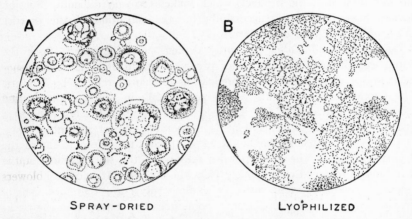

SPRAY-DRIED   LYOPHILIZED

FIG. 390. Appearance of particles of eggs dehydrated by two different methods. (After Shaw, Vorkoeper, and Dyche, 1946.)

of frozen egg in a vacuum yields a product composed of platelike particles of irregular outline (Fig. 390).

In the belt-drying method, the liquid egg is thinly applied to a metal belt which revolves in a chamber ventilated by filtered air at a temperature of about 60° C. One revolution requires approximately 2 hours. The egg material is then automatically scraped off the conveyor. Drying is completed in 2 or 3 hours in a cabinet where the temperature is between 38° C. and 43° C. The final moisture content is from 3 to 8 per cent (*Swenson, 1941*).

When whole egg or yolk is dried by the pan method in cabinets, a crust usually forms over the surface of the liquid material and impedes the evaporation of moisture from the underlying portion. The liquid in the pans must therefore be of proper depth if dehydration is not to be delayed. About 6 mm. is considered the optimum depth, since a thinner layer does not hasten evaporation sufficiently to be justified. At a temperature of 40° C. to 45° C., drying to a water content of 5 per cent may be accomplished in about 20 hours (*Bumazhnov, 1944c*).

PHYSICOCHEMICAL CHANGES IN DRIED EGGS

Unless carefully controlled, the process of drying may destroy many of the properties of the liquid egg. The heat necessary to evaporate moisture is also a powerful denaturing agent; and, as evaporation of water proceeds, more heat becomes available for absorption by the egg material. The result may be irreversible changes in the constituents of the egg, especially the proteins. These changes, furthermore, are not instantaneous and may occur slowly at a temperature lower than that at which the egg is dried. For this reason, once it is initiated, deterioration of the egg powder may continue during storage.

Destruction of some of the essential properties of the egg during dehydration is revealed by low solubility of the dried product, with accompanying loss of viscosity and foaming properties; by a gradually increasing acidity upon storage; and by changes in flavor and color, sometimes not apparent until the powder has been held for a time. The moisture content of the dried egg determines to a large extent many of the characteristics of the product.

Solubility. When the dried egg is reconstituted, some of the egg's colloidal characteristics are lost if the proteins are unable to reabsorb water. Among the physicochemical changes observable in reconstituted dried eggs are a decrease in viscosity and an increase in surface tension, sometimes to the extents shown in the table (*Bumazhnov, 1944a*). In dried eggs of

	VISCOSITY (poise)	SURFACE TENSION (erg^3/sq. cm.)
Before drying	0.155	52.1
After drying	0.015	59.8

poor solubility, the yolk emulsion fails to reconstitute properly. Small fat globules coalesce into larger ones containing air and rise to the top of the liquid (*Bate-Smith, Brooks, and Hawthorne, 1943*).

The solubility of dried-egg products may be rapidly lowered if the powder is not cooled as quickly as possible after dehydration. The deleterious effect of heat continues, for example, when the dehydrated material is allowed to accumulate at the bottom of the drying chamber. Dried whole egg may lose 10 per cent of its solubility in 1 hour at 80° C., and 28 per cent in 4 hours (Fig. 391). Dried albumen better withstands prolonged heating. At the above temperature, its solubility

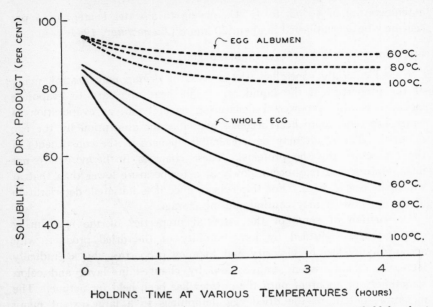

FIG. 391. Decrease in solubility of dried egg albumen and whole egg held for short periods of time at high temperatures. (After Bumazhnov, 1944b.)

may be lowered by 5 per cent in 1 hour and by 8 per cent in 4 hours (cf. Fig. 391).

During storage, the solubility of dried eggs continually decreases at a rate that depends upon the initial water content of the egg powder and upon the holding temperature. Relatively rapid loss of solubility, accompanied by deterioration in other respects, is associated with high moisture content. A small difference in the percentage of water in the dried product may markedly affect the rate at which solubility decreases. If dried whole egg contains, for example, 2 per cent water, its KCl value (or solubility in 10 per cent potassium chloride solution) may decrease from 80 per cent to 50 per cent in 10 days of storage at 43° C.; but, if the water content is 7 per cent, the solubility may fall to 20 per cent (Fig. 392-A). In dried albumen, the loss in solubility similarly depends upon the percentage of water (if this is over 3 per cent), as indicated by the tabulated data for samples of albumen held at 50° C. (*Stewart, Best, and Lowe, 1943*).

WATER CONTENT (per cent)	SOLUBILITY (per cent)	
	INITIAL	AT 15 DAYS
3.0	87	88
8.3	85	43

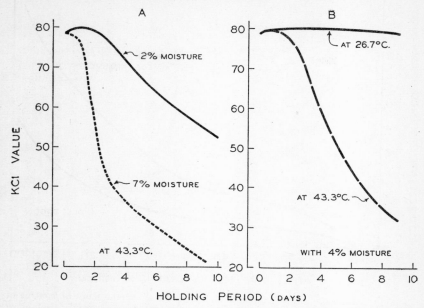

Fɪɢ. 392. Changes in KCl (solubility) value in egg powders of various moisture contents when held at different temperatures. (After White and Thistle, 1943.)

A, effect of moisture content. *B,* effect of temperature.

Unless dried-egg powder is stored in a waterproof container, its water content, and consequently its rate of deterioration, may be influenced by the relative humidity of the atmosphere. Moisture is absorbed from the environment until equilibrium is attained (Fig. 393).

The vapor pressure of water over dried-egg powders is proportional to the temperature and to the moisture content of the egg material, as shown in Fig. 394. In dried eggs containing less than 1 per cent water, the rate of change in vapor pressure, with change in moisture content, is very high at any given temperature. In egg powders of extremely low water content, the ratio of the heat of adsorption of water to the heat of condensation of water vapor at the same temperature also decreases very sharply with each increase in the percentage of water in the dried egg. In powders containing more than 1 per cent water, the value of the ratio changes at a slower rate as the moisture content of the dried-egg material increases (Fig. 395). Figure 396 shows the direct relationship between the specific heat and the moisture content of egg powder, when water is reabsorbed by powder dried in a vacuum to a water content of 0.15 per cent.

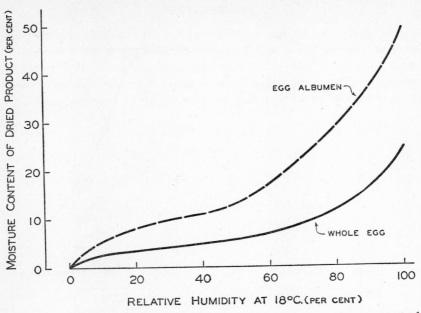

FIG. 393. Influence of the relative humidity of the air on the moisture content of dried albumen and whole egg held at 18° C. (After Bumazhnov, 1944b.)

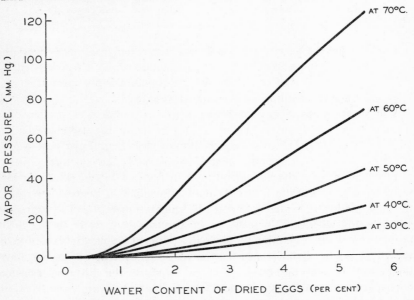

FIG. 394. Relationship (at various temperatures) between the vapor pressure of water over spray-dried whole eggs and the water content of the egg powder. (After Makower, 1945.)

748

The higher the storage temperature, the more quickly solubility decreases, a fact which is well demonstrated when egg powders of identical water content are held at different temperatures. In 10 days at 27° C., whole-egg powder containing 4 per cent moisture may exhibit a barely perceptible change in solubility, whereas, in the same period at 43° C.,

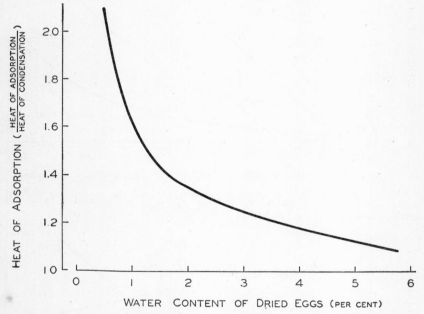

FIG. 395. Relationship between the water content and the heat of adsorption of spray-dried whole eggs. (After Makower, 1945.)

its original solubility of 80 per cent may fall to 30 per cent (cf. Fig. 392-*B*).

The decrease in solubility proceeds somewhat differently in each type of dried product—yolk, whole egg, and albumen. Initially, the solubility of whole egg is relatively high, in comparison with that of yolk and albumen, presumably because of the protective action which the albumen proteins exert upon the fats of the yolk. However, a fairly rapid loss of solubility begins immediately after processing, although in a few days the rate of decrease slackens. In dried albumen, the solubility remains unchanged for a short time and then declines; and, in dried yolk, it diminishes steadily from the beginning (Fig. 397-*A*).

The moisture content of unfermented dried albumen is dependent to

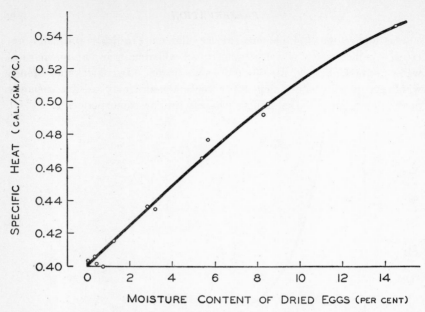

FIG. 396. Relationship between the moisture content and the specific heat of dried egg (between 27° and 66° C.). (After Stitt and Kennedy, 1945.)

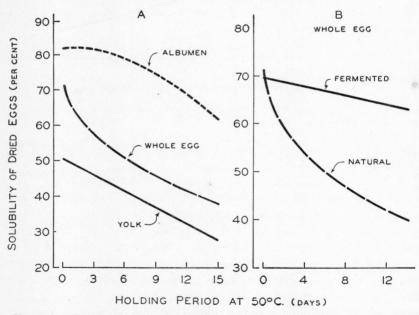

FIG. 397. Course of change in solubility of dried-egg products held at 50° C. (After Stewart, Best, and Lowe, 1943.)

A, egg components. *B,* natural and fermented whole egg.

some extent upon the thickness of the flakes. The rate at which solubility is lost therefore varies in flakes of different sizes. Loss of solubility is demonstrably less rapid in thin flakes, obtained from albumen layers ⅛ inch in depth, than in flakes of the size that results when the drying pans are filled to the depth of ¾ inch (Fig. 398).

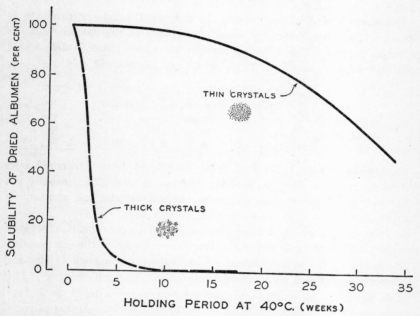

FIG. 398. Changes in solubility of dried albumen prepared as thick and thin flakes and stored at 40° C. (After Stewart and Kline, 1941.)

The dehydrated product prepared from fermented albumen remains relatively soluble during storage. Fully fermented dried albumen is 98 per cent soluble after it has been held 61 weeks at 40° C.; dried natural albumen of similar water content is only 48 per cent soluble (*Stewart and Kline, 1941*). The removal of glucose by bacterial action is responsible for the better stability of the fermented albumen. This fact has been demonstrated not only by the solubility loss occurring when 0.4 per cent glucose is replaced in fermented albumen before dehydration (*Stewart, Best, and Lowe, 1943*), but also by the stability achieved when fermentation is accomplished by the action of yeast rather than bacteria (*Hawthorne and Brooks, 1944*). If only 0.1 per cent of glucose remains in dried albumen, solubility changes will occur (*Stewart and Kline, 1941*).

The ability of dried whole egg to retain its solubility during storage is also markedly improved if glucose is removed by fermentation prior to dehydration (cf. Fig. 397-*B*). Similarly, the solubility of fermented yolk may decrease by only 11 per cent during 2 weeks at 47° C., whereas the solubility of unfermented yolk may decrease by 40 per cent (*Bate-Smith and Hawthorne, 1945*).

The development of insolubility in untreated dried-egg products is paradoxically accompanied by the disappearance of the glucose which is naturally present. The following changes have been noted in dehydrated eggs held 14 days at 47° C. (*Bate-Smith and Hawthorne, 1945*):

	GLUCOSE (gram/100 cc.)		SOLUBILITY (per cent)	
	INITIAL	FINAL	INITIAL	FINAL
Whole egg	0.47	0.09	97	55
Albumen	0.47	0.09	99	86
Yolk	0.23	0.10	80	39

Eventually, most of the free sugar disappears from desiccated eggs during storage. In albumen, however, the loss in solubility is usually not observable until some time after the glucose content has measurably decreased.

It is possible that the loss of glucose is due to condensation between the reducing group of the sugar and free amino groups of the egg proteins. The change in the sugar content of unfermented dried whole egg and albumen is paralleled by a decrease in amino nitrogen content. This fact is indicated by analyses of samples before and after 5 to 7 days' storage at 47° C. (*Bate-Smith and Hawthorne, 1945*). Amino nitrogen

	SUGAR (gram/100 cc.)		AMINO NITROGEN (gram/100 cc.)	
	INITIAL	FINAL	INITIAL	FINAL
Whole egg	0.45	0.22	0.447	0.320
Albumen	0.42	0.25	0.549	0.427

decreases, during storage, in mixtures of glucose and ovalbumin, ovoglobulin, ovolivetin, or ovovitellin. It may therefore be assumed that the reaction of proteins with glucose is not confined to any particular protein constituent of the egg (*Olcott and Dutton, 1945*).

Synthetically prepared glucose-amine condensation products are highly fluorescent (*Olcott and Dutton, 1945*). The existence of condensation products in stored dried eggs is further indicated, therefore, by the fact that the fluorescence of salt extracts of egg products increases as the holding period is extended. Also, the fluorescence parallels the loss of solubility and the occurrence of various other changes. The data in the table indicate how fluorescence readings may rise during storage at 24° C. (*Pearce, 1943*). Rapid development of fluorescence, as of insolubility,

FLUORESCENCE READING

	INITIAL	AT 1 MONTH	AT 2 MONTHS
Yolk	10.0	15.0	20.4
Albumen	10.1	20.5	32.2
Mixed yolk and albumen	14.8	16.5	23.5

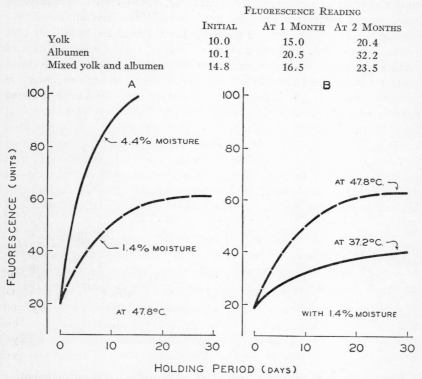

Fig. 399.   Changes in fluorescence values of dried whole eggs during storage. (After Thistle, White, Reid, and Woodcock, 1944.)

A, effect of moisture content of egg powder.   *B*, effect of holding temperature.

is associated with high moisture content in the egg product, and with high holding temperatures.   Figure 399-*A* shows how high is the fluorescence reading of dried whole egg containing 4.4 per cent of water when compared with that of a powder containing only 1.4 per cent water.   The latter powder is even less fluorescent when it is held at about 37° C. rather than at 47° C. (cf. Fig. 399-*B*).

A condensation between reducing groups of glucose and free amino groups of proteins would cause loss of solubility only indirectly.   It is possible that the products of the primary reaction are involved in further reactions which are directly responsible for insolubility; but the nature of the secondary reactions is as yet unknown.   Of course, condensation cannot occur when the natural free glucose of the egg is eliminated before dehydration, and the prevention of insolubility by removal of glucose is thus explained (*Bate-Smith and Hawthorne, 1945*). In all probability, the moisture content of egg products influences the development of insolubility merely by providing a vehicle for chemical change.

The loss of solubility in dried eggs may be retarded by the addition of sucrose or lactose, in concentrations of 10 to 15 per cent, before dehydration (*Brooks and Hawthorne, 1943a*). The reason for the protective action of these sugars is not clear; upon storage, the amino nitrogen content of the egg decreases in the presence of either sugar, and—at least in the presence of sucrose—the glucose content also diminishes. Lactose, by virtue of its reducing group, might prevent the condensation of dextrose and amino groups by entering, itself, into combination with proteins. It must be assumed that this reaction, if it occurs, leads to others whose end products are soluble. Sucrose possibly inhibits the secondary reactions which produce loss of solubility (*Bate-Smith and Hawthorne, 1945*).

The addition of lactose gives better results than the addition of sucrose. The superiority of lactose is shown by the following comparison of the solubility changes in sucrose-dried egg and in lactose-dried egg, during 18 days of storage at 47° C. (*Bate-Smith and Hawthorne, 1945*):

	SOLUBILITY (per cent)	
SUGAR ADDED	INITIAL	FINAL
Sucrose	100	78
Lactose	100	98

The addition of certain amino acids to liquid egg before dehydration also retards development of insolubility, apparently because the glucose of the egg reacts with these substances in preference to those naturally present. The table compares the solubility of untreated dried egg, before and after 14 days of storage at 47° C., with that of samples containing glycine and cysteine hydrochloride (*Bate-Smith and Hawthorne, 1945*).

	SOLUBILITY (per cent)	
ACID ADDED	INITIAL	FINAL
None	94	46
Glycine (1 per cent)	98	79
Cysteine (0.35 per cent)	97	77

Glycine or cysteine hydrochloride is effective also in preserving the solubility of dried albumen. Figure 400 shows the changes in the solubility (in water) of an untreated sample, compared with that of samples to which glycine (0.23 per cent) or cysteine hydrochloride (0.48 per cent) was added before drying.

It has also been observed that loss of solubility is delayed if the pH of the liquid egg is adjusted to approximately pH 6.0 by the addition of lactic acid (*Stewart, Best, and Lowe, 1943; Bate-Smith and Hawthorne, 1945*). The reason for this effect is not clear, since glucose and amino

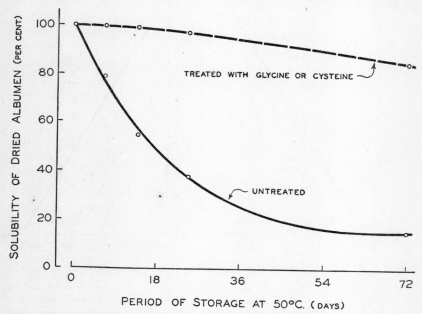

FIG. 400. Changes in solubility of dried egg albumen during storage, as affected by the addition of glycine or cysteine hydrochloride to albumen prior to dehydration. (After Kline and Fox, 1946.)

nitrogen disappear in the presence of lactic acid as rapidly as in its absence.

Aerating Power. In general, the initial aerating power of dried eggs is best in products of good solubility (Fig. 401-*A*). The foaming properties of an egg powder are determined to a considerable extent by drying conditions. Spray-dried whole eggs of almost identical water content may have entirely different aerating power, as measured by cake volume

DRYING CONDITIONS	WATER (per cent)	CAKE VOLUME (cubic centimeters)
Proper	2.01	444
Adverse	2.08	317

(*Woodcock and Reid, 1943*).   In dried eggs containing more than 3.5 per cent water, there is little apparent correlation between whipping quality and moisture content.   In powders with less than 3.5 per cent of water, the beating properties are inferior, probably because of dehydration under poor conditions (*Hawthorne, 1943*).

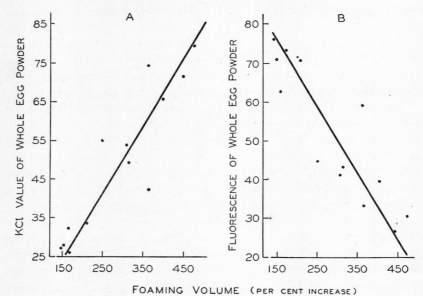

Fig. 401.   Relationship between the foaming volume of whole egg powder and its fluorometric reading and KCl value (solubility in 10 per cent potassium chloride). (After Reid and Pearce, 1945.)

A, relationship between foaming volume and KCl value.   *B*, relationship between foaming volume and fluorescence.

The foaming properties of egg products decrease as insolubility develops (*Bate-Smith, Brooks, and Hawthorne, 1943*).   There is also an inverse linear relationship between aerating power and fluorescence value (cf. Fig. 401-*B*).

Lactose and sucrose, by preventing the reactions that result in loss of solubility, also prevent loss of aerating power.   Sucrose, although a better preservative of whipping qualities during short holding periods, is

	FOAM VOLUME (cubic centimeters)		
SUGAR ADDED	INITIAL	AT 7 WEEKS	AT 16 WEEKS
None	170	110	
Lactose (10 per cent)	305	250	250
Sucrose (10 per cent)	330	300	110

less effective than lactose when storage time is extended. The manner in which the type of sugar may influence the rate at which foaming properties disappear during storage at 37° C. is shown by the accompanying data (*Brooks and Hawthorne, 1943a*).

Changes in Reaction. Except for fermented dried albumen, which is made acid by bacterial action, freshly dehydrated eggs are more alkaline than the liquid egg from which they are prepared. During the drying process, the high temperatures which cause the evaporation of water also break down the carbonates of the egg and drive off carbon dioxide. The original pH value of the egg material, ordinarily within the range of 7.2 to 7.6, may thus be raised to the range of 7.6 to 8.6, or even higher.

The extent to which alkalinity increases during dehydration can be reduced by the preliminary addition of 10 per cent lactic acid to liquid egg. The increase in pH, after reconstitution, is inversely proportional to the amount of adjustment made in the pH of the liquid egg. For example, if the pH value of the liquid egg is reduced only to 7.0, that of the reconstituted egg is about 8.1; but adjustment to pH 6.5 results in a subsequent increase only to pH 7.1 (*Stewart, Best, and Lowe, 1943*).

During storage, the dried egg becomes increasingly acid, for more than one reason. When glucose condensation with proteins occurs, an increase in free acidic groups results (*Bate-Smith and Hawthorne, 1945*). In addition, free fatty acids are liberated during storage at the relatively low temperatures of 15° C. to 20° C., although less readily in egg powders of low moisture content. In dehydrated eggs, containing various percentages of water and stored 40 weeks at 20° C., the free fatty acids (expressed as cubic centimeters of 0.05 N sodium ethoxide per gram of ether extract) may increase as follows, when their original value ranges from 1.3 to 1.6 cc. (*Brooks and Hawthorne, 1943b*):

WATER CONTENT (per cent)	FREE FATTY ACIDS
2.6	1.80
4.5	2.40
8.5	4.60

When stored at high temperatures, dried eggs of high water content show a decrease in pH which is too large to be caused by the increase in free fatty acids alone. It is possible that this relatively greater acidity is due to an increase in orthophosphates and acid-soluble phosphorus. In freshly dehydrated eggs, these substances, respectively, are present in the amounts of about 0.084 mg. per cubic centimeter and 0.115 mg. per cubic centimeter. During storage for 24 weeks at 37° C., they increase to the following levels in egg products containing various amounts of water (*Brooks, 1943*):

WATER (per cent)	ORTHOPHOSPHATES (mg./cc.)	ACID-SOLUBLE PHOSPHORUS (mg./cc.)
2.5	0.104	0.144
8.4	0.146	0.240
11.0	0.190	0.319

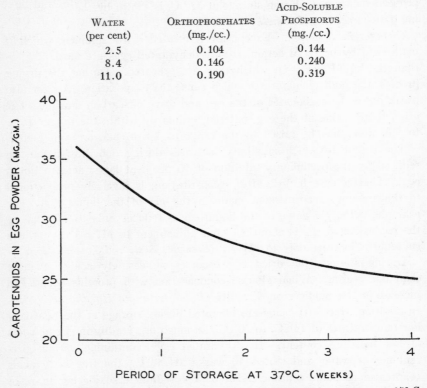

FIG. 402. Change in carotenoid content of egg powder during storage at 37° C. The change is identical in egg powders of moisture content from 0.7 to 4.3 per cent. (After Dutton and Edwards, 1946.)

Changes in Color. Dried eggs sometimes grow undesirably darker during storage. When they do so, their color passes through a sequence of changes from yellow to brown. Unfermented dried albumen, in fact, may eventually become black (*Stewart and Kline, 1941*).

The color changes in dried eggs are of two types. The carotenoid pigments gradually decrease in amount (Fig. 402), probably because of oxidation. In 6 months of storage at 37° C., as much as 27 per cent of the carotenoids may disappear (*Dutton and Edwards, 1945*). Concurrently, brown substances develop. The latter are complex in nature and are formed by reactions which are as yet incompletely understood. The fluorescent products of glucose-amine condensation are brown (*Olcott and Dutton, 1945*) and therefore may contribute to discoloration in stored dried eggs. In addition, there is evidence that brown, fluorescing compounds arise from a similar condensation between aldehydes and lipid amines, perhaps cephalin-amino groups. Possibly aldehydes result from the oxidation of fatty acids, which may also polymerize and yield brown materials (*Edwards and Dutton, 1945; Dutton and Edwards, 1945*).

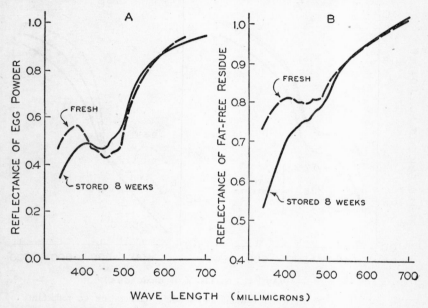

FIG. 403. Reflectance of *A*, egg powder, and *B*, its fat-free residue, before and after 8 weeks of storage. (After Dutton and Edwards, 1946.)

The changes in the color of dried eggs during storage may be demonstrated by the altered intensity of reflected monochromatic light, expressed as the ratio of intensity of light reflected from the egg product to the intensity of the light reflected from a standard magnesium oxide surface. In a fresh sample of egg powder of low water content, the reflectance curve exhibits minima, characteristic of carotenoid pigments, at various wave lengths between 430 and 480 mμ. (Fig. 403-*A*). After storage for 8 weeks at 37° C., reflectance increases in the region of carotenoid absorption and decreases at the short wave-length end of the spectrum. After almost complete removal of the carotenoid pigments, the reflectance curve of the fat-free residue of egg powder also shows a pronounced decrease in reflectance at the short wave-length end of the spectrum (Fig. 403-*B*). It is thus indicated that the formation of ether-insoluble brown materials (glucose-protein condensation products) is probably responsible for part of the decrease in the reflectance of the original powder (*Dutton and Edwards, 1946*).

The rate of carotenoid destruction is accelerated by increased holding temperatures, but it is not greatly affected by the moisture content of

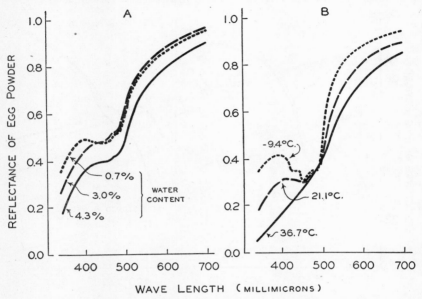

FIG. 404. Influence of water content and holding temperature on reflectance of egg powder. (After Dutton and Edwards, 1946.)

A, reflectance of egg powders of various moisture contents, stored 8 weeks at 37° C. *B*, reflectance of egg powders, all containing 5 per cent moisture, stored 9 months at various temperatures.

the egg powder. After 8 weeks of storage at 37° C., the carotenoid absorption region may be distinguishable in the reflectance curves of samples containing 0.7, 3.0, and 4.3 per cent water, although it is partially obscured by the effects of simultaneous formation of brown substances (Fig. 404-*A*). The latter appear more rapidly in egg products of greater water content, and in those held at higher temperatures. In egg powder with a fairly high percentage of moisture, the reflectance curves may exhibit little destruction of carotenoids after 9 months of storage at −9.4° C.; but, after storage at 21.1° C. and 36.7° C., the disappearance of the carotenoid minima and the development of brown absorption are both very marked (cf. Fig. 404-*B*).

Removal of glucose from albumen by fermentation before dehydration completely prevents darkening of the dried product during storage (*Stewart and Kline, 1941*). However, color changes in dried whole egg are not entirely eliminated by the addition of sugars, such as sucrose and lactose, prior to the drying process (*Brooks and Hawthorne, 1943b*).

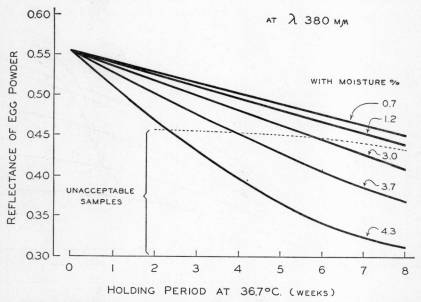

FIG. 405. Changes in the reflectance (at 380 mμ) of stored egg powders containing various percentages of water. Development of unacceptable flavor is indicated. (After Dutton and Edwards, 1946.)

Changes in Flavor. During storage, various unpleasant tastes may develop in dried eggs, simultaneously with loss of solubility and changes in color. The flavor and appearance of egg products are closely associated (*Tracy, Sheuring, and Hoskisson, 1944*).

A noticeable burnt flavor in whole egg powders may be the result either of overheating during drying or of storage at high temperatures. When dried eggs have been held at temperatures in the vicinity of 15° C., the ordinary storage or "cardboard" taste is often detectable. In products of low water content, held at comparable temperatures for long periods of time, the taste is often described as "fishy" (*Bate-Smith, Brooks, and Hawthorne, 1943*). An acid, "cheesy" flavor has been noted in powders of low *p*H and very low glucose content (*Fryd and Hanson,*

1945) and has been ascribed to the action of bacteria that produce lactic acid (*Brooks and Hawthorne, 1943b*).

In general, good flavor in dried eggs is associated with high glucose content. Fermentation, by removing glucose, may decrease the palatability of dried albumen.

Loss of palatability, associated as it is with the disappearance of glucose, accompanies the formation of brown substances and occurs more

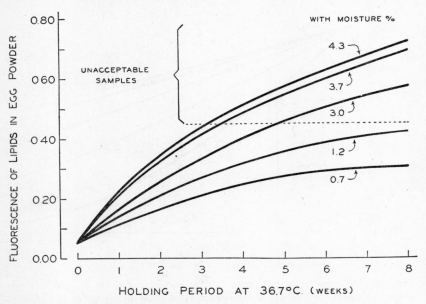

Fig. 406. Changes in the fluorescence values of lipid substances in stored egg powders containing various percentages of water. Development of unacceptable flavor is indicated. (After Dutton and Edwards, 1946.)

rapidly in egg powders of higher moisture content. In 2 to 6 weeks of storage at about 37° C., dried eggs containing 3.0 to 4.3 per cent water become unacceptable in flavor and show a marked decrease in light reflectance at 380 mμ; by comparison, powders containing only 0.7 to 1.2 per cent water deteriorate little, either in flavor or color (Fig. 405).

For some time, the fluorescence associated with the browning reaction has been employed as an index of palatability. The degree of fluorescence shown by salt solution extracts of defatted egg powders gives a certain indication of the extent of flavor deterioration. Lipid fluorescence is considered a more accurate criterion of palatability than the fluorescence of salt-solution extracts (*Boggs, Dutton, Edwards, and Fevold, 1946*). The formation of fluorescent lipid substances proceeds at a rate proportional to the moisture content of the dried egg (Fig. 406).

The fluorescent lipid substances responsible for loss of solubility are apparently products of the oxidation of yolk phospholipids (*Fevold, Edwards, Dimick, and Boggs, 1946*). Loss in palatability can be greatly retarded if dried eggs are stored in an atmosphere from which oxygen

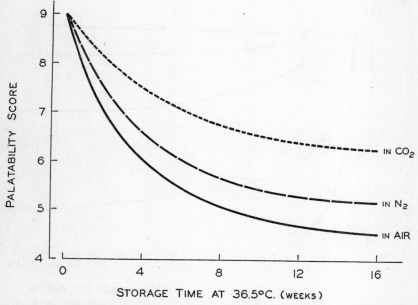

FIG. 407. Changes in the palatability of egg powders stored in different atmospheres. (After Boggs and Fevold, 1946.)

is excluded. The merits of carbon dioxide and nitrogen as storage atmospheres, as compared with normal air, are shown in Fig. 407. The relative superiority of carbon dioxide over nitrogen appears to be due in part to the acidic nature of the former gas. Acidification of liquid egg to pH 5.5 or pH 6.0, prior to dehydration, improves palatability; if the pH is adjusted much below this range, the taste of the dried product is disagreeable (*Stewart, Best, and Lowe, 1943*). Either the acidity must be neutralized upon reconstitution of the dried egg, or dry sodium bicarbonate, equivalent to the acid, must be added to the egg powder after dehydration and before packaging (*Boggs and Fevold, 1946*).

MICROFLORAL CONTENT OF DRIED EGGS

During the drying process, the bacterial content of liquid egg is much reduced. More bacteria are destroyed as the drying temperature is raised. The numbers of surviving bacteria may be decreased three- to fivefold by increasing the outlet temperature of the spray drier by 4° C. to 15° C. Also, the more slowly the egg powder is cooled, the more

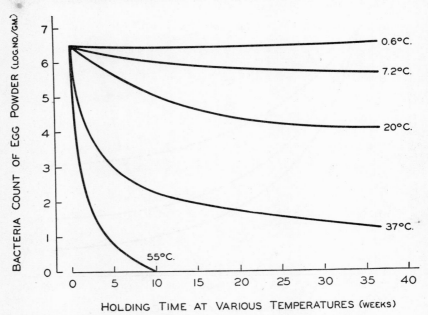

Fig. 408. Changes in the bacterial content of egg powders stored at various temperatures. (After Gibbons and Fulton, 1943.)

bacteria are destroyed (*Gibbons and Fulton, 1943*). Since dehydration under conditions favorable to the death of bacteria would result in egg products of very poor quality, microorganisms are generally present in freshly dried eggs.

During storage, the bacterial content of egg powder usually decreases, at a rate that depends chiefly upon the holding temperature. High temperatures, such as 55° C., result in the fairly prompt death of nearly all bacteria; but, at 0.6° C., little change may occur during periods as long as 36 weeks (Fig. 408).

The survival of bacteria in stored dried eggs seems to be little affected by the egg's moisture content, if this is within the usual range. If the egg powder absorbs a limited amount of water from the atmosphere, the bacterial count decreases. The death rate of bacteria apparently is increased by the chemical changes responsible for the loss of solubility in dried eggs containing undesirably high percentages of water. On the other hand, if the relative humidity of the storage atmosphere is so high that the dried egg absorbs water in excess of 20 per cent of its

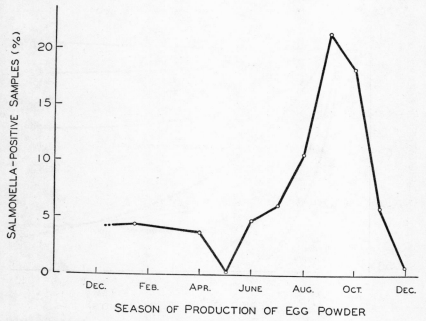

FIG. 409. Seasonal changes in the incidence of *Salmonella* organisms in dried whole egg powders. (After Gibbons and Moore, 1944a.)

weight, bacterial life is supported, and multiplication occurs (*Stuart, Hall, and Dicks, 1942*).

Mold growth may occur in dried-egg powders. It has been found at temperatures from 7° C. to 32° C., in products with moisture content from approximately 3 per cent to 8.5 per cent (*Stuart, Hall, and Dicks, 1942; Gibbons and Fulton, 1943*).

Several species of *Salmonellae* have been detected in dried eggs. Most of the same species have been found infecting fowl, and it is therefore possible that egg powders containing *Salmonellae* are derived from contaminated shell eggs (see Chapter 8, "Biological Properties"). The drying process reduces the numbers of *Salmonellae* in liquid egg 65 to 99 per cent. The presence of these organisms in dehydrated eggs probably indicates high contamination of the liquid egg. The incidence of *Salmonellae* in egg powders, although it varies in different geographic regions, usually increases between June and September, perhaps because of warm weather (Fig. 409). In stored dried eggs, the death of *Salmonellae,* as of other bacteria, occurs at a rate dependent upon the

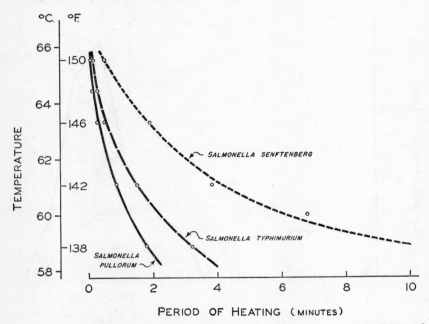

FIG. 410. Heating period required at various temperatures to accomplish total destruction of three species of *Salmonella* in liquid whole eggs. (After Winter, Stewart, McFarlane, and Solowey, 1946.)

holding temperature. Although there is some question as to how many of these organisms constitute an infective dose for man, some individuals are more susceptible than others. Ordinary cooking procedures often destroy *Salmonellae* completely. Dried-egg powders contaminated with *Salmonellae* may be a health hazard, however, if they are incorporated into foods that are consumed after being heated only slightly, or not at all (*Gibbons and Moore, 1944a, 1944b*). Different species of *Salmonellae* vary in their resistance to heat, as shown in Fig. 410. Pasteurization of the liquid egg for about 7 minutes at 60° C. is sufficient completely to destroy the most resistant species, *Salmonella senftenberg*. The practice of pasteurizing liquid egg prior to dehydration is coming into increasing use. Holding the egg material at the temperature of 60° C. for 30 minutes reduces the viable count by 98 to 99 per cent and destroys not only *Salmonellae,* but also *Staphylococci* and coliform organisms, without injuring the quality of the egg powder (*Gibbons, Fulton, and Reid, 1946*). Ordinarily, *Escherichia coli* tends to remain viable in unpasteurized dried eggs (*Stuart, Hall, and Dicks, 1942*).

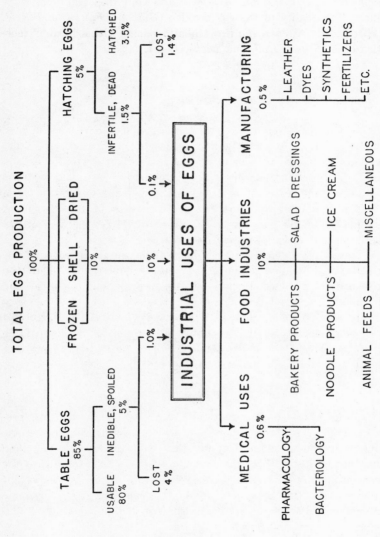

Sources and estimated industrial uses of hens' eggs in the United States, in 1945.

Chapter Eleven

INDUSTRIAL USES

The avian egg is a valuable raw material for many
industrial products.

In the modern industrial and scientific era, new uses have been created for many materials. As investigation reveals its distinctive properties, the egg becomes valuable to an increasingly large number of industries.

The industrial usefulness of the egg is due largely to its colloidal nature. Its ability to form and stabilize emulsions and the coagulability of its proteins give it special importance for the makers of food products. Other properties enable the egg to fill many diversified needs, both in manufacturing and in various applied arts and sciences.

In the United States, about 10 per cent of the annual supply of hens' eggs, chiefly in frozen or dried form, is used industrially. The food industries absorb the largest share of these eggs, pharmaceutical and research laboratories a smaller proportion. Inedible market and hatchery discards are used to some extent as raw material for the manufacture of a number of products. Of the remaining 90 per cent of the annual egg production, 5 per cent is reserved for hatching purposes, and 85 per cent is marketed as shell eggs for table use.

FOODSTUFF INDUSTRIES

Eggs are incorporated into many different foodstuffs that are prepared commercially in larger quantities each year as housewives become increasingly dependent upon manufactured food products rather than upon homemade items. To the makers of these food products, the physicochemical properties of the egg constitute the most important reason for its use. To the general public, however, the egg is desirable in foods not only because it adds nourishment, but also because it improves taste, texture, and appearance.

BAKERY PRODUCTS

The baking industry in the United States has grown rapidly in the past and is still expanding as bakers experiment with a greater variety of products to tempt the public palate. Eggs are an essential ingredient of most baked goods, and at present bakers alone use nearly 7 per cent of the country's total egg supply, or about 4 billion eggs annually.

Types of Eggs Used

Eggs are available in several forms—as whole eggs in the shell (fresh or storage), and as frozen or dried whole eggs, whites, or yolks. Frozen

Food Products	Shell Eggs: Fresh, storage	Frozen Eggs: Whole	Frozen Eggs: Albumen	Frozen Eggs: Plain yolk	Frozen Eggs: Salt yolk	Frozen Eggs: Sugar yolk	Frozen Eggs: Glycerine yolk	Desiccated Eggs: Whole	Desiccated Eggs: Albumen	Desiccated Eggs: Yolk
Biscuits, cookies	●	●	●	●		●	●			
Cakes, dark	●	●		●		●	●			
Cakes, white	●		●							
Candies, confections			●						●	
Custards	●			●		●				
Doughnuts	●	●		●		●	●	●		●
Egg noodles	●	●		●				●		●
Food beverages	●	●	●			●		●		●
Ice cream	●	●	●	●		●		●		●
Icings	●	●	●						●	
Macaroni	●	●		●				●		●
Mayonnaise	●			●	●	●	●			
Pancake & other prep flours								●		●
Pies	●	●						●		
Prepared puddings	●		●					●		●
Salad dressing	●	●		●	●	●	●			

FIG. 411. Use of various forms of eggs in food products. (After Ovson, 1933, 1938.)

yolk may be obtained as plain yolk, or as yolk containing salt, glycerin (5 to 8 per cent), or sugar (10 per cent). These ingredients serve to some extent as preservatives, but they are added chiefly as stabilizers, to prevent loss of consistency when the yolk is thawed (see Chapter 10, "Preservation"). Figure 411 shows some of the food products that are made with eggs in these different forms. The relative extent to which each type of egg is used by all industry is shown in the following table.

Fresh or storage shell eggs are most convenient for baking on a small scale or in the home, but they constitute only 1 to 5 per cent of the total purchased by commercial bakers. The baking industry makes extensive use of eggs in other forms, especially frozen eggs. In cooking

KIND OF EGGS	AMOUNT USED (per cent)
Shell eggs (fresh or storage)	91.36
Frozen eggs:	
Whole	3.51
Albumen	1.81
Plain yolk	0.83
Salted yolk	0.38
Sugared yolk ⎱ Glycerin yolk ⎰	0.52
Desiccated eggs:	
Whole	0.03
Albumen	0.08
Yolk	1.48

properties, frozen eggs compare most favorably with fresh eggs and yet have all the time- and space-saving advantages of being already broken, or even separated. The baker thaws only what he needs for each day. In addition, frozen eggs must meet certain specified standards and therefore yield uniformly good results. Some bakers claim that the glycerin yolks aid in keeping moisture in baked goods.

Of the total supply of the various forms of frozen eggs, the baking industry uses the following percentages (*Ovson, 1938*):

TYPE OF FROZEN EGG	AMOUNT USED (per cent of supply)
Mixed whole	50
Albumen	50
Sugared yolk	50
Glycerin yolk	50
Plain yolk	25

Dried eggs are less popular with bakers than frozen ones (*Watts and Elliott, 1941*), but are often used to improve the consistency of pie fillings (*Snyder, 1930*). Desiccation sometimes denatures the proteins of the albumen, and the egg thus loses some of its desirable properties. Dried eggs, upon reconstitution, are of much less uniform quality than frozen ones and are more perishable. As desiccating processes are improved, dried eggs will no doubt be more extensively used by the baking industry (*Jordan and Sisson, 1943b; Ary and Jordan, 1945*).

It has been suggested that the standard value of 13.5 gm. (0.48 oz.) of spray-dried whole egg powder be adopted as the minimum weight equivalent to an average liquid egg (*Jordan and Sisson, 1943a*).

Cakes

Of all the products made by bakers, cakes are the most important. Cookies, crackers, biscuits, doughnuts, and puff pastries are also generally considered to be cake products. Some 4 to 5 billion pounds of cake products are eaten annually in the United States, and at least 25 per cent of this amount is made commercially (*Bailey and LeClerc, 1935*).

There is an almost countless variety of cakes. A particular kind of cake may be so popular in a country as to become a national dish. The *Kuchen* of Germany and the *gateau* of France are examples. The proportion of egg varies in different kinds of cake. In the three principal types of light cakes baked in the United States, eggs form the percentages shown in the table (*Bailey and LeClerc, 1935*).

	ALBUMEN (per cent)	YOLK (per cent)	WHOLE EGGS (per cent)
Angel-food cake	41.3	....	....
Gold cake	....	17.7	....
Pound cake	....	....	24.8

Although flour is the basic component of cakes, eggs aid in moistening the batter. The fat content of the yolk has shortening value (*Bailey and LeClerc, 1935*). The egg also acts as a binder to retain the air cells formed during beating, mixing, and baking, and thus it exerts a certain leavening power (*Jordan and Pettijohn, 1946*). The leavening property of eggs made them especially important before the widespread use of baking powder, and, for this reason, old-fashioned cake recipes call for more eggs than modern ones.

In some cakes, such as angel-food and sponge cake, the cellular structure is due largely to the ability of albumen to form foam when beaten. Albumen is a colloid of low surface tension and high viscosity. Whipping exerts a shearing stress which introduces air bubbles. Around these, the albumen forms walls, liquid at first, but progressively more solid as the protein particles become oriented. Beating must stop before too much air has been entrapped, since the protein walls of the bubbles must not become so thin and fragile that they lose their elasticity. Baking expands the air bubbles, coagulates the egg proteins, and sets the dough in a permanent state of lightness (*Barmore, 1934*). White cakes, made with egg albumen, but without egg yolk, regain their original volume and cellular structure if suspended in water after being compressed to one-sixth of their thickness. Cakes containing neither yolk nor albumen swell to a moderate degree but disintegrate (Fig. 412).

The tenderness, volume, elasticity, and compressibility of the finished cake depends upon the quality of the egg albumen foam. Temperature, and the speed and method of whipping the albumen, can affect the lightness of the foam to some extent; but the best cakes can be obtained only when egg whites of good quality are used. Although the liquid layer of the albumen whips more quickly than the dense layer, the beating properties of the albumen are ruined by the type of liquefaction which eventually occurs as the egg ages (*Barmore, 1936*).

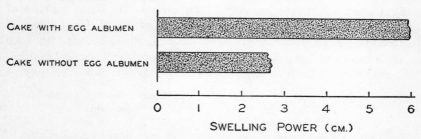

CAKE WITH EGG ALBUMEN

CAKE WITHOUT EGG ALBUMEN

0 1 2 3 4 5 6

SWELLING POWER (CM.)

Fig. 412. Reconstitution (after compression) of the cellular structure of cake containing egg albumen compared with that of cake made without egg albumen. (After Glabau and Kepes, 1935.)

Experiments have been made to determine how the various layers of frozen albumen compare with the entire albumen of fresh eggs, when used in cakes (*Miller and Vail, 1943*). The results indicate that the frozen liquid portion may be slightly better than the frozen dense layer, but that both compare favorably with fresh albumen. The similar quality of cakes made with these three forms of albumen is indicated in the table.

	COMPRESSIBILITY (square centimeters)	ELASTICITY (millimeters)	TENDERNESS (grams)
Fresh albumen	9.3	2.7	63.9
Frozen dense layer	8.4	2.8	76.1
Frozen liquid layer	10.0	2.6	56.8

Most cakes require frosting as a final touch to add tempting appearance. Egg albumen is an essential ingredient of several types of frosting and ornamental icing, both cooked and uncooked. A few recipes require the whole egg.

Other Products

Eggs are important not only in commercial baking, but also as ingredients of packaged cake flours and mixes of various types for home use.

The utilization of dried whole eggs in waffle and pancake flours, ice cream mixes, and other preparations has been increasing rapidly in the past few years.

Ether extracts of egg yolks have been experimentally added to bread flour (*Freilich and Frey, 1941*). Bread made from the flour was increased in volume and improved in texture and color.

ICE CREAM

Eggs are used in most commercial ice creams. In some varieties of ice cream, eggs constitute only a small proportion of the total solid ingredients. Their percentage is somewhat higher in other varieties, such as the Vienna and Neapolitan types, which are made from cooked custard-like mixtures. The tabulated data (*Heller, 1927*) show the average

	CONTENT OF EGG SOLIDS (per cent)
Average commercial brand	0.25–0.50
Vienna ice cream	0.60–1.24
Neapolitan (New York) ice cream	1.42–2.90

percentage of egg solids in the different types of ice cream. In home-made ice creams, the content of egg solids may be as high as 2 per cent, especially if fresh whole eggs are used.

The addition of even a small percentage of egg solids to the total solids of an ice cream mix increases the smoothness of the final product. The smoothness of ice cream is affected by many factors, but it depends largely on the size and distribution of ice crystals. Large crystals cause coarse texture, which is undesirable (*Arbuckle, 1940*). Egg yolks markedly improve the whipping quality of the ice cream mix, thereby shortening the time required for freezing; as a result, the size of the crystals is materially reduced. The following table (*Price, 1931*) compares the time necessary to complete the whipping of mixtures with and without egg yolk:

YOLK CONTENT	TIME (minutes)
None	13
Sugared frozen	9
Dried	6

The effect of egg yolk is due to the action of the lecithoprotein complex. The lecithoprotein of the yolk forms an adsorption film around the fat globules in the mixture; the fat globules then adhere more tenaciously to the other solids. The clumping of the globules is thus decreased, and the whipping quality of the mix is improved. Egg albumen, without the yolk, is never a constituent of commercial ice cream, since it has little or no effect on whipping quality (*Button, 1929*).

Several forms of eggs are utilized in commercial ice creams, the choice depending largely on cost. To produce ice cream of good quality, the relative amount of egg which must be used in the mix varies with the different types of eggs (*Horrall, 1934*), as shown in the table. Custard

Type of Egg	Proportion of Egg in Mix (per cent)
Dried yolks	0.1–0.5
Dried whole eggs	0.2–0.6
Frozen yolks	0.6–0.8
Frozen whole eggs	1.0–2.0
Fresh whole eggs	1.0–2.0

ice creams, of course, require a somewhat greater egg content, and their flavor is distinctive because of the presence of the eggs.

Approximately 30 per cent of all ice cream manufacturers prefer dried yolk powder to any other form of eggs. About 10 per cent of the manufacturers use frozen sugared yolks or glycerin-preserved yolks. Not more than 1 per cent favor the use of whole eggs in any form (*LeClerc and Bailey, 1940*).

The eggs are added to the mix prior to its pasteurization. If fresh eggs or yolks are used, they are beaten first. Frozen eggs must be thawed at 2° C. to 10° C. before they can be added. Dried eggs may be introduced into the cold mix directly, or they may first be dissolved in cold skim milk, or mingled with part of the sugar.

There are one or two disadvantages to the use of eggs for increasing the total solids content of ice cream. Eggs are of relatively high cost; also, if present in ice cream in an excessive amount, they produce an undesirably heavy consistency. In addition, some individuals object to the flavor of the egg in ice cream.

MACARONI AND NOODLES

For many centuries, macaroni and noodles have been popular foods in Europe, especially in Germany and Italy. In the United States, they were never in great demand until after the first world war, but their production then increased phenomenally. Since 1920, America has been an exporter of both macaroni and noodles (*LeClerc, 1933*), and in 1929, for example, produced 553 million pounds of noodles (*LeClerc and Bailey, 1940*).

Noodle products are made entirely from wheat-flour dough and egg. Water and salt may be added, although these two ingredients are not considered absolutely necessary. Needless to say, eggs are extremely important to the noodle industry. If they are omitted from the dough, the product can no longer be considered noodles, but resembles spaghetti.

Macaroni, like spaghetti, seldom contains eggs in any form. However, some manufacturers have found that an enriched macaroni product results when eggs are added to the dough (*LeClerc, 1933*).

In noodles, egg yolks alone are usually used, in either the fresh, frozen, or dried state; the whole egg is also used, to some extent. The United States food laws require that the moisture content of noodles should not exceed 13 per cent, and that the egg solids content should not be less than 5.5 per cent, by weight. In order to insure this ratio, the following approximate amounts of various forms of eggs are required for each 100 lb. of flour (*Anonymous, 1929b; LeClerc, 1933*):

	WEIGHT (pounds/100 lb.)
Liquid whole egg	20.0
Liquid yolk	17.5
Dried whole egg	5.7
Dried yolk	5.3

To meet legal standards, noodles must contain minimum quantities of certain substances, as follows (*Buchanan, 1924*):

	CONTENT (per cent)
Moisture	8.3
Lipoids	4.4
Lipoid phosphoric acid	0.1
Nitrogen	2.4
Water-soluble nitrogen	0.5
Egg solids (calculated)	5.9

MAYONNAISE AND SALAD DRESSING

The United States is the only country that produces mayonnaise and salad dressings on a large commercial scale. Here, in a representative year (1937), between 15 and 20 million pounds of mayonnaise and salad dressing were made (*LeClerc and Bailey, 1940*). There have been at least 558 different brands on the American market (*Snyder, 1931a*).

Mayonnaise and salad dressing are semisolid emulsions of edible vegetable oil, egg yolk, vinegar, and various seasonings. Ordinarily, dispersions of such mutually insoluble substances as these tend to separate quickly; but the egg yolk acts as a stabilizing agent. The lecithoprotein complex of the yolk maintains the emulsification of the other constituents. Neither lecithin nor lecithin-free yolk is satisfactory alone (*Harris, Epstein, and Cahn, 1941*).

Experiments have been made to determine which of the yolk's constituents is the most effective emulsifying agent in mayonnaise (*Sell, Olsen, and Kremers, 1935*). The results indicate that lecithin may entirely destroy mayonnaise consistency if added in excessive amount, but that the lecithoprotein complex definitely improves it. As the table shows, when mayonnaise is emulsified with pure lecithoprotein and egg oil, the consistency is much greater than when salted yolk is used. (Consistency is expressed as the time required for a 500-gm. plunger to penetrate the samples in a Gardner mobilometer.)

EMULSIFYING AGENT	RELATIVE CONSISTENCY (seconds)
Salted yolk (control sample)	15
Lecithoprotein and egg oil	900+

In mayonnaise and salad dressing, egg albumen has no emulsifying power, and most manufacturers use egg yolk only. Those who make their products with whole eggs consider the albumen as useless extra weight. However, it has been found (*Clickner, 1936*) that the addition of whey powder makes egg albumen as effective an emulsifier as egg yolk. This discovery may possibly lead to an increased demand for whole eggs and egg albumen.

Egg yolk may constitute as much as 11 per cent, by volume, of mayonnaise; in salad dressing, the percentage is always considerably lower (*LeClerc and Bailey, 1940*). Epstein, Reynolds, and Harris (*1937*) analyzed several commercial brands of mayonnaise and found the following approximate average composition:

INGREDIENTS	AMOUNTS (per cent)
Oil	79.0
Total moisture	11.6
Egg yolk	9.0
Acetic acid	0.4

Because of the complex colloidal nature of egg yolk, mayonnaise and salad dressing must be mixed at certain temperatures and be otherwise specially handled. Egg yolk of the highest quality is essential. Eggs laid in the spring and summer seem to produce a more stable, stiffer mayonnaise product than those laid at other times of the year, perhaps because the solids content of spring and summer eggs is higher. Yolks destined for use in mayonnaise and salad dressing must contain at least 43 per cent of solids for the best results (*Snyder, 1931a*).

Frozen salted yolks constitute well over 50 per cent of all eggs used in the manufacture of salad dressings. In addition, some frozen plain yolks are utilized, and also a few fresh refrigerated eggs and dried products. Very few glycerin-preserved yolks find their way into mayonnaise.

In the United States, mayonnaise, like other food products, must conform to government regulations and standards. It must not contain artificial coloring, and its content of vegetable oil and of egg yolk must exceed certain specified minimum percentages.

OTHER USES OF EGGS IN FOOD MATERIALS

In addition to the foods described above, there are many others in which the egg is an important ingredient. As methods for freezing and dehydrating eggs are improved, there will undoubtedly be an even more extensive utilization of processed eggs by the food industries. Probably many new uses for eggs in food products will be developed in the future.

PUDDING AND CUSTARD POWDERS

Packaged mixes are convenient for the quick preparation of puddings and custards. Over 9 million eggs are used yearly for this purpose (*LeClerc and Bailey, 1940*). Almost all of these are in the form of dried egg albumen, which has good foaming and whipping qualities (*Hinks, 1923*).

INFANT FOODS

In the United States, milk, of which there is a plentiful supply, has always been considered the most valuable infant food. In many countries, milk is so scarce that other foods must be found that will supply infants with the same nutritional elements. Experiments in China have resulted in the development of a soybean egg powder which is an adequate substitute for milk as a food for small children. Soybean egg powder, as formulated by Reid (*1934*), contains egg yolk, sucrose, sodium chloride, and calcium lactate, in addition to soybean milk.

CONFECTIONERY

The use of eggs in candies is very old. Eggs are required in the recipes of a cookbook, dated 1733, which contains "all the receipts of the late ingenious Mrs. Earle, Confectioner to their late majesties King William and Queen Anne." The earliest French bonbons, called *pâte de guimauve*, consisted of marshmallow root, sugar, and egg albumen.

It is the albumen that is used in candy. Most confectioners prefer commercial dried albumen, although frozen albumen is used to make

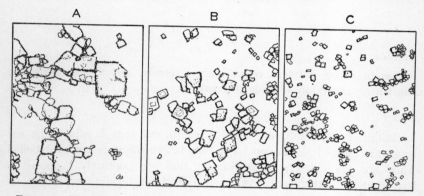

Fig. 413. The effect of egg albumen on the crystallization of sugar from syrups. (After Swanson, 1929.)

A, crystals from fondant candy made of sugar and water only. *B,* crystals from fondant candy made by cooking and beating the syrup and adding egg albumen. *C,* crystals from divinity candy made by whipping hot syrup into stiffly beaten egg albumen until crystallization occurs.

marshmallows. In 1935, the candy industry used 120,000 frozen and 24,000 dried egg whites (*LeClerc and Bailey, 1940*). Many candy manufacturers use fresh eggs, but this entails separating the yolks and finding some way to dispose of them.

Eggs affect the crystallization of sugar in candy and make a much smoother, creamier product (Fig. 413). Probably 90 per cent of all cream centers contain egg albumen. Egg is essential to nougatine, marshmallow whips, and other similar confections (*Snyder, 1931b*).

JAMS

In England, dried or fresh eggs are used in the manufacture of a very popular variety of jam called "fruit curd." To meet accepted standards, commercially prepared fruit curd must contain certain percentages of eggs and other ingredients (*Bagnall and Smith, 1945*). These are shown in the table.

Ingredient	Amount (per cent)
Fat	3.0
Citric acid	0.3
Oil of lemon	0.1
or	
Oil of orange	0.2
Dried egg	2.0
or	
Fresh egg	6.0

SAUSAGE

Certain fancy sausages contain eggs, cereals, and various seasoning and curing agents, in addition to meat.

BALUT

"Balut" is the Philippine word for fertilized eggs in which the embryos are allowed to develop for a time before the eggs are hard boiled. They are considered a great delicacy and are popularly supposed to have a definite therapeutic value as a food for invalids, especially the tuberculous.

In the Philippines, balut is prepared commercially from duck eggs and is sold on the market in two forms—*balut mamatong,* which has been incubated for 14 days, and *balut sa puti,* which has been incubated for 17 or 18 days (*Santos and Pidlaoan, 1931*).

MAPLE SYRUP

Eggs, when used in the production of maple syrup (*Bryan, 1910*), do not enter into the final product but serve merely to clarify the sap. As the syrup boils down in an evaporating vat, a continuous scum rises to the surface and must be cleared away. The addition of egg, especially of egg albumen, aids in bringing the scum to the surface. This use of egg white, however, is not universal among maple syrup producers and, in fact, is now rather rare.

BEVERAGES

In fruit beverages, wines, beers, and liquors, egg albumen is sometimes used as a clarifying agent. For this purpose, a 2 per cent solution of dried albumen is added to the fruit juice or liquor. The whole mixture is then heated to a temperature of 38° C. to 80° C. The albumen coagulates and settles out, carrying with it the finely divided particles, which, if left, would cause cloudiness.

Eggs also find their way into many other beverages in a variety of forms. The beverage-powder industry buys 5 million eggs a year, mostly for the yolks. Beverage powders are widely used in hospitals, school cafeterias, and other institutions, because their nutritive value is high, and they can be prepared quickly.

Fresh eggs are used extensively in drinks, such as milk shakes and wines, to add flavor and nutriment. Egg cognac is a favorite liqueur of many connoisseurs. It is most often prepared at home by the addition of powdered sugar and 3 egg yolks to each 100 gm. of cognac. On a commercial scale, producers add from 250 to 300 gm. of yolk to each liter of cognac. Analyses of a homemade egg cognac and of several commercial brands (*Feder, 1913*) showed the following weights of egg yolk in 100-gm. samples:

	EGG YOLK (grams/100 gm.)
Homemade	25.6
Commercial brands	16.8–23.1

EGG SUBSTITUTES

Eggs are often the most expensive ingredient of foodstuffs, and many commercial food manufacturers have tried to find substitutes that would perform the same tasks at lower cost. However, the colloidal complexities of the egg have defied imitation, and so-called substitutes can make no claim to possessing all the qualities of eggs.

Many of the egg substitutes—or "egg savers," as they are called—which appeared on the market in the early part of the twentieth century were complete frauds. In the United States, their sale has now been banned by the Federal Food and Drug Act; but it is interesting to note the great variety of substances foisted on the unsuspecting public as equivalents of eggs. Most of these products bore no resemblance to eggs, other than a rich yellow color, to simulate that of egg yolk.

The typical egg substitute of this fraudulent type was usually made with a starch base (cornstarch, tapioca starch, rice starch, or potato starch), to which was added some albuminous compound (casein, gelatin, albumin, or gums), and a vegetable or coal tar dye (*McCann, 1918*). These substitutes were harmless, and indeed they had a certain food value. One-fourth of a teaspoon of such a substitute, however, could hardly be considered equal to 2 natural eggs; yet that was the claim made by one manufacturer.

Analyses of these egg substitutes showed them to be very low in fat and protein, which are present in eggs in considerable amounts. In a series of analyses (*LaWall, 1918*), the following percentages of these important food substances were found in egg substitutes. One German

	AMOUNT (per cent)
Fat	0.20–29.0
Protein	0.70–33.0

egg substitute at the time of World War I consisted of nothing more than corn meal, colored yellow (*Markus, 1916*).

In a class with these egg savers were the custard powders which were advertised as being as nourishing as custard made from eggs. Chemical analysis of such a custard powder (*Anonymous, 1893*) is given below, to show how false was this claim.

INGREDIENT	AMOUNT (per cent)
Starch	86.0
Water	12.0
Albuminous compounds	0.6
Soluble coloring matter	1.0
Ash	0.4

There are many legitimate products on the market which are not disguised to resemble eggs, nor are they intended to replace eggs entirely in foodstuffs; yet through their use, bakers, mayonnaise manufacturers, confectioners, and others are enabled to decrease the number of eggs in their products, and thus to reduce costs. To this class of egg substitute belong substances that aid in the emulsification of salad oils (*Barthels, 1939*) and in the production of foam in batters for baked goods (*Bollmann and Rewald, 1929; May, 1939; Bishop, 1940; and others*).

A product called "dried grain-germ egg" has been made in Germany from grain embryos. It is said that 20 gm. of this food is equivalent to 1 hen's egg (*Grandel, 1940*).

DETERMINATION OF EGG CONTENT

As mentioned above, it has been a great temptation for food manufacturers to substitute cheaper materials for eggs in an attempt to cut costs. It has therefore been necessary to establish certain legal standards of minimum egg content for baked goods, ice cream, macaroni and noodles, salad dressings, and other foodstuffs. Tests for determining the egg content of foods must not only indicate the presence or absence of eggs but must also measure the egg content quantitatively.

Since about 1890, food chemists of Europe and the United States have been searching for rapid and accurate methods for determining the egg content of foods. Most modern tests are based upon the principle first applied by Bein (*1890*) and Winchelhaus (*1890*), who estimated the amount of egg yolk by ascertaining the quantity of lecithin-phosphoric acid in an alcohol-ether extract of the food sample. Other investigators varied the technique. For example, Juckenack and Pasternack (*1904*) extracted the lecithin with hot absolute alcohol. Still others have taken advantage of the presence of cholesterol in egg yolk in determining the egg content of a food. Some tests are specific for egg albumen. In discussing the detection of eggs, it is convenient to consider the various foodstuffs separately.

Bakery Products. The presence and the amount of eggs in baked goods can be determined by the alcohol-ether extraction of the lecithin-phosphoric acid from samples of the product. The residue remaining after the evaporation of this extract is dissolved in aqueous potassium hydroxide, evaporated again, and then dissolved in dilute nitric acid, so that the phosphoric acid content can be determined. For every egg present in the cake, 0.13 gm. of phosphoric acid can be recovered by this method (*Buogo, 1934*). Grossfeld (*1940*) describes a method of extracting the egg fats from baked goods and comparing them with the known percentages of these fats in eggs, in order to determine the number of eggs present in any given bakery product.

Sometimes it is advisable to test the egg powders used by bakers and other food manufacturers for the presence of cornstarch, which may be added to give bulk to the powders. Samples are centrifuged with carbon tetrachloride and ether until the egg proteins have collected at the top and the starch has settled to the bottom. The starch can then be recovered, dried, and weighed (*Comte, 1929*).

Ice Cream. Egg, or custard, ice creams are required by law to contain certain specified percentages of egg. When ice cream began to increase in popularity, the need for methods of determining its egg content grew in proportion.

Guarnieri (*1929*), in testing ice cream for its egg content, made use of the alcohol-ether extraction of lecithin-phosphoric acid. His test depends on the calculation of the iodine numbers of the extracts of the samples. The greater the percentage of egg in the ice cream, the higher is the iodine number, as shown in the table.

Egg Content (per cent)	Iodine Number
5	35.8
10	39.8

Later, N. Smith (*1930*) found that the lipoid phosphorus content of ice cream mixes was correlated as follows with the yolk content:

Yolk Content (per cent)	Lipoid Phosphorus (milligrams/100 gm.)
None	2.50
0.25	4.25
0.50	5.84
1.0	6.53
1.0	9.23
3.0	13.87

Colorimetric tests are also used for determining both the phosphoric acid content (*Dessirier, 1938*) and the cholesterol content (*Lampert, 1930*) of ice cream, and thus, indirectly, the egg content.

Lindner (*1939*) has suggested using the refraction of fats extracted from ice cream to ascertain the egg content of each liter of product. The number of eggs is equal to $0.018(R - 44)F$, where R represents the refractometer reading and F is the fat content of 1 liter of the ice cream. This formula is accurate only when no other fats except those of eggs and milk are present.

To calculate the egg content of ice cream, Horrall (*1935*) devised the following formula:

$$\% \text{ eggs} = \frac{\% \text{ lecithin in mix} - \left(\dfrac{\% \text{ fat in mix} \times \% \text{ lecithin in fat}}{100}\right)}{\% \text{ lecithin in eggs}} \times 100.$$

Macaroni and Noodles. Tests for the egg content of macaroni and noodles have been made in Europe for many years and have become increasingly important in the United States as larger quantities of these products are manufactured.

The quantitative determination of the amount of lecithin-phosphoric acid in samples of macaroni and noodles is made by techniques similar to those described for other foods. However, flour and semolina contain appreciable amounts of lecithin, for which it is necessary to correct. Flours vary in their lecithin content, so that it is often difficult to make corrections. Estimates of the amount of lecithin in average flours vary from 0.0228 per cent (*Fiehe, 1931*) to 0.055 per cent (*Haenni, 1941*).

These percentages are great enough to affect the accuracy of the test for egg yolk.

In an attempt to eliminate the error due to the presence of lecithin in flour, Popp (*1908*) and Cappenberg (*1909*) devised techniques based on the determination of the cholesterol content of the yolk. These tests were not widely used, however, until new techniques were suggested by Settimj (*1929*), Tillmans, Riffart, and Kühn (*1930*), Soldi and Testori (*1931*), and Haenni (*1941*). The cholesterol is extracted with ether or xylene and saponified. The addition of acid gives a characteristic color, according to the amount of cholesterol present. The average cholesterol content of the yolk of an egg is 0.25 to 0.26 gm., as determined colorimetrically. The cholesterol content of each kilogram of food, divided by this value, therefore gives the number of eggs per kilogram of food (*Schmidt-Hebbel and Bollmann, 1941; Koehn and Collatz, 1944*).

It has also become necessary to distinguish between egg lecithin and soybean lecithin, which is finding its way into noodle products. It has been suggested that these two forms of lecithin can be differentiated by the fact that soybean lecithin fluoresces to a much greater extent than either egg lecithin or the lecithin of wheat flour (*Winston and Jacobs, 1945*).

Macaroni and noodles often contain whole eggs, instead of egg yolk alone. Many methods for ascertaining the egg content of these products depend upon this fact and are specific for egg albumen. When using these methods, no correction factor is needed, because flour and semolina do not contain substances that can be confused with the substances in egg white. Nitric acid (*Schmid, 1912*) or acetic acid (*Martin, 1921*) may be used to precipitate the albumen from aqueous solutions of the samples to be tested. In the method employed by Vautier (*1922*), the precipitate (obtained by treatment with magnesium sulfate and heat) is burned; the weight lost during combustion is equivalent to the amount of albumen present. Farcy (*1914*) used heat as a precipitating agent and calculated the number of eggs from the increase in nitrogen in the precipitate over the amount to be expected if flour alone were present. Gothe (*1915*) proposed the use of specific sera to distinguish albumen from yolk, in food products. Schaffer and Gury (*1916*) added sodium hydroxide and copper sulfate to the filtrate obtained from test samples and noted the time required for the resulting blue color to disappear. The presence of yolks or whole eggs is indicated accordingly, as shown in the table. A method devised by Hertwig (*1923*) differentiates between

Egg Content	Time for Clearing (minutes)
No eggs	21–23
3 yolks per kilogram	31–36
3 whole eggs per kilogram	49–52

products made with egg yolk, and those made with whole eggs, according to the proportional amounts of fat, lecithin, phosphoric acid, and water-soluble nitrogen present. When the sample contains whole egg, the amount of water-soluble protein nitrogen is relatively high.

Mayonnaise. Cahn and Epstein (*1943*) worked out the following formula for determining the amount of yolk, of any solids content, in mayonnaise:

$$\text{Per cent yolk}_{(g)} = \frac{9899P}{g - 12.2} - \frac{69N}{g - 12.2},$$

where g represents the solids content of the yolk, P the per cent of phosphorus pentoxide and N the per cent of nitrogen in the sample.

Eggs as Adulterants. Chemical tests are useful not only in determining the egg content of foods, but also in detecting the presence of the egg as an adulterant. Coffee beans, for example, are sometimes glazed with a coating of egg albumen, which, in this case, is an adulterant. It is possible to recover a considerable amount of nitrogen from water in which glazed beans have been washed, whereas less than 0.01 gm. of nitrogen per 100 gm. of the sample is found if the beans are unglazed.

ANIMAL FEEDS

Eggs have generally been considered too expensive to be fed to animals. Some dried eggs are incorporated into feeds for birds, turtles, goldfish, and other pets, because pet owners are willing to pay fairly high prices. About 400,000 eggs are used for this purpose each year. Feeding eggs to large animals, however, is impractical, although experiments have shown that eggs are as nutritionally valuable for animals as for human beings. Rarkin (*1903*) reported that a 132-lb. calf gained weight at the average rate of 3.4 lb. per day, when fed for 24 days on skim milk and eggs, in the ratio of 6 eggs to 14 liters of milk. This rate of gain is slightly better than the normal for young calves.

Incubator discards—infertile eggs, and eggs with dead embryos—have also come into use as animal feed. These eggs, as well as others unfit for human consumption, can be eaten by some animals if decomposition has not progressed too far. Eggs with certain rots, and the refuse from egg-breaking plants and hatcheries, are often sold locally as hog feed. They are valuable for this purpose because of their high protein content (*Anonymous, 1941*) and are of satisfactory digestibility (*William, McCay, Salmon, and Krider, 1942*). In addition, incubator discards make good feed for fur-bearing animals, such as fox and mink. Excellent pelts are developed by animals fed hatchery refuse.

Experiments indicate that the cooked and dried contents of incubated, fertile eggs are an excellent source of protein for growing chicks, when substituted for the proteins usually fed in a starting and growing diet (*Hammond, Fritz, Nestler, and Titus, 1944; Kennard and Chamberlin, 1944*).

Since eggshells contain approximately 93 per cent calcium carbonate, they are as good a source of this mineral for chicks and laying hens as the oyster shells and limestone generally used for this purpose. Before eggshells can be fed to chickens, however, they must be processed to remove as much of the adhering whites as possible. They must also be sterilized, since several diseases common to poultry are known to be carried by eggs (see Chapter 8, "Biological Properties"). Finally, the shells are crushed. Fifteen pounds of crushed eggshell in every 500 lb. of mash supplies hens with adequate amounts of calcium carbonate (*Wilcke, 1940*). The only objection to the use of eggshells in this way is the fact that they are usually more expensive than oyster shells or limestone.

MEDICINE AND MEDICAL RESEARCH

The use of eggs for research in human and veterinary medicine, and in the preparation of serums, vaccines, and pharmaceutical products, is of tremendous importance, although the actual number of eggs involved may seem small in comparison to the numbers used by industries.

Therapeutic Uses and Pharmaceutical Preparations

Eggs have a definite place in preventive medicine and are of therapeutic value in the treatment of many dietary deficiency diseases (see Chapter 9, "Food Value"). Eggs, however, have other medical uses.

Oil of egg is used by druggists in the preparation of certain ointments and emulsions. The oil is obtained from yolks either by pressure, or by the use of solvents. When the pressure method is to be employed, the yolks of fresh eggs are separated from the albumen and heated until most of the water has evaporated. The dried mass is then placed in bags and the oil is pressed out between hot plates; finally, the oil is filtered. Paladino and Toso (*1896*) were able to obtain 25 to 35 per cent of the total egg oil in this way.

The use of fat solvents for extracting the oil from hard-boiled eggs is more effective than the pressure method. The yield of egg oil depends upon the solvent employed. Jean (*1903*) experimented with various solvents on aliquots of a single sample of yolk; the approximate percentages of the total egg oil obtained with each solvent is shown below.

Solvent	Yield (per cent)
Petroleum ether	48
Carbon bisulfide	50
Carbon tetrachloride	50
Ether	51
Chloroform	58

The emulsifying action of yolk and the clarifying properties of albumen are sometimes useful to pharmacists in the preparation of medicinal prescriptions. Egg yolk has also been used as an emulsifier of turpentine and water for enemas (*Simmonds, 1889*).

Egg albumen is sometimes injected with cocaine and cocaine substitutes, as it has the effect of prolonging anesthesia, especially of the cornea (*Stephany and Matschulan, 1937; Matschulan and Amsler, 1938*). It apparently increases the absorption of cocaine (*Rosenkranz, 1938*).

The egg, especially the albumen, has been used as a remedy for burns. It is also a well-known antidote for certain poisons, especially arsenic compounds and salts of heavy metals. As an antidote, egg albumen prevents absorption of the poison by coating the mucous membrane of the stomach and inactivates the poison by combining with it chemically (*Muspratt, 1852*). Yolks are sometimes used by the peasants of Europe as an ointment for cuts, scratches, and bruises. Raw egg, added to various drinks, has been said to relieve soreness and to clear congestion of the throat (*Simmonds, 1889*). Lampe (*1933*) reports a modern use of the old folk custom of healing open wounds and ulcers by covering them with the membranes from eggs boiled 5 minutes.

Egg extract (lecithin and lutein) has been put up in ampoules for hypodermic injection (*Vita and Bracaloni, 1936*).

Culture Media

Bacteriological laboratories use large quantities of eggs each year in the preparation of culture media (see Chapter 8, "Biological Properties"). Because of its high nutritive value, the egg is a good enriching agent. Egg yolk is an excellent source of lecithin, protein, and phosphoric acid

(*Besredka and Jupille, 1914*), and egg white is able to supply all the nitrogen required by some bacteria (*Pozerski, 1936*). Eggs are also used in preferential media for isolating particular organisms (*Capaldi, 1896*). The purification of sewage by the activated sludge method is indicated by the reduced ability of the remaining bacteria to utilize the coagulated protein of egg albumen (*Courmont and Rochaix, 1920*).

As an ingredient for culture media the egg has the advantage of being easy to obtain. Media containing eggs are quickly prepared and therefore are valuable for comparative studies. At first, the sterilization and preservation of such culture media was difficult, due to factors introduced by the colloidal properties of eggs, but these problems have been largely solved.

Serology

In Wassermann tests, the tubes containing the highest dilutions of serum may give falsely positive reactions; so, also, may all dilutions of spinal fluid, which contains very little protein. Egg albumen very successfully compensates for this protein deficiency. The test becomes slightly less sensitive after the addition of egg white, but this disadvantage is offset by the greater reliability of the positive reactions (*Boerner and Lukens, 1941*).

Artificial Insemination

A limited use for eggs has been found in the artificial insemination of livestock, particularly cattle. By means of artificial insemination, it is possible for a superior bull to sire numerous calves and thus hasten the improvement of the herd. In order to impregnate a large number of cows, bull semen must be diluted; but the diluent must in no way injure the spermatozoa. Diluents containing egg yolk are the most satisfactory, since they permit the spermatozoa to remain motile for a week or more (in unpreserved semen, the spermatozoa lose their motility within 24 hours). Fresh yolks are carefully separated from the albumen, which is mildly spermicidal, and are added to a salt solution. Originally, the yolk was mixed with an equal quantity of a solution of 0.2 per cent potassium phosphate and 2 per cent dibasic sodium phosphate, and four or five parts of this mixture were then added to one part of semen (*Phillips and Lardy, 1940*). Figure 414 shows how long the spermatozoa of several species of animal retain their motility in this yolk-phosphate buffer. The optimum pH of the solution is about 6.75.

A yolk-citrate diluent gives more satisfactory results than the yolk-phosphate diluent (*Salisbury, Fuller, and Willett, 1941*). It preserves the fertility of the semen over a longer storage period. The citrate also disperses the fat globules of the yolk, so that it is possible to see the spermatozoa clearly under the microscope when testing for semen quality; this is impossible when phosphate salts are used. Further research has

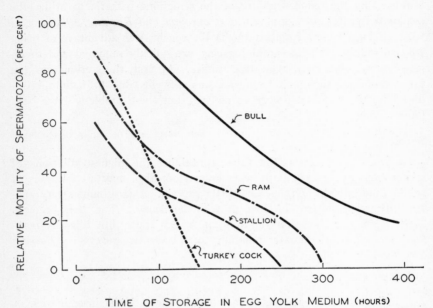

FIG. 414. Duration of motility of spermatozoa in yolk-phosphate medium. (After Lardy and Phillips, 1939.)

shown that the addition of glucose to the yolk-citrate diluent increases the viability of bovine spermatozoa (*Salisbury and Vandemark, 1945*).

Use of the Developing Chick Embryo for Medical Research

The developing chick embryo has long been used in the academic study of embryology, histology, genetics, and pathology. It is also valuable experimental material for the study of human and animal diseases.

STUDY OF PATHOLOGICAL ORGANISMS AND TUMORS

The developing chick embryo is now used for the culture of various viruses, bacteria, and protozoa which have never been grown successfully on artificial media, and which formerly required the living tissues of the animals they commonly infect. The expense of supplying animals for laboratory research on these organisms was sometimes prohibitive. After it was discovered that the relatively inexpensive chick embryo is satisfactory as a research animal, the identification, classification, and detailed study of many pathogenic organisms were greatly simplified.

It has been known for many years that malignant tumor cells may be grown in the developing egg (*Murphy, 1912*). This knowledge has now been utilized to advantage in cancer research (*Taylor, Thacker, and Pennington, 1942*). Not only is the incubating egg used for tissue culture; so, too, are media composed of reconstituted, frozen-dried embryo juice, or ground embryos (*Hetherington and Craig, 1939, 1940*).

DIAGNOSIS

The incubated egg is of value in the diagnosis of such diseases as spinal meningitis (*Blattner, Heys, and Hartmann, 1943*). Evidence of meningitis may appear in chick embryos inoculated with suspected spinal fluid in which no organisms can be found. Chick embryos are also useful in differential diagnosis, since various diseases which cannot be distinguished by their symptoms often produce characteristic gross and microscopic changes in embryonic chick tissues (*Brandly, 1939*).

VACCINES

Fertile eggs have become very important in the production of immunizing sera and vaccines against human and animal diseases. In 1939, when encephalomyelitis ("sleeping sickness") became epidemic among horses in the United States, laboratories producing the specific vaccine purchased, for culture purposes, 1,750,000 fertile eggs incubated for 10 to 12 days. Ordinarily, the production of this particular vaccine requires fewer eggs, annually, than this number. Laboratories also supply chick vaccines for other diseases, such as laryngotracheitis, lymphogranuloma venereum, measles, and various pox diseases, including smallpox. New chick vaccines are being developed constantly.

Chick vaccines have a number of advantages over ordinary vaccines. They can be produced in larger quantities at lower cost. Many chick vaccines cause less pronounced reactions and are more effective as immunizers. The superior effectiveness of the chick vaccine against equine encephalomyelitis, for example, is shown by the fact that it leaves no more than 0.5 per cent of vaccinated horses unprotected, whereas the disease is contracted by 1.5 per cent of horses inoculated with vaccine produced in equine brain tissue (*Clarke, 1940*).

In addition, the production of vaccines in the chick embryo is simplified because the chick embryo is less likely than a mature animal to have a prior infection at the time it is inoculated.

TESTING GERMICIDES

Incubated eggs have proved useful in testing the efficacy of germicides. In treating wounds, the toxicity of the antiseptic agent for tissues, as well as for pathogenic organisms, must be considered. The effectiveness of a germicide against microorganisms grown *in vitro* can be determined in the laboratory, but its practical value can be revealed only when it is applied to animal infections. The chick embryo is excellent test material, because its tissues are more or less susceptible to the same toxins as human tissues. A good indication of the safety and effectiveness of a germicide may be obtained by infecting incubated eggs with *Staphylococcus aureus,* injecting 0.2 cc. of germicide, and examining for bacterial growth and general condition of the embryo 6 days later (*Green and Birkeland, 1942*). This procedure is reliable, inexpensive, and easily performed.

Dunham and MacNeal (*1942, 1943*) developed a somewhat similar method of testing the effectiveness of antiseptics against viruses. The antiseptic is mixed with vaccinia virus, and the mixture is injected into the developing egg. Vaccinia was chosen as a test virus because more is known of it than of any other virus.

TESTING MILK

Another use for incubated eggs is in testing milk for the presence of such pathogenic bacteria as *Brucella abortus,* which causes undulant fever. This organism is very difficult to isolate, but the incubated egg has been found to be a good growth medium for it (*Metzger, Beaudette, and Stokes, 1939*). *Brucella* may be grown in the chick embryo at less cost, and more rapidly, than in other media. It can also be obtained in pure culture.

MANUFACTURING

Eggs possess certain physical and chemical properties that make them useful as basic ingredients of many manufactured goods. Much experimentation has been done to discover new industrial uses for the entire egg and for its component parts. The versatile egg has been utilized in such varied processes as the production of photographic plates and the spinning of synthetic fibers. In some instances, eggs have been employed extensively for specialized purposes, only to be replaced later by other materials. However, new uses have been suggested and tried, and others are continually being developed (Fig. 415).

In the United States, about 6.5 per cent of the annual egg supply is classified as inedible (see frontispiece of this chapter). In this category are spoiled eggs from cold storage plants, and infertile eggs, or eggs with dead embryos, from hatcheries. Only slightly more than 1 per cent of all inedible eggs are reclaimed by industry; the remainder, unfortunately,

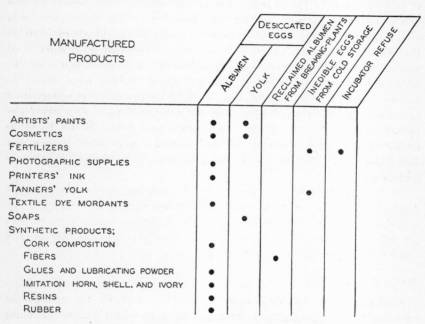

FIG. 415. Use of various forms of eggs in manufacturing.

is discarded. In the future, new ways of using inedible eggs may possibly eliminate this waste.

The Leather Industry

Many inedible eggs from various sources are utilized in leather tanning, in the form of "tanners' yolk," which, in spite of its name, usually consists of both yolk and albumen. It is estimated that between 5,500,000 and 6,500,000 lb. of tanners' yolk are produced annually in the United States (*Anonymous, 1941*). Tanners' yolk is essential in fat-liquoring, a process designed to incorporate a certain amount of oil into the dried hides to give them added softness, elasticity, and strength, and to enhance the grain and flesh surfaces.

There are several types of tanners' yolk, and the composition of each is variable within certain fairly wide limits. Liquid tanners' yolk consists of thoroughly mixed whole eggs, to which a suitable preservative has been added. In salted tanners' yolk, the preservative is common salt, in proportions varying from one-sixth to one-fourth the weight of the eggs; the usual ratio is 20 lb. of salt to 100 lb. of egg. Since salt-free eggs with low ash content are desirable for certain types of leather, other preservatives are also used, such as borax, boric acid, sodium benzoate, and sodium fluoride. In order to insure against the resale of tanners' yolk for human consumption, it is sometimes denatured by the addition of substances, such as birch tar oil, which render it inedible but do not alter its usefulness in the leather industry (*Rogers, 1916*).

Tanners have made considerable effort to standardize the composition of tanners' yolk, in order to guarantee an oil content of at least 9 per cent, and a salt content of not more than 18 per cent. When preservatives other than salt are used, the oil content is higher. The composition of a typical commercial brand (*Wilson, 1927*) of tanners' yolk is as follows:

	AMOUNT (per cent)
Water	63
Salt	17
Oil	10
Other egg constituents	10

The emulsification of the fat liquor is necessary, in order to obtain the desired absorption and uniform distribution of oil throughout the layers of the hide. Tanners' yolk acts as a stabilizer in the fat-liquor emulsion. If the liquor is unstable, the oil precipitates on the surface of the leather, giving a smeary, dirty appearance. The addition of egg prevents this precipitation, and produces a drier, less greasy surface. Tanners' yolk also alters the distribution of oil between the grain and flesh sides of the hide, so that the flesh side takes up more oil than usual, and a better surface is produced for coloring and finishing (*Merrill, 1928*). This is especially important with suède and white leather.

The protein of the egg is now thought to be the most important constituent of tanners' yolk, rather than lecithin and other phosphatides (*McLaughlin and Theis, 1945*). This conclusion is supported by the fact that albumen and whole egg very markedly increase the stability of the emulsion, whereas the effect of egg yolk alone is almost imperceptible (Fig. 416). Whole egg is improved as a stabilizer if it is allowed to stand for several weeks and thicken.

The *p*H of tanners' yolk is important. The absorption of oil by the leather is increased when the *p*H of the fat liquor is between 4.5 and 6.75, but it decreases if the *p*H is higher (Fig. 417).

The addition of tanners' yolk to fat liquor also improves the firmness of the leather. Theis and Hunt (*1932*) found that the addition of 25

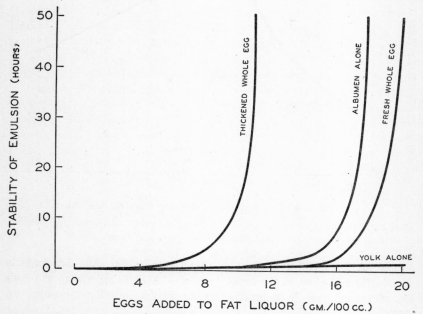

FIG. 416. The stability of fat-liquor emulsion, as influenced by the addition of various quantities of yolk, albumen, and whole egg (fresh and thickened). (After Wilson, 1927.)

per cent of tanners' yolk increased the average tensile strength of leather from 200 lb. per square inch to 275 lb. and increased the average tear strength from 80 lb. per square inch to 100 lb. (Fig. 418). When more than 25 per cent of egg is added, however, the strength of the leather decreases.

Eggs are also used in the finishing of certain types of leather, particularly glazed colored stock. The albumen alone is utilized. It acts as a fixative for dyestuffs. Heating and drying coagulate the albumen to some extent, so that it forms an insoluble film on the surface of the leather. Albumen has additional uses in seasoning goldbeaters' skin, drum heads, and banjo heads (*Anonymous, 1941*).

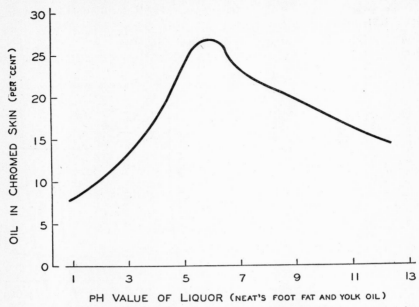

F<small>IG</small>. 417. Effect of *p*H of fat-liquor (consisting of neat's foot fat and egg yolk oil) on total oil taken up by chromed skin. (After Theis and Hunt, 1931.)

Furs. A few inedible eggs are also used by furriers. Albumen imparts extra luster and gloss to the surface of furs. Yolk is applied to the flesh side of tanned furs in a process similar to fat-liquoring (*Anonymous, 1941*).

Artists' Materials

Egg albumen and yolk oil both possess qualities that make them useful as painting media when they are mixed with resins and pigments. The oil of the yolk is a diluting agent for the pigments, and the proteins of the albumen coagulate upon exposure to light and air. Yolk oil imparts certain desirable and pleasing effects to the surface of paintings. Egg media may be applied to paper, cardboard, wood, canvas, and even plaster walls.

All the early literature on art contains references to the use of the whole egg or its component parts in artists' materials. Each artist had his favorite formula for the mixing of egg paints and guarded it as a valuable trade secret. Many famous works of art (including Ghirlandaio's *Old Man with a Child,* which hangs in the Louvre) were done with egg oil paints. It is possible that Hendrik Bloemaert used them in 1632

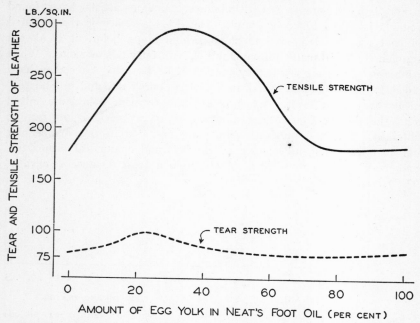

Fig. 418. Effect of egg-yolk content of neat's foot oil on the tear and tensile strengths of leather. (After Theis and Hunt, 1932.)

when he painted his picture of a woman with a basket of eggs, reproduced in part in Fig. 419. Since that time, egg oil has been largely superseded by linseed, poppy seed, and other vegetable oils, which have better drying qualities; but a few traditional formulas have come down to modern painters.

The principal painting medium requiring eggs is tempera. This is usually applied to a rigid panel previously prepared with a coat of absorbent gesso. Egg tempera was used by European artists of all periods and was known to the early Egyptians. In the Middle Ages, it was widely employed in the illumination of manuscripts.

Tempera paints may be prepared from either albumen or yolk. The most widely known formula calls for the separation of the yolk and albumen of a very fresh egg. The vitelline membrane is then punctured, and the yolk material is caught and mixed with an equal quantity of pigment paste prepared from pigment ground in water. Egg tempera must be very fresh in order to adhere properly, so only small quantities can be mixed at one time.

For tempera prepared with albumen, modern artists have found commercial dried albumen useful. The dried albumen is reconstituted in cold water and allowed to stand several hours before it is filtered through cloth and made into an emulsion with fatty oils or resins. When fresh egg whites are used, they must first be beaten. After they have stood for a few hours, the clear liquid is poured off, and ground pigments are added to it. When commercial tempera paints are mixed with the thinner part of egg albumen, especially vivid colors can be obtained. Unusually brilliant paints are prepared in this manner for use on advertising and educational displays and anatomical demonstration models, which must be seen well from a great distance (*Congdon, 1932*).

Fɪɢ. 419. Pen-and-ink drawing from Hendrik Bloemaert's painting, which was possibly made with egg yolk medium.

Pictures made with egg paints may be either as transparent as water colors or as opaque as oils. In fact, the addition of oil to egg paints tends to add certain characteristics of oil paints, without the loss of any of the water-color properties. The use of oil with egg paints is very ancient, but none of the traditional recipes is now available. Modern artists have devised a number of satisfactory formulas for mixing egg with oil to form the proper emulsion. Most of these call for either yolk or whole egg, and water, stand oil, and damar varnish. Egg and oil in approximately equal volumes are mixed slowly, and the volume of the mixture is then doubled by the addition of water.

The egg is important in other art media besides tempera. Egg albumen, with gum or hydromel, has been used since early times to obtain brilliant, glazed colors. Egg albumen size has also been employed by gilders to attach gold leaf to picture frames, leather, etc.

Egg yolk is incorporated in grounds applied to textiles, woods, and other painting surfaces, to make them tighter, less absorbent, and more luminous. The egg dries hard and becomes almost waterproof.

Secco painting with egg yolk on finished, dried lime plaster walls was formerly practiced. The egg yolk and lime combine solidly and permanently, unless the mural is subjected to undue weathering. Egg yolk, well diluted with water, is an excellent sizing for walls that are to be decorated with egg paint murals.

Artists have also used the eggshell as a painting surface. In past centuries, wealthy young men of Venice paid fabulous prices for their miniature portraits delicately painted on eggshells. Drinking cups have been made of ostrich and emu eggshells, beautifully mounted in silver (*Simmonds, 1889*).

Dyes

The dyeing industry uses small quantities of eggs as mordants, or fixing agents. A mordant must attach itself firmly to the fiber to be dyed or printed and must also take up the dye well. The egg can perform both of these essential functions (*Eitner, 1909*).

Albumen colors are used for textiles because they are bright, and also because they are held to the fiber mechanically, by virtue of the coagulated albumen in them (*Anonymous, 1941*).

Photography

At one time, eggs were extensively used by photographers, before they were replaced by other substances that do the same work more efficiently.

Albumen was used to sensitize glass negatives as early as 1848. A Frenchman, Niépce de Saint-Victor, published the first account of preparing glass negatives by coating them thinly with egg albumen. Later, photographers improved upon his method by adding a coat of collodion over the albumen.

In 1853, a Boston photographer first used egg albumen to sensitize paper for photographic prints, which he called "crystallotypes." These prints were notable for their glossy surface. Unfortunately, photographic paper prepared with albumen soon loses its sensitivity, so that it is impractical to prepare much at one time. However, by 1863, an American paper company was importing 15,000 reams of paper stock annually from Germany and France for the manufacture of albumen paper. The paper was sized with common salt and albumen and sold dry to photographers. To sensitize the paper for use, it was necessary to soak it, albumen side down, in a silver nitrate solution.

By 1890, gelatin-bromide paper for photographic prints was becoming popular. The advantages of the new product spelled the doom of albumen paper. Since 1895, scarcely any photographers have used albumen paper.

PRINTING

Egg albumen is still employed as a vehicle for ammonium or potassium dichromate in sensitizing zinc or aluminum plates for lithographic and offset printing, such as is used for the printing of maps. Fresh albumen is superior to the dried product for this purpose, and it has been found that plates sensitized with albumen at a relative humidity of 50 to 55 per cent give the clearest image (*Mayer, 1932*).

Cosmetics

Most women are well acquainted with the role eggs play in cosmetics. Egg shampoos and facials are offered by many beauty shops, and skin creams and packaged shampoos containing eggs are available.

In the past, it was claimed that egg yolk was valuable as a tissue nutrient for the skin. Egg oil was used in Russia during the nineteenth century in the manufacture of Kazan soap, noted for its excellence as a skin treatment. It was too expensive for the general public, but many wealthy women used it (*Simmonds, 1889*). Egg yolk was also an ingredient of certain soaps made in Germany. When the egg is incorporated into soaps, skin creams, lotions, or shampoos, preservatives must be added, generally glycols and sulfonated castor or corn oils (*Beckert, Kritchevsky, and Harris, 1933*).

Fresh eggs are used both for egg mask facials and for shampoos. Largely because of their expense and the inconvenience involved, these treatments have never gained great popularity in commercial beauty parlors and are chiefly used at home. Egg masks supposedly aid in tightening the skin and smoothing out wrinkles. It is claimed that egg shampoos impart a smooth glossiness to the hair (*Niles, 1938*), and that they are superior to soap shampoos for hair that has been subjected to the alkalis used in permanent waving. Since egg yolk is a natural emulsifying agent, it is probably of value in removing excess oil and dirt (*Snyder, 1933b*).

Fertile eggs incubated for a short time have been used in the preparation of a skin nutrient (*Humblet, Teyssier, and deValkener, 1939*).

Synthetic Products

The production of synthetic materials of all kinds is one of the most important modern industries. Eggs are used in the manufacture of several such materials.

RUBBER

Egg albumen has been employed in the vulcanization of rubber (*Dippel, 1898*). Low grades of rubber, in particular, are improved when egg albumen paste, prepared with calcium or magnesium oxide, is introduced and rendered insoluble by treatment with certain agents (*Esch, 1912*).

In addition to its use with natural rubber, egg aids in the emulsification of the constituents of synthetic methyl rubber (*I. G. Farbenindustrie, 1931*).

Diesser (*1912*) made a product in which egg albumen was combined with certain chemicals to form a satisfactory substitute for elastic.

RESINS

Egg albumen is used in synthetic phenol-formaldehyde resins as a modifier, to reduce brittleness (*Tarasov, 1916*). The phosphatides of egg yolk are used for their oils. Such articles as phonograph records are made from plastics containing eggs (*Frank, 1934*).

CORK COMPOSITION

Egg albumen is mixed with granular cork and dispersed rubber, packed into a mold while wet, and subjected to heat to coagulate the albumen and coalesce the rubber. This renders the product nonreactive and waterproof (*Levin, 1939*).

IMITATION HORN, SHELL, AND IVORY

Dried albumen, and also pulverized eggshells, are used in the manufacture of imitation horn, shell, and ivory. Albumen powder is mixed with water and poured into molds, to which increasing pressure is applied for several minutes. The pressed albumen is then chemically treated and dried for several days. Any deformities in its surface are smoothed out. Tortoise-shell designs can be applied to this substance by adding tea powder or Japanese ink to the albumen before it is processed. Other colors and effects are also possible (*Sikata and Huzii, 1940*).

GLUES AND LUBRICATING POWDERS

The adhesive qualities of egg albumen were known to the ancient Chinese, who mended pottery with powdered glass combined with albumen. In modern times, albumen glue has been used for cementing cork pads into bottle caps. In biological laboratories, microtome sections of tissues are fixed to glass slides with a mixture of equal parts of egg albumen and glycerin, to which a suitable preservative has been added (*Wilson, 1923, p. 18*).

A graphite lubricating powder is made of finely powdered graphite, mixed with egg albumen and coagulated by heat. However, this preparation is no better than graphite and tallow mixtures, which are cheaper.

ARTIFICIAL FIBER

Theoretically, at least, all known proteins are capable of being drawn into fibers. Many fibrous products are manufactured from proteins that are the surplus or waste materials of agriculture and industry. Such a waste protein is ovalbumin from the albumen of inedible eggs. Egg white may be reclaimed from broken or inedible eggs by a patented process (*Tranin, 1938*). The albumen is separated from shell material by centrifuging. It is next subjected to the filtering action of a solution of gelatin and chlorine to purify and sterilize the albumen. The albumen is then acidified with citric or tartaric acid, in order to effect the separation of foreign matter, which is skimmed off.

The process whereby ovalbumin is elongated into threads was developed by Lundgren (*1941*). Briefly, the technique is as follows. A 3 per cent solution of twice-crystallized ovalbumin is combined with an equal amount of a 3 per cent solution of sodium alkyl benzene-sulfonate at a pH of 6.5. This complex of protein and detergent is then precipitated with a saturated solution of magnesium sulfate, and the resulting precipitate is pulled into fiber. The detergent must be extracted from the fiber. This is accomplished by soaking the fibers in a 60 per cent solution of acetone. The pure protein product thus obtained can be elongated from 300 to 700 per cent in an atmosphere of live steam. The thread is then annealed at lower temperatures and allowed to equilibrate for a day at a relative humidity of 65 per cent and a temperature of 21° C. The final product is shown in Fig. 420.

In molecular orientation, tensile strength, and moisture resistance, ovalbumin fibers are equal to natural protein fibers. Tensile strengths of more than 70,000 lb. per square inch have been achieved. There is a direct relationship between the tensile strength and the amount of elonga-

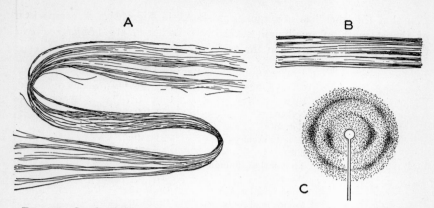

Fɪɢ. 420. Synthetic fibers from the egg protein, ovalbumin. (After Lundgren and O'Connell, 1944.)

A, a strand of fibers prepared from ovalbumin. *B,* a spun thread, containing 100 filaments fused together; magnified about 100 times. *C,* X-ray diffraction pattern of stretched, oriented ovalbumin. (For X-ray diffraction pattern of unoriented ovalbumin, see Fig. 252, in Chapter 7, "Physicochemical Properties.")

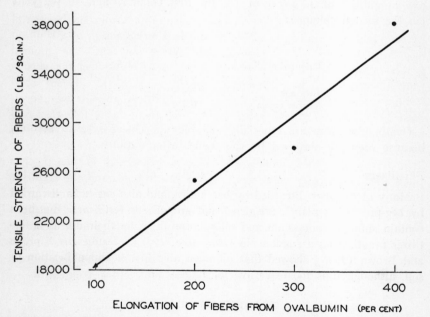

Fɪɢ. 421. Effect of elongation on tensile strength of fibers made from oval-bumin. (After Palmer and Galvin, 1943.)

tion that the fiber has undergone, as shown in Fig. 421. A comparison
of the strength of ovalbumin thread with other fibers (*Lundgren and
O'Connell, 1944*) is given in the table.

FIBER	BREAKING STRENGTH (1000 lb./sq. in.)
Flax	Up to 156
Nylon	72–100
Cotton	40–111
Acetate rayon	23–110
Silk	46– 74
Technical ovalbumin	20– 70
Wool	17– 25
Commercial casein	Up to 10
Commercial soybean	Up to 10

Besides having a tensile strength comparable to other fibers, ovalbumin
fibers retain moisture almost as well as wool, and better than rayon
(*Lundgren and O'Connell, 1944*). The following table indicates the
percentage of water, on a dry basis, held in various threads at the relative humidities of 30 per cent (in the first column) and 70 per cent
(in the second column):

	MOISTURE RETENTION (per cent)	
Oriented ovalbumin	7.3	13.5
Wool	9.0	17.0
Raw silk	7.1	13.3
Viscose rayon	6.7	12.9

Ovalbumin fibers are especially suitable for use as surgical sutures,
because they are absorbed by the tissues after a time.

Fertilizers

Many of the eggs discarded by hatcheries, and also eggshells discarded
by egg-processing plants, are dried and ground for fertilizer. Eggshells
contain sufficient potassium and phosphorus to make their use as a fertilizer practical on a small scale (*Browning, 1917*). Studies by Lipman
and Brown (*1908*) showed that albumen aids in the ammonification of
soil, although it is too expensive for large-scale use.

TRADITIONAL USES

The avian egg has been a familiar object to man for many centuries
of his history, and it has always been a rather mysterious symbol of life
and resurrection. It is natural, therefore, that the egg should play a
significant role in the ceremonies associated with various religious beliefs.

Today, much of the meaning of these ceremonies has been forgotten; but the traditional use of the egg remains, at least in some portions of the world.

Among many groups of people, chickens are kept only for purposes of magic and divination. The Polynesians and Melanesians, for example, never eat chicken meat or eggs, but they believe that eggs, when carefully gathered and broken ceremonially, have the power to augur future events. Eggs have a part in Malaysian entertainments in Borneo, and in other parts of the East Indies.

In the days of the early Greeks, Persians, and Romans, and, before that time, in ancient Egypt, eggs were a symbol of new life. Colored eggs were given away at ceremonies celebrating the beginning of the new year (*Wigent, 1929*), and again at spring festivals held in recognition of the renewal of life. The Chinese also made use of eggs in symbolic spring rites, long before the Christian era.

Popular Arts

Man has apparently always had a desire to beautify his possessions by the addition of color and decorative design. In the past, traditional designs and colors have come to be associated with many objects, including the egg. Novel uses have also been found for the egg in modern popular art.

Eggshell Mosaics. The production of mosaics, by inlaying small pieces of glass, tile, enamel, or a variety of other substances, is an ancient craft that has frequently reached a high degree of artistry. It is also practiced as a hobby. Among the materials that have been used for this purpose is colored eggshell (*Anonymous, 1936b*), which has been found adaptable for making many novel designs (Fig. 422). The shells are broken into pieces from ⅛ to ¼ inch in diameter, dyed, and glued to a base in the desired arrangement.

Easter Eggs. The custom of coloring eggs at Easter time has a long history. In the fourth century, the use of eggs for food during Lent was forbidden by the Church, but the eggs were saved. On Holy Saturday, they were dyed yellow, violet, or red, and were then taken to church to be blessed by the priest. On Easter Sunday they were distributed to friends. In France, during the reigns of Louis XIV and Louis XV, it was the custom for the king to present his courtiers with gilded eggs on Easter (*Simmonds, 1889*). Decorated Easter eggs that had been blessed by a priest were believed, in some countries, to have special power as good luck charms.

In Poland, Germany, and other European countries, certain religious groups decorate Easter eggs with great elaborateness (Fig. 423). In

Moravia, young girls apply complex and original designs to eggs by means of wax and dyes, much as batik work is done, and then present the eggs as love tokens (*Snyder, 1933a*).

Colored Easter eggs are still popular in many countries, although much of the significance of the religious symbolism has been lost. In the United States, dyes and stencils for decorating Easter eggs are made

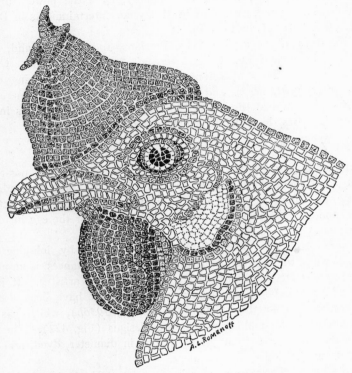

Fig. 422. Suggested motif for a mosaic made of colored pieces of eggshell.

commercially and appear on the market every year, before Easter.

Other Art Forms. In Malaya, eggs are dyed, and then decorated with floral designs by scaling the color away to expose the white of the shell, where desired (*Simmonds, 1889*).

Comical figures of animals and people can be made by attaching whole eggshells to each other, and adding arms, legs, and other appropriate appendages (Fig. 424).

In 1905, on the island of Formosa, as many as 160,000 duck eggs were used for hatching ducklings, which were stuffed and sold as toys and Christmas decorations, chiefly in the United States (*Kikuma, 1917*).

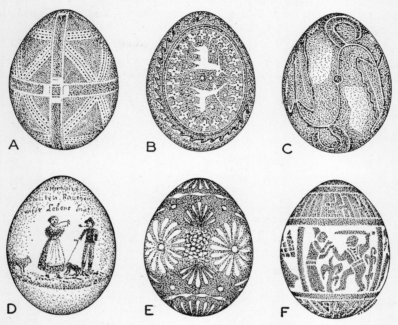

FIG. 423. Decorated Easter eggs from various European countries and Africa.

A and *B*, made in Poland (after Roscoe, 1938). *C*, made in Rumania (from Hamburg Ethnological Museum). *D* and *E*, made in Germany, in Baden and Brandenburg provinces, respectively (from Berlin Ethnological Museum). *F*, made in Sudan, Africa (after Espasa, 1925).

FIG. 424. Fanciful creatures made from empty eggshells. (After Espasa, 1925.)

BIBLIOGRAPHY[1]

ABDERHALDEN, E., and EBSTEIN, E. (1906), Z. physiol. Chem., 48:531–534 (333).
ABDERHALDEN, E., and HUNTER, A. (1906), Z. physiol. Chem., 48:505–512 (333).
ABDERHALDEN, E., and PETTIBONE, C. J. V. (1912), Z. physiol. Chem., 81:458–472 (585).
ABDERHALDEN, E., and PREGL, F. (1905), Z. physiol. Chem., 46:24–30 (333).
ABELIN, I. (1935), Schweiz. med. Wochschr., 65:728 (568).
ABELS, J. C. (1936), J. Am. Chem. Soc., 58:2609–2610 (355).
ABERCROMBIE, R. G. (1931), Naturalist, pp. 105–108 (96).
ABIKO, A. (1935), J. Oriental Med., 23:19 (479).
ABRAHAM, E. P. (1939), Biochem. J., 33:622–630 (502).
ABRAMSON, H. A. (1932), J. Gen. Physiol., 15:575–603 (473).
ADAIR, G. S., and ADAIR, M. E. (1936), Proc. Roy. Soc. (London), 120(A):422–446 (429).
ADAIR, G. S., and MORAN, T. (1935), Dept. Sci. Ind. Research (Brit.), Food Invest. Repts., 1934:29–30 (442).
ADAIR, G. S., and ROBINSON, M. E. (1931), J. Physiol., 72:2P–3P (442, 469).
ADLER, E., and EULER, H. VON (1934), Z. physiol. Chem., 225:41–45 (347).
AGGAZZOTTI, A. (1913), Arch. ital. biol., 59:287–304 (379).
AKAGI, S. (1930), J. Biochem. (Japan), 11:423–425 (445).
ALBANESE, A. A., HOLT, L. E., BRUMBACK, J. E., HAYES, M., KADJI, C., and WANGERIN, D. M. (1941), Proc. Soc. Exptl. Biol. Med., 48:728–730 (591).
ALBANESE, A. A., HOLT, L. E., BRUMBACK, J. E., KADJI, C., FRANKSTON, J. E., and WANGERIN, D. M. (1943), Proc. Soc. Exptl. Biol. Med., 52:18–20 (591).
ALBANESE, A. A., HOLT, L. E., FRANKSTON, J. E., KADJI, C., BRUMBACK, J. E., and WANGERIN, D. M. (1943), Proc. Soc. Exptl. Biol. Med., 52:209–211 (591).
ALBANESE, A. A., RANDALL, R. McI., and HOLT, L. E. (1943), Science, 97:312–313 (591).
ALBERTUS MAGNUS (1250), De animalibus, Hermann Stadler, Munster i.W., Aschendorff, 1916–20, 2 vol., 1664 pp. (287).
ALBRIGHT, W. P., and THOMPSON, R. B. (1935), Poultry Sci., 14:373–375 (128, 648, 669).
ALDERTON, G., and FEVOLD, H. L. (1946), J. Biol. Chem., 164:1–5 (501).
ALDERTON, G., LEWIS, J. C., and FEVOLD, H. L. (1945), Science, 101:151–152 (506).
ALDERTON, G., WARD, W. H., and FEVOLD, H. L. (1945), J. Biol. Chem., 157:43–58 (502).
ALDERTON, G., WARD, W. H., and FEVOLD, H. L. (1946), Arch. Biochem., 11:9–13 (507).
ALDROVANDI, D' (1642), Bononiae, 387 (306).
ALEXANDER, H. L., SHIRLEY, K., and ALLEN, D. (1936), J. Clin. Invest., 15:163–167 (558).
ALMQUIST, H. J. (1933), Univ. California Agr. Expt. Sta. Bull., 561:1–31 (149, 160, 301, 302, 423, 636, 637, 667).
ALMQUIST, H. J. (1934), Poultry Sci., 13:375 (160, 224, 328).
ALMQUIST, H. J. (1936), Poultry Sci., 15:460–461 (219).
ALMQUIST, H. J., and BURMESTER, B. R. (1934), Poultry Sci., 13:116–122 (150, 271, 321, 354).
ALMQUIST, H. J., and GIVENS, J. W. (1935), Poultry Sci., 14:182, 190 (358, 361).
ALMQUIST, H. J., GIVENS, J. W., and KLOSE, A. (1934), Ind. Eng. Chem., 26:847–848 (418, 637).

[1] The italicized figures in parentheses refer to pages of the text.

ALMQUIST, H. J., and GREENBERG, D. M. (1934), *J. Biol. Chem.*, **105**:519–522 (*452, 453, 454*).

ALMQUIST, H. J., and HOLST, W. F. (1931), *Hilgardia*, **6**:61–72 (*166*).

ALMQUIST, H. J., and LORENZ, F. W. (1932a), *U. S. Egg Poultry Mag.*, **38(4)**:20–23 (*682*).

ALMQUIST, H. J., and LORENZ, F. W. (1932b), *U. S. Egg Poultry Mag.*, **38(5)**:48–49, 60 (*136, 143*).

ALMQUIST, H. J., and LORENZ, F. W. (1933a), *Poultry Sci.*, **12**:83–89 (*138, 139, 319, 321*).

ALMQUIST, H. J., and LORENZ, F. W. (1933b), *U. S. Egg Poultry Mag.*, **39(4)**:28–30, 52 (*667*).

ALMQUIST, H. J., and LORENZ, F. W. (1935), *Poultry Sci.*, **14**:340–341 (*682*).

ALMQUIST, H. J., LORENZ, F. W., and BURMESTER, B. R. (1932), *Ind. Eng. Chem., Anal. Ed.*, **4**:305–306 (*414, 415, 416*).

ALMQUIST, H. J., LORENZ, F. W., and BURMESTER, B. R. (1934), *J. Biol. Chem.*, **106**:365–371 (*132, 342*).

ALMQUIST, H. J., NELSON, B. O., and LORENZ, F. W. (1934), *U. S. Egg Poultry Mag.*, **40(12)**:13–16 (*695*).

ALMQUIST, H. J., and STOKSTAD, E. L. R. (1936), *J. Nutrition*, **12**:329–335 (*622*).

ALMY, L. H., MACOMBER, H. I., and HEPBURN, J. S. (1922), *J. Ind. Eng. Chem.*, **14**:525–527 (*712, 713*).

ALTMANN, M., and HUTT, F. B. (1938), *Endocrinology*, **23**:793–799 (*517*).

ALVAREZ, W. C., and HINSHAW, H. C. (1935), *J. Am. Med. Assoc.*, **104**:2053–2058 (*580*).

AMAR, J. (1924), *Compt. rend. acad. sci.*, **178**:803–805 (*401, 407, 408*).

AMBROSIO, A. D' (1933), *Giorn. chim. ind. applicata*, **15**:231–233 (*358, 361*).

AMTHOR, C., and ZINK, J. (1897), *Z. anal. Chem.*, **36**:1–17 (*483*).

ANDERSON, C. F., and PLATT, A. E. (1936), *J. Dept. Agr. S. Australia*, **39**:1342–1350 (*688*).

ANDROSS, M. (1940), *Chemistry & Industry*, **18**:449–454 (*580, 582*).

ANNETTS, M. (1936), *Biochem. J.*, **30**:1807–1814 (*536*).

ANONYMOUS (1893), *Food and Sanitation*, (Nov. 25) **3**:361 (*782*).

ANONYMOUS (1897), *Landw. Zentr.* (*Posen*), **25(34)**:209 (*697*).

ANONYMOUS (1899), *Queensland Agr. J.*, **4(3)**:418–419 (*698*).

ANONYMOUS (1918), *J. Soc. Chem. Ind.* (*London*), **37**:225 (*314*).

ANONYMOUS (1929a), *Can. Nat. Poultry Rec. Assoc., Blue Book*, pp. 67, 69, 71 (*31*).

ANONYMOUS (1929b), *Macaroni J.*, **10(12)**:36 (*776*).

ANONYMOUS (1936a), *Rev. avicole*, p. 109 (*257*).

ANONYMOUS (1936b), *U. S. Egg Poultry Mag.*, **42**:591 (*805*).

ANONYMOUS (1937a), *Poultry Item*, **39(7)**:22–23 (*268, 275*).

ANONYMOUS (1937b), *Poultry Item*, **39(8)**:12 (*263, 268, 275*).

ANONYMOUS (1939), *Harper Adams Utility Poultry J.*, **24**:445–455 (*57, 58*).

ANONYMOUS (1941), *U. S. Dept. Agr. Circ.*, **583**:1–91 (*787, 793, 795, 796, 799*).

ANSON, M. L. (1939), *J. Gen. Physiol.*, **23**:239–246 (*457*).

ANSON, M. L. (1940a), *J. Gen. Physiol.*, **23**:321–331 (*475, 682*).

ANSON, M. L. (1940b), *J. Biol. Chem.*, **135**:797–798 (*476*).

ANSON, M. L. (1941), *J. Gen. Physiol.*, **24**:399–421 (*475*).

AOKI, M. (1925), *J. Biochem.* (*Japan*), **5**:71–76 (*352*).

ARBUCKLE, W. S. (1940), *Missouri Agr. Expt. Sta. Research Bull.*, **320**:1–48 (*774*).

ARCHIBALD, R. M. (1944), *J. Biol. Chem.*, **154**:643–656 (*539*).

ARLOING, F., and LANGERON, L. (1925), *Compt. rend. soc. biol.*, **93**:1305–1307 (*556*).

ARMIT, H. W. (1910), *Z. Immunitätsforsch.*, **6**:703–726 (*551*).

ARNOLD, L. (1929), *J. Hyg.*, **29**:82–116 (*567*).

ARNOLD, V. (1911), *Z. physiol. Chem.*, **70**:300–309 (*475*).

ARNOW, L. E. (1935), *J. Biol. Chem.*, 110:43–59 (*455, 460, 463*).

ARY, J. E., and JORDAN, R. (1945), *Food Research*, 10:476–484 (*771*).

ASCHAM, L. (1935), *J. Nutrition*, 10:337–342 (*605*).

ASHBY, G. K. (1946), *Proc. Soc. Exptl. Biol. Med.*, 61:13–15 (*569*).

ASMUNDSON, V. S. (1931a), *Poultry Sci.*, 10:157–165 (*46, 84, 85, 212, 213*).

ASMUNDSON, V. S. (1931b), *Sci. Agr.*, 11:590–606 (*220, 222, 223, 228*).

ASMUNDSON, V. S. (1931c), *Sci. Agr.*, 11:662–672 (*88, 89, 105*).

ASMUNDSON, V. S. (1931d), *Sci. Agr.*, 11:672–680 (*84, 115*).

ASMUNDSON, V. S. (1931e), *Sci. Agr.*, 11:775–785 (*264, 265, 277, 279, 282, 293, 294, 295, 296, 297*).

ASMUNDSON, V. S. (1933a), *Proc. World's Poultry Congr.* (*Rome*), 5(2):344–348 (*117*).

ASMUNDSON, V. S. (1933b), *Zool. Anz.*, 104:209–217 (*225, 273, 288, 290, 291, 294*).

ASMUNDSON, V. S. (1938), *J. Agr. Research*, 56:387–393 (*18*).

ASMUNDSON, V. S. (1939), *Poultry Sci.*, 18:138–145 (*295, 297*).

ASMUNDSON, V. S., ALMQUIST, H. J., and KLOSE, A. A. (1936), *J. Nutrition*, 12:1–14 (*45*).

ASMUNDSON, V. S., and BAKER, G. A. (1940), *Poultry Sci.*, 19:227–232 (*108, 382*).

ASMUNDSON, V. S., BAKER, G. A., and EMLEN, J. T. (1943), *Auk*, 60:34–44 (*144, 382*).

ASMUNDSON, V. S., and BIELY, J. (1928), *Poultry Sci.*, 7:293–299 (*34*).

ASMUNDSON, V. S., and BIELY, J. (1930), *Sci. Agr.*, 10:497–507 (*34*).

ASMUNDSON, V. S., and BURMESTER, B. R. (1936), *J. Exptl. Zoöl.*, 72:225–246 (*220, 221, 223*).

ASMUNDSON, V. S., and BURMESTER, B. R. (1938), *Poultry Sci.*, 17:126–130 (*223*).

ASMUNDSON, V. S., GUNN, C. A., and KLOSE, A. A. (1937), *Poultry Sci.*, 16:194–206 (*212, 219*).

ASMUNDSON, V. S., and JERVIS, J. G. (1933), *J. Exptl. Zoöl.*, 65:395–420 (*219, 223*).

ASMUNDSON, V. S., and LLOYD, W. E. (1935), *Poultry Sci.*, 14:259–266 (*12*).

ASMUNDSON, V. S., LORENZ, F. W., and MOSES, B. D. (1946), *Poultry Sci.*, 25:346–354 (*51*).

ASMUNDSON, V. S., and PINSKY, P. (1935), *Poultry Sci.*, 14:99–104 (*121, 125, 213, 214, 226*).

ASMUNDSON, V. S., and WOLFE, M. J. (1935), *Proc. Soc. Exptl. Biol. Med.*, 32:1107–1109 (*212*).

ASTBURY, W. T., DICKINSON, S., and BAILEY, K. (1935), *Biochem. J.*, 29:2351–2360 (*479*).

ASTON, T. F., and STEVENS, W. H. (1924), Brit. Pat. 242,780 (*712*).

ATANASOFF, J. V., and WILCKE, H. L. (1937), *J. Agr. Research*, 54:701–709 (*374, 375*).

ATWELL, B. D., and CRAWFORD, G. H. (1867), U. S. Pat. 65,988 (*716*).

ATWOOD, H. (1914), *West Virginia Agr. Expt. Sta. Bull.*, 145:71–102 (*70*).

ATWOOD, H. (1923), *West Virginia Agr. Expt. Sta. Bull.*, 182:1–16 (*70*).

ATWOOD, H. (1926), *West Virginia Agr. Expt. Sta. Bull.*, 201:1–30 (*78*).

ATWOOD, H. (1927), *Poultry Sci.*, 6:108–109 (*79, 80*).

ATWOOD, H. (1929a), *West Virginia Agr. Expt. Sta. Bull.*, 223:1–11 (*25, 231*).

ATWOOD, H. (1929b), *Poultry Sci.*, 8:137–140 (*24, 78, 79, 231*).

ATWOOD, H., and CLARK, T. B. (1929), *Poultry Sci.*, 8:193–197 (*46, 47, 83, 84*).

ATWOOD, H., and CLARK, T. B. (1930a), *Proc. Poultry Sci. Assoc.*, 22:52–54 (*77*).

ATWOOD, H., and CLARK, T. B. (1930b), *West Virginia Agr. Expt. Sta. Bull.*, 233:1–19 (*33*).

ATWOOD, H., and WEAKLEY, C. E. (1917), *West Virginia Agr. Expt. Sta. Bull.*, 166:1–35 (*78, 116*).

ATWOOD, H., and WEAKLEY, C. E. (1924), *West Virginia Agr. Expt. Sta. Bull.*, 185:1–15 (*378*).

AUFRECHT, S., and SIMON, F. (1908), *Deut. med. Wochschr.*, 34:2308–2310 (*579, 580*).

AUSTIN, E. H. (1908), *Farm Poultry*, 19:347 (*7*).

AVENATI-BASSI, B., and BRAVO, G. A. (1925), *Boll. uffic. e tec regia staz. sper. ind. pelli e mat. concianti, Napoli-Torino*, 3:24–37 (*314*).

AVERILL, C. K. (1933), *Condor*, 35:93–97 (*6*).

AXELSSON, J. (1932), *Kgl. Fysiograf. Sällskap. Lund. Handl.*, 43(4):1–196 (*97, 99, 100, 101*).

AXELSSON, J. (1934), *Ann. Agr. Coll. Sweden*, 1:69–207 (*64, 71*).

AYKROYD, W. R., and ROSCOE, M. H. (1929), *Biochem. J.*, 23:483–497 (*626*).

BABIN, F. P. (1938), *Kholodil'naya Prom.*, 16:1–19 (*376*).

BAER, K. E. VON (1845), *Über doppelleibige Missgeburten oder organische Verdoppelungen in Wirbelthieren*, St. Petersburg, 116 pp. (*265, 266, 285*).

BAGNALL, D. J. T., and SMITH, A. (1945), *Analyst*, 70:211 (*780*).

BAILEY, C. H. (1914), *Proc. Soc. Exptl. Biol. Med.*, 12:68–70 (*568*).

BAILEY, L. H., and LECLERC, J. A. (1935), *Cereal Chem.*, 12:175–212 (772).

BAILEY, M. I. (1935), *Ind. Eng. Chem.*, 27:973–976 (*394, 395, 396*).

BAILEY, M. I. (1936), *Ind. Eng. Chem., Anal. Ed.*, 7:385–386 (*415, 416*).

BAILLY, —— (1838), *Mém. soc. roy. sci.*, pp. 226–228 (*307*).

BAIN, W. (1912), *Lancet*, 90(I):918–921 (*570*).

BAINBRIDGE, F. A. (1911), *J. Hyg.*, 11:341–355 (*500*).

BAIRD, J. C., and PRENTICE, J. H. (1930), *Analyst*, 55:20–23 (*680*).

BALDES, E. J. (1934), *Proc. Roy. Soc. (London)*, 114(B):436–440 (*420, 424*).

BALL, C. D., HARDT, C. R., and DUDDLES, W. J. (1943), *J. Biol. Chem.*, 151:163–169 (*476*).

BALL, E. D., and ALDER, B. (1917), *Utah Agr. Expt. Sta. Bull.*, 149:1–71 (*16*).

BALLAND, M. (1881), *Compt. rend. acad. sci.*, 93:550–551 (*354*).

BALLS, A. K., and HOOVER, S. R. (1940), *Ind. Eng. Chem.*, 32:594–596 (*682*).

BALLS, A. K., and SWENSON, T. L. (1934a), *Ind. Eng. Chem.*, 26:570–572 (*518*).

BALLS, A. K., and SWENSON, T. L. (1934b), *J. Biol. Chem.*, 106:409–419 (*518, 585*).

BALLS, A. K., and SWENSON, T. L. (1936a), *U. S. Pat.* 2,054,213 (*743*).

BALLS, A. K., and SWENSON, T. L. (1936b), *Food Research*, 1:319–324 (*743*).

BANCROFT, W. P., and RUTZLER, J. E. (1931), *J. Phys. Chem.*, 35:144–161 (*400, 461*).

BANDEMER, S. L., SCHAIBLE, P. J., and DAVIDSON, J. A. (1946), *Poultry Sci.*, 25:446–450 (*671*).

BARBIERI, A. N. (1907), *Compt. rend. acad. sci.*, 145:133–135 (*338*).

BARFURTH, D. (1896), *Arch. Entwicklungsmech. Organ.*, 2:303–351 (*122*).

BARKER, H. A. (1933a), *J. Biol. Chem.*, 103:1–12 (*467, 471*).

BARKER, H. A. (1933b), *J. Gen. Physiol.*, 17:21–34 (*460, 469*).

BARKER, H. A. (1934), *J. Biol. Chem.*, 104:667–673 (*454, 470*).

BARLOW, E. H. (1902), Brit. Pat. 11,054 (*712*).

BARMORE, M. A. (1934), *Colorado Agr. Expt. Sta. Tech. Bull.*, 9:1–58 (*393, 394, 395, 397, 398, 772*).

BARMORE, M. A. (1936), *Colorado Agr. Expt. Sta. Tech. Bull.*, 15:1–54 (*394, 398, 773*).

BARNES, R. H., MAACK, J. E., KNIGHTS, M. J., and BURR, G. O. (1945), *Cereal Chem.*, 22:273–286 (*592, 593*).

BARNUM, G. L. (1935), *J. Nutrition*, 9:621–635 (*622*).

BARRAL, ——, and TRÉSORIER, —— (1907), French Pat. 375,652 (*699*).

BARROW, A. (1890), *Eggs*, D. Lothrop Co., Boston, 159 pp. (*313*).

BARTELMEZ, G. W. (1912), *J. Morphol.*, 23:269–329 (*122, 123, 198, 207, 214, 215*).

BARTELMEZ, G. W. (1918), *Biol. Bull.*, 35:319–361 (*123, 137*).

BARTELMEZ, G. W., and RIDDLE, O. (1924), *Am. J. Anat.*, 33:57–66 (*122*).

BARTHELS, W. (1939), Belgian Pat. 437,430 (*782*).

BARTHOLIN, T. (1661), *Misc. nat. cur.*, decur. I, ann. I, obs. 36:104 (*290*).

BASKETT, R. G., DRYDEN, W. H., and HALE, R. W. (1937), *J. Ministry Agr. Northern Ireland*, 5:132–142 (*370, 382*).

BASTEI, P. (1926), *Rev. sud-americana endocrinol. immunol. químioterap.*, 9:1281–1286 (*558*).

BATEMAN, J. B. (1932), *J. Exptl. Biol.*, 9:322–331 (*420*).

BATEMAN, J. B., and CHAMBERS, L. A. (1939), *J. Chem. Phys.*, 7:244–250 (*439, 440*).

BATEMAN, W. G. (1916), *J. Biol. Chem.*, 26:263–291 (*585, 587, 588, 589, 594, 595, 596, 597*).

BATEN, W. D., and HENDERSON, E. W. (1941), *Poultry Sci.*, 20:556–564 (*108*).

BATES, R. W., LAHR, E. L., and RIDDLE, O. (1935), *Am. J. Physiol.*, 111:361–368 (*216*).

BATE-SMITH, E. C. (1936), *Dept. Sci. Ind. Research (Brit.), Food Invest. Repts.*, 1935:35 (*682*).

BATE-SMITH, E. C. (1937), *Dept. Sci. Ind. Research (Brit.), Food Invest. Repts.*, 1936:58–59 (*707*).

BATE-SMITH, E. C., BROOKS, J., and HAWTHORNE, J. R. (1943), *J. Soc. Chem. Ind. (London)*, 62:97–100 (*745, 756, 761*).

BATE-SMITH, E. C., and HAWTHORNE, J. R. (1945), *J. Soc. Chem. Ind. (London)*, 64:297–302 (*752, 754, 755, 757*).

BATESON, W. (1902), *Evolution Comm. Roy. Soc., Rept.*, 1:87–124 (*29*).

BAUDRIMONT, A., and ST. ANGE, M. (1847), *Ann. chim. phys.*, (Ser. 3) 21:195–295 (*312, 380, 406*).

BAUER, R. W. (1895), *Biol. Zentr.*, 15:448 (*257, 280*).

BAUERFEIND, J. C., and NORRIS, L. C. (1939), *Poultry Sci.*, 18:400 (*624*).

BAUMANN, C. A., SEMB, J., HOLMES, C. E., and HALPIN, J. G. (1939), *Poultry Sci.*, 18:48–53 (*610*).

BAUMGARTEN, —— (1884), *Zentr. klin. Med.*, 2:25–30 (*492*).

BEACH, J. R., and DAVIS, D. E. (1927), *Hilgardia*, 2:411–424 (*249, 510*).

BEADLE, B. W., CONRAD, R. M., and SCOTT, H. M. (1938), *Poultry Sci.*, 17:498–504 (*221*).

BEAUDETTE, F. R. (1925), *J. Am. Vet. Med. Assoc.*, 67:741–745 (*510*).

BEAUDETTE, F. R. (1926), *J. Am. Vet. Med. Assoc.*, 68:644–652 (*511*).

BÉCHAMP, M. A. (1873), *Compt. rend. acad. sci.*, 77:1525–1529 (*327*).

BECKERT, C. J., KRITCHEVSKY, W., and HARRIS, B. R. (1933), U. S. Pat. 1,924,972 (*800*).

BEHRE, E. H., and RIDDLE, O. (1919), *Am. J. Physiol.*, 50:364–376 (*222, 323*).

BEILINSSON, A. (1929), *Biochem. Z.*, 213:399–405 (*461*).

BEIN, S. (1890), *Ber. deut. chem. Ges.*, 23:423–424 (*783*).

BELL, M. E. (1933), *Biochem. J.*, 27:1430–1437 (*588*).

BELL, P. H. (1938), *Ind. Eng. Chem., Anal. Ed.*, 10:579–582 (*358*).

BELLER, K., and HENNINGER, E. (1930), *Arch. Geflügelkunde*, 4:453–562 (*510*).

BELLER, K., and ZEKI, M. (1934), *Deut. tierärztl. Wochschr.*, 42:273–275 (*514*).

BELLINI, A. (1907), *Arch. fisiol.*, 4:123–132 (*403, 412, 413*).

BENDIRE, C. (1892–95), *Life Histories of North American Birds*, Smithsonian Institution, Washington, Vols. 1–2, 446 and 518 pp. (*1*).

BENESCH, R., BARRON, N. S., and MAWSON, C. A. (1944), *Nature*, 153:138–139 (*226, 274*).

BENFORD, F., and HOWE, R. F. (1931), *Trans. Illum. Eng. Soc.*, 26:292 (*386, 387*).

BENJAMIN, E. W. (1914), *Cornell Univ. Agr. Expt. Sta. Bull.*, 353:1–46 (*370*).

BENJAMIN, E. W. (1920), *Cornell Univ. Agr. Expt. Sta. Mem.*, 31:195–312 (*70, 71, 92, 93, 94, 96, 99, 100, 102*).

BENJAMIN, H. R., and HESS, A. F. (1933), *J. Biol. Chem.*, 103:629–641 (*236*).

BENNETTS, H. W. (1931), *Australian Vet. J.*, 7:27–31 (*687*).

BENNION, E. B., HAWTHORNE, J. R., and BATE-SMITH, E. C. (1942), *J. Soc. Chem. Ind. (London)*, 61:31–34 (*392*).

BENNION, N. L., and WARREN, D. C. (1933a), *Poultry Sci.*, **12**:69–82 (*83, 86, 119, 213, 222*).

BENNION, N. L., and WARREN, D. C. (1933b), *Poultry Sci.*, **12**:362–367 (*77, 78, 79, 81*).

BENOIT, J. (1932), *Arch. zool. exptl. et gén. Supplementaire*, **73**:1–112 (*176*).

BENOIT, J. (1935a), *Compt. rend. soc. biol.*, **120**:133–136 (*51*).

BENOIT, J. (1935b), *Compt. rend. soc. biol.*, **120**:905–908 (*216*).

BENOIT, J., and COURRIER, R. (1933), *Compt. rend. soc. biol.*, **114**:1335–1338 (*296*).

BENOIT, J., GRANGAUS, R., and SARFATI, S. (1941), *Bull. histol. appl. physiol. et path. et tech. microscop.*, **18**:173–184 (*51, 213*).

BERARD, J. E. (1821), *Ann. chim.*, **16**:152–183, 225–252 (*707*).

BERG, L. R. (1945), *Poultry Sci.*, **24**:555–563 (*104, 157, 158*).

BERG, R. (1925), *Die Nahrungs- und Genussmittel*, E. Pahl, Dresden, 67 pp. (*358*).

BERG, W. N. (1909), *Am. J. Physiol.*, **23**:420–459 (*520*).

BERGH, A. A. H. VAN DEN, and GROTEPASS, W. (1936), *Compt. rend. soc. biol.*, **121**:1253–1258 (*347*).

BERGMANN, M., and NIEMANN, C. (1937), *J. Biol. Chem.*, **118**:301–314 (*334, 435*).

BERGTOLD, W. H. (1929), *Auk*, **46**:466–473 (*107*).

BERKESY, L., and GÖNCZI, K. (1933a), *Arch. exptl. Path. Pharmakol.*, **171**:260–268 (*607*).

BERKESY, L., and GÖNCZI, K. (1933b), *Magyar Orvosi Arch.*, **34**:402–409 (*361*).

BERNARD, C. (1850), *Compt. rend. soc. biol.*, **1**:9–10 (*280*).

BERNARD, R., and GENEST, P. (1945), *Science*, **101**:617–618 (*226*).

BERNHART, F. W. (1939), *J. Biol. Chem.*, **128**:289–295 (*457, 463*).

BERNHART, F. W. (1940), *J. Biol. Chem.*, **132**:189–193 (*334, 435*).

BERNHART, F. W., and ARNOW, L. E. (1939), *J. Phys. Chem.*, **43**:733–736 (*463*).

BERRIER, H. (1939), *Compt. rend. soc. biol.*, **131**:943–944 (*351*).

BERRY, L. N. (1938), *New Mexico Agr. Expt. Sta. Bull.*, **255**:1–15 (*350*).

BERRY, L. N., and WALKER, A. L. (1927), *New Mexico Agr. Expt. Sta. Bull.*, **158**:1–18 (*19*).

BERRYMAN, G. H., and CHATFIELD, C. (1943), *J. Nutrition*, **25**:23–37 (*598*).

BERTHELOT, M., and ANDRÉ, G. (1890), *Compt. rend. acad. sci.*, **110**:925–934 (*419*).

BERTRAND, G. (1903), *Bull. soc. chim. France*, (Ser. 3) **29**:790–794 (*358*).

BERTRAND, G., and AGULHON, H. (1913), *Compt. rend. acad. sci.*, **156**:2027–2029 (*358*).

BERTRAND, G., and MEDIGRECEANU, F. (1913), *Ann. inst. Pasteur*, **27**:1–11 (*358*).

BERTRAND, G., and VLADESCO, R. (1922), *Bull. soc. chim. France*, (Ser. 4) **31**:268–272 (*358*).

BESREDKA, A. (1921), *Ann. inst. Pasteur*, **35**:291–293 (*507*).

BESREDKA, A., and JUPILLE, F. (1914), *Ann. inst. Pasteur*, **27**:1008–1017 (*507, 789*).

BEST, C. H., CHANNON, H. J., and RIDOUT, J. H. (1934), *J. Physiol.*, **81**:409–421 (*568*).

BEST, C. H., FERGUSON, G. C., and HERSHEY, J. M. (1933), *J. Physiol.*, **79**:94–102 (*568*).

BEST, C. H., and HUNTSMAN, M. E. (1932), *J. Physiol.*, **75**:405–412 (*568*).

BETHKE, R. M., KENNARD, D. C., and SASSAMAN, H. L. (1927), *J. Biol. Chem.*, **72**:695–706 (*618*).

BETHKE, R. M., RECORD, P. R., KICK, C. H., and KENNARD, D. C. (1936), *Poultry Sci.*, **15**:326–335 (*618*).

BETHKE, R. M., RECORD, P. R., and WILDER, F. W. (1936), *J. Nutrition*, **12**:309–320 (*623*).

BETHKE, R. M., RECORD, P. R., WILDER, O. H. M., and KICK, C. H. (1936), *Poultry Sci.*, **15**:336–344 (*617, 620*).

BIALASZEWICZ, K. (1912), *Arch. Entwicklungsmech. Organ.*, **34**:489–540 (*420, 672*).

BIALASZEWICZ, K. (1926), *Prace Inst. Nenck*, 3(52), p. 17 (*411*).

BIASOTTI, M. (1932), *Arch. intern. pharmacodynamie*, **42**:305–310 (*344*).

BIDAULT, C., and BLAIGNAN, S. (1927), *Compt. rend. soc. biol.*, **96**:837–839 (*428*).

BIERRY, H., and GOUZON, B. (1939), *Ann. faculté sci. Marseille*, **12**:221–289 (*348*).

BILEK, F. (1936), *Proc. World's Poultry Congr.* (*Leipzig*), **6(1)**:233–236 (*350*).

BIRCKNER, V. (1919), *J. Biol. Chem.*, **38**:191–203 (*358*).

BIRD, H. R. (1943), *U. S. Egg Poultry Mag.*, **49**:402–404, 419–422, 427 (*610, 621*).

BIRD, S. (1937), *Sci. Agr.*, **17**:359–375 (*48*).

BISBEY, B., APPLEBY, V., WEIS, A., and COVER, S. (1934), *Missouri Agr. Expt. Sta. Research Bull.*, **205**:1–32 (*609, 612, 617*).

BISBEY, B., COVER, S., APPLEBY, V., and WEIS, A. (1933), *Missouri Agr. Expt. Sta. Bull.*, **340** (Ann. Rept., 1933):60–61 (*612*).

BISHOP, J. L. (1940), Australian Pat. 110,660 (*782*).

BISHOP, W. B. S. (1929a), *Med. J. Australia*, **1929(I)**:96–99 (*358*).

BISHOP, W. B. S. (1929b), *Med. J. Australia*, **1929(I)**:792–806 (*358*).

BISHOP, W. B. S., and COOKSEY, T. (1929), *Med. J. Australia*, **1929(II)**:660–662 (*358*).

BISSONNETTE, T. H. (1931), *Physiol. Zoöl.*, **4**:542–574 (*51*).

BISSONNETTE, T. H. (1932a), *Science*, **75**:18–19 (*49, 51*).

BISSONNETTE, T. H. (1932b), *Science*, **76**:253–256 (*51*).

BISSONNETTE, T. H. (1933a), *Poultry Sci.*, **12**:396–399 (*51*).

BISSONNETTE, T. H. (1933b), *Quart. Rev. Biol.*, **8**:201–208 (*51*).

BISSONNETTE, T. H., and CHAPNICK, M. H. (1930), *Am. J. Anat.*, **45**:307–343 (*51*).

BIZARRO, A. H. (1913), *J. Physiol.*, **46**:267–284 (*587*).

BLACK, D. J. G., and TYLER, C. (1944), *Nature*, **153**:682–683 (*166*).

BLACKWOOD, J. H., and WISHART, G. M. (1934), *Biochem. J.*, **28**:550–558 (*528, 529, 533, 534*).

BLAKESLEE, A. F., HARRIS, J. A., WARNER, D. E., and KIRKPATRICK, W. F. (1917), *Storrs* (*Connecticut*) *Agr. Expt. Sta. Bull.*, **92**:95–194 (*59*).

BLAKESLEE, A. F., and WARNER, D. E. (1915), *Am. Naturalist*, **49**:360–368 (*132*).

BLANCHETIÈRE, A. (1927), *Compt. rend. acad. sci.*, **185**:1321–1323 (*527*).

BLASIUS, R. (1867), *Z. wiss. Zoöl.*, **17**:480–524 (*164, 165, 185*).

BLATTNER, R. J., HEYS, F. M., and HARTMANN, A. F. (1943), *Arch. Pathol.*, **36**:262–268 (*791*).

BLOCH, B., and PRIETO, J. A. G. (1929), *Ann. dermatol. syphilig.*, **10**:461–480 (*558*).

BLOCK, R. J. (1934), *J. Biol. Chem.*, **105**:455–461 (*333*).

BLOCK, R. J. (1943), *Yale J. Biol. Med.*, **15**:723–728 (*591*).

BLOCK, R. J., and BOLLING, D. (1944), *J. Am. Dietet. Assoc.*, **20**:69–76 (*591*).

BLOMBERG, C. G. (1932), *Food Industries*, **4**:100–102 (*740*).

BLOOM, W., BLOOM, M. A., and McLEAN, F. C. (1941), *Anat. Record*, **81**:443–475 (*238*).

BLOOR, W. R. (1943), *Biochemistry of the Fatty Acids*, Reinhold Publishing Corp., New York, 387 pp. (*341*).

BLUM, F. (1897), *Z. physiol. Chem.*, **22**:127–131 (*479*).

BOAS, M. A. (1924), *Biochem. J.*, **18**:422–424 (*629*).

BOAS, M. A. (1927), *Biochem. J.*, **21**:712–724 (*587, 630*).

BOERNER, F., and LUKENS, M. (1941), *Am. J. Clinical Path.*, **11**:71–74 (*789*).

BOGGS, M. M., DUTTON, H. J., EDWARDS, B. G., and FEVOLD, H. L. (1946), *Ind. Eng. Chem.*, **38**:1082–1084 (*763*).

BOGGS, M. M., and FEVOLD, H. L. (1946), *Ind. Eng. Chem.*, **38**:1075–1079 (*763*).

BOHR, C., and HASSELBALCH, K. A. (1903), *Skand. Arch. Physiol.*, **14**:398–429 (*376*).

BOHREN, B. B., THOMPSON, C. R., and CARRICK, C. W. (1945), *Poultry Sci.*, **24**:356–362 (*132*).

BOISSEVAIN, C. H., and SCHULTZ, H. W. (1938), *Am. Rev. Tuberc.*, **38**:624–628 (*508*).

BOLEY, L. E., and GRAHAM, R. (1939), *J. Am. Vet. Med. Assoc.*, **95**:545–550 (*252*).

BOLLMANN, H., and REWALD, B. (1929), Brit. Pat. 328,075 (*782*).

BOLTON, W., and COMMON, R. H. (1941), *Nature*, **148**:373 (*129*).

BOND, M. (1922), *Biochem. J.*, **16**:479–481 (*588*).

BONNANI, A. (1912), *Boll. accad. med. Roma*, **38**:22 (*214*).

BONNET, H., and LEBLOIS, C. (1939), *Compt. rend. soc. biol.*, **130**:630–631 (*510*).

BONNET, R. (1883), *Deut. Z. Tiermed.*, **9**:239–252 (*292, 306*).

BOOHER, L. E., and HARTZLER, E. R. (1939), *U. S. Dept. Agr. Tech. Bull.*, **707**:1–20 (*623*).

BOOHER, L. E., and MARSH, R. L. (1941), *U. S. Dept. Agr. Tech. Bull.*, **802**:1–30 (*612, 614*).

BOOTH, N. (1930), *Biochem. J.*, **24**:158–168 (*431, 457, 467*).

BORNSTEIN, S. (1942), *N. Y. State J. Med.*, **42**:163–166 (*512*).

BORNSTEIN, S., and SAPHRA, I. (1942), *J. Infectious Diseases*, **71**:55–56 (*514*).

BORNSTEIN, S., and SCHWARZ, H. (1942), *Am. J. Med. Sci.*, **204**:546–550 (*513*).

BORSOOK, H., MACFADYEN, D. A., and WASTENEYS, H. (1930), *J. Gen. Physiol.*, **13**:295–306 (*525*).

BORSOOK, H., and WASTENEYS, H. (1924), *J. Biol. Chem.*, **62**:633–639 (*525*).

BORSOOK, H., and WASTENEYS, H. (1925), *J. Biol. Chem.*, **63**:563–574 (*524*).

BOSCH, E., GYÖRGY, P., and WITEBSKY, E. (1931), *Klin. Wochschr.*, **10**:2264–2265 (*558*).

BOSCH, E., GYÖRGY, P., and WITEBSKY, E. (1932), *Z. Kinderheilk.*, **53**:394–410 (*558*).

BOUCHER, R. B. (1941), *U. S. Egg Poultry Mag.*, **47**:470–473, 512 (*610*).

BOYD, J. D., DRAIN, C. L., and NELSON, M. V. (1929), *Am. J. Diseases Children*, **38**:721–725 (*621*).

BRADLEY, O. C. (1928), *J. Anat.*, **62**:339–345 (*190, 191, 192, 193, 194, 195, 196*).

BRAGA, A. (1939), *Biol. Med. (Rio de Janeiro)*, **5**(13):29–35 (*514*).

BRAMBELL, F. W. R. (1926), *Trans. Roy. Soc. (London)*, **214**(B):113–151 (*201, 203, 209*).

BRANDLY, C. A. (1939), *Proc. World's Poultry Congr. (Cleveland)*, **7**:246–250 (*791*).

BRANDLY, C. A., MOSES, H. E., and JUNGHERR, E. L. (1946), *Poultry Sci.*, **25**:397–398 (*563*).

BRANION, H. D., DRAKE, T. G. H., and TISDALL, F. F. (1934a), *U. S. Egg Poultry Mag.*, **40**(7):20–22, 54, 56–57 (*617, 618, 619*).

BRANION, H. D., DRAKE, T. G. H., and TISDALL, F. F. (1934b), *U. S. Egg Poultry Mag.*, **40**(8):22–23, 58 (*617*).

BRANION, H. D., DRAKE, T. G. H., and TISDALL, F. F. (1934c), *U. S. Egg Poultry Mag.*, **40**(9):22–23, 52, 54 (*620*).

BRANION, H. D., DRAKE, T. G. H., and TISDALL, F. F. (1935), *J. Can. Med. Assoc.*, **32**:9–12 (*617*).

BRANION, H. D., and SMITH, J. B. (1932), *Poultry Sci.*, **11**:261–265 (*618*).

BRAUN, M. (1901), *Centr. Bakt. Parasitenk.*, **29**:13–15 (*305*).

BRIDGMAN, P. W. (1914), *J. Biol. Chem.*, **19**:511–512 (*401, 457, 464*).

BRIGGS, D. R. (1928), *J. Am. Chem. Soc.*, **50**:2358–2363 (*448, 450*).

BRIGGS, D. R. (1935), *J. Phys. Chem.*, **39**:983–995 (*435*).

BRIGGS, F. A. (1938), *J. Infectious Diseases*, **63**:103–112 (*563, 564*).

BRITES, G. (1933), *Folia anat. Univ. Conimbrigensis*, **8**(3):1–3 (*288, 293*).

BROCA, P. (1861), *Compt. rend. soc. biol.*, (Ser. 3) **3**:154–161 (*277*).

BRODY, S. (1921), *J. Gen. Physiol.*, **3**:431–437 (*217*).

BRODY, S. (1945), *Bioenergetics and Growth*, Reinhold Publishing Corp., New York, 1023 pp. (*598, 599, 600*).

BRODY, S., FUNK, E. M., and KEMPSTER, H. L. (1938), *Univ. Missouri Agr. Expt. Sta. Research Bull.*, **278**:1–59 (*39*).

BRODY, S., HENDERSON, E. A., and KEMPSTER, H. L. (1923), *J. Gen. Physiol.*, **4**:41–45 (*11*).

BRONFENBRENNER, J., ANDREWS, V. L., and SCOTT, Z. R. (1915), *J. Am. Med. Assoc.*, **64**:1306–1307 (*558*).

Brooks, J. (1937), *Dept. Sci. Ind. Research (Brit.)*, *Food Invest. Repts.*, **1936**:49 (*347*).

Brooks, J. (1943), *J. Soc. Chem. Ind. (London)*, **62**:137–139 (*758*).

Brooks, J., and Hawthorne, J. R. (1943a), *J. Soc. Chem. Ind. (London)*, **62**:165–167 (*754, 757*).

Brooks, J., and Hawthorne, J. R. (1943b), *J. Soc. Chem. Ind. (London)*, **62**:181–185 (*757, 761, 762*).

Brooks, J., and Pace, J. (1938a), *Dept. Sci. Ind. Research (Brit.)*, *Food Invest Repts.*, **1937**:27–30 (*378, 387, 423*).

Brooks, J., and Pace, J. (1938b), *Proc. Roy. Soc. (London)*, **126(B)**:196–210 (*378, 387, 411, 681*).

Brooks, J., and Pace, J. (1939), *Dept. Sci. Ind. Research (Brit.)*, *Food Invest. Repts.*, **1938**:37–38 (*423*).

Brown, A. J., and Millar, E. T. (1906), *J. Chem. Soc.*, **89**:145–155 (*533*).

Brown, E. G., Combs, G. R., and Wright, E. (1940), *J. Am. Med. Assoc.*, **114**:642–644 (*512*).

Brown, J. M. (1939), *Farmer and Stockbreeder*, **53**:746 (*257, 291*).

Brown, J. T. (1910), *Encyclopedia of Poultry*, Southwood, London, Vol. I, 268 pp. (*267*).

Brown, W. L. (1934), *Georgia Agr. Expt. Sta. Bull.*, **183**:1–8 (*127*).

Brown, W. L. (1938), *J. Biol. Chem.*, **122**:655–659 (*350*).

Browning, P. E. (1917), *Ind. Eng. Chem.*, **9**:1043 (*804*).

Brownlee, D. S., and James, L. H. (1939), *Proc. World's Poultry Congr. (Cleveland)*, **7**:488–492 (*726, 735, 737*).

Bruckner, J. H. (1935), *The Optimum Environmental Conditions for Winter Egg Production*, Inaug. Diss., Cornell University, Ithaca, New York, 218 pp. (*48*).

Brull, L. (1933), *Compt. rend. soc. biol.*, **113**:67–69 (*564*).

Brunett, E. L. (1925), *Cornell Univ. Vet. Coll. Rept.*, **1923–24**:70–73 (*246*).

Bryan, A. H. (1910), *U. S. Dept. Agr., Bur. Chem. Bull.*, **134**:110 (*780*).

Bryant, R. L., and Sharp, P. F. (1934), *J. Agr. Research*, **48**:67–89 (*169*).

Buchanan, R. (1924), *J. Assoc. Offic. Agr. Chemists*, **7**:407–424 (*776*).

Buckner, G. D., Insko, W. M., and Harms, A. (1943), *Poultry Sci.*, **22**:95 (*377*).

Buckner, G. D., Insko, W. M., Martin, J. H., and Harms, A. (1939), *Proc. World's Poultry Congr. (Cleveland)*, **7**:534–535 (*341*).

Buckner, G. D., and Martin, J. H. (1920), *J. Biol. Chem.*, **41**:195–203 (*44, 154, 354*).

Buckner, G. D., and Martin, J. H. (1928), *Am. J. Physiol.*, **89**:164–169 (*197*).

Buckner, G. D., Martin, J. H., and Hull, F. E. (1930), *Am. J. Physiol.*, **93**:86–89 (*236, 237*).

Buckner, G. D., Martin, J. H., and Insko, W. M. (1930), *Proc. World's Poultry Congr. (London)*, **4**:357–361 (*225, 226*).

Buckner, G. D., Martin, J. H., and Peter, A. M. (1923), *Kentucky Agr. Expt. Sta. Research Bull.*, **250**:333–367 (*44, 359*).

Buckner, G. D., Martin, J. H., and Peter, A. M. (1925), *Am. J. Physiol.*, **71**:349–354 (*185, 196, 197, 225*).

Buckner, G. D., Martin, J. H., and Peter, A. M. (1928), *J. Agr. Research*, **36**:263–268 (*44, 121*).

Buckner, G. D., Martin, J. H., Pierce, W. G., and Peter, A. M. (1922), *J. Biol. Chem.*, **51**:51–54 (*275*).

Bujard, E. (1917), *Arch. sci. phys. et nat.*, (Ser. 4) **44**:483–486 (*290*).

Bull, H. B. (1932), *Z. physik. Chem.*, **161(A)**:192–194 (*442*).

Bull, H. B. (1938a), *J. Biol. Chem.*, **123**:17–30 (*455, 462*).

Bull, H. B. (1938b), *J. Biol. Chem.*, **125**:585–598 (*440, 470*).

BULL, H. B. (1938c), *Cold Spring Harbor Symposia Quant. Biol.*, **6**:140–149 (*457, 458, 466, 474, 475, 477*).

BULL, H. B. (1940a), *J. Biol. Chem.*, **133**:39–49 (*443, 464, 465, 474*).

BULL, H. B. (1940b), *Trans. Faraday Soc.*, **36**:80–84 (*444*).

BULL, H. B. (1941a), *J. Biol. Chem.*, **137**:143–151 (*334, 429, 435, 437*).

BULL, H. B. (1941b), *Advances in Enzymol.*, **1**:1–42, Interscience Publishers, Inc., New York, 417 pp. (*457*).

BULL, H. B. (1944), *J. Am. Chem. Soc.*, **66**:1499–1507 (*441*).

BULL, H. B., and FRAMPTON, V. L. (1936), *J. Am. Chem. Soc.*, **58**:594–596 (*484, 485*).

BULL, H. B., and NEURATH, H. J., (1937), *J. Biol. Chem.*, **118**:163–175 (*457, 462, 463, 467*).

BUMAZHNOV, A. D. (1944a), *U. S. Egg Poultry Mag.*, **50**:198–202, 231, 234–236 (*745*).

BUMAZHNOV, A. D. (1944b), *U. S. Egg Poultry Mag.*, **50**:259–262, 277, 282, 284–285 (*746, 748*).

BUMAZHNOV, A. D. (1944c), *U. S. Egg Poultry Mag.*, **50**:356–358, 374, 381–384 (*745*).

BUNGE, G. (1885), *Z. physiol. Chem.*, **9**:49–59 (*528*).

BUNKFELDT, R., and STEENBOCK, H. (1943), *J. Nutrition*, **25**:309–317 (*601*).

BUNYEA, H. (1942a), *U. S. Dept. Agr. Yearbook, Separate*, **1891**:931–943 (*509*).

BUNYEA, H. (1942b), *U. S. Dept. Agr. Yearbook, Separate*, **1891**:993–1006 (*494, 510*).

BUNYEA, H., and HALL, W. J. (1930), *J. Agr. Research*, **40**:209–223 (*509*).

BUOGO, G. (1934), *Industria chimica*, **9**:1634–1636 (*783*).

BURGER, J. W. (1943), *J. Exptl. Zoöl.*, **94**:161–168 (*51*).

BURK, N. F. (1937), *J. Biol. Chem.*, **120**:63–83 (*474*).

BURK, N. F., and GREENBERG, D. M. (1930), *J. Biol. Chem.*, **87**:197–238 (*474*).

BURMESTER, B. R. (1940), *J. Exptl. Zoöl.*, **84**:445–500 (*220, 221, 222, 223, 224, 225, 229*).

BURMESTER, B. R., and CARD, L. E. (1938), *Poultry Sci.*, **17**:235–239 (*304*).

BURMESTER, B. R., and CARD, L. E. (1939), *Poultry Sci.*, **18**:138–145 (*137, 260, 261, 297*).

BURMESTER, B. R., SCOTT, H. M., and CARD, L. E. (1940), *Poultry Sci.*, **19**:299–302 (*221*).

BURNS, D. (1916), *Biochem. J.*, **10**:263–279 (*352*).

BURROWS, W. H., and BYERLY, T. C. (1938), *Poultry Sci.*, **17**:324–330 (*29*).

BURROWS, W. H., and BYERLY, T. C. (1940), *Poultry Sci.*, **19**:346 (*252*).

BURROWS, W. H., and BYERLY, T. C. (1942), *Poultry Sci.*, **21**:416–421 (*252*).

BURROWS, W. H., and FRAPS, R. M. (1942), *Endocrinology*, **30**:702–705 (*252*).

BUSHNELL, L. D., and MAURER, O. (1914), *Kansas Agr. Expt. Sta. Bull.*, **201**:749–777 (*493, 494*).

BUSS, W. J. (1919), *Ohio Agr. Expt. Sta. Monthly Bull.*, **4(3)**:79–82 (*19, 21*).

BUSTER, M. W. (1927), *California Agr. Expt. Sta. Bull.*, **424**:1–21 (*22*).

BUTTON, F. C. (1929), *Ice Cream Trade J.*, **25(8)**:37–38 (*775*).

BYERLY, T. C. (1941), *Univ. Maryland Agr. Expt. Sta. Tech. Bull.*, **A-1**:1–29 (*39*).

BYERLY, T. C., and MOORE, O. K. (1941), *Poultry Sci.*, **20**:387–390 (*27, 49*).

BYERLY, T. C., TITUS, H. W., and ELLIS, N. R. (1933), *J. Agr. Research*, **46**:1–22 (*335*).

BYWATERS, H. W. (1909), *Biochem. Z.*, **15**:322–349 (*330*).

CAHN, F. J., and EPSTEIN, A. K., (1943) *Ind. Eng. Chem., Anal. Ed.*, **15**:281–282 (*786*).

CALLENBACH, E. W. (1934), *Poultry Sci.*, **13**:267–273 (*76*).

CALLENBACH, E. W., NICHOLAS, J. E., and MURPHY, R. R. (1944), *Pennsylvania Agr. Expt. Sta. Bull.*, **461**:1–12 (*23*).

CALVERT, F. C. (1873), *Compt. rend. acad. sci.*, **67**:1024–1026 (*698, 699, 707*).

CALVERY, H. O. (1929), *J. Biol. Chem.*, **83**:649–656 (*331*).

CALVERY, H. O. (1932), *J. Biol. Chem.*, **94**:613–634 (*331, 332, 333, 435*).

CALVERY, H. O. (1933a), *J. Biol. Chem.*, **100**:183–186 (*328, 331, 332*).

CALVERY, H. O. (1933b), *J. Biol. Chem.*, **102**:73–89 (*526, 527, 532, 533, 535, 536*).

CALVERY, H. O. (1935), *J. Biol. Chem.*, **112**:171–174 (*527*).

CALVERY, H. O. (1942), *Food Research*, **7**:313–331 (*364*).

CALVERY, H. O., BLOCK, W. D., and SCHOCK, E. D. (1936), *J. Biol. Chem.*, **113**:21–25 (*482, 519, 527, 528*).

CALVERY, H. O., and TITUS, H. W. (1934), *J. Biol. Chem.*, **105**:683–689 (*331, 333, 335*).

CALVERY, H. O., and WHITE, A. (1932), *J. Biol. Chem.*, **94**:635–639 (*331, 332, 333*).

CAMPANINI, —— (1908), *J. Roy. Soc. Arts*, **56**:886 (*712*).

CAMUS, M. L. (1904), *Compt. rend. soc. biol.*, **57**:90–93 (*383*).

CANNAN, R. L., KIBRICK, A. C., and PALMER, A. H. (1941), *Ann. N. Y. Acad. Sci.*, **41**:243–266 (*431*).

CANNON, W. B. (1904), *Am. J. Physiol.*, **12**:387–418 (*585, 587*).

CAPALDI, A. (1896), *Centr. Bakt. Parasitenk.*, **20**:800–803 (*507, 789*).

CAPPENBERG, H. (1909), *Chem. Ztg.*, **33**:985 (*785*).

CAPRARO, V., and FORNAROLI, P. (1939a), *Arch. sci. biol.* (*Italy*), **25**:117–125 (*426*).

CAPRARO, V., and FORNAROLI, P. (1939b), *Arch. sci. biol.* (*Italy*), **25**:279–291 (*352*).

CARD, L. E. (1917), *Storrs* (*Connecticut*) *Agr. Expt. Sta. Bull.*, **91**:41–90 (*48*).

CARD, L. E., and SLOAN, H. J. (1935), *Poultry Sci.*, **14**:300–301 (*142*).

CARO, L. DE, and LOCATELLI, A. (1936), *Quaderni nutriz.*, **3**:187–191 (*614*).

CARPENTER, N. K. (1906), *Condor*, **8**:57 (*259*).

CARPENTER, N. K. (1907), *Condor*, **9**:198–199 (*275*).

CARPIAUX, E. (1903a), *Bull. agr.* (*Brussels*), **19**(2):200–212 (*317, 320*).

CARPIAUX, E. (1903b), *Chem. zentr.*, **74**:58 (*377*).

CARPIAUX, E. (1908), *Bull. acad. Belg. cl. sci.*, pp. 283–295 (*410, 411*).

CARR, R. H. (1939), *Poultry Sci.*, **18**:225–231 (*370*).

CASPARIS, H., SHIPLEY, P. G., and KRAMER, B. (1923), *J. Am. Med. Assoc.*, **81**:818–819 (*616*).

CASTER, A. B., and ST. JOHN, J. L. (1944), *Arch. Biochem.*, **4**:51–58 (*402*).

CATTANEO, G. (1877), *Atti soc. ital. sci. nat.*, **20**:89–102 (*684*).

CHAHOVITH, X., and ARNOVLEVITCH, V. (1927), *Compt. rend. soc. biol.*, **96**:16–18 (*565*).

CHAIKOFF, I. L., LORENZ, F. W. and ENTENMAN, C. (1941), *Endocrinology*, **28**:597–602 (*180, 184, 234, 235, 236, 241*).

CHAMBERS, A. L., and FLOSDORF, E. W. (1936), *J. Biol. Chem.*, **114**:75–83 (*457, 464*).

CHAMPY-COUJARD, C. (1945), *Compt. rend. soc. biol.*, **139**:398–399 (*212*).

CHAPMAN, F. M. (1937), *Handbook of Birds of Eastern North America*, D. Appleton-Century Co., New York, 581 pp. (*6*).

CHARGAFF, E. (1942a), *J. Biol. Chem.*, **142**:491–504 (*332*).

CHARGAFF, E. (1942b), *J. Biol. Chem.*, **142**:505–512 (*205, 363*).

CHARLES, E., and HOGBEN, L. (1933), *Quart. J. Exptl. Physiol.*, **23**:343–349 (*236*).

CHATFIELD, C., and ADAMS, G. (1940), *U. S. Dept. Agr. Circ.*, **549**:1–91 (*598, 599, 600*).

CHATIN, J. (1883), *Compt. rend. soc. biol.*, (Ser. 7) **5**:384–386 (*307*).

CHATIN, J. (1887), *Compt. rend. soc. biol.*, (Ser. 8) **4**:466 (*285*).

CHELDELIN, V. H., and WILLIAMS, R. J. (1942), *Univ. Texas Pub.*, **4237**:105–124 (*623, 627, 628, 629, 632*).

CHERRINGTON, V. A., GILDOW, E. M., and MOORE, P. (1937), *Poultry Sci.*, **16**:226–231 (*511*).

CHIBNALL, A. C., REES, M. W., and WILLIAMS, E. E. (1943), *Biochem. J.*, **37**:372–388 (*333*).

CHICK, H., and LUBRZYNSKA, E. (1914), *Biochem. J.*, **8**:59–69 (*429, 443*).

CHICK, H., and MARTIN, C. J. (1910), *J. Physiol.*, **40**:404–430 (*459*).

CHICK, H., and MARTIN, C. J. (1912), *J. Physiol.*, **45**:261–306 (*458*).

CHICK, H., and MARTIN, C. J. (1913), *Biochem. J.*, **7**:92–96 (*429*).

CHICK, H., and ROSCOE, M. H. (1929), *Biochem. J.*, **23**:498–503 (*623*).

CHIDESTER, F. E. (1915), *Am. Naturalist*, **49**:49–51 (*263, 264, 286*).

CHIN, T. C. (1933), *Chem. Ind.* (*China*), **8**:145–155 (*699*).

CHIRAY, M. (1907), *Jahresber. Fortschr. Tierchem.*, **36**:805 (*563*).

CHOBAUT, A. (1897), *Feuille jeun. Naturalistes*, (Ser. 3) **27**:215 (*292, 293*).

CHOMKOVIC, G. (1928), *Ann. Czechoslov. Acad. Agr.*, **3**:1–62 (*205*).

CHOU, C. Y., and WU, H. (1934), *Chinese J. Physiol.*, **8**:145–153 (*433*).

CHRISTENSEN, K. M. (1916), U. S. Pat. 1,184,621 (*712*).

CHRZASZCZ, T. (1935), *Wiadomości Farm.*, **62**:481–483 (*84, 358*).

CLAIREMONT, V. (1914), U. S. Pat. 1,092,897 (*712, 716*).

CLARENBURG, A., and DORNICKX, C. G. J. (1932), *Z. Hyg. Infektionskrankh.*, **114**:31–41 (*514*).

CLARENBURG, A., and POT, A. W. (1935), *Tijdschr. Diergeneesk.*, **62**:1240–1246 (*512*).

CLARK, G. A., GAUNT, R. T., and TIMPERLEY, W. A. (1938), *Quart. J. Exptl. Physiol.*, **28**:149–154 (*566*).

CLARK, G. L., and MANN, W. A. (1922), *J. Biol. Chem.*, **52**:157–182 (*438*).

CLARK, G. L., and SHENK, J. H. (1937a), *Radiology*, **28**:58–67 (*428, 456, 473*).

CLARK, G. L., and SHENK, J. H. (1937b), *Radiology*, **28**:357–361 (*479*).

CLARK, G. W. (1925), *J. Biol. Chem.*, **65**:597–600 (*411*).

CLARK, J. H. (1925), *Am. J. Physiol.*, **73**:647–660 (*462*).

CLARK, J. H. (1935), *J. Gen. Physiol.*, **19**:199–210 (*457, 458, 463*).

CLARK, J. H. (1943), *J. Gen. Physiol.*, **27**:101–111 (*457, 461, 467, 470*).

CLARK, J. H. (1945), *J. Gen. Physiol.*, **28**:539–545 (*461*).

CLARK, L. B., LEONARD, S. L., and BUMP, G. (1937), *Science*, **85**:339–340 (*50*).

CLARK, L. N. (1915), *J. Biol. Chem.*, **22**:485–491 (*46*).

CLARK, T. B. (1932), *Poultry Sci.*, **11**:176 (*284*).

CLARK, T. B. (1940), *Poultry Sci.*, **19**:61–66 (*11, 64, 68, 70*).

CLARKE, M. C. (1940), *U. S. Egg Poultry Mag.*, **46**:644–646, 697–698 (*791*).

CLAYTON, M. M. (1930), *J. Nutrition*, **3**:23–38 (*583, 592*).

CLEYER, A. (1682), *Misc. Curiosa, Live Ephem. ined. phys. germ. Acad.*, obs. **XVI**:36 (*254, 287, 306, 307*).

CLICKNER, F. H. (1936), U. S. Pat. 2,030,964 (*777*).

CLINTON, H. F. (1927), *J. Dept. Agr. Victoria*, **25**:739–741 (*254, 257, 258, 284, 291, 292*).

COHEN, S. J. (1920), *J. Lab. Clin. Med.*, **5**:285–294 (*565*).

COHN, A. H., HENRY, J. L., and PRENTIS, A. M. (1925), *J. Biol. Chem.*, **63**:721–766 (*435*).

COHN, E. J. (1936), *Chem. Rev.*, **19**:241–273 (*432*).

COHN, E. J. (1938), *Cold Spring Harbor Symposia Quant. Biol.*, **6**:8–20 (*451, 452*).

COHN, E. J. (1939), *Bull. N. Y. Acad. Med.*, **15**:639–667 (*431, 433*).

COHN, E. W., and WHITE, A. (1935), *J. Biol. Chem.*, **109**:169–175 (*518, 519, 529, 530, 531*).

COLE, A. G. (1933), *Proc. Soc. Exptl. Biol. Med.*, **30**:1162–1164 (*428*).

COLE, A. G. (1938), *Arch. Path.*, **26**:96–101 (*544, 545*).

COLE, L. J. (1933), *Auk*, **50**:284–296 (*50*).

COLE, L. J., and HALPIN, J. G. (1916), *J. Am. Assoc. Instr. Invest. Poultry Husb.*, **3**:7–8 (*56*).

COLE, L. J., and HUTT, F. B. (1928), *Poultry Sci.*, **7**:60–66 (*46*).

COLE, R. K. (1938), *Anat. Record*, **71**:349–361 (*142, 189, 219, 646*).

COLE, V. V., and CURTIS, G. M. (1935), *J. Nutrition*, **10**:493–506 (*607*).

COLEMAN, E. S. (1914), U. S. Pat. 1,120,029 (*712*).

COLES, R. (1936), *J. Ministry Agr. (Engl.)*, **43**:317–332 (*385*).

COLES, R. (1938), *J. Ministry Agr. (Engl.)*, **44**:1204–1213 (*156*).

COLLIER, P. (1892), *New York Agr. Expt. Sta. Bull.*, **38**:1–10 (*154*).

COMMAILLE (1874), *Centr. agr. Chem.*, **4**:419 (*314*).

COMMON, R. H. (1933), *J. Agr. Sci.*, **23**:555–569 (*359*).

COMMON, R. H. (1936a), *J. Agr. Sci.*, **26**:85–100 (*238, 240*).

COMMON, R. H. (1936b), *J. Agr. Sci.*, **26**:492–508 (*226*).

COMMON, R. H. (1938), *J. Agr. Sci.*, **28**:347–366 (*195, 196*).

COMMON, R. H. (1941), *J. Agr. Sci.*, **31**:412–414 (*226*).

COMMON, R. H. (1943), *J. Agr. Sci.*, **33**:213–220 (*238*).

COMMON, R. H., and HALE, R. H. (1941), *J. Agr. Sci.*, **31**:415–437 (*238, 240*).

COMTE, —— (1929), *Ann. fals. et fraudes*, **22**:600 (*783*).

CONGDON, E. D. (1932), *Anat. Record*, **51**:327–329 (*798*).

CONRAD, R. M. (1939), *Poultry Sci.*, **18**:327–329 (*154, 236*).

CONRAD, R. M., and PHILLIPS, R. E. (1938), *Poultry Sci.*, **17**:143–146 (*218, 219, 220*).

CONRAD, R. M., and SCOTT, H. M. (1938), *Physiol. Revs.*, **18**:481–494 (*202, 216, 217, 220, 223, 224, 225*).

CONRAD, R. M., and SCOTT, H. M. (1939), *Proc. World's Poultry Congr. (Cleveland)*, **7**:528–530 (*682*).

CONRAD, R. M., and SCOTT, H. M. (1942a), *Poultry Sci.*, **21**:77–80 (*646*).

CONRAD, R. M., and SCOTT, H. M. (1942b), *Poultry Sci.*, **21**:81–85 (*196, 219, 242*).

CONRAD, R. M., and WARREN, D. C. (1939), *Poultry Sci.*, **18**:220–224 (*134*).

CONRAD, R. M., and WARREN, D. C. (1940), *Poultry Sci.*, **19**:9–17 (*218, 279, 282*).

CONTINI, M. (1934), *Arch. fisiopatol.*, **2**:1 (*569*).

CORIN, G., and BÉRARD, E. (1888), *Bull. acad. roy. Belg.*, **15**:643–662 (*458*).

CORIN, G., and BÉRARD, E. (1889), *Jahresber. Fortschr. Tierchem.*, **18**:13–14 (*327*).

CORPER, H. J., and COHN, M. L. (1933), *Am. J. Hyg.*, **18**:1–25 (*507*).

CORRAN, J. W., and LEWIS, W. C. M. (1924), *Biochem. J.*, **18**:1364–1370 (*399*).

CORRELL, J. T., and HUGHES, J. S. (1933), *J. Biol. Chem.*, **103**:511–514 (*237*).

COSTE, J. (1847), *Histoire générale et particulière du développement des corps organisés* V. Masson, Paris, T. 1., 602 pp. (*149, 214, 222*).

COURMONT, P., and ROCHAIX, A. (1920), *Compt. rend. acad. sci.*, **170**:967–970 (*789*).

COWARD, K. H., and MORGAN, B. G. E. (1935), *Brit. Med. J.*, **2**:1041–1044 (*614*).

CRAIG, W. (1911), *J. Morphol.*, **22**:299–305 (*6*).

CRAMER, W. (1908), *J. Physiol.*, **37**:146–157 (*586*).

CRAMMER, J. L., and NEUBERGER, A. (1943), *Biochem. J.*, **37**:302–310 (*455, 456, 472, 473*).

CRAVENS, W. W., SEBESTA, E. E., HALPIN, J. G., and HART, E. B. (1943), *Poultry Sci.*, **22**:94–95 (*46*).

CRAVENS, W. W., SEBESTA, E. E., HALPIN, J. G., and HART, E. B. (1946), *Poultry Sci.*, **25**:80–82 (*46*).

CREW, F. A. E. (1923a), *J. Heredity*, **14**:361–362 (*249*).

CREW, F. A. E. (1923b), *Proc. Roy. Soc. (London)*, **95**(B):256–278 (*248*).

CREW, F. A. E., and HUXLEY, J. S. (1923), *Vet. J.*, **79**:343–348 (*46*).

CROSS, L. J. (1912), *A Study of the Relation of the Chemical Composition of Hens' Eggs to the Vitality of the Young Chick*, Inaug. Diss., Cornell University, Ithaca, New York, 16 pp. (*342*).

CRUICKSHANK, E. M. (1934), *Biochem. J.*, **28**:965–977 (*337, 339, 342*).

CRUICKSHANK, E. M. (1939), *Proc. World's Poultry Congr. (Cleveland)*, **7**:539–542 (*342, 619, 648*).

CRUICKSHANK, E. M. (1941), *Nutrition Abstracts & Revs.*, **10**:645–657 (*360, 610, 612*).

CRUICKSHANK, E. M., KODICEK, E., and WANG, Y. L. (1945), *J. Soc. Chem. Ind.* (*London*), 64:15–17 (*615, 616, 620, 624, 626, 627*).

CRUICKSHANK, E. M., and MOORE, T. (1937), *Biochem. J.*, 31:179–187 (*611*).

CUBIN, H. K. (1929), *Biochem. J.*, 23:25–30 (*457, 458, 479*).

CULBERTSON, J. T. (1932), *J. Immunol.*, 23:439–453 (*548, 549, 550*).

CULBERTSON, J. T. (1934), *J. Immunol.*, 26:350–351 (*549*).

CUNNINGHAM, C. H. (1944), *J. Am. Vet. Med. Assoc.*, 105:410–416 (*509*).

CUNNINGHAM, C. H., and STUART, H. W. (1944), *Am. J. Vet. Research*, 5:142–146 (*508*).

CUNNINGHAM, I. J. (1931), *Biochem. J.*, 25:1267–1294 (*360, 604*).

CURTIS, G. M. (1920), *Use of Artificial Light to Increase Winter Egg Production*, Reliable Poultry Journal Publishing Co., Quincy, Illinois, 112 pp. (*49*).

CURTIS, M. R. (1910), *Maine Agr. Expt. Sta. Bull.*, 176:1–20 (*179, 182, 183, 215*).

CURTIS, M. R. (1914a), *Maine Agr. Expt. Sta. Bull.*, 228:105–136 (*78, 79, 81, 88, 91*).

CURTIS, M. R. (1914b), *Arch. Entwicklungsmech. Organ.*, 39:217–327 (*115, 118, 119*).

CURTIS, M. R. (1914c), *Biol. Bull.*, 26:55–83 (*277, 278, 283, 284*).

CURTIS, M. R. (1915a), *J. Agr. Research*, 3:375–385 (*218, 276, 278, 279, 280, 281, 282, 285, 286*).

CURTIS, M. R. (1915b), *J. Agr. Research*, 5:397–404 (*246*).

CURTIS, M. R. (1915c), *Biol. Bull.*, 28:154–158 (*251, 252*).

CURTIS, M. R. (1916) *Biol. Bull.*, 31:181–213 (*233, 254, 264, 265, 288, 290, 291, 292, 293, 294, 299, 300*).

CUSHNY, A. R. (1902), *Am. J. Physiol.*, 6:xviii–xix (*183, 193*).

CZADEK, O. VON (1916), *Z. landw. Versuchsw. Deut. Oesterr.*, 19:440–444 (*314, 317, 320*).

DAGGS, R. G. (1931), *J. Nutrition*, 4:443–467 (*585, 592*).

DAHLE, D. (1936), *J. Assoc. Offic. Agr. Chemists*, 19:228–232 (*358*).

DAKIN, H. D., and DALE, H. H. (1919), *Biochem. J.*, 13:248–257 (*481, 544*).

DALE, H. H., and HARTLEY, P. (1916), *Biochem. J.*, 10:408–433 (*428, 544, 545*).

DALLING, T., and WARRACK, G. H. (1932), *J. Path. Bact.*, 35:655 (*511*).

DAM, H. (1928), *Biochem. Z.*, 194:188–196 (*342*).

DAM, H. (1929), *Biochem. Z.*, 215:475–492 (*342*).

DANIELS, A. L., and EVERSON, G. J. (1935), *J. Nutrition*, 9:191–203 (*607*).

DANIELS, A. L., and EVERSON, G. J. (1936), *J. Nutrition*, 11:327–341 (*603*).

DANIELS, A. L., and WRIGHT, O. E. (1934), *J. Nutrition*, 8:125–138 (*605, 606*).

DANILOVA, A. K., and NEFEDOVA, W. A. (1935), *Zentr. Tierernähr.*, 7:532–542 (*314, 317, 320*).

DARBY, C. W., and STAFSETH, H. J. (1942), *Proc. U. S. Livestock Sanitary Assoc.*, 46:189–201 (*514*).

DARESTE, C. (1877), *Recherches sur la production artificielle des monstruosités ou essais de tératogénie expérimentale*, C. Reinwald, Paris, 304 pp. (*285, 300*).

DARIN, O. (1936), *Rass. clin. terap. sci. affini*, 35:115–137 (*569*).

DARWIN, C. (1868), *The Variation of Animals and Plants under Domestication*, John Murray, London, Vols. 1–2, pp. 494 and 568. (*95*).

DAUPHINEE, J. A., and HUNTER, A. (1930), *Biochem. J.*, 24:1128–1147 (*529, 532, 533*).

DAVAINE, C. (1860), *Compt. rend. soc. biol.*, (Ser. 3) 2:183–268 (*254, 255, 264, 280, 285, 287, 291, 293, 299, 300, 305, 307*).

DAVIDSON, J. A., McCRARY, C. M., and CARD, C. G. (1946), *Michigan Agr. Expt. Sta. Quart. Bull.*, 28(4):281–286 (*19*).

DAVIES, G. (1937), *Dept. Sci. Ind. Research* (*Brit.*), *Food Invest. Repts.*, 1936:51–52 (*669*).

DAVIS, B. D., and COHN, E. J. (1939), *Ann. N. Y. Acad. Sci.*, 39:209–212 (*448*).

DAVIS, D. E. (1942a), *Anat. Record*, 82:153–165 (*210, 244, 248*).

DAVIS, D. E. (1942b), *Anat. Record*, 82:297–307 (*182, 210, 244*).

DAVIS, D. E. (1942c), *Auk*, 59:549–554 (7).

DAVIS, M. J., and METZ, H. A. (1916), Brit. Pat. 105, 840 (*712*).

DAVY, J. (1860), *Proc. Roy. Soc. (London)*, 10:31 (*425*).

DAVY, J. (1863), *Edinburgh New Philos. J.*, 18:249–258 (*400, 410*).

DAY, H. G. (1942), *Federation Proc.*, 1:188–189 (*606*).

DE, S. S. (1940), *Indian J. Med. Research*, 27:793–806 (*569*).

DELEZENNE, C., and POZERSKI, E. (1903), *Compt. rend. soc. biol.*, 55:935–937 (*518*).

DENTON, C. A. (1940), *Poultry Sci.*, 19:281–285 (*129, 350*).

DENTON, C. A., CABELL, C. A., BASTRON, H., and DAVIS, R. (1944), *J. Nutrition*, 28:421–426 (*615, 620, 626*).

DEOBALD, H. J., CHRISTIANSEN, J. B., HART, E. B., and HALPIN, J. G. (1938), *Poultry Sci.*, 17:114–119 (*236, 240*).

DEOBALD, H. J., LEASE, E. J., HART, E. B., and HALPIN, J. G. (1936), *Poultry Sci.*, 15:179–185 (*238, 359*).

DERRIEN, E. (1924), *Compt. rend. soc. biol.*, 91:634–636 (*348*).

DESAI, S. B. (1930), *Biochem. J.*, 24:1897–1904 (*526*).

DESSIRIER, L. (1938), *Ann. fals. et fraudes*, 31:464–467 (*784*).

DEUEL, H. J., HRUBETZ, M. C., MATTSON, F. H., MOREHOUSE, M. G., and RICHARDSON, A. (1943), *J. Nutrition*, 26:673–685 (*349, 611*).

DEVANEY, G. M., MUNSELL, H. E., and TITUS, H. W. (1933), *Poultry Sci.*, 12:215–222 (*617, 619*).

DEVANEY, G. M., MUNSELL, H. E., and TITUS, H. W. (1936), *Poultry Sci.*, 15:149–153 (*617*).

DEVANEY, G. M., TITUS, H. W., and NESTLER, R. B. (1935), *J. Agr. Research*, 50:853–860 (*612*).

DICKIE, G. (1849), *Ann. Mag. Nat. Hist.*, 2(2):169–176 (*171*).

DIENES, L. (1927), *J. Immunol.*, 14:61–75 (*547*).

DIENES, L. (1929), *J. Immunol.*, 17:531–538 (*547*).

DIENES, L. (1930a), *Proc. Soc. Exptl. Biol. Med.*, 27:690–692 (*547*).

DIENES, L. (1930b), *J. Immunol.*, 18:279–283 (*547*).

DIENES, L. (1931a), *J. Immunol.*, 20:221–238 (*547*).

DIENES, L. (1931b), *J. Immunol.*, 20:333–345 (*547*).

DIENES, L., and SCHOENHEIT, E. Q. (1927), *J. Immunol.*, 14:9–42 (*546, 547*).

DIESSER, G. (1912), Swiss Pat. 62,592 (*801*).

DINGEMANS, J. J. J. (1932), *Chem. Weekblad*, 29:138–140 (*419*).

DIPPEL, F. (1898), German Pat. 106,730 (*801*).

DIZMANG, V. R., and SUNDERLIN, G. (1933), *U. S. Egg Poultry Mag.*, 39(11):18–19 (*396*).

DOAN, R. L. (1934), *U. S. Egg Poultry Mag.*, 40(10):19–23, 53–55 (*635*).

DOMM, L. V. (1929), *Arch. Entwicklungsmech. Organ.*, 119:171–187 (*177*).

DOMM, L. V., and DAVIS, D. E. (1941), *Proc. Soc. Exptl. Biol. Med.*, 48:665–668 (*176*).

DONHOFFER, C. (1933), *Biochem. J.*, 27:806–817 (*343*).

DONNALLY, H. H. (1930), *J. Immunol.*, 19:15–40 (*557*).

DORSET, M. (1902), *Am. Med.*, 3:555–556 (*507*).

DOUGHERTY, J. E. (1922), *California Agr. Expt. Sta. Circ.*, 254:1–8 (*49*).

DOUGHERTY, J. E. (1926), *Am. J. Physiol.*, 76:265–267 (*633*).

DOW, R. B. (1940), *Food Manuf.*, 15:207–210 (*401*).

DOYLE, J. D., and DOYLE, M. K. (1935), Brit. Pat. 463,403 (*712*).

DOYLE, T. M. (1926), *J. Comp. Path. Therap.*, 39:137–140 (*510*).

DREA, W. F. (1935), *J. Nutrition*, **10**:351–355 (*358*).

DREW, G. A. (1907), *Science*, **26**:119–120 (*25*).

DREYER, G., and HANSSEN, O. (1907), *Compt. rend. acad. sci.*, **145**:234–236 (*463*).

DRYDEN, J. (1899), *Utah Agr. Expt. Sta. Bull.*, **67**:121–160 (*65*).

DRYDEN, J. (1905), *Utah Agr. Expt. Sta. Bull.*, **92**:174–175 (*127*).

DRYDEN, J. (1921), *Oregon Agr. Expt. Sta. Bull.*, **180**:1–96 (*12, 59*).

DU, S. D., and WILLIAMS, C. M. (1930), *China Med. J.*, **44**:565–567 (*305*).

DUERDEN, J. E. (1918), *J. Heredity*, **9**:243–245 (*101*).

DUGÈS, A. (1839), *Traité physiol. comparée*, **3**:318 (*278*).

DUGGAN, R. E. (1946), *J. Assoc. Offic. Agr. Chemists*, **29**:76–84 (*726*).

DUHAMEL, M. (1826), *Rec. trav. soc. sci. Lille*, **1823–1824**:273 (*307*).

DUMAS, J. B., and CAHOURS, L. (1842), *Ann. chim. et phys.*, (Ser. 3) **6**:385–448 (*325*).

DUNHAM, H. H., and RIDDLE, O. (1942), *Physiol. Zoöl.*, **15**:383–394 (*218*).

DUNHAM, W. B., and MACNEAL, W. J. (1942), *J. Bact.*, **44**:413–424 (*792*).

DUNHAM, W. B., and MACNEAL, W. J. (1943), *J. Lab. Clin. Med.*, **28**:947–953 (*792*).

DUNKERLY, J. S. (1930), *Proc. World's Poultry Congr.* (*London*), **4**:46–72 (*55*).

DUNN, L. C. (1923), *Storrs* (*Connecticut*) *Agr. Expt. Sta. Bull.*, **111**:135–172 (*55, 56, 71*).

DUNN, L. C. (1924a), *Storrs* (*Connecticut*) *Agr. Expt. Sta. Bull.*, **117**:17–88 (*11, 28, 55*).

DUNN, L. C. (1924b), *Storrs* (*Connecticut*) *Agr. Expt. Sta. Bull.*, **118**:89–140 (*11, 28, 55*).

DUNN, L. C. (1924c), *Storrs* (*Connecticut*) *Agr. Expt. Sta. Bull.*, **122**:215–277 (*11, 28, 55*).

DUNN, L. C. (1927), *Storrs* (*Connecticut*) *Agr. Expt. Sta. Bull.*, **147**:241–282 (*11, 16*).

DUNN, L. C., and SCHNEIDER, M. (1923), *Poultry Sci.*, **2**:90–92 (*109, 379*).

DURANT, A. J., and McDOUGLE, H. C. (1943), in *Diseases of Poultry*, edited by Biester, H. E., and Davies, L., Iowa State College, Ames, Iowa, 1005 pp. (*307*).

DÜRIGEN, B. (1922), *Die Geflügelzucht*, Paul Parey, Berlin, 552 pp. (*265*).

DURME, M. VAN (1914), *Arch. biol.* (*Liége*), **29**:71–200 (*200*).

DURSKI, S. P. VON (1907), *Die pathologischen Veränderungen des Eies und Eileiters bei den Vögeln*, Inaug. Diss., Bern, 35 pp. (*254, 307*).

DUTTON, H. J., and EDWARDS, B. G. (1945), *Ind. Eng. Chem.*, **37**:1123–1126 (*758, 759*).

DUTTON, H. J., and EDWARDS, B. G. (1946), *Ind. Eng. Chem.*, **38**:347–350 (*758, 759, 760, 761, 762*).

DUVAL, M. (1884), *Ann. sci. nat. Zool.*, **18**:1–208 (*124*).

DZIERJGOWSKI, S. K. (1901), *Arch. sci. biol.* (*St. Petersburg*), **8**:429–440 (*562*).

EAKIN, R. E., McKINLEY, W. A., and WILLIAMS, R. J. (1940), *Science*, **92**:224–225 (*631*).

EAKIN, R. E., SNELL, E. E., and WILLIAMS, R. J. (1940), *J. Biol. Chem.*, **136**:801–802 (*506, 631*).

EAKIN, R. E., SNELL, E. E., and WILLIAMS, R. J. (1941), *J. Biol. Chem.*, **140**:535–543 (*631*).

EBBECKE, U., and HAUBRICH, R. (1937), *Arch. ges. Physiol.* (*Pflüger's*), **238**:429–430 (*404*).

EBBELL, H. (1935), *Arch. Geflügelkunde*, **9**:77–79 (*51*).

EBER, A. (1932), *Z. Fleisch- u. Milchhyg.*, **42**:297–300 (*515*).

EDWARDS, B. G., and DUTTON, J. H. (1945), *Ind. Eng. Chem.*, **37**:1121–1122 (*759*).

EDWARDS, C. L. (1902), *Am. J. Physiol.*, **6**:351–397 (*123*).

EICHHOLZ, A. (1898), *J. Physiol.*, **23**:163–177 (*327*).

EITNER, W. (1909), *Gerber*, **35**:156 (*799*).

ELLINGER, P., and KOSCHARA, W. (1933), *Ber. deut. chem. Ges.*, **66B**:808–813 (*347*).

ELLIS, N. R., MILLER, D., TITUS, H. W., and BYERLY, T. C. (1933), *J. Nutrition*, **6**:243–262 (*612, 623*).

ELSHOLTII, J. S. (1675), *Misc. nat. cur.*, decur. I., ann. VI and VII, obs. **80**:115 (*287*).

ELVEHJEM, C. A., KEMMERER, A. R., HART, E. B., and HALPIN, J. G. (1929), *J. Biol. Chem.*, 85:89–96 (*358, 363*).

ELVEHJEM, C. A., and SHERMAN, W. C. (1932), *J. Biol. Chem.*, 98:309–319 (*604*).

ENGEL, R. W. (1943), *J. Nutrition*, 25:441–446 (*632*).

ENGEL, R. W., PHILLIPS, P. H., and HALPIN, J. G. (1940), *Poultry Sci.*, 19:135–142 (*624*).

ENGLER, H. (1936), *Proc. World's Poultry Congr.* (*Leipzig*), 6:249–255 (*40*).

ENTENMAN, C., LORENZ, F. W., and CHAIKOFF, I. L. (1938), *J. Biol. Chem.*, 126:133–139 (*241*).

EPSTEIN, A. K., HARRIS, B. R., and KATZMAN, M. (1943), *Proc. Soc. Exptl. Biol. Med.*, 53:238–241 (*721*).

EPSTEIN, A. K., REYNOLDS, M. C., and HARRIS, B. R. (1937), *Mayonnaise and Other Salad Dressings*, The Emulsol Corp., Chicago, 68 pp. (*777*).

ERIKSON, S. E., BOYDEN, R. E., INSKO, W. M., and MARTIN, J. H. (1938), *Kentucky Agr. Expt. Sta. Bull.*, 378:1–24 (*359, 671*).

ERIKSON, S. E., BOYDEN, R. E., MARTIN, J. H., and INSKO, W. M. (1938), *Kentucky Agr. Expt. Sta. Bull.*, 382:113–124 (*359, 671*).

ERIKSON, S. E., and INSKO, W. M. (1934), *Kentucky Agr. Expt. Sta. Ann. Rept.*, 46:53–55 (*358*).

ERLANDSEN, A. (1907), *Z. physiol. Chem.*, 51:71–155 (*338*).

ERMOLENKO, N. F., and GUTERMAN, V. Y. (1938), *Kolloid. Zhur.*, 4:85–91 (*398*).

ERRERA, J., and HIRSHBERG, Y. (1933), *Biochem. J.*, 27:764–770 (*432*).

ESCH, W. (1912), German Pat. 273,482 (*801*).

ESPASA, H. DE J. (1925), *Enciclopedia Universal Ilustrada*, 28:593–603 (*807*).

EVANS, H. M., and BURR, G. O. (1927), *Univ. Calif. Mem.*, 8:1–176 (*621*).

EVANS, R. J. (1942), *U. S. Egg Poultry Mag.*, 48:596–599 (*713*).

EVANS, R. J., and CARVER, J. S. (1942), *U. S. Egg Poultry Mag.*, 48:546–549 (*713, 714*).

EVANS, R. J., CARVER, J. S., and BRANT, A. W. (1944a), *Poultry Sci.*, 23:9–15 (*240*).

EVANS, R. J., CARVER, J. S., and BRANT, A. W. (1944b), *Poultry Sci.*, 23:36–42 (*154, 238*).

EVERSON, G. J., and DANIELS, A. L. (1934), *J. Nutrition*, 8:497–502 (*607*).

EWELL, A. W. (1930), *Refrig. Eng.*, 20:358–360 (*711*).

EWELL, A. W. (1936), *Ice and Refrig.*, 91:295–296 (*710, 711*).

EWELL, A. W. (1938), *Food Research*, 3:101–108 (*711*).

FABRICIUS, H. (1600), *De formatione Ovi et Pulli*, translated by H. B. Adelmann, Cornell University Press, Ithaca, New York, 1942, 883 pp. (*304*).

FALCK-LOENDAHL, C. (1924), Danish Pat. 33,257 (*697*).

FALTA, W. (1906a), *Deut. Arch. klin. Medi.*, 86:517–564 (*594, 595*).

FALTA, W. (1906b), *Verhandl. Ges. deut. Naturforsch. Ärtze*, 77(2):40–41 (*596*).

FALTA, W., and NOEGGERATH, C. T. (1906), *Beitr. chem. Physiol. Pathol.*, 7:313–322 (*595*).

FARCY, L. (1914), *Ann. fals. et fraudes*, 7:183–187 (*785*).

FARINAS, E. C. (1930), *Philippine J. Agr.*, 1:311–365 (*511*).

FARRAR, G. E., and GOLDHAMER, S. M. (1935), *J. Nutrition*, 10:241–254 (*605*).

FARRELL, J. S. (1928), *Am. J. Physiol.*, 85:672–684 (*585*).

FARRIES, E. H. M., and BELL, A. F. (1930), *Ann. Botany*, 44:423–455 (*508*).

FEDER, E. (1913), *Z. Untersuch. Nahr. u. Genussm.*, 25:277–285 (*781*).

FEINBERG, J. G., HUGHES, J. S., and SCOTT, H. M. (1937), *Poultry Sci.*, 16:132–134 (*237, 239, 359*).

FELLENBERG, T. VON (1923), *Biochem. Z.*, 139:371–451 (*358*).

FELLENBERG, T. VON (1930), *Biochem. Z.*, 218:300–317 (*358*).

FELLNER, O. O. (1925), *Klin. Wochschr.*, 4:1651–1652 (*516*).

FERDINANDOFF, V. V. (1931), *Practical Handbook of Poultry Industry* (in Russian), Moscow, 157 pp. (*144, 160, 169, 268, 314, 383*).

FÉRÉ, C. (1897), *Bull. mêm. soc. med. hôp. Paris*, (Ser. 3) **14**:608–617 (*84*).

FÉRÉ, C. (1898a), *J. anat. phys.*, **34**:123–127 (*69, 73, 78, 278*).

FÉRÉ, C. (1898b), *Compt. rend. soc. biol.*, (Ser. 10) **5**:922–924 (*256*).

FÉRÉ, C. (1899), *Compt. rend. soc. biol.*, (Ser. 11) **1**:921–922 (*280*).

FÉRÉ, C. (1902), *Compt. rend. soc. biol.*, **54**:348–349 (*290*).

FERNAU, A., and PAULI, W. (1915), *Biochem. Z.*, **70**:426–441 (*457, 463*).

FERRY, R. M., and LEVY, A. H. (1934), *J. Biol. Chem.*, **105**:xxvii–xxviii (*556*).

FEVOLD, H. L., EDWARDS, B. G., DIMICK, A. N., and BOGGS, M. M. (1946), *Ind. Eng. Chem.*, **38**:1079–1082 (*763*).

FIEHE, J. (1931), *Z. Untersuch. Lebensm.*, **61**:428 (*784*).

FINDLAY, G. M., and STERN, R. O. (1929), *Arch. Disease Childhood*, **4(19)**:1–11 (*629*).

FINLAY, G. F. (1925), *Brit. J. Exptl. Biol.*, **2**:439–468 (*176*).

FINNEGAN, W. J. (1941), *Ice and Refrig.*, **101**:403–408 (*732*).

FIORI, A. (1930), *Biochim. e terap. sper.*, **17**:267–271 (*569*).

FISCHER, H., and KÖGL, F. (1923), *Z. physiol. Chem.*, **131**:241–261 (*226, 348*).

FISCHER, H., and KÖGL, F. (1924), *Z. physiol. Chem.*, **138**:262–275 (*348*).

FISCHER, H., and MÜLLER, R. (1925), *Z. physiol. Chem.*, **142**:120–140 (*348*).

FISCHER, N. C. (1935), U. S. Pat. 1,966,801 (*743*).

FITCH, C. P., and LUBBEHUSEN, R. E. (1928), *J. Am. Vet. Med. Assoc.*, **72**:636–649 (*510*).

FITCH, C. P., LUBBEHUSEN, R. E., and DIKMANS, R. N. (1924), *J. Am. Vet. Med. Assoc.*, **66**:43–53 (*36, 510*).

FIXSEN, M. A. B. (1931), *Biochem. J.*, **25**:596–605 (*629, 630*).

FLEMING, A. (1922), *Proc. Roy. Soc.* (*London*), **93(B)**:306–317 (*501, 502, 503, 506*).

FLEMING, A., and ALLISON, V. D. (1922a), *Brit. J. Exptl. Path.*, **3**:252–260 (*502, 503*).

FLEMING, A., and ALLISON, V. D. (1922b), *Proc. Roy. Soc.* (*London*), **94(B)**:142–151 (*502, 503*).

FLEMING, A., and ALLISON, V. D. (1924), *Lancet*, **206(I)**:1303–1306 (*502, 503, 504, 506, 683*).

FLEMING, A., and ALLISON, V. D. (1927), *Brit. J. Exptl. Path.*, **8**:214–218 (*502, 504, 505*).

FLEURENT, E., and LÉVI, L. (1920), *Bull. soc. chim. France*, **27**:441–442 (*358*).

FLOERICKE, K. (1909), *Jahrbuch der Vogelkunde, 1908*, Kosmos Verlag, Stuttgart, 130 pp. (*7*).

FLOSDORF, E. W., and CHAMBERS, L. A. (1933), *J. Am. Chem. Soc.*, **55**:3051–3052 (*464*).

FLOSDORF, E. W., and CHAMBERS, L. A. (1935), *J. Immunol.*, **28**:297–310 (*556*).

FLOURENS, M. J. P. (1835), *Compt. rend. acad. sci.*, **1**:182–183 (*285*).

FOLIN, O., and DENIS, W. (1912), *J. Biol. Chem.*, **12**:245–251 (*333*).

FORBES, E. B., BEEGLE, F. M., and MENSCHING, J. E. (1913), *Ohio Agr. Expt. Sta. Bull.*, **255**:211–231 (*314, 410*).

FOSBINDER, R. J., and LESSIG, A. E. (1933), *J. Franklin Inst.*, **215**:579–591 (*440, 450*).

FOSTER, M., and BALFOUR, F. M. (1874), *Elements of Embryology*, Macmillan and Co., London, 272 pp. (*1, 134, 135*).

FOWLER, S., and EDSER, E. (1926), Brit. Pat. 274,200 (*712*).

FOX, W. S. (1899), *Zoologist*, **3**:23–26 (*6*).

FRAMPTON, V. L., and ROMANOFF, A. L. (1947), *Arch. Biochem.*, **13**:315–321 (*328, 448*).

FRANK, F. (1934), U. S. Pat. 1,977,940 (*801*).

FRANK, P. (1911), *J. Biol. Chem.*, **9**:463–470 (*587*).

FRANKE, K. W., and TULLY, W. C. (1935), *Poultry Sci.*, **14**:273–279 (*364*).

FRÄNKEL, S., and JELLINEK, C. (1927), *Biochem. Z.*, **185**:392–399 (*343, 344*).

FRANKENBERG, G. VON (1928), *Zool. Anz.*, **78**:323–329 (*288, 292*).

FRAPS, R. M. (1940), *Poultry Sci.*, **19**:348 (*212*).

FRAPS, R. M. (1942), *Anat. Record*, **84**:521 (*218*).

FRAPS, R. M. (1946), *Anat. Record*, **96**:573 (*231*).

FRAPS, R. M., and DURY, A. (1943), *Proc. Soc. Exptl. Biol. Med.*, **52**:346–349 (*218*).

FRAPS, R. M., HERTZ, R., and SEBRELL, W. H. (1943), *Proc. Soc. Exptl. Biol. Med.*, **52**:140–142 (*196*).

FRAPS, R. M., and NEHER, B. H. (1945), *Endocrinology*, **37**:407–414 (*218*).

FRAPS, R. M., OLSEN, M. W., and NEHER, B. H. (1942), *Proc. Soc. Exptl. Biol. Med.*, **50**:308–312 (*216*).

FRAPS, R. M., and RILEY, G. M. (1942), *Proc. Soc. Exptl. Biol. Med.*, **49**:253–257 (*218*).

FRAPS, R. M., RILEY, G. M., and OLSEN, M. W. (1942), *Proc. Soc. Exptl. Biol. Med.*, **50**:313–317 (*216, 218*).

FRASER, D. T., JUKES, T. H., BRANION, H. D., and HALPERN, K. C. (1934), *J. Immunol.*, **26**:437–446 (*562, 563*).

FREILICH, J., and FREY, C. N. (1941), U. S. Pat. 2,243,860 (*774*).

FRENCH, M. H. (1935a), *Ann. Rept. Dept. Vet. Sci. Animal Husbandry, Tanganyika Terr.*, **1935**:25–27 (*103*).

FRENCH, M. H. (1935b), *Ann. Rept. Dept. Vet. Sci. Animal Husbandry, Tanganyika Terr.*, **1935**:40–42 (*712*).

FREUNDLICH, H., and NEUKIRCHER, H. (1926), *Kolloid-Z.*, **38**:180–181 (*444*).

FREUNDLICH, H., and SEIFREZ, W. (1923), *Z. physik. Chem.*, **104**:233–261 (*405*).

FRIEDBERGER, E. (1926), *Klin. Wochschr.*, **5**:1966–1967 (*581*).

FRIEDBERGER, E., and ABRAHAM, A. (1929), *Deut. med. Wochschr.*, **55**:396–397 (*589, 590*).

FRIEDBERGER, E., and ABRAHAM, A. (1930), *Z. ges. exptl. Med.*, **72**:490–513 (*581, 587, 589, 590, 614*).

FRIEDBERGER, E., ABRAHAM, A., and SEIDENBERG, S. (1928), *Deut. med. Wochschr.*, **54**:2092–2093 (*588*).

FRIEDBERGER, E., and HODER, F. (1932), *Z. Immunitätsforsch.*, **74**:429–447 (*501*).

FRIEDBERGER, E., and SEIDENBERG, S. (1927), *Deut. med. Wochschr.*, **53**:1507–1509 (*581*).

FRIEDGOOD, H. B., and UOTILA, U. U. (1941), *Endocrinology*, **29**:47–58 (*248*).

FRIESE, W. (1923), *Z. Untersuch. Lebensm.*, **46**:33–37 (*114*).

FRISCH, J., PAULI, W., and VALKÓ, E. (1925), *Biochem. Z.*, **164**:401–436 (*470*).

FROMME, W. (1933), *Deut. med. Wochschr.*, **59**:655–656 (*512*).

FROMME, W. (1934a), *Deut. med. Wochschr.*, **60**:1969–1970 (*512*).

FROMME, W. (1934b), *Arch. Hyg. u. Bakt.*, **113**:29–45 (*512*).

FRONDA, F. M. (1924), *Philippine Agr.*, **13**:99–100 (*279*).

FRONDA, F. M. (1935), *Philippine Agr.*, **24**:229–238 (*48*).

FRONDA, F. M., and CLEMENTE, D. D. (1934), *Philippine Agr.*, **23**:187–196 (*65, 136, 368*).

FRONDA, F. M., and CLEMENTE, D. D. (1936a), *Philippine Agr.*, **24**:635–648 (*656, 658*).

FRONDA, F. M., and CLEMENTE, D. D. (1936b), *Philippine Agr.*, **25**:191–206 (*657*).

FRONDA, F. M., CLEMENTE, D. D., and BASIO, E. (1935), *Philippine Agr.*, **24**:49–58 (*369*).

FRYD, C. F. M., and HANSON, S. W. F. (1945), *J. Soc. Chem. Ind. (London)*, **64**:55–56 (*761*).

FRYKLIND, K. E. (1901), Brit. Pat. 9898 (*712*).

FUJII, N. (1924), *J. Biochem. (Japan)*, **3**:394–406 (*484*).

FUNK, E. M. (1935), *Univ. Missouri Coll. Agr. Expt. Sta. Bull.*, **350**:1–15 (*693*).

FUNK, E. M. (1937), *Univ. Missouri Coll. Agr. Expt. Sta. Bull.*, **384**:1–12 (*693*).

FUNK, E. M. (1938), *Univ. Missouri Coll. Agr. Expt. Sta. Bull.*, **394**:1–15 (*691*).

FUNK, E. M. (1943a), *Univ. Missouri Agr. Expt. Sta. Research Bull.*, **362**:1–38 (*715*).

FUNK, E. M. (1943b), *Univ. Missouri Agr. Expt. Sta. Research Bull.*, **364**:1–28 (*715*).

FUNK, E. M., and KEMPSTER, H. L. (1934), *Univ. Missouri Agr. Expt. Sta. Bull.*, **332**:1–15 (*79*).

FURREG, E. (1931), *Biol. Zentr.*, **51**:162–173 (*229, 348*).

FÜRTH, E., and KLEIN, K. (1933), *Veröffentl. Geb. Med. Verwalt.*, **39**:363–376 (*512*).

FÜRTH, J. (1925), *J. Immunol.*, **10**:777–789 (*553*).

FÜRTH, O., and DEUTSCHBERGER, O. (1927), *Biochem. Z.*, **186**:139–154 (*333*).

FÜRTH, R. (1923), *Ann. Physik*, **70**:63–80 (*413, 451*).

GADOW, H. VON (and SELENKA, E.) (1891), in H. G. Bronn's *Klassen und Ordnungen des Thierreichs*, Bd. 6, Abt. 4, C. F. Winter, Leipzig, 1008 pp. (*165, 172*).

GAGE, S. H., and FISH, P. A. (1924), *Am. J. Anat.*, **34**:1–85 (*129, 130, 202*).

GAGE, S. H., and GAGE, S. P. (1908), *Science*, **28**:494–495 (*129, 350*).

GAGGERMEIER, G. (1932), *Arch. Geflügelkunde*, **6**:105–109 (*659*).

GALLUP, W. D., and NORRIS, L. C. (1939), *Poultry Sci.*, **18**:83–88 (*45, 358, 361, 364*).

GANE, R. (1934), *Dept. Sci. Ind. Research (Brit.), Food Invest. Repts.*, **1933**:128–131 (*403*).

GANE, R. (1935), *Dept. Sci. Ind. Research (Brit.), Food Invest. Repts.*, **1934**:57–58 (*677, 678*).

GANE, R. (1937), *Dept. Sci. Ind. Research (Brit.), Food Invest. Repts.*, **1936**:44–46 (*375*).

GARREAU, Y., and MARINESCO, N. (1929), *Compt. rend. acad. sci.*, **189**:331–333 (*451*).

GARRIGON, —— (1907), French Pat. 394,455 (*712*).

GAUGER, H. C., and GREAVES, R. E. (1946), *Poultry Sci.*, **25**:119–123 (*511*).

GAUJOUX, E., and KRIJANOWSKY, A. (1932), *Compt. rend. soc. biol.*, **110**:1083–1084 (*341*).

GAUTIER, A. (1874), *Compt. rend. acad. sci.*, **79**:227–229 (*327*).

GAUTIER, A. (1900), *Compt. rend. acad. sci.*, **130**:289–291 (*358*).

GELARIE, A. J. (1944), *J. Lab. Clin. Med.*, **29**:532–533 (*508*).

GENEST, P., and BERNARD, R. (1945), *Rev. can. biol.*, **4**:172–192 (*103, 104, 156, 157, 273*).

GENGOU, —— (1902), *Ann. inst. Pasteur*, **16**:734–755 (*540*).

GERBER, L., and CARR, R. H. (1930), *J. Nutrition*, **3**:245–256 (*335*).

GEYELIN, G. R. (1865), *Poultry Breeding in a Commercial Point of View*, Simkin and Marshal, London, 95 pp. (*180*).

GIACOMINI, E. (1893), *Mon. zool. ital.*, **4**:202–265 (*185, 193, 194*).

GIBBONS, N. E., and FULTON, C. O. (1943), *Can. J. Research*, **21(D)**:332–339 (*723, 764, 765*).

GIBBONS, N. E., FULTON, C. O., and HOPKINS, J. W. (1942), *Can. J. Research*, **20(D)**:306–319 (*713*).

GIBBONS, N. E., FULTON, C. O., and REID, M. (1946), *Can. J. Research*, **24(F)**:327–337 (*767*).

GIBBONS, N. E., and MOORE, R. L. (1944a), *Can. J. Research*, **22(F)**:48–53 (*765, 766*).

GIBBONS, N. E., and MOORE, R. L. (1944b), *Can. J. Research*, **22(F)**:54–63 (*766*).

GIBBONS, N. E., and MOORE, R. L. (1946), *Poultry Sci.*, **25**:115–118 (*514*).

GIERSBERG, H. (1921), *Biol. Zentr.*, **41**:252–268 (*193, 222*).

GIERSBERG, H. (1922), *Z. wiss. Zoöl.*, **120**:1–97 (*173, 185, 187, 191, 194, 195, 226, 227*).

GILBERT, A. G. (1891), *Can. Exptl. Farm Rept.*, pp. 221–234 (*65*).

GILLAM, A. E., and HEILBRON, I. M. (1935), *Biochem. J.*, **29**:1064–1067 (*348*).

GISH, C. L., PAYNE, L. F., and PETERSON, W. J. (1940), *Poultry Sci.*, **19**:154–156 (*129*).

GIVENS, J. W., ALMQUIST, H. J., and STOKSTAD, E. L. R. (1935), *Ind. Eng. Chem.*, **27**:972–973 (*386*).

GLABAU, C. A., and KEPES, E. (1935), *Cereal Chem.*, **12**:108–120 (*408, 773*).

GLASER, O. (1913), *Biol. Bull.*, **24**:175–186 (*283*).

GLOMSET, D. A., and BOLLMAN, J. L. (1943), *Gastroenterology*, **1**:776–783 (*568*).

GOBLEY, —— (1846), *Recherches chimiques sur le jaune d'œuf*, DeFain and Thunot, Paris, 40 pp. (*325, 330*).

GOBLEY, —— (1847), *J. pharm. chim.*, (Ser. 3) **11**:409–417 (*317*).

GODFREY, A. B., and TITUS, H. W. (1934), *Poultry Sci.*, **13**:56–60 (*48, 87*).

GODSTON, J. (1934), *Food Industries*, **6**:201–203 (*735*).

GOLDSMITH, J. B. (1928), *J. Morphol.*, **46**:275–315 (*199, 200, 203, 208, 209*).

GOODALE, H. D. (1915), *J. Am. Assoc. Instr. Invest. Poultry Husb.*, **1**:18–19 (*24*).

GOODALE, H. D. (1918), *J. Agr. Research*, **12**:547–574 (*18, 25, 27, 80*).

GOODALE, H. D. (1927), *Proc. World's Poultry Congr.* (*Ottawa*), **3**:87–92 (*55*).

GOODALE, H. D., SANBORN, R., and WHITE, D. (1920), *Massachusetts Agr. Expt. Sta. Bull.*, **199**:93–116 (*29, 30*).

GORDON, C. D. (1940), *Poultry Sci.*, **19**:349 (*140*).

GORESLINE, H. E., and MCFARLANE, V. H. (1944), *U. S. Egg Poultry Mag.*, **50**:116–119, 140–144 (*719*).

GORTER, E. (1934), *Am. J. Diseases Children*, **47**:947–957 (*439, 440*).

GORTER, E., and GRENDEL, F. (1926), *Trans. Faraday Soc.*, **22**:477–483 (*440*).

GORTER, E., and PHILIPPI, G. T. (1934), *Proc. Acad. Sci. Amsterdam*, **37**:788–793 (*440*).

GOSS, L. J. (1941), *Cornell Univ. Vet. College Rept.*, **1939–40**:103–113 (*246, 247, 248*).

GOTHE, F. (1915), *Z. Untersuch. Nahr. u. Genussm.*, **30**:389–399 (*785*).

GOUZON, B. (1934), *Compt. rend. soc. biol.*, **116**:925–926 (*347, 419*).

GRAHAM, J. C. (1930), *Proc. Am. Poultry Sci. Assoc.*, **22**:48–51 (*51*).

GRALÉN, N., and SVEDBERG, T. (1939), *Nature*, **143**:519–520 (*437*).

GRANDEL, F. (1940), *Z. Volksernähr.*, **15**:354–356 (*782*).

GRANT, E. A., DOW, R. B., and FRANKS, W. P. (1941), *Science*, **94**:616 (*401, 457, 464*).

GRAY, J. T. (1887), U. S. Pat. 358,656 (*716*).

GREEN, T. W., and BIRKELAND, J. M. (1942), *Proc. Soc. Exptl. Biol. Med.*, **51**:55–56 (*792*).

GREENSMITH, H. W. (1934), Brit. Pat. 409, 623 (*712*).

GREENSTEIN, J. P. (1938), *J. Biol. Chem.*, **125**:501–513 (*457, 476*).

GREENWOOD, A. W. (1925), *Brit. J. Exptl. Biol.*, **2**:469–492 (*176*).

GREENWOOD, A. W. (1936), *Proc. World's Poultry Congr.* (*Leipzig*), **6**:265–269 (*30*).

GRESHOFF, M., SACK, J., ECK, J. J. VAN, and BOSZ, Q. (1903), *Chem. Ztg.*, **27**:499–501 (*314, 320*).

GRINI, O. (1939), *Proc. World's Poultry Congr.* (*Cleveland*), **7**:496–498 (*659*).

GROB, D. (1943), *J. Gen. Physiol.*, **26**:431–442 (*500, 501*).

GROEBBELS, F. (1927), *J. Ornithol.*, **75**:225–235 (*323*).

GROEBBELS, F., and MÖBERT, F. (1927), *J. Ornithol.*, **75**:376–384 (*114*).

GROH, J. (1926), *Biochem. Z.*, **173**:249–257 (*435*).

GROLLMAN, A. (1931), *Biochem. Z.*, **238**:408–417 (*426*).

GRÖNBERG, A., and LUNDBERG, A. (1928), *Acta Med. Scand.*, **69**:99–118 (*570*).

GRÖNBERG, A., and LUNDBERG, A. (1929), *Acta Med. Scand.*, **72**:291–307 (*570*).

GRONCHI, V. (1932), *Boll. soc. ital. biol. sper.*, **7**:1297–1301 (*569*).

GROSS, A. (1899), *Zur Kenntnis des Ovovitellins*, Diss. Med., Strassburg, 32 pp. (*326*).

GROSSFELD, J. (1933a), *Arch. Geflügelkunde*, **7**:369–374 (*108*).

GROSSFELD, J. (1933b), *Z. Untersuch. Lebensm.*, **65**:311–314 (*337*).

GROSSFELD, J. (1938), *Handbuch der Eierkunde*, Julius Springer, Berlin, 375 pp. (*2*).

GROSSFELD, J. (1940), *Z. Untersuch. Lebensm.*, **80**:1–12 (*783*).

GROSSFELD, J., and KANITZ, H. R. (1937), *Z. Untersuch. Lebensm.*, **74**:471–477 (*129*).

GRUENWALD, P. (1942), *J. Morphol.*, **71**:299–305 (*176*).

GRULEE, C. G. (1920), *Am. J. Diseases Children*, **20**:15–17 (*558*).

GRULEE, C. G., and BONAR, B. E. (1921), *Am. J. Diseases Children*, **21**:89–95 (*558*).

GRUVEL, A. (1902), *Proc.-verb. soc. sci. phys. nat.*, *Bordeaux, 1901–02*, pp. 72–76 (*291, 292*).

GUARNIERI, P. (1929), *Ind. olii minerali e grassi.*, 9:141 (*783*).

GUERRANT, N. B., KOHLER, E., HUNTER, J. E., and MURPHY, R. R. (1935), *J. Nutrition*, 10:167–168 (*619, 620*).

GUERRANT, N. B., and RUDY, W. J. (1939), *Proc. Soc. Exptl. Biol. Med.*, 40:166–169 (*585*).

GUGGENHEIM, A. (1929), *Z. Immunitätsforsch.*, 61:361–380 (*559, 560, 561*).

GUNDERSON, M. F., and GUNDERSON, S. D. (1945), *U. S. Egg Poultry Mag.*, 51:533, 560–562 (*722*).

GUNN, T. E. (1912), *Proc. Zool. Soc. London*, pp. 63–79 (*176*).

GÜNTHER, F. C. (1772), *Sammlung von Nestern und Eyern verschiedener Vögel*, Nürnberg (*88*).

GURNEY, J. H. (1899), *Ibis*, 5:19–42 (*10, 11*).

GUTOWSKA, M. S. (1931), *Quart. J. Exptl. Physiol.*, 21:197–216 (*46*).

GUTOWSKA, M. S., and MITCHELL, C. A. (1945), *Poultry Sci.*, 24:159–167 (*226*).

GUTOWSKA, M. S., and PARKHURST, R. T. (1942a), *Poultry Sci.*, 21:277–287 (*373*).

GUTOWSKA, M. S., and PARKHURST, R. T. (1942b), *Poultry Sci.*, 21:321–328 (*155*).

GUTOWSKA, M. S., PARKHURST, R. T., PARROTT, E. M., and VERBURG, R. M. (1943), *Poultry Sci.*, 22:195–204 (*196, 226*).

GUTTERIDGE, H. S., BIRD, S., MacGREGOR, H. I., and PRATT, J. M. (1944), *Sci. Agr.*, 25:31–42 (*49*).

GUTTERIDGE, H. S., and PRATT, J. M. (1946), *Poultry Sci.*, 25:89–91 (*156*).

GUTTERIDGE, H. S., PRATT, J. M., and O'NEIL, J. B. (1944), *Sci. Agr.*, 24:240–250 (*22*).

GYÖRGY, P., MELVILLE, D. B., BURK, D., and duVIGNEAUD, V. (1940), *Science*, 91:243–245 (*631*).

GYÖRGY, P., MORO, E., and WITEBSKY, E. (1930), *Klin. Wochschr.*, 9:1012–1017 (*557*).

GYÖRGY, P., and ROSE, C. S. (1941), *Science*, 94:261–262 (*631, 632*).

GYÖRGY, P., and ROSE, C. S. (1942), *Proc. Soc. Exptl. Biol. Med.*, 49:294–298 (*352, 629*).

GYÖRGY, P., and ROSE, C. S. (1943), *Proc. Soc. Exptl. Biol. Med.*, 53:55–57 (*631, 632*).

GYÖRGY, P., ROSE, C. S., EAKIN, R. E., SNELL, E. E., and WILLIAMS, R. J. (1941), *Science*, 93:477–478 (*631*).

GYÖRGY, P., ROSE, C. S., and TOMARELLI, R. (1942), *J. Biol. Chem.*, 144:169–173 (*631*).

HAAS, A. R. C. (1918), *J. Biol. Chem.*, 35:119–125 (*454*).

HADLEY, P., and CALDWELL, D. W. (1916), *Rhode Island Agr. Expt. Sta. Bull.*, 164:1–70 (*492, 493, 494*).

HADLEY, P., and CALDWELL, D. W. (1920), *Rhode Island Agr. Expt. Sta. Bull.*, 181:5–64 (*67, 68, 69, 70, 73, 77*).

HAENNI, E. B. (1941), *J. Assoc. Offic. Agr. Chemists*, 24:119–147 (*784, 785*).

HAHN, J. (1930), *Zool. Anz.*, 89:259–264 (*292, 293*).

HAHN, L., and HEVESY, G. (1937), *Nature*, 140:1059–1060 (*241*).

HAINES, R. B. (1930), *Dept. Sci. Ind. Research (Brit.)*, *Food Invest. Repts.*, 1929:39 (*660*).

HAINES, R. B. (1932), *Biochem. J.*, 26:323–336 (*500*).

HAINES, R. B. (1934), *Biol. Rev. Biol. Proc. Cambridge Phil. Soc.*, 9:235–261 (*500*).

HAINES, R. B. (1935), *Dept. Sci. Ind. Research (Brit.)*, *Food Invest. Repts.*, 1934:44–48 (*711*).

HAINES, R. B. (1937), *Dept. Sci. Ind. Research (Brit.)*, *Food Invest., Spec. Repts.*, 45:1–85 (*500*).

HAINES, R. B. (1938), *J. Hyg.*, 38:338–355 (*492, 683, 687, 689*).

HAINES, R. B. (1939), *Dept. Sci. Ind. Research (Brit.)*, *Food Invest., Spec. Repts.*, 47:1–65 (*492, 684, 689, 691, 705, 710*).

HAINES, R. B. (1940), *Chemistry & Industry*, 18:391–396 (*667*).

HAINES, R. B., and MORAN, T. (1940), *J. Hyg.*, 40:453–461 (*169, 496, 497, 498, 691*).

HAITINGER, M. (1928), *Mitt. staatl. tech. Versuchsamtes*, 17:147–156 (*659*).

HALE, H. P. (1933), *Proc. Roy. Soc.* (*London*), 112(B):473–479 (*420, 424*).

HALL, G. O. (1926), *The Relation of the Internal Anatomy of Fowls to Intensity, Cycle, and Annual Egg Production*, Inaug. Diss., Cornell University, Ithaca, New York, 76 pp. (*180, 184*).

HALL, G. O. (1935), *Poultry Sci.*, 14:323–329 (*52, 53*).

HALL, G. O. (1938), *J. Heredity*, 29:51–53 (*11*).

HALL, G. O. (1939), *Poultry Sci.*, 18:282–287 (*117, 140*).

HALL, G. O. (1944), *Poultry Sci.*, 23:259–265 (*99, 100*).

HALL, G. O. (1945a), *Poultry Sci.*, 24:451–458 (*306*).

HALL, G. O. (1945b), *Poultry Sci.*, 24:496–498 (*700, 701*).

HALL, G. O. (1946), *Poultry Sci.*, 25:3–12 (*51*).

HALL, G. O., and MARBLE, D. R. (1931), *Poultry Sci.*, 10:194–203 (*12*).

HALL, G. O., and ROMANOFF, A. L. (1943), *Poultry Sci.*, 22:396–397 (*380*).

HALLAUER, C. (1936), *Z. Hyg. Infektionskrankh.*, 118:605–614 (*563*).

HALLER, A. VON (1768), *Opera minora Anat.*, Lausannae, 3:121 (*289*).

HALNAN, E. T. (1936), *Proc. World's Poultry Congr.* (*Leipzig*), 6:53–64 (*44*).

HALNAN, E. T., and DAY, H. D. (1935), *J. Ministry Agr.* (*Engl.*), 42:236–245 (*139, 638*).

HALNAN, E. T., and MORAN, T. (1937), *Dept. Sci. Ind. Research* (*Brit.*), *Food Invest. Repts.*, 1936:44–45 (*140, 391*).

HAMMOND, J. C., FRITZ, J. C., NESTLER, R. B., and TITUS, H. W. (1944), *Poultry Sci.*, 23:217–220 (*787*).

HAND, D. B. (1935a), *J. Biol. Chem.*, 108:703–707 (*454*).

HAND, D. B. (1935b), *J. Gen. Physiol.*, 18:847–852 (*455, 464, 465*).

HANKE, —— (1908), in *Gefiederte Welt*, quoted by K. FLOERICKE, 1909, *Jahrbuch der Vogelkunde, 1908*, Kosmos Verlag, Stuttgart, 130 pp. (*7*).

HANNAS, R. R. (1920), *New Jersey Agr. Expt. Sta. Hints to Poultrymen*, 8(10):1–4 (*29*).

HANNING, F. M. (1945), *Iowa State Coll. J. Sci.*, 20:10–12 (*396*).

HANSEMANN, D. (1913), *Arch. Entwicklungsmech. Organ.*, 35:223–235 (*181*).

HANSEN, C. H. (1933), *Proc. World's Poultry Congr.* (*Rome*), 5(2):587–593 (*221*).

HARGITT, C. W. (1899), *Zool. Bull.*, 2:225–229 (*265, 274, 286, 287, 290, 292*).

HARGITT, C. W. (1912), *Am. Naturalist*, 46:556–560 (*263, 287, 288, 290, 294*).

HARLAND, H. C. (1927), *J. Genetics*, 18:55–62 (*7*).

HARPER, E. H. (1904), *Am. J. Anat.*, 3:349–386 (*207, 215*).

HARPER, J. A., and MARBLE, D. R. (1945), *Poultry Sci.*, 24:56–60 (*88*).

HARRIS, B. R., EPSTEIN, A. K., and CAHN, F. J. (1941), *Oil & Soap*, 18:179–182 (*777*).

HARRIS, J. A. (1926), *Poultry Sci.*, 6:1–8 (*36*).

HARRIS, J. A. (1927), *Poultry Sci.*, 6:215–224 (*36, 37*).

HART, E. B., STEENBOCK, H., LEPKOVSKY, S., KLETZIEN, S. W. F., HALPIN, J. G., and JOHNSON, O. N. (1925), *J. Biol. Chem.*, 65:579–595 (*617*).

HART, E. B., STEENBOCK, H., WADDELL, J., and ELVEHJEM, C. A. (1928), *J. Biol. Chem.*, 77:797–812 (*604*).

HARTE, R. A. (1945), *Science*, 102:563–564 (*519, 530*).

HARTLEY, G., and LUSHBAUGH, C. C. (1942), *Am. J. Path.*, 18:323–331 (*565*).

HARTUNG, C. (1902), *Z. Biol.*, 43:195–212 (*360*).

HARVEY, E. N., and DANIELLI, J. F. (1936), *J. Cellular Comp. Physiol.*, 8:31–36 (*401*).

HARVEY, W. (1651), *Exerc. de Generatione Animalium*, XIII:55 (*287*).

HARVEY, W. (1737), *Exerc. de Generatione Animalium*, XXIV:98 (*300*).

HARVIE-BROWN, J. A. (1868), *Zoologist*, **3**:1098 (*278*).

HARVIE-BROWN, J. A., and BUNYARD, P. F. (1910), *Brit. Birds*, **3**:252–255 (*6*).

HASTERLIK, A. (1916), *Z. Fleisch- u. Milchhyg.*, **27**:84–87 (*98*).

HAUGE, S. M., and CARRICK, C. W. (1925), *J. Biol. Chem.*, **64**:111–112 (*633*).

HAUGE, S. M., and ZSCHEILE, F. P. (1942), *Science*, **96**:536 (*610, 615, 616*).

HAUGH, R. R. (1933), *U. S. Egg Poultry Mag.*, **39**(3):27, 49 (*390*).

HAUGH, R. R. (1937), *U. S. Egg Poultry Mag.*, **43**:552–555, 572–573 (*665*).

HAUGH, R. R. (1939), *Proc. World's Poultry Congr. (Cleveland)*, **7**:525–528 (*663*).

HAUROWITZ, F., TUNCA, M., SCHWERIN, P., and GÖKSU, V. (1945), *J. Biol. Chem.*, **157**: 621–625 (*531, 532*).

HAUROWITZ, F., TUNCA, M., and YURD, N. (1943), *Tib Fakültesi (Istanbul)*, **6**:3231–3233 (*535*).

HAUSER, E. A., and SWEARINGEN, L. E. (1941), *J. Phys. Chem.*, **45**:644–659 (*438, 439*).

HAUSMANN, W. (1899), *Z. physiol. Chem.*, **27**:95–108 (*330*).

HAUSMANN, W. (1900), *Z. physiol. Chem.*, **29**:136–145 (*330*).

HAWTHORNE, J. R. (1943), *J. Soc. Chem. Ind. (London)*, **62**:135–137 (*756*).

HAWTHORNE, J. R., and BROOKS, J. (1944), *J. Soc. Chem. Ind. (London)*, **63**:232–234 (*751*).

HAYCRAFT, J. B., and DUGGAN, C. W. (1890), *Am. J. Physiol.*, **24**:288–306 (*399, 400*).

HAYDEN, C. E., and SAMPSON, J. (1931), *Cornell Univ. Vet. Coll. Rept.*, **1929–1930**:174–182 (*236, 237, 242*).

HAYS, F. A. (1924), *Am. Naturalist*, **58**:43–59 (*55, 56*).

HAYS, F. A. (1924–25), *Poultry Sci.*, **4**:43–50 (*18, 27, 80*).

HAYS, F. A. (1927), *Proc. World's Poultry Congr. (Ottawa)*, **3**:92–95 (*18*).

HAYS, F. A. (1929a), *J. Agr. Research*, **38**:511–519 (*74*).

HAYS, F. A. (1929b), *Massachusetts Agr. Expt. Sta. Bull.*, **258**:255–302 (*55*).

HAYS, F. A. (1930a), *Proc. Poultry Sci. Assoc.*, **22**:16–19 (*70*).

HAYS, F. A. (1930b), *Proc. World's Poultry Congr. (London)*, **4**:134 (*80*).

HAYS, F. A. (1932a), *Massachusetts Agr. Expt. Sta. Bull.*, **288**:1–8 (*75*).

HAYS, F. A. (1932b), *Massachusetts Agr. Expt. Sta. Bull.*, **289**:1–12 (*18*).

HAYS, F. A. (1934), *Massachusetts Agr. Expt. Sta. Bull.*, **312**:1–8 (*21, 55, 56, 72*).

HAYS, F. A. (1935), *Massachusetts Agr. Expt. Sta. Bull.*, **316**:1–15 (*55, 56*).

HAYS, F. A. (1937a), *Massachusetts Agr. Expt. Sta. Bull.*, **344**:1–28 (*71, 97, 100, 103, 104*).

HAYS, F. A. (1937b), *Poultry Sci.*, **16**:353 (*153*).

HAYS, F. A. (1938), *J. Agr. Research*, **57**:575–581 (*24*).

HAYS, F. A. (1939), *Massachusetts Agr. Expt. Sta. Bull.*, **364**:1–16 (*32*).

HAYS, F. A. (1940), *J. Heredity*, **31**:476 (*71*).

HAYS, F. A. (1941), *Poultry Sci.*, **20**:217–220 (*72*).

HAYS, F. A. (1944a), *Massachusetts Agr. Expt. Sta. Bull.*, **411**:1–16 (*70*).

HAYS, F. A. (1944b), *Poultry Sci.*, **23**:310–313 (*18*).

HAYS, F. A., and SANBORN, R. (1926), *Massachusetts Agr. Expt. Sta. Tech. Bull.*, **7**:53–83 (*29*).

HAYS, F. A., and SANBORN, R. (1930), *Massachusetts Agr. Expt. Sta. Bull.*, **264**:71–85 (*30*).

HAYS, F. A., and SANBORN, R. (1934), *Massachusetts Agr. Expt. Sta. Bull.*, **305**:55–56 (*54*).

HAYS, F. A., SANBORN, R., and JAMES, L. L. (1924), *Massachusetts Agr. Expt. Sta. Bull.*, **220**:44–53 (*19*).

HAYS, F. A., and SUMBARDO, A. H. (1927), *Poultry Sci.*, **6**:196–200 (*145, 147, 369*).

HEALY, D. J., and PETER, A. M. (1925), *Am. J. Physiol.*, **74**:363–368 (*411*).

HEFFTER, A. (1907), *Mediz. Naturw. Arch.*, **1**:81–104 (*475*).

HEIDELBERGER, M. (1938), *J. Am. Chem. Soc.*, **60**:242–244 (*549*).

HEIDELBERGER, M., DAVIS, B., and TREFFERS, H. P. (1941), *J. Am. Chem. Soc.*, **63**:498–503 (*479*).

HEIDELBERGER, M., and KENDALL, F. E. (1935), *J. Exptl. Med.*, **62**:697–720 (*547, 548, 549, 550, 551, 552*).

HEIMAN, V. (1935), *Poultry Sci.*, **14**:137–146 (*350*).

HEIMAN, V., and CARVER, J. S. (1935), *U. S. Egg Poultry Mag.*, **41(8)**:40–41 (*126, 131*).

HEIMAN, V., and CARVER, J. S. (1936a), *Poultry Sci.*, **15**:141–148 (*644, 645*).

HEIMAN, V., and CARVER, J. S. (1936b), *U. S. Egg Poultry Mag.*, **42**:426–429 (*644*).

HEIMAN, V., and WILHELM, L. A. (1937a), *J. Agr. Research*, **54**:551–557 (*644*).

HEIMAN, V., and WILHELM, L. A. (1937b), *Poultry Sci.*, **16**:400–403 (*128*).

HEINROTH, O. (1922), *J. Ornithol.*, **70**:172–285 (*61*).

HEKTOEN, L., and COLE, A. G. (1928), *J. Infectious Diseases*, **42**:1–24 (*327, 541, 542*).

HEKTOEN, L., and COLE, A. G. (1929), *J. Infectious Diseases*, **44**:165–168 (*556*).

HEKTOEN, L., and COLE, A. G. (1932), *J. Infectious Diseases*, **50**:171–176 (*547*).

HEKTOEN, L., and COLE, A. G. (1937), *Proc. Soc. Exptl. Biol. Med.*, **36**:97–99 (*543*).

HELLER, B., & Co., (1927), *Heller's Guide for Ice Cream Makers*, B. Heller & Co., Chicago, 312 pp. (*774*).

HELLER, V. G., PAUL, H., and THOMPSON, R. B. (1934), *J. Biol. Chem.*, **106**:357–364 (*236, 239, 241*).

HELLERMAN, L., CHINARD, F. P., and DEITZ, V. R. (1943), *J. Biol. Chem.*, **147**:443–462 (*476*).

HELLERMAN, L., CHINARD, F. P., and RAMSDELL, P. A. (1941), *J. Am. Chem. Soc.*, **63**:2551–2553 (*471, 476*).

HENDERSON, D. C. (1945), *U. S. Egg Poultry Mag.*, **51**:122 (*43*).

HENDERSON, E. W., and WILCKE, H. L. (1933), *Poultry Sci.*, **12**:266–273 (*130, 132*).

HENDRICKS, W. A. (1934), *Poultry Sci.*, **13**:290–294 (*83*).

HENDRICKSON, N., and SWAN, G. C. (1918), *J. Ind. Eng. Chem.*, **10**:614–617 (*670, 706*).

HENDRIX, B. M., and DENNIS, J. (1938), *J. Biol. Chem.*, **126**:315–322 (*477*).

HENDRIX, B. M., and DENNIS, J. (1943), *Arch. Biochem.*, **2**:371–380 (*478*).

HENDRIX, B. M., and WILSON, V. (1928), *J. Biol. Chem.*, **79**:389–403 (*431, 467*).

HENNEGUY, L. F., (1911), *Compt. rend. soc. biol.*, **70**:779–780 (*257, 288, 293*).

HENNERTY, M. (1930), *Proc. World's Poultry Congr.* (London), **4**:881–887 (*53*).

HENRIQUES, V., and HANSEN, C. (1903), *Skand. Arch. Physiol.*, **14**:390–397 (*342, 485*).

HENRY, K. M., KON, S. K., and THOMPSON, S. Y. (1940), *Biochem. J.*, **34**:998–1001 (*588*).

HENRY, M. (1928), *Bull. acad. vét. France*, **1**:157–158 (*305, 306*).

HENRY, W. C., and BARBOUR, D. (1933), *Ind. Eng. Chem.*, **25**:1054–1058 (*393, 394, 395, 396*).

HEPBURN, J. S., FEGLEY, N. A., SOHN, K. S., and COX, J. (1933), *Am. J. Pharm.*, **105**:547–550 (*600*).

HEPBURN, J. S., and KATZ, A. B. (1927), *J. Franklin Inst.*, **203**:835–841 (*314, 317, 320, 483, 599, 600*).

HEPBURN, J. S., and MIRAGLIA, P. R. (1937), *J. Franklin Inst.*, **223**:375–377 (*314, 317, 320, 483, 599, 600*).

HERCUS, C. E., and ROBERTS, K. C. (1927), *J. Hyg.*, **26**:49–83 (*358, 360, 361, 607*).

HERLITZKA, A. (1907), *Arch. ital. biol.*, **48**:169–189 (*470*).

HERLITZKA, A. (1910), *Z. Chem. u. Ind. Kolloide*, **7**:251–256 (*414, 454*).

HERRAIZ, M. L., and ALVAREZ HERRERO, H. G. DE (1943), *Rev. asoc. argentina dietol.* **1**:46–48 (*624*).

HERRICK, E. H. (1944), *Poultry Sci.*, **23**:65–66 (*219, 243*).

HERRICK, F. H. (1899), *Am. Naturalist*, **33**:409–414 (*287, 289, 291, 292, 293*).

HERRICK, F. H. (1907), *Science*, **25**:725–726 (*5*).

HERRING, V., quoted by BRODY, S. (1945), *Bioenergetics and Growth*, Reinhold Publishing Corp., New York, 1023 pp. (*598, 599, 600*).

HERRIOTT, R. M. (1936), *J. Gen. Physiol.*, **19**:283–299 (*476*).

HERSHEY, J. M. (1930), *Am. J. Physiol.*, **93**:657–658 (*568*).

HERSHEY, J. M., and SOSKIN, S. (1931), *Am. J. Physiol.*, **98**:74–85 (*568*).

HERTWIG, R. (1923), *J. Assoc. Offic. Agr. Chemists*, **6**:508–510 (*785*).

HERTZ, R., FRAPS, R. M., and SEBRELL, W. H. (1943), *Proc. Soc. Exptl. Biol. Med.*, **52**: 142–144 (*218*).

HERVEY, G. W., and DECKER, M. (1926), *Poultry Sci.*, **5**:149–151 (*22, 75*).

HERZOG, R. O. (1907), *Z. Elektrochem.*, **13**:533–541 (*436*).

HERZOG, R. O., and GONELL, H. W. (1925), *Kolloid-Z.*, **36** (suppl.—*Zsigmondy Festschrift*): 44–48 (*165*).

HERZOG, R. O., and MARGOLIS, M. (1909), *Z. physiol. Chem.*, **60**:298–305 (*529, 533*).

HESS, A. F. (1923a), *Proc. Soc. Exptl. Biol. Med.*, **20**:369–370 (*616*).

HESS, A. F. (1923b), *J. Am. Med. Assoc.*, **81**:15–17 (*616*).

HESS, A. F., and WEINSTOCK, M. (1924), *Proc. Soc. Exptl. Biol. Med.*, **21**:441–442 (*620*).

HESSELVIK, L. (1938), *Z. physiol. Chem.*, **254**:144–146 (*433*).

HETHERINGTON, D. C., and CRAIG, J. S. (1939), *Proc. Soc. Exptl. Biol. Med.*, **42**:831–834 (*791*).

HETHERINGTON, D. C., and CRAIG, J. S. (1940), *Proc. Soc. Exptl. Biol. Med.*, **44**:282–285 (*791*).

HEUSER, G. F., and NORRIS, L. C. (1934), *Proc. World's Poultry Congr. (Rome)*, **5(2)**:551–558 (*74, 75*).

HEUSER, G. F., and NORRIS, L. C. (1946), *Poultry Sci.*, **25**:173–179 (*373*).

HEVESY, G., and HAHN, L. (1938), *Kgl. Danske Videnskab. Selskab, Biol. Medd.*, **14**:1–39 (*363*).

HEWITT, E. A. (1931), *Vet. Med.*, **26**:1–8 (*274*).

HEWITT, E. A. (1939), *J. Am. Vet. Med. Assoc.*, **95**:201–210 (*273, 275, 307*).

HEWITT, L. F. (1927), *Biochem. J.*, **21**:216–224 (*452*).

HEWLETT, R. T. (1892), *J. Physiol.*, **13**:493–512 (*327*).

HEYWANG, B. W. (1938), *Poultry Sci.*, **17**:240–247 (*231*).

HEYWANG, B. W. (1940), *Poultry Sci.*, **19**:29–34 (*40, 83*).

HEYWANG, B. W. (1946), *Poultry Sci.*, **25**:215–222 (*86, 119*).

HEYWANG, B. W., and TITUS, H. W. (1941), *Poultry Sci.*, **20**:483–489 (*643*).

HICKS, C. S., and HOLDEN, H. F. (1934), *Australian J. Exptl. Biol. Med. Sci.*, **12**:91–97 (*455, 472*).

HIDEN, R. B. (1921), *Am. Naturalist*, **55**:373–377 (*250*).

HIGGINSON, T. W. (1863), *Out-door Papers*, Ticknor and Fields, Boston, 370 pp. (*1*).

HILL, A. V. (1930a), *Proc. Roy. Soc. (London)*, **106(B)**:477–505 (*402*).

HILL, A. V. (1930b), *Trans. Faraday Soc.*, **26**:667–678 (*420*).

HILL, R. (1930), *Proc. Roy. Soc. (London)*, **107(B)**:205–214 (*604*).

HINKS, E. (1923), *Analyst*, **48**:542 (*778*).

HINSHAW, W. R., JONES, E. E., and GRAYBILL, H. W. (1931), *Poultry Sci.*, **10**:375–382 (*34, 35*).

HINSHAW, W. R., UPP, C. W., and MOORE, J. M. (1926), *J. Am. Vet. Med. Assoc.*, **68**: 631–641 (*509*).

HIS, W. (1868), *Untersuchungen über die erste Anlage des Wirbelthierleibes. Die erste Entwickelung des Hühnchens im Eis.* F. C. W. Vogel, Leipzig, 237 pp. (*180*).

HOAGLAND, R., and LEE, A. R. (1924), *J. Agr. Research*, **28**:461–472 (*623, 624*).

HOFFMAN, H. A. (1932), *Science,* **76**:489–490 *(247).*

HOFFMANN, J. (1943), *Chem. Ztg.,* **57**:49–52 *(358).*

HOFFMANN, P. (1901), *Z. Anal. Chem.,* **40**:450–459 *(360).*

HOLDEFLEISS, P. (1911), *Ber. phys. Lab. Versuchanst. landwirt. Inst. Univ. Halle,* **20**:93–111 *(100).*

HOLDEN, H. F., and FREEMAN, M. (1930), *Australian J. Exptl. Biol. Med. Sci.,* **7**:13–26 *(452, 470, 471).*

HOLFORD, F. E. (1930), *J. Immunol.,* **19**:177–216 *(557).*

HOLL, M. (1890), *Sitzber. Akad. Wiss. Wien. Math. naturw. Klasse,* **99**:311–370 *(215).*

HOLLAND, G., HINSBERG, K., KOHLS, G., and NICKEL, V. (1934), *Z. ges. exptl. Med.,* **93**:62–68 *(517).*

HOLLANDE, A. C., and GATÉ, J. (1918), *Compt. rend. soc. biol.,* **81**:148–151 *(540).*

HOLLANDER, F. d' (1902), *Anat. Anz.,* **21**:168–171 *(199).*

HOLLANDER, F. d' (1905), *Arch. Anat. Micr.,* **7**:117–180 *(215).*

HOLMAN, R. T., TAYLOR, M. W., and RUSSELL, W. C. (1945), *J. Nutrition,* **29**:277–281 *(43).*

HOLMBERG, C. G. (1939), *Biochem. J.,* **33**:1901–1906 *(606).*

HOLST, W. F., and ALMQUIST, H. J. (1931a), *Hilgardia,* **6**:45–48 *(415).*

HOLST, W. F., and ALMQUIST, H. J. (1931b), *Hilgardia,* **6**:49–60 *(139, 674).*

HOLST, W. F., ALMQUIST, H. J., and LORENZ, F. W. (1932a), *Poultry Sci.,* **11**:144–149 *(658).*

HOLST, W. F., ALMQUIST, H. J., and LORENZ, F. W. (1932b), *U. S. Egg Poultry Mag.,* **38(7)**:50–53, 57 *(150).*

HOLT, L. E., ALBANESE, A. A., BRUMBACK, J. E., KADJI, C., and WANGERIN, D. M. (1941), *Proc. Soc. Exptl. Biol. Med.,* **48**:726–728 *(591).*

HOLT, L. E., ALBANESE, A. A., SHETTLES, L. B., KADJI, C., and WANGERIN, D. M. (1942), *Federation Proc.,* **1(2)**:116–117 *(591).*

HOLZAPFEL, L. (1938), *Kolloid-Z.,* **85**:272–278 *(436).*

HOOKER, S. B., and BOYD, W. C. (1934), *J. Immunol.,* **26**:469–479 *(546).*

HOOVER, S. R. (1938), *J. Assoc. Offic. Agr. Chemists,* **21**:496–502 *(646, 647).*

HOOVER, S. R. (1939), *U. S. Egg Poultry Mag.,* **45**:162–163 *(714).*

HOPKINS, F. G. (1900), *J. Physiol.,* **25**:306–330 *(477).*

HOPKINS, F. G. (1930), *Nature,* **126**:328–330 *(457, 461).*

HOPKINS, F. G., and PINKUS, S. N. (1898), *J. Physiol.,* **23**:130–136 *(428).*

HOROWITZ, A. (1903), *Baumgarten's Jahresber.,* **19**:984–985 *(250, 495).*

HOROWITZ-VLASOVA, L. M., BALIN, E. O., and NOVOTELNOV, N. V. (1932), *Schriften zentral. biochem. Forsch. Inst. Nahr. u. Genussmittelind. (U.S.S.R.),* **2**:3–9 *(493).*

HORRALL, B. E. (1934), *Ice Cream Field,* **24(6)**:24 *(775).*

HORRALL, B. E. (1935), *Indiana Agr. Expt. Sta. Bull.,* **401**:1–31 *(784).*

HORST, M. G. TER (1936), *Rec. trav. chim.,* **55**:33–42 *(440).*

HOSOI, K., ALVAREZ, W. C., and MANN, F. C. (1928), *Arch. Int. Med.,* **41**:112–126 *(580).*

HOUGHTON, C. E. (1935), *Agr. Gazette, Miscellaneous Publication,* **3011**:283–291 *(53).*

HOUGHTON, H. W., and WEBER, F. C. (1914), *Biochem. Bull.,* **3**:447 *(685).*

HOUSSET, —— (1785), *Neuchatel,* p. 72 *(290).*

HOWARD, H. E. (1914), *The British Warblers,* R. H. Porter, London, Vol. 1, 203 pp. *(5).*

HOWARD, H. E. (1920), *Territory in Bird Life,* John Murray, London, 308 pp. *(5).*

HOXIE, W. (1887), *Ornithol. Oologist,* **12**:207 *(107).*

HOXIE, W. (1890), *Ornithol. Oologist,* **15**:165–166 *(107).*

HUANG, T. C., and WU, H. (1930), *Chinese J. Physiol.,* **4**:221–230 *(474).*

HUEPER, W. C. (1942), *Am. J. Path.,* **18**:895–933 *(564).*

HUEPPE, F. (1888), *Centr. Bakt. Parasitenk.*, **4**:80–81 *(507)*.

HÜFNER, C. G. (1892), *Arch. Anat. Physiol. (Physiol. Abt.)*, pp. 467–479 *(383, 385)*.

HUGHES, J. S. (1936), *Kansas Agr. Expt. Sta. Biennial Rept.*, **1934–1936**:84–85 *(221)*.

HUGHES, J. S., and PAYNE, L. F. (1937), *Poultry Sci.*, **16**:135–138 *(127, 349, 610)*.

HUGHES, J. S., PAYNE, L. F., TITUS, R. W., and MOORE, J. M. (1925), *J. Biol. Chem.*, **66**:595–600 *(617)*.

HUGHES, J. S., and SCOTT, H. M. (1936), *Poultry Sci.*, **15**:349–351 *(328)*.

HUGHES, J. S., SCOTT, H. M., and ANTELYES, J. (1936), *Ind. Eng. Chem., Anal. Ed.*, **8**:310–311 *(518)*.

HUGHES, J. S., TITUS, R. W., and SMITS, B. L. (1927), *Science*, **65**:264 *(225)*.

HUGOUNENQ, L. (1906), *Compt. rend. acad. sci.*, **142**:173–175 *(333)*.

HUGOUNENQ, L., and MOREL, A. (1905), *Compt. rend. acad. sci.*, **140**:1065–1067 *(528)*.

HUGOUNENQ, L., and MOREL, A. (1907), *Bull. soc. chim. France*, (Ser. 4) **1**:154–165 *(333)*.

HÜLPHERS, G. (1939), *Skand. Vet. Tid.*, **29**:1213–1226 *(510, 515)*.

HULSEBOSCH, C. J. VAN L. (1927), *Pharm. Weekblad*, **64**:325 *(348)*.

HUMBLET, J., TEYSSIER, M., and VALKENER, L. DE (1939), French Pat. 842,774 *(800)*.

HUNSCHER, H. A. (1930), *J. Biol. Chem.*, **86**:37–57 *(621)*.

HUNT, C. H., WINTER, A. R., and BETHKE, R. M. (1939), *Poultry Sci.*, **18**:330–336 *(626)*.

HUNTER, J. A., VAN WAGENEN, A., and HALL, G. O. (1936), *Poultry Sci.*, **15**:115–118 *(126, 140, 141, 643, 644, 645)*.

HUNTER, R. F. (1923), *Chem. News*, **127**:134–135 *(422, 423)*.

HURST, C. C. (1905), *Evolution Comm. Roy. Soc., Rept.*, **2**:131–154 *(29, 100)*.

HURST, C. C. (1921), *Proc. World's Poultry Congr. (The Hague)*, **1**:3–20 *(18, 29, 71)*.

HUTT, F. B. (1938), *Am. Naturalist*, **72**:268–284 *(57)*.

HUTT, F. B. (1939), *Poultry Sci.*, **18**:276–278 *(274)*.

HUTT, F. B. (1946), *Auk*, **63**:171–174 *(307)*.

HUTT, F. B., and BOZIVICH, H. (1946), *Poultry Sci.*, **25**:554–561 *(72)*.

HUTT, F. B., COLE, R. K., and BRUCKNER, J. H. (1941), *Poultry Sci.*, **20**:514–526 *(57, 247)*.

HUTT, F. B., and GRUSSENDORF, D. T. (1933), *J. Exptl. Zoöl.*, **65**:199–214 *(213)*.

HUXLEY, J. S. (1927), *J. Linnean Soc. London, Zoöl.*, **36**:457–466 *(61, 63)*.

IDZUMI, S. (1924), *Mitt. Med. Fac. Univ. Tokyo*, **32**:197–216 *(343)*.

I. G. FARBENINDUSTRIE AKT.-GES. (1931), French Pat. 720,189 *(801)*.

ILJIN, M. D. (1917), *Studies on Development of the Chicken Embryo* (in Russian), Petrograd, 115 pp. *(317, 320, 358, 411)*.

INGERSOLL, A. M. (1910), *Condor*, **12**:15–17 *(94, 259, 275)*.

INGERSOLL, E. (1897), *Harper's Mag.*, **96**:40–57 *(7, 63)*.

ISAKI, T. (1930), *Z. physiol. Chem.*, **188**:189–192 *(411)*.

ISCOVESCO, H. (1910), *Compt. rend. soc. biol.*, **69**:622–624 *(427)*.

IVANOVA, S. (1935), *Arch. exptl. Path. Pharmakol.*, **179**:349–359 *(51)*.

IWAKURA, N. (1939), *Japan. J. Med. Sci. VIII. Internal Med. Pediat. Psychiat. 5, No. 2, Proc. Jap. Soc. Internal Med.*, **34**:181–182 *(558)*.

IWATA, M., SUSUKI, S., and IWATA, K. (1941), *Bull. Inst. Phys. Chem. Research (Tokyo)*, **20**:25–40 *(517)*.

JAAP, R. G., PENQUITE, R., and THOMPSON, R. B. (1943), *Poultry Sci.*, **22**:11–19 *(75)*.

JACKSON, S. H., DRAKE, T. G. H., SLINGER, S. J., EVANS, E. V., and POCOCK, R. (1946), *J. Nutrition*, **32**:567–581 *(626)*.

JACOBSEN, N. P. (1910), Danish Pat. 14,902 (*712*).

JACOBY, E. (1910), German Pat. 251,281 (*712*).

JADASSOHN, W. (1926), *Schweiz. med. Wochschr.*, **56**:667–670 (*559*).

JAKOB, H. (1916), *Tijdschr. Diergeneesk.*, **43**:411–428 (*264*).

JANKE, A., and JIRAK, L. (1934), *Biochem. Z.*, **271**:309–323 (*415, 671*).

JARVIS, L. G. (1899), *Ontario Agr. Coll. Expt. Farms, Ann. Rept.*, **1898**:130–134 (*697*).

JASCHIK, A., and KIESELBACH, J. (1931), *Z. Untersuch. Lebensm.*, **62**:572–575 (*358*).

JEAN, F. (1903), *Ann. chim. anal.*, **8**:51–53 (*788*).

JEANS, P. C., STEARNS, G., GOFF, E. A., MCKINLEY, J. B., and OELKE, M. J. (1933), *Am. J. Diseases Children*, **46**:69–89 (*602*).

JEFFREY, F. P. (1941), *Poultry Sci.*, **20**:298–301 (*645*).

JEFFREY, F. P. (1945), *Poultry Sci.*, **24**:241–244 (*301*).

JEFFREY, F. P., and DARAGO, V. (1940), *New Jersey Agr. Expt. Sta. Bull.*, **682**:1–16 (*658*).

JEFFREY, F. P., and PINO, J. (1943), *Poultry Sci.*, **22**:230–234 (*304*).

JEFFREY, F. P., and PLATT, S. P. (1941), *New Jersey Agr. Expt. Sta. Bull.*, **687**:1–23 (*19, 20*).

JENKINS, M. K., HEPBURN, J. S., SWAN, C., and SHERWOOD, C. M. (1920), *Ice and Refrig.*, **58**:140–147 (*691*).

JENKINS, M. K., and PENNINGTON, M. E. (1919), *U. S. Dept. Agr. Bull.*, **775**:1–36 (*670*).

JERNE, H. (1911), Brit. Pat. 2145 (*712*).

JESSEN-HANSEN, H. (1927), *Compt. rend. trav. lab. Carlsberg*, **16**(10):1–20 (*453*).

JOHLIN, J. M. (1933), *J. Gen. Physiol.*, **16**:605–613 (*420*).

JOHNS, C. K., and BÉRARD, H. L. (1946), *Sci. Agr.*, **26**:34–42 (*736*).

JOHNSON, E. P., and ANDERSON, G. W. (1933), *J. Am. Vet. Med. Assoc.*, **82**:258–259 (*510*).

JONES, D. B., GERSDORFF, C. E. F., and MOELLER, O. (1924), *J. Biol. Chem.*, **62**:183–195 (*333*).

JONES, D. B., MURPHY, J. C., and MOELLER, O. (1925), *Am. J. Physiol.*, **71**:265–273 (*615*).

JONES, D. B., and WIDNESS, K. D. (1946), *J. Nutrition*, **31**:675–683 (*593*).

JONES, H. I., and DUBOIS, R. (1920), *Ind. Eng. Chem.*, **12**:751–757 (*700, 714*).

JONES, I. D., and GORTNER, R. A. (1932), *J. Phys. Chem.*, **36**:387–436 (*401*).

JORDAN, R., and PETTIJOHN, M. S. (1946), *Cereal Chem.*, **23**:265–277 (*772*).

JORDAN, R., and SISSON, M. S. (1943a), *U. S. Egg Poultry Mag.*, **49**:168–169, 184 (*771*).

JORDAN, R., and SISSON, M. S. (1943b), *U. S. Egg Poultry Mag.*, **49**:218–221 (*771*).

JOSEPHS, H. W. (1939), *Bull. Johns Hopkins Hosp.*, **65**:167–195 (*605*).

JUCKENACK, A., and PASTERNACK, R. (1904), *Z. Untersuch. Nahr. u. Genussm.*, **8**:94–100 (*783*).

JUKES, T. H. (1933), *J. Biol. Chem.*, **103**:425–437 (*332, 333, 343*).

JUKES, T. H. (1941), *J. Nutrition*, **21**:193–200 (*627*).

JUKES, T. H., FRASER, D. T., and ORR, M. D. (1934), *J. Immunol.*, **26**:353–360 (*562*).

JUKES, T. H., and KAY, H. D. (1932a), *J. Nutrition*, **5**:81–101 (*336, 430*).

JUKES, T. H., and KAY, H. D. (1932b), *J. Biol. Chem.*, **98**:783–788 (*333*).

JUKES, T. H., and KAY, H. D. (1932c), *J. Exptl. Med.*, **56**:469–482 (*562*).

JULL, M. A. (1924a), *Poultry Sci.*, **3**:77–88 (*115, 116, 118*).

JULL, M. A. (1924b), *Poultry Sci.*, **3**:153–167, 170–172 (*74, 76*).

JULL, M. A. (1933), *J. Heredity*, **24**:93–101 (*55, 56*).

JUNG, G. S. (1671), *Misc. nat. cur.*, decur. I, ann. II, obs. **250**:348 (*287*).

JUNGHERR, E., and CLANCY, C. F. (1939), *J. Infectious Diseases*, **64**:1–17 (*514*).

JUNGHERR, E., and MINARD, E. L. (1942), *J. Am. Vet. Med. Assoc.*, **100**(778):38–46 (*511*).

JUSTOW, N. (1927), *Anat. Anz.*, **64**:184–186 (*290, 293*).

KABLE, G. W., Fox, F. E., and LUNN, A. G. (1928), *Oregon Agr. Expt. Sta. Bull.*, **231:** 1–37 (*51*).

KAHANE, E., and LÉVY, J. (1944), *Compt. rend. acad. sci.*, **219**:431–433 (*537*).

KAR, A. B. (1947), *Anat. Record*, **97**:175–195 (*182*).

KARADY, S., and BROWNE, J. S. L. (1939), *J. Immunol.*, **37**:463–468 (*553*).

KARRER, P., and SCHÖPP, K. (1934), *Helv. Chim. Acta*, **17**:735–737 (*347, 487*).

KATO, K., and KO, J. (1938), *J. Coll. Agr., Tokyo Imp. Univ.*, **3(2)**:139–153 (*166, 368, 369, 382*).

KAUCHER, M., GALBRAITH, H., BUTTON, V., and WILLIAMS, H. H. (1943), *Arch. Biochem.*, **3**:203–215 (*340*).

KAUFFMANN, F. (1934), *Zentr. Bakt. Parasitenk.*, **132**:337–342 (*514*).

KAUFFMANN, F., and SILBERSTEIN, W. (1934), *Zentr. Bakt. Parasitenk.*, **132**:431–437 (*513*).

KAUFMAN, L., and BACZKOWSKA, H. (1938), *Mém. inst. nat. polon. écon. rurale Pulawy*, **17**:176–187 (*130*).

KAUPP, B. F. (1918), *The Anatomy of the Domestic Fowl*, W. B. Saunders Co., Philadelphia, 373 pp. (*183, 185*).

KAUPP, B. F., and DEARSTYNE, R. S. (1927), *North Carolina Agr. Expt. Sta. Bull.*, **29**:1–44 (*509*).

KAWAMOTO, T. (1936), *Japan. Z. Microbiol. u. Path.*, **30**:1108–1113 (*508*).

KAY, H. D., and MARSHALL, P. G. (1928), *Biochem. J.*, **22**:1264–1269 (*326, 330, 333, 433, 454*).

KEHOE, R. A., CHOLAK, J., and LARGENT, E. J. (1944), *J. Am. Water Works Assoc.*, **36:** 637–644 (*363*).

KEILIN, D., and MANN, T. (1939), *Nature*, **144**:442–443 (*606*).

KEKWICK, R. A., and CANNAN, R. K. (1936a), *Biochem. J.*, **30**:227–234 (*431, 432*).

KEKWICK, R. A., and CANNAN, R. K. (1936b), *Biochem. J.*, **30**:235–240 (*479*).

KELLAWAY, C., and WILLIAMS, F. E. (1933), *Australian J. Exptl. Biol. Med. Sci.*, **11**:81–94 (*569*).

KELLY, A. (1901), *Jenaische Z. Natur.*, **35**:429–494 (*161, 382*).

KEMPSTER, H. L. (1924), *Missouri Agr. Expt. Sta. Bull.*, **225**:1–16 (*42*).

KEMPSTER, H. L. (1927), *Proc. World's Poultry Congr.* (*Ottawa*), **3**:143–147 (*22, 23, 24, 75*).

KEMPSTER, H. L., and HENDERSON, E. W. (1922), *Missouri Agr. Expt. Sta. Bull.*, **197**:1–95 (*19*).

KENDEIGH, S. C. (1941), *Ecology*, **22**:237–248 (*51, 87, 213*).

KENNARD, D. C. (1925), *Poultry Sci.*, **4**:109–117 (*372*).

KENNARD, D. C., and CHAMBERLIN, V. D. (1931), *Ohio Agr. Expt. Sta. Bull.*, **476**:1–22 (*51*).

KENNARD, D. C., and CHAMBERLIN, V. D. (1944), *Ohio Agr. Expt. Sta. Bimonthly Bull.*, **29(228)**:186–187 (*787*).

KENSLER, C. J., WADSWORTH, C., SUGIURA, K., RHOADS, C. P., DITTMER, K., and DU-VIGNEAUD, V. (1943), *Cancer Research*, **3**:823–824 (*632*).

KENTON, H. B. (1938), *J. Infectious Diseases*, **62**:48–51 (*563*).

KHAW, O. K. (1930), *China Med. J.*, **44**:922–923 (*305*).

KIKUMA, L. (1917), *Agr. Expt. Sta., Govt. of Formosa, Japan, Bull.*, **114** (*806*).

KILGORE, L. B. (1933), *U. S. Egg Poultry Mag.*, **39(3)**:42–45, 63 (*398*).

KING, A., and MUKHERJEE, L. N. (1940), *J. Soc. Chem. Ind.* (*London*), **59**:185–191 (*399*).

KING, E. J. (1931), *Biochem. J.*, **25**:799–811 (*537*).

KING, E. J. (1934), *Biochem. J.*, **28**:476–481 (*538, 539*).

KINGHORNE, J. W. (1920), *U. S. Dept. Agr., Farmer's Bull.*, **1109**:1–7 (*700*).

KIONKA, H. (1894), *Anat. Hefte*, **3**:391–445 (*122*).

KIRKPATRICK, W. F., and CARD, L. E. (1917), *Storrs (Connecticut) Agr. Expt. Sta. Bull.*, 89:255–302 (*29*).

KITT, M. (1897), *Chem. Ztg.*, 21:303–304 (*483*).

KLEIN, W. (1933), *Arch. Geflügelkunde*, 7:65–74 (*31, 45*).

KLIMMER, M. (1930), *Berlin. tierärztl. Wochschr.*, 46:702–710 (*515*).

KLIMMER, M. (1931), *Münch. med. Wochschr.*, 78:1212 (*510*).

KLIMMER, M. (1932), *Berlin. tierärztl. Wochschr.*, 48:737–739 (*510*).

KLINE, R. W., and FOX, S. W. (1946), *Iowa State Coll. J. Sci.*, 20:265–267 (*755*).

KLOSE, A. A., and ALMQUIST, H. J. (1937), *Poultry Sci.*, 16:173–174 (*144, 172, 223, 228, 347*).

KLOSE, A. A., JONES, G. I., and FEVOLD, H. L. (1943), *Ind. Eng. Chem.*, 35:1203–1205 (*614, 615, 620, 624, 626, 627, 628*).

KNECHT, E. (1920), *J. Soc. Dyers Colourists*, 36:195–198 (*408*).

KNIERIEM, W. VON (1885), *Z. Biol.*, 21:67–139 (*585*).

KNOWLES, H. R., HART, E. B., and HALPIN, J. G. (1935), *Poultry Sci.*, 14:83–89 (*236*).

KNOX, C. W. (1930a), *Iowa Agr. Expt. Sta. Research Bull.*, 119:311–332 (*21, 22*).

KNOX, C. W. (1930b), *Iowa Agr. Expt. Sta. Research Bull.*, 128:237–252 (*23*).

KNOX, C. W., and BITTENBENDER, H. A. (1927), *Iowa Agr. Expt. Sta. Research Bull.*, 103:51–64 (*19*).

KNOX, C. W., and GODFREY, A. B. (1934), *Poultry Sci.*, 13:18–22 (*140*).

KNOX, C. W., and GODFREY, A. B. (1938), *Poultry Sci.*, 17:159–162 (*140, 141, 142*).

KNOX, C. W., and GODFREY, A. B. (1940), *Poultry Sci.*, 19:291–294 (*140, 141*).

KNOX, C. W., JULL, M. A., and QUINN, J. P. (1935), *J. Agr. Research*, 50:573–589 (*18*).

KNOX, C. W., and OLSEN, M. W. (1938), *Poultry Sci.*, 17:193–199 (*57*).

KOCH, M., and RABINOWITSCH, L. (1907), *Arch. path. Anat. (Virchow's)*, 190:246–541 (*494*).

KOCH, W. (1934), *Klin. Wochenschr.*, 13:1647–1648 (*212, 217*).

KOEHN, R. C., and COLLATZ, F. A. (1944), *J. Assoc. Offic. Agr. Chemists*, 27:451–455 (*785*).

KOENIG, M. C., KRAMER, M. M., and PAYNE, L. F. (1935), *Poultry Sci.*, 14:178–182 (*611*).

KOGA, T. (1923), *Biochem. Z.*, 141:430–446 (*517, 518*).

KOGAN, I. M. (1940), *Voprosy Pitaniya*, 9(4):13–19 (*358*).

KÖGL, F., and TÖNNIS, B. (1936), *Z. physiol. Chem.*, 242:43–73 (*631*).

KOHMURA, T. (1931), *J. Heredity*, 22:77–80 (*66*).

KOJO, K. (1911), *Z. physiol. Chem.*, 75:1–12 (*317, 320, 352*).

KOMORI, Y. (1926), *J. Biochem. (Japan)*, 6:1–20 (*333*).

KON, S. K. (1940), *Chemistry & Industry*, 18:360–363 (*585*).

KONDO, K., and IWAMAE, H. (1936), *J. Agr. Chem. Soc. Japan*, 13:554–557 (*454*).

KÖNIG, J., and FARWICK, B. (1876), *Z. Biol.*, 12:497–512 (*314*).

KÖNIG, J., and KRAUCH, C. (1878), *Chemische Zusammensetzung der Menschlichen Nahrung- und Genussmittel*, fourth edition (1903), Julius Springer, Berlin, Vol. 1, 1535 pp. (*314, 317*).

KONOPACKA, B. (1933), *Arch. Biol.*, 44:251–305 (*202, 203*).

KOPEĆ, S. (1922), *Mém. inst. nat. polon. écon. rurale Pulawy*, 3:328–342 (*97, 100*).

KOPEĆ, S. (1924), *Mém. inst. nat. polon. écon. rurale Pulawy*, 5:294–327 (*71*).

KOPEĆ, S. (1926a), *J. Genetics*, 16:269–286 (*99*).

KOPEĆ, S. (1926b), *Mém. inst. nat. polon. écon. rurale Pulawy*, 7:158–179 (*100*).

KOPEĆ, S. (1927), *Arch. Geflügelkunde*, 1:344–357 (*99, 101*).

KOPEĆ, S., and GREENWOOD, A. W. (1929), *Arch. Entwicklungsmech. Organ.*, 121:87–95 (*516*).

KOSIN, I. L. (1944), *Poultry Sci.*, 23:266–269 (*121, 122*).

KOSIN, I. L. (1945), *Anat. Record*, 91:245–251 *(122)*.

KOTLAN, A., and CHANDLER, W. L. (1925), *J. Am. Vet. Med. Assoc.*, 67:756–763 *(305)*.

KRIEGER, C. H., BUNKFELDT, R., THOMPSON, C. R., and STEENBOCK, H. (1941), *J. Nutrition*, 21:213–220 *(602)*.

KRIZENECKY, J. (1934), *Věstník Českoslov. Akad. Zemědělské*, 10:581–586 *(117)*.

KRONMANN, C. A. (1899), Brit. Pat. 1322 *(698)*.

KRUKENBERG, W. (1883), *Verhandl. phys. med. Ges. Wurzburg*, 17:109–127 *(226)*.

KRUYT, H. R., and DEJONG, J. R. (1934), *Kolloid-Beihefte*, 40:55–86 *(444, 445)*.

KRZHISHKOWSKII, K. N. (1933), *Physiology of Domesticated Birds* (in Russian), Leningrad, 296 pp. *(10)*.

KUHN, R. (1933), *J. Soc. Chem. Ind. (London)*, 52:981–986 *(609)*.

KUHN, R., and BROCKMANN, H. (1932), *Z. physiol. Chem.*, 206:41–64 *(346, 486, 487)*.

KUHN, R., and DESNUELLE, P. (1938), *Z. physiol. Chem.*, 251:14–18 *(475)*.

KUHN, R., GYÖRGY, P., and WAGNER-JAUREGG, T. (1933), *Ber. deut. chem. Ges.*, 66B:317–320 *(136, 347)*.

KUHN, R., and LEDERER, E. (1933), *Ber. deut. chem. Ges.*, 66B:488–495 *(350)*.

KUHN, R., and WAGNER-JAUREGG, T. (1933), *Ber. deut. chem. Ges.*, 66B:1577–1582 *(487, 489)*.

KUHN, R., WINTERSTEIN, A., and LEDERER, E. (1931), *Z. physiol. Chem.*, 197:141–160 *(345, 487)*.

KUMMERLÖWE, H. (1931), *Zool. Anz.*, 95:103–105 *(265, 266, 292)*.

KUNSTLER, J. (1907), *Bibliographie anat.*, 16:262–272 *(263, 268, 287, 290, 302)*.

KUPSCH, W. (1934), *Arch. Geflügelkunde*, 8:97–115 *(132)*.

KYES, P., and POTTER, T. S. (1934), *Anat. Record*, 60:377–379 *(238)*.

LABOULBÈNE, A. (1859), *Compt. rend. soc. biol.*, (Ser. 3) 1:161–164 *(307)*.

LAGRANGE, E. (1935), *Compt. rend. soc. biol.*, 120:846–848 *(688)*.

LAKELA, O. (1931), *Poultry Sci.*, 11:181–184 *(305)*.

LAMPE, R. (1933), *Zentr. Chir.*, 60:2887–2890 *(788)*.

LAMPERT, L. M. (1930), *Ind. Eng. Chem., Anal. Ed.*, 2:159–162 *(784)*.

LAMSON, G. H. (1909), *Storrs (Connecticut) Agr. Expt. Sta. Bull.*, 55:203–214 *(494, 495)*.

LANDOIS, H. (1865), *Z. wiss. Zoöl.*, 15:1–31 *(225)*.

LANDOIS, H. (1878), *Zool. Garten*, 19:17–24 *(262, 263, 265, 266, 267)*.

LANDOIS, H. (1882), *Humboldt*, 1:22–24 *(305)*.

LANDOIS, H. (1892), *Jahresber. westfäl. Prov.-ver. Wiss. Kunst*, 20:34–35 *(292)*.

LANDSTEINER, K., LONGSWORTH, L. G., and SCHEER, J. VAN DER (1938), *Science*, 88:83–85 *(448)*.

LANDSTEINER, K., and SCHEER, J. VAN DER (1940), *J. Exptl. Med.*, 71:445–454 *(545, 546)*.

LANDY, M., DICKEN, D. M., BICKING, M. M., and MITCHELL, W. R. (1942), *Proc. Soc. Exptl. Biol. Med.*, 49:441–444 *(506, 508)*.

LANGE, F. E. M., and NORD, F. F. (1935), *Biochem. Z.*, 278:173–190 *(442)*.

LANGE, R. (1907), *Arch. Hyg.*, 62:201–215 *(495, 497)*.

LANGSTEIN, L. (1901), *Beitr. chem. Physiol. Path.*, 1:83–104 *(327)*.

LANGSTEIN, L. (1903), *Beitr. chem. Physiol. Path.*, 3:510–513 *(330)*.

LANGSTEIN, L. (1907), *Biochem. Z.*, 5:410–412 *(330)*.

LANGWORTHY, C. F. (1901), *U. S. Dept. Agr., Farmer's Bull.*, 128:1–31 *(314, 317, 320, 354)*.

LANGWORTHY, C. F. (1917), *U. S. Dept. Agr. Bull.*, 471:1–30 *(312, 599, 600)*.

LANGWORTHY, C. F., and BAROTT, H. G. (1921), *J. Biol. Chem.*, 46:xlix–l *(376)*.

LANGWORTHY, C. F., and HOLMES, A. D. (1917), *U. S. Dept. Agr. Bull.*, 507:1–19 *(597)*.

LAPLAND, M. (1924), *Proc. World's Poultry Congr. (Barcelona)*, 2:391–395 *(48)*.

LARDY, H. A., and PHILLIPS, P. H. (1939), *Proc. Am. Soc. Animal Prod.*, **32**:219–221 (*790*).

LARIONOV, W. T. (1941), *Compt. rend. acad. sci. U.R.S.S.*, **32**:227–229 (*30*).

LARosa, W. (1927), *Chemist-Analyst*, **16**, Nos. 2, 3 (*428*).

LASCHTSCHENKO, P. (1909), *Z. Hyg. Infektionskrankh.*, **64**:419–427 (*499, 501*).

LASKOWSKI, M. (1933), *Biochem. Z.*, **260**:230–240 (*237*).

LASKOWSKI, M. (1934), *Biochem. Z.*, **273**:284–290 (*226, 237*).

LATASTE, F. (1924), *Proc. verb. soc. linn.*, **76**:224–238 (*149*).

LAUPRECHT, E. (1939), *Proc. World's Poultry Congr.* (*Cleveland*), **7**:148–151 (*64*).

LAURENCE, W. L. (1944), *Science*, **99**:392–393 (*506*).

LAURIE, D. F. (1912), *Dept. Agr. So. Australia Bull.*, **72**:1–12 (*248, 249*).

LAWALL, C. H. (1918), *Pennsylvania Dept. Agr. Bull.*, **1(7)**:1–20 (*782*).

LAWATSCHEK, R. (1914), *Prager med. Wochschr.*, **39**:185–189 (*558, 586*).

LAWRENCE, J. V., and RIDDLE, O. (1916), *Am. J. Physiol.*, **41**:430–437 (*237*).

LEAGUE OF NATIONS (1935), *Health Organization Rept. on the Physiological Bases of Nutrition*, Geneva, 19 pp. (*610*).

LEAGUE OF NATIONS (1938), *Quart. Bull. Health Organization*, **7**(June):460–502 (*610*).

LEASE, J. G. (1936), *Z. Vitaminforsch.*, **5**:110–118 (*631*).

LEASE, J. G. (1937), *Poultry Sci.*, **16**:374–377 (*630*).

LEBBIN, G. (1900), *Z. öffentl. Chem.*, **6**:148–149 (*312, 314, 317, 320*).

LEBBIN, G. (1902), *Therap. Monatsh.*, **15**:552–553 (*579, 593, 597*).

LEBLOND, C. (1834), *L'Institut*, **II**:266 (*307*).

LECLERC, J. A. (1933), *Cereal Chem.*, **10**:383–419 (*775, 776*).

LECLERC, J. A., and BAILEY, L. H. (1940), *Cereal Chem.*, **17**:279–312 (*775, 776, 777, 778, 779*).

LEDERER, J. (1939), *Compt. rend. soc. biol.*, **139**:491–493 (*606*).

LEE, C. D., HOLM, G., and MURRAY, C. (1936), *J. Am. Vet. Med. Assoc.*, **89**(N.S. 42):65–76 (*511*).

LEE, C. E., HAMILTON, S. W., and HENRY, C. L. (1936), *Poultry Sci.*, **15**:307–310 (*45*).

LEE, D. H. K., ROBINSON, K. W., YEATES, N. T. M., and SCOTT, M. I. R. (1945), *Poultry Sci.*, **24**:195–207 (*657*).

LEE, J.-S., and TSAI, C. (1942), *Quart. J. Exptl. Physiol.*, **31**:281–297 (*569, 570*).

LEE, W. Y., and WU, H. (1932), *Chinese J. Physiol.*, **6**:307–318 (*435, 474*).

LEHMAN, J. C. (1864), *Arch. path. Anat.* (*Virchow's*), **30**:593–598 (*563*).

LEICHSENRING, J. M., and FLOR, I. H. (1932), *J. Nutrition*, **5**:141–146 (*605*).

LEITCH, I. (1937), *Nutrition Abstracts & Revs.*, **6**:553–578 (*602*).

LEMBERG, R. (1931), *Ann.*, **488**:74–90 (*348*).

LENTI, C. (1939), *Arch. sci. biol.*, **25**:1–6 (*604*).

LENTI, C. (1945), *Boll. soc. ital. biol. sper.*, **20**:531–532 (*458*).

LEONE, S. T. (1936), *Boll. chim. farm.*, **75**:303 (*358, 607*).

LEONG, P. C. (1940), *J. Malaya Branch Brit. Med. Assoc.*, **4**:66–107 (*623*).

LEPESCHKIN, W. W. (1922), *Biochem. J.*, **16**:678–701 (*427*).

LEPKOVSKY, S., TAYLOR, L. W., JUKES, T. H., and ALMQUIST, H. J. (1938), *Hilgardia*, **11**:559–591 (*625, 626*).

LEPPER, E. H., and MARTIN, C. J. (1927), *Biochem. J.*, **21**:356–361 (*410*).

LEPPER, H. A., BARTRAM, M. T., and HILLIG, F. (1944), *J. Assoc. Offic. Agr. Chemists*, **27**:204–223 (*728*).

LERNER, I. M. (1942), *J. Agr. Research*, **64**:333–338 (*59*).

LERNER, I. M. (1946), *Poultry Sci.*, **25**:392–394 (*305*).

LERNER, I. M., and SMITH, W. R. (1942), *Poultry Sci.*, **21**:473 (*303, 305*).

LERNER, I. M., and TAYLOR, L. W. (1937), *J. Agr. Research*, **55**:703–712 (*18*).

LERNER, I. M., and TAYLOR, L. W. (1940), *Poultry Sci.*, **19**:187–190 (*52, 59*).

LERNER, I. M., and TAYLOR, L. W. (1943), *Am. Naturalist*, **77**:119–132 (*18*).

LESCARDÉ, F. (1908), *L'œuf de poule, sa conservation par le froid*, H. Dunod and E. Pinat, Paris, 132 pp. (*707*).

LESHER, S. W., and KENDEIGH, S. C. (1941), *Wilson Bull.*, **53**:169–180 (*30*).

LESNÉ, E., and CLEMENT, R. (1932), *Compt. rend. soc. biol.*, **107**:1533–1534 (*620*).

LESNÉ, E., ZIZINE, P., and BRISKAS, S. (1938a), *Compt. rend. soc. biol.*, **128**:935–936 (363).

LESNÉ, E., ZIZINE, P., and BRISKAS, S. (1938b), *Compt. rend. soc. biol.*, **128**:937–939 (363).

LEVENE, P. A. (1901), *Med. News.*, **79**:981–982 (*540*).

LEVENE, P. A. (1916), *J. Biol. Chem.*, **24**:69–89 (*338, 339, 340, 483*).

LEVENE, P. A., and ALSBERG, C. L. (1906), *J. Biol. Chem.*, **2**:127–133 (*333*).

LEVENE, P. A., and BEATTY, W. A. (1907a), *Biochem. Z.*, **4**:299–304 (*533*).

LEVENE, P. A., and BEATTY, W. A. (1907b), *Biochem. Z.*, **4**:305–311 (*333*).

LEVENE, P. A., and MORI, T. (1929), *J. Biol. Chem.*, **84**:49–61 (*344*).

LEVENE, P. A., and ROTHEN, A. (1929), *J. Biol. Chem.*, **84**:63–68 (*344*).

LEVENE, P. A., and WEST, C. J. (1916), *J. Biol. Chem.*, **24**:111–116 (*339, 340*).

LEVENE, P. A., and WEST, C. J. (1917), *J. Biol. Chem.*, **31**:649–654 (*338, 341*).

LEVENE, P. A., and WEST, C. J. (1918), *J. Biol. Chem.*, **35**:285–290 (*483*).

LEVEQUE, P. F., and PONSCARME, L. J. (1913), *Ann. école nat. agr. Grignon*, **4**:38–42 (*342*).

LEVI, W. M. (1941), *The Pigeon*, R. L. Bryan Co., Columbia, South Carolina, 512 pp. (*257, 259, 262, 274, 280, 285, 292, 306*).

LEVIN, B.-S. (1935), *Compt. rend. soc. biol.*, **119**:80–82 (*569, 570*).

LEVIN, B.-S. (1936a), *Compt. rend. soc. biol.*, **121**:535–536 (*570*).

LEVIN, B.-S. (1936b), *Compt. rend. soc. biol.*, **121**:1093–1095 (*570*).

LEVIN, M. (1939), U. S. Pat. 2,155,429 (*801*).

LEVINE, M., and ANDERSON, D. Q. (1930), *J. Bact.*, **19**:55–56 (*689*).

LEVITES, S. J. (1909), *Biochem. Z.*, **20**:220–223 (*597*).

LEVY, A. J., and GEORGIAKAKIS, N. (1934), *Arch. maladies app. digest. et maladies nutrition*, **24**:785–793 (*341*).

LEWIN, L., MIETHE, A., and STENGER, E. (1908), *Arch. ges. Physiol. (Pflüger's)*, **124**:585–590 (*418*).

LEWIS, H. R., HANNAS, R. R., and WENE, E. H. (1919), *New Jersey Agr. Expt. Sta. Bull.*, **338**:5–96 (*41, 49*).

LEWIS, J. H., and WELLS, H. G. (1927), *J. Infectious Diseases*, **40**:316–325 (*542, 543*).

LEWIS, P. S. (1926), *Biochem. J.*, **20**:978–983 (*458*).

LEWITH, S. (1890), *Arch. exptl. Path. Pharmakol.*, **26**:341–354 (*459*).

LHAMON, L. (1919), *J. Home Econ.*, **11**:108–115 (*583*).

LICHTENSTEIN, S. (1932), *Z. Tuberk.*, **64**:245–251 (*510*).

LIEBERMANN, C. (1878), *Ber. deut. chem. Ges.*, **11**:606–610 (*226*).

LIEBERMANN, L. (1888), *Arch. ges. Physiol. (Pflüger's)*, **43**:71–151 (*202, 326, 328, 330, 390*).

LIÉGEOIS, M. (1859), *Compt. rend. soc. biol.*, (Ser. 3) **1**:254–255 (*301*).

LILLIE, R. S. (1907), *Am. J. Physiol.*, **20**:127–179 (*438*).

LINDERSTRÖM-LANG, K. (1939), *Ann. Rev. Biochem.*, **8**:37–58 (*532*).

LINDNER, E. (1939), *Kisérletügyi Közlemények.*, **42**:43–45 (*784*).

LINDOW, C. W., ELVEHJEM, C. A., and PETERSON, W. H. (1929), *J. Biol. Chem.*, **82**:465–471 (*358, 604, 606*).

LINDWALL, V. (1881a), *Jahresber. Fortschr. Tierchem.*, **11**:38–39 (*330*).

LINDWALL, V. (1881b), *Upsala Läkarenfören. Förh.*, **16**:546–554 (*328*).

LINTON, E. (1887), *Proc. U. S. Natl. Museum*, **10**:367–369 (*305*).

LIPCSCY, L. (1930), Brit. Pats. 364,128, and 364,129 (*712*).

LIPMAN, J. G., and BROWN, P. E. (1908), *New Jersey Agr. Expt. Sta. Rept.*, **29**:129–136 (*804*).

LIPPINCOTT, W. A. (1921), *J. Am. Assoc. Instr. Invest. Poultry Husb.*, **7**:73–74 (*76*).

LIPPINCOTT, W. A., PARKER, S. L., and SCHAUMBURG, L. M. (1925), *Poultry Sci.*, **4**:127–140 (*76*).

LITTLE, A. (1924), *Rhodesia Agr. J.*, **21**:750–753 (*282*).

LIU, S. C. (1933), *Chinese J. Physiol.*, **7**:107–116 (*464, 466*).

LIVERANI, E. (1934), *Compt. rend. soc. biol.*, **115**:133–134 (*510*).

LOEB, J. (1924), *J. Gen. Physiol.*, **6**:307–328 (*431*).

LOESCHKE, A. (1931), *Z. physiol. Chem.*, **199**:125–128 (*358*).

LOGES, G., and PINGEL, T. (1900), *Sächs. landw. Z.*, **22**:409–411 (*360*).

LOISEL, G. (1900), *Compt. rend. soc. biol.*, **52**:757–759 (*320*).

LONDON, E. S., and SULIMA, A. T. (1905), *Z. physiol. Chem.*, **46**:209–235 (*585, 587*).

LONGSWORTH, L. G. (1941), *Ann. N. Y. Acad. Sci.*, **41**:267–285 (*435, 449, 473*).

LONGSWORTH, L. G., CANNAN, R. K., and MacINNES, D. A. (1940), *J. Am. Chem. Soc.*, **62**:2580–2590 (*327, 328, 433, 446, 447, 449*).

LOOS, P. A. (1881), *Z. wiss. Zoöl.*, **35**:478–504 (*193*).

LORENZ, F. W. (1939), *Poultry Sci.*, **18**:295–300 (*350, 667, 668*).

LORENZ, F. W., and ALMQUIST, H. J. (1934), *Ind. Eng. Chem.*, **26**:1311–1313 (*668*).

LORENZ, F. W., and ALMQUIST, H. J. (1936), *Poultry Sci.*, **15**:14–18 (*141, 645*).

LORENZ, F. W., ENTENMAN, C., and CHAIKOFF, I. L. (1938), *J. Biol. Chem.*, **122**:619–633 (*237, 241*).

LORENZ, F. W., and NEWLON, W. E. (1944), *Poultry Sci.*, **23**:193–198 (*636*).

LORENZ, F. W., PERLMAN, I., and CHAIKOFF, I. L. (1943), *Am. J. Physiol.*, **138**:318–327 (*240, 242, 363*).

LORENZ, F. W., and TAYLOR, L. W. (1940), *J. Agr. Research*, **61**:293–301 (*141*).

LORENZ, F. W., TAYLOR, L. W., and ALMQUIST, H. J. (1934), *Poultry Sci.*, **13**:14–17 (*139*).

LOUGHLIN, W. J. (1932), *Biochem. J.*, **26**:1557–1565 (*451, 464*).

LOUGHLIN, W. J. (1933), *Biochem. J.*, **27**:99–105 (*431*).

LOUGHLIN, W. J., and LEWIS, W. C. M. (1932), *Biochem. J.*, **26**:476–487 (*466*).

LÖWENSTEIN, E. (1925), *Z. Tuberk.*, **41**:18–25 (*515*).

LOY, W. C. (1944), *U. S. Egg Poultry Mag.*, **50**:537–539, 569–570 (*720, 725*).

LUCAS, A. M. (1946), *Am. J. Anat.*, **79**:431–471 (*304*).

LUND, W. A., HEIMAN, V., and WILHELM, L. A. (1938), *Poultry Sci.*, **17**:372–376 (*370, 371, 372, 382*).

LUNDGREN, H. P. (1941), *J. Am. Chem. Soc.*, **63**:2854–2855 (*802*).

LUNDGREN, H. P., ELAM, D. W., and O'CONNELL, R. A. (1943), *J. Biol. Chem.*, **149**:183–193 (*457, 474*).

LUNDGREN, H. P., and O'CONNELL, R. A. (1944), *Ind. Eng. Chem.*, **36**:370–374 (*803, 804*).

LYONS, M. (1939), *Arkansas Agr. Expt. Sta. Bull.*, **374**:1–18 (*156, 358*).

LYONS, M., and INSKO, W. M. (1937a), *Science*, **86**:328 (*358, 364*).

LYONS, M., and INSKO, W. M. (1937b), *Kentucky Agr. Expt. Sta. Bull.*, **371**:63–75 (*358, 364*).

McALDOWIE, M. (1886), *J. Anat. Phys.*, **20**:225–237 (*95*).

McBAIN, J. W., DAWSON, C. R., and BARKER, H. A. (1934), *J. Am. Chem. Soc.*, **56**:1021–1027 (*435, 436*).

McBAIN, J. W., and LEYDA, F. A. (1938), *J. Am. Chem. Soc.*, **60**:2998–3002 (*437*).

McCAMMON, R. B., PITTMAN, M. S., and WILHELM, L. A. (1934), *Poultry Sci.*, **13**:95–101 (*648, 668, 669*).

McCann, A. W. (1918), *Pennsylvania Dept. Agr.*, **16**(7 and 8):21 (*781*).

McCharles, C. H., and Mulvaney, H. A. (1937), U. S. Pat. 2,087,985 (*743*).

McClary, C. F., and Upp, C. W. (1939), *Poultry Sci.*, **18**:210–219 (*511*).

McCollum, E. V. (1909), *Am. J. Physiol.*, **25**:120–141 (*589*).

McCollum, E. V., Halpin, J. G., and Drescher, A. H. (1912), *J. Biol. Chem.*, **13**:219–224 (*342, 485*).

McFarlane, J., Dunbar, V. E., Borsook, H., and Wasteneys, H. (1927), *J. Gen. Physiol.*, **10**:437–450 (*521, 522, 525, 526*).

McFarlane, V. H., Watson, A. J., and Goresline, H. E. (1945), *U. S. Egg Poultry Mag.*, **51**:250–257, 270–273, 275–277, 279–281, 282–286 (*721*).

McFarlane, W. D., Fulmer, H. L., and Jukes, T. H. (1930), *Biochem. J.*, **24**:1611–1631 (*333, 335, 360*).

McHargue, J. S. (1924), *J. Agr. Research*, **27**:417–424 (*358*).

McHargue, J. S. (1925), *Am. J. Physiol.*, **72**:583–594 (*358, 604*).

McHenry, E. W., and Gavin, G. (1941), *J. Biol. Chem.*, **138**:471–475 (*628*).

Machida, S., and Sasaki, T. (1937), *J. Agr. Chem. Soc. Japan*, **13**:305–308 (*633*).

McLaughlin, G. D., and Theis, E. R. (1945), *The Chemistry of Leather Manufacture*, Reinhold Publishing Corp., New York, 800 pp. (*794*).

McLean, H. (1909), *Z. physiol. Chem.*, **59**:223–229 (*338*).

McNally, E. H. (1943), *Poultry Sci.*, **22**:40–43 (*135*).

Macowan, M. M. (1932), *Quart. J. Exptl. Physiol.*, **21**:383–391 (*236*).

MacPherson, C. F. C., and Heidelberger, M. (1940), *Proc. Soc. Exptl. Biol. Med.*, **43**:646–647 (*554*).

MacPherson, C. F. C., and Heidelberger, M. (1945a), *J. Am. Chem. Soc.*, **67**:574–577 (*477*).

MacPherson, C. F. C., and Heidelberger, M. (1945b), *J. Am. Chem. Soc.*, **67**:585–591 (*555*).

MacPherson, C. F. C., Heidelberger, M., and Moore, D. H. (1945), *J. Am. Chem. Soc.*, **67**:578–585 (*473*).

MacPherson, C. F. C., Moore, D. H., and Longsworth, L. G. (1944), *J. Biol. Chem.*, **156**:381–382 (*447, 449*).

M'Williams, J. M. (1927), *Scottish Naturalist*, **166**:108–110 (*259, 262, 268, 275*).

Macy, I. G., Hunscher, H. A., McCosh, S. S., and Nims, B. (1930), *J. Biol. Chem.*, **86**:59–74 (*621*).

Macy, R. W. (1934), *Univ. Minnesota Agr. Expt. Sta. Tech. Bull.*, **98**:1–71 (*35, 36, 252, 305, 306*).

Macy, R. W. (1940), *Proc. Minnesota Acad. Sci.*, **8**:39–41 (*35*).

Maie, S. (1922), *Biochem. Z.*, **132**:311–324 (*552*).

Maile, W. C. D., and Scott, K. J. L. (1935), *Lancet*, **228**(1):21–23 (*579*).

Makower, B. (1945), *Ind. Eng. Chem.*, **37**:1018–1022 (*748, 749*).

Mallmann, W. L., and Churchill, E. (1942), *Ice and Refrig.*, **102**:303–305 (*726*).

Mallmann, W. L., and Davidson, J. A. (1944), *U. S. Egg Poultry Mag.*, **50**:169–171, 189–191 (*713, 714*).

Mallmann, W. L., and Michael, C. E. (1940), *Michigan State Coll. Agr. Expt. Sta. Tech. Bull.*, **174**:1–34 (*689, 705, 706*).

Manen, E. van, and Rimington, C. (1935), *Onderstepoort J. Vet. Sci.*, **5**:329–344 (*518*).

Mankin, W. R. (1928), *Med. J. Australia*, **1928**(II):87 (*358*).

Mann, T., and Keilin, D. (1940), *Nature*, **146**:164–165 (*226*).

Mansfeld, M. (1901), *Oesterr. chem. Ztg.*, **4**:442–443 (*314*).

Manville, I. A. (1926), *Am. J. Hyg.*, **6**:238–253 (*615*).

MANWARING, W. H., MARINO, H. D., McCLEAVE, T. C., and BOONE, T. H. (1927), *Proc. Soc. Exptl. Biol. Med.*, **24**:650–651 (*540*).

MARBLE, D. R. (1930a), *Cornell Univ. Agr. Expt. Sta. Bull.*, **503**:1–42 (*30*).

MARBLE, D. R. (1930b), *Poultry Sci.*, **9**:257–265 (*77, 78, 82*).

MARBLE, D. R. (1943), *Poultry Sci.*, **22**:61–71 (*92, 93, 94, 95*).

MARCHLEWSKI, L., and WIERZUCHOWSKA, J. (1928), *Bull. intern. acad. polon. sci.*, **1928A**: 471–478 (*456*).

MARKS, M. B. (1936), *Am. J. Digestive Diseases Nutrition*, **3**:41–44 (*559*).

MARKUS, A. (1916), *Chem. Ztg.*, **40**:397–398 (*782*).

MARLATT, A. L., and CLOW, B. (1927), *Wisconsin Agr. Expt. Sta. Bull.*, **396**:1–46 (*620*).

MARLOW, H. W., and KING, H. H. (1936), *Poultry Sci.*, **15**:377–380 (*333, 603*).

MARLOW, H. W., and RICHERT, D. (1940), *Endocrinology*, **26**:531–534 (*352, 516*).

MARRACK, J., and HEWITT, L. F. (1929), *Biochem. J.*, **23**:1079–1089 (*435*).

MARSHAK, A. (1944), *J. Gen. Physiol.*, **28**:95–102 (*566*).

MARSHALL, W., and CRUICKSHANK, D. B. (1938), *J. Agr. Sci.*, **28**:24–42 (*109, 168, 172, 225*).

MARSTON, H. R. (1926), *Australian J. Exptl. Biol. Med. Sci.*, **3**:233–236 (*533*).

MARTIN, E. (1921), *Mon. sci.*, **11**:151–152 (*785*).

MARTIN, J. H., ERIKSON, S. E., and INSKO, W. M. (1930), *Kentucky Agr. Expt. Sta. Bull.* **304**:187–218 (*46*).

MARZA, V. D., and MARZA, E. (1932), *Bull. histol. appl. physiol. et path. et tech. microscop.*, **9**:313–340 (*341*).

MARZA, V. D., and MARZA, E. (1935), *Quart. J. Micr. Sci.*, **78**:134–189 (*201, 202, 203, 208 210, 212*).

MASAI, Y., and FUKUTOMI, T. (1923), *J. Biochem. (Japan)*, **2**:271–277 (*356*).

MASAMUNE, H., and HOSHINO, S. (1936), *J. Biochem. (Japan)*, **24**:219–224 (*344*).

MASSENGALE, O. N., and PLATT, C. S. (1930), *Poultry Sci.*, **9**:240–246 (*83*).

MASTIN, H., and REES, H. G. (1926), *Biochem. J.*, **20**:759–762 (*478*).

MASUDA, Y., and HORI, T. (1937), *J. Agr. Chem. Soc. Japan*, **13**:200–205 (*340*).

MASUROVSKY, B. I. (1942), *Food Industries*, **14(4)**:65 (*717*).

MATHIEU, ——, and URBAIN, —— (1908), quoted by LESCARDÉ, F. (1908), *L'œuf de poule, sa conservation par le froid*, H. Dunod and E. Pinat, Paris, 132 pp. (*422*).

MATSCHULAN, G., and AMSLER, C. (1938), *Arch. exptl. Path. Pharmakol.*, **190**:560–564 (*788*).

MATSUDA, M. (1926), *Japan Med. World*, **6**:335–345 (*561*).

MATSUDA, T. (1931), *Japan. J. Gastroenterol.*, **3**:14–17 (*563*).

MATTIKOW, M. (1932), *Poultry Sci.*, **11**:83–93 (*489*).

MAUGHAN, G. H., and MAUGHAN, E. (1933), *Science*, **77**:198 (*617*).

MAURER, O. (1911), *Kansas Agr. Expt. Sta. Bull.*, **180**:333–396 (*492, 493, 494*).

MAW, A. J. G. (1934), *Poultry Sci.*, **13**:131–134 (*46, 83, 214*).

MAW, A. J. G., and MAW, W. A. (1928), *Sci. Agr.*, **9**:201–208 (*23*).

MAW, A. J. G., and MAW, W. A. (1932), *Sci. Agr.*, **12**:281–286 (*74*).

MAY, F. E. (1939), Brit. Pat. 503,297 (*782*).

MAY, H. G. (1924), *Rhode Island Agr. Expt. Sta. Bull.*, **197**:1–48 (*494*).

MAYER, G. R. (1932), *Penrose's Annual*, **34**:86–88 (*800*).

MAYHEW, R. L. (1934), *Poultry Sci.*, **13**:148–154 (*35*).

MAYNEORD, W. V. (1927), *Brit. J. Radiol.*, **23**:19–33 (*161*).

MAZZA, F. P. (1930), *Arch. sci. biol. (Italy)*, **15**:12–26 (*453, 455*).

MEHARLISCU, N. (1933), *Arch. Geflügelkunde*, **7**:320 (*148*).

MELDRUM, R. U., and ROUGHTON, F. J. W. (1933), *J. Physiol.*, **80**:113–142 (*226*).

MELLANBY, E. (1921), *Med. Research Council (London), Spec. Rept. Ser.*, **61**:65 (*616*).

MELLANBY, M., and KILLICK, E. M. (1926), *J. Physiol.*, **61**:xxiii (*616*).

MENDEL, L. B., and LEAVENWORTH, C. S. (1908), *Am. J. Physiol.*, **21**:77–84 (*352*).

MENDEL, L. B., and LEWIS, R. C. (1913), *J. Biol. Chem.*, **16**:55–77 (*586, 596*).

MENSCHICK, W., and PAGE, I. H. (1932), *Z. physiol. Chem.*, **211**:246–252 (*352*).

MERRILL, H. B. (1928), *Ind. Eng. Chem.*, **20**:654–656 (*794*).

MERZ, W. (1930), *Z. physiol. Chem.*, **193**:59–87 (*340*).

MESZAROS, G. (1934), *Z. Untersuch. Lebensm.*, **68**:548–553 (*117*).

METZGER, H. J., BEAUDETTE, F. R., and STOKES, F. R. (1939), *J. Am. Vet. Med. Assoc.*, **95**:158 (*792*).

MEULEN, H. TER (1932), *Nature*, **130**:966 (*358*).

MEURS, G. J. VAN (1923), *Rec. trav. chim.*, **42**:800–803 (*314, 317, 320*).

MEYER, K. (1944), *Science*, **99**:391–392 (*506*).

MEYER, K., and HAHNEL, E. (1946), *J. Biol. Chem.*, **163**:723–732 (*502, 503, 506*).

MEYER, K., HAHNEL, E., and STEINBERG, A. (1946), *J. Biol. Chem.*, **163**:733–740 (*502*).

MEYER, K., THOMPSON, R., PALMER, J. W., and KHORAZO, D. (1934), *Science*, **79**:61 (*503*).

MIDELTON, W. J. (1907), *Brit. Med. J.*, **2422**:1302 (*589*).

MIESCHER, F. (1871), *Med.-chem. Untersuch.*, **2**:502–509 (*528*).

MILBY, T. T., and THOMPSON, R. B. (1941), *Proc. Oklahoma Acad. Sci.*, **1941**:41–44 (*50*).

MILES, A. A., and HALNAN, E. T. (1937), *J. Hyg.*, **37**:79–97 (*494, 687*).

MILESI, C. (1898), *Jahresber. Fortschr. Tierchem.*, **28**:38–39 (*330*).

MILLER, E. L., and VAIL, G. E. (1943), *Cereal Chem.*, **20**:528–535 (*773*).

MILLER, J. (1927), *Analyst*, **52**:457–458 (*699*).

MILLER, M. W., and BEARSE, G. E. (1934), *Washington Agr. Expt. Sta. Tech. Bull.*, **306**:5–20 (*44*).

MILLER, R. C., FORBES, E. B., and SMYTHE, C. V. (1929), *J. Nutrition*, **1**:217–232 (*605*).

MILLER, R. J., FOWLER, H. L., BERGEIM, O., REHFUSS, M. E., and HAWK, P. B. (1919), *Am. J. Physiol.*, **49**:254–270 (*580, 585, 589*).

MILLER, W. S., CHESLEY, K. G., ANDERSON, H. V., and THEIS, E. R. (1932), *J. Am. Leather Chem. Assoc.*, **27**:174–182 (*473*).

MILLS, B. J. B. (1891), Brit. Pat. 17,717 (*712*).

MIRSKY, A. E. (1941), *J. Gen. Physiol.*, **24**:709–723 (*475, 476*).

MIRSKY, A. E., and ANSON, M. L. (1935), *J. Gen. Physiol.*, **18**:307–323 (*475, 476*).

MIRSKY, A. E., and ANSON, M. L. (1936a), *J. Gen. Physiol.*, **19**:427–438 (*476*).

MIRSKY, A. E., and ANSON, M. L. (1936b), *J. Gen. Physiol.*, **19**:451–459 (*476*).

MIRSKY, A. E., and PAULING, L. (1936), *Proc. Natl. Acad. Sci. U. S.*, **22**:439–447 (*457, 474*).

MITCHELL, H. H., and CARMAN, G. G. (1924), *J. Biol. Chem.*, **60**:613–620 (*593, 594*).

MITCHELL, H. H., and CARMAN, G. G. (1926), *J. Biol. Chem.*, **68**:183–215 (*594*).

MITCHELL, H. S. (1925), *Am. J. Physiol.*, **74**:359–362 (*588*).

MITCHELL, L. C. (1932), *J. Assoc. Offic. Agr. Chemists*, **15**:310–326 (*317, 320*).

MITCHELL, L. C. (1940), *J. Assoc. Offic. Agr. Chemists*, **23**:285–288 (*685, 686*).

MITCHELL, L. C., and HORWITZ, W. (1941), *J. Assoc. Offic. Agr.Chemists*,**24**:319–326(*686*).

MIYAMORI, S. (1934), *Nagoya J. Med. Sci.*, **8**:176 (*341*).

MONTEVERDE, J. J., and SIMEONE, D. H. (1944a), *Rev. Facultad Agron. Vet. (Univ. Buenos Aires)*, **1(10)**:1–30 (*514*).

MONTEVERDE, J. J., and SIMEONE, D. H. (1944b), *Rev. Facultad Agron. Vet. (Univ. Buenos Aires)*, **2(3)**:49–59 (*514*).

MONTIUS, C. (1757), *Bononiae sci. art. Inst. Commentarii*, **4**:330 (*306*).

MOORE, B., and PARKER, W. H. (1902), *Am. J. Physiol.*, **7**:261–293 (*426*).

MOORE, E. N. (1940), *J. Am. Vet. Med. Assoc.*, **96**:727–732 *(247)*.

MOORE, E. N., and MARTEN, E. A. (1944), *Am. J. Vet. Research*, **5**:256–261 *(247)*.

MORAN, T. (1924), *Proc. Intern. Congr. Refrig.*, **4(1)**:122–137 *(731)*.

MORAN, T. (1925), *Proc. Roy. Soc. (London)*, **98(B)**:436–456 *(134, 377, 408, 409, 424, 430)*.

MORAN, T. (1935a), *Dept. Sci. Ind. Research (Brit.), Food Invest. Repts.*, **1934**:52–53 *(402)*.

MORAN, T. (1935b), *Proc. Roy. Soc. (London)*, **118(B)**:548–559 *(402, 442)*.

MORAN, T. (1936a), *Dept. Sci. Ind. Research (Brit.), Food Invest. Repts.*, **1935**:36–40 *(405, 424, 708)*.

MORAN, T. (1936b), *Dept. Sci. Ind. Research (Brit.), Food Invest. Repts.*, **1935**:40–41 *(674, 703)*.

MORAN, T. (1936c), *J. Exptl. Biol.*, **13**:41–47 *(390, 677)*.

MORAN, T. (1937a), *Dept. Sci. Ind. Research (Brit.), Food Invest. Repts.*, **1936**:46–49 *(665, 667, 681, 707)*.

MORAN, T. (1937b), *Dept. Sci. Ind. Research (Brit.), Food Invest. Repts.*, **1936**:52–53 *(666)*.

MORAN, T. (1937c), *Dept. Sci. Ind. Research (Brit.), Food Invest. Repts.*, **1936**:53–55 *(710)*.

MORAN, T. (1937d), *J. Soc. Chem. Ind. (London)*, **56**:96–101 *(321, 704, 707, 708, 710)*.

MORAN, T., and HAINES, R. B. (1939), *Dept. Sci. Ind. Research (Brit.), Food Invest. Repts.*, **1938**:39–41 *(383)*.

MORAN, T., and HALE, H. P. (1936), *J. Exptl. Biol.*, **13**:35–40 *(135,142,143, 146, 147,170, 326, 328, 329, 389, 414, 420, 424)*.

MORAN, T., and PIQUÉ, J. (1926), *Dept. Sci. Ind. Research (Brit.), Food Invest. Spec. Repts.*, **26**:1–80 *(148, 656, 685, 688, 692, 699, 700, 706)*.

MORGAN, C. L. (1932), *Poultry Sci.*, **11**:172–175 *(372)*.

MORGAN, W. A., and WOODROOF, J. G. (1927), *Georgia Agr. Expt. Sta. Bull.*, **147**:210–215 *(127)*.

MÖRNER, C. T. (1912), *Z. physiol. Chem.*, **80**:430–473 *(323, 453)*.

MÖRNER, K. A. H. (1901–02), *Z. physiol. Chem.*, **34**:207–338 *(330)*.

MORRELL, C. A., BORSOOK, H., and WASTENEYS, H. (1927), *J. Gen. Physiol.*, **8**:601–617 *(522, 523, 524)*.

MORRIS, J. P. (1940), *J. Malaya Branch Brit. Med. Assoc.*, **4**:114–117 *(358)*.

MOTA, F. E. DE LA (1917), U. S. Pat. 1,229,592 *(712)*.

MOUQUET, A. (1924), *Rec. méd. vét.*, **100**:181–185 *(7)*.

MOXON, A. L. (1937), *South Dakota Agr. Expt. Sta. Bull.*, **311**:1–91 *(358)*.

MOXON, A. L., and POLEY, W. E. (1938), *Poultry Sci.*, **17**:77–80 *(358, 362)*.

MULDER, G. J. (1838), *Ann.*, **28**:73–82 *(434)*.

MÜLLER, R. (1933), *Münch. med. Wochschr.*, **80**:1771–1772 *(513)*.

MUNKS, B., ROBINSON, A., BEACH, E. F., and WILLIAMS, H. H. (1945), *Poultry Sci.*, **24**: 459–464 *(42)*.

MUNRO, S. S. (1938), *Poultry Sci.*, **17**:17–27 *(140)*.

MUNRO, S. S. (1940), *Sci. Agr.*, **21**:53–62 *(369)*.

MUNRO, S. S., and ROBERTSON, G. (1935), *U. S. Egg Poultry Mag.*, **41(12)**:48, 50 *(390)*.

MURISIER, P. (1928), *Bull. soc. Vaud. sci. nat.*, **56**:517–523 *(248)*.

MURLIN, J. R., EDWARDS, L. E., and HAWLEY, E. E. (1944), *J. Biol. Chem.*, **156**:785–786 *(593, 594)*.

MURLIN, J. R., NASSET, E. S., and MARSH, M. E. (1938), *J. Nutrition*, **16**:249–269 *(594)*.

MURPHY, J. B. (1912), *J. Am. Med. Assoc.*, **59**:874–875 *(791)*.

MURPHY, J. C., and JONES, D. B. (1924), *J. Agr. Research*, **29**:253–257 *(609)*.

MURRAY, H. A. (1925), *J. Gen. Physiol.*, **9**:1–37 *(109)*.

MURTHY, G. N. (1937), *Indian J. Med. Research*, **24**:1083–1092 *(624)*.

MUSPRATT, S. (1852), *Quart. J. Chem. Soc. (London)*, **4**:178–180 *(788)*.

MUSSEHL, F. E., and HALBERSLEBEN, D. L. (1923), *J. Agr. Research*, **23**:717–720 (*369, 382*).

MYERS, W. G., and FRANCE, W. G. (1940), *J. Phys. Chem.*, **44**:1113–1126 (*432*).

NALBANDOV, A. V. (1942), *J. Heredity*, **33**:53–54 (*285*).

NALBANDOV, A. V., and CARD, L. E. (1944), *Poultry Sci.*, **23**:170–180 (*303, 304, 305*).

NATHUSIUS, W. VON (1867), *Arch. mikroskop. Anat.*, **45**:654–692 (*1, 304*).

NATHUSIUS, W. VON (1868), *Z. wiss. Zoöl.*, **18**:225–270 (*162, 163, 168, 169, 170, 271, 328*).

NATHUSIUS, W. VON (1869), *Z. wiss. Zoöl.*, **19**:322–348 (*170, 262, 263, 268, 269, 272, 274, 281, 302*).

NATHUSIUS, W. VON (1871a), *J. Ornithol.*, **19**:241–260 (*150, 164*).

NATHUSIUS, W. VON (1871b), *Z. wiss. Zoöl.*, **21**:330–355 (*161, 169, 170*).

NATHUSIUS, W. VON (1874), *J. Ornithol.*, **22**:1–26 (*164*).

NATHUSIUS, W. VON (1878), *Z. wiss. Zoöl.*, **30** (suppl.):69–77 (*170*).

NATHUSIUS, W. VON (1879), *J. Ornithol.*, **27**:225–261 (*100*).

NATHUSIUS, W. VON (1882a), *J. Ornithol.*, **30**:129–161 (*150*).

NATHUSIUS, W. VON (1882b), *J. Ornithol.*, **30**:255–314 (*160*).

NATHUSIUS, W. VON (1885), *J. Ornithol.*, **33**:165–178 (*150*).

NATHUSIUS, W. VON (1893), *Z. wiss. Zoöl.*, **55**:576–584 (*224, 225*).

NATHUSIUS, W. VON (1894), *Zool. Anz.*, **17**:440–445, 449–452 (*171, 173, 229*).

NATHUSIUS, W. VON (1895), *Arch. mikroskop. Anat.*, **45**:654–692 (*1, 280, 302, 307*).

NEDZEL, A. J., and ARNOLD, L. (1931a), *Proc. Soc. Exptl. Biol. Med.*, **28**:358–360 (*567*).

NEDZEL, A. J., and ARNOLD, L. (1931b), *Proc. Soc. Exptl. Biol. Med.*, **28**:360–361 (*567*).

NEDZEL, A. J., and ARNOLD, L. (1931c), *Proc. Soc. Exptl. Biol. Med.*, **28**:364–366 (*567*).

NEEDHAM, J. (1924), *Biochem. J.*, **18**:1371–1380 (*352*).

NEEDHAM, J. (1927), *Brit. J. Exptl. Biol.*, **5**:6–26 (*343*).

NEEDHAM, J. (1931a), *Chemical Embryology*, Cambridge University Press, 3 vols., 2021 pp. (*1, 100, 334, 352, 410*).

NEEDHAM, J. (1931b), *J. Exptl. Biol.*, **8**:330–344 (*135, 389, 677*).

NEEDHAM, J., and SMITH, M. (1931), *J. Exptl. Biol.*, **8**:286–292 (*401, 424*).

NEEDHAM, J., STEPHENSON, M., and NEEDHAM, D. M. (1931), *J. Exptl. Biol.*, **8**:319–329 (*352, 377*).

NEHER, B. H., and FRAPS, R. M. (1946), *J. Exptl. Zoöl.*, **101**:83–90 (*216*).

NEUBERGER, A. (1938), *Biochem. J.*, **32**:1435–1451 (*344, 486*).

NEURATH, H. (1936), *J. Phys. Chem.*, **40**:361–368 (*441, 450*).

NEURATH, H., and BULL, H. B. (1936), *J. Biol. Chem.*, **115**:519–528 (*466, 468*).

NEWTON, A. (1893), *A Dictionary of Birds*, Adam and Charles Black, London, 1088 pp. (*102*).

NICHOLAS, J. E. (1936), *Refrig. Eng.*, **32**:213–215 (*693*).

NICHOLAS, J. E., CALLENBACH, E. W., and MURPHY, R. R. (1944), *Pennsylvania Agr. Expt. Sta. Bull.*, **462**:1–24 (*49*).

NICHOLS, J. B. (1930), *J. Am. Chem. Soc.*, **52**:5176–5187 (*436, 450, 466*).

NICKLÈS, J. (1856), *Compt. rend. acad. sci.*, **43**:885 (*358*).

NIELSEN, F. A., and GARNATZ, G. F. (1940), *Proc. Inst. Food Technol.*, **1**:289–294 (*736*).

NILAKANTAN, P. (1940), *Quart. J. Indian Inst. Sci.*, **3**(2):190 pp. (*161*).

NILES, K. B. (1938), *U. S. Egg Poultry Mag.*, **44**:676–679, 705 (*800*).

NOETHER, P. (1931), *Arch. Exptl. Path. Pharmakol.*, **160**:369–374 (*46*).

NORRIS, L. C., and BAUERNFEIND, J. C. (1940), *Food Research*, **5**:521–532 (*347, 624, 625, 626*).

NORRIS, L. C., HEUSER, G. F., RINGROSE, A. T., and WILGUS, H. S. (1934), *Poultry Sci.*, **13**:308–309 *(154)*.

NORTH, M. O. (1935), *U. S. Egg Poultry Mag.*, **41(1)**:18–19 *(643)*.

NORTHRUP, E. F. (1913), *J. Franklin Inst.*, **175**:413–419 *(376)*.

NOTTBOHM, F. E., and MAYER, F. (1933), *Z. Untersuch. Lebensm.*, **66**:585–592 *(340, 352)*.

NOVAK, J., and DUSCHAK, F. (1923), *Z. ges. Anat.*, **69**:483–492 *(209)*.

NOWINSKI, W. W., and FERRANDO, R. J. (1942), *Rev. soc. argentina biol.*, **18**:397–403 *(633)*.

NUTTING, G. C., SENTI, F. R., and COPLEY, M. J. (1944), *Science*, **99**:328–329 *(479)*.

OBERMAYER, F., and PICK, E. P. (1902), *Wiener Klin. Rundschau.*, **16**:277–279 *(327)*.

O'DONOGHUE, C. H. (1911), *Anat. Anz.*, **37**:530–536 *(290)*.

OELLACHER, J. (1872), *Z. wiss. Zoöl.*, **22**:181–234 *(122)*.

OGLE, R. C. (1938), *Cornell Univ.*, *N. Y. State Coll. Agr.*, *Ext. Bull.*, **394**:1–32 *(246)*.

OGLE, R. C., and LAMOREUX, W. F. (1942), *Cornell Univ.*, *N. Y. State Coll. Agr.*, *Ext. Bull.*, **544**:1–4 *(50)*.

OKEY, R., and STEWART, D. (1933), *J. Biol. Chem.*, **99**:717–727 *(567)*.

OKEY, R., and YOKELA, E. (1936), *J. Nutrition*, **11**:463–470 *(568)*.

OLCOTT, H. S., and DUTTON, H. J. (1945), *Ind. Eng. Chem.*, **37**:1119–1121 *(752, 758)*.

OLSEN, M. W. (1942a), *J. Morphol.*, **70**:513–533 *(200)*.

OLSEN, M. W. (1942b), *Poultry Sci.*, **21**:497–499 *(129)*.

OLSEN, M. W., and BYERLY, T. C. (1932), *Poultry Sci.*, **11**:266–271 *(232, 233)*.

OLSEN, M. W., and FRAPS, R. M. (1944), *J. Morphol.*, **74**:297–309 *(200)*.

OLSEN, M. W., and KNOX, C. W. (1940), *Poultry Sci.*, **19**:254–257 *(72)*.

OLSON, C., and BULLIS, K. L. (1942), *Massachusetts Agr. Expt. Sta. Bull.*, **391**:1–56 *(244)*.

OLSSON, N. (1934), *Studies on the Specific Gravity of Hens' Eggs*, Otto Harrasowitz, Leipzig, 16 pp. *(368, 387)*.

OLSSON, N. (1936), *Proc. World's Poultry Congr.* *(Leipzig)*, **6(1)**:310–320 *(115, 116, 117, 121, 148, 150, 151, 157, 382, 387)*.

ONCLEY, J. L. (1942), *Chem. Rev.*, **30**:433–450 *(451, 452)*.

ONCLEY, J. L., FERRY, J. D., and SHACK, J. (1940), *Ann. N. Y. Acad. Sci.*, **40**:371–388 *(452)*.

O'NEIL, J. B. (1946), *Poultry Sci.*, **25**:83–84 *(46)*.

ONOE, T. (1936), *J. Biochem.* *(Japan)*, **24**:1–8 *(326, 453)*.

OPIE, E. L. (1924), *J. Exptl. Med.*, **39**:659–675 *(563, 565)*.

ORENT, E. R., and McCOLLUM, E. V. (1931), *J. Biol. Chem.*, **92**:651–678 *(607)*.

ORRU, A. (1933a), *Arch. sci. biol.* *(Italy)*, **19**:69–70 *(425)*.

ORRU, A. (1933b), *Arch. sci. biol.* *(Italy)*, **19**:71–75 *(425)*.

ORRU, A. (1933c), *Boll. soc. ital. biol. sper.*, **8**:284–286 *(425)*.

ORRU, A. (1933d), *Boll. soc. ital. biol. sper.*, **8**:668–669 *(390)*.

ORRU, A. (1935), *Atti accad. Lincei*, **22**:458–463 *(413)*.

ORRU, A. (1936), *Atti accad. Lincei*, **23**:959–965 *(413)*.

ORRU, A. (1939), *Arch. sci. biol.* *(Italy)*, **25**:292–308 *(390)*.

ORRU, A. (1940a), *Arch. sci. biol.* *(Italy)*, **26**:32–50 *(390)*.

ORRU, A. (1940b), *Ricerca sci.*, **11**:215–217 *(390)*.

OSBORNE, T. B. (1899), *J. Am. Chem. Soc.*, **21**:477–485 *(330)*.

OSBORNE, T. B., and CAMPBELL, F. (1900), *J. Am. Chem. Soc.*, **22**:422–450 *(325, 327, 330)*.

OSBORNE, T. B., and HARRIS, I. F. (1903), *J. Am. Chem. Soc.*, **25**:323–353 *(332)*.

OSBORNE, T. B., and JONES, D. B. (1909), *Am. J. Physiol.*, **24**:153–160 *(333)*.

OSBORNE, T. B., JONES, D. B., and LEAVENWORTH, C. S. (1909), *Am. J. Physiol.*, **24**:252–262 *(333)*.

OSBORNE, T. B., LEAVENWORTH, C. S., and BRAUTLECHT, C. A. (1908), *Am. J. Physiol.*, **23**:180–200 *(333)*.

OSBORNE, T. B., and MENDEL, L. B. (1913), *J. Biol. Chem.*, **15**:311–326 *(594, 595)*.

OSBORNE, T. B., and MENDEL, L. B. (1915), *J. Biol. Chem.*, **20**:351–378 *(592, 594)*.

OSBORNE, W. A. (1931), *Australian J. Exptl. Biol. Med. Sci.*, **8**:239–240 *(380)*.

OSBORNE, W. A., and KINKAID, H. E. (1914), *Biochem. J.*, **8**:28–29 *(390, 425)*.

OSTWALD, W. (1913), *Kolloid-Z.*, **12**:213–222 *(404)*.

OVSON, L. D. (1933), *Food Industries*, **5**:502–504 *(770)*.

OVSON, L. D. (1938), *Quick Frozen Foods*, **1**(3):15–18, 40 *(770, 771)*.

OWEN, E. C. (1939), *Biochem. J.*, **33**:22–26 *(602)*.

OWEN, R. (1866), *On the Anatomy of Vertebrates*, Vol. 2, *Birds and Mammals*, Longmans, Green, and Co., London, 592 pp. *(1)*.

OZAWA, E. (1936), *Japan. J. Exptl. Med.*, **14**:115–146 *(562)*.

PACE, J. (1937), *Dept. Sci. Ind. Research (Brit.)*, *Food Invest. Repts.*, **1936**:50–51 *(518)*.

PAGNIEZ, P., VALLERY-RADOT, P., and HAGUENAU, J. (1921), *Bull. mêm. soc. méd. hôp. Paris*, **45**:1077 *(558)*.

PALADINO, G. (1891), *J. Microgr.*, **15**:79–84 *(202)*.

PALADINO, P., and TOSO, D. (1896), *Analyst*, **21**:161 *(788)*.

PALIT, C. C., and DHAR, N. R. (1930a), *J. Phys. Chem.*, **34**:711–723 *(422)*.

PALIT, C. C., and DHAR, N. R. (1930b), *J. Phys. Chem.*, **34**:993–1005 *(422)*.

PALMER, K. J., and GALVIN, J. A. (1943), *J. Am. Chem. Soc.*, **65**:2187–2190 *(803)*.

PALMER, L. S., and KEMPSTER, H. L. (1919), *J. Biol. Chem.*, **39**:313–330 *(131)*.

PANORMOFF, A. A. (1905a), *J. Russ. Phys. Chem. Soc.*, **37**:915–923 *(326)*.

PANORMOFF, A. A. (1905b), *J. Russ. Phys. Chem. Soc.*, **37**:923–930 *(452)*.

PANUM, P. L. (1860), *Untersuchungen über die Entstehung der Missbildungen zunächst in den Eiern der Vögel*, G. Reimer, Berlin, 260 pp. *(263, 264, 280, 285, 286, 291, 292, 293, 307)*.

PAPPENHEIMER, A. (1940), *J. Exptl. Med.*, **71**:263–269 *(551)*.

PARKE, J. L. (1867), *Med.-chem. Untersuch.*, **2**:209–214 *(483)*.

PARKE, J. L. (1868), *Z. Chem.*, **11**:157–158 *(317)*.

PARKER, G. H. (1906), *Am. Naturalist*, **40**:13–25 *(255, 277, 279, 280, 282, 287, 290, 292, 293, 299, 300, 306)*.

PARKER, J. E., and BARTON, O. A. (1945), *North Dakota Agr. Expt. Sta. Bimonthly Bull.*, **8**(2):3–5 *(29)*.

PARKER, J. E., and KEMPSTER, H. L. (1940), *Poultry Sci.*, **19**:157–158 *(264, 265)*.

PARKER, S. L. (1927a), *Poultry Sci.*, **6**:259–273 *(126)*.

PARKER, S. L. (1927b), *Proc. World's Poultry Congr. (Ottawa)*, **3**:402–405 *(667)*.

PARKER, S. L., GOSSMAN, S. S., and LIPPINCOTT, W. A. (1926), *Poultry Sci.*, **5**:131–145 *(127, 128)*.

PARKHURST, R. T. (1924), *Idaho Agr. Expt. Sta. Bull.*, **134**:1–8 *(42)*.

PARKHURST, R. T. (1926), *Poultry Sci.*, **5**:121–126 *(74)*.

PARKHURST, R. T. (1933), *Poultry Sci.*, **12**:97–111 *(87)*.

PARKHURST, R. T. (1934), *Poultry Sci.*, **13**:317 *(85)*.

PARONA, C., and GRASSI, B. (1877), *Atti soc. ital. sci. nat. Milano*, **20**:103–124 *(268, 287, 299, 300)*.

PARSONS, C. H., and MINK, L. D. (1937), *U. S. Egg Poultry Mag.*, **43**:484–491, 509–512 *(645)*.

PARSONS, H. T. (1931a), *J. Biol. Chem.*, **90**:351–367 *(586)*.

PARSONS, H. T. (1931b), *J. Biol. Chem.*, **92**:lxiv–lxv *(629, 630)*.

PARSONS, H. T., JANSSEN, P., and SCHOENLEBER, F. (1934), *J. Biol. Chem.*, **105**:lxvii *(630)*.

PARSONS, H. T., and KELLY, E. (1933a), *Am. J. Physiol.*, **104**:150–164 (*630*).

PARSONS, H. T., and KELLY, E. (1933b), *J. Biol. Chem.*, **100**:645–652 (*630*).

PARSONS, H. T., and LEASE, J. G. (1934), *J. Nutrition*, **8**:57–67 (*630*).

PATTERSON, C. T. (1916), *J. Am. Assoc. Instr. Invest. Poultry Husb.*, **3**:16–20 (*24*).

PATTERSON, J. T. (1910), *J. Morphol.*, **21**:101–134 (*217, 232*).

PATTERSON, J. T. (1911), *Am. Naturalist*, **45**:54–59 (*268, 288, 293, 294*).

PAUL, H. (1934), *Proc. Oklahoma Acad. Sci.*, **14**:32–34 (*237*).

PAULI, W. (1928), *Biochem. Z.*, **202**:337–364 (*431*).

PAULI, W., and KÖLBL, W. (1935), *Kolloid-Beihefte*, **41**:417–460 (*453, 467*).

PAULI, W., and WEISSBROD, J. (1935), *Kolloid-Beihefte*, **42**:429–462 (*457, 461*).

PAVARINO, G. L. (1929), *Ann. chim. applicata*, **19**:266–272 (*687*).

PAVESI, P. (1893), *Boll. soc. zool. ital.*, **2**:101–109 (*306*).

PAVLOV, V. A., and ISAKOVA-KEO, M. M. (1929), *Biochem. Z.*, **216**:19–27 (*412*).

PAYAWAL, S. R., LOWE, B., and STEWART, G. F. (1946), *Food Research*, **11**:246–260 (*403, 404*).

PAYEN, A. (1863), *Précis des substances alimentaires*, Paris, 488 pp. (*314*).

PAYNE, L. F. (1925), *Poultry Sci.*, **4**:102–108 (*350*).

PAYNE, L. F., and HUGHES, J. S. (1933), *Kansas Agr. Expt. Sta. Bull.*, **34**:1–64 (*45*).

PCHELIN, V. A. (1940), *J. Phys. Chem. (U.S.S.R.)*, **14**:1085–1102 (*450*).

PEANO, E., and PISSARO, I. (1935), *Rev. sud-americana endocrinol. immunol. quîmioterap.*, **18**:85–106 (*84*).

PEARCE, J. A. (1943), *Can. J. Research*, **21(D)**:98–107 (*753*).

PEARCE, J. A., REID, M., METCALFE, B., and TESSIER, H. (1946), *Can. J. Research*, **24(F)**:215–223 (*739*).

PEARCE, R. M. (1912), *J. Exptl. Med.*, **16**:349–362 (*563*).

PEARL, R. (1909), *J. Exptl. Zoöl.*, **6**:339–359 (*73, 88, 94, 95, 262*).

PEARL, R. (1910), *Zool. Anz.*, **35**:417–423 (*277, 284*).

PEARL, R. (1912a), *Am. Naturalist*, **46**:697–711 (*18*).

PEARL, R. (1912b), *J. Exptl. Zoöl.*, **13**:153–268 (*52*).

PEARL, R. (1916), *J. Biol. Chem.*, **24**:123–135 (*46*).

PEARL, R. (1917), *Genetics*, **2**:417–432 (*12*).

PEARL, R., and CURTIS, M. R. (1912), *J. Exptl. Zoöl.*, **12**:99–132 (*181, 220*).

PEARL, R., and CURTIS, M. R. (1914), *J. Exptl. Zoöl.*, **17**:395–424 (*215, 223*).

PEARL, R., and CURTIS, M. R. (1916), *J. Agr. Research*, **6**:977–1042 (*66, 258, 260, 261, 262, 274, 276, 295, 296, 297, 298, 299*).

PEARL, R., and SCHOPPE, W. F. (1921), *J. Exptl. Zoöl.*, **34**:101–118 (*180, 181, 213*).

PEARL, R., and SURFACE, F. M. (1914a), *J. Biol. Chem.*, **19**:263–278 (*214, 218*).

PEARL, R., and SURFACE, F. M. (1914b), *U. S. Dept. Agr., Bur. Animal Industry, Bull.*, **110(3)**:171–241 (*87, 90, 91, 92, 107, 108*)

PELLEGRINO, L. P. (1927), *Ann. igiene*, **37**:28–31 (*305*).

PENIONZHKEVICH, E. E. (1933), *Science of the Egg* (in Ukrainian), Kamenets Podolsk, U.S.S.R., 183 pp. (*166*).

PENIONZHKEVICH, E. E. (1936), *Progress Zoötech. Studies* (in Russian), **2**:449–464 (*383, 388*).

PENNINGTON, D., SNELL, E. E., and EAKIN, R. S. (1942), *J. Am. Chem. Soc.*, **64**:469 (*631*).

PENNINGTON, M. E. (1909–1910), *J. Biol. Chem.*, **7**:109–132 (*317, 483, 492, 494*).

PENNINGTON, M. E. (1941), *U. S. Egg Poultry Mag.*, **48**:91–95, 127–128 (*719, 725, 731, 737*).

PENNINGTON, M. E., and HORNE, G. A. (1924), *Proc. Intern. Congr. Refrig.*, **4**:200 (*711*).

PENNINGTON, M. E., JENKINS, M. K., and BETTS, H. M. P. (1918), *U. S. Dept. Agr. Bull.*, 565:1–20 *(638)*.

PENNINGTON, M. E., JENKINS, M. K., ST. JOHN, E. A., and HICKS, W. B. (1914), *U. S. Dept. Agr. Bull.*, 51:1–77 *(688)*.

PENNINGTON, M. E., JENKINS, M. K., and STOCKING, W. A. (1916), *U. S. Dept. Agr. Bull.*, 224:1–99 *(726)*.

PENNINGTON, M. E., and PIERCE, H. C. (1910), *U. S. Dept. Agr. Yearbook*, 1910:461–476 *(692)*.

PENNINGTON, M. E., and ROBERTSON, H. C. (1912), *U. S. Dept. Agr., Bur. Chem., Circ.* 104:1–8 *(517, 518)*.

PENNISTON, V., and HEDRICK, L. R. (1944), *U. S. Egg Poultry Mag.*, 50:26–27, 47–48 *(722)*.

PENQUITE, R., and THOMPSON, R. B. (1933), *Poultry Sci.*, 12:201–205 *(50)*.

PERLMAN, I., and CHAIKOFF, I. L. (1939), *J. Biol. Chem.*, 127:211–220 *(568)*.

PERNOT, E. F. (1909), *Oregon Agr. Coll. Expt. Sta. Bull.*, 103:1–16 *(495)*.

PEROV, S. S. (1935), *Trudy Lab. Izucheniyu Belka i Belkovogo Obmena Organizme, Vsesoyuz. Akad. Sel'sko-Khoz. Nauk im. V. I. Lenina*, 8:64–74 *(427)*.

PEROV, S. S., and DOLINOV, K. (1932), *Trudy Lab. Izucheniyu Belka i Belkovogo Obmena Organizme, Vsesoyuz. Akad. Sel'sko-Khoz. Nauk im. V. I. Lenina*, 3:3–20 *(412)*.

PERRAULT, C. (1666–1699), *Mém. acad. sci. Paris*, 10:559–561 *(287)*.

PERRAULT, C. (1675), *Hist. acad. roy. sci. Paris*, 3:329–333 *(306)*.

PETER, P. N., and BELL, R. W. (1930), *Ind. Eng. Chem.*, 22:1124–1128 *(401, 438)*.

PETERSEN, C. F., LAMPMAN, C. E., and STAMBERG, O. E. (1947), *Poultry Sci.*, 26:180–186 *(626)*.

PETERSON, W. J., DEARSTYNE, R. S., COMSTOCK, R. E., and WELDON, V. (1945), *Ind. Eng. Chem., Anal. Ed.*, 17:370–371 *(625)*.

PETERSON, W. J., HUGHES, J. S., and PAYNE, L. F. (1939), *Kansas Agr. Expt. Sta. Tech. Bull.*, 46:1–74 *(126, 345, 346, 349, 350, 487)*.

PETERSON, W. J., and PARRISH, D. B. (1939), *Poultry Sci.*, 18:54–58 *(237)*.

PETIT, A. (1871), *Compt. rend. acad. sci.*, 62:503–506 *(226, 407)*.

PETROV, S. G. (1935), *Poultry Sci.*, 14:330–339 *(54)*.

PFEIFFER, C. A., and GARDNER, W. U. (1938), *Endocrinology*, 23:485–491 *(238)*.

PHILIPPI, G. T. (1936), *On the Nature of Proteins*, Thesis, University of Leyden *(440)*.

PHILLIPS, C. L. (1887), *Auk.*, 4:346 *(7)*.

PHILLIPS, P. H., HALPIN, J. G., and HART, E. B. (1935), *J. Nutrition*, 10:93–98 *(358, 360)*.

PHILLIPS, P. H., and LARDY, H. A. (1940), *J. Dairy Sci.*, 23:399–404 *(789)*.

PHILLIPS, R. E. (1943), *Poultry Sci.*, 22:368–373 *(46, 212, 244, 249)*.

PHILLIPS, R. E., and WARREN, D. C. (1937), *J. Exptl. Zoöl.*, 76:117–136 *(214, 215)*.

PHILLIPS, R. E., and WILLIAMS, C. S. (1944), *Poultry Sci.*, 23:110–113 *(368, 369)*.

PHILPOTT, D. (1933), *Proc. World's Poultry Congr. (Rome)*, 5(2):479–490 *(118, 120)*.

PICARD, W. K. (1928), *Dept. Agr. Ind. Commerce, Dutch E. Indies, Bull.*, 65:1–46 *(511)*.

PICCHI, C. (1911), *Brit. Birds*, 5:45–49 *(176)*.

PICHARD, P. (1898), *Compt. rend. acad. sci.*, 126:1882–1885 *(358)*.

PICK, E. W. (1911), *Poultry World*, 7:495 *(293)*.

PIERCE, H. C. (1909), *Iowa State Coll. Expt. Sta. Press Bull.*, 17:1–3 *(700)*.

PIETTRE, M. (1936), *Compt. rend. acad. sci.*, 202:699–701 *(326)*.

PINOY, P. E., and FABIANI, G. (1937), *Compt. rend. soc. biol.*, 124:562–563 *(553)*.

PIORKOWSKI, M. (1895), *Arch. Hyg.*, 25:145–153 *(495)*.

PLATT, C. S. (1927), *Poultry Sci.*, 6:285–289 *(32)*.

PLATT, C. S. (1936), *Poultry Sci.*, 15:249–251 *(57)*.

PLIMMER, R. H. A. (1908), *J. Chem. Soc.*, **93**:1500–1506 (*326, 330*).

PLIMMER, R. H. A. (1921), *Analyses and Energy Values of Foods*, H. M. Stationery Office, London, 255 pp. (*317, 320, 599, 600*).

PLIMMER, R. H. A., and BAYLISS, W. M. (1905), *J. Physiol.*, **33**:439–461 (*533*).

PLIMMER, R. H. A., and LOWNDES, J. (1924), *Biochem. J.*, **18**:1163–1169 (*353*).

PLIMMER, R. H. A., and PAGE, H. J. (1913), *Biochem. J.*, **7**:157–174 (*352*).

PLIMMER, R. H. A., and ROSEDALE, J. L. (1925), *Biochem. J.*, **19**:1004–1021 (*333*).

PLIMMER, R. H. A., and SCOTT, F. H. (1909), *J. Physiol.*, **38**:247–253 (*355, 356*).

POLECK, T. (1850), *Ann. Physik*, **79**:155–161 (*358, 411*).

POLSON, A. (1939a), *Kolloid-Z.*, **87**:149–181 (*436*).

POLSON, A. (1939b), *Kolloid-Z.*, **88**:51–61 (*436, 474*).

POMMERENKE, W. T., SLAVIN, H. B., KARIHER, D. N., and WHIPPLE, G. H. (1935), *J. Exptl. Med.*, **61**:261–282 (*587*).

POPP, G. (1908), *Z. öffentl. Chem.*, **14**:453–563 (*785*).

POPPE, K. (1910), *Arb. kaiserl. Gesundh.* (*Berlin*), **34(2)**:186–221 (*492, 495, 497*).

PORTER, T. (1941), *J. Nutrition*, **21**:101–113 (*605*).

POSIN, D. Q. (1942), *Proc. Montana Acad. Sci.*, **3** and **4**:10–15 (*364*).

POSTERNAK, S., and POSTERNAK, T. (1927), *Compt. rend. acad. sci.*, **184**:909–911 (*534*).

POZERSKI, E. (1936), *Compt. rend. soc. biol.*, **122**:909–911 (*789*).

POZERSKI, E., and GUÉLIN, MME. (1938), *Compt. rend. soc. biol.*, **128**:504–506 (*518*).

PRENTICE, J. H., BASKETT, R. G., and ROBERTSON, G. S. (1930), *Proc. World's Poultry Congr.* (*London*), **4**:242–251 (*22*).

PRENZLAU, L. (1910), Brit. Pat. 15,309 (*697*).

PRESS, R. (1941), *Nature*, **148**:753 (*358*).

PRÉVOST, J. L., and MORIN, A. (1846), *J. pharm. chim.*, **9**:249–256 (*314*).

PRICE, C. W. (1933), *Biochem. J.*, **27**:1789–1792 (*484*).

PRICE, C. W., and LEWIS, W. C. M. (1933), *Trans. Faraday Soc.*, **29**:775–787 (*484*).

PRICE, H. I., and LEWIS, W. C. M. (1929), *Biochem. J.*, **23**:1030–1043 (*484, 485, 486*).

PRICE, W. V. (1931), *J. Dairy Sci.*, **14**:221 (*774*).

PRIDEAUX, E. B. R., and WOODS, D. E. (1932), *Proc. Roy. Soc.* (*London*), **110(B)**:353–360 (*431*).

PRITZKER, I. Y. (1940), *Compt. rend. acad. sci. U. R. S. S.*, **6**:41–44 (*79*).

PRITZKER, I. Y. (1941), *Poultry Sci.*, **20**:99–101 (*172, 336*).

PUCHER, G. W. (1927), *Proc. Soc. Exptl. Biol. Med.*, **25**:72–73 (*390, 671*).

PUNNETT, R. C. (1923), *Heredity in Poultry*, Macmillan and Co., Limited, London, 204 pp. (*29, 100*).

PUNNETT, R. C. (1933), *J. Genetics*, **27**:465–470 (*100*).

PUNNETT, R. C., and BAILEY, P. G. (1920), *J. Genetics*, **10**:277–292 (*29, 101*).

PURJESZ, B., BERKESY, L., and GÖNCZI, K. (1933), *Arch. exptl. Path. Pharmakol.*, **173**:553–557 (*358, 360*).

PURJESZ, B., BERKESY, L., GÖNCZI, K., and KOVÁCS-OSKOLAS, M. (1934), *Arch. exptl. Path. Pharmakol.*, **176**:578–582 (*358, 360*).

PURJESZ, B., and WEISS, S. (1925), *Wien. Arch. inn. Med.*, **10**:377–392 (*518*).

PURKINJE, J. E. (1825), *Symbolae ad ovi avium historium ante incubationem*, Typ. Universitatis, Vratislaviae, 22 pp. (*232*).

PURVIS, M. (1921), *Breeder's Gaz.*, **79**:815 (*65*).

PUTZEYS, P., and BROSTEAUX, J. (1935), *Trans. Faraday Soc.*, **31**:1314–1325 (*455*).

PYKE, M. (1937), *Biochem. J.*, **31**:1958–1963 (*623*).

PYKE, M. (1939), *J. Soc. Chem. Ind.* (*London*), **58**:338–340 (*623*).

PYKE, W. E., and JOHNSON, G. (1941), *Poultry Sci.*, **20**:125–138 (*641*).

Quinn, J. P., and Godfrey, A. B. (1940), *Poultry Sci.*, **19**:359–360 (*303, 304*).
Quisenberry, J. E. (1915–16), *Missouri State Poultry Board, Bienn. Rept.*, pp. 59–60 (*43*).

Rabaud, E. (1902), *Feuille jeun. Naturalistes*, **32**:201–202 (*293*).
Radeff, T. (1935), *Arch. Geflügelkunde*, **9**:111–114 (*350*).
Raebiger, H. (1929), *Beitr. Klin. Tuberk.*, **71**:209–215 (*510*).
Rahn, O. (1932), *Physiology of Bacteria*, P. Blakiston's Son and Co., Philadelphia, 438 pp. (*459*).
Rajewski, B. (1929), *Strahlentherapie*, **33**:362–374 (*463*).
Rajewski, B. (1930), *Biochem. Z.*, **227**:272–285 (*457, 463*).
Rakuzin, M. A. (1915), *J. Russ. Phys. Chem. Soc.*, **47**:1050–1054 (*320*).
Rakuzin, M. A. (1916), *J. Russ. Phys. Chem. Soc., Proc.*, **48**:1335 (*454*).
Rakuzin, M. A., Brando, E. M., and Pekarskaja, G. F. (1915), *J. Russ. Phys. Chem. Soc.*, **47**:2051–2056 (*422*).
Rakuzin, M. A., and Flieher, G. D. (1923), *Chem. Ztg.*, **47**:66 (*406, 429, 430*).
Ramon, G. (1928), *Compt. rend. soc. biol.*, **99**:1476–1478 (*563*).
Rantasalo, V. (1929), *Acta Paediat.*, **8**:1–11 (*563*).
Rarkin, M. (1903), *J. soc. agr. suisse romande*, **44**:252–257 (*786*).
Ratner, B. (1928), *Am. J. Diseases Children*, **36**:277–278 (*558*).
Ratner, B., and Gruehl, H. L. (1931), *J. Exptl. Med.*, **53**:677–686 (*557*).
Ratner, B., Jackson, H. C., and Gruehl, H. L. (1927a), *J. Immunol.*, **14**:249–265 (*557*).
Ratner, B., Jackson, H. C., and Gruehl, H. L. (1927b), *J. Immunol.*, **14**:291–302 (*557*).
Ratner, B., Jackson, H. C., and Gruehl, H. L. (1927c), *J. Immunol.*, **14**:303–319 (*557*).
Rayer, —— (1849), *Compt. rend. soc. biol.*, (Ser. 1) **1**:9–10, 123 (*254, 278, 289, 292*).
Réaumur, R. A. F. de (1749), *Art de faire éclore et d'élever en toute saison des oiseaux domestiques, de toute espèces, soit par le moyen de la chaleur du fumier, soit par le moyen de celle du feu ordinaire*, Paris, Vol. 1, 364 pp.; vol. 2, 427 pp. (*29*).
Reder, R. (1938), *Poultry Sci.*, **17**:521–522 (*618*).
Reder, R. (1942), *Poultry Sci.*, **21**:528–531 (*618*).
Reed, F. D., and Martin, C. L. (1933), *Poultry Sci.*, **12**:90 (*248*).
Reedman, E. J., and Hopkins, J. W. (1942), *Can. J. Research*, **20(D)**:297–305 (*714*).
Reese, A. M. (1919), *J. Am. Assoc. Instr. Invest. Poultry Husb.*, **5**:22 (*250*).
Reichard, J. C. (1910), *Pharm. Ztg.*, **55(16)**:158–160 (*422*).
Reichenow, A. (1870), *J. Ornithol.*, **18**:385–392 (*89*).
Reid, E. (1934), *Chinese J. Physiol.*, **8**:53–64 (*778*).
Reid, M., and Pearce, J. A. (1945), *Can. J. Research*, **23(F)**:239–242 (*756*).
Reinhardt, C. (1899), German Pat. 112,892 (*712*).
Remotti, E. (1929), *Ricerche morfol.*, **9**:1–42 (*142*).
Rettger, L. F. (1913), *Storrs (Connecticut) Agr. Expt. Sta. Bull.*, **75**:191–213 (*250, 492, 493, 494, 495, 682*).
Rettger, L. F. (1914), *J. Exptl. Med.*, **19**:552–561 (*509*).
Rettger, L. F., Hull, T. G., and Sturges, W. S. (1916), *J. Exptl. Med.*, **23**:475–489 (*513*).
Rettger, L. F., Kirkpatrick, W. F., and Card, L. E. (1919), *Storrs (Connecticut) Agr. Expt. Sta. Bull.*, **101**:75–88 (*249, 495, 509*).
Rettger, L. F., Kirkpatrick, W. F., and Jones, R. E. (1914), *Storrs (Connecticut) Agr. Expt. Sta. Bull.*, **77**:263–309 (*244*).
Rettger, L. F., and Scoville, M. M. (1920), *J. Infectious Diseases*, **26**:217–229 (*511*).
Rettger, L. F., Slanetz, C. A., and McAlpine, J. C. (1926), *Storrs (Connecticut) Agr. Expt. Sta. Bull.*, **140**:99–110 (*513*).
Rettger, L. F., and Sperry, J. A. (1912), *J. Med. Research*, **21**:55–64 (*501*).

RETTGER, L. F., and STONEBURN, F. H. (1909), *Storrs (Connecticut) Agr. Expt. Sta. Bull.*, **60**:32–57 (*494, 509*).

RETTGER, L. F., and STONEBURN, F. H. (1911), *Storrs (Connecticut) Agr. Expt. Sta. Bull.*, **68**:278–301 (*509*).

RHOADS, C. P., and ABELS, J. C. (1943), *J. Am. Med. Assoc.*, **121**:1261–1263 (*632*).

RIBOULLEAU, J. (1938), *Compt. rend. soc. biol.*, **129**:914–916 (*352*).

RIBOULLEAU, J. (1940–41), *J. pharm. chim.*, **1**:661–662 (*352*).

RICE, F. E., and YOUNG, D. L. (1928), *Poultry Sci.*, **7**:116–118 (*413, 420*).

RICE, J. E., NIXON, C., and ROGERS, C. A. (1908), *Cornell Univ. Agr. Expt. Sta. Bull.*, **258**:1–68 (*30*).

RICHARDSON, K. C. (1935), *Trans. Roy. Soc. (London)*, **225(B)**:149–195 (*196, 222, 225, 226, 228, 229*).

RIDDLE, O. (1911), *J. Morphol.*, **22**:455–491 (*129*).

RIDDLE, O. (1921a), *Proc. Soc. Exptl. Biol. Med.*, **19**:12–14 (*257*).

RIDDLE, O. (1921b), *Science*, **54**:664–666 (*252*).

RIDDLE, O. (1923), *Am. J. Physiol.*, **66**:309–321 (*9, 252*).

RIDDLE, O. (1925a), *Am. J. Physiol.*, **73**:5–16 (*26*).

RIDDLE, O. (1925b), *Anat. Record*, **30**:365–382 (*176*).

RIDDLE, O. (1927), *Proc. Am. Phil. Soc.*, **66**:497–509 (*236*).

RIDDLE, O. (1931), *Am. J. Physiol.*, **97**:581–587 (*6, 19, 51*).

RIDDLE, O. (1938a), *Sci. Month.*, **47**:97–113 (*213, 245*).

RIDDLE, O. (1938b), *Carnegie Inst. Wash., Pub.*, **501**:259–273 (*235*).

RIDDLE, O. (1942a), *Cold Spring Harbor Symposia Quant. Biol.*, **10**:7–14 (*234*).

RIDDLE, O. (1942b), *Endocrinology*, **31**:498–506 (*235*).

RIDDLE, O., and ANDERSON, C. E. (1918), *Am. J. Physiol.*, **47**:92–102 (*214*).

RIDDLE, O., and BASSET, G. C. (1916), *Am. J. Physiol.*, **41**:425–429 (*214*).

RIDDLE, O., BATES, R. W., and DYKSHORN, S. W. (1933), *Am. J. Physiol.*, **105**:191–216 (*246*).

RIDDLE, O., BATES, R. W., and LAHR, E. L. (1935), *Am. J. Physiol.*, **111**:352–360 (*30, 246*).

RIDDLE, O., BATES, R. W., WELLS, B. B., MILLER, R. A., LAHR, E. L., SMITH, G. C., DUNHAM, H. H., and OPDYKE, D. F. (1941), *Carnegie Inst. Wash., Year Book*, **40**:234–245 (*218*).

RIDDLE, O., and BURNS, F. H. (1927), *Am. J. Physiol.*, **81**:711–724 (*235*).

RIDDLE, O., and DOTTI, L. B. (1936), *Science*, **84**:557–559 (*245*).

RIDDLE, O., HOLLANDER, W. F., McDONALD, M. R., LAHR, E. L., and SMITH, G. C. (1945), *Carnegie Inst. Wash., Year Book*, **44**:139–146 (*213*).

RIDDLE, O., and HONEYWELL, H. E. (1924), *Am. J. Physiol.*, **67**:333–336 (*235*).

RIDDLE, O., and LAHR, E. L. (1944), *Endocrinology*, **35**:255–260 (*30*).

RIDDLE, O., and McDONALD, M. R. (1945), *Endocrinology*, **36**:48–52 (*245*).

RIDDLE, O., RAUCH, V. M., and SMITH, G. C. (1944), *Anat. Record*, **90**:295–305 (*238*).

RIDDLE, O., RAUCH, V. M., and SMITH, C. C. (1945), *Endocrinology*, **36**:41–47 (*245*).

RIDDLE, O., and REINHART, W. H. (1926), *Am. J. Physiol.*, **76**:660–676 (*235*).

RIDDLE, O., and SCHOOLEY, J. P. (1944), *J. Washington Acad. Sci.*, **34**:341–346 (*244*).

RIDDLE, O., SMITH, G. C., and BENEDICT, F. G. (1933), *Am. J. Physiol.*, **105**:428–433 (*236*).

RIDGEWAY, R. (1912), *Color Standards and Color Nomenclature*, published by the author, Washington, 53 plates, 43 pp. (*97, 126, 128*).

RIEMENSCHNEIDER, R. W., ELLIS, N. R., and TITUS, H. W. (1938), *J. Biol. Chem.*, **126**: 255–263 (*337, 339, 340*).

RIEVEL, H. (1939), *Proc. World's Poultry Congr. (Cleveland)*, **7**:478–483 (*495*).

RILEY, G. M., and FRAPS, R. M. (1942), *Endocrinology*, **30**:537–541 (*212*).

RINGOEN, A. R. (1940), *J. Exptl. Zoöl.*, **83**:379–386 *(212, 242)*.

RINGOEN, A. R. (1942), *Am. J. Anat.*, **71**:99–116 *(51)*.

RINGOEN, A. R. (1945), *J. Morphol.*, **77**:265–283 *(238)*.

RINGOEN, A. R., and KIRSCHBAUM, A. (1939), *J. Exptl. Zoöl.*, **80**:173–191 *(213)*.

RINGROSE, R. C., and MORGAN, C. L. (1939), *Poultry Sci.*, **18**:125–128 *(142)*.

RIVALIEZ, P. (1683), *Acta eruditorum*, **1683**:220–221 *(287)*.

RIZZO, A. (1899), *Ricerche lab. anat. norm. univ. Roma*, **7**:171–177 *(109, 166, 167)*.

ROBERTS, E., and CARD, L. E. (1929), *Poultry Sci.*, **8**:271–272 *(288)*.

ROBERTS, E., and CARD, L. E. (1934), *Proc. World's Poultry Congr. (Rome)*, **5(2)**:353–358 *(29)*.

ROBERTS, I. P. (1891), *Cornell Univ. Agr. Expt. Sta. Bull.*, **37**:376–377 *(697)*.

ROBERTSON, G. S. (1931), *Can. Nat. Poultry Rec. Assoc. Blue Book*, 68 *(76, 82)*.

ROBERTSON, T. B. (1909–1910), *J. Biol. Chem.*, **7**:359–364 *(454)*.

ROBERTSON, T. B. (1916), *J. Biol. Chem.*, **25**:647–661 *(598)*.

ROBERTSON, T. B., and CUTLER, E. (1916), *J. Biol. Chem.*, **25**:663–667 *(598)*.

ROBINSON, T. W., and WOODSIDE, G. (1937), *J. Cellular Comp. Physiol.*, **9**:241–260 *(352)*.

ROBINSON, W. E. (1930), *American Poultry Record*, Poultry Item, Sellersville, Pa., 239 pp. *(25)*.

ROCHLINA, M. (1934a), *Bull. soc. chim. biol.*, **16**:1645–1651 *(237)*.

ROCHLINA, M. (1934b), *Bull. soc. chim. biol.*, **16**:1652–1662 *(241)*.

ROEPKE, R. R., and BUSHNELL, L. D. (1936), *J. Immunol.*, **30**:109–113 *(242)*.

ROEPKE, R. R., and HUGHES, J. S. (1935), *J. Biol. Chem.*, **108**:79–83 *(237)*.

ROGERS, C. A. (1912), *Proc. Intern. Assoc. Instr. Invest. Poultry Husb.*, **1**:77–81 *(129)*.

ROGERS, J. S. (1916), *J. Am. Leather Chem. Assoc.*, **11**:412–425 *(794)*.

ROHONYI, H. (1912), *Biochem. Z.*, **44**:165–179 *(521)*.

ROLLMAN, H. (1934), German Pat. 604,386 *(712)*.

ROMANKEWITSCH, N. A. (1932), *Z. Zellforsch. u. mikroscop. Anat.*, **16**:208–216 *(144, 146, 147)*.

ROMANKEWITSCH, N. A. (1934), *Z. Zellforsch. u. mikroscop. Anat.*, **21**:110–118 *(171, 172)*.

ROMANOFF, A. L. (1926), *Comparative Results of Different Animal Protein Feeds in Egg Production*, Inaug. Diss., Cornell University, Ithaca, New York, 133 pp. *(41, 83)*.

ROMANOFF, A. L. (1929a), *Biol. Bull.*, **56**:351–356 *(150, 151, 152, 153, 157, 159, 270, 370, 371, 372, 373)*.

ROMANOFF, A. L. (1929b), *Science*, **70**:314 *(222, 321)*.

ROMANOFF, A. L. (1931), *Biochem. J.*, **25**:994–996 *(181, 185, 203, 206, 318)*.

ROMANOFF, A. L. (1932), *Biol. Bull.*, **62**:54–62 *(336, 483)*.

ROMANOFF, A. L. (1933), *Poultry Sci.*, **12**:305–309 *(176)*.

ROMANOFF, A. L. (1940), *Food Research*, **5**:291–306 *(379, 406, 655, 656, 670, 671)*.

ROMANOFF, A. L. (1941), *J. Cellular Comp. Physiol.*, **18**:199–214 *(377)*.

ROMANOFF, A. L. (1943a), *Anat. Record*, **85**:261–267 *(125, 133, 181, 202, 204, 205)*.

ROMANOFF, A. L. (1943b), *Food Research*, **8**:212–223 *(166, 378, 383, 384, 385, 388, 655)*.

ROMANOFF, A. L. (1943c), *Food Research*, **8**:286–291 *(138, 321, 400, 406, 411, 413, 414, 644)*.

ROMANOFF, A. L. (1944a), *Biodynamica*, **4**:329–358 *(426)*.

ROMANOFF, A. L. (1944b), U. S. Pat. 2,362,774 *(376, 640, 660)*.

ROMANOFF, A. L. (1944c), *Biol. Bull.*, **87**:223–226 *(411)*.

ROMANOFF, A. L. (1948), *Poultry Sci.*, **27**:369–370 *(701)*.

ROMANOFF, A. L. (unpublished) *(66, 67, 90, 91, 114, 144, 150, 159, 166, 167, 276, 279, 285, 315, 316, 320, 368, 369, 371, 376, 382, 383, 384, 395, 408, 415, 640, 661)*.

ROMANOFF, A. L., and BLESS, A. A. (1942), *Proc. Natl. Acad. Sci. U. S.*, **28**:306–311 *(425)*.

ROMANOFF, A. L., BUMP, G., and HOLM, E. (1938), *N. Y. State Conservation Dept. Bull.*, **2**:1–44 (*150*).

ROMANOFF, A. L., and COTTRELL, C. L. (1939), *Proc. Soc. Exptl. Biol. Med.*, **42**:298–301 (*376*).

ROMANOFF, A. L., and FRANK, K. (1941), *Proc. Soc. Exptl. Biol. Med.*, **47**:527–530 (*376, 413*).

ROMANOFF, A. L., and GROVER, H. J. (1936), *J. Cellular Comp. Physiol.*, **7**:425–431 (*412, 413*).

ROMANOFF, A. L., and HALL, G. O. (1944), *Food Research*, **9**:218–220 (*376, 659*).

ROMANOFF, A. L., and HUTT, F. B. (1945), *Anat. Record*, **91**:143–154 (*65, 257, 288, 289, 293, 294, 302*).

ROMANOFF, A. L., and KOSHKIN, S. I. (unpublished) (*108, 109*).

ROMANOFF, A. L., and ROMANOFF, A. J. (1929), *Biol. Bull.*, **57**:300–306 (*411*).

ROMANOFF, A. L., and ROMANOFF, A. J. (1933), *Cornell Univ. Agr. Expt. Sta. Mem.*, **150**: 1–36 (*411*).

ROMANOFF, A. L., and ROMANOFF, A. J. (1944), *Food Research*, **9**:358–366 (*392, 393, 395, 716*).

ROMANOFF, A. L., and SULLIVAN, R. A. (1937), *Ind. Eng. Chem.*, **29**:117–120 (*137, 321, 414, 416, 417, 673*).

ROMANOFF, A. L., and YUSHOK, W. D. (unpublished) (*412*).

ROMIJN, C., and ROOS, J. (1938), *J. Physiol.*, **94**:365–379 (*148, 149*).

RONA, P., and FISCHGOLD, H. (1933), *Biochem. Z.*, **261**:366–373 (*528*).

ROSCOE, C. S. (1938), *Poultry*, **1938** (Oct. 25):457, 463 (*807*).

ROSE, B. (1941), *J. Immunol.*, **42**:161–180 (*553*).

ROSE, H. (1850), *Ann. Physik*, **79**:398–428 (*358, 411*).

ROSE, M. S. (1926), *J. Biol. Chem.*, **67**:xx–xxi (*584*).

ROSE, M. S., and BORGESON, G. M. (1935), *Nutrition Abstracts & Revs.*, **5**:474 (*584, 606*).

ROSE, M. S., and MCCOLLUM, E. L. (1928), *J. Biol. Chem.*, **78**:549–555 (*584*).

ROSE, M. S., and MACLEOD, M. (1922), *J. Biol. Chem.*, **50**:83–88 (*585*).

ROSE, M. S., and MACLEOD, M. (1923), *J. Biol. Chem.*, **58**:369–371 (*585, 587*).

ROSE, M. S., and MACLEOD, G. (1928a), *J. Nutrition*, **1**:29–38 (*584*).

ROSE, M. S., and MACLEOD, G. (1928b), *Proc. Soc. Exptl. Biol. Med.*, **25**:697–698 (*584*).

ROSE, M. S. and VAHLTEICH, E. (1938), *J. Am. Dietet. Assoc.*, **14**:593–614 (*610*).

ROSE, M. S., VAHLTEICH, E., and MACLEOD, G. (1934), *J. Biol. Chem.*, **104**:217–229 (*604, 605*).

ROSE, W. C., HAINES, W. J., and JOHNSON, J. E. (1942), *J. Biol. Chem.*, **146**:683–684 (*591*).

ROSE, W. C., HAINES, W. J., JOHNSON, J. E., and WARNER, D. T. (1943), *J. Biol. Chem.*, **148**:457–458 (*591*).

ROSENBERGER, F. (1908), *Z. physiol. Chem.*, **56**:464–467 (*352*).

ROSENBERGER, I. (1932), *Wien. med. Wochschr.*, **82**:1208–1209 (*607*).

ROSENKRANTZ, J. A., and BRUGER, M. (1946), *Arch. Path.*, **42**:81–87 (*568*).

ROSENKRANZ, S. (1938), *Arch. exptl. Path. Pharmakol.*, **189**:555–556 (*788*).

ROSNER, L. (1940), *J. Biol. Chem.*, **132**:657–662 (*476*).

ROSS, W. F., and TRACY, A. H. (1942), *J. Biol. Chem.*, **145**:19–26 (*534, 535*).

ROSSER, F. T., WHITE, W. H., WOODCOCK, A. H., and FLETCHER, D. A. (1942), *Can. J. Research*, **20**(D):57–70 (*692, 713*).

ROTHCHILD, I., and FRAPS, R. M. (1944), *Proc. Soc. Exptl. Biol. Med.*, **56**:79–82 (*210, 216, 244*).

ROTHCHILD, I., and FRAPS, R. M. (1945), *Endocrinology*, **37**:415–430 (*248*).

ROUX, C. (1923), *Ann. soc. linn. Lyon*, **69**:245–246 (*267*).

ROWAN, W. (1928), *Nature*, **122**:11–12 (*49*).

ROWAN, W. (1930), *Trans. Roy. Soc. Can.*, Sect. V., (3) **24**:157–164 (*210, 211*).

ROWAN, W. (1931), *Proc. Boston Soc. Nat. Hist.*, **39**:151–208 (*51*).

ROWAN, W. (1936), *Proc. World's Poultry Congr.* (*Leipzig*), **6**:142–152 (*49, 51*).

ROWAN, W. (1938), *Biol. Rev.*, **13**:374–402 (*213*).

ROY, A. C., and CHOPRA, R. H. (1941), *Indian J. Med. Research*, **29**:773–781 (*569, 570*).

RUBIN, M., and BIRD, H. R. (1942), *Univ. Maryland Agr. Expt. Sta. Tech. Bull.*, **A–12**: 339–350 (*45*).

RUDD, G. V. (1924a), *Australian J. Exptl. Biol. Med. Sci.*, **1**:179–185 (*526*).

RUDD, G. V. (1924b), *Australian J. Exptl. Biol. Med. Sci.*, **1**:187–190 (*331*).

RUDOLPH, W. (1938), *Z. Untersuch. Lebensm.*, **75**:428–430 (*412*).

RULLMAN, W. (1916), *Centr. Bakt. Parasitenk.*, **45**:219–230 (*518*).

RUNNELLS, R. A., and VAN ROEKEL, H. (1927), *Poultry Sci.*, **6**:141–147 (*509*).

RUOTSALAINEN, A. (1928), *Z. Kinderheilk.*, **46**:370–383 (*589*).

RUSSELL, W. C. (1934), *New Jersey Agr.*, **16(2)**:6 (*612*).

RUSSELL, W. C., HOWARD, C. H., and HESS, A. F. (1930), *Science*, **72**:506–507 (*237*).

RUSSELL, W. C., and McDONALD, F. G. (1929), *J. Biol. Chem.*, **84**:463–474 (*240, 359*).

RUSSELL, W. C., and TAYLOR, M. W. (1935a), *J. Nutrition*, **10**:613–623 (*611*).

RUSSELL, W. C., and TAYLOR, M. W. (1935b), *New Jersey Agr.*, **17(4)**:3 (*612*).

RUSSELL, W. C., TAYLOR, M. W., and WALKER, H. A. (1941), *Poultry Sci.*, **20**:372–378 (*43, 342*).

RYDER, J. A. (1893), *Proc. Am. Phil. Soc.*, **31(141)**:203–209 (*88*).

SACCHI, M. (1887), *Atti soc. ital. sci. nat. Milano*, **30**:273–309 (*183, 185*).

SAENZ, A., and COSTIL, L. (1932), *Compt. rend. soc. biol.*, **110**:1189–1190 (*508*).

SAHA, K. C., and MAZUMDER, P. (1945), *Ann. Biochem. Exptl. Med.*, **5**:33–38 (*355, 604*).

SAIJA, G. (1899), quoted by RIZZO, A. (1899), *Ricerche lab. anat. norm. univ. Roma*, **7**:171–177 (*109*).

ST. JOHN, J. L. (1931), *J. Am. Chem. Soc.*, **53**:4014–4019 (*401, 402, 419*).

ST. JOHN, J. L. (1936), *Proc. World's Poultry Congr.* (*Leipzig*), **6(1)**:285–289 (*641*).

ST. JOHN, J. L., and CASTER, A. B. (1944), *Arch. Biochem.*, **4**:45–49 (*402*).

ST. JOHN, J. L., and FLOR, I. H. (1931), *Poultry Sci.*, **10**:71–82 (*393, 395, 641*).

ST. JOHN, J. L., and GREEN, E. L. (1930), *J. Rheol.*, **1**:484–504 (*406*).

SAINZ, P. A., and RODRIGUES, A. (1943), *Inform. Méd.* (*Cuba*), **7(1)**:8–31 (*569*).

SAKURAJI, K. (1917), *J. Tokyo Med. Soc.*, **31**:1 (*343*).

SALISBURY, G. W., FULLER, H. K., and WILLETT, E. L. (1941), *J. Dairy Sci.*, **24**:905–910 (*789*).

SALISBURY, G. W., and VANDEMARK, N. L. (1945), *Am. J. Physiol.*, **143**:692–697 (*790*).

SALKOWSKI, E. (1911), *Biochem. Z.*, **32**:335–361 (*352*).

SALMON, W. D., and GOODMAN, J. G. (1934), *J. Nutrition*, **8**:1–24 (*629*).

SALVATORI, A. (1936a), *Riv. biol.*, **21**:16–25 (*387*).

SALVATORI, A. (1936b), *Riv. biol.*, **21**:76–80 (*388*).

SANCTIS, A. G. DE (1922), *Arch. Pediat.*, **39**:104–106 (*589*).

SANTOS, F. O., and PIDLAOAN, N. (1931), *Philippine Agr.*, **19**:659–664 (*780*).

SAVAGE, W. G. (1932), *J. Prev. Med.*, **6**:425–451 (*512, 513*).

SCHADE, A. L., and CAROLINE, L. (1944), *Science*, **100**:14–15 (*501*).

SCHAFER, E. A. (1907), *Nature*, **77**:159–163 (*49*).

SCHAFFER, F., and GURY, E. (1916), *Mitt. Lebensm. Hyg.*, **7**:217–222 (*785*).

SCHAIBLE, P. J., and BANDEMER, S. L. (1946a), *Poultry Sci.*, **25**:451–452 (*672*).

SCHAIBLE, P. J., and BANDEMER, S. L. (1946b), *Poultry Sci.*, **25**:456–459 (*668*).

SCHAIBLE, P. J., and BANDEMER, S. L. (1947), *Poultry Sci.*, **26**:207–209 (*666, 667*).

SCHAIBLE, P. J., BANDEMER, S. L., and DAVIDSON, J. A. (1946), *Poultry Sci.*, **25**:440–445 (*668*).

SCHAIBLE, P. J., and CARD, C. G. (1943), *Food Industries*, **15**:67–68 (*730*).

SCHAIBLE, P. J., DAVIDSON, J. A., and BANDEMER, S. L. (1944), *Poultry Sci.*, **23**:441–445 (*360, 672*).

SCHAIBLE, P. M., DAVIDSON, J. A., and MOORE, J. M. (1936), *Poultry Sci.*, **15**:298–303 (*300, 301*).

SCHAIBLE, P. J., MOORE, L. A., and MOORE, J. M. (1933), *Poultry Sci.*, **12**:334 (*667*).

SCHAIBLE, P. J., MOORE, L. A., and MOORE, J. M. (1934), *Science*, **79**:372 (*350*).

SCHALM, O. W. (1937), *J. Infectious Diseases*, **61**:208–216 (*511*).

SCHANTZ, E. J., BOUTWELL, R. K., ELVEHJEM, C. A., and HART, E. B. (1940), *J. Dairy Sci.*, **23**:1201–1204 (*598*).

SCHARRER, K., and SCHROPP, W. (1932), *Z. Tierernähr.*, **4**:249–264 (*358, 360, 361*).

SCHENCK, E. G. (1932), *Z. physiol. Chem.*, **211**:111–153 (*326*).

SCHEUNERT, A., and WAGNER, E. (1928), *Deut. med. Wochschr.*, **53**:1258–1260 (*581*).

SCHEUNERT, A., and WAGNER, E. (1929), *Deut. med. Wochschr.*, **55**:395–396 (*581, 589*).

SCHMID, A. (1912), *Chem. Ztg.*, **36**:796 (*785*).

SCHMIDT, J. (1932), *Berlin. tierärztl. Wochschr.*, **48**:161–166 (*84, 364*).

SCHMIDT-HEBBEL, H., and BOLLMANN, I. O. (1941), *Rev. med. y aliment. (Santiago, Chile)*, **4**:256–259 (*785*).

SCHMITT, F. O., OLSON, A. R., and JOHNSON, C. H. (1928), *Proc. Soc. Exptl. Biol. Med.*, **25**:718–720 (*401*).

SCHNEIDER, H. A., ASCHAM, J. E., PLATZ, B. R., and STEENBOCK, H. (1939), *J. Nutrition*, **18**:99–104 (*629*).

SCHNEITER, R., BARTRAM, M. T., and LEPPER, H. A. (1943), *J. Assoc. Offic. Agr. Chemists*, **26**:172–182 (*725, 734, 738*).

SCHÖNBERG, A., and GHONEIM, A. (1946), *Nature*, **157**:77–78 (*23*, 47).

SCHÖNWETTER, M. (1930), *Beitr. Fortpfl. biol. Vögel*, **6**:185–193 (*150*).

SCHÖNWETTER, M. (1932), *Ool. Rec.*, **12**:83–86 (*107, 109*).

SCHOOL, H. (1893), *Arch. Hyg.*, **17**:535–551 (*411*).

SCHOOLEY, J. P., and RIDDLE, O. (1944), *Physiol. Zoöl.*, **16**:187–193 (*26*).

SCHRENK, W. G., CHAPIN, D. S., and CONRAD, R. M. (1944), *Ind. Eng. Chem., Anal. Ed.*, **16**:632–634 (*615*).

SCHROEDER, C. H. (1933), *U. S. Egg Poultry Mag.*, **39(6)**:44–46, 50 (*136*).

SCHULTZE, H. E. (1943), *Z. physiol. Chem.*, **279**:87–93 (*518*).

SCHULZ, F. N., and ZSIGMONDY, R. (1903), *Beitr. chem. physiol. Path.*, **3**:137–160 (*327, 401, 443*).

SCHUMACHER, S. (1896), *Zool. Anz.*, **19**:366–368 (*287, 293, 300*).

SCOTT, D. A. (1934), *Biochem. J.*, **28**:1592–1602 (*606*).

SCOTT, H. M. (1940), *Am. Naturalist*, **74**:185–188 (*262*).

SCOTT, H. M. (1941), *Poultry Sci.*, **20**:472 (*646*).

SCOTT, H. M., and HUANG, W. L. (1941), *Poultry Sci.*, **20**:402–405 (*219, 220*).

SCOTT, H. M., HUGHES, J. S., and WARREN, D. C. (1937), *Poultry Sci.*, **16**:53–61 (*221*).

SCOTT, H. M., JUNGHERR, E., and MATTERSON, L. D. (1944a), *Proc. Soc. Exptl. Biol. Med.*, **57**:7–10 (*34, 46*).

SCOTT, H. M., JUNGHERR, E., and MATTERSON, L. D. (1944b), *Poultry Sci.*, **23**:446–453 (*46*).

SCOTT, H. M., and PAYNE, L. F. (1937), *Poultry Sci.*, **16**:90–96 (*50*).

SCOTT, H. M., and WARREN, D. C. (1936), *Poultry Sci.*, **15**:381–389 (*25, 50*).

SCOTT, W. J., and GILLESPIE, J. M. (1943), *J. Council Sci. Ind. Research*, 16:15–17 (727).

SCOTT, W. J., and GILLESPIE, J. M. (1944), *J. Council Sci. Ind. Research*, 17:299–304 (727).

SCOTT, W. M. (1930), *Brit. Med. J.*, 1930(2):56–58 (512).

SCOTT, W. M. (1932), *J. Path. Bact.*, 35:655 (511, 512).

SCOULAR, F. I. (1939), *J. Nutrition*, 17:103–113 (606).

SCRIMSHAW, N. W., HUTT, F. B., SCRIMSHAW, M. W., and SULLIVAN, C. R. (1945), *J. Nutrition*, 30:375–383 (623).

SCRIMSHAW, N. W., THOMAS, W. P., McKIBBEN, J. W., SULLIVAN, C. R., and WELLS, K. C. (1944), *J. Nutrition*, 28:235–239 (624).

SEASTONE, C. V. (1938), *J. Gen. Physiol.*, 21:621–629 (440).

SEEBOHM, H. (1896), *Coloured Figures of the Eggs of British Birds*, Pawson and Brailsford, Sheffield, England, 304 pp. (1, 89, 97, 173).

SELIGMANN, E., and HERTZ, J. J. (1944), *Ann. Internal Med.*, 20:743–751 (512, 514).

SELL, H. M., OLSEN, A. G., and KREMERS, R. E. (1935), *Ind. Eng. Chem.*, 27:1222–1223 (777).

SENDJU, Y. (1927), *J. Biochem. (Japan)*, 7:181–189 (352).

SENG, H. (1913), *Z. Immunitätsforsch.*, 20:355–366 (559, 560).

SEREBROVSKY, R., and SEREBROVSKY, A. (1926), *Mem. Anikowo Genetical Sta.*, 132 pp. (89).

SERONO, C., and MONTEZEMOLO, R. (1943), *Rass. clin. terap. sci. affini*, 42:1–9 (622).

SERONO, C., and PALOZZI, A. (1911), *Arch. farmacol. sper.*, 11:553–570 (483).

SERRES, M. (1860), *Mém. acad. sci. (Paris)*, 25:92 (285).

SETTIMJ, M. (1929), *Ann. chim. applicata*, 19:182–188 (785).

SEVAG, M. G., and SEASTONE, C. V. (1934), *Z. Immunitätsforsch.*, 83:464–471 (556).

SHANNON, W. R. (1922), *Am. J. Diseases Children.*, 23:392–405 (558).

SHARMA, B. M. (1932), *Z. Immunitätsforsch.*, 77:79–100 (553, 557, 692).

SHARP, P. F. (1929a), *Science*, 69:278–280 (680, 707).

SHARP, P. F. (1929b), *U. S. Egg Poultry Mag.*, 35(6):14–17, 64 (641).

SHARP, P. F. (1932), *Ind. Eng. Chem.*, 24:941–946 (714).

SHARP, P. F. (1934), *U. S. Egg Poultry Mag.*, 40(11):33–37 (646).

SHARP, P. F. (1937), *Food Research*, 2:477–498 (682).

SHARP, P. F., and POWELL, C. K. (1927), *Proc. World's Poultry Congr. (Ottawa)*, 3:399–402 (679).

SHARP, P. F., and POWELL, C. K. (1930), *Ind. Eng. Chem.*, 22:908–910 (663, 674).

SHARP, P. F., and POWELL, C. K. (1931), *Ind. Eng. Chem.*, 23:196–199 (680).

SHARP, P. F., and STEWART, G. F. (1931), *U. S. Egg Poultry Mag.*, 37(6):30–32, 63, 65–66, 68 (707, 709).

SHARP, P. F., and STEWART, G. F. (1936), *Cornell Univ. Agr. Expt. Sta. Mem.*, 191:1–11 (684, 688, 703, 704).

SHARP, P. F., STEWART, G. F., and HUTTAR, J. C. (1936), *Cornell Univ. Agr. Expt. Sta. Mem.*, 189:1–26 (669).

SHARP, P. F., and WHITAKER, R. (1927), *J. Bact.*, 14:17–46 (501, 683).

SHARPE, J. S. (1924), *Biochem. J.*, 18:151–152 (352).

SHAW, T. M., VORKOEPER, A. R., and DYCHE, J. K. (1946), *Food Research*, 11:187–194 (744).

SHEARD, C., and HIGGINS, G. M. (1930), *J. Exptl. Zoöl.*, 57:205–222 (387, 388).

SHERMAN, H. C. (1941), *Chemistry of Foods and Nutrition*, sixth edition, The Macmillan Co., New York, 611 pp. (602, 605).

SHERMAN, H. E., and WANG, T. C. (1929), *Philippine J. Sci.*, 38:69–79 (600).

SHERMAN, W. C., ELVEHJEM, C. A., and HART, E. B. (1934), *J. Biol. Chem.*, **107**:289–295 *(604, 605)*.

SHERWOOD, R. M., and FRAPS, G. S. (1932), *Texas Agr. Expt. Sta. Bull.*, **468**:1–19 *(45, 612)*.

SHERWOOD, R. M., and FRAPS, G. S. (1934), *Texas Agr. Expt. Sta. Bull.*, **493**:1–22 *(613)*.

SHERWOOD, R. M., and FRAPS, G. S. (1935), *Texas Agr. Expt. Sta. Bull.*, **514**:1–21 *(610, 611)*.

SHIBATA, S. (1933), *Proc. World's Poultry Congr.* (*Rome*), **5(2)**:475–478 *(217)*.

SHIBATA, S. (1936), *Proc. World's Poultry Congr.* (*Leipzig*), **6**:202–203 *(317, 320)*.

SHIKINAMI, Y. (1928), *Tôhoku J. Exptl. Med.*, **10**:1–25 *(517)*.

SHKLYER, N. (1937), *Ukrainian Biochem. J.*, **10**:406–418 *(410)*.

SHOHL, A. T. (1939), *Mineral Metabolism*, Reinhold Publishing Corp., New York, 384 pp. *(602, 603)*.

SHOHL, A. T., and SATO, A. (1923), *J. Biol. Chem.*, **58**:257–266 *(603)*.

SHORE, A., WILSON, H., and STUECK, G. (1935), *J. Biol. Chem.*, **112**:407–413 *(480)*.

SHUTT, F. T. (1927), *Can. Dept. Agr. Circ.*, **31**:1–2 *(699, 700)*.

SHUTT, W. J. (1934), *Trans. Faraday Soc.*, **30**:893–898 *(451)*.

SIBLEY, W. K. (1890), *J. Comp. Med. Vet. Arch.*, **11**:317–334 *(510)*.

SIEKE, F. (1943), *Arch. Hyg. u. Bakt.*, **129**:108–115 *(377)*.

SIEVERS, O. (1938), *Acta Path. Microbiol. Scand. Suppl.*, **37**:458–475 *(561)*.

SIEVERS, O. (1939), *Acta Path. Microbiol. Scand.*, **16**:44–98 *(541, 542, 560)*.

SIKATA, M., and HUZII, S. (1940), *Imp. Zoötech. Expt. Sta. (Chiba, Japan) Research Bull.*, **43**:1–31 *(801)*.

SIMEONE, D. H. (1944), *Rev. facultad agron. y vet., Univ. Buenos Aires*, **2(2)**:15–47 *(513)*.

SIMMONDS, P. L. (1889), *Popular Sci. Monthly*, **35**:92–105 *(287, 788, 799, 800, 805, 806)*.

SIMPSON, B. W., and STRAND, R. (1930), *New Zealand J. Agr.*, **40**:403–406 *(360)*.

SIMPSON, S. (1920), *Proc. Soc. Exptl. Biol. Med.*, **17**:87–88 *(46)*.

SIMPSON, S. (1923), *Quart. J. Exptl. Physiol.*, **13**:181–189 *(46)*.

SJÖGREN, B., and SVEDBERG, T. (1930), *J. Am. Chem. Soc.*, **52**:5187–5192 *(435, 436, 437, 455, 472)*.

SJOLLEMA, B., and DONATH, W. F. (1940), *Biochem. J.*, **34**:736–748 *(346, 349, 612)*.

SJÖQVIST, J. (1895), *Skand. Arch. Physiol.*, **6**:255–261 *(451)*.

SKOGLUND, W. C., and TOMHAVE, A. E. (1941), *Poultry Sci.*, **20**:322–326 *(141)*.

SKRAUP, Z. H., and HÜMMELBERGER, F. (1909), *Monatsh. Chem.*, **30**:125–145 *(333)*.

SLUKES, T. M. (1902), U. S. Pat. 699,258 *(697)*.

SMAKULA, A. (1934), *Z. angew. Chem.*, **47**:657–665 *(488)*.

SMITH, A. J. M. (1930), *Dept. Sci. Ind. Research (Brit.), Food Invest. Repts.*, **1929**:74–80 *(378, 380, 656)*.

SMITH, A. J. M. (1931), *Dept. Sci. Ind. Research (Brit.), Food Invest. Repts.*, **1930**:80–102 *(378, 379, 655, 656)*.

SMITH, A. J. M. (1932), *Dept. Sci. Ind. Research (Brit.), Food Invest. Repts.*, **1931**:148–182 *(407, 420, 426, 674, 675, 676, 677)*.

SMITH, A. J. M. (1933), *Dept. Sci. Ind. Research (Brit.), Food Invest. Repts.*, **1932**:118–120 *(380, 402)*.

SMITH, A. J. M. (1934), *J. Exptl. Biol.*, **11**:228–242 *(402, 403, 420, 421, 422)*.

SMITH, A. J. M. (1935a), *Dept. Sci. Ind. Research (Brit.), Food Invest. Repts.*, **1934**:53–56 *(403, 404, 405, 407, 674)*.

SMITH, A. J. M. (1935b), *Dept. Sci. Ind. Research (Brit.), Food Invest. Repts.*, **1934**:211–214 *(704)*.

SMITH, C. W. (1930), *Nebraska Agr. Expt. Sta. Bull.*, **247**:1–34 *(48, 51)*.

SMITH, E. C. (1935), *Dept. Sci. Ind. Research (Brit.)*, *Food Invest. Repts.*, **1934**:50–52 (*681*).

SMITH, E. E. (1928), *Aluminum Compounds in Food*, Paul B. Hoeber, Inc., New York, 378 pp. (*358*).

SMITH, E. R. B. (1935), *J. Biol. Chem.*, **108**:187–194 (*433*).

SMITH, E. R. B. (1936), *J. Biol. Chem.*, **113**:473–478 (*433, 434*).

SMITH, M. (1931), *J. Exptl. Biol.*, **8**:312–318 (*377, 378, 678, 679*).

SMITH, M., and SHEPHERD, J. (1931), *J. Exptl. Biol.*, **8**:293–311 (*421, 425, 675, 676*).

SMITH, M. I. (1939), *Proc. Pacific Sci. Congr.*, **6**:167–176 (*566*).

SMITH, M. I. (1941), *J. Am. Med. Assoc.*, **116**:562–566 (*362, 566*).

SMITH, M. I., and STOHLMAN, E. F. (1940), *J. Pharmacol.*, **70**:270–278 (*566*).

SMITH, N. C. (1930), *J. Assoc. Offic. Agr. Chemists*, **13**:272–291 (*784*).

SMITH, R. M. (1943), *Arkansas Agr. Expt. Sta. Bull.*, **431**:1–26 (*46*).

SMYTH, F. S., and BAIN, K. (1931), *J. Allergy*, **2**:282–284 (*557*).

SMYTH, F. S., and STALLINGS, M. (1931), *J. Allergy*, **3**:16–19 (*558*).

SNELL, E. E., ALINE, E., COUCH, J. R., and PEARSON, P. B. (1941), *J. Nutrition*, **21**:201–205 (*627, 628*).

SNELL, E. E., EAKIN, R. E., and WILLIAMS, R. J. (1940), *J. Am. Chem. Soc.*, **62**:175–178 (*629*).

SNELL, E. E., and QUARLES, E. (1941), *J. Nutrition*, **22**:483–489 (*608, 627, 628*).

SNELL, E. E., and STRONG, F. M. (1939), *Ind. Eng. Chem., Anal. Ed.*, **11**:346–350 (*347, 624*).

SNELL, H. M., OLSEN, A. G., and KREMERS, R. E. (1935), *Ind. Eng. Chem.*, **27**:1222–1223 (*399*).

SNYDER, C. G. (1930), *U. S. Egg Poultry Mag.*, **36(7)**:14–21, 80 (*771*).

SNYDER, C. G. (1931a), *U. S. Egg Poultry Mag.*, **37(12)**:46–48, 75 (*776, 778*).

SNYDER, C. G. (1931b), *U. S. Egg Poultry Mag.*, **37(5)**:16–21, 76, 78 (*779*).

SNYDER, C. G. (1933a), *U. S. Egg Poultry Mag.*, **39(4)**:49 (*806*).

SNYDER, C. G. (1933b), *U. S. Egg Poultry Mag.*, **39(9)**:20–21, 51–53 (*800*).

SNYDER, E. S. (1945), *Ontario Dept. Agr. Bull.*, **446**:1–47 (*131*).

SOLDI, A., and TESTORI, S. (1931), *Ann. chim. applicata*, **21**:338–343 (*785*).

SONNENBRODT, —— (1908), *Arch. mikroscop. Anat. Entwicklungsmech.*, **72**:415–480 (*122, 215*).

SORBY, H. C. (1875), *Proc. Zool. Soc. London*, pp. 351–365 (*226, 348*).

SÖRENSEN, M. (1934), *Compt. rend. trav. lab. Carlsberg*, **20(3)**:1–19 (*327, 332, 344*).

SÖRENSEN, M., and SÖRENSEN, S. P. L. (1925), *Compt. rend. trav. lab. Carlsberg*, **15(9)**:1–26 (*457, 469, 475*).

SÖRENSEN, S. P. L. (1917), *Compt. rend. trav. lab. Carlsberg*, **12**:262–372 (*435, 438, 468*).

SÖRENSEN, S. P. L. (1925), *J. Am. Chem. Soc.*, **47**:457–469 (*428*).

SÖRENSEN, S. P. L., and HÖYRUP, M. (1916), *Compt. rend. trav. lab. Carlsberg*, **12**:12–67 (*428*).

SÖRENSEN, S. P. L., and HÖYRUP, M. (1917), *Compt. rend. trav. lab. Carlsberg*, **12**:164–212 (*429, 441*).

SÖRENSEN, S. P. L., LINDERSTRÖM-LANG, K., and LUND, E. (1927), *J. Gen. Physiol.*, **8**:543–601 (432).

SOWELL, D. F., and MORGAN, C. L. (1936), *Poultry Sci.*, **15**:219–222 (*142*).

SPAMER, C. O. (1931), *U. S. Egg Poultry Mag.*, **37(2)**:48–51, 58, 60, 62 (*699, 712*).

SPANSWICK, M. P. (1930), *Am. J. Pub. Health*, **20**:73–74 (*721*).

SPIEGEL-ADOLF, M. (1928), *Klin. Wochschr.*, **7**:1592–1596 (*472*).

SPIEGEL-ADOLF, M., and KRUMPEL, O. (1929), *Biochem. Z.*, **208**:49–59 (*462, 463, 464*).

SPIES, T. D. (1946), *Ann. N. Y. Acad. Sci.*, **48**:313–342 (*632*).

SPIES, T. D., BEAN, W. B., and ASHE, W. E. (*1939*), *J. Am. Med. Assoc.*, **112**:2414–2415 (*629*).

SPOHN, A. A., and RIDDLE, O. (1916), *Am. J. Physiol.*, **41**:397–408 (*132, 133, 206, 317, 318, 355*).

STAFSETH, H. J., BIGGAR, R. J., THOMPSON, W. W., and NEU, L. (1934), *J. Am. Vet. Med. Assoc.*, **85**:342–359 (*510*).

STAIR, R., and COBLENTZ, W. W. (1935), *J. Research Natl. Bur. Standards*, **15**:295–316 (*388, 389, 418, 419*).

STANLEY, A., and WITSCHI, E. (1940), *Anat. Record*, **76**:339–342 (*176*).

STARKENSTEIN, E. (1911), *Biochem. Z.*, **30**:56–98 (*352*).

STARY, Z., and ANDRATSCHKE, I. (1925), *Z. physiol. Chem.*, **148**:83–98 (*331*).

STASSANO, H., and BILLON, F. (1902), *Compt. rend. soc. biol.*, **54**:167–169 (*570*).

STEAD, T. (1882), Brit. Pat. 4910 (*712*).

STEARNS, G., JEANS, P. C., and VANDECAR, V. (1936), *J. Pediat.*, **9**:1–10 (*621*).

STEDMAN, H. L., and MENDEL, L. B. (1926), *Am. J. Physiol.*, **77**:199–210 (*462*).

STEENBOCK, H., and BLACK, A. (1925), *J. Biol. Chem.*, **64**:263–298 (*618*).

STEENBOCK, H., and SCOTT, H. T. (1932), *Wisconsin Agr. Expt. Sta. Bull.*, **421**(Ann. Rept. 1930–1931):122 (*617*).

STEGGERDA, M. (1928), *J. Exptl. Zoöl.*, **51**:403–416 (*213*).

STEGGERDA, M. (1931), *Poultry Sci.*, **10**:98–103 (*213*).

STEGGERDA, M., and HOLLANDER, W. F. (1944), *Poultry Sci.*, **23**:459–461 (*270, 271, 275*).

STEINER, A. (1941), *J. Am. Med. Assoc.*, **116**:2752 (*589*).

STENQVIST, F. (1928), *Deut. med. Wochschr.*, **54**:1920–1923 (*581, 582*).

STEPANEK, O. (1905), *Centr. Physiol.*, **18**:188–205 (*352*).

STEPHANY, A., and MATSCHULAN, G. (1937), *Arch. exptl. Path. Pharmakol.*, **187**:234–236 (*788*).

STEPP, W., FEULGEN, R., and VOIT, K. (1927), *Biochem. Z.*, **181**:284–288 (*352*).

STERN, M., and THIERFELDER, H. (1907), *Z. physiol. Chem.*, **53**:370–385 (*338, 343, 485*).

STEWART, G. F. (1935), *Poultry Sci.*, **14**:24–32 (*144, 160*).

STEWART, G. F. (1936), *Poultry Sci.*, **15**:119–124 (*371, 372*).

STEWART, G. F., BEST, L. R., and LOWE, B. (1943), *Proc. Inst. Food Technol.*, **1943**:77–89 (*746, 750, 751, 755, 757, 763*).

STEWART, G. F., GANS, A. R., and SHARP, P. F. (1932a), *U. S. Egg Poultry Mag.*, **38(5)**: 31–34 (*639*).

STEWART, G. F., GANS, A. R., and SHARP, P. F. (1932b), *U. S. Egg Poultry Mag.*, **38(7)**: 17–22 (*636, 639*).

STEWART, G. F., GANS, A. R., and SHARP, P. F. (1933a), *U. S. Egg Poultry Mag.*, **39(1)**: 47–50 (*642*).

STEWART, G. F., GANS, A. R., and SHARP, P. F. (1933b), *U. S. Egg Poultry Mag.*, **39(2)**: 37–39, 60–61 (*639*).

STEWART, G. F., and KLINE, R. W. (1941), *Proc. Inst. Food Technol.*, **1941**:48–56 (*741, 751, 758, 761*).

STEWART, J. H., and ATWOOD, H. (1903), *West Virginia Agr. Expt. Sta. Bull.*, **88**:147–162 (*127*).

STIEBELING, H. K. (1936), *Bur. Labor Stat., U. S. Dept. Labor, Monthly Labor Rev.*, **43**:14–23 (*598*).

STIEBELING, H. K. (1939), *Year Book of Agriculture, U. S. Dept. Agr.*, **1939**:321–340 (*598*).

STIEBELING, H. K. (1942), *Federation Proc.*, **1**:327–330 (*577*).

STIEVE, H. (1918), *Arch. Entwicklungsmech. Organ.*, **44**:530–588 (*27, 48, 217, 232, 248*).

STIEVE, H. (1922), *Arch. Entwicklungsmech. Organ.*, **50**:607–617 (*188*).

STILES, G. W., and BATES, C. (1912), *U. S. Dept. Agr., Bur. Chem., Bull.*, **158**:1–36 (*492*).

STITT, G., and KENNEDY, E. K. (1945), *Food Research*, **10**:426–436 (*750*).

STODDARD, K. L. (1931), *The Bobwhite Quail: Its Habits, Preservation, and Increase*, Charles Scribner's Sons, New York, 559 pp. (*9*).

STRAIN, H. H. (1935), *J. Biol. Chem.*, **111**:85–93 (*346*).

STRAIN, H. H. (1936), *Science*, **83**:241–242 (*345*).

STRAUB, J. (1932), *Debreceni Szemle*, **6**:129–131 (*361, 362*).

STRAUB, J. (1933), *Z. Untersuch. Lebensm.*, **65**:97–100 (*358, 361*).

STRAUB, J., and DONCK, C. M. (1934), *Chem. Weekblad*, **31**:461–465 (*411, 422*).

STRAUB, J., and HOOGERDUYN, M. J. J. (1929), *Rec. trav. chim.*, **48**:49–82 (*388, 390, 420, 424, 425*).

STRAUB, J., and KABOS, W. J. (1938), *Chem. Weekblad*, **35**:739–741 (*659*).

STRITAR, J., DACK, G. M., and JUNGEWAELTER, F. G. (1936), *Food Research*, **1**:237–246 (*514*).

STUART, L. S., and GORESLINE, H. E. (1942a), *J. Bact.*, **44**:541–549 (*740, 741, 742*).

STUART, L. S., and GORESLINE, H. E. (1942b), *J. Bact.*, **44**:625–632 (*741, 742*).

STUART, L. S., HALL, H. H., and DICKS, E. (1942), *U. S. Egg Poultry Mag.*, **48**:629–633, 658 (*765, 767*).

STUART, L. S., and McNALLY, E. H. (1943), *U. S. Egg Poultry Mag.*, **49**:28–31, 45–47 (*492, 496, 499, 683*).

STURKIE, P. D. (1946), *Poultry Sci.*, **25**:369–372 (*226*).

STURM, —— (1910), *Arch. wiss. u. prakt. Tierheilk.*, **36**:177–207 (*183, 185*).

STÜTZER, A. (1882), *Repert. anal. Chem.*, **2**:161–169 (*317*).

SUBIRANA, I. (1917), Swiss Pat. 74,124 (*712*).

SUEYOSHI, Y., and KAWAI, K. (1932), *J. Biochem. (Japan)*, **15**:277–283 (*484, 485*).

SUGIMOTO, T. (1913), *Arch. exptl. Path. Pharmakol.*, **74**:14–26 (*518*).

SUGITA, T. (1940), *J. Tokyo Med. Assoc.*, **54**:241–254 (*569*).

SUMNER, E. (1938), *J. Nutrition*, **16**:129–139 (*583, 593, 594*).

SUMNER, E., and MURLIN, J. R. (1938), *J. Nutrition*, **16**:141–152 (*594*).

SUMNER, E., PIERCE, H. B., and MURLIN, J. R. (1938), *J. Nutrition*, **16**:37–56 (*594*).

SUMULONG, M. D. (1925), *Philippine J. Sci.*, **28**:549–557 (*257, 279, 280*).

SUOMALAINEN, P. (1939), *Ann. Acad. Sci. Fennicae*, **A53(8)**:1–13 (*633*).

SUPINO, F. (1897), *Feuille jeun. Naturalistes*, (Ser. 3) **27**:201 (*287*).

SURFACE, F. M. (1912), *Maine Agr. Expt. Sta. Bull.*, **206**:395–430 (*183, 185, 186, 187, 190, 191, 192, 193, 194*).

SUSSMAN, H., DAVIDSON, A., and WALZER, M. (1928), *Arch. Internal Med.*, **42**:409–414 (*558, 586*).

SVEDBERG, T. (1930), *Kolloid-Z.*, **51**:10–24 (*437*).

SVEDBERG, T. (1931), *Nature*, **128**:999–1000 (*427*).

SVEDBERG, T., and ERIKSSON, I.-B. (1933), *Biochem. Z.*, **258**:1–3 (*536*).

SVEDBERG, T., and NICHOLS, J. G. (1926), *J. Am. Chem. Soc.*, **48**:3081–3092 (*434, 435*).

SVEDBERG, T., and PEDERSON, K. (1940), *The Ultracentrifuge*, translated from the German, Oxford University Press, London, 154 pp. (*435*).

SVEDBERG, T., and TISELIUS, A. (1926), *J. Am. Chem. Soc.*, **48**:2272–2278 (*455*).

SVETOSAROV, E., and STREIGH, G. (1940), *Compt. rend. acad. sci. U.R.S.S.*, **27**:398–401 (*23, 213*).

SVOBODA, H. (1902), *Oesterr. Chem.-Ztg.*, **5(21)**:483–484 (*697*).

SWANSON, E. L. (1929), *The Effect of Egg Albumen on the Crystallization of Sugar from Sirups*, Inaugur. Diss., Iowa State College, Ames, Iowa, 58 pp. (*779*).

SWENSEN, A. D., FIEGER, E. A., and UPP, C. W. (1942), *Poultry Sci.*, **21**:374–378 (*129, 667*).

SWENSON, T. L. (1938), U. S. Pat. 2,110,613 (742).

SWENSON, T. L. (1939), *U. S. Dept. Agr., Bur. Agr. Chem. Eng., Bull.*, MC-58:1–27 (713, 714, 715).

SWENSON, T. L. (1941), *U. S. Egg Poultry Mag.*, 47:407–409, 448 (742, 744).

SWENSON, T. L., and JAMES, L. H. (1932), *U. S. Egg Poultry Mag.*, 38(11):14–16, 58 (370, 374).

SWENSON, T. L., and JAMES, L. H. (1935), *U. S. Egg Poultry Mag.*, 41(3):16–19 (735).

SWENSON, T. L., SLOCUM, R. R., and JAMES, L. H. (1936), *U. S. Egg Poultry Mag.*, 42: 297–298, 310–311 (713).

SWIFT, C. H. (1914), *Am. J. Anat.*, 15:483–516 (199, 200).

SWIFT, C. H. (1915), *Am. J. Anat.*, 18:441–470 (199).

SWINGLE, D. B., and POOLE, G. E. (1923), *Univ. Montana Agr. Expt. Sta. Bull.*, 155:1–30 (699, 700).

SYDENSTRICKER, V. P., SINGAL, S. A., BRIGGS, A. P., DeVAUGHN, N. M., and ISBELL, H. (1942), *J. Am. Med. Assoc.*, 118:1199–1200 (632).

SZIELASKO, A. (1904), *Untersuchungen über die Gestalt und die Bildung der Vogeleier*, Inaug. Diss., Königsberg, 30 pp. (108).

SZIELASKO, A. (1905), *J. Ornithol.*, 53:273–297 (88).

SZIELASKO, A. (1913), *J. Ornithol.*, 61:52–117 (170).

SZUMAN, J. G. (1925), *Compt. rend. acad. sci.*, 181:257–259 (145, 149).

SZUMAN, J. G. (1926), *Compt. rend. soc. biol.*, 95:1314–1315 (254, 257, 258).

TABAKOFF, J. J. (1939), *Proc. World's Poultry Congr. (Cleveland)*, 7:151–153 (64).

TAKAHASHI, R. (1937), *Mitt. med. Akad. Kioto*, 14:159–172 (352).

TALENTI, M. (1935), *Arch. farmacol. sper.*, 59:287–290 (352).

TALLARICO, G. (1933), *Proc. World's Poultry Congr. (Rome)*, 5(2):581–586 (583).

TAMMANN, G. (1889), *Z. physiol. Chem.*, 12:322–326 (358).

TANGL, F. (1903), *Arch. ges. Physiol. (Pflüger's)*, 93:327–376 (598, 599, 600).

TANGL, H. (1939), *Magyar Orvosi Arch.*, 40:435–438 (83).

TAPERNOUX, A. (1930), *Compt. rend. soc. biol.*, 105:405–407 (348).

TARASOV, K. I. (1916), U. S. Pat. 1,187,869 (801).

TARCHANOFF, J. R. (1884), *Arch. ges. Physiol. (Pflüger's)*, 33:303–378 (114, 323, 400, 410).

TAYLOR, A., THACKER, J., and PENNINGTON, D. (1942), *Science*, 96:342–343 (791).

TAYLOR, A. G. (1930), *Proc. World's Poultry Congr. (London)*, 4:785–791 (53).

TAYLOR, G. L., ADAIR, G. S., and ADAIR, M. E. (1932), *J. Hyg.*, 32:340–348 (435).

TAYLOR, L. W., and LERNER, I. M. (1939), *J. Agr. Research*, 58:383–396 (153).

TAYLOR, L. W., and LERNER, I. M. (1941), *J. Heredity*, 32:33–36 (103).

TAYLOR, L. W., and MARTIN, J. H. (1928), *Poultry Sci.*, 8:39–44 (117, 153, 368).

TAYLOR, M. W., and RUSSELL, W. C. (1935), *J. Agr. Research*, 51:663–667 (237).

TAYLOR, M. W., and RUSSELL, W. C. (1943), *Poultry Sci.*, 22:334–336 (31).

TEISLER, E. (1902), *J. soc. cent. agr. Belg.*, 49:10 (699).

TEIXEIRA E SILVA, H. M. (1946), *Bol. ind. animal (São Paulo)*, 8(3):91–101 (483).

TEN BROECK, C. (1914), *J. Biol. Chem.*, 17:369–375 (556).

TENCONI, A. (1934), *Pensiero Medico*, 23:226–232 (358).

TEPLEY, L. J., STRONG, F. M., and ELVEHJEM, C. A. (1942), *J. Nutrition*, 23:417–423 (627).

TERMOHLEN, W. D., WARREN, E. L., and WARREN, C. C. (1938), *U. S. Dept. Agr., Marketing Inf. Ser.*, PSM-1:1–80 (742, 743).

TERROINE, E. F., and BELIN, P. (1927), *Bull. soc. chim. biol.*, 9:1074–1083 (342)

TEZNER, O. (1935), *Klin. Wochschr.*, 14:539–541 (557).

THAYER, S. A., MCKEE, R. W., BINKLEY, S. B., MACCORQUODALE, D. W., and DOISY, E. A. (1939), *Proc. Soc. Exptl. Biol. Med.*, **41**:194–197 (*622*).

THEIS, E. R., and HUNT, F. S. (1931), *Ind. Eng. Chem.*, **23**:50–53 (*796*).

THEIS, E. R., and HUNT, F. S. (1932), *Ind. Eng. Chem.*, **24**:799–802 (*795, 797*).

THEORELL, H. (1938), *Biochem. Z.*, **298**:242–267 (*489*).

THIERIOT, J. H. (1897), *U. S. Consular Repts.*, **1897** (Dec.):563–564 (*699, 712*).

THISTLE, M. W., WHITE, W. H., REID, M., and WOODCOCK, A. H. (1944), *Can. J. Research*, **22(F)**:80–86 (*753*).

THOMAS, A. W., and BAILEY, M. I. (1933), *Ind. Eng. Chem.*, **25**:669–674 (*409, 731, 732, 733*).

THOMAS, B. H., and QUACKENBUSH, F. W. (1933), *Iowa Agr. Expt. Sta. Rept.*, **1933**:27 (*611*).

THOMAS, B. H., and WILKE, H. L. (1940), *Iowa Agr. Expt. Sta., Ann. Rept.*, pp. 111–112 (*620*).

THOMPSON, D'A. W. (1943), *On Growth and Form*, Cambridge University Press, 1116 pp. (*88*).

THOMPSON, R. B., and ALBRIGHT, W. P. (1934), *Oklahoma A. and M. Coll. Agr. Expt. Sta. Rept.*, **1932–1934**:114–117 (*668*).

THOMPSON, R. B., ALBRIGHT, W. P., SCHNETZLER, E. E., and HELLER, V. G. (1932), *Oklahoma A. and M. Coll. Agr. Expt. Sta. Rept.*, **1930–1932**:128–136 (*129, 648*).

THOMPSON, W. C. (1942), *New Jersey Agr. Expt. Sta. Bull.*, **700**:1–12 (*11*).

THOMPSON, W. C., PHILPOTT, D., and PAGE, H. C. (1936), *New Jersey Agr. Expt. Sta. Bull.*, **614**:1–31 (*21, 59*).

THOMPSON, W. L. (1944), *Bull. Univ. Pittsburgh*, **40**:311–317 (*517, 518*).

THOMSON, A. (1844), *London and Edinburgh Mon. J. Med. Sci.*, **7**:581 (*285*).

THOMSON, A. (1859), *Cyclopaedia of Anatomy and Physiology*, edited by R. B. Todd. Longmans, Brown, Green, Longmans, and Roberts, London, Vol. 5, 890 pp. (*124, 146*).

THOMSON, R. T., and SORLEY, J. (1924), *Analyst*, **49**:327–332 (*317, 320, 483, 671, 698*).

THUNBERG, T. (1902), *Skand. Arch. Physiol.*, **13**:99–106 (*383*).

THUNBERG, T. (1941), *Kgl. Fysiograf. Sällskap. Lund, Förh.*, **11**:126–128 (*352*).

TILLMANS, J., RIFFART, H., and KÜHN, A. (1930), *Z. Untersuch. Lebensm.*, **60**:361–389 (*785*).

TIMPERLEY, W. A., NAISH, A. E., and CLARK, G. A. (1936), *Lancet*, **231**:1142–1149 (*566*).

TINKLER, C. K., and SOAR, M. C. (1920), *Biochem. J.*, **14**:114–119 (*408*).

TISELIUS, A. (1930), *The Moving Boundary Method of Studying the Electrophoresis of Proteins*, Inaug. Diss., Upsala, Sweden, 107 pp. (*432*).

TISELIUS, A. (1937), *Trans. Faraday Soc.*, **33**:524–531 (*446, 448*).

TISELIUS, A., and ERIKSSON-QUENSEL, I.-B. (1939), *Biochem. J.*, **33**:1752–1756 (*327, 525, 527*).

TISELIUS, A., and GROSS, D. (1934), *Kolloid-Z.*, **66**:11–20 (*436*).

TISHIMA, —— (1934), quoted by KATO, K., and KO, J. (1938), *J. Coll. Agr., Tohoku Imp. Univ.*, **3(2)**:139–153 (*167*).

TISSIER, H. (1926), *Compt. rend. soc. biol.*, **94**:446–447 (*688*).

TISSIER, H., and LAGRANGE, E. (1925), *Compt. rend. soc. biol.*, **93**:426–428 (*688*).

TITTSLER, R. P. (1927), *Pennsylvania Agr. Expt. Sta. Bull.*, **213**:20–21 (*509*).

TITUS, H. W., FRITZ, J. C., and KAUFFMAN, W. R. (1938), *Poultry Sci.*, **17**:38–45 (*128*).

TITUS, H. W., and NESTLER, R. B. (1935), *Poultry Sci.*, **14**:90–98 (*618, 619*).

TIXIER, R. (1945), *Bull. soc. chim. biol.*, **27**:627–631 (*348*).

TODD, W. R., ELVEHJEM, C. A., and HART, E. B. (1934), *Am. J. Physiol.*, **107**:146–156 (*606*).

TODRICK, A., and WALKER, E. (1937), *Biochem. J.*, **31**:292–296 (*476*).

TOMITA, M. (1921), *Biochem. Z.*, **116**:12–14 (*352*).

TOMKINS, R. G. (1937a), *Dept. Sci. Ind. Research (Brit.), Food Invest. Repts.*, **1936**:53 (*492, 689*).

TOMKINS, R. G. (1937b), *Dept. Sci. Ind. Research (Brit.), Food Invest. Repts.*, **1936**:149–151 (*688, 705*).

TOMPKINS, E. H. (1936a), *Am. J. Syphilis, Gonorrhea, Venereal Diseases*, **20**:22–36 (*570, 571*).

TOMPKINS, E. H. (1936b), *Am. Rev. Tuberc.*, **33**:625–638 (*571*).

TOMPKINS, E. H. (1943), *Arch. Path.*, **35**:695–712 (*570*).

TORRANCE, W. (1923), *J. S. African Chem. Inst.*, **6**:11–17 (*354*).

TORRISI, D. (1935a), *Boll. soc. ital. biol. sper.*, **10**:443–445 (*572*).

TORRISI, D. (1935b), *Boll. soc. ital. biol. sper.*, **10**:445–447 (*572*).

TRACY, P. H., SHEURING, J., and HOSKISSON, W. A. (1944), *Food Research*, **9**:126–131(*761*).

TRANIN, S. (1938), U. S. Pat. 2,119,657 (*802*).

TREDERN, L. S. (1808), *Ovi Avium Historiae et Incubationis Prodromus*, Diss., Jenae, Etzdorf, 16 pp. (*138*).

TROSSARELLI, L., and MASSANO, A. (1938), *Giorno batteriol. immunol.*, **20**:1059–1071 (*682*).

TSCHERMAK, A. VON (1915), *Biol. Zentr.*, **35**:46–63 (*100*).

TSO, E. (1925a), *China Med. J.*, **39**:136–140 (*116, 598, 600*).

TSO, E. (1925b), *Proc. Soc. Exptl. Biol. Med.*, **22**:263–265 (*616*).

TSO, E. (1926a), *Biochem. J.*, **20**:17–22 (*616, 624, 698*).

TSO, E. (1926b), *Am. J. Physiol.*, **77**:192–198 (*603, 620*).

TSUJI, K. (1922), *Endocrinology*, **6**:587 (*517*).

TUKKER, J. G. (1930), *Proc. World's Poultry Congr. (London)*, **4**:900–906 (*64, 70, 77*).

TULLY, W. C., and FRANKE, K. W. (1935), *Poultry Sci.*, **14**:280–284 (*44*).

TURNER, A. W. (1927), *Australian J. Exptl. Biol. Med. Sci.*, **4**:57–60 (*689*).

TURNER, C. W., IRWIN, M. R., and REINEKE, E. P. (1945), *Poultry Sci.*, **24**:171–180 (*46*).

TURNER, C. W., KEMPSTER, H. L., HALL, N. M., and REINEKE, E. P. (1945), *Poultry Sci.*, **24**:522–533 (*46*).

TURPIN, G. M. (1918), *Iowa Agr. Expt. Sta. Bull.*, **178**:211–232 (*9*).

TYLER, C. (1946), *J. Agr. Sci.*, **36**:111–116 (*155*).

UBER, F. M., HAYASHI, T., and ELLS, V. R. (1941), *Science*, **93**:22–23 (*391*).

UHLENHUTH, S. (1900), *Deut. med. Wochschr.*, **26**:734–735 (*540*).

UHLENHUTH, S. (1901), *Deut. med. Wochschr.*, **27**:260–261 (*540*).

ULPIANI, C. (1901), *Atti accad. Lincei*, **10**:368–375 (*483*).

UNDERHILL, F. P., PETERMAN, F. I., GROSS, E. G., and KRAUSE, A. C. (1929), *Am. J. Physiol.*, **90**:72–75 (*358*).

UPP, C. W. (1934), *Louisiana Agr. Expt. Sta. Circ.*, **13**:1–8 (*667*).

UPP, C. W., and THOMPSON, R. B. (1927), *Oklahoma Agr. Expt. Sta. Bull.*, **167**:1–36 (*19, 20, 76*).

URBAIN, O. M., and MILLER, J. N. (1930), *Ind. Eng. Chem.*, **22**:355–356 (*731, 733*).

URBAIN, W. M., WOOD, I. H., and SIMMONS, R. W. (1942), *Ind. Eng. Chem., Anal. Ed.*, **14**:231–233 (*415, 728*).

URNER, F. G. (1921), *Am. Poultry Advocate*, **29**:114 (*712*).

UTESCHER, E. (1898), German Pat. 98,231 (*697*).

UTESCHER, E. (1921), German Pat. 328,423 (*699*).

VAČEK, T. (1935), *Arch. Geflügelkunde*, **9**:17–22 (*46*).

VACIRCA, F., and BERTOLA, A. (1937), *Biochim. e terap. sper.*, **24**:91–98 (*569*).

VALENCIENNES, A. (1856), *Compt. rend. acad. sci.*, **42**:3–6 (*280, 283*).

VALENCIENNES, A., and FRÉMY, E. (1854), *J. pharm. chim.*, (Ser. 3) **26**:5–16 (*326*).

VAN ALSTYNE, E. V. N., and GRANT, P. A. (1911), *J. Med. Research*, **25**:399–408 (*563*).

VAN SLYKE, D. D. (1911), *J. Biol. Chem.*, **10**:15–55 (*330*).

VAN SLYKE, D. D. (1915), *J. Biol. Chem.*, **22**:281–285 (*330*).

VAN WAGENEN, A., and HALL, G. O. (1936), *Poultry Sci.*, **15**:405–410 (*130, 642, 646*).

VAN WAGENEN, A., HALL, G. O., and WILGUS, H. S. (1937), *J. Agr. Research*, **54**:767–777 (*140, 646*).

VAN WAGENEN, A., and WILGUS, H. S. (1935), *J. Agr. Research*, **51**:1129–1137 (*139, 646, 647*).

VAUGHAN, V. C., CUMMING, J. G., and McGLUMPHY, C. B. (1911), *Z. Immunitätsforsch.*, **9**:16–28 (*563, 564*).

VAUGHAN, V. C., CUMMING, J. G., and WRIGHT, J. H. (1911), *Z. Immunitätsforsch.*, **9**: 458–489 (*563, 564, 565*).

VAUGHAN, V. C., and WHEELER, S. M. (1907), *J. Infectious Diseases*, **4**:476–508 (*540*).

VAUTIER, E. (1922), *Mitt. Gebiete Lebensm. Hyg.*, **13**:63–66 (*785*).

VAUTIN, C. T. J., and WHIFFEN, W. G. (1926), Brit. Pat. 274,164 (*697*).

VECCHI, A. (1933), *Proc. World's Poultry Congr.* (*Rome*), **5(2)**:502–511 (*361*).

VERGE, J. (1934), *Rec. méd. vét.*, **110**:705–723 (*515*).

VERNON, H. M. (1904), *J. Physiol.*, **31**:346–358 (*518, 585, 587*).

VEZZANI, V. (1939), *Proc. World's Poultry Congr.* (*Cleveland*), **7**:117–119 (*48*).

VEZZANI, V., DEVALLE, T., MEYNIER, E., and SIMONETTA-CUIZZA, R. (1936), *Proc. World's Poultry Congr.* (*Leipzig*), **6(1)**:358–361 (*607*).

VICKERY, H. B., and SHORE, A. (1932), *Biochem. J.*, **26**:1101–1106 (*333*).

VILLÈLE, A. DE (1903), *Rev. agr. Réunion*, **9**:372–376 (*699*).

VILTER, F. W., and SCHMIDT, O. (1933), *Z. Untersuch. Lebensm.*, **65**:649 (*692*).

VINSON, A. E. (1908), *Arizona Agr. Expt. Sta. Bull.*, **60**:431–433 (*697*).

VITA, G., and BRACALONI, L. (1933), *J. pharm. chim.*, **18**:104–108 (*483*).

VITA, G., and BRACALONI, L. (1936), *J. pharm. chim.*, **24**:558–563 (*788*).

VLÈS, F., and HEINTZ, E. (1935), *Compt. rend. acad. sci.*, **200**:1927–1929 (*456*).

VOELKER, O. (1940), *J. Ornithol.*, **88**:604–611 (*348*).

VOIT, C., HERMANN, E., FORSTER, H., FEDER, H., and STUMPF, H. (1877), *Z. Biol.*, **13**: 518–526 (*411*).

VONDELL, J. H. (1933), *U. S. Egg Poultry Mag.*, **39(4)**:18–19 (*648*).

VONDELL, J. H., and PUTNAM, J. N. (1945), *Poultry Sci.*, **24**:284–286 (*648*).

VOSSELER, J. (1908), *Pflanzer*, **4(9)**:129–136 (*697*).

WALKER, E. (1925), *Biochem. J.*, **19**:1082–1084 (*476*).

WALTHER, A. R. (1914), *Landw. Jahrb.*, **46**:89–104 (*65*).

WALZER, M. (1927), *J. Immunol.*, **14**:143–174 (*586*).

WAN, S., WU, H., and CHANG, C. Y. (1935), *Chinese Med. J.*, **49**:1264–1266 (*584*).

WANG, C. C. (1929), *U. S. Egg Poultry Mag.*, **35(10)**:20–23, 74, 76 (*698*).

WANG, C. C., KERN, R., and KAUCHER, M. (1930), *Am. J. Diseases Children*, **39**:768–773 (*602*).

WANG, C.-F., and WU, H. (1940), *Chinese J. Physiol.*, **15**:67–82 (*462*).

WARNER, D. E., and EDMOND, H. D. (1917), *J. Biol. Chem.*, **31**:281–294 (*236, 237*).

WARNER, D. E., and KIRKPATRICK, W. F. (1916), *J. Heredity*, **7**:128–131 (*256, 257, 258, 260, 295*).

WARNER, R. C. (1942), *J. Biol. Chem.*, **142**:741–756 (*480, 481*).

WARNER, R. C., and CANNAN, R. K. (1942), *J. Biol. Chem.*, **142**:725–739 (*480*).

WARRACK, G. H., and DALLING, T. (1933), *Vet. J.*, **89**:483–487 (*511*).

WARREN, D. C. (1927), *Poultry Sci.*, **7**:1–8 (*57*).

WARREN, D. C. (1930a), *Poultry Sci.*, **9**:184–193 (*24, 27, 48*).

WARREN, D. C. (1930b), *Proc. World's Poultry Congr.* (*London*), **4**:146–152 (*56, 57*).

WARREN, D. C. (1930c), *Kansas Agr. Expt. Sta. Bull.*, **252**:1–54 (*100*).

WARREN, D. C. (1939), *J. Agr. Research*, **59**:441–452 (*85*).

WARREN, D. C., and CONRAD, R. M. (1939), *J. Agr. Research*, **58**:875–893 (*129*).

WARREN, D. C., and CONRAD, R. M. (1942), *Poultry Sci.*, **21**:515–520 (*227, 228*).

WARREN, D. C., and SCHNEPEL, R. L. (1940), *Poultry Sci.*, **19**:67–72 (*153, 155*).

WARREN, D. C., and SCOTT, H. M. (1935a), *Poultry Sci.*, **14**:195–207 (*185, 214, 216, 229, 232, 279, 282, 294*).

WARREN, D. C., and SCOTT, H. M. (1935b), *J. Agr. Research*, **51**:565–572 (*229, 230, 231, 232*).

WARREN, D. C., and SCOTT, H. M. (1936), *J. Exptl. Zoöl.*, **74**:137–156 (*213, 216, 217*).

WASTENEYS, H., and BORSOOK, H. (1924), *J. Biol. Chem.*, **62**:675–686 (*523, 524*).

WASTENEYS, H., and BORSOOK, H. (1925), *J. Biol. Chem.*, **63**:575–578 (*533*).

WATERS, N. F. (1931), *Rhode Island Agr. Expt. Sta. Bull.*, **228**:1–105 (*75*).

WATERS, N. F. (1934), *Iowa State Coll. J. Sci.*, **7**:367–384 (*75*).

WATERS, N. F. (1937), *Poultry Sci.*, **16**:305–313 (*32*).

WATERS, N. F. (1941), *Poultry Sci.*, **20**:14–27 (*71*).

WATERS, N. F. (1945a), *Poultry Sci.*, **24**:81–82 (*72*).

WATERS, N. F. (1945b), *Poultry Sci.*, **24**:226–233 (*511*.)

WATERS, N. F. (1945c), *Poultry Sci.*, **24**:318–323 (*71*).

WATT, J. (1945), *U. S. Public Health Service, Public Health Repts.*, **60**:835–839 (*514*).

WATTS, B. M., and ELLIOTT, C. (1941), *Cereal Chem.*, **18**:1–9 (*771*).

WEAVER, C. H. (1930), *Proc. World's Poultry Congr.* (*London*), **4**:379–390 (*37, 38*).

WEDEMANN, W. (1940), *Z. Fleisch- u. Milchhyg.*, **50**:277–279 (*715*).

WEECH, A. A. (1942), *Bull. Johns Hopkins Hosp.*, **70**:157–176 (*587*).

WEECH, A. A., and GOETSCH, E. (1939), *Bull. Johns Hopkins Hosp.*, **64**:425–433 (*587*).

WEHNER, A. (1930), *Deut. landsw. Geflügel-Ztg.*, quoted by GAGGERMEIER, G. (1932), *Arch. Geflügelkunde*, **6**:105–109 (*659*).

WEIANT, A. S. (1917), *U. S. Dept. Agr., Farmer's Bull.*, **791**:1–26 (*7, 9*).

WEIMER, B. R. (1918), *J. Am. Assoc. Inst. Invest. Poultry Husb.*, **4**:78–79 (*265, 266, 286*).

WEINSTEIN, P. (1933), *Z. Untersuch. Lebensm.*, **66**:48–59 (*314, 317, 320, 420*).

WELCH, A. DE M. (1936), *Proc. Soc. Exptl. Biol. Med.*, **35**:107–108 (*568*).

WELLS, H. G. (1911), *J. Infectious Diseases*, **9**:147–171 (*327, 540, 542, 551, 561*).

WELMANNS-KÖLN, P. (1903), *Pharm. Ztg.*, pp. 665–667, 804 (*314*).

WENDT, G. VON (1925), "Mineralstoffwechsel," in *Handbuch der Biochemie*, edited by K. Oppenheimer, second edition, G. Fischer, Jena, **8**:183–255 (*603*).

WENZEL, K. (1908), *Ornit. Monatschr.*, **33**:462–472, 494–501 (*7*).

WESSELMANN, A. (1936), *Vet. Bull.*, **6**:544 (*511*).

WEST, P. M., and WOGLOM, W. H. (1941), *Science*, **93**:525–527 (*632*).

WESTGATE, J. M. (1937), *Hawaii Agr. Expt. Sta. Ann. Rept.*, **1935**:23–24 (*358*).

WESTON, W. A. R. D., and HALNAN, E. T. (1927), *Poultry Sci.*, **6**:251–258 (*689*).

WHEELER, W. P. (1890), *N. Y. State Agr. Expt. Sta. Rept.*, pp. 122–140 (*697*).

WHETHAM, E. O. (1933), *J. Agr. Sci.*, **23**:383–418 (*14*).

WHIPPLE, G. H., and ROBSCHEIT-ROBBINS, F. S. (1930), *Am. J. Physiol.*, **92**:388–399 (*605*).

WHITE, G. F. (1919), *Science*, **49**:362 (*508*).

WHITE, W. H., and THISTLE, M. W. (1943), *Can. J. Research*, **21(D)**:211–222 (*747*).

WHITSON, D., TITUS, H. W., and BIRD, H. R. (1946), *Poultry Sci.*, **25**:143–147 (*47*).

WICKE, W. (1856), *Ann. Chem. Pharm.*, **97**:350 (*354*).

WICKE, W. (1858), *Naumannia*, **8**:393–397 (*347*).

WICKE, W. (1863), *Ann. Chem. Pharm.*, **125**:78–80 (*354*).

WICKMANN, H. (1896), *J. Ornithol.*, **44**:81–92 (*232, 233*).

WIGENT, Z. (1929), *Am. Poultry J.*, **60**:243 (*805*).

WIIDIK, R. (1937), *Z. Fleisch- u. Milchhyg.*, **48**:81–83, 106–110 (*688*).

WIJK, P. J., and UBBELS, P. (1930), *Proc. World's Poultry Congr. (London)*, **4**:99–110 (77).

WILCKE, H. L. (1936), *Iowa Agr. Expt. Sta. Research Bull.*, **194**:77–103 (*374*).

WILCKE, H. L. (1939), *Poultry Sci.*, **18**:236–243 (*87*).

WILCKE, H. L. (1940), *U. S. Egg Poultry Mag.*, **46(10)**:617–618 (*787*).

WILDER, O. H. M., BETHKE, R. M., and RECORD, P. R. (1933), *J. Nutrition*, **6**:407–412 (*461, 607*).

WILEY, H. W. (1908), *U. S. Dept. Agr., Bur. Chem., Bull.*, **115**:1–117 (*134, 665, 670*).

WILHELM, L. A. (1939), *Proc. World's Poultry Congr. (Cleveland)*, **7**:521–524 (*664*).

WILHELM, L. A. (1940), *Poultry Sci.*, **19**:246–253 (*153, 154, 157*).

WILHELM, L. A., and HEIMAN, V. (1936), *U. S. Egg Poultry Mag.*, **42**:750–751, 761 (*643*).

WILHELM, L. A., and HEIMAN, V. (1938), *State Coll. Washington Agr. Exp. Sta. Tech. Paper*, **358**:1–19 (*644, 645*).

WILKINS, R. H. (1915), *A Study of the Reproductive System of the Female Domestic Fowl*, Diss., Cornell University, Ithaca, New York, 64 pp. (*180, 184*).

WILLARD, J. T., and SHAW, R. H. (1909), *Kansas Agr. Expt. Sta. Bull.*, **159**:143–177 (*312, 317, 320*).

WILLHAM, O. S. (1931), *Panhandle (Oklahoma) Agr. Expt. Sta. Bull.*, **28**:1–16 (*51*).

WILLIAM, J. P., McCAY, C. M., SALMON, O. N., and KRIDER, J. L. (1942), *J. Animal Sci.*, **1**:38–40 (*787*).

WILLIAMS, J. W., and WATSON, C. C. (1937), *Nature*, **139**:506–507 (*461*).

WILLIS, H. (1927), Australian Pat. 10,250 (*712*).

WILLSTÄTTER, R., and ESCHER, H. H. (1912), *Z. physiol. Chem.*, **76**:214–225 (*487*).

WILM, —— VON (1895), *Arch. Hyg.*, **23**:145–169 (*495*).

WILSON, J. A. (1923), *The Chemistry of Leather Manufacture*, Chemical Catalog Co., Inc., New York, 350 pp. (*802*).

WILSON, J. A. (1927), *J. Am. Leather Chem. Assoc.*, **22**:559–565 (*794, 795*).

WILSON, J. W., and LEDUC, E. H. (1947), *Anat. Record*, **97**:470–494 (*571*).

WILSON, S. J., and WALZER, M. (1935), *Am. J. Diseases Children*, **50**:49–54 (*558, 586*).

WINCHELHAUS, —— (1890), *Analyst*, **15**:116 (*783*).

WINCHESTER, C. F. (1939), *Endocrinology*, **24**:697–701 (*46*).

WINCHESTER, C. F. (1940), *Poultry Sci.*, **19**:239–245 (*242*).

WINNEK, P. S., and SMITH, A. H. (1937), *J. Biol. Chem.*, **121**:345–352 (*607*).

WINSTON, J. J., and JACOBS, B. R. (1945), *J. Assoc. Offic. Agr. Chemists*, **28**:607–616 (*785*).

WINTER, A. R., STEWART, G. F., McFARLANE, V. H., and SOLOWEY, M. (1946), *Am. J. Pub. Health*, **36**:451–460 (*766*).

WINTON, B. (1929), *Poultry Sci.*, **8**:103–111 (*37, 38*).

WISE, W. L. (1901), Brit. Pat. 8937 (*699*).

WITTICH, W. VON (1851), *Z. wiss. Zoöl.*, **3**:213–219 (*146, 149, 168, 495*).

WITZ, G. (1876), *Dinglers Polytech. J.*, **219**:93 (*406, 407*).

WLADIMIROFF, G. E. (1926), *Biochem. Z.*, **177**:280–297 (*412*).

WOITKEWITSCH, A. A. (1940), *Compt. rend. acad. sci. U.R.S.S.*, **N.S.27**:294–297 (*31*).

WOLF, C. G. L. (1912), *Biochem. Z.*, **40**:234–276 (*585*).

WOLFF, L. K., VORSTMANN, N. J. M., and SCHOENMAKER, P. (1923), *Chem. Weekblad* **20**:193–195 (*358*).

WOLFHÜGEL, K. (1906), *Z. Infektionskrankh.*, **1**:21–25 (*305*).

WOLLMAN, E., and BARDACH, M. (1935), *Compt. rend. soc. biol.*, **118**:1425–1427 (*552*).

WOODCOCK, A. H., and REID, M. (1943), *Can. J. Research*, **21(D)**:389–393 (*743, 756*).

WOODGER, J. H. (1925), *Quart. J. Microscop. Sci.*, **69**:445–462 (*199*).

WOODROOF, J. G. (1942), *Ice and Refrig.*, **102**:62–64 (*734*).

WOODS, E., and CLOW, B. (1927), *Wisconsin Agr. Expt. Sta. Bull.*, **396**:46 (*620*).

WOOLLEY, D. W. (1942), *Proc. Soc. Exptl. Biol. Med.*, **49**:540–541 (*628*).

WOOLLEY, D. W., and LONGSWORTH, L. G. (1942), *J. Biol. Chem.*, **142**:285–290 (*631*).

WORINGER, P. (1932a), *Z. Kinderheilk.*, **52**:586–593 (*557*).

WORINGER, P. (1932b), *Bull. soc. péd.*, **30**:419–423 (*557*).

WORINGER, P. (1932c), *Compt. rend. soc. biol.*, **111**:723–725 (*557*).

WORINGER, P. (1932d), *Compt. rend. soc. biol.*, **111**:726–728 (*557*).

WORINGER, P. (1933), *Ann. inst. Pasteur*, **50**:270–281 (*557*).

WORMS, W. (1903), *J. Russ. Phys. Chem. Soc.*, **35**:835–844 (*326*).

WORMS, W. (1906), *J. Russ. Phys. Chem. Soc.*, **38**:597–607 (*452*).

WORTH, C. B. (1940), *Auk*, **57**:44–60 (*108*).

WRIGHT, A., and WHIPPLE, G. H. (1934), *J. Exptl. Med.*, **59**:411–425 (*567, 568*).

WU, C., and NOYES, B. (1928), *China Med. J.*, **42**:209 (*305*).

WU, H. (1927), *Chinese J. Physiol.*, **1**:81–87 (*457, 464*).

WU, H. (1929), *Chinese J. Physiol.*, **3**:1–6 (*457*).

WU, H. (1931), *Chinese J. Physiol.*, **5**:321–344 (*457, 474*).

WU, H., and LING, S. M. (1927), *Chinese J. Physiol.*, **1**:407–430 (*400, 477*).

WU, H., and LIU, S. C. (1931), *Proc. Soc. Exptl. Biol. Med.*, **28**:782–784 (*464*).

WU, H., LIU, S. C., and CHOU, C. Y. (1931), *Chinese J. Physiol.*, **5**:309–320 (*431*).

WU, H., TENBROECK, C., and LI, C. P. (1927), *Chinese J. Physiol.*, **1**:277–296 (*553*).

WU, H., and WANG, C. F. (1938), *J. Biol. Chem.*, **123**:439–442 (*462*).

WU, H., and WU, D. Y. (1925), *J. Biol. Chem.*, **64**:369–378 (*478*).

WURTZ, R. (1890), *Semaine méd.*, **10**:21 (*499*).

YOKOYAMA, Y. (1934), *Proc. Imp. Acad. Tokyo*, **10**:582–585 (*340*).

YOSINAGA, T. (1936), *J. Biochem. (Japan)*, **24**:21–30 (*538*).

YOUNG, E. G. (1922a), *Proc. Roy. Soc. (London)*, **93(B)**:15–35 (*452*).

YOUNG, E. G. (1922b), *Proc. Roy. Soc. (London)*, **93(B)**:235–248 (*462, 469*).

YOUNG, E. G. (1937), *J. Biol. Chem.*, **120**:1–9 (*333*).

YOUNG, E. G., and MACDONALD, I. G. (1927), *Trans. Roy. Soc. Can.*, **21**:385–393 (*520, 587*).

ZÄCH, C. (1929), *Mitt. Lebensm. Hyg.*, **20**:209–215 (*659*).

ZAGAEVSKY, J. S., and LUTIKOVA, P. O. (1937), *Voprosy Pitaniya*, **6(3)**:84–88 (*493, 682*).

ZAGAEVSKY, J. S., and LUTIKOVA, P. O. (1944a), *U. S. Egg Poultry Mag.*, **50**:17–20, 43–46 (*721, 722, 724*).

ZAGAEVSKY, J. S., and LUTIKOVA, P. O. (1944b), *U. S. Egg Poultry Mag.*, **50**:75–77, 88–90 (*497, 498, 499*).

ZAGAEVSKY, J. S., and LUTIKOVA, P. O. (1944c), *U. S. Egg Poultry Mag.*, **50**:121–123 (*495, 497*).

ZAITSCHEK, A. (1934), *Biedermanns Zentr., B., Tierernähr.*, **6**:102–111 (*45*).

ZANDER, D. V., LERNER, M., and TAYLOR, L. W. (1942), *Poultry Sci.*, **21**:455–461 (*12*).

ZANETTI, C. U. (1897), *Ann. chim. farm.*, **26**:529–534 *(330)*.

ZAVADOVSKI, B., TITAIEV, A., and FAIERNIARK, S. (1929), *Endokrinologie*, **5**:416–425 *(30)*.

ZAWADZKI, B. (1933), *Protoplasma*, **19**:485–509 *(404)*.

ZAWADZKI, B. (1935), *Compt. rend. soc. biol.*, **118**:1099–1101 *(404)*.

ZÖLLNER, I. (1932), *Biedermanns Zentr., B., Tierernähr.*, **4**:369–376 *(41)*.

ZSIGMONDY, R. (1901), *Z. Anal. Chem.*, **40**:697–719 *(401)*.

ZUCKERKANDL, F., and MESSINER-KLEBERMASS, L. (1931), *Biochem. Z.*, **236**:19–28 *(344)*.

INDEX

Glycerophosphoric acid in yolk phospholipids, 339

Glycine, addition of, to liquid egg, 754–755

in egg proteins, 333, 334

Glycogen in egg, 343

Glycolipids of yolk, 341. *See also* Ovokerasin, Ovophrenosin.

Glycoproteins of albumen, 326, 332, 343, 344. *See also* Ovomucin, Ovomucoid.

Gold number, of albumen, 401

of albumen proteins, 443

Goldfinch (green-backed) egg, markings on, 275

Golgi bodies, 199, 201, 202, 209

Goose, egg-laying of, 8

Goose egg, air cell in, 148

albumen, composition of, general chemical, 320

energy value of, 600

species specificity of, 541, 542

composition of, general chemical, 314

proportional, 114

energy value of, 599

ovalbumin, electrophoretic mobility of, 448

species specificity of, 545–546

ovomucoid, species specificity of, 543

shell, density of, 382

mammillary layer of, structure of, 164

minerals of, 354

oöporphyrin in, 348

pores in, 166

water in, 323

weight of, 114

yolk, absence of Forssman antigen from, 561

chemical composition of, 317

energy value of, 600

species specificity of, 559, 560

yolk oil, physicochemical properties of, 483

Goose (*Anser cinereus*) egg, mammilla in shell of, 164

Goose (*Anser cygnoides*) egg, mammilla in shell of, 164

Goose (*Anser segetum*) egg, mammilla in shell of, 164

Goose (Canada) egg, dimensions of, 106

weight of, 62, 64

Goose (*Cereopsis novae hollandiae*) egg, mammilla in shell of, 164

Goose (domestic) egg, mammilla in shell of, 164

Goose (Embden) egg, dimensions of, 106

weight of, 62, 64

Gossypol and yolk color, 129, 350, 667

"Grass egg," 129

Grebe (American eared) egg, dwarf, 259

Grebe (great crested) egg, shape of, 89

Grebe egg ovomucoid, specific rotation of, 454

Greenshank egg albumen, species specificity of, 541, 542

Grouse, breeding season of, 50

egg-laying of, 8

Grouse egg, air cell in, 148

shell, density of, 382

Grouse (ruffed) egg, dimensions of, 106

weight of, 62

Guanidine in egg, 352

Guillemot egg, markings on, 97

shape of, 90

Guinea fowl, egg-laying of, 8

Guinea fowl egg, albumen, composition of, general chemical, 320

energy value of, 600

composition of, general chemical, 314

proportional, 114

dimensions of, 106

energy value of, 599

ovalbumin, electrophoretic mobility of, 448

species specificity of, 545–546

ovomucoid, species specificity of, 543

specific rotation of, 454

water in, distribution of, 323

weight of, 62, 114

yolk, chemical composition of, 317

energy value of, 600

Guinea fowl (pearl) egg, ovalbumin, species specificity of, 544–545

ovoconalbumin, species specificity of, 544